ECOLOGY

Second Edition

ECOLOGY

Second Edition

ROBERT E. RICKLEFS
University of Pennsylvania

NELSON

Thomas Nelson and Sons Ltd
Lincoln Way   Windmill Road
Sunbury-on-Thames
Middlesex   TW16   7HP
P.O. Box 73146   Nairobi   Kenya
P.O. Box 943   95 Church Street   Kingston   Jamaica

Thomas Nelson (Australia) PTY Ltd
19-39 Jeffcott Street   West Melbourne   Victoria 3003

Thomas Nelson and Sons (Canada) Ltd
81 Curlew Drive   Don Mills   Ontario

Thomas Nelson (Nigeria) Ltd
8 Ilupeju Bypass   PMB 21303   Ikeja   Lagos

First published in Great Britain 1973 by Thomas Nelson and Sons Ltd
Second edition first published in Great Britain 1980 by Thomas Nelson and
Sons Ltd

ISBN 0 17 761086 7 (boards)
     0 17 771088 8 (paper)

Printed in the United States of America

Drawings by Joel Ito and John Woolsey

Contents

Contents   vii

Preface

Almost six years will have passed between the first and second editions of *Ecology*. These years have seen substantial progress in many aspects of ecology, among them the application of evolutionary principles to interpreting life history patterns and the development of theory about community organization. In revising *Ecology,* I have tried to reflect this progress and similar progress in other studies and to bring coverage of the literature up to date for ecology as a whole. I have also tried to provide more balance in examples and in application of principles, giving equal emphasis to aquatic and terrestrial systems and to studies of plants, invertebrates, and vertebrates.

Many of the changes the reader will find in the book have resulted from changes in my own perception of ecology. Six years of teaching, reading, and research have broadened my outlook and given me time to think about the relationships among the many subdisciplines of ecology and between ecology and allied fields of genetics, evolution, behavior, and comparative anatomy and physiology. But my belief remains firm in the principles that evolution is the most important unifying concept in ecology and that ecological systems are best understood in terms of interactions among their components. These principles are revealed most clearly at the organism and population levels of organization, yet one sees a strong movement to apply them to studies of communities and ecosystems.

In appearance, organization, and tone, *Ecology* has changed little. It is still intended for courses in general ecology at the college level. I have continued to try to convey to the reader both the important findings and theories of ecology and the fascinating diversity of patterns in the natural world, which ecologists seek to explain. The coverage remains broad, although I have alloted less space to some topics, notably ethology, while enlarging the discussion of others, particularly life history patterns, quantitative genetics, community ecology, and nutrient cycling. In addition, I have written new chapters on soils, extinction, and niche theory.

For me, *Ecology* is a continuing experience and a significant part of my professional life. I have been immensely pleased by the response of students and colleagues to the book, and I am grateful for the many comments and suggestions provided by its readers. I hope that *Ecology* will continue to be as readable, as stimulating to students, and as helpful to teachers as it has been gratifying to me.

Preface to the First Edition

The array of complex ecological interactions—among individuals, among populations, and between life forms and the physical and chemical environment—may be understood in terms of a few basic principles that govern these interactions. I have written this book primarily to show how fundamental ecological principles elucidate the meaning of patterns in nature.

We study ecology for two reasons: to gain the intellectual gratification that comes from understanding natural patterns and processes, and to apply that understanding to environmental problems that confront mankind. The recent emphasis that ecologists have placed on practical applications is timely and appropriate, but it also tends to narrow our view of ecology as a science. The kinds of knowledge that we draw upon to solve environmental problems represent only a portion of the science of ecology. This book tries to bring into balance the inevitable tendency toward specialization in science, and it tries to keep in view the sensible relationship between ecology the science and ecology the solver of environmental problems. My aim is to impart understanding of, and provoke curiosity about, ecology the science. Where applications demonstrate a particular point better than do natural ecosystems, I will discuss the applications. Readers will quickly appreciate the links between natural systems and practical problems. Thorough understanding of principles of structure and function in natural communities is, I think, the best preparation for tackling those problems.

The scope of *Ecology* is wider than that of most other texts; I have included sections on the physiology and behavior of organisms, on genetics, and on evolutionary biology. I feel that such breadth is important because the structure and function of biological communities are based on the attributes of organisms that form the communities. These attributes, in turn, are evolved characteristics that are determined by the interaction of the organism with its environment. An understanding of evolution and adaptation is essential to, and inseparable from, the study of all ecological relationships.

I have written this book primarily for readers beginning their study of ecology at the college level, but the book will also serve as a source of material and ideas for those whose study is more advanced and for those engaged in the practice of ecology as a profession. The reader does not need a strong background in biology or other sciences to understand what appears here. The first part of the book is not meant to be rigorous and demanding, and it leans heavily on detailed examples to make its points. As the narrative progresses, subjects are presented more quantitatively, and more diversified knowledge is summarized and synthesized. The organization of the book, though clearly proceeding from organism level to population and community levels of ecological interaction, is not rigid; from time to time, related subjects are discussed in detail to show that ecology is intimately tied to all science and that boundaries should not be stringently drawn around disciplines.

Although this book is intended as an introduction, I have not shied away from presenting a mathematical treatment of population processes. I have tried to explain the mathematics thoroughly, including those algebraic steps that are not always evident, and I hope that readers who feel uncomfortable with mathematics will try to work through the mathematical presentations to their own satisfaction. None of the derivations in the book is difficult. I have included the mathematical treatment to demonstrate ways in which ideas can be described and organized quantitatively, and to demonstrate that one may gain insights that are not always intuitively obvious. Some acquaintance with mathematical description and mathematical modeling of ecological processes is required to evaluate contemporary thought in ecology.

I hope this book will provide a broad but sound introduction to a fascinating and essential area of human thought. If it encourages critical evaluation of our thinking about ecology and stimulates some readers to contribute further to our knowledge and understanding of ecology, I will be all the more gratified.

Part 1

Introduction

Ecology—Rationale and Approach

Potassium cycles in forests, genetic equilibrium between selection and mutation, energy budgets of pond organisms, pollination systems of plants, mating calls of toads, predator-prey population oscillations—all these fall within the realm of ecology. They represent a small sample of what ecologists have learned during the past century, knowledge that is rapidly being unified by the discovery of basic principles. Hence the fascination and challenge of the study of ecology.

But ecology has another side as well. Man clearly has brought upon himself a bewildering variety of ecological crises by his efforts to manipulate the environment to his own ends and by the unforeseen detrimental effects of his material affluence upon the environment. Faced with these crises, should ecologists continue to expand our understanding of the natural world, or should they apply this understanding to the ecological problems that directly confront mankind? To forsake the intense pursuit of ecological theory merely because our immediate problems require the application of knowledge would be a grave error of shortsightedness. Experience has shown that solutions to problems are often extracted from or based on knowledge originally acquired in studies that had no obvious application. So, too, have many advances in theoretical ecology come from applied research: Studies on the control of agricultural pests, for example, have greatly helped us to understand population dynamics.

Another problem in this issue of theoretical versus applied ecological study is that theoretical knowledge is sometimes applied inappropriately or overlooked altogether. What little we know about soil nutrients should have warned that when applied to tropical regions, farming techniques developed for temperate areas could only lead to disappointing and often disastrous results. Time and time again, tropical land has been cleared for crops in much the same way as in Iowa or Ohio, exposing the bare soil to heavy rains that carry away nutrients and to sunlight that turns some kinds of soil into a hard material more suitable for masonry than farming. Our failures, while enlightening, still have not taught us enough about soils and plant nutrition in the tropics to devise foolproof agricultural programs. But what is the best way to apply knowledge to a practical problem? To develop sound farming techniques for the tropics, should we now set up a large series of experimental plots to test every conceivable combination of agricultural methods, or should we study basic principles of plant productivity and nutrient cycling and extrapolate from this the general knowledge to develop suitable agricultural practices?

A more detailed example may clarify somewhat the problem of the relationship between theoretical and applied ecology. Just after World War II, a control program at Clear Lake, California, was directed toward eliminating a certain kind of fly called a midge (Hunt and Bischoff 1960). Because basic principles of ecology were ignored, or not even recognized, the program ended in disaster. Clear Lake is a beautiful body of water, twenty miles long, that had become an important resort area well known for its fishing and water skiing. For many years the lake was regarded as an idyllic vacation spot, but during the summer months, immense clouds of midges were increasingly attracted to the lights of houses near the edge of the lake. Though they did not pose a threat to anyone's health, the midges were considered very annoying. Midges are closely related to mosquitoes but are not bloodsucking, and they are bothersome only because of their presence in large numbers. It is ironic that the problem was caused by human activities. Nutrients draining into the lake in the runoff from farm lands and in raw sewage dumped directly into the lake had changed the bottom sediments to favor the growth of the midge larvae.

For many years, researchers failed to find a suitable control program for the midges, but by the end of World War II, many such pesticides as DDD and DDT became commercially available. In 1949, DDD was applied to the entire lake at a level of less than two-hundredths of a part per million of insecticide in the water. This measure was successful, and for the next two or three years few midges were reported. After 1951, however, the midges began to increase in numbers; in 1954, DDD was again applied, resulting in a substantial killing off of the bothersome insects. Later that year, residents began to notice large numbers of dead birds, mostly western grebes, along the shore of the lake. No one saw any reason to connect these deaths with the midge control program.

Populations of the midges recovered more rapidly after the second application than after the first, and a third treatment was required in 1957. It was less successful, and biologists suspected that the midges had become resistant to the insecticide. Later that year, numbers of western grebes were found dead along the lake shore. This time, however, two dying birds were sent to the Bureau of Chemistry of the California Department of Agriculture. Their analyses showed that the concentration of DDD residues in the fat of the dead grebes was almost one hundred thousand times the concentration of the original application of insecticide to the lake. It was later discovered that edible fish had also concentrated the poisons to the extent that many species were no longer safe to eat. The midge control program had been a death warrant for Clear Lake. Its effect had been to breed a DDD-resistant strain of midge, which is still a local nuisance, to eliminate the western grebe from the lake, to reduce populations of other water birds, and to place in jeopardy the lake's most important asset, its fishery.

The results of the midge control program came as a complete surprise. Its designers had not counted on the ability of the midge to evolve resistance to a very toxic insecticide, nor had they foreseen that the residues of DDD would accumulate in the bodies of animals and be passed from prey to predator up the food chain, concentrated at each step until they became lethal. Such unforeseen consequences have plagued nearly all our efforts to bring under control the complicated system of checks and balances that maintains the stability of the natural world. As the impact of our activities expands, management of our environment will become more difficult and complex, and reliance upon goal-oriented research in ecology will become even riskier. Understanding general principles and being able to extrapolate from them will become increasingly important.

There is, I believe, a more subtle and in some ways more compelling reason to study ecology. Such study involves the process of scientific inquiry, one of the integral features in the development of civilization. Museums record and preserve civilization in works of art, but modern civilization, transcending traditional art forms, embraces the wonders of industry and science. Civilization also has fostered skyscraper cities, pollution-belching industries, and modern war machines—monuments to money and personal gain. But with the advance of technology has come intellectual advance in science and expansion of our basic knowledge and understanding of the natural world. Good and bad, this is all part of civilization—our heritage—and the basis of what we contribute to the future.

We commonly make the mistake of equating science with those who participate in it, and vice versa. Science embodies rigorous and demanding inquiry; scientists are merely people who apply these methods. Scientific inquiry does not automatically lead to scientific progress. A well-tuned piano is a precise instrument, but the presence of a piano does not assure that great music will be composed. Composition is an endeavor of the human mind, not of the qualities of a particular instrument. A scientist demonstrates artistic qualities when he identifies important problems, develops the proper techniques to observe and experiment with natural phenomena, and interprets and synthesizes the knowledge produced by his investigations. Scientific inquiry—on whatever subject—may have an aesthetic quality that is strictly attributable to the mind of the individual scientist. I hope that this book will awaken in some readers an appreciation for scientific inquiry as an intellectual process.

Civilization is the creation of optimistic minds. Think of the tremendous optimism that urged Charles Darwin through twenty-five years of searching for shreds of evidence to support his theory of evolution before bringing it to public light. What confidence in the future must lead scientists to explore the Universe and contemplate the birth and death of stars. The intellectual contributions of Newton and Einstein, apart from their application to mechanics and nuclear energy, are as much a part of our civilization as are "Mona Lisa" and the cathedral at Chartres. The concept of evolution has had as great an influence on modern thinking as almost any idea. But what about the future of scientific endeavor? Can we be optimistic? In particular, how much will we ever be able to know about the natural world? Many persons, including some scientists, are pessimistic, pointing to current crises and to the distinct possibility that man may not survive even to the end of this century. Others deplore the immense allocation of human and material resources to academic endeavors while so many people live in poverty and misery. An even more basically pessimistic viewpoint was revealed by geneticist Bentley Glass, who said that "... the laws of life are based on similarities, finite in number and comprehensible to us in the main now. For all time to come, these have been discovered, here and now, in our lifetime."

Many scientists have also suggested that some natural phenomena are completely beyond the realm of man's comprehension. J. B. S. Haldane, one of the greatest scientific minds of the century, has said: "My own suspicion is that the universe is not only queerer than we suppose, but queerer than we can suppose." My preference is for optimism. How else could society have reached our present condition, unless at times in the past people believed that *today* would actually exist? The pessimism that constrains man to live day by day mires the development of the mind.

To approach a body of knowledge so vast and complex as that of ecology is particularly difficult through a book or course that proceeds along a linear sequence, chapter by chapter, week by week. We cannot organize ecology in a single dimension, but some attempt to do so must be made so as to give the information and concepts presented in this book a logical order. Any of several approaches may intelligibly organize a body of knowledge and understanding. One way is to reconstruct the path of intellectual development, progressing from facts and simple concepts to the modern frontiers of an expanding science, attaining progressively more complex understanding of the natural world along the way. A second approach, one that is equally valid and frequently employed, is to order the present knowledge of the field into some sort of logical structure.

During our own intellectual development, we initially try to define natural surroundings in terms that are readily understood and coped with by naming organisms and describing their structure. Only after we are fully at ease with a system do we pose questions of it to understand its function. Some ideas are best introduced after certain basic information is well in mind. My teaching experience has shown that familiarity with organisms and environments on a casual and descriptive basis provides a useful foundation for understanding more subtle ecological concepts.

The conceptual development of an individual during his lifetime often parallels the historical development of thought in a particular discipline. Ecology is no exception. Paintings by Cro-Magnon man in caves in the south of France attest to his interest in the natural world. Much of this interest, of course, stemmed from his having to kill animals for food and clothing. Many contemporary primitive societies have names for a wide variety of organisms. Natives in certain areas of New Guinea have local names for several hundred species of birds (Diamond 1966); birds are an important part of their diet. The Tzeltal language of the Indians of Chiapas, Mexico, contains hundreds of names for local species of plants (Berlin *et al.* 1974). As one might expect, the degree to which species are distinguished depends on the importance of the organisms in daily life. Those with high cultural significance, intensively cultivated, or otherwise used for food, are sometimes overdifferentiated, names being given to several varieties of the same species. Conversely, plants or animals with little or no cultural significance are often underdifferentiated, one name being sufficient for several species.

The Greeks were the first to study biology systematically. Aristotle named many hundreds of organisms and grouped them according to a scheme of classification. He had recognized that although species differ, some share many attributes and can be grouped together at various levels of similarity. The Greeks also wrote about the form and structure of organisms, their morphology and anatomy. Biology did not progress much past that point in classical times, and whatever knowledge was accumulated was largely lost to the Western world. Interest in the study of natural surroundings was revived in Europe in the seventeenth and eighteenth centuries, and progressed rapidly through periods of activity that initially were directed toward classification, placing organisms into some sort of natural and logical order. The natural history of organisms, the patterns of their lives in nature, began to occupy the minds of biologists during the latter part of the eighteenth century, and has continued to do so to the present.

The knowledge acquired by the study of natural history in the past century led to new questions about the interrelationships between organisms and their environment. Charles Darwin, whom most would agree was the first great ecologist, synthesized observations on nature into the unifying principle embodied in his theory of evolution through natural selection. But Darwin's great work raised more questions than it was able to answer. Although descriptive natural history dominated ecology until the early part of this century, interest rapidly spread to genetics, population dynamics, and energetics, which today provide the basis for a modern synthesis of ecology.

New concepts are built largely upon prior knowledge and understanding. Equally important, all disciplines develop through interaction with others. Physiology, the study of organism function, and genetics, the study of inheritance, have had a strong impact on the development of ecology. As I mentioned earlier, such applied fields as agriculture and forestry have also made important contributions. Stimulation has not come entirely from other areas of biology, or for that matter, exclusively from the sciences. While physics, chemistry, mathematics, and statistics were fashioning invaluable tools for the study of ecology, the social sciences were also contributing concepts that have become central ecological ideas. Darwin himself readily acknowledged that the ideas of Thomas Malthus on population and economics greatly influenced the development of his theory of organic evolution. Even now, parallels are drawn between ecological systems and economic systems based on free enterprise. Many points of similarity, particularly competition and exploitation, were originally formulated by economists.

The development of a field proceeds largely through interaction with other fields, which frequently provide analytical tools and stimulating ideas. For this reason, the historical development of ecology does not follow a linear sequence of thought, but has been linked to a variety of disciplines for varying periods. Although our understanding of ecology is enriched by the study of its history, this approach does not of itself provide a sufficient organizational basis for a modern ecology textbook.

I considered arranging the present knowledge of ecology into a logical order based on some hierarchical scheme of classification; the use of increasing or decreasing orders of structural complexity was compelling. One such scheme might start with the cell, which is the simplest unit that exhibits most of the properties of life, and pass progressively to the organism, which may be viewed as a collection of cells, to the population, a collection of similar organisms, to the biological community, an assemblage of many different kinds of organisms, and finally to the ecosystem, the biological community and the physical realm together. Suitable arguments can be made for starting at any one of these levels and working in either or both directions.

One philosophy of scientific endeavor, that of reductionism, maintains that to understand a particular phenomenon, one must dissect it into its components. Accordingly, one would begin at a particular level and work down to progressively finer levels of analysis, for example, from the organism to the cell and further. This approach leads to the difficulty that no phenomenon can be understood absolutely, but only in terms of some lower level of organization. Taking reductionism to its logical extent could eventually lead one into the realm of worlds that Haldane supposed may be queerer than we can ever know.

Conversely, it may be possible to understand a level of organization from knowing the structure and function of the level just below it. This deductive process, though seemingly the opposite of reductionism, shares the implicit assumption that the whole is equal to the sum of its parts. But various levels of organization may have unique properties that are not obvious from our understanding of lower levels of organization. If we knew all that was to be known about cells, could we predict everything that we know about organisms? A moment's reflection will show that we cannot. For example, although we understand the structure of the genetic material, and how the hereditary "blueprint" is translated into the structure and function of cells, we cannot predict from this information exactly how the creatures that inhabit Earth should appear. The whole seems equal to more than the sum of its parts—each level of organization possesses a degree of specificity that is unique.

Whether we focus our inquiry up or down through this hierarchy of organization determines the kind of question we ask. Consider, for example, the southerly migration of birds in the autumn. How do birds migrate? "How" questions are usually answered by reductionist approaches: in the present example perhaps leading to an investigation of the cellular mechanisms of muscle contraction, the biochemistry of energy metabolism, and the neurophysiology of coordination and orientation. Why do birds migrate? "Why" questions often lead to a consideration of the external relations of the system: in this case, those environmental factors that stimulate the initiation of migratory behavior or to the significance of migration to the survival and well-being of the organism. Our answer might be based on experiments with caged birds that become restless when the daylight period reaches a certain length corresponding to their autumn migration period. Or our answer might be that there are no insects in New York during the winter, hence insectivorous birds must fly to more southerly latitudes where their prey remain active through the year (many seed-eating species do not migrate). Investigations of phenomena on any level of organization may require diverse approaches, depending on the kind of questions asked.

My approach to ecology in this book recognizes all levels and approaches as facets of a complex science. At the same time, I have tried to provide a consistent theme based on how organisms respond to their environments and how these responses determine the structure and function of ecological communities. This book expresses a primarily mechanistic viewpoint rather than a descriptive one; that is, its purpose is to convey understanding rather than merely to recount phenomena or patterns in nature. The single most important mechanistic generalization in ecology is that of natural selection and evolution, the great unifying principle of biology that gives pattern and meaning to the bewildering complexity and diversity of life. This book stresses an evolutionary viewpoint because of the basic importance of evolution to understanding life. I cannot overemphasize that a book does not substitute for one's direct experience with nature. Throughout this book I have tried to convey as much feeling as is possible on a printed page for organisms and their environments. However satisfactory this may be, I should hope that from time to time you will venture into the field to discover for yourself many of the principles and phenomena discussed here.

2

The Order of the Natural World

The English word "ecology" is taken from the Greek "oikos," meaning house, the immediate environment of man. The literal meaning of the word has been expanded so that "ecology" now denotes the study of the natural environment, particularly the interrelationships between organisms and their surroundings. Although coined by the German zoologist Ernst Haeckel in the 1860s, "ecology" was brought into general use by the Danish botanist Eugen Warming in the 1890s.

The environment of each organism consists of different physical characteristics and different biological interactions, even though several organisms may live in a single location. A squirrel and an oak tree inhabit the same place, bathed by the same sunlight and drenched by the same rains; but the world of the oak includes the subterranean realm of its roots, which is completely foreign to the squirrel; the world of the squirrel is decisively affected by the bobcat, which merely passes the oak casually in pursuit of its prey.

The environment of the organism depends both on the organism and on the habitat in which it lives. In the desert, hot, dry air wrings moisture from the body; in the humid forest, predators and parasites may be more important. Diverse aspects of the environment mold the life strategy of the organism in different ways.

Ecologists are interested in patterns of nature beyond those embodied in organisms: the diversity and complexity of species assemblages; the flow of energy and nutrients through a community. In a sense, ecologists strive to answer questions raised by the observations of early natural historians and to explain the patterns they discovered. To understand the aims of ecology, one must see patterns in the natural world; questions about such patterns form the basis of all scientific endeavor.

What better place to experience nature than in the tropical rain forest, where life is most luxurious and diverse? Nowhere else is one so acutely aware of nature. At night, especially, the vast numbers of active creatures make the life of man's greatest cities seem paltry by comparison. In the pervasive blackness, we hear occasional frog calls and the monotonous, piercing whines of cicadas, but we dimly sense the teeming activity that actually surrounds us. In the Panamanian rain forest, the delicate first light of a November day is shattered by the explosive cries of a nearby troop of howler monkeys. The howling subsides, but it is presently answered by that of another troop farther away, and then another and another. We can easily locate the nearby troops of monkeys and so, we would guess, can the howlers themselves. Are we witnessing a morning proclamation of territory by each group?

It is now light enough so that we can begin to wend our way along the narrow forest trails, being careful not to brush against the long sharp spines that viciously ornament the trunk of the black palm; we make our way around the giant buttresses that spread out from the base of trees over the forest floor. There is a faint snapping sound, and our eyes quickly search for some small animal moving in the bushes. Another snap. A small object catapults in front of us and strikes the ground nearby. Finally we locate the source—a small bush whose fruits have opened up and are shooting their seeds over the forest floor. As the fruit, which resembles a three-sided pea pod, dries out, the edges of the pod come together, squeezing the seeds with greater and greater force until the stalk of the seed finally breaks and the seed is ejected from the pod (Figure 2-1).

The drab browns and greens of the inner rain forest are occasionally broken by the flashing of iridescent butterfly wings. A blue morpho butterfly streaks by us, just out of reach. How striking it is that these butterflies should be so conspicuous. How different from the moths that are attracted at night to the lights around buildings; they possess every conceivable device for being unobtrusive. Their brown and gray colors look like dead leaves and bark. Many of the moths conceal their legs, normally a sure betrayal of any animal, beneath their wings, or they have modified them to resemble bark or lichens. One species characteristically protrudes one leg or another from underneath its regular outline to break up the symmetry so characteristic of animal forms. Others have wings that although intact, often have the broken and contorted appearance of dead leaves. The hind edge of the wing of another species bears the unmistakable picture of a rolled-up leaf complete with its shadow. How perfectly these moths must blend with their back-

Figure 2-1   Opening of the fruit of *Rinorea sylvatica* (violet family), showing how the seeds are shot out as the pod dries out.

ground when they are resting on leaves or the bark of trees. How have these moths come to be that way, and why are other kinds of butterflies and moths so conspicuous?

We are constantly impressed by a sense of purpose in the adaptations of organisms. As we walk through the forest we wonder about the purpose of nature. Whose purpose? What purpose? Our thoughts are suddenly interrupted as we step, as if going through a door into another room, into a large clearing in the forest where we are blinded momentarily by the sun. A recent treefall caused this break in the forest canopy. This giant did not exit gracefully, but took with it at least a dozen other trees that either had been directly in its path or had been bound to it by vines. The clearing gives us our first glimpse of the sky in several hours. A long graceful ribbon of hawks silently glides southward many hundreds of feet above us. Their flight seems effortless—none of the birds is moving its wings. The narrow ribbon is caught by a small updraft, sending it spiraling upward in wide, lazy circles until it breaks away again toward South America, seemingly impatient to be gone. Are these the same birds that we saw earlier in the fall, gliding along the ridges of the Appalachians in Pennsylvania?

We look again at the fallen tree in the clearing, hoping to glimpse some of the plants and animals that usually live in the sunlight on top of the forest and rarely venture to its lower depths. The tree has been down too long, and most of the animals have died or moved back into the forest canopy. But we do see several orchids that are entirely new to us, some small and delicate, others larger and more brightly displaying their beauty. The clearing will soon be choked by the seedlings and saplings of the regenerating forest.

We move on, and soon we are enclosed once more in the dim eerie green-brown light that enshrouds the inner part of the forest. There is some activity around a pile of dung off to the side of the trail. Several small dung beetles are busy parceling off bits of this treasure, rolling it up into balls to push across the forest floor to provide food for their young at some distant spot. The beetles push their little ball of dung backward, which they are obliged to do because of the shapes of their bodies and legs—at least I think that is the reason; perhaps they just cannot face the task. One beetle is trying to push a dung ball down a gentle slope. Occasionally, the ball gets away from it and starts to roll; but rather than lose the ball, the beetle hangs on with its hind legs and rolls over and over several turns down the trail, it and the ball together.

By the time we return to our house, our eyes have experienced more than our minds can sort out in a day, and we are tired after the long walk. But a shower and a good meal start us on an evening of conversation that takes us back over the day's journey to recount our observations and to ask questions—mostly to ask questions—far into the night, as seems always to be the outcome of our forays into the untouched world of the tropical rain forest.

The natural world is diverse and complex.

The naturalist stands in awe of the diversity of forms in nature and of the intricate interactions of these forms with one another and with their environments. In our walk through the tropical forest, we passed several hundred varieties of trees, and yet to untrained eyes they all seemed to be "doing" about the same thing. They were using the energy of sunlight assimilated by their green leaves to convert carbon dioxide from the air, and water and minerals from the soil, into the organic molecules that make up their structure. Why should there be so many kinds of plants when one seemingly could perform the same functions as all the others? We know that vast forests consisting of one or, at most, a very few species of coniferous trees stretch across the middle latitudes of Canada.

There are hundreds of species of butterflies and moths in the forest. Their appearance varies so much that one wonders if there are as many different environments within one forest as there are kinds of butterflies and moths. In spite of how diverse these butterflies and moths look to us and, presumably, to their natural predators, they are remarkably similar in their feeding habits. As adults, they all have long tubular snouts, which normally lie curled beneath their heads, but which can extend to the depths of tubular flowers for feeding. And as larvae, or caterpillars, they all eat green vegetation. How do these species interact with one another ecologically? How are the resources of the forest parceled out among them?

Many species are mutually bound to others for their survival. The delicately colored harlequin beetle, for example, carries on its back a small community of mites and pseudoscorpions that feed upon the mites (Figure 2-2). I have heard the story two different ways; I am not sure which is true, but both are equally fascinating. Some say the mites feed on the blood of their host, and the pseudoscorpions, which live under the wings of the beetle, feed on the mites. So far as I know, this kind of pseudoscorpion is found nowhere

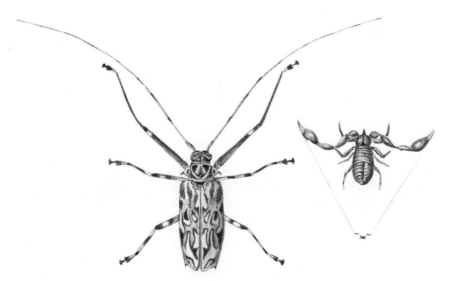

Figure 2-2   A harlequin beetle and its pseudoscorpion traveling companion, much enlarged.

else in nature. Others maintain that the mites perform a beneficial function by scouring fungi from the delicate membranous wings of the harlequin beetle. (Fungi seem to grow on almost everything in the tropics.) If this were the case, the mites would perform a beneficial function for the beetle, hindered considerably by the predacious pseudoscorpions. This explanation is strengthened by the fact that the forewings of the beetles have numerous small pits, just large enough to protect an individual mite from the pincers of the pseudoscorpion.

The natural world is dynamic, but it is also stable and self-replenishing.

The large tree that falls in a forest may be several hundred, or even several thousand, years old when it finally topples, but its life is exceptionally long compared to most living things. Death comes in many ways to organisms. Some fall victim to predators and parasites, while others die of exposure to the physical environment. A cold snap in spring, or a pond's drying up late in the summer, can bring death to scores of organisms. The natural community whose presence seems so stable is actually undergoing constant turnover, with replacement by new individuals, just as the organic structure of our body is continually replaced during our lifetime. In spite of its dynamic aspect, the natural world is also measurably stable—an equilibrium is maintained. The forest we walk through today is very much as it was five, ten, or even a hundred years ago. The same kinds of plants and animals persist to the present, though many of these are not the same individuals that were here earlier.

The dead bodies of organisms and the wastes of biological processes do not pile up. They are broken down and their component parts are recycled by the community. The dead leaves rustling under our feet cover the decomposed remains of other leaves in the soil beneath them. Soil organisms transform their elegant shapes into an amorphous mass of decaying and decomposing plant tissues, finally reducing them to the mineral elements from which they were once, in part, synthesized.

Populations of organisms are also continually replaced. An insect may lay thousands of eggs each year, and some marine organisms shed millions of eggs into the water—more than necessary to compensate losses of individuals from populations. Yet in spite of tremendous potential for population growth, populations remain within rather narrow limits because of a series of checks and balances. The frailty of these delicate balances is usually not fully appreciated until they are upset, often at the hand of man.

A spectacular case of unrestrained growth occurred in the early part of this century in the population of mule deer inhabiting the Kaibab Plateau of northern Arizona, just north of the Grand Canyon (Rasmussen 1941). The plateau was set aside as a game preserve in 1906, particularly for the maintenance of a large deer population. A program of predator removal was initiated to protect the deer population; during the next twenty-five years, more than 6,000 mountain lions, wolves, coyotes, and bobcats were killed in the area. The wolf population was exterminated.

Following the disappearance of the predators, the deer population, which had an estimated 4,000 individuals in 1906, began to increase rapidly. By 1923, there were between 60,000 and 70,000 deer within the game preserve. The natural browse of the area could not support such a large population. Even by 1918, wardens noticed that the vegetation used by the deer as food had been badly overgrazed. As the population increased, the food shortage worsened, and in September, 1923, one investigator estimated that between 30,000 and 40,000 deer were on the verge of starvation. In 1925, almost two-thirds of the total herd died. Only 20,000 deer remained in 1931, and by 1939, the population had dwindled to 10,000, about twice its level at the beginning of the control program.

Although the direct causes of the population increase and subsequent crash were not determined by critical study, this case, and others like it, indicate that predator populations are an integral part of the system of checks and balances that operates to keep an animal population at a level suitable for the amount of food available in its area. When the natural control system is altered, many years must pass before the population is again restored to a balanced state.

Man is a part of the natural world.

We are becoming increasingly aware that man has a deeply felt impact on the natural world, but we must also keep in mind that we ourselves are a part of the natural world. Our anatomy, physiology, and behavior testify to our animal nature. Stripped of our culture, we are not very impressive compared to other organisms. Beside other mammals, we are neither large nor small, strong nor weak, fast nor slow; rather, we have capitalized on being generalists. We have developed new ways of coping with our environment through inventiveness, learning, and communication. It is man's ability to innovate and make rapid changes, as embodied in technology, that sometimes gives him a degree of control over nature. Man can produce changes in the environment faster than most organisms can cope with them. Some of these changes have conquered diseases and controlled insect pests, and they have been a great benefit to man. But some technological developments have been mixed blessings, effecting unforeseen and disastrous consequences along with the achievement of their goal; and others, less harmful, can most charitably be described as shortsighted.

The days when a white whale might get the better of a sea captain have disappeared under the prows of fast-moving ships equipped with sonar, harpoon guns, and floating slaughterhouses. This extravagant technology, combined with greed, will shortly result in the extinction of most kinds of whales hunted for commercial purposes. The whaling industry has employed myopic practices, ironically cutting short its own existence. The problem lies not in lacking the knowledge to manage the herds properly (I am purposely evading the moral question of whether we have any right to hunt these magnificent animals), but rather in the political and economic nature of man. The regulation of an industry, practiced by many nations in international waters, raises several political problems that can be solved only by nations voluntarily restricting their own participation in hunting. Some do; but because others do not, the whales are doomed.

Local political interests can also obstruct the intelligent enforcement of control programs. When the mule deer population of the Kaibab Plateau became excessive, the U.S. Forest Service sought a program to remove some of the deer by hunting. This was protested for many years by the state of Arizona, which challenged the federal government's right to remove deer from state lands. Arizona eventually lost the case, and the deer were removed, but even now the deer population is split by political boundaries. The Kaibab Plateau lies partly within Grand Canyon National Park, in which no hunting of any kind is allowed, and partly within the Kaibab National Forest, which is open to hunting. Only the deer fail to make the distinction.

All our activities affect the natural world, and as our population continues to grow and our material productivity continues to increase, our impact will become progressively more damaging. At the same time, we cannot live apart from the natural world. We rely on it completely to provide us with food and most of our materials for shelter and clothing, as well as for recreation. Our survival depends on maintaining the natural world in a healthy state. Are we ruining the natural world by producing changes faster than organisms can cope with them? Are we ruining it for ourselves, producing changes faster than we, too, can cope with them? How can man, who seems to have partially escaped control by natural forces, redesign his own activities to fit into the natural system? These questions are as much a part of ecology as man is a part of the natural world.

None of these questions can be answered, none of these problems solved, without first understanding how the natural world functions. The emphasis of this book is placed not so much on how to cope with our environmental crises, but rather on developing a fundamental understanding of ecological processes.

3

How We Perceive and Understand Our Environment

As we embark on a new discipline, we should take stock of what we already know that might be relevant to our study; we should also determine how our thinking processes might affect our approach to new phenomena. We are already familiar with some basic aspects of the environment, such as the variation in natural communities over the surface of the Earth. We understand how the structure and function of an organism enable it to survive and reproduce in the environment in which it lives. But we must also accept that our ability to perceive the environment through our senses and to conceptualize it in our minds is severely limited.

The organism is the fundamental unit of ecology.

The organism is a well-defined entity, having a physical boundary that separates it from the outside world. The organism is maintained by a system of internal controls that keep it in an intimate and dynamic relationship with the environment. Many of the properties of organisms also apply to the cells that make up the body of the organism, but the organism has special properties of organization and integration that cells do not exhibit. Populations and communities superficially appear to be made up of organisms in the same way organisms are composed of cells, but populations and communities do not exhibit the level of organization and integration that is characteristic of organisms. For our purposes, biological structure and function at all levels are most readily understood in terms of the whole organism; having muscle cells makes no sense unless an organism moves. The population is a collection of similar entities having few properties other than those embodied in its constituent organisms. Social behavior does confer structure on a population, but such behavior is most readily interpreted as the response of organisms to their social environment. Communities embrace diverse species that may be analogous to the variety of cell types found in organisms, but communities lack the discreteness and organization of organisms.

Most of the organisms we encounter are clearly recognizable as individual entities, but in some species it is difficult to distinguish where one organism ends and another begins. For example, in colonies of some corals and tunicates, individuals are fused together and share a common blood supply. Are we to call these individuals organisms or not? With regard to many physiological functions, the entire colony is like one organism; reproduction, however, accomplished by an outgrowth of cells that eventually gives rise to a new individual, is strictly a function of the individual and not of the colony. All of the cells that constitute each offspring are derived from a single parent individual.

In colonies of bees, wasps, ants, and termites—the social insects—individuals are more readily distinguished than in corals and tunicate colonies, but the social insect colony is highly integrated. The division of labor is developed in most social insects to the extent that most of the individuals in the colony never reproduce during their lifetime and perform a purely supporting function for reproductive individuals. Moreover, the individuals in each colony usually are produced from a single mating, so they are genetically closely related. The individual in the insect colony has many of the properties of cells in organisms, particularly with the division of labor among the various units of the colony. For this reason, insect colonies have frequently been referred to as "superorganisms."

Organisms fall into natural groups called species.

Organisms can be grouped according to their similarities into basic kinds, or species. All humans, for example, are alike in certain basic respects, but different from all other kinds of organisms. We recognize that category of organisms called humans as a species. But there are problems with this concept, as well. We could say that all humans with blue eyes are alike in eye color, and differ from all other humans. Are we to call those individuals with blue eyes a different species from those with brown eyes? Does any one level of similarity provide a more useful criterion for species distinctions than any other? For humans, this question poses no great problem because the variation among humans is negligible compared to the differences that separate humans from their closest relatives, the great apes. For many kinds of organisms, such distinctions are not always easy to make. Taxonomists, the people who classify organisms, recognize many tens of thousands of kinds of flies, some distinguished only by the most minute details of their anatomy. Although we have a basic notion of "kind," difficult taxonomic problems continually try our understanding of the species concept.

Organisms are well designed with respect to their environments.

Organisms seem to have been designed in a purposeful manner to meet the requirements of their environments. Their structures are adapted to the various functions they perform for the organism. Birds have wings to fly. The purpose of the eye is to see. Teeth enable organisms to eat. The sharp, pointed teeth of the lion are well suited for tearing apart animal flesh, and the flat, broad teeth of the cow are ideal for grinding its plant food. Such adaptations suggest that some external force has imposed designs on organisms.

The leaves of most kinds of trees exhibit a consistent pattern, that of flatness. How does this flatness arise? We could determine how cell growth organizes the leaf bud into a structure that we recognize as the leaf, but we would merely have described its development. Would knowing the chemical constituents of leaves allow us to predict the shape of the leaf? Does anything about the organism, the tree itself, imply that its leaves should be flat? Why should deciduous trees not have needles, or for that matter, a big blob of leaf tissue at the end of each twig? Why should there be leaves at all? The only meaningful answer to these questions is that the flat shape of the leaf must serve some purpose. Flatness makes the leaf an ideal organ to intercept light, the source of energy for the photosynthetic process of the tree, and for gas and heat exchange with the air. A flat object has a large surface and requires relatively little material for its construction.

The design of a particular characteristic of the organism often affects other adaptations. That is, the parts of the organism constitute a well-integrated whole. For example, the horse's legs are designed for the rapid running speed it needs to escape predators in its open grassy habitat. The design for rapid motion required the evolution of a foot longer and more simply constructed than that of other mammals (Figure 3-1). But changes in bone structure, particularly in the hind legs, have caused the legs to be relatively stiff. Most mammals use their hind legs for scratching at external parasites, but the hind legs of the horse can only minimally provide this function. As a result, at least two other parts of the body have been redesigned to help the horse combat parasites: the long swishing tail, which can cover a large portion of the hind quarters of the horse; and a loose skin that the horse can shake very thoroughly without moving its body.

D'Arcy Thompson pointed out long ago in *On Growth and Form* that many problems involved in the design of organisms are similar to those encountered in the design of bridges or other structures. For example, the

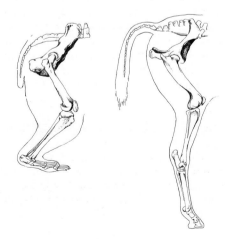

Figure 3-1   Hind limbs of a bear (left) and a horse, showing the elongated and simplified structure of the horse's foot for rapid motion (after Dunbar 1960).

struts of an airplane's wing closely resemble the structure of the bones in a bird's wing. Both are designed to achieve the maximum possible strength and rigidity using the least material. The skeletal structure of the extinct bison pictured in Figure 3-2 must have supported a very heavy suspended mass of viscera between two sets of supports, the front and the hind legs. The design of the bison is like the combination of a cantilever and a suspension bridge. The engineering solution to the bridge problem involves the use of two types of supporting members: one to bear the full weight of the structure, supporting it from below; and another to suspend parts of the structure from above. The bison was similar. The bones of the legs and the vertical extensions of the vertebrae supported the full weight of the bison, while the ligaments, tendons, and muscles connecting the vertebrae to the rib cage supported the weight of the viscera from above. The massive head and neck musculature of the bison counterbalanced the viscera in much the same way that the two ends of a cantilevered bridge balance each other. The bison, however, carried the cantilever design one step further by solving the engineer's nightmare of making the entire structure extremely flexible.

We frequently interpret the coloration of animals in terms of the reasons we sometimes camouflage man-made objects. Particularly during times of war, ability to escape detection can be a tremendous advantage. The environmental context—the pattern and coloration of the background—determines the kind of camouflage to be used: One would not have much luck wearing green and brown splotched clothing against the white background of winter snow. The presence of an enemy or of predators in the natural world is also a crucial part of the environment; these are the agents

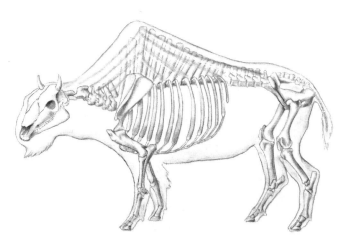

Figure 3-2   Skeleton of an extinct bison. The stout bones of the legs and projections of the vertebrae supported most of the weight of the bison, while the ligaments connecting the bones suspended the weight of the head and the viscera from above (after Thompson 1961).

that judge how well the design of the organism is fulfilling its purpose. In the absence of visually hunting predators, coloration to match the background has no purpose whatsoever.

The natural world can be conceived as a set of patterns.

We constantly recognize patterns in our surroundings, organization, and interrelationships in the complexity around us. We are able to perceive these patterns because of the predictability of their elements. We can anticipate nature in various ways, but most efficiently by the generalization of past experience. For example, experience with the movement of objects thrown into the air enables us to predict their trajectories with accuracy. A good outfielder can predict where a baseball will land long before it begins to drop.

Our lives are organized around patterns of our environment. Only if nature is predictable can we respond properly to it. Birds live in the woods, fish inhabit the sea. Without ever having been in a particular forest, we could expect to find birds rather than fish simply on the basis of past experience with birds and fish and forests and seas. By experiencing the unnaturalness of surrealism, we realize how completely our minds are bound up in the various patterns we recognize in nature. As ecologist G. Evelyn Hutchinson (1953) pointed out, ". . . (if) we imagine ourselves encountering in the middle of a desert a rock crystal carving of a sewing machine associated with a dead fish to which postage stamps are stuck, we may suspect that we have entered a region of the imagination in which ordinary concepts have become completely disordered."

Patterns have two sources of predictability, one achieved through observation, and the second by understanding the mechanism that produces the pattern. In the first case, predictions are based on extrapolating observations to new but similar situations. We do this when we predict the flight path of a ball. But by applying the laws of motion, we could have predicted the trajectory of the ball without any previous experience with the phenomenon, knowing only its initial speed and direction. Similarly, by applying the laws of probability we can predict the frequency distribution of random events without having to perform a series of trials.

In the development of science, empirically observed patterns almost always preceded the discovery of the causative principles that produce the patterns. After detailed observations, the German astronomer Johannes Kepler discovered that the time required for a planet to revolve around the Sun is inversely related to its distance from the Sun. Only later did the English physicist Isaac Newton formulate laws of motion whose predictive powers are so great that they made possible the detection of unseen planets on the basis of just their gravitational effects on the motion of some of the known planets. The same has been true in the biological sciences. Early naturalists knew that organisms could be classified into a regular hierarchy of similarities. But only after Charles Darwin proposed his theory of evolution was the basis for these patterns understood.

Our understanding of nature is limited by our perception of our surroundings.

We are very much limited by our senses. Overwhelmed by the sights and sounds in our surroundings, we often forget that other organisms perceive the world differently from the way we do. We mostly see and hear the world around us. Smells drift by largely unnoticed. Our sense of taste is dull. We also become accustomed to interpreting natural phenomena in terms of surroundings that we are familiar with. Islanders know water better than natives of Kansas, for example.

Our limited perception of nature can encumber our ability to understand natural phenomena. For years, most ecologists were trained and worked in Europe and temperate North America where seasonal fluctuations in temperature are a major aspect of the environment. When naturalists visited the tropics, they were impressed by the constant year-round temperature, and they assumed that biological communities are less variable in tropical regions than in temperate regions. Only recently have ecologists bothered to count individuals of tropical species over long periods. Surprisingly, they have found that these populations undergo marked seasonal fluctuations in the tropics, and additionally that they can vary considerably from year to year. Temperature defines the seasons in a temperate climate and forms a familiar pattern. But the patterns are different in the tropics, where the seasons are marked by wet and dry periods, and where rainfall is notoriously unpredictable at many times of the year.

Our interpretation of the natural world is also plagued by the problem of scale in time and space. A flea can jump a hundred times its length. We immediately react by translating that distance to a familiar scale—comparing flea lengths to human lengths. The comparable human jump would clear two football fields. How incredibly strong fleas are! But the flea is not actually performing athletic miracles. As size changes, the relationships between distance, power, and time change accordingly. Consider the rate at which flying organisms can beat their wings. Try to move your arms up and down as rapidly as possible. How fast? Two or three times per second. Most large birds can flap their wings between two and twenty times per second, and some small insects up to 500 times per second. This may seem amazing, but as the size scale changes, so does the time scale. You can demonstrate this very simply for yourself. Tie a weight to a string, and start it swinging like a pendulum. As you shorten the length of the string, you will notice that the weight swings back and forth much faster. In fact, the rate of swinging varies inversely with the length of the string. Halve the length of the pendulum, and its swinging frequency doubles. The wings of most small insects are less than a centimeter long, so we should not be surprised that an insect can beat its wings more than a hundred times faster than we can flap our arms.

Many biological phenomena occur on scales of time and space that our minds can easily comprehend, partly because our senses have been greatly extended by technology. Our sensory perception and our ability to conceptualize patterns are presumably adapted for coping with surroundings similar to the biological patterns that we try to understand. Biologists do not suffer so much the hair-pulling frustration of the physicist confronted with phenomena that are infinitely larger and smaller, and occur infinitely before and after, our own scale of existence. There is also a philosophical question that asks whether we can ever know and understand the world as it truly is, or whether we are limited to understanding what we perceive as the world. What matters, of course, is that we can conceptualize a consistent system that behaves as we predict it should.

Scientists look at the natural world from many different angles, depending on their training and temperament, and on the problems they study. All perspectives and approaches are valid to the extent that they can help us to understand the natural world. No one approach is inherently superior to another. Scientists use various approaches—empirical observation, experimentation, and theorizing—each of which is an integral part of scientific inquiry and is necessary to further our understanding of the natural world. The population biologist L. B. Slobodkin has said that "... in one sense, the distinction between theoretician, laboratory worker, and field worker is that the theoretician deals with all conceivable worlds while the laboratory worker deals with all possible worlds and the field worker is confined to the real world. The laboratory ecologist must ask the theoretician if the possible world is an interesting one and must ask the field worker if it is at all related to the real one."

Through scientific inquiry, we slowly begin to see how fragments of understanding fit into a larger setting. A question is the starting point of any inquiry. Phenomena—patterns in nature—prompt us to inquire how or why a pattern came to be. We form several hypotheses to answer our question, and then test each hypothesis by suitable experiments or observations. If the results are consistent with one of the hypotheses, we may begin to generalize our understanding of a phenomenon and make valid predictions based on this understanding. Experimental results provide new observations, which in turn may prompt us to ask new questions. Scientific inquiry is thus self-perpetuating; once a new inquiry is begun, it snowballs, and questions lead to new questions.

Scientific endeavor is like a selection process, separating good and bad hypotheses, incorporating some ideas into our understanding, and rejecting others. Because hypotheses can be ambiguous and observations imprecise, scientific knowledge must be continually re-evaluated in the context of technological and conceptual advances. Theories are not rigid and unbending. Newton's laws of motion explain everyday phenomena perfectly, at least within the limits of observation, but modifications proposed by Albert Einstein in his theories of relativity were needed to account for the behavior of objects moving at nearly the speed of light. Charles Darwin would probably be both amazed and pleased by present-day modifications to, and elaborations on, his theory of evolution.

As we strive to understand the natural world in terms of causative or mechanistic theories such as those of Newton and Darwin, should we ignore such empirical theories as those of Kepler? Should Darwin's theory of evolution have been rejected because the genetical basis of inheritance was not understood? This would be analogous to rejecting Newton's laws of motion because we do not understand what gravity is. We do tend to be more receptive to causative than empirical theories. To search for Newtonian solutions to problems, especially to overemphasize quantitative aspects of natural patterns, may impede the progress of science. The deliberate rejection of empirical theories has often led to superficial research on easily handled problems, where results are easily obtained without raising controversial issues.

Few fields or disciplines are ever ripe for a major theoretical synthesis. This has occurred few times in ecology: the theory of evolution by natural selection proposed in the mid-1800s; the development of the community concept just after the turn of the century; the rigorous mathematical formulation of natural selection theory and the development of population genetics in the 1920s and early 1930s; the elucidation of general properties of competition and predator-prey relationships at about the same time; the translation of ecological phenomena into energetic terms, giving rise to the ecosystem concept, during the 1940s and 1950s; the marriage of evolution and ecology during the 1960s. These syntheses have made the greatest contributions to our understanding of ecology, and have continued to stimulate further inquiry into natural phenomena.

Part 2

The Physical Environment

4

Life and the Physical Environment

We often contrast the living and the nonliving as opposites: biological versus physical and chemical, organic versus inorganic, biotic versus abiotic, animate versus inanimate, active versus passive. While these two great realms of the natural world are almost always readily distinguishable and separable, they do not exist one apart from the other. The dependence of life upon the physical world is obvious. The impact of living beings on the physical world is more subtle, but this impact is equally important to the continued existence of life on Earth. Soils, the atmosphere, lakes and oceans, and many sediments turned to stone by geological forces owe their characteristics in part to the activities of plants and animals.

Life has unique properties not shared by physical systems.

All forms of life have many properties in common that set organisms apart from stones and other inanimate objects. Motion and reproduction are the two most evident of these properties, even though many plants may be said to move very little indeed and one might describe the growth of crystals as a kind of reproduction. Exceptional cases that appear to fall on the wrong side of the great fence separating the biotic and the abiotic are less worrisome than they seem at first. Motion is merely an expression of a more fundamental property of life, the ability to perform work directed toward a predetermined goal. Reproduction represents, above all, the emancipation of the specific quality of biological structure and function from determination by simple physical structures like crystals. One might compare the genetic material passed from generation to generation by reproduction to language in its abstraction and specificity.

Although living beings are distinct from inanimate objects, life also must be viewed as an elaboration of the physical world, not its alternative. Organisms function within constraints set by physical laws. They are like internal combustion engines transforming energy to perform work. The Earth is also a giant heat machine, utilizing the energy in sunlight to drive the winds and ocean currents. But here lies the difference between physical and biological systems. In physical systems, energy transformations act to even out differences in energy level throughout the system, always following the path of least resistance. In biological systems, whether energy transformation is directed toward pursuing prey, producing seeds, keeping warm, or maintaining such basic body functions as breathing, blood circulation, and salt balance, it is used purposefully by the organism to maintain itself *out* of equilibrium with the physical forces of gravity, heat flow, diffusion, and chemical reaction. In a sense, this is the secret of life. A boulder rolling down a steep slope releases energy during its descent, but no useful work is performed. The source of energy, gravity in this case, is external, and as soon as the boulder comes to rest in the valley below, it is once more brought into equilibrium with the forces in its physical environment.

A bird in flight must constantly expend energy to maintain itself aloft against the pull of gravity. The bird's source of energy is internal, being the food that it has assimilated into its body, and the work performed serves a purpose useful to the bird in its pursuit of prey, escape from predators, or migration. To be able to act against external physical forces is the one common property of all living forms, the source of animation that distinguishes the living from the nonliving. Bird flight may be a supreme expression of animation, but plants just as surely perform work to counter physical forces when they absorb soil minerals into their roots or synthesize the highly complex carbohydrates and proteins that make up their structure.

Physical forces in the environment could not be held at bay without the expenditure of energy to perform work. The ultimate source of energy for life in the physical world is light from the Sun. Plants have special pigments, among them chlorophyll, that absorb light and capture its energy. That energy is then converted to food energy during the manufacture of sugars from simple inorganic compounds—carbon dioxide and water. The energy-trapping process is called *photosynthesis,* literally, a putting together with light. Energy locked up in the chemical bonds of sugars, and thence proteins and fats, is used by plants and animals, which either eat plants or eat other animals that eat plants, and so on, to perform the work required of an animate existence.

The biological and physical worlds are interdependent.

Life is totally dependent on the physical world. On one hand, organisms receive their nourishment from the physical world, and on the other, the distributions of plants and animals are limited by their tolerance of the physical environment. The heat and dryness of deserts prevent the occurrence of most life forms, just as the bitter cold of polar regions prevents the establishment of all but the most hardy organisms. Form and function are also brought under the yoke of the physical world (Alexander 1968, 1971, Gans 1974, Hochachka and Somero 1973). The viscosity and density of water require that fish be streamlined according to rigid hydrodynamic rules if they are to be swift. The concentration of oxygen in the atmosphere, at 21 per cent, places upper bounds on the metabolic rates of organisms. Similarly, the ability of plants and animals to dissipate body heat—accomplished by the purely physical means of evaporative cooling, conduction, and radiation of heat from the body surface to the surroundings—limits their rate of activity and their safe exposure to direct sunlight (Gates 1963).

The activities of organisms also affect the physical world, sometimes in a profound manner. The oxygen that we take for granted with every breath was produced largely by the photosynthetic activities of green plants (Berkner and Marshall 1964, 1965, Cloud 1968, Van Valen 1971). Before green plants evolved in primitive seas, the atmosphere of the Earth was composed mostly of methane (CH_4), ammonia (NH_3), water vapor (H_2O), and hydrogen (H_2). As early aquatic plants began to utilize sunlight as a source of energy, they began to liberate oxygen, some of which escaped from the oceans and accumulated in the atmosphere. Over the past two billion years, the span of life on Earth, most of the hydrogen once contained in the Earth's primitive atmosphere has escaped into outer space. Plants have assimilated the carbon contained in atmospheric methane and the nitrogen in ammonia, and their place in the atmosphere has been partly taken by oxygen released during photosynthesis.

Plants play an equally influential role in the development of soil properties (Crocker 1952). Plant roots find their way into tiny crevices and pulverize rock as they grow and expand. Bacteria and fungi hasten the weathering of rock by chemical means. Fungi secrete acids to dissolve minerals out of unaltered rock, thereby weakening the crystalline structure of the rock and speeding its decomposition. Rotting plant detritus also releases acids that do the work of chemical decay, while fragments of detritus alter the physical structure of the soil. Animals, by burrowing, trampling, and defecating, also play a part in the development of soil.

The role of plants and animals in maintaining soil characteristics is shown most dramatically when communities are disturbed. The development of the Dust Bowl in the midwestern part of the United States during the 1930s provides a vivid example. The Dust Bowl area is normally dry and windy, but the root systems of the natural vegetation, mostly perennial grasses, are extensive enough to hold the soil in place. When the prairies were converted to agriculture, the perennial grasses were replaced by annual crops, with less extensive root systems, that were plowed up each year. A series of dry years reduced crop growth and turned the soil surface into fine dust. The result, shown in Figure 4-1, is legendary.

Plants also influence movement of water. Rain does not accumulate where it falls. If it did, New York State would be under 200 feet of water within a lifetime. Some water flows over the soil surface or through the underlying earth to enter rivers, lakes, and, eventually, the ocean. The remainder escapes by evaporation from the ground surface and vegetation. The leaf area of an eastern deciduous forest is, on the average, about four times the area of the ground surface; that is, there are about four acres of leaf surface per acre of forest floor. Plants are thus the major pathway of evaporation. When a forest is cut, most of the water that normally would have evaporated from the leaves flows instead into rivers. The consequences of clear-cutting without provision for extensive replanting are flooding, increased erosion and the silt

Figure 4-1   The Dust Bowl area of the midwestern United States. Wind erosion begins when soils are plowed but there is insufficient rain for crop growth. Above, a winter-wheat crop failure in Finney County, Kansas, has resulted in soil blowing (March 1954). Below, wind-blown soil particles—a dust storm— approaches Springfield, Colorado, in May 1937, during the height of the Dust Bowl tragedy. Dust storms completely destroyed some farming areas (photographs courtesy of the U.S. Soil Conservation Service).

deposition that accompanies it, and removal of mineral nutrients from the denuded soil. As we have already seen (page 29), vegetation has a profound influence on the local heat budget of the atmosphere and hence on local precipitation. By fostering increased rainfall, the development of vegetation, particularly in dry climates, is self-reinforcing.

The interdependence of the physical and biological realms is the basis of the *ecosystem concept* in ecology. In spite of the ecosystem's being the most encompassing unit of ecology, the term itself was not used until 1935, when it was coined by the English botanist A. G. Tansley. The ecosystem, he wrote, includes ". . . not only the organism-complex, but also the whole complex of physical factors forming what we call the environment of the biome—the habitat factors in the widest sense. Though the organism may claim our primary interest, when we are trying to think fundamentally we cannot separate them from their special environment, with which they form a physical system."

The biotic and abiotic parts of the ecosystem are linked by a constant exchange of material through *nutrient cycles* driven by energy from the Sun. The basic pattern of energy and material flux in the ecosystem is shown in Figure 4-2. Plants manufacture organic compounds, utilizing energy obtained from sunlight and nutrients from soil and water. The plants use these compounds as a source of building materials for their tissues and as a source of energy for their maintenance functions. To release stored chemical energy, plants metabolize organic compounds, breaking them apart into their

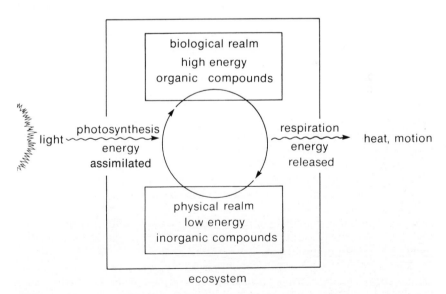

Figure 4-2 Schematic diagram of the flow of energy through the ecosystem and the cycling of chemical nutrients within the ecosystem. The biological and physical realms, represented by organic and inorganic compounds, together comprise the ecosystem.

original inorganic constituents—carbon dioxide, water, nitrates, phosphates, and so on—and thus completing the nutrient cycle.

All ecosystems are variations on a common theme. Forests, deserts, marshes, oceans, and streams all share the thermodynamic properties of ecosystems outlined in Figure 4-2, even though their structure and the details of their function differ greatly. It is this diversity that captures the imagination of those who study ecology, who strive to understand how fundamental principles in the physical and biological worlds interact with the heterogeneity of the Earth to produce the variety of ecosystems that surround us.

The Earth is a giant heat machine.

A clue to diversity in the biological world comes from the study of variation in the physical world itself, which can be considered as a model of more complicated biological systems. The surface of the Earth, its waters, and the atmosphere above it behave as a giant heat machine, obeying the same thermodynamic rules as do ecosystems and exhibiting a similar variety in time and space.

The simplest abstraction of a thermodynamic system might be a uniform hunk of rock in outer space. Lifeless and motionless, this alien world intercepts light energy emanating from the Sun, or some more distant star, and radiates energy into the black depths of space. Our rock behaves as a simple ecosystem: energy is assimilated, energy is transformed, energy is lost from the system. When the rock absorbs sunlight, the molecules in the rock are caused to move more rapidly, that is, the rock heats up. The energy in light is transformed to heat energy. But the hotter an object, the more rapidly it loses heat to its surroundings, in this case empty space. A law of thermodynamics states that the distribution of energy in the Universe tends to become more even with time; in other words, that energy moves from points of high concentration to points of low concentration. As our rock heats up, it re-radiates the energy it received from the Sun at a proportionately greater rate. When the rate of energy radiation has reached a level equal to the rate at which sunlight is received, the net energy balance of the rock is zero (energy loss equals energy gain), and the rock attains a constant steady-state temperature. And so it is with any thermodynamic model, including an ecosystem. Animals and plants liberate energy as heat by respiration at approximately the same rate that plants assimilate energy as light by photosynthesis. Just as the temperature of the rock measures the energy stored in the system at any moment, the chemical energy in plants, animals, and organic detritus measures the energy stored in the ecosystem. Both will tend to achieve a steady state if left unperturbed.

The surface of the Earth is much more complex than a homogeneous rock in space. As the Earth's surface varies from bare rock to forested soil, open ocean, and frozen lake, its ability to absorb sunlight varies as well, thus creating differential heating and cooling. As with our rock in space, the heat energy absorbed by the Earth is eventually radiated back into space, but not before undergoing further transformations that perform the work of evaporating water and contributing to the circulation of the atmosphere and oceans. All these factors result in tremendous varieties of physical conditions over the surface of the Earth, which, in turn, have fostered the diversification of ecosystems that we discover about us.

Variation in solar radiation with latitude creates major global patterns in temperature and rainfall.

The Earth's climate tends to be cold and dry toward the poles and hot and wet toward the Equator. Although this oversimplification has as many exceptions as a weather report, climate nonetheless does exhibit broadly defined patterns (Barry and Chorley 1970, Flohn 1968, Lowry 1969, Trewartha 1954).

Global variation in climate is determined largely by the position of the Sun relative to the surface of the Earth. The Sun exerts its greatest warming effect on the atmosphere, oceans, and land when it is directly overhead. The Sun's warmth is diminished when it lies close to the horizon and its rays strike the surface at an oblique angle. Not only does a beam of sunlight spread over a greater area when the Sun is low, it also travels a longer path through the atmosphere, where much of its light energy is either reflected or absorbed by the atmosphere and re-radiated into space as heat. The Sun's highest position each day varies from directly overhead in the tropics to near the horizon in polar regions; hence the warming effect of the Sun increases from the poles to the Equator. This uneven distribution of the Sun's energy over the surface of the Earth creates major geographical patterns in temperature, rainfall, and wind.

Warming air expands, becomes less dense, and thus tends to rise. Its ability to hold water vapor increases, and evaporation is accelerated. The rate of evaporation from a wet surface nearly doubles with each ten degrees Celsius* rise in temperature. The Sun heats the atmosphere most intensely in the tropics. The warmed surface air picks up water vapor and begins to rise. As the moisture-laden air gains altitude and cools, its water vapor condenses into thick clouds that drench the tropical landscape with rain. Daily cycles of

---

*Although we are accustomed to using Fahrenheit (F) temperatures in the United States, the Celsius (C) scale of temperature (sometimes referred to as the centigrade scale), is used throughout most of the world and almost exclusively in scientific work. The Celsius scale is much more convenient to ecologists than the Fahrenheit scale because it forms the basis of the commonly used heat scale of calories. The Celsius scale divides the temperature range into 100 degrees between the freezing point of water (zero degrees Celsius, or 0 C) and the boiling point (100 C). On the Fahrenheit scale, the corresponding temperatures are 32 F and 212 F, delimiting a 180-degree range between the freezing and boiling points of water. Each degree Celsius is therefore equal to 1.8 degrees Fahrenheit. Convenient conversion formulae are $C = 5/9$ $(F - 32)$ and, conversely, $F = 9/5 \, C + 32$. Some familiar bench marks on the Celsius scale are: 0 C (32 F), freezing point of water; 10 C (50 F), a cool day; 20 C (68 F), a mild day; 30 C (86 F), a warm day; 40 C (104 F), a very hot day. A very cold day is about -18 C (0 F). Room temperatures usually range between 20 and 25 C; normal body temperature of man is 37 C.

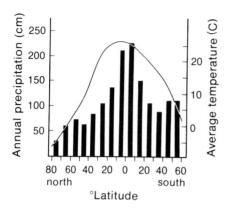

Figure 4-3   Average annual precipitation (vertical bars) and temperature (solid line) for 10° latitudinal belts within continental land masses. The figure represents averages for many localities, which obscures the great variation within each latitudinal belt (from data in Clayton 1944).

heating and cooling cause most tropical rain to fall during the afternoon and evening; in temperate areas, as well, summer thunder showers, resulting from strong vertical currents of warm, moist air, occur most often late in the day.

Because warm tropical air can hold much more water than temperate or arctic air, annual precipitation is greatest in tropical regions (Figure 4-3). The tropics are so wet not because more water occurs in tropical latitudes than elsewhere, but because water is cycled more rapidly through the tropical atmosphere. Cycles of evaporation and precipitation are driven by the Sun, and it is the source of energy, not the quantity of water, that primarily determines latitudinal patterns in rainfall. The distribution of continental land masses exerts a secondary effect. Rainfall is more plentiful in the Southern Hemisphere because oceans and lakes cover a greater proportion of its surface (81 per cent, compared with 61 per cent of the Northern Hemisphere, List 1966). Water evaporates more readily from exposed surfaces of water than from soil and vegetation.

Winds are driven by energy from the Sun, just as the cycling of water in the atmosphere. Indeed, the two cannot be separated, and wind patterns exert a strong influence on precipitation. The mass of warm air that rises in the tropics eventually spreads to the north and south in the upper layers of the atmosphere. It is replaced from below by surface-level air from subtropical latitudes (Figure 4-4). The tropical air mass that rose under the warming sun cools as its heat is radiated back into space. By the time it has extended to 30° north and south of the solar equator,* the cooled air mass becomes so dense that it begins to sink back to the surface. It is relatively dry because condensation has removed much of its water, which fell as rain over the tropical regions where the air current originated. Its capacity to evaporate

*The solar equator is the parallel of latitude that lies directly beneath the Sun. The position of the solar equator varies seasonally from 23° north latitude on June 21 (the summer solstice) to 23° south latitude on December 21 (the winter solstice). The solar equator coincides with the Earth's geographical equator at the equinoxes (March 21 and September 21).

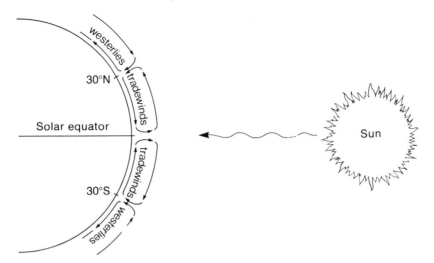

Figure 4-4   Simplified diagram of convection currents driven by the Sun's energy in the atmosphere of tropical and subtropical latitudes.

and hold water increases further as it sinks and warms. As the air strikes the Earth's surface in subtropical latitudes and spreads to the north and south, it draws moisture from the land, creating zones of arid climate.

Convection currents in the atmosphere, driven by the Sun's energy, redistribute heat and moisture about the surface of the Earth. The region of rising air in the tropics—the doldrums—is one of high rainfall. Conversely, descending air robs the land of water, which is transported elsewhere by wind currents. The tradewinds, blowing steadily toward the Equator from the dry horse latitudes, carry moisture picked up along the way into the tropics. Just as warm tropical air rises, cold air masses over the north and south polar regions descend and flow along the surface toward lower latitudes. When cold air meets a warmer air mass moving poleward across temperate latitudes, the warm moist air rises above the denser polar air and cools, bringing precipitation.

Precipitation is distributed over the surface of the Earth in such a way that most wet regions occur close to the Equator, and the major deserts occupy a belt centered about 30° latitude north and south of the Equator (Figure 4-5). Great names in deserts—the Arabian, Sahara, Kalahari, and Namib of Africa, the Atacama of South America, the Mohave and Sonoran of North America, and the Australian—all belong to regions within these belts.

Exceptions to this pattern are caused by major land masses. Mountains force air upward, causing it to cool and lose its moisture as precipitation on the windward side of the range. As the air descends the leeward slopes of the mountains and travels across the lowlands beyond, it picks up moisture and creates arid environments called *rain shadows* (Figure 4-6). The Great Basin deserts of the western United States and the Gobi Desert of Asia lie in the rain shadows of extensive mountain ranges.

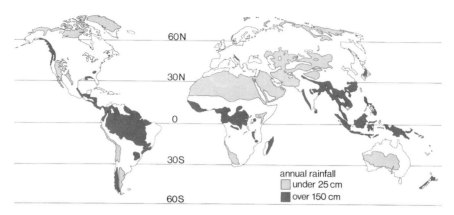

Figure 4-5   Distribution of the major deserts (regions with less than 25 centimeters [10 inches] annual precipitation) and wet areas (having more than 150
centimeters [80 inches] annual precipitation) (after Espenshade 1971).

The interior of a continent is usually drier than its coasts simply because
the interior is farther removed from the major site of water evaporation, the
surface of the ocean. Furthermore, coastal (maritime) climates are less variable than interior (continental) climates because the tremendous heat-
storage capacity of water reduces temperature fluctuations. For example, the
difference between the hottest and coldest mean monthly temperature near

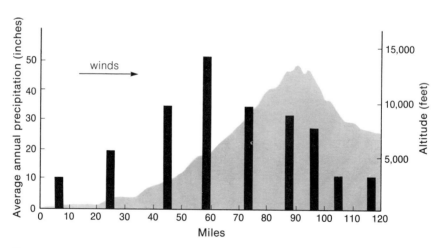

Figure 4-6   The influence of the Sierra Nevada mountain range on local
precipitation and in causing a rain shadow to the east. Weather comes predominantly from the west (left) across the Central Valley of California. As moisture-
laden air is deflected upwards by the mountains, it cools and its moisture
condenses, resulting in heavy precipitation on the western slope of the
mountains. As the air rushes down the eastern slope, it warms and begins to pick
up moisture, creating arid conditions in the Great Basin.

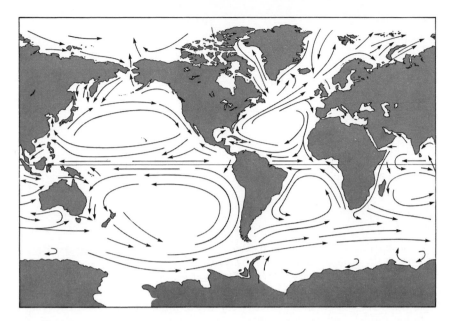

Figure 4-7   The major ocean currents. Water movement generally proceeds clockwise in the Northern Hemisphere and counterclockwise in the Southern Hemisphere (after Duxbury 1971).

the Pacific coast of the United States at Portland, Oregon, is 16 C (28 F). Farther inland, this range increases to 18 C (33 F) at Spokane, Washington; 26 C (47 F) at Helena, Montana; and 33 C (60 F) at Bismarck, North Dakota.

Ocean currents also play a major role in transferring heat over the surface of the Earth. In large ocean basins, cold water tends to move toward the tropics along the western coasts of the continents, and warm water tends to move toward temperate latitudes along the eastern coasts of continents (Figure 4-7). The cold Humboldt Current moving north from the Antarctic Ocean along the coasts of Chile and Peru is partly responsible for the presence of cool deserts along the west coast of South America right to the Equator, though these regions also lie within the rain shadow of the Andes Mountains. Conversely, the warm Gulf Stream, emanating from the Gulf of Mexico, carries a mild climate far to the north into western Europe.

The changing seasons.

Although we may characterize a region's climate as hot or cold and wet or dry, regular cycles of change are as important aspects of climate as long term averages of temperature and precipitation. Periodic cycles in climate are based upon cyclical astronomical events: the rotation of the Earth upon its axis causes daily periodicity in the environment; the revolution of the Moon around the Earth determines the periodicity of the tides; the revolution of the Earth around the Sun brings seasonal change.

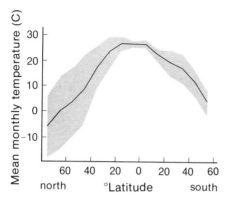

Figure 4-8   Annual range of mean monthly temperatures as a function of latitude. Note that seasonal variation is reduced in the Southern Hemisphere where there is a high ratio of water to land. The temperature range at a station may vary considerably from the mean for the latitudinal belt in which it is located (data compiled from Clayton and Clayton 1947).

The Earth's Equator is tilted slightly with respect to the path the Earth follows in its orbit around the Sun. As a result, the Northern Hemisphere receives more solar energy than the Southern Hemisphere during the northern summer, less during the northern winter. The seasonal change in temperature increases with distance from the Equator (Figure 4-8). At high latitudes in the Northern Hemisphere, mean monthly temperatures vary by an average of 30 C (54 F), with extremes of more than 50 C (90 F) annually; the mean temperatures of the warmest and coldest months in the tropics differ by as little as two or three degrees.

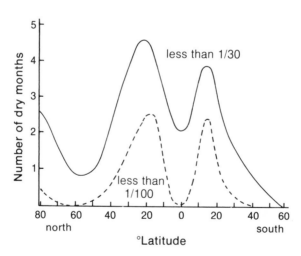

Figure 4-9   Seasonal distribution of rainfall as a function of latitude. The length and severity of the dry season are indicated by the number of months with rainfall less than a certain fraction (1/30 and 1/100) of the yearly total. If rainfall were evenly distributed throughout the year, 1/12 of the total would fall during each month (data compiled from Clayton and Clayton 1947).

Latitudinal patterns in rainfall seasonality are complicated by belts of wet and dry climate that move north and south with the changing seasons. Annual variation in rainfall is greatest in broad latitudinal belts lying about 20° north and south of the Equator (Figure 4-9). As the seasons change, these regions are alternately crossed by the solar equator, bringing heavy rains, and by the subtropical high-pressure belts, bringing clear skies.

Panama, at 9°N, lies within the wet tropics, but even there the seasonal movement of the solar equator profoundly influences the climate. The major tropical belt of high rainfall remains south of Panama during most of our winter, but it lies directly over Panama during the summer. Hence the winter is dry and windy, the summer humid and rainy. Panama's climate is wetter on the northern (Caribbean) side of the Isthmus, the direction of prevailing winds, than on the southern (Pacific) side. This rain-shadow effect is more pronounced in nearby western Costa Rica, where a high mountain range intercepts moisture coming from the Caribbean side of the Isthmus. The Pacific lowlands are so dry during the winter months that most trees lose their leaves. The tinder-dry forest and bare branches contrast sharply with the wet, lush, more typically tropical forest during the wet season (Figure 4-10).

Figure 4-10   Kiawe forest on the island of Maui, Hawaii, during the peak of the dry season. Note the complete absence of leaves at this time. The grasses of the forest understory are tinder-dry (photograph courtesy of the U.S. Forest Service).

Farther to the north, at 30°N in central Mexico, rainfall comes only during the summer when the solar equator reaches its most northward limit (Figure 4-11). During the rest of the year this region falls within the dry, subtropical high-pressure belt. The influence of the solar equator, bringing summer rainfall, extends into the Sonoran Desert of southern Arizona and New Mexico. This area also receives moisture during the winter from the Pacific Ocean, carried by the southwesterly winds emanating from the subtropical high-pressure belt farther south. Southern California is beyond the summer rainfall belt and has a winter-rainfall–summer-drought climate, often referred to as a Mediterranean climate.

The seas are warmed by the Sun just as the continents and the atmosphere are, but their great mass of water acts like a heat sink to dampen daily and seasonal fluctuations in temperature. Large seasonal changes in temperature are more often caused by seasonal movements of water masses of different temperature than by local heating and cooling. During the Panamanian dry season, roughly January to April, steady winds blowing in a southwesterly direction create strong *upwelling* currents in the Pacific Ocean along the southern and western coasts of Central America. During the upwelling period, warm surface water is blown away from the coast and cooler water moves upward from deeper regions to replace it. As a result, the annual range of water temperature on the Pacific coast of Panama is about three times that of the Caribbean coast (Rubinoff 1968).

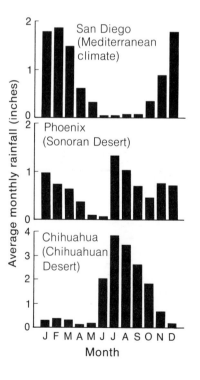

Figure 4-11   Seasonal occurrence of rainfall at three localities in western North America showing the summer rainy season of the Chihuahuan Desert, the winter rain—summer drought of the Pacific Coast (Mediterranean-climate type), and the combined climate pattern of the Sonoran Desert (rainfall data from Clayton 1944).

Small temperate-zone lakes are more sensitive than oceans to the changing seasons. Temperature cycles are an important force in the nutrient budgets of lakes because changes in temperature gradients from surface to bottom cause vertical mixing of the water twice each year, during the spring and the fall (Hutchinson 1957). In winter, the *temperature profile* of the lake is inverted, with the coldest water (0 C) at the surface, just beneath the ice. Because the density of water increases between freezing and 4 C, the warmer water sinks, and temperature increases (up to 4 C) with depth. In early spring, the Sun warms the surface of the water gradually. Until surface temperature reaches 4 C, surface water tends to sink into the cooler layers immediately below. This minor vertical mixing creates a uniform temperature distribution throughout the water column. Without thermal layering to impede mixing, surface winds cause deep vertical movement of water in early spring *(spring turnover),* bringing nutrients from regions of decomposition in the bottom sediments to the surface.

As the Sun rises higher each day and the air above the lake warms, surface water warms up faster than deeper water, creating a sharp zone of temperature change, called the *thermocline,* across which water does not mix. The warm surface water literally floats on the cooler water below. The depth of the thermocline varies with local wind patterns and with the depth and turbidity of the lake. Thermoclines occur anywhere between five and twenty meters below the surface; lakes less than five meters deep usually lack stratification. Thoreau's Walden Pond, in Concord, Massachusetts, develops a sharp thermocline between six and ten meters depth (Deevey 1942). In August, water temperatures decrease from 25 C at the top of the thermocline to 5 C at its bottom.

The thermocline demarcates an upper layer of warm water (the *epilimnion*) and a deep layer of cold water (the *hypolimnion*). Most of the primary production of the lake occurs in the epilimnion where sunlight is intense. Photosynthesis supplements mixing of oxygen at the lake surface to keep the epilimnion well aerated and thus suitable for animal life, but plants often deplete dissolved mineral nutrients, and thereby curtail their own productivity. The hypolimnion is cut off from the surface of the lake by sharp temperature *stratification,* and, being frequently below the *euphotic* zone of photosynthesis, animals and bacteria deplete the hypolimnion of oxygen, creating anaerobic conditions.

During the fall, surface layers of the lake cool more rapidly than deeper layers and, becoming heavier than the underlying water, begin to sink. This vertical mixing *(fall overturn),* persists into late fall, until the temperature at the lake surface drops below 4 C, and winter stratification ensues. Fall overturn produces greater vertical mixing of water than spring overturn because temperature differences in the lake are greater during summer stratification than during winter stratification. Fall overturn speeds the movement of oxygen to deep waters and rushes nutrients to the surface. Where the hypolimnion becomes fairly warm in midsummer, deep vertical mixing may take place in late summer when temperatures are still warm enough for plant growth. Infusion of nutrients into surface waters at this time often causes a later burst of phytoplankton population increase—the *fall bloom.* In deep cold lakes, vertical mixing does not penetrate to all depths until late fall or early winter when water temperatures are too cold to support plant growth.

Topographic and geologic influences superimpose local variation on global patterns.

Variation in topography and geology can create variation in the environment within regions of uniform climate. In hilly areas, the slope of the land and its exposure to the Sun influence the temperature and moisture content of the soil. Soils on steep slopes are well drained, often creating conditions of moisture stress for plants when the soils of nearby lowlands are saturated with water. In arid regions, stream bottomlands and seasonally dry river beds often support well-developed forests, which contrast sharply with the surrounding desert vegetation. Plant communities on shady and sunny sides of mountains and valleys frequently differ in accordance with the temperature and moisture regimes of each exposure. South-facing slopes are exposed to the direct-heating effect of the Sun, which limits vegetation to shrubby, drought-resistant *(xeric)* forms. The corresponding north-facing slopes are relatively cool and wet and harbor moisture-requiring *(mesic)* vegetation (Figure 4-12).

Figure 4-12   The effect of exposure on the vegetation of a series of mountain ridges near Aspen, Colorado. The north-facing (left-facing) slopes are cool and moist, permitting the development of spruce forest. Shrubby, drought-resistant vegetation grows on the south-facing slopes.

Figure 4-13   Even in the tropics one may find communities dominated by cold temperatures. Biological communities stop abruptly at the snow line, at about 5,000 meters elevation in central Peru.

Air temperature decreases with altitude by about 6 C for each 1,000-meter increase in elevation (Holdridge 1967). Even in the tropics, if one could climb high enough, one would eventually encounter freezing temperatures and perpetual snow. Where the temperature at sea level is 30 C, freezing temperatures would be reached at about 5,000 meters (16,000 feet). This, indeed, is the approximate altitude of the snow line in the Andes of central Peru (Figure 4-13).

A 6 C drop in temperature corresponds, in temperate latitudes, to an 800-kilometer (500-mile) increase in latitude. In many respects, the climate and vegetation of high altitudes resembles that of sea level localities at higher latitudes (Billings and Mooney 1968). Despite their similarities, however, alpine environments are usually less seasonal than their low-elevation counterparts at higher latitudes, even though average temperature and annual rainfall may be similar. Temperatures in tropical montane environments remain nearly constant over the year, and the occurrence of frost-free conditions at high altitudes allows many tropical plants and animals to live in the cool environments found there.

In the mountains of the southwestern United States, changes in plant communities with elevation create more or less distinct belts of vegetation, referred to as *life zones* by the early naturalist C. H. Merriam (1894). Merriam's scheme of classification included five broad zones that he named, from south to north (or low to high elevation): Lower Sonoran, Upper Sonoran, Transition, Canadian (or Hudsonian), and Arctic-Alpine.

At low elevations in the southest, one encounters a cactus and desert-shrub association characteristic of the Sonoran Desert of northern Mexico and southern Arizona (Figure 4-14). In the woodlands along stream beds, plants and animals have a distinctly tropical flavor. Many hummingbirds and flycatchers, ring-tailed cats, jaguars, and peccaries make their only temperate-zone appearances in this area. At 2,500 meters (8,200 feet) higher, in the Alpine Zone, we find a landscape resembling the tundra of northern Canada and Alaska. By climbing 2,500 meters, we experience changes in climate and vegetation that would require a journey to the north of 2,000 kilometers (1,250 miles), or more, at sea level.

Local variation in the bedrock underlying a region promotes the differentiation of soil types and enhances biotic heterogeneity. In the northern Appalachian Mountains and in mountains near the Pacific coast of the United States, outcrops of serpentine (a kind of igneous rock) produce soils with so much magnesium that species of plants characteristic of surrounding soil types cannot grow (Whittaker 1954, Walker 1954). Serpentine *barrens,* as they are called, are usually dominated by a sparse covering of grasses and herbs, many of which are distinct *endemics* (species found nowhere else) that have evolved a high tolerance for magnesium (Figure 4-15). Depending on the composition of the bedrock and the rate of weathering, granite, shale, and sandstone also can produce a barren type of vegetation. The extensive pine barrens of southern New Jersey occur on a large outcrop of sand, which produces a dry, acid, infertile soil capable of supporting no more than knee-high pygmy forests of pines (McPhee 1968, McCormick 1970). Physical characteristics of the soil and of the underlying rock also influence drainage and the ability of the soil to hold moisture. The extensive pine forests found on the coastal plain of the southeastern United States grow on sandy soils that drain too well to support the growth of most broad-leaved trees. Climate, too, plays an important role in the weathering of rock and the formation of soils; in temperate and arctic regions of the Northern Hemisphere, glacial activity during the last 100,000 years has influenced soil characteristics over vast areas (Bunting 1967). We shall consider the influence of climate on the development of soil properties in more detail in the next chapter.

Integrated descriptions of climate stress the interaction of temperature and availability of water.

We find it easier to dissect climate into its component properties of temperature, humidity, precipitation, wind, and solar radiation, than to appreciate at once all the implications of these factors for the ecosystem. We must, however, understand *interactions* among climate factors because these factors are clearly interdependent in their effect on life. For example, seasonal rainfall promotes plant growth more strongly during warm months than during cold months. Wind movement and solar radiation interact with temperature to determine thermal stress; temperature and humidity together influence water balance.

Hudsonian Zone
Elevation 8,500 feet

Alpine Zone
Elevation 11,000 feet

Upper Sonoran Zone
Elevation 5,000 feet

Transition Zone
Elevation 6,500 feet

Lower Sonoran Zone
Elevation 3,000 feet

Upper Sonoran Zone
Elevation 4,000 feet

Figure 4-14   Vegetation types corresponding to different elevations in the mountains of southeastern Arizona. Lower Sonoran vegetation is mostly saguaro cactus, small desert trees such as paloverde and mesquite, numerous annual and perennial shrubs, and small succulent cacti. Agave, ocotillo, and grasses are conspicuous elements of the Upper Sonoran Zone, with oaks appearing toward its upper edge. Large trees are predominant at higher elevations: ponderosa pine in the Transition Zone; spruce and fir in the Hudsonian Zone. These gradually give way to bushes, willows, herbs, and lichens in the Alpine Zone above the treeline (courtesy of the U.S. Soil Conservation Service, U.S. Forest Service, W. J. Smith, and R. H. Whittaker, from Whittaker and Niering 1965).

Figure 4-15   A small serpentine barren in eastern Pennsylvania. The soils surrounding the barren support oak-hickory-beech forest.

The *climograph* portrays seasonal changes in temperature and rainfall simultaneously. The climograph has a rainfall scale (horizontal axis) and a temperature scale (vertical axis); each month is plotted on the graph according to its average temperature and rainfall. Seasonal progression of climate is portrayed on the climograph by following the points for each month of the year in succession (Figure 4-16). The horizontal spread of months represents seasonal variation in rainfall; the vertical spread, variation in temperature.

The climograph permits a visual comparison of climates at different localities. We note immediately that the seasons in Panama bring marked variation in rainfall but little change in temperature, whereas the reverse characterizes New York City. During no months are the climates of the two localities similar, although July and August conditions in New York approach the April climate in Panama. Moving east to west across the United States from Cincinnati, Ohio, to Winnemucca, Nevada, climate becomes more arid but temperatures remain within the same range. The change in vegetation from deciduous, broad-leaved forest in Ohio, to short-grass prairie in Wyoming and desert shrubs in Nevada is thus dependent upon the water relations of different plant forms rather than temperature tolerance. Although San Diego's weather in January resembles that of Lincoln, Nebraska, in April, the overall climates differ as much as their vegetation. San Diego's Mediterranean-type climate, with hot, dry summers and cool, moist winters, favors slow-growing, drought-resistant shrubs (chaparral), while Lincoln is (or rather was) surrounded by tall-grass prairie. Summer rainfall in the Great

Plains states supports greater plant productivity than the winter rainfall of the West Coast because water is abundant on the prairies during the warm summer months, the best growing season. The winters, however, are too cold and dry to support shrubby, evergreen vegetation.

Although the climograph is useful for comparing localities, it fails to combine the effects of temperature and rainfall in any biologically meaningful way, and it does not show the cumulative effects of weather upon the environment. For example, during dry seasons both evaporation and *transpiration* (evaporation of water from leaves) remove water from the soil. If rainfall is insufficient to balance evaporation and transpiration losses, the water deficit in the soil steadily increases, perhaps for months at a time. In other words, soil water reflects last month's rainfall as well as more recent input to the soil.

In 1948, geographer C. W. Thornthwaite published a method for utilizing climate data to estimate the seasonal availability of water in the soil. He compared the rate at which water is drawn from the soil by plants and by direct evaporation with the rate at which it is restored by precipitation. The

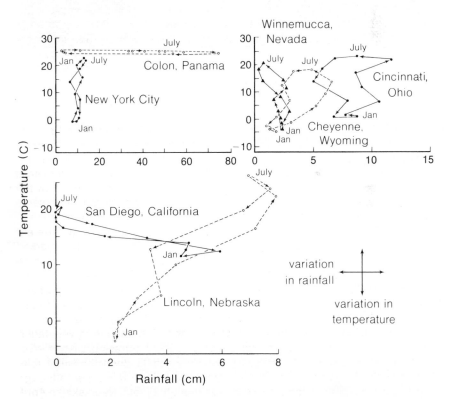

Figure 4-16   Climographs of representative localities in North America. Each point represents the mean temperature and rainfall for one month. Lines connecting the months at each locality indicate seasonal change in climate (data from Clayton and Clayton 1947).

sum of evaporation and transpiration is the total *evapotranspiration* of the habitat. Evaporation and transpiration increase with temperature by a factor of nearly two for each 10 C rise in temperature, other things being equal, although the character of the soil and vegetation cover also influences water loss from the soil. In natural environments, evapotranspiration is at times limited by the availability of water in the soil. Potential evapotranspiration, which represents the amount of water that would be drawn from the soil if soil moisture were not limiting, can be calculated from temperature and precipitation (Ward 1967). When precipitation input to the soil (rainfall minus surface runoff) exceeds potential evapotranspiration at all seasons, the soil will remain saturated with water throughout the year, as at Brevard, North Carolina (Figure 4-17). At Bar Harbor, Maine, precipitation falls below potential evapotranspiration during the warm summer months, and the soil is depleted of water during late summer and early fall. Canton, Mississippi, receives rainfall similar to that of Bar Harbor, but the hotter climate of Mississippi increases the potential evaporation of water from its soils, and serious water deficits are incurred during the summer months. Manhattan, Kansas, receives much less rain than Canton, Mississippi, but because the rainfall is concentrated during the summer period of maximum potential evapotranspiration, soil-water deficits are no more serious than in Mississippi. On the other hand, Grand Junction, Colorado, has a dry climate where soils are depleted of water most of the year and rarely become saturated. Plant productivity is correspondingly low.

Because potential evapotranspiration increases with temperature, temperature and water stress go hand in hand. Thus, boreal regions receiving 25 to 50 centimeters of precipitation each year have a more favorable water budget for plant production than tropical regions with similar levels of precipitation.

Thornthwaite's analysis may lack the detail to predict local variation in soil moisture and plant production, but his graphs indicate the relative length and severity of seasonal drought. By keeping a running seasonal balance of gains and losses of water, one can appreciate the cumulative effects of climate on soil moisture.

Irregular fluctuations in the environment are superimposed on periodic cycles.

Major geographic and seasonal trends in climate can be related to physical factors whose influences gather at a particular point in time and space. The average, or expected, conditions at this point are understandable in terms of simple principles governing physical forces acting in a heterogeneous physical world. But nature has an unexpected element as well.

Everyone knows that the weather is difficult to predict far in advance. We often remark that a year was particularly dry or cold compared with others. Almost all climatic factors are unpredictable to some extent. Rainfall is most variable where it is sparsest. At a given locality, mean monthly precipitation varies most during the driest season. Year-to-year variation in temperature on

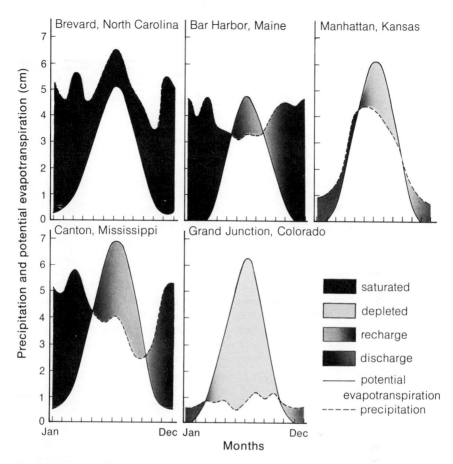

Figure 4-17   The relationship of precipitation and potential evapotranspiration to changes in the availability of soil moisture. When evapotranspiration exceeds precipitation, water is withdrawn from the soil until the deficit exceeds ten centimeters (four inches), the average amount of moisture that soils can hold (after Thornthwaite 1948).

a particular date is greatest where the annual fluctuation in temperature is greatest. The most extreme conditions are also generally the rarest, but these may be critically important in shaping the adaptations of organisms, which must be prepared for all possible contingencies. Some events—earthquakes, tornadoes, volcanic eruptions, and hurricanes—are truly disasters for life, beyond the response mechanisms of organisms, and their general effect on the biological community is totally destructive. Even some of the most constant environments do not escape occasional disastrous conditions. The rich Peruvian fishing industry, as well as some of the world's largest seabird colonies, thrives on the abundant fish in the rich waters of the Humboldt Current, an antarctic mass of cold water that flows up the western coast of

South America and finally veers offshore at Ecuador, where warm tropical inshore waters prevail. Each year a warm countercurrent known as El Niño (referring to the Christ child because it appears about Christmastime) moves down the coast of Peru, occasionally forcing the cold Humboldt Current offshore and taking with it the food supply of millions of birds. The ornithologist Robert Cushman Murphy (1936:102) described the effects of El Niño:

> The immediate result of an advance of El Niño is to raise the temperature of the littoral ocean water by five or more degrees Centigrade. The normal plankton of the cool Humboldt Current waters next succumbs, perhaps because of the increased temperature, perhaps in part because of a different composition of salts in the water. The common schooling fish leave the region or die, and less familiar species, such as flying fish, dolphins, and other tropical types, invade the shore waters and even enter harbors. Later, if the incursion of tropical waters is marked and widespread, disease attacks the population of cormorants, boobies, pelicans, and other guano birds belonging to the normal Humboldt Current fauna. Carcasses drift ashore in vast numbers, and the survivors of such species are driven southward.

El Niño also can have a profound effect on the terrestrial environment. In 1925, the current extended more than 1,500 miles south to Arica, Chile, where during March alone the rainfall was ten times the total of the preceding decade and brought a flush of biological activity to the usually lifeless Atacama Desert.

5

Soil Development

We take the dirt under our feet for granted, unwisely it would seem, because most of the vital mineral exchange between the biosphere and the inorganic world occurs in the soil. Plants obtain water and nutrients from the soil. When they die, they return to the soil where they are decomposed and their mineral nutrients released. Organisms responsible for decomposition—the myriad bacteria and fungi, the minute arthropods and worms, the termites and millipedes—abound in the surface layers of the soil where dead organic matter is freshest. Their activities contribute to the development of soil properties from above, while physical and chemical decomposition of the bedrock contribute to the soil from below.

As with climate, soil formation is determined by physical and chemical processes whose results are as varied as the conditions under which they occur. Soil characteristics vary greatly over the world and both influence and reflect the distribution of vegetation types (Bunting 1967, Eyre 1968). Five factors are largely responsible for variation in soils: climate, parent material, vegetation, local topography, and, to some extent, age (Brady 1974). In general, the decomposition and weathering of parent rock and the addition of organic material to the soil proceed most rapidly in warm, wet climates. As a result, the influence of parent rock on the structure and composition of soil *decreases* with increasing rainfall, temperature, and age.

Once formed, soils remain in a dynamic state. Some minerals are removed by ground water; others blow in as dust or are released by the decomposition of underlying rock layers. Although soil is in a constant state of flux, soils of most regions attain characteristic steady-state properties. In dry areas, rainfall is so sparse that chemical weathering of bedrock or other parent material is slow, and plant production is so low that little organic detritus is added to the soil. Soils of arid regions are typically shallow, and bedrock lies close to the surface (Figure 5-1). Weathering may extend to the depth of only one foot; such soils can be shallower or even absent where erosion removes weathered rock and organic detritus as rapidly as they are formed. The faces of cliffs and rocks in the upper regions of intertidal zones at the edge of the sea are extreme examples of sites where soil formation is prevented by erosion. Soil development is also stopped short on alluvial

Figure 5-1   Profile of a poorly developed soil in Logan County, Kansas, illustrating shallow soil depth and absence of soil zonation (courtesy of the U.S. Soil Conservation Service).

deposits, where the weathering process does not have a chance to work owing to the fresh layers of silt deposited each year by floodwaters. At the other extreme, parent material is most highly weathered in parts of the humid tropics, where chemical alteration of the parent material may extend to depths of 200 feet or more. Most temperate soils are intermediate in depth, usually extending to a few feet (Mohr and Van Baren 1954, Leopold *et al.* 1964).

Soil horizons reflect the changing influence of soil-forming factors with depth.

Where a recent roadcut or excavation exposes the soil in cross section, one is often struck by the presence of distinct layers, called *horizons* (Figure 5-2). Soil horizons have been described with complex and sometimes conflicting terminology by soil classifiers. A generalized, and somewhat simplified, soil profile has four major divisions: O, A, B, and C horizons, with two

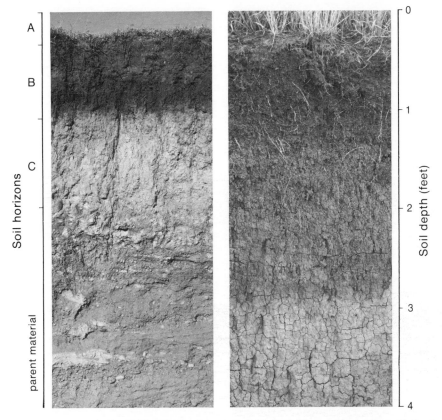

Figure 5-2   Soil profiles from the central United States illustrating distinct horizons. The profile at left, from eastern Colorado, is weathered to a depth of about two feet where the subsoil contacts the original parental material, consisting of loosely aggregated, calcium-rich, wind-deposited sediments (loess). The A_1 and A_2 horizons are not clearly distinguished except that the latter is somewhat lighter-colored. The B horizon contains a dark band of redeposited organic materials which were leached from the uppermost layers of the soil. The C horizon is light-colored and has been leached of much of its calcium. Some of the calcium has been redeposited at the base of the C horizon and at greater depths in the parent material. The profile at right is that of a typical prairie soil from Nebraska. Rainfall is sufficient to leach readily soluble ions completely from the soil. Hence there are no B layers of redeposition, as in the drier Colorado soil at left, and the profile is more homogeneous. The A horizon is weakly subdivided into a darker upper layer and lighter lower layer. The weathered soil lies upon a parent material composed of loess, the wind-blown remnants of glacial activity. The depth scale, in feet, at right, applies to both profiles; the soil horizons, at left, apply only to the left-hand profile (courtesy of the U.S. Soil Conservation Service).

subdivisions of the A horizon. Arrayed in order descending from the surface of the soil, the horizons and their predominant characteristics are:

O   primarily dead organic litter. Most soil organisms are found in this layer.

A_1 a layer rich in humus, consisting of partly decomposed organic material mixed with mineral soil.

A_2 a region of extensive *leaching* (or *eluviation*) of minerals from the soil. Because minerals are dissolved by water (mobilized) in this layer, plant roots are concentrated here where the minerals are most readily available.

B   a region of little organic material whose chemical composition closely resembles that of the underlying rock. Clay minerals and oxides* of aluminum and iron leached out of the overlying A_2 horizon are sometimes deposited here (*illuviation*).

C   primarily weakly weathered material, which closely resembles the parent rock. Calcium and magnesium carbonates accumulate in this layer, often forming hard, impenetrable layers within the C horizon.

The soil horizons demonstrate the decreasing influence of climate and the increasing influence of bedrock with increasing depth. Soil formation is greatly complicated, however, by the movement of mineral elements upward and downward through the soil profile. Before considering these processes in detail, we shall examine the initial weathering of the bedrock and how it influences soil characteristics.

Weathering is the physical and chemical breakdown of rock material near the Earth's surface.

Decomposition of rock at the bottom of a soil profile, or on a newly exposed rock surface, is fostered by the action of both physical and chemical agents. Repeated freezing and thawing of water in crevices breaks up rock into smaller pieces and exposes new surfaces to chemical action. Initial chemical alteration of the rock occurs when water dissolves some of the more soluble minerals, particularly sodium chloride (NaCl) and calcium sulfate ($CaSO_4$), and leaches them from the soil profile. Other minerals, particularly the oxides of titanium, aluminum, iron, and silicon, do not dissolve readily and are thus resistant to leaching under most conditions.

The weathering of granite exemplifies some basic processes of soil formation. Granite is an igneous rock formed when the less dense molten material deep within the earth rose to the surface, cooled, and crystallized. Granite consists chiefly of three minerals: feldspar, mica, and quartz. Feldspar, which consists of aluminosilicates of potassium ($K_2O \cdot Al_2O_3 \cdot 6\,SiO_2$), weathers rapidly owing to the removal of potassium (K) in the pres-

*Oxides are compounds consisting of oxygen and one or more other elements.

ence of carbonic acid (Eyre 1968, Buol *et al.* 1973). (Carbonic acid [H_2CO_3], formed when carbon dioxide dissolves in water, is always present in rain water.) The remainder of the feldspar mineral is reorganized with water to form one of several types of silicate clays, such as kaolinite ($Al_2O_3 \cdot 2 SiO_2 \cdot 2 H_2O$), depending on weathering conditions. As a general class of materials, clays perform an extremely important function in the soil, that of providing sites for ion exchange between the soil and plants. We shall look in detail at the role of clay particles later.

The mica grains in granite are composed of aluminosilicates of potassium, magnesium (Mg), and iron (Fe). When granite weathers, potassium and magnesium are removed rapidly, and the remaining oxides of iron, aluminum, and silicon form clay particles. Quartz, a form of silica (SiO_2), is relatively insoluble in acidic water and, therefore, remains more or less unaltered in the soil as sand grains. Changes in chemical composition of granite as it weathers from rock to soil are summarized in Figure 5-3. Calcium, magnesium, sodium, and potassium disappear quickly, while aluminum, silicon, and iron remain.

Removal of minerals from weathered granite rock varies greatly with climate. Similar parent materials in localities with progressively higher temperature and greater rainfall (for example, Massachusetts, Virginia, and British Guiana) exhibit greater loss of total rock volume and increased removal of specific elements, particularly silicon and potassium (Figure 5-4).

The decomposition of granite illustrates some of the principal chemical influences on soil formation, but weathering follows quite different courses on different types of bedrock. Pure quartz sand (silica) and pure limestone (calcium carbonate) do not produce clay readily because they lack iron and aluminum oxides; soil formation thus proceeds slowly unless other materials are mixed into the bedrock. Limestone frequently has a high percentage of clay particles originally derived from eroded soils. When such limestones weather, the calcium carbonate is readily leached, leaving a soil of high clay content. In general, the composition of the bedrock and its initial weathering

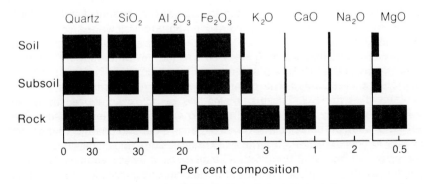

Figure 5-3   Percentage composition of soil layers and parent rock (granite) for each mineral in a soil profile from British Guiana (after Bunting 1967).

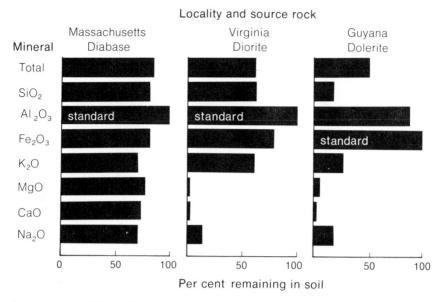

Figure 5-4   Differential removal of minerals from granitic rocks as a result of weathering in Massachusetts, Virginia, and British Guiana. Values are compared to either aluminum or iron oxides, which are assumed to be the most stable components of the mineral soil (after Russell 1961).

determine the relative amounts of clay and sand in derived soils. These qualities in turn influence the availability of mineral ions in the soil and the capacity of the soil to hold water.

The clay and humus content of soil determines its cation-exchange capacity.

Plants can obtain minerals from the soil in the form of dissolved ions, which are electrically charged atoms or compounds, whose solubility is determined by their electrostatic attraction to water molecules. Because ions dissolve in water, they would quickly be washed out of the soil if they were not strongly attracted to stable soil particles. Clay and humus particles, separately or associated in a complex, are large enough to form a stable component of the soil. These particles and complexes, referred to as *micelles,* actively play a role in the flux of mineral ions in the ecosystem. The surface of each particle has numerous negative electrical charges that attract positive ions (*cations*), such as calcium, magnesium, and potassium, and so retain them in the soil (Figure 5-5). The quantity of these negatively charged attachment sites in the soil is the *cation-exchange capacity* of the soil.

The role of the clay and humus particles in soil chemistry is not, however, so simple. The bonds between the mineral ions and the micelle are relatively

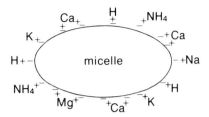

Figure 5-5   Schematic representation of a clay or humus particle (micelle) with hydrogen ions and mineral ions attracted by negative charges at its surface (after Eyre 1968).

weak, so they constantly break and re-form. When a potassium ion (K^+) dissociates from a micelle, its place may be taken by any other ion that is close by. Some ions cling more strongly to micelles than others. In order of decreasing tenacity, the common ions are hydrogen (H^+), calcium (CA^{++}), magnesium (Mg^{++}), potassium (K^+), and sodium (Na^+). Hydrogen ions thus tend to displace calcium and all other ions on the micelle. If ions were not added to or removed from the soil, the relative proportions of mineral ions associated with clay-humus particles would reach a steady state. But carbonic acid in rain water and organic acids produced by the decomposition of organic detritus continually add hydrogen ions to the upper layers of the soil; the hydrogen ions readily displace other minerals, which are then washed out of the soil and into ground water. The influx of hydrogen ions in water percolating through the soil is largely responsible for the mobility of ions in the soil and the differentiation of layers in the soil profile, as we shall see below.

Water retention is related to the size of soil particles.

In addition to determining the clay content of the soil, the mineral composition of the bedrock determines the size and abundance of sand grains and silt particles, collectively known as the *skeleton* of the soil. The materials of the soil skeleton are chemically inert, but they influence the physical structure of the soil and its water-holding capacity.

Water is sticky. The capacity of water molecules to cling to each other and to surfaces they touch underlies the familiar phenomena of surface tension and the rise of water against gravity in capillary tubes. Water clings tightly to surfaces of the soil skeleton. Because total surface area of particles in the soil increases as particle size decreases, silty soils hold more water than coarse sands, through which water drains quickly.

Water capacity is not equivalent to water availability. Plant roots easily take up water that clings loosely to soil particles by surface tension, but water near the surface of sand and silt particles is bound tightly to the soil particles by stronger forces. Soil scientists measure the strength with which the cells of root hairs can absorb water from the soil in terms of equivalents of atmospheric pressure. Capillary attraction holds water in the soil with a force equivalent to a pressure of one-tenth to one-fifth atmosphere, about 1.5 to 3.0 pounds per square inch. (Sea level atmospheric pressure is 14.7 psi.) Water attracted to soil particles with less force than one-tenth atmosphere (water in the middle of large interstices between soil particles, hence at great distance from their surfaces) drains out of the soil, under the pull of gravity, into the ground water in the crevices of the bedrock below. The amount of water held against gravity by forces of attraction greater than one-tenth to one-fifth atmosphere is called the *field capacity* of the soil.

A force equivalent to one-tenth atmosphere is sufficient to raise a column of water about three feet above the water surface. We know that plant roots can exert a much greater pull on water in the soil, because water rises in the tallest trees to leaves more than 300 feet above the ground. In fact, plants can exert a pull of about fifteen atmospheres on soil water and, therefore, can take up water held in the soil by forces weaker than fifteen atmospheres. Water close to the surface of soil particles is bound by forces of attraction that often exceed thirty atmospheres. Water held by forces greater than fifteen atmospheres is unavailable to plants, and the amount of water held by such forces is called the *wilting coefficient* or *wilting point* of the soil. (The wilting point is usually determined by using sunflower plants as the standard.) Once plants under drought stress have taken up all the water in the soil held by forces weaker than fifteen atmospheres, they can no longer obtain water, and they wilt, even though water remains in the soil.

As soil water is depleted, the remainder is held by increasingly stronger forces, on average, because a greater proportion of the water is situated close to the surfaces of soil particles. This relationship is shown in Figure 5-6 for a typical soil with a more or less even distribution of soil particle sizes from clay (up to 0.002 mm) through silt (0.002–0.05 mm) to sand (0.05–2.0 mm). Such soils are called *loams*. When saturated, this soil holds about 45 grams of water per 100 grams of dry soil (45 per cent water). The field capacity is about 32 per cent, and the wilting coefficient about 7 per cent. The available water is the field capacity minus the wilting coefficient, or 25 per cent, but water is more readily obtained by plants when the soil moisture is closer to the field capacity.

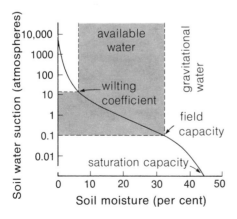

Figure 5-6   Relationship between water content of a loam soil and the average force of attraction of the soil water to soil particles (soil-water suction). The difference between the soil water content at field capacity (0.1 atmospheres) and the wilting coefficient (15 atm) is the water available to plants (after Brady 1974).

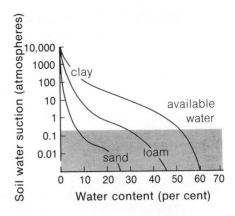

Figure 5-7 Relationship between soil-water content and soil-water suction for typical sand, loam, and clay soils. The available water (field capacity minus wilting coefficient) is greatest for the loam. Water availability is less for sandy soils because large pore spaces allow too free drainage, and for clay-based soils because the high particle surface area holds water too tightly (after Brady 1974).

In soils with predominantly smaller particles, the surface area of the soil skeleton is relatively large, and both the wilting coefficient and field capacity are higher; a correspondingly larger proportion of soil water is held by forces greater than fifteen atmospheres. Soils with predominantly larger skeletal particles have less surface area and larger interstices between particles. A larger proportion of the soil water is held loosely and is thus available to plants, but these soils have a lower field capacity. Availability of water to plants is maximum in soils with a variety of particle sizes between sand and clay (Figure 5-7).

Podsolization occurs when clay particles are broken down under acid-soil conditions.

Under mild, temperate conditions of temperature and rainfall, sand grains and clay particles are resistant to weathering and form stable components of the soil skeleton. Under conditions of high soil acidity, however, clay particles break down in the A horizon of the soil profile, and their soluble ions are transported downward and deposited in lower horizons. This process, known as *podsolization,* reduces the ion-exchange capacity, and, therefore, the fertility of the upper layers of the soil, by reducing the clay content.

Acid conditions are found primarily in cold regions where coniferous trees dominate the forests. The slow decomposition of acidic plant litter under conifer forests produces organic acids, which increase the acidity of the soil and promote breakdown of clay particles. In addition, rainfall usually exceeds evaporation during the year in regions of podsolization. Under these moist conditions, water continually moves downward through the soil profile, so there is little upward transport of new clay materials from the weathered bedrock below.

Podsolization advances farthest under spruce and fir forests in New England and the Great Lakes region and across a wide belt of southern and western Canada. A typical profile of a podsolized soil (Figure 5-8) has striking

Figure 5-8   Profile of a podsolized soil in Plymouth County, Massachusetts. The light-colored, eluviated A_2 horizon and the dark-colored, illuviated B_1 horizon immediately below it form distinct bands. Note the general absence of roots in the A_2 horizon compared with the lower B_1 horizon.

bands corresponding to regions of leaching (*eluviated* horizons) and redeposition (*illuviated* horizons). The topmost layers of the profile (O and A_1) are dark and rich in organic matter. These are underlain by a light-colored A_2 horizon, which has been leached of most of its clay content. As a result, the A_2 consists mainly of sandy skeletal material that holds neither water nor nutrients well. One usually finds a dark-colored band of deposition immediately under the eluviated A_2 horizon. This is the uppermost layer of the B horizon, where iron and aluminum oxides are redeposited, giving the layer a dark color. Other, more mobile minerals may accumulate to some extent in lower parts of the B horizon, which then grades almost imperceptibly into a C horizon and the parent material.

Laterization occurs when silica is leached from soil under alkaline conditions.

Whereas acidic conditions foster the breakdown and eluviation of clay minerals (iron and aluminum oxides), a basic or alkaline soil reaction facilitates the removal of silica (SiO_2) from the soil. Leaching of silica, called *laterization,* occurs primarily in tropical regions (Eyre 1968), where it is localized in the soils under humid forests (Sanchez 1973). Soil scientists do not fully understand why many tropical soils have low acidity (high alkalinity).

It is known, however, that plant detritus decays rapidly in the tropics, owing to high temperatures and moisture. Humic acids, therefore, do not persist in tropical soils so long as they do in cooler regions. Furthermore, decomposition of plant litter is accomplished primarily by bacteria, which produce no acid. The plant litter of cool forests is decomposed in part by fungi, which themselves produce acids to aid the chemical breakdown of organic detritus (Harley 1972).

Laterization has an effect on the soil profile opposite to that of podsolization. Removal of silica from the top layers of the soil increases the proportion of iron and aluminum oxides, which give tropical soils their characteristic red color. If laterization proceeds far enough, all the silica disappears from the soil, including that in clay particles, leaving behind a material called *laterite,* which is more like concrete than soil.

Laterization normally does not alter soil completely in undisturbed tropical forests. Organic humus particles, which accumulate in the upper layers of the soil, maintain a soil structure and cation-exchange capacity suitable for root growth and plant nutrition. Strongly laterized layers may, however, form deeper in the soil profile.

Disturbance of tropical forests can have disastrous effects where soil is prone to laterization. Removal of trees for agriculture or housing exposes the soil to the drying effects of the Sun. Evaporation of water from the ground surface frequently reverses the usual downward flow of water through the soil profile. Iron and aluminum oxides are then brought to the surface where they cement soil particles into a substance so hard that it is at best suitable for masonry. A completely laterized soil is nearly impervious to water and thus promotes surface runoff and erosion. Such disturbed soils are, of course, useless for agriculture, and their hardness and low water content slows the regeneration of natural vegetation.

Soils can become calcified and salinized in arid regions.

Under arid conditions, where evaporation exceeds rainfall (see Figure 4-17), water does not percolate completely through the soil. Calcium carbonate, dissolved in the topmost layers of the soil profile after a rainfall, is often redeposited in these same layers when water evaporates from the soil, or it may be transported downward to the lower limit of water penetration. The results of this process, called *calcification,* can be seen in the left-hand soil profile in Figure 5-2, in which a narrow, diffuse band of calcium carbonate (light-colored) has been redeposited about two feet below the soil surface. This horizon marks the lower limit of water percolation, immediately below which one finds relatively unweathered parent material. The depth of water penetration, hence the depth of the calcified layer, becomes less and less as rainfall diminishes.

Where ground water occurs close to the soil surface, dissolved minerals are drawn to the surface by evaporation and the upward pull of capillary movement. Evaporating water then leaves the minerals behind at the surface, sometimes forming thick crusts called *caliches* (Figure 5-9), which inhibit

Figure 5-9   An alkaline area devoid of plants in Chouteau County, Montana, where rising ground water has deposited a crust of calcium carbonate.

plant growth (Magistad 1945). A more serious reduction of soil fertility occurs when soluble neutral salts, such as sodium chloride (table salt, NaCl) and calcium sulfate ($CaSO_4$), accumulate in the soil and on its surface. This process, referred to as *salinization,* occurs where soil drainage is impeded and surface evaporation far exceeds percolation. Neutral salts reduce plant growth owing to the high salt concentration of soil water when it is available (saline soil conditions). When a salinized soil is relatively free of neutral salts but contains abundant sodium ion adsorbed to cation-exchange sites on soil micelles (sodic conditions), soil fertility is reduced twice over, first by increasing the alkalinity of soil and second by sodium toxicity (Russell 1961).

In many desert basins, ground water is close enough to the surface to be drawn upward by evaporation. The resulting caliche and salt deposits form the "dry lakes" that are widespread in the Mohave Desert and Great Basin of the western United States. Standing water on the surface in such regions is usually so full of dissolved minerals that it is undrinkable. (For many early pioneers who crossed the deserts, the choice between dying of thirst or alkali poisoning must have been difficult.)

Irrigation can, indeed, make a desert bloom. Dry soils become highly fertile when they are irrigated because of the high concentration of adsorbed mineral ions in the upper layers of the profile. But the rich agricultural returns are often cut short by speeded salinization of the soil. Irrigation water is ordinarily obtained from rivers that, in dry regions, are usually loaded with silt and dissolved salts. Most water added to the soil by irrigation eventually evaporates, leaving behind, near the soil surface, the salts it carried. The ultimate result is similar to what happens when water naturally enters the soil profile from underlying ground water. Salts accumulate rapidly, and the soil soon becomes too alkaline for agriculture.

The development of vegetation and soil go hand in hand.

The initial weathering of bedrock and the secondary alteration of the soil profile by podsolization, laterization, and salinization primarily influence the inorganic composition of the soil. Yet many important characteristics of the soil, including its humus content and the availability of nitrogen and phosphorus, are determined largely by vegetation (Crocker 1952, Jenny 1958). Soil changes that can follow removal of vegetation in the tropics show dramatically the role of vegetation in maintaining a steady state in the soil. Denudation quickly alters the movement of water through the soil and rapidly changes patterns of leaching and deposition.

Vegetation exerts its most dramatic influence on the development of soils where the underlying parent material is freshly exposed. Primary soil development occurs where geologic agents remove layers of existing soil or add sediments over the top of existing soil horizons. Since the recession of the glaciers from the Great Lakes region 10,000 to 12,000 years ago, the surface level of the Great Lakes has periodically lowered, leaving behind a chronological series of sand dunes at the southern end of Lake Michigan. The value of these dunes to the study of ecological processes was first recognized by the pioneering plant ecologist Henry C. Cowles, who, in 1899, described changes in vegetation observed on progressively older dunes. A newly formed dune consists largely of sand (silica). Water percolates rapidly through the dune and, because clays are absent, quickly leaches out any mineral nutrients. The dune environment excludes all but the most hardy plants. Marram grass (genus *Ammophila*) colonizes the sand at an early stage by sending out rhizomes (horizontal roots) from plants growing in better soil at the edge of the dunes (Figure 5-10).

Once grasses become established, they stabilize the dune and begin to add organic detritus to the dune surface. By building up the humus content of the sand, dune grasses encourage true soil development. Grasses and shrubs dominate the first century of plant succession on dunes. These are followed by the establishment of pine and its rapid replacement by black oak at an age of 150 to 200 years.

Changes brought about by vegetation, and the ultimate attainment of a steady state in the soil, were described by Olson (1958). Olson found that the humus added to the soil provided sites for cation exchange, just as the clay particles do in clay-based soils. Silt and clay particles eventually are added to the soil by wind deposition. The cation-exchange capacity of the soil increases rapidly for 500 to 1,000 years after dune stabilization, then levels off.

Figure 5-10   Marram grass growing on dunes in Indiana Dunes State Park, Indiana. At left, plants are seen extending out over fresh sand. At right, sand has been removed to expose the underground rhizomes by which the grass spreads.

Litter continues to accumulate on the forest floor, and humic acids eventually make the soil acidic. Hydrogen ions replace other cations (calcium, potassium, magnesium, etc.), until they occupy almost half the ion-exchange sites in the soil. As a result, soil fertility declines slowly, then levels off about 4,000 years after dune stabilization.

Crocker and Major (1955) have described soil development on areas bared by receding glaciers at Glacier Bay, Alaska. The retreat of the edge of the glacier had been recorded for a century, thus Crocker and Major knew the exact age of each of their study sites. Unlike the Lake Michigan sand dunes, sediments left behind by receding glaciers contained abundant calcium and clay. Vegetation established itself rapidly, and decaying plant detritus changed the hydrogen-ion concentration of the soil from slightly alkaline to slightly acid in twenty years. Each species of plant influenced soil acidity differently, however. Alder thickets acidified the soil rapidly, but willow and cottonwood did so slowly. In any case, increasing soil acidity accelerated the removal of calcium, while accumulating detritus steadily added to the inorganic nitrogen content of the soil. These changes in turn influenced the suitability of the habitat for different species of plants and fostered further changes in vegetation, eventually leading to spruce forest. The relationship between vegetation and soil properties during the early development of soil on newly exposed sites dramatizes the more general interactive roles of the physical and biological world and the dynamic nature of the ecosystem itself.

We have seen how variation in climate and geology affect soil properties. Geographical and temporal variations in all aspects of the physical world broadly influence the structure and functioning of the entire ecosystem, determining not only physical and chemical properties of the soils, but also levels of organic production, paths of energy flow and nutrient cycling, and the adaptations of plants and animals that give each habitat its characteristic appearance. In the next chapter, we shall examine the structure and appearance of terrestrial and aquatic communities as they are influenced by major patterns of climate and topography. This variety in the physical and biological world is derived from relatively few basic processes combined in different ways according to the particular nature of the environment, just as a few musicians can blend their individual contributions into an infinite variety of sounds.

6

The Diversity of Natural Communities

If man is anything, he is orderly. Natural history, and later ecology, grew out of schemes of classification, systems by which animals and plants could be given names in an orderly fashion based upon their similarities. After European botanists had named most of the local plant species by the end of the last century, they redirected much of their energy toward systems of classification for entire communities of plants. Most of these schemes were based on vegetation structure—height of vegetation, leaf or needle structure, deciduousness, and dominant plant form. Of course, these properties of plants are adaptations to the physical environment in which they live, and we should not be surprised, therefore, to note the close correspondence between vegetation zones and climate.

Major vegetation types are clearly discernible: tall forests, shrubland, and prairie are distinct; so are coniferous and broad-leaved forest types (*e.g.,* Figure 6-1). The problem with classifications of vegetation, or with similar schemes for aquatic communities, is that many intermediates occur. In fact, most biological communities intergrade, sometimes almost imperceptibly, as the physical environment changes from one locality to the next. One is tempted to deal with these intermediates by making finer distinctions between communities. This practice can, however, lead to a bewildering variety of names for plant communities. The somewhat conservative scheme of botanist A. W. Küchler (1964, 1967) lists 116 vegetation types in the United States alone.

Classifications of plant communities are most useful if they relate vegetation types to environment: temperature, rainfall, soil, and topography. In this chapter, we shall examine the influence of environment on vegetation structure and survey the diversity of natural communities by way of a photographic essay.

Structural schemes of classification are based on plant form.

The earliest traditions of vegetation classification were based on attempts to describe the most important plants of each major association (see

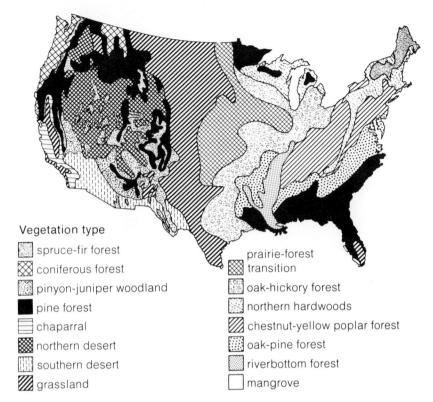

Vegetation type

- spruce-fir forest
- coniferous forest
- pinyon-juniper woodland
- pine forest
- chaparral
- northern desert
- southern desert
- grassland
- prairie-forest transition
- oak-hickory forest
- northern hardwoods
- chestnut-yellow poplar forest
- oak-pine forest
- riverbottom forest
- mangrove

Figure 6-1   Vegetation map of the United States (after Fowells 1965).

Schimwell 1971, Mueller-Dombois and Ellenberg 1974). These classifications embraced complete analysis of the *species* composition of communities (*floristic* analysis) on one hand, and description of plant *forms,* regardless of the particular species, on the other hand. Floristic analysis proved useful in restricted areas where botanists knew all the species and where minor differences between communities involved the replacement of species by similar ones with slightly different ecological requirements. But floristic analysis is completely unworkable on a global scale because biogeographical barriers restrict the distributions of individual species. Forests in Europe and the United States that are functionally similar have few species in common. Floristic differences between vegetational counterparts in California and Australia would be greater. The tropics pose still further problems for floristic analysis because of their great species diversity. Few botanists can recognize a majority of the hundreds of kinds of trees in a tropical forest, and to make matters more difficult, many species can be distinguished only when they are in flower or fruit.

Difficulties of floristic analysis for world-wide vegetation classification are partly overcome by analysis of form and function of plants rather than

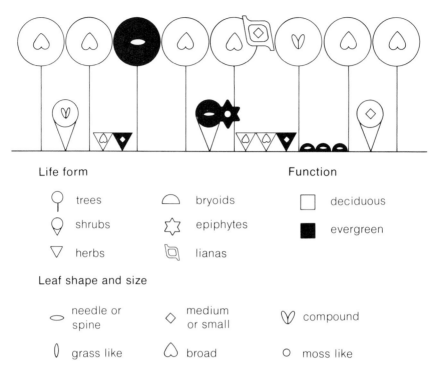

Life form

○̣ trees   ◠ bryoids

◯̣ shrubs   ☆ epiphytes

▽ herbs   ◌ lianas

Function

▢ deciduous

◼ evergreen

Leaf shape and size

◠ needle or spine   ◇ medium or small   ◊ compound

◗ grass like   ◠ broad   ○ moss like

Figure 6-2   Symbolic representation of a forest by the Dansereau method of vegetation classification (after Dansereau 1957).

their scientific names. Numerous sets of symbols were devised to describe such characteristics as plant size, life form, leaf shape, size, and texture, and per cent of ground coverage. Küchler (1949) worked out a system of letter and number symbols that could be combined into a formula describing the characteristics of any given plant formation. Thus M6iCXE5cD3i6H2pLlc represents an oak-yew woodland and E4hcD2rGH2rLlc(b), a madrone-holly scrub. But you have to know the players to follow the game. A similar symbolic method of description, devised by Pierre Dansereau (1957), portrays vegetation formations by use of lollipop- and ice-cream-cone–shaped figures with internal shading and symbols varying according to the nature of the plant (Figure 6-2). Although they emphasize plant form rather than floristics, the classifications of Küchler and Dansereau are primarily descriptive. They are too complex to be used as a hierarchical system, although they have found application where classification of only the predominant features of a plant formation is desired.

In 1903, the Danish botanist Christen Raunkiaer proposed to classify plants according to the position of their buds (regenerating parts), and found that the occurrence of his major categories corresponded closely to climatic

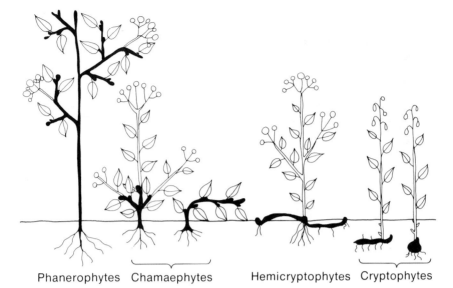

Phanerophytes Chamaephytes   Hemicryptophytes Cryptophytes

Figure 6-3   Diagrammatic representation of Raunkiaer's life forms. Unshaded parts of the plant die back during unfavorable seasons, while the solid black portions persist and give rise to the following year's growth. Proceeding from left to right, the buds are progressively better protected (after Raunkiaer 1937).

conditions (Raunkiaer 1934). He distinguished five principal life forms (Figure 6-3):

phanerophytes (from the Greek *phaneros,* visible) carry their buds on the tips of branches, exposed to extremes of climate. Most trees and large shrubs are phanerophytes. As one might expect, this plant form dominates in moist, warm environments where buds require little protection.

chamaephytes (from the Greek *chamai,* on the ground, dwarf) comprise small shrubs and herbs which grow close to the ground (prostrate life form). Proximity to the soil protects the bud. In regions of heavy snowfall, the buds are protected beneath the snow from extreme air temperatures. Chamaephytes are most frequent in cool, dry climates.

hemicryptophytes (from the Greek *kryptos,* hidden) persist through the extreme environmental conditions of the winter months by dying back to ground level where the regenerating bud is protected by soil and withered leaves. This growth form is characteristic of cold, moist zones.

cryptophytes are further protected from freezing and desiccation by having their buds completely buried beneath the soil. The bulbs of irises and

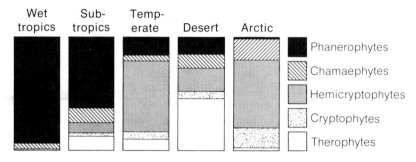

Figure 6-4   Proportion of plant life forms, classified according to Raunkiaer (1934, 1937), in various climatic regions (after compilations of Richards 1952, Dansereau 1957, Daubenmire 1968).

daffodils are representative of the regenerating buds of cryptophyte plants. Like hemicryptophytes, cryptophytes are also found in cold, moist climates.

therophytes (from the Greek *theros,* summer) die during the unfavorable season of the year and do not have persistent buds. Therophytes are regenerated solely by seeds, which resist extreme cold and drought. The therophyte form includes most annual plants and occurs most abundantly in deserts and grasslands.

The proportional occurrence of Raunkiaer's life forms in the floras of various climatic regions is summarized in Figure 6-4. Life form and climate go closely together. Phanerophytes dominate vegetation forms in warm, moist environments. They are progressively replaced by chamaephytes, hemicryptophytes, and cryptophytes in temperate and arctic regions. Deserts are distinctive in having a large proportion of therophytes.

The Holdridge classification defines plant communities by temperature and rainfall.

Botanist L. R. Holdridge (1967) proposed a classification of the world's plant formations based solely on climate (Figure 6-5). Holdridge considered temperature and rainfall to prevail over other environmental factors in determining vegetation, although soils and exposure may strongly influence plant communities within each climate zone.

Holdridge's scheme classifies climate according to the biological effects of temperature and rainfall on vegetation. As in Thornthwaite's analysis of climate (page 47), temperature and rainfall are seen as interacting to define humidity provinces. The dividing lines between humidity provinces are determined by critical ratios of potential evapotranspiration to precipitation. Insofar as potential evapotranspiration is a function of temperature, the humidity provinces relate temperature and rainfall to the water relations of

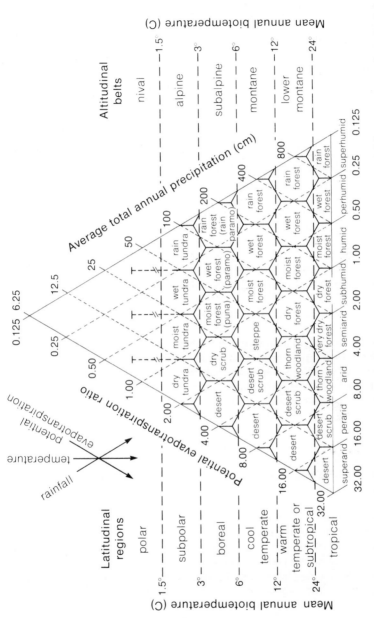

Figure 6-5   The Holdridge scheme for the classification of plant formations. *Mean annual biotemperature* is calculated from monthly mean temperatures after converting means below freezing to 0 C. The *potential evapotranspiration ratio* is the potential evapotranspiration divided by the precipitation; the ratio increases from humid to arid regions (after Holdridge 1967).

plants in a way that is meaningful. Holdridge's formula indicates, for example, that the availability of moisture to plants in wet tundra, with an annual rainfall of 25 centimeters (cm) and an average temperature near freezing, is similar to that in a wet tropical forest, with 400 cm precipitation and an average temperature of 27 C.

Holdridge related differences between plant formations to proportional differences between their climates. The temperature or rainfall of each zone is either twice or one half that of the adjoining zone. Thus 25 cm annual precipitation can make as big a difference in the vegetation of arid regions as 250 cm does between plant formations in the humid tropics. It seems intuitively reasonable that a little rainfall should stimulate desert annuals much more than it would rain-forest trees. Experiments under controlled conditions largely support this notion.

Holdridge also constructed his temperature scale with biological considerations in mind. He assumed that biological activity ceases below 0 C. The temperature of any month whose mean was below freezing was set at 0 C for calculating mean annual temperature. Furthermore, because small increases in temperature affect biological systems more at low temperatures than at high temperatures, Holdridge set the temperature boundaries of his life zones at 1.5, 3, 6, 12, and 24 C, each temperature being twice the previous one. A factorial scale of temperature is consistent with increases in rate of evaporation and rate of biological activity in relation to increasing temperature.

Simple climate classifications, like the Holdridge scheme, are far from ideal. In regions with similar mean rainfall and temperature, differing seasonal patterns of precipitation and temperature can create differences in vegetation structure. Topography, soil, and fire can also influence the development of vegetation types. It is fair to say, however, that climatic schemes of life zone classification do emphasize the pervasive influence of temperature and moisture on plant formations.

The structure of natural communities reflects climate.

If a random sample of terrestrial localities is placed on a graph according to the mean annual temperature and rainfall of each locality, most points fall within a triangular area whose three corners represent warm-moist, warm-dry, and cool-dry environments (Figure 6-6). Cold regions with high rainfall are conspicuously absent; water does not evaporate rapidly at low temperature, and the atmosphere in cold regions holds little water vapor. But owing to the depressing effect of low temperature on evaporation, a little water goes a long way. In the tropics, 50 cm (20 in) of rainfall can support little more than a desert scrub-type vegetation, but the same 50 cm permit the development of a passable coniferous forest in northern Canada.

Plant ecologist R. H. Whittaker (1975) has combined several structural classifications of plant communities into one scheme, which he has transposed onto a graph of temperature and rainfall (Figure 6-7). Within the tropical and subtropical realms, with mean temperatures between 20 and

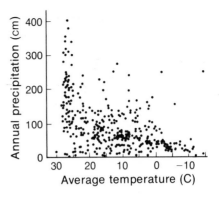

Figure 6-6 Average annual temperature and rainfall for a large sample of localities more or less evenly distributed over the land area of the Earth. Most of the points can be enclosed by a triangular region that includes the total range of possible climates on Earth, excluding high mountains (data from Clayton 1944).

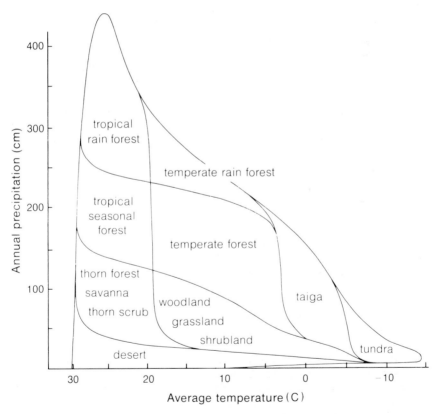

Figure 6-7   Whittaker's classification of vegetation types superimposed on the range of terrestrial climates. In climates intermediate between those of forested and desert regions, fire, soil, and climate seasonality determine whether woodland, grassland, or shrubland develops (from Whittaker 1975).

30 C, vegetation types grade from true rain forest, which is wet throughout the year, to desert. Intermediate climates support seasonal forests, in which some or all trees lose their leaves during the dry season (see Figure 4-10), and short, dry forests or scrublands with many thorntrees. As aridity increases, shrubs appear farther apart, exposing large patches of bare ground, a habitat characteristic most highly developed in true deserts.

The range of plant communities in temperate areas follows the same pattern as tropical communities, with the same basic vegetation types distinguishable in both. In colder climates, however, the range of precipitation from one locality to another is so narrow that vegetation types are poorly differentiated on the basis of climate. Where mean annual temperatures are below −5 C, Whittaker lumps all plant associations into one type—tundra. The whole scale of moisture gradients represented in the tropics is compressed into a narrow band in the arctic. Which is tundra: Rain forest or desert? Water abounds on moist tundra, but because it is frozen most of the year and permanently frozen a few feet below the soil surface, it is mostly unavailable to plants.

Toward the drier end of the rainfall spectrum within each temperature range, fire plays a distinct role in shaping the form of plant communities (Borchert 1950, Daubenmire 1968). For example, in the African savannas and midwestern American prairies, frequent fires kill the seedlings of trees and prevent the establishment of tall forests, for which favorable conditions otherwise exist. Burning favors the perennial grasses with extensive root systems that can survive fires. After an area has burned over, the grass roots send up fresh shoots and quickly revegetate the surface. In the absence of frequent fires, tree seedlings can become established and eventually shade out the prairie vegetation.

As in all classifications, exceptions to the scheme outlined in Figure 6-7 appear frequently. Boundaries between vegetation types are at best fuzzy. All plant forms do not respond to climate in the same way. For example, some species of Australian eucalyptus trees form forests under climate conditions that support only shrubland or grassland on other continents. Plant communities are affected by factors other than temperature and rainfall. We have already examined the influence of topography, soils, fire, and seasonal variation, but the point deserves emphasis.

Part 3

Adaptation

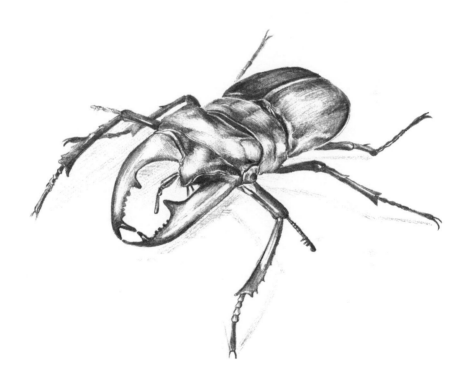

7

Environment and Adaptation: Some Examples

All species encounter a variety of environmental patterns deriving from many sources. One is exposure to physical and chemical factors in the environment. A second is exposure to predators, parasites, prey, and other species with which an individual interacts. A third source of environmental pattern is exposure to individuals of the same species, through sexual behavior, family ties, and social interaction. All these patterns in the environment mold the adaptations that are the structure and functioning of the organism.

In this chapter, we shall continue to examine environmental patterns, but more closely, by considering three species whose environments differ greatly. The first, the giant red velvet mite, lives in the harsh environment of the Mojave Desert of southern California; its activities are closely geared to physical factors in its environment, particularly to temperature and to rainfall. The next species, the chestnut-headed oropendola, a large relative of our blackbirds and grackles, lives in the lowland tropical forest of Panama. The oropendola is partner to a complicated arrangement evolved among a diverse group of interacting species. Unlike the harsh desert environment, the climate in wet tropical areas is highly favorable for the development of life; biological factors are an all-important part of the environment, and physical factors are relegated to less important roles as agents of natural selection. Our third example, that of the baboon on the savannas of Africa, is striking in its complex social life. As for all animals and plants, the physical and biological aspects of the baboon's environment have laid the foundations of its adaptations, but the rich social environment of the baboon has additionally shaped the behavior of the species.

In the harsh desert environment, the giant red velvet mite spends only a few hours aboveground each year.

The Mojave Desert of southern California is a forsaken area with little rain, searing summer heat, and chilling winter cold. These conditions are so forbidding that, except for a few struggling plants, the desert appears to be nearly devoid of life for most of the year. But the desert's silence and apparent sterility are occasionally broken during the milder days of winter by swarms of insects and other creatures that appear on the surface or fly above it for a few hours, and then disappear as mysteriously as they came. One of the more conspicuous of these creatures is the giant red velvet mite (Figure 7-1) whose scientific name, *Dinothrombium pandorae,* paints a vivid picture of this close relative of spiders. The generic name, *Dinothrombium,* is derived from the Greek *deinos,* meaning terrible, and *thrombos,* a lump or a clot. This particular species is named after the mythological Pandora, who was sent by Zeus to bring evil to the human race, as a counterbalance to the gift of fire that Prometheus, disobeying Zeus, had presented. For her dowry Pandora carried a box containing all the ills humans might suffer. When she opened the box out of curiosity, the ills escaped and have plagued us ever since.

Several years ago, biologists Lloyd Tevis and Irwin Newell (1962) began a study of the behavior of the mite in relation to the physical conditions of its environment. They found that the mites spend most of the year in burrows dug in the sand. The particular conditions that are favorable for the emergence of the giant red velvet mite occur infrequently in the Mojave Desert. During four years of observation, adult mites appeared aboveground only ten times, always during the cooler months of December, January, or February, when they can tolerate the temperatures on the desert's surface. Tevis and Newell could predict from their observations that an emergence would occur on the first sunny day after a rain of more than three-tenths of an inch, provided that air temperatures were moderate. An individual mite appeared only once each year.

On the day of a major emergence, the mites come out of their burrows between 9 and 10 A.M., and by late morning one may find thousands of mites scurrying across the desert sands in all directions. At midday, between 11:30 and 12:30, the mites dig back into the sand and wait until the following year before they emerge again.

The mites do not leave their burrows to terrorize unsuspecting desert travelers. During their two- or three-hour stay aboveground each year, they must perform two important functions: feeding and mating. On the same day the mites emerge, large swarms of termites appear, flying over the desert sand, their own emergence presumably triggered by the same physical factors that cause the mites to leave their burrows. It is upon these termites that the mites feed. A flying termite cannot, of course, be captured by the flightless mites; a mite must locate its prey after the termite drops to the ground and sheds its wings, but before it burrows into the sand. All this happens very quickly, giving the mites about an hour to find their prey.

Because the mites are solitary, they must mate during their brief period aboveground each year. Courtship behavior of the giant red velvet mite is similar to that of spiders and their relatives. The males walk nervously around and over a feeding female, tapping and stroking her, and cover the sand around her with loosely spun webs. Males court *feeding* females for two reasons. First, the females have ravenous appetites (they have not eaten for a year, after all) and would probably be just as likely to devour a male mite as a termite. Second, and perhaps more important, females can produce eggs only if they have had a meal. Thus, by mating with a feeding female, the male guarantees that his efforts to reproduce will not be wasted.

Figure 7-1   Adult giant red velvet mites on the surface of the ground (left) and in a vertical burrow (right, a cutaway view of the ventral side of the mite). (Photographs by Philip L. Boyd, Deep Canyon Research Center, courtesy of Lloyd Tevis, Jr.)

About midday, after feeding and mating have taken place, the mites congregate in troughs on the windward sides of sand dunes, where surface temperatures and the size of the sand particles are just right (less than one-half millimeter in diameter), and re-enter the sand almost simultaneously. The mites continue to dig their new burrows on the first day until the coolness of the late winter afternoon slows their activity. Burrowing continues on subsequent days when the sand becomes warm enough, until the burrows are completed. During the rest of the year, the adult mite spends its time moving up and down in its burrow to follow the movement of the preferred temperature zone as the surface of the sand is heated and cooled each day.

Eggs are laid during the early spring. They soon hatch, and the young mites crawl to the surface of the desert to search for a host to attach themselves to, usually a grasshopper. While they are growing, the young mites remain with their host, obtaining their nourishment from its body fluids. When they are full grown, the mites drop off their unwilling hosts, and seek suitable spots to dig their own burrows in the sand, thus renewing the life cycle of the giant red velvet mite.

Oropendolas and cowbird parasites form the basis for a complex interrelationship of species in the humid tropics.

The tropical surroundings of the chestnut-headed oropendola in Panama are a far cry from the rigorous environment of the giant red velvet mite. In the tropics air temperature varies little during the year, and abundant rainfall maintains the lush vegetation. Life abounds; diverse animals and plants are intricately interwoven into a rich fabric of biological interactions. In the harsh environments of deserts and the polar regions such interactions are noticeably simplified. In the tropics we are not so aware of adaptations to the physical environment, rather we are impressed by adaptations to biological environments.

The situation we are going to examine involves two birds, the chestnut-headed oropendola and its brood parasite the giant cowbird (Figure 7-2). As we shall see, two insects—a bot fly and a wasp—also play an integral part in the interaction between the two birds. The biological sleuthing that uncovered this story was performed by Neal Smith (1968), a staff biologist at the Smithsonian Tropical Research Institute in the Panama Canal Zone. Smith's insight, inventiveness, and perseverance proved a good match for the complexity of the oropendola-cowbird relationship.

Brood parasites (cowbirds and cuckoos are familiar ones) have been known for a long time. They are so named because the female lays her eggs in the nest of another species, the host, and the young are raised by the foster parents. Naturalists used to believe that the presence of a brood parasite in a nest always reduced the survival of the host young because parasites compete for the food brought in by the host parents. As one would expect, many potential hosts can detect the presence of parasite eggs in their nests and eject them (Rothstein 1971, 1975). To counter this defense, many brood parasite species have evolved an elaborate egg mimicry to fool the host species into accepting the alien egg as one of its own (Wickler 1968).

Smith was aware that nesting colonies of the chestnut-headed oropendola, usually consisting of anywhere from ten to a hundred nests, were often parasitized by the giant cowbird. He observed a curious phenomenon. Examining the eggshells that had been thrown out of the nests by the females after the young had hatched, Smith noted that under some colonies the eggshells of the cowbirds were distinctly different from those of the oropendolas, whereas in other colonies the eggshells of the two species so closely resembled each other that they could be distinguished only on the basis of shell thickness. Thus it appeared that in some colonies the cowbirds had evolved to mimic the eggs of their host, while in others they had not. These promising observations were to form the basis of a detailed study of brood parasitism.

Smith's first problem was to devise methods of studying bird nests suspended from the outlying limbs of large trees, twenty to sixty feet above his head. One generally avoids climbing trees in the tropics because of all the other creatures that might be climbing the same tree. And even if one could reach the top of a tree without an unpleasant encounter, climbing out on the small branches from which the nests are treacherously suspended would be challenging even to a circus acrobat. To remove the nests, examine and manipulate their contents, and then replace the nests in their original position seemed impossible. The nest is constructed in the form of a long, pensile bag made of interwoven grasses and small vines. It should be feasible, Smith reasoned, to cut or rip the nests down with the aid of long poles and then reattach them to the original site. Most ecologists would have difficulty imagining themselves standing atop a 15-foot ladder, balancing 48 feet of extendable aluminum poles with instruments at the tip controlled from below by long ropes. But with this apparatus, Smith performed the delicate task of lowering a nest of fragile eggs fifty feet to the ground, examining the con-

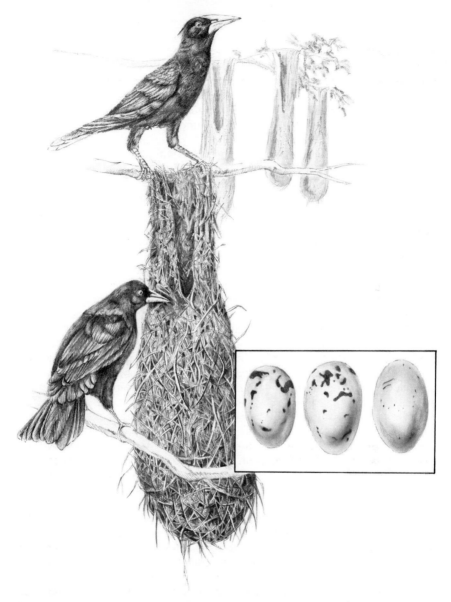

Figure 7-2   Female chestnut-headed oropendola at her nest (above) and giant cowbird (below). In the inset, the oropendola egg (left) is compared with a mimic cowbird egg (center) and nonmimetic cowbird egg (right).

tents, and replacing it—all of this carried out at night, under the duress of incessant mosquito attack, so as not to provoke the adult birds to abandon their nests.

Using a formidable, Rube Goldberg array of pincers and flaps connected by lever and pulley, Smith was able to replace the nests in their original positions with a sticky variety of contact tape or contact glue. In the most highly developed version of the nest-replacement protocol, he stapled rat snaptraps to the end of the nest. These clung tenaciously to any tree limb they were pressed against.

Having mastered the technique of working with the nests, Smith set out to determine whether the presence or absence of egg mimicry in the cowbirds elicited different oropendola behavior toward foreign eggs. A number of objects—including mimic cowbird eggs, nonmimic cowbird eggs, other kinds of eggs, and a variety of objects only remotely, if at all, resembling eggs—were put into nests in both kinds of oropendola colonies, those that tolerated nonmimic eggs and those that did not. Smith found, as he had suspected, that in oropendola colonies where the cowbird eggs closely mimicked those of their hosts, the oropendolas removed virtually everything from their nests except their own eggs and very closely matching cowbird eggs. These oropendolas were discriminators and tried, often in vain, to discover and eject the cowbird eggs. In oropendola colonies where cowbirds were poor egg mimics, the oropendolas were willing to accept all sorts of foreign objects. Smith had thus identified two types of colonies, one in which the oropendolas discriminated against cowbirds and tried to eject everything from their nests but their own eggs, the other in which oropendolas were nondiscriminators and accepted notably different eggs and other materials as well.

In what other ways did the two colonies differ? In the nondiscriminator colonies, were the cowbirds actually beneficial to the oropendolas? Why else would their eggs be tolerated by the oropendolas? While making an extensive survey of the oropendolas in the Panama Canal Zone and nearby Panama, Smith found that in nondiscriminator colonies, young oropendolas were often infested with the larvae of a species of bot fly. The parasites sometimes killed the nestling oropendolas, and frequently so weakened them that their chances of surviving to adulthood were slim. In discriminator colonies these parasites were rarely present. Here was a major difference between the two types of colonies. Was it possible that the role of the cowbird in the colonies was linked to bot-fly parasitism?

Smith examined oropendola young in nondiscriminator colonies (susceptible to bot-fly parasitism), and discovered that the incidence of bot flies was higher in nests that did not contain cowbirds than in nests that did, as shown by the following data:

	Number of nestling oropendolas in nests	
	With cowbirds	Without cowbirds
With bot-fly parasites	57	382
With no parasites	619	42

Further observations plainly showed that the nestling cowbirds would snap at anything small that moved within the nest, including adult bot flies, and, furthermore, they would remove bot-fly larvae from the skin of the nestling oropendolas. This behavior on the part of the young cowbird benefited the oropendola and accounted for the acceptance of the brood parasites in nondiscriminator colonies.

The cowbird young are well suited to groom their nest mates. They hatch five to seven days before the oropendola young, and develop precociously. Their eyes are open within 48 hours after hatching, whereas the eyes of oropendola nestlings open six to nine days after hatching. Also, the cowbird young are born with a thick covering of down, absent in the oropendola young, that presumably deters bot flies from laying their eggs on the skin of the young cowbirds. By the time the oropendolas hatch, the cowbirds are sufficiently developed to groom the oropendola young. In discriminator colonies, which are not troubled by bot-fly parasitism, cowbird young perform no such useful function for the oropendolas; and because they compete with the oropendola young for food, they are detrimental to the productivity of the colony.

The role of the cowbird in this story is now evident, but we have not yet determined why bot flies are present in some colonies and absent from others. Smith noted that all the discriminator colonies, and none of the nondiscriminator colonies, were built near the nests of wasps or bees, whose occupants swarm in large numbers around their nests and virtually fill the air throughout the oropendola colony. Wasps and bees presumably prevent the bot flies from entering the colonies. Occasionally, Smith found the detached wings of bot flies beneath the nests of discriminator colonies. By hanging up rolls of fly paper in the two types of colonies, Smith found that adult bot flies rarely enter the area around discriminator nests. But the protection is not perfect. Some of the nests at the periphery of the discriminator colonies, and thus at some distance from the wasp and bee nests, are occasionally parasitized by bot flies. Because the protective influence of the wasp extends over a limited distance, discriminator colonies tend to be much more compact than nondiscriminator colonies, with nests placed close together around the wasp nests at the center.

Oropendolas and cowbirds are not bothered by the same wasps and bees that viciously attack other intruders at their nests and hives. The birds could benefit the wasps inasmuch as their defenses against nest predators may also protect the wasp nests. In addition, the oropendolas have a characteristic, strong odor that may act as a chemical signal to the insects to suppress their normal defense reactions. Clearly, the oropendola story may someday chronical subtle mutual interdependences of which we now have only tantalizing hints.

The nesting success of the oropendolas in the nondiscriminator colonies shows the advantage of having cowbirds as nest mates (Table 7-1). Relatively few young are raised from the nests of nondiscriminating oropendolas, less than one half of a young per nest on the average, owing to predation, starvation, and abandonment of nests. Nondiscriminator broods with cowbirds nonetheless produce more than twice as many host young as nests

TABLE 7-1

Fledging success of oropendolas in discriminator and nondiscriminator colonies tabulated according to the presence or absence of cowbird nest mates. Fledging success is expressed as the number of host young leaving the nest (from Smith 1968).

		Fledging success	
Number of young in nest		Discriminator colonies	Nondiscriminator colonies
Oropendola	Cowbird		
2	0	0.53	0.19
3	0	0.55	0.19
2	1	0.28	0.53
2	2	0.20	0.43

without parasites. The difference in success between nondiscriminator nests and discriminator nests, both with no cowbirds, may be attributed to the detrimental effects of bot-fly parasitism in the nondiscriminator nests.

We might expect discriminator and nondiscriminator nests, both with cowbirds, to produce similar numbers of host young. Neither is beset by bot-fly parasitism, and both have the same number of mouths to feed. But in fact, cowbirds reduce success more in discriminator colonies, where they are disadvantageous to the host, than in nondiscriminator colonies, where they confer some advantage. In discriminator colonies, a cowbird chick reduces the number of host young fledged by about one half. In nondiscriminator colonies, however, a brood parasite more than doubles the nesting success of the host. These facts suggest that if there were no bot-fly parasitism, nondiscriminator colonies would be the more successful, perhaps because they can select colony sites without regard to the presence or absence of wasp nests. Furthermore nondiscriminator colonies breed earlier in the year than discriminator colonies because they do not have to wait for wasp nests to become active. More food may be available earlier in the year.

For the oropendola, the presence or absence of wasps in the vicinity of the colony completely alters the role of the cowbird as a factor of the environment. Accordingly, the behavior of the oropendolas toward the cowbirds and their eggs also varies between the two types of colonies, and in turn affects the environment that molds the behavior of the cowbird. In discriminator colonies, adult oropendolas not only eject the eggs of cowbirds from their nests if they can distinguish them, but they also chase adult cowbirds out of the colony. The oropendolas in nondiscriminating colonies are indifferent to cowbirds. Behavior of the cowbird in the two types of colonies reflects this difference. In discriminator colonies, female cowbirds are cautious and always enter the colony singly. Behavioral adaptation has gone so far that the cowbirds mimic the behavior of the discriminator oropendolas; they often gather nest-building materials and act as if they are beginning to build a nest, a most uncharacteristic behavior of brood parasites. On the other hand, cowbirds that parasitize nondiscriminator colonies are often gregarious and enter the colonies in small groups. They behave aggressively toward the oropendolas and sometimes even chase them from their nests.

It is not sufficient for a successful egg mimic merely to produce an egg that is indistinguishable from host eggs. It must also lay the egg at the proper time. A female oropendola normally lays her two eggs on consecutive days. She is sensitive to the appearance of eggs in the nest before or after her laying period, and will frequently desert her nest if it contains more than three eggs. Even when confronted by perfect egg mimicry, a discriminating oropendola can be fooled only if a single egg is laid by the cowbird soon after the first oropendola egg has been laid. If the cowbird lays its egg a day too early or too late, it reveals its presence and the oropendola will abandon the nest and start another. In nondiscriminator colonies, however, neither the number of eggs laid by the cowbird nor their appearance matters to the oropendola. Commonly, a female cowbird lays two to five eggs over several days in the nest of a single nondiscriminating oropendola. Smith refers to these birds as dumpers.

The interaction between the oropendola and the cowbird shows how the pattern of the environment determines the adaptations of the organism and how two or more kinds of organisms can be major factors in each other's environments (Table 7-2). The oropendola, the cowbird, the bot fly, and the wasp are all mutually important to each other and affect the evolution of one another. This relationship differs sharply from the relationship of an organism to the physical environment, which neither evolves nor responds by adaptation to changes in the biotic environment. Whereas the physical environment is passive, the biotic environment is responsive. It continually readjusts to evolutionary changes in any one of its living components. We might expect evolution in a biotically dominated environment to differ from evolution in a physically dominated environment. Different populations, like the oropendola and the cowbird, may co-evolve with respect to each other to form a mutually beneficial relationship that could not be achieved between a population and its physical environment.

TABLE 7-2

Biological attributes of discriminator and nondiscriminator colonies of the chestnut-headed oropendola (from Smith 1968).

	Colony type	
	Discriminator	Nondiscriminator
Wasp nests	Present	Absent
Possibility of bot-fly parasitism	Slight or absent	Heavy
Effect of cowbird on oropendola	Disadvantageous	Advantageous
Foreign objects in nest	Rejected	Accepted
Cowbird eggs	Mimetic	Nonmimetic
Cowbird eggs per nest	One	Several
Cowbird behavior in colony	Timid	Aggressive
Colony structure	Compact	Open
Nesting season	Late	Early

The social environment of an organism consists of neighboring individuals of the same species who provide the selective force, the environmental pattern, for the co-evolution of the roles of individuals in that population. Such adaptations are clearly expressed in the social behavior of baboons.

The behavior of the baboon has evolved in a social context.

Social behavior has developed to different degrees in various animals, but nowhere has its study excited so much interest as in the subhuman primates, the closest relatives of man. The social group consists of individuals of the same species, many of whom are closely related, and upon whom the individual depends in part for his survival. The extent to which some animals have evolved in a social context is demonstrated by their utter dependence upon the group for survival. Such is the case among the baboons of East Africa. In groups, these animals are avoided by all predators except lions; individual baboons, despite their strength and fearsome teeth, have very poor prospects. As a consequence of living in a social group, the individual adapts to an environment determined largely by the behavior of its cohort. Whereas individual antagonism and avoidance are the general rules among nonsocial animals, group cohesiveness is absolutely necessary for the survival of such social animals as the baboon. This cohesiveness is promoted by the evolution of a high level of organization within the baboon troop.

Anthropologists S. L. Washburn and Irven DeVore (1961) studied baboons intensively at two localities in Kenya: a small park near Nairobi and the Amboseli Reserve at the foot of Mount Kilimanjaro. Months of careful observation revealed details of the social interactions among the baboons. For example, the troop, moving in an open savanna, has a spatial structure that reflects the roles of individuals in the troop (Figure 7-3). When resting in trees, baboons are normally safe from predators, but when they move across open grasslands without the immediate safety of trees, members of a troop assume an organized, defensive pattern. The less dominant adult males and some of the older juveniles are found toward the periphery of the group. Females without infants and some of the older juveniles stay closer to the center. Females with infants and young juveniles, the most vulnerable members of the troop, move at the center of the troop with the dominant males. When the troop is threatened by a predator, the dominant males move into position between the center of the troop and the threat, while the rest of the troop seeks safety behind them. The importance of trees as potential escape routes is so great that baboons are not often found in areas lacking suitable trees, even if the habitat is otherwise perfectly suitable.

The spatial organization of the baboon troop as a defense strategy often is supplemented by the presence of other nonpredatory animals, particularly gazelles. While the gazelles gain a degree of protection from predators by their association with baboons, the baboons benefit from the gazelles' keen hearing and sense of smell by early detection of predators.

Interactions among individuals within the troop are many and diverse. The baboon has a strong tendency to stay with other baboons. Even within

Figure 7-3   A moving troop of baboons. Females with infants riding on their backs are near the center of the troop, close to some large males. Small juveniles and young males are closer to the periphery of the troop.

the troop, individuals form small, persistent friendship groups, within which there is much grooming and playing. Most baboons, particularly adult females, spend many hours each day grooming others that present themselves in the proper manner. To groom, one baboon parts the hair of the other with its hands and removes dirt, lice, ticks, and the like with its hands and mouth. Grooming and play additionally strengthen the social bonds within the troop and promote its stability and cohesiveness.

Care of infant baboons is left to their mothers. The fathers' attention is directed toward the well-being of the troop as a whole. After an infant has been weaned, its strong bond with its mother is replaced by looser associations within a play group of other juveniles. As young baboons reach adulthood, they enter the dominance hierarchy of the troop, in which their position is determined by fighting, bluffing, and the relative rank of those with whom they associate most frequently. Several males may band together and enhance each other's position in the hierarchy, although individually they hold lower ranks. The position of an individual in the social hierarchy determines to some degree the amount of grooming it receives and its precedence over other baboons at a food and water source. Rank also extends privileges in mating with females.

The female is sexually receptive for about one week each month. At the beginning of her estrus cycle, a female often mates with the older juveniles and some of the subordinate males. But when she becomes fully fertile a female mates only with the dominant males, who consequently sire most of the offspring in the troop. During the mating period, a male and female form what is called a consort pair for a few hours to several days; all other social functions are disrupted for them. Paired baboons generally move at the periphery of the troop, and the male can be extremely aggressive toward others. Baboons appear not to form long-lasting pair bonds, although lack of close association within the troop need not imply a weak social bond.

The baboon must cope with a wide range of behavioral interactions during its lifetime. The relationship between mother and infant, among individuals in a friendship or play group, among males in a dominance hierarchy or among those individuals that cooperate to defend their dominance rank, between the male and female in a consort pair—all produce a complex social environment.

Social adaptation does not occur independently of the physical and biotic environments (Crook 1970, Kummer 1971). Baboon troops in different habitats exhibit considerable variation in their social organization. For example, in the open savannas, an average-sized baboon troop, perhaps forty individuals near Nairobi and eighty individuals in the Amboseli, may occupy a home range commonly of two to six square miles, but sometimes as large as fifteen square miles (DeVore and Washburn 1964). In Uganda, the baboons inhabit forest or mixed forest-grassland areas where troops of forty to eighty usually range over less than two square miles (Rowell 1967). In the Uganda forests, the preferred food of the baboon, which is fruit, is much more plentiful than in the open savannas of Kenya. Thus the troops can fill their food needs in a smaller area.

The social organization within the troops of the two areas also differs in several important aspects. In the open savannas of Kenya, antagonistic interactions between individuals are more frequent than in Uganda. In the forest, an individual can go out of sight behind a bush until his antagonist "forgets" about the dispute between them. No such social escape exists on the open savannas. Forest baboons can fill their stomachs in an hour's feeding each day, whereas baboons of the open grasslands spend most of their time in search of food. Whereas grassland troops are often on the move, forest baboons are frequently engaged in grooming, playing, sitting, and drowsing. Because grassland baboons forage almost continuously, there is less social interaction among individuals, and that which does occur is more frequently antagonistic than among forest dwellers.

The hamadryas baboons inhabiting the arid grassland at the southern edge of the Danakil Desert in Ethiopia are organized at three social levels: single males with several females and their offspring, persistent bands of several of these single-male family groups, and a larger troop composed of many bands (Kummer 1968; Figure 7-4). During the day, the hamadryas disperse to feed as small bands or single-male groups. At dusk, they congregate in large numbers to sleep on cliffs, where they are protected from nocturnal predators. Before breaking up in the morning, the troop congregates near the roosting cliffs for a morning social hour, filled with grooming and, among infant and juvenile males, play. The gelada baboon, a cliff-roosting species of the Ethiopian highlands, forages in large groups when food is abundant but disperses into single-male groups during seasons of scarcity (Crook 1966). In contrast to hamadryas, geladas do not appear to have distinct bands within the troop. The patas monkey of the dry grasslands of Uganda has an entirely different social structure, based solely on single-male groups that seldom meet and are rarely on friendly terms (Hall 1965). These studies, and others describing additional variations on the basic single-male groups, multi-male bands, and the larger troops of savanna-

Figure 7-4   Above: A troop of hamadryas baboons departing from the roosting cliff. Smaller bands are not evident during this early morning progression. Below: A single-male group resting at mid-morning after splitting from the rest of the troop to feed (after Kummer 1971, courtesy of H. Kummer).

dwelling species, emphasize the role of the environment in determining social organization. The particular social structure itself influences adaptations of behavior that strengthen social bonds and both organize and facilitate social interactions.

8

Natural Selection and Design in Nature

The adaptations of the velvet mite, oropendola, and baboon are well suited to their particular environments. Every detail of their morphology, physiology, and behavior seems capable of meeting the challenge of their surroundings. The close correspondence between organism and environment is no accident. Only those individuals that are well suited to the environment survive and produce offspring. The inherited traits which they pass on to their progeny are preserved. Unsuccessful individuals do not survive and reproduce, hence their less suitable traits are eliminated from the population. This process, which Charles Darwin called *natural selection,* allows the population to respond to its environment over periods of many generations and slowly refines the adaptations of organisms to fit the requirements of the environment.

Natural selection can be outlined in general terms and by specific example as follows:

(a) The reproductive potential of populations is great, but

(a) Rabbits should cover the Earth, but

(b) populations tend to remain constant in size, because

(b) they don't, because

(c) populations suffer high mortality.

(c) many are caught by predators.

(d) Individuals vary within populations, leading to

(d) Some rabbits run faster than others,

(e) differential survival of individuals.

(e) and escape from predators.

(f) Traits of individuals are inherited by their offspring.

(f) So do their young.

(g) The composition of the population changes by the elimination of unfit individuals

(g) Populations of rabbits, as a whole, tend to run faster than their predecessors.

As it is described above, natural selection occurs because of three properties of organisms and their relationship to the environment: (1) genetic variability, (2) inheritance of traits, and (3) influence of the environment on survival and reproduction. Design is not inherent to the process of natural selection. The environment itself is the template for the design we see in organisms. Selection merely acts as an agent for realizing that pattern. Whether a rabbit runs slowly or rapidly is inconsequential to natural selection; only the influence of its swiftness on the number of offspring that it leaves is important.

Inheritance and genetic variation within populations are facts of genetics obvious to anyone who has noticed both the variability of physique, facial appearance, eye color, and hair among humans and the tendency of related individuals to share many of these traits. The relationship between particular genetic traits, on one hand, and survival and reproduction, on the other, is less obvious. By our own technological devices we have surrounded ourselves with an environment largely of our own making, one that seems to have little influence on physique, eye color, blood type, baldness, and other genetic traits. To be sure, some inherited diseases—such as retinoblastoma, achondroplastic dwarfism, aniridia, and sickle-cell anemia—have a profound effect on survival. These are deleterious, often fatal traits. They occur rarely and produce major disruption of body function. But they hardly seem representative of the traits that might have fostered eagles from primitive single-celled organisms balanced uncertainly on the fence between the coming kingdoms of plants and animals a billion years ago.

Plant and animal breeders have known for centuries that by carefully selecting breeding lines for a desired trait, the appearance of population could be altered toward a desired end: longer wool, increased egg and milk production, sweeter fruit. Selection was practiced on domesticated plants and animals long before its role in shaping biological communities was appreciated.

The demonstration of natural selection acting to produce evolutionary changes in natural populations did not come until more than half a century after Charles Darwin originally proposed the mechanism. The first evidence that a change in the environment could select a new trait in a population came from early programs designed to control insect pests. Agricultural researchers had found that cyanide gas could be used to control populations of scale insects on citrus crops in southern California. As early as 1914, however, populations in some groves had become tolerant of the gas, and fumigation was no longer effective. Laboratory experiments demonstrated not only that tolerance was inherited, but also that some individuals naturally resisted cyanide poisoning in areas where the fumigant had never been used (Quayle 1938, Dickson 1940). This resistance was caused by a mutation—an error in the genetic code—which appeared rarely and sporadically in individual scale insects. In the absence of cyanide, the mutation for cyanide resistance would be of no value to the organism and could be detrimental if it altered normal body functions. Where the gas treatment was used to control scale insects, it resulted in survival of individuals carrying the resistant trait, while killing all others. The more recent development of a wide variety of insecticides has further revealed the presence of resistant traits among individuals of numerous pest species (Crow 1957). More than performing their intended function, pesticides have often selected resistant populations that are all the more difficult to control (Debach 1974).

Studies on industrial melanism demonstrated natural selection in action.

The English have always been avid butterfly and moth collectors, and such enthusiasts look carefully for rare variant forms. Early in the last century, occasional dark (or melanistic) specimens of the common peppered moth (*Biston betularia*) were collected. Over the succeeding hundred years, the dark form, referred to as *carbonaria,* became increasingly common in some industrial areas, until at present it makes up nearly 100 per cent of some populations. The phenomenon aroused considerable interest among geneticists, who showed by cross-mating light and dark forms of the moth that melanism is a simple inherited trait.

In the early 1950s, H. B. D. Kettlewell, an English physician who had been practicing medicine for fifteen years and who was also an amateur butterfly and moth collector, changed the pattern of his life to pursue the study of industrial melanism (Kettlewell 1959). Several facts about melanism were already known before Kettlewell began his studies: (1) the melanistic trait is an inherited characteristic, hence the widespread occurrence of melanism had been the result of genetic changes in populations; (2) the earliest records of *carbonaria* were from forests near heavily industrialized regions of England; (3) the dark form occurs most frequently in populations near modern industrial centers (Figure 8-1); (4) where there is relatively little industrialization, the light form of the moth still prevails. It was also known that dark forms have similarly appeared in many other moths and other insects. Melanism is not unique to the peppered moth.

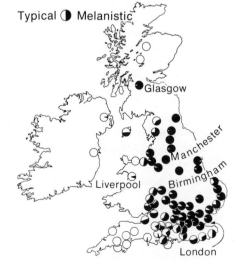

Typical ◑ Melanistic

Glasgow

Manchester

Liverpool   Birmingham

London

Figure 8-1   The frequency of melanistic individuals in populations of the peppered moth (*Biston betularia*) in various localities in the British Isles. The map is based on more than 20,000 records, from 83 centers, collected during 1952 to 1956 (after Kettlewell 1958).

Kettlewell knew that the peppered moth inhabits dense woods and rests on tree trunks during the day. He reasoned that where melanistic individuals had become common, the environment must have been altered in some way to give the dark form a greater survival advantage than the light form. Could natural selection have led to the replacement of the "typical" light form by *carbonaria*? To test this hypothesis, Kettlewell had to find some measure of fitness other than the relative evolutionary success of the two forms.

To determine whether *carbonaria* had a greater fitness than typical peppered moths in areas where melanism occurred, Kettlewell chose the mark-release-recapture method. Large numbers of individuals of both forms were to be marked with cellulose paint and then released in a suitable forest. The area would be thoroughly trapped for moths, and the number of marked individuals of each type that were recaptured would be noted. Any difference in the survival of the two forms would appear in the percentages of the light and dark forms recaptured. Kettlewell did not design his experiment to detect differential survival of the larvae of the two forms of *Biston*. Nor could it detect differences in the fecundity of adult moths. Thus, if the results of the experiment turned out negative—that is, if the two forms had similar survival rates as adults—Kettlewell could not have flatly rejected the original hypothesis that the two forms had different fitnesses. Even if predators were detecting and killing equal numbers of light and dark moths, a dark moth might bear more eggs than a light one, or its offspring may have better survived the caterpillar stage. Kettlewell was banking heavily on measurements of adult survival to reveal differences in fitness between the light and dark forms.

Because large numbers of individuals had to be released to guarantee even a small number of recaptures, Kettlewell collected and raised more than 3,000 caterpillars to provide adult moths for his experiments. Two patches of woods, one in a polluted area and one in an unpolluted area, were chosen for the experiment. Adult moths of both forms were marked with a dot of cellulose paint and then released. The mark was placed on the underside of the wing so that it would not call the attention of predators to a moth resting on a tree trunk. Moths were recaptured by attracting them to a mercury vapor lamp in the center of the woods and to caged virgin females around the edge of the woods. (Only males could be used in the study because females are attracted neither to lights nor to virgin females.)

In one experiment, Kettlewell (1955, 1956) released 201 typicals and 601 *carbonaria* in a polluted woods near Birmingham. The results were as follows:

Birmingham (industrial area)	Typicals	Melanics
Number of moths released	201	601
Number of moths recaptured	34	205
Per cent recaptured	16.0	34.1

These figures indicate that the dark form survived better than the light.

Although consistent with Kettlewell's original hypothesis, these results could be interpreted otherwise: as differential attraction of the two forms to the traps used or as differential dispersion of the two forms away from the point of release. Variables besides differential mortality had to be accounted for.

To test the hypothesis of natural selection unequivocally, Kettlewell ran a control experiment in an unpolluted forest near Dorset with the following results: of 496 marked typicals released, 62 (12.5 per cent) were recaptured; of 473 marked melanics, 30 (6.3 per cent) were recaptured. Thus, in the unpolluted forest, light adults had a higher survival rate than dark adults. If typicals and melanics were differently attracted to light traps or dispersed from a release point at different rates, the level of pollution in the forest should not have influenced the result of the experiment. Only differential fitness could account for a reversal in the relative rate of recapture of typical and melanistic forms in polluted and unpolluted forests. The differences in relative survival of the two forms in the different environments confirmed Kettlewell's hypothesis and established natural selection as responsible for the high frequency of *carbonaria* in industrial areas.

Having demonstrated that natural selection had been responsible for the replacement of typical by melanistic forms of the peppered moth in industrial regions, Kettlewell then sought to determine the specific agent of selection. He reasoned that in industrial areas pollution had darkened the trunks of trees so much that typicals stood out against them and were easily found by predators. Any aberrant dark forms would be better camouflaged against the darkened tree trunks, and their coloration would confer survival value to them. Eventually, differential survival of dark and light forms would lead to changes in their relative frequency in a population. To test this hypothesis, Kettlewell had to determine whether tree trunks are, in fact, darker in areas where the melanic form was prevalent, and then to demonstrate that camouflage is important to the survival of the moths.

A clean handkerchief rubbed against a tree trunk can satisfy even the most doubting critic that the trunks of trees in polluted areas are darker than those in nonpolluted areas. And as one might have expected, the *carbonaria* stands out against tree trunks in unpolluted woods, whereas typicals are more conspicuous in polluted settings (Figure 8-2).

Kettlewell was certain that conspicuousness of light-colored forms resting on darkened backgrounds must have greatly increased predation on them by visually oriented animals. The dark forms must have a similar disadvantage in unpolluted woods. To test this, Kettlewell placed equal numbers of the light and dark forms on tree trunks in polluted and unpolluted woods and watched them carefully at some distance from behind a blind. (A blind is a tentlike structure intended to conceal an observer from his subjects, more appropriately called a "hide" by the English.) He quickly discovered that several species of birds regularly searched the tree trunks for moths and other insects and that these birds more readily found the moth that contrasted with its background than the moth that blended with the bark. Kettlewell tabulated the following instances of predation:

	Individuals taken by birds	
	Typicals	Melanics
Unpolluted woods	26	164
Polluted woods	43	15

Figure 8-2   Typical and melanistic forms of the peppered moth at rest on a lichen-covered tree trunk in unpolluted countryside (left) and on a soot-covered tree trunk near Birmingham (from the experiments of H. B. D. Kettlewell).

These data are fully consistent with the results of the mark-release-recapture experiments. Together they clearly demonstrate the operation of natural selection that, over a long period, resulted in genetic changes in populations of the peppered moth in polluted areas. Many decades were required for the replacement of one form by the other. The agents of selection were insectivorous birds whose ability to find the moths depended on the coloration of the moth with respect to its background. The evolution of industrial melanism shows clearly how the interaction between the organism and its environment determines the organism's fitness.

The theory of natural selection enables us to predict genetic changes in a population from known changes in the environment. If pollution were to be controlled in industrialized areas, and if this allowed forests to revert to their natural state, we would predict that the frequency of the light form of the peppered moth would begin to increase. In fact, smoke-control programs were started in Manchester in 1952. Collections of the peppered moth over the last twenty years in the Manchester area do show a statistically significant increase in the proportion of the light form in the population (Cook, Askew, and Bishop 1970).

Natural selection tends to maximize the reproductive rate of the individual.

Any trait, or combination of traits, that enables an individual to leave more progeny than other individuals leave will be selected and incorporated into the population. Most traits have optimum expressions. Individuals that vary to either side of the optimum are the less fit. The snail whose shell is too thin is easily preyed upon; the snail whose shell is too thick is encumbered. The area of a bird's wing must conform to aerodynamic principles: too large or too small is not good enough a thousand feet up. In a classic paper on selection, H. C. Bumpus (1898) observed that English sparrows killed by a severe winter storm tended to be the largest and smallest individuals; a larger proportion of birds of intermediate size survived (Johnston *et al.* 1972, O'Donald 1973). If this were not generally true, populations would continually evolve to produce larger sparrows, faster rabbits, darker moths, and so on.

Whereas most traits have optimum values in a particular environment, one would expect that among individuals with different reproductive rates, which directly determine number of progeny left to the future, selection should always favor the most fecund. Indeed, the English ornithologist David Lack (1954) has argued that one should see in nature adaptations to maximize reproductive rate: "If one type of individual lays more eggs than another and the difference is hereditary, then the more fecund type must come to predominate over the other (even if there is over-population)—unless for some reason the individuals laying more eggs leave fewer, not more, eventual descendants." Thus, the number of offspring per clutch or litter should be the maximum number that the parents can adequately nourish or otherwise provide for.

To test his hypothesis, Lack began a series of studies on the number of eggs laid in a nest by birds (clutch size). Would individuals with the highest reproductive potential—those laying the largest number of eggs—actually be the most fecund? Well, not all the young of a brood necessarily survive. Many starve or die from other causes before leaving their nest, and most fail to reach reproductive maturity. Lack suspected that the number of eggs laid by a species had evolved to correspond to the maximum number of young parents could feed under normal conditions.

The common swift feeds by taking insects on the wing. The survival of swift young, which usually number two or three per nest, varies according to the feeding conditions for adults (Lack and Lack 1951). In years of fair weather, broods of three fledged the most young, an average of 2.3. (A bird is fledged when its plumage is developed and it is able to fly.) Broods of one and two were less productive. When cold and rainy weather impaired foraging conditions for the adults, broods of one, two, and three young all produced about one fledged young per nest. Poor feeding resulted in slow growth and starvation, particularly in large broods. Even if young of large broods successfully left the nest, they were frequently underweight and had poor prospects of surviving the first few months of independence.

Unfortunately, young birds usually disperse too far after fledging to measure their survival after they have left the nest. In a study on the European starling (Lack 1948), nestlings were individually identified with numbered aluminum leg bands in several thousand broods, the size of which varied between one and ten young at the time of leaving the nest. The number of young subsequently recovered by trapping or shooting formed the basis for estimates of reproductive success. Lack included only recoveries of birds more than three months after fledging, reasoning that by that time the young were fully developed and variation in nutritional state at fledging should have exerted its effect on the survival of young after leaving the nest. Less than two per cent of all the young banded were later recovered, but these were sufficient to demonstrate that the number of recoveries per nest increased with increasing brood size up to four and five but not beyond, perhaps decreasing for the largest broods (Table 8-1). In this study, Lack showed that an intermediate brood size (five) is the most productive in the long run. The weights of nestlings in broods of different size (Table 8-2) revealed that the failure of the larger broods to produce more offspring than broods of inter- mediate size is due largely to undernourishment of the young. Beyond a brood size of five, the adults could not fully satisfy the food requirements of the young.

Lack's hypothesis and his observations on the reproductive success of broods of different size explain the evolutionary optimization of brood size on one level only. On another level, we must ask what determines the rate at which adults can gather food for their young. One factor must be the effec- tiveness of the adaptations of adults as predators to find and capture prey, compared with the adaptations of the prey organisms to escape detection and capture by the predators (Ricklefs 1970a). It is also clear that the number of offspring raised from a given amount of food depends on the efficiency

TABLE 8-1

Reproductive success of the European starling with respect to brood size, demonstrating that the maximum productivity corresponds to the modal brood size (from Lack 1948).

Brood size	Number of broods	Number of young	Recovered more than three months after fledging		
			Number	Per cent of individuals	Per cent per brood
1	65	65	0	—	—
2	164	328	6	1.8	3.7
3	426	1,278	26	2.0	6.1
4	989	3,956	82	2.1	8.3
5	1,235	6,175	128	2.1	10.4
6	526	3,156	53	1.7	10.1
7	93	651	10	1.5	10.2
8	15	120	1	0.8	6.4
9	2	18	0	—	—
10	1	10	0	—	—

TABLE 8-2

Weights of nestling European starlings from broods of different size, demonstrating undernourishment of the young in larger broods (from Lack 1948).

Age of nestling (days)	Weights of nestling (grams) in broods of		
	2	5	7
6	48.3	45.1	38.2
15	88.0	77.6	71.4

with which the young utilize food for growth: With higher food requirements of individual young, fewer can be raised. Furthermore, where many individuals gather food in the same area, food resources are divided among them. In more dense populations, less food is available to each individual, hence fewer young can be raised.

No general theory, such as Lack's, can survive without stimulating criticism and alternative theories. Lack first presented his ideas in 1947 and then expanded upon them considerably in 1954 in his book, *The Natural Regulation of Animal Numbers.* New ideas and more data have followed: Some contradict the original theory and some elaborate on it (Klomp 1970). As a result, our knowledge about reproductive rates has become increasingly complex. Perhaps no one general theory can be sufficiently robust to explain the variation in reproductive rates among all species. In particular, the number of eggs laid by quail and ducks, which do not feed their young, cannot be determined by the ability of the parents to provide food. Here the physiological strain of egg laying or the inability of adults to incubate large numbers of eggs may set an upper limit to clutch size.

The number of young produced in a single brood is only one term in the equation for the individual's evolutionary fitness. Others include the number of broods that can be raised each breeding season, the survival of young to sexual maturity, and the life span of the adults. To the extent that a change in one term can influence the others, selection may favor some compromise in which the number of young raised per brood is somewhat less than the maximum possible at any one time. Ecologist George Williams (1966a, b) has suggested that the increased effort required to raise a larger brood might reduce the survival of the adults, perhaps because they would have less time to watch for predators or groom themselves properly, or because they become physiologically exhausted. Under these circumstances, selection may favor individuals that reduce their reproductive effort, and hence the size of their broods, but that more than compensate for this loss of fecundity by extending their reproductive life spans. The wandering albatross, among the most long-lived of animals, lays only one egg every other year, but may live to be forty or fifty years old.

The salmon has taken an approach to life opposite to that of the albatross. While most large fish lay a moderate number of eggs each year for several years, many species of salmon spawn only once, producing eggs at the expense of other tissues and preserving only those organs needed to migrate from the sea to their spawning grounds at the headwaters of rivers. Most of their digestive organs and some muscle are converted to eggs. Unable to assimilate food, they die of starvation soon after spawning.

The number of eggs laid by fish and other organisms generally decreases as the intensity of parental care for the young increases (for example, see Thorson 1950). There is a trade-off between the number of eggs laid by the adult and the survival of progeny to reproductive maturity, although the environmental conditions that foster the evolution of these varied combinations of fecundity and parental care are not well understood. It is important to realize that although natural selection tends to maximize the fitness of individuals in a population, we have no generalized view of the precise nature of the adaptations to bring about this result. Nothing inherent in natural selection can lead us to predict the existence of a specific set of adaptations. As Darwin remarked in *On the Origin of Species,* ". . . maternal love and maternal hatred is all the same to the inexorable principle of natural selection." That is, the patterns observed in nature reflect patterns in the environment. Natural selection is the translator, not the designer of that pattern.

Man applies selection to cultivated plants and domesticated animals.

Man has consciously created by "artificial" selection a variety of domesticated animals and plants tailored to his particular needs. Cows have been selected to give more milk, chickens to lay more eggs, sheep to yield more wool, cereal crops to produce more grain—all these changes directed toward some desired goal. Can we not conclude that selection itself has been made goal oriented, or purposeful? Why should we not believe that natural selection is purposefully directed in the same way that artificial selection is an extension of man's design?

What we call man's "purpose" is expressed by his providing particular environments for the animals and plants he raises. This new environment is designed to increase the fitness of desired traits so that those traits are incorporated into the population by selection. If by controlled breeding programs man greatly increases the relative fecundity of cows with high milk production by preventing inferior producers from reproducing, genetic factors for high milk production will be passed on to future generations, while genetic factors for low production are eliminated. In the environment of the milk barn, low production drastically curtails fecundity.

The environments of tropical forests and the arctic tundra have characteristic patterns that select for widely differing adaptations among the organisms that live in these environments. If man consciously were to change, say, the environmental temperature for a population, he would not change the process of natural selection, but merely the adaptations that it produced. The following detailed example demonstrates this point.

Fruit flies of the genus *Drosophila* are frequently used for experiments on selection because they are easily raised in the laboratory, exhibit many variable morphological traits that are amenable to selection, and have a short life span, about two weeks, so that selection can be applied to many generations of flies within time limits determined by the patience of the investigators. The experiments we shall consider were performed with *D. melanogaster* by K. Mather and B. J. Harrison (1949) in the Department of Genetics at the University of Birmingham, England, between 1942 and 1946. A single trait was chosen for selection: the number of bristles on the ventral surface of the fourth and fifth abdominal segments. These bristles normally number about forty per individual, with some variation in wild populations of flies; the investigators were interested in attempting to both increase and decrease the number of bristles.

The procedure followed to select for an increased or decreased number of bristles is illustrated in Figure 8-3. To begin the experiment, two pairs of flies were allowed to mate and lay eggs in a small culture bottle, with a layer of food on the bottom. The adults were removed from the bottle just before the first of their progency began to emerge as adults from the pupal stage of development (about two weeks). As the progeny emerged, they were removed from the culture bottles and the sexes were kept separate to assure the virginity of the females. Twenty offspring of each sex were chosen at random and the number of bristles on each were counted. For the high line, Mather and Harrison selected the two individuals of each sex with the largest number of bristles and mated these to produce the next generation. The four flies were first placed together in a small vial for one to three days to mate and they were then transferred to larger culture bottles where they laid their eggs, starting the next generation. The investigators followed a comparable procedure for the low line.

In each successive generation only one-tenth of the flies in each group of progeny, those with the extreme numbers of bristles, were selected and allowed to reproduce. This selection regime produced dramatic results (Figure 8-4). In twenty generations the number of bristles per individual in the high line increased from 36 at the beginning of the experiments to almost 56. Selection in the low line produced a somewhat slower, but nonetheless significant decrease in the number of bristles.

Clearly, Mather and Harrison were able to change the genetic composition of populations of fruit flies according to their own design. They did not, however, alter the process of selection itself, because the genetic changes they produced were brought about through the differential reproduction of individuals with different genetic constitutions, just as Darwin envisioned the mechanism of natural selection. In the high line, for example, only those individuals with the highest number of bristles were allowed to mate and reproduce. Thus the genetic factors that tend to produce high numbers of bristles were transmitted into future populations at a greater rate than those that result in lower bristle numbers. Only the environment, not the mechanism of selection, had been changed.

By applying artificial selection, man alters the pattern of the environment in such a way that selection produces desired changes in populations. Selection itself does not produce design, rather design is conferred on the organism by the pattern of the environment acting through selection. In the peppered moth, certain patterns in the forest—the background color of the bark and the hunting characteristics of visually oriented bird predators—

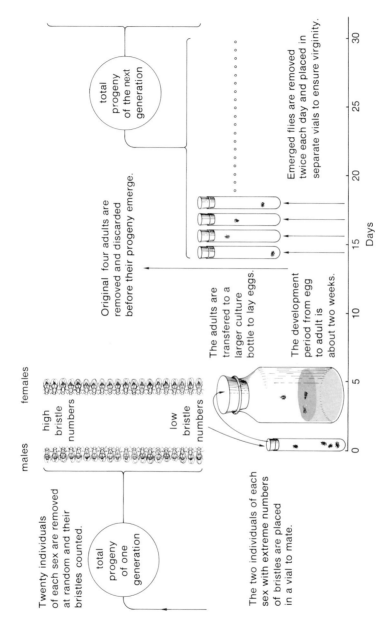

Figure 8-3   Schematic diagram of Mather and Harrison's (1949) method for artificial selection of high and low numbers of bristles on the fourth and fifth abdominal segments of the fruit fly, *Drosophila melanogaster*. The details for the high- and low-selected lines (high and low numbers of bristles) are identical. Each generation of selection required about thirty days. The initial population for the first generation of selection was formed by interbreeding two wild strains, Oregon and Samarkand.

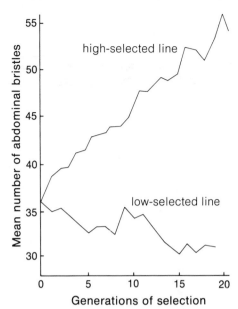

Figure 8-4   The results of artificial selection for high and low numbers of abdominal bristles in the fruit fly, *Drosophila melanogaster* (after Mather and Harrison 1949).

provide the evolutionary "reasons" for the adaptations of wing color in the moths, the "purpose" of the design. In experiments with artificial selection, such as Mather and Harrison's, man has merely changed the pattern of the environment, and through the process of natural selection, the design of the organism is changed accordingly.

9

The Nature of Adaptation

The diverse environments of plants and animals have resulted in a bewildering variety of adaptations to better suit the organism to these conditions. Virtually all imaginable adaptations exist somewhere in the living world. Where they do not, but would nonetheless be useful means to mankind's ends, we have applied artificial selection to domesticated and cultivated lines to achieve a desired result. The variety of adaptations and their susceptibility to selective pressures may give the false impression that the properties of plants and animals are infinitely flexible, like clay in an artist's hand. But, of course, adaptation does not proceed without limit. Constraints are imposed because the organism must function as a whole, because of the nature of the mechanism of inheritance, because of compromises required by properties of living systems that are beyond the reach of natural selection, because of the adaptations of other organisms, and because of random and historical elements in the circumstances within which adaptations evolve.

The adaptations of an organism must be integrated.

We tend to regard the adaptations of a particular organ in the context of its immediate environment. The shape of a bird's beak corresponds closely to the type of food that it eats; the structure of the eye is adapted to perceive patterns of light; the thickness of fur is seen to vary with regard to the temperature of the surrounding air. Paramount to the environment of the organ is the rest of the organism to which it is attached. The structure of the horse's leg, which enables its bearer to run so swiftly, has clearly been selected by swift predators. But the legs of mammals have functions besides running—grooming, scratching, and fighting, to name three. Owing to its structure, the horse's leg has lost much of the flexibility that the legs of other mammals possess, and the foot can no longer reach other parts of the body. Because scratching is important to the horse's survival and well-being, the modification of the leg was accompanied by the evolution of a long, highly movable, brushlike tail to swish at insects and a loose skin that can be shaken by underlying muscles.

104

While it seems absurd to imagine an organism that, for example, could create sound but could not hear, the interaction between these functions may be quite subtle, particularly when the organism must sort from the myriad noises around it the particular sounds to which it must respond. Most bats emit very high pitched sounds, far above our hearing range, and locate food by the echo that is bounced back from flying insects. Not only have the ears and faces of some species of bat become grotesquely (or exquisitely, depending on how you feel about bats) modified to determine the direction of a returning echo, but a bat's hearing is more sensitive to the characteristic sound frequencies and temporal patterns of echoes produced by its own voice than by the voices of other species of bats (Grinnell 1970). This specialization, and many similar ones, emphasizes that the organism is the basic unit of adaptation and that its parts do not evolve independently of one another.

The integrity of organismic function and the basic conservatism of the adaptation process underlie a long-standing controversy over the loss or degeneration of eyes in some cave-dwelling organisms (Barr 1968; Figure 9-1). Although this phenomenon has been referred to as regressive evolution, changes leading to a loss of vision seemingly must have conferred some advantage to the individual in the cave environment if these changes have

Figure 9-1   The northern cavefish (*Amblyopsis spelaea*) inhabits subterranean streams in Indiana and Kentucky (from Barr 1968, courtesy of T. C. Barr).

been favored by natural selection. A nineteenth century writer went so far as to propose that the presence of eyeless animals in caves resulted from individuals with more or less degenerate eyes being unable to find their way out of caves after having fallen or having been washed into them accidentally. More recent theories attribute the advantage of eyelessness to economy in growth and development or to beneficial effects of genetic traits for reduced vision expressed in other parts of the body. Whatever the underlying basis, the degeneration of organs useless to the cave organism should be thought of as progressive in that it allows the organism to become better adapted to the cave environment.

Natural selection acts in the present, but adaptations reflect the past.

The evolutionary fitness of that combination of hereditary traits embodied in an organism derives from the interaction of the organism with its environment during its lifetime. Because evolutionary changes follow the replacement of individuals in populations, new adaptations may require many generations to become predominant. A sudden change in the environment may immediately and completely alter the relative fitnesses of different genetic traits and change the course of evolution, but a population will continue to reflect earlier conditions until complete evolutionary adjustment to the new environment has occurred. Thus the degree to which a population is well adapted to its present environment depends partly on how rapidly the environment changes and partly on how rapidly evolutionary change occurs. Populations of bacteria can adapt to an antibiotic in a matter of days. Populations of moths may require several years to respond to the effects of pollution in their environments. A long-lived, slowly reproducing organism, such as man, may take many millennia to make an evolutionary transition of the same scale.

Natural selection is opportunistic.

Given the variation present in a population, natural selection favors traits that are best suited to the maintenance and reproduction of the organism, and hence the traits generally best suited to the environment. When the environment changes, the adaptive response of a population is limited to selection among genetic traits already present in the population. That is to say, environmental change does not stimulate production of the genetic variation needed to produce particular evolutionary responses. The ultimate source of all variation in hereditary material is mutation, the spontaneous and random change in the complex molecules of the cell nucleus that encode the genetic information. Because mutation occurs continuously, genetic variants are always present in the population, and it is among them that natural selection chooses.

Prior to the industrial revolution in England, when forests throughout the country were pollution-free, black forms of the peppered moth were occasionally collected. These variants bore the observable expression of the genetic trait *carbonaria,* which occurred in the population with very low frequency. As woodlands surrounding industrial areas became polluted with soot, dark forms of the peppered moth became prevalent in those areas (see Chapter 8, particularly Figure 8-2 on page 96). In some localities, however, particularly in the south and in a few pockets in the northwest of England, another melanistic form, *insularia,* which differs from *carbonaria* in having white speckling on the wings, became prevalent. The two dark forms have different genetic bases but similarly increase the evolutionary fitness of the individual in polluted woods. One cannot predict beforehand the precise form that adaptation to a particular environmental change will take. Furthermore, a single evolutionary problem may have more than one genetic solution in a species.

Many seemingly imperfect adaptations attest to the fact that the environment cannot stimulate suitable changes in the hereditary material upon which natural selection acts. The defenses against malaria that have evolved in some human populations in tropical areas provide a good example of this. The best known of these defenses is a hereditary change in the blood protein, hemoglobin, known as the sickle-cell trait. The trait (hemoglobin type S), common in many parts of tropical Africa, confers protection from malaria when the gene is inherited from only one parent (Allison 1956, Ingram 1963, Cavalli-Sforza and Bodmer 1971). If the hemoglobin trait is inherited from both parents, the red blood cells can become distorted into a sickle shape, causing a deadly form of anemia. In spite of this negative effect on individual fitness, the sickle-cell trait has increased to a high frequency in many areas because it confers immunity from the malaria parasite to more individuals than it kills by producing anemia. This adaptation, therefore, must be considered successful—but only because no better evolutionary alternative has appeared. It is noteworthy that in parts of western Africa, another blood "defect" (hemoglobin C), which provides effective protection against malaria but which produces a milder form of anemia than hemoglobin S, has nearly replaced the sickle-cell trait.

The opportunism of evolution can be demonstrated particularly well by the ultimate fates of various lines of fish adapted to obtain oxygen directly from air. The fossil record shows that all air-breathing vertebrates, the amphibians, reptiles, mammals, and birds, were derived some 350 million years ago from a group of air-breathing fish, the lungfish, descendants of which have persisted to the present. But other kinds of fish, particularly among groups that inhabit stagnant water, also are able to obtain oxygen directly from air, some being so dependent on this trait that they will drown if kept underwater (Schmidt-Nielsen 1975). Some of these breathe through modified gills, others directly across the skin and the linings of the mouth, throat, and stomach, the latter literally gulping air in order to breathe. Why were the lungfish the antecedents of terrestrial vertebrates? Why not one of the other air breathers? The answer is simply that truly terrestrial vertebrates consume oxygen at a far greater rate than any air-breathing fish, including the lungfish, but only the lungfish possessed a breathing apparatus that could be modified to provide greatly increased ventilation. The volume of the "lung" of the lungfish could be rapidly changed by contraction and relaxation of the body-wall muscles of the thorax, its surface area greatly increased by infolding of the lung wall. A moment's reflection will convince you that these modifications could not be achieved by the gills, skin, stomach, and so on, employed for respiration by other air breathers. One must remember, however, that during their evolution, lungfish were not destined to be terrestrial vertebrates, they were destined to be lungfish, and many still are. That they were also the progenitors of birds and mammals was a fortuitous and happy accident. But that lungfish and not other air-breathing forms had this honor is a logical consequence of their particular adaptations for gas exchange with the atmosphere.

Are all species well adapted?

Because natural selection tends to weed out unfit hereditary traits, populations should be well adapted to their environments. In a very general sense, this is true. Populations are evolutionarily successful, by definition, if they persist. Except where the environment is changing, adaptations, by and large, represent the best genetic possibilities for a population. We often hear people say that the rat, the cockroach, and the dandelion are better adapted than the wolf or the extinct passenger pigeon, which could not cope with the changes man created in the environment. But evolution is not forward looking. Natural selection cannot anticipate such ravages upon the environment as are caused by man. Thus, adaptability to new conditions is not a fair measure of evolutionary adaptation. The wolf was, and still is, well adapted to its existence in wild habitats of North America. Before the appearance of European man in the Western Hemisphere, it was one of the most widespread and abundant of large predatory mammals. But the wolf is shy of man. Its wild habitats are disappearing rapidly, and its body cannot withstand man's bullets. Would we say then that coyotes, which are probably more abundant now than they ever have been in spite of persistent efforts to exterminate them, are better adapted than wolves? Yes, but only because their fearlessness and habit of scavenging for food make them well suited to the conditions being created by man. Wolves remain well adapted to the particular environments in which they evolved.

We cannot judge adaptation by the geographical range or numerical abundance of a particular species. All species are more or less equally well adapted to their own particular environments simply because they persist in them. Species do, however, become extinct, while the populations and distributions of others expand. To the extent that adaptability, the capacity for evolutionary response to environmental change, is responsible for the persistence of a population, it must be a criterion for judging the relative degree of adaptedness of different species. At present, however, we have little more than a rudimentary understanding of the basis for adaptability and why it should differ among species.

Adaptation can never be perfected.

Because natural selection continually favors more fit individuals over those that are less fit, one would expect the average fitness of a population to increase indefinitely if the environment remained unchanged. The effects of natural selection are not unrestrained, however, and several factors continually act to balance the fitness gains of a population resulting from selection, and thereby limit the perfection of adaptation.

A basic genetic force opposing the perfection of adaptation is the spontaneous appearance of unfit hereditary traits in populations. These may be introduced by random mutation of the genetic material or by the immigration of individuals into a population from areas where the environment selects different traits. For example, the interbreeding of peppered moths from unpolluted woods with populations in polluted woods would increase the number of typical, or light-colored, forms, thereby reducing the average fitness of the population in the polluted woods. In a sense, mutation and immigration of genes (gene flow) are like the noise in a radio receiver. No matter how closely the radio is tuned to a station, some static may persist; in the same way, mutation and gene flow occur regardless of how closely a population is adapted to its environment.

As we have seen, evolution is opportunistically based on the variation present in populations. We can be quite sure, for example, that mice could not rapidly evolve the ability to fly, even if it should be advantageous for them to do so. The present adaptations of populations determine the evolutionary pathways that are open. This restriction of evolutionary opportunity greatly constrains the specific adaptations to solve a particular evolutionary problem. Both the gazelle and the ostrich inhabit the savannas of southern Africa and are adapted for rapid movement to escape predators. But the gazelle and the ostrich have solved the problem differently, because the forelimbs (wings) of the ostrich, or of its ancestral species, were not suited for terrestrial locomotion.

Adaptation is also constrained by physical, chemical, and mathematical properties of the natural world that are inaccessible to evolutionary change. An organism cannot adapt to a cold environment by changing the rate of conductance of heat through air or water. To tolerate the cold, it must manipulate such characteristics as the insulative value of fur, which is amenable to evolutionary adjustment. Aquatic vertebrates have adapted to highly saline environments, not by changing the rate of diffusion of salt through water, but by changing the biological properties of their cell membranes to reduce permeability to salt and to effect active secretion of salt from inside the cell.

Detailed analysis of physical constraints to adaptation have led to some surprising discoveries. A locust's jump, for example, has been analyzed and timed down to its last detail (Alexander 1968). Calculations based on the laws of motion and on measurements of the ability of muscle to generate power by contraction proved that locusts could not possibly jump as far as they do, though locusts were apparently unimpressed by the calculations. This contradiction was finally reconciled by the discovery that the leg muscle not only contracts very rapidly but that, just before the locust jumps, the muscle is stretched like a rubber band to give the extra boost needed to account for the biophysicists' calculations.

Adaptation also leads to compromise because organisms have at their disposal limited time, energy, and body constituents to allocate to the many functions they need to survive and reproduce. Bats could not evolve an increased proportion of flight muscles to improve flight capabilities without decreasing some other component, perhaps the weight of their digestive organs. Plants cannot increase their seed crop without reducing their annual growth rate. The bright plumage of some birds may make them more attractive to mates, but it also makes them more conspicuous to predators. Heavy armor reduces mobility. One seemingly cannot have the best of two worlds. The compromise achieved must be a delicate balance among the various functions performed by an organism.

Perhaps the most formidable obstacle to the perfection of adaptation is environmental change, which forces populations to continually readjust their adaptations to new environmental conditions. Major changes in the physical and chemical environment are perhaps less disruptive than changes in the biotic environment, because populations may shift their geographical ranges to match changes in climate. Most climates now present on earth have existed for long periods, although perhaps not in the same location. Changes in the biological environment of a species pose more persistent evolutionary problems. A new invader of a habitat or an evolutionary change in a resident will alter the environment for other species. These, in turn, respond to the new challenge by new adaptations of their own, often leading to a stalemate. The cowbird or cuckoo discovers (in an evolutionary sense) that it can improve its fitness by laying its eggs in the nests of other birds. The hosts counter this evolutionary move with the evolution of discrimination against foreign objects in their nests. The brood parasite returns the play by evolving egg mimicry to fool the host. Mimicry, in turn, may be countered by the host's evolving aggressive behavior toward adult brood parasites. And on it goes, this game of counteradaptation, with populations continually striving to foil the evolutionary strategies of other populations that compete with or exploit them.

Who measures fitness?

The evolutionary success of a trait is directly related to its survival value, or fitness, compared with other expressions of the same character: long versus short limbs, brown versus black fur, cold versus warm temperature adaptation. The competition between such traits for their persistence in the population is what Darwin referred to as the struggle for existence. But what determines the relative fitness of traits? What natural agent selects among them? Predators quite literally select against moths whose coloration does not match their resting backgrounds, but can one say that large seeds select birds with beaks large enough to crack them? This could not be in the seed's best interest, but surely as there are large seeds in the world, there are birds with large, powerful beaks, not to mention squirrels, seed-eating insects, seed molds, and others bent on destroying the seed for their own benefit. Come to think of it, selective predation of conspicuous moths could not serve the best interests of the predator because selection favoring cryptic forms ultimately makes the prey more difficult to find.

Selection is not the strategy of selective agents in the environment. Rather, it is the strategy of traits competing with others in the population. The fitness of an individual in the Darwinian sense is meaningful only in comparison with that of other individuals. Of course, competition is not limited to individuals of the same species but extends to individuals of other species with similar ecological requirements. Competition within and between species can have greatly different outcomes. The first results in the evolutionary replacement of traits within a population through the selective death of individuals, whereas the second can cause the replacement of entire populations by those more fit. Although fitness may be expressed absolutely in terms of progeny, it assumes evolutionary meaning only in relation to the fitness of competing genotypes. The outcome of competition is the most critical yardstick by which fitness is measured. It is one of the most basic, and yet most elusive of ecological principles, one which we shall touch on many times again in this book.

Cultural change is analogous to genetic evolution.

Our typically human characteristics are not all inherited. Years of struggle to master mathematics or a foreign language drive home the importance of learning to human development. Language, the ability to use tools, the knowledge accumulated during a lifetime are as much a part of the human organism as its structure and physiology. Just as physical traits are passed from generation to generation through the genetic material, cultural traits are also passed on through teaching and learning. Anyone who has studied anthropology or linguistics will recognize that cultural characteristics change in an evolution-like manner, much as genetic information changes through the process of evolution. These are parallel mechanisms, differing between species only in emphasis. In simple organisms the genetic line predominates, but in organisms with progressively higher levels of complexity, one finds increasing reliance on cultural information and learning. To specify genetically the complex social interactions of many higher vertebrates and the language of man would surely be impossible. Of course, not all learning is the child of culture. Organisms are impressionable to a remarkable degree, and their development frequently reflects the particular environment of each individual. This is not the learning of culture as such, because the environment exists apart from the population and its lessons are assimilated independently by each organism.

One interesting example from England illustrates particularly well the incorporation of a new cultural trait into a population. Some years ago, when milk was still delivered to the doorstep each morning in bottles capped with tin foil, residents of one locality became annoyed when some sort of animal began to peck open the bottle caps and take some of the cream from the top (milk was not homogenized in those days). The culprit turned out to be the great tit, a small bird closely related to the chickadees and titmice of North America. In a few years, bottle opening had spread throughout the neighborhood and eventually to a large part of England (Fisher and Hinde 1949, Hinde and Fisher 1952). The habit spread too rapidly to have been a genetically inherited trait; individuals had learned the behavior by watching others.

New cultural traits in most animals probably originate as accidental extensions of normal behavior (Alcock 1972). Undoubtedly, one bird must have accidently, or out of curiosity, pecked on the top of a milk bottle one morning and found its contents tasty. Pecking is not novel behavior for titmice. In their search for food they frequently hammer at the bark of trees and at pine cones to uncover insects that might be within. And so the habit of stealing the cream from bottled milk may have developed as the application of a normal behavior pattern to a novel source of food—the cultural analogue of genetic mutation. Learning situations are common among animals, especially birds and mammals, and there are many examples of the spread of new habits through a population by learning and imitation (Strum 1975, Thorpe 1963).

Learning is cultural opportunism, enabling individuals to take the greatest advantage of their surroundings. Opportunism occurs, too, in developmental flexibility, where the growing organism adjusts itself to its surroundings, but developmental responses cannot be passed on to other individuals in the population.

Learning may also play an important role in the cultural transmission of such species-specific traits as song and other vocalizations. In some birds, song is innate and no learning is required for its full development, but many birds must hear the song of other individuals before they themselves can produce the characteristic song of the population (Marler 1970). We must realize, however, that the innate ability and propensity to learn species-specific characteristics is itself a genetically determined trait with adaptive value.

Genetics and culture can be closely related. Organisms that learn species-specific traits may also be genetically programmed within narrow limits for the specific kinds of characteristics that they will actually learn. Acceptable cultural information is specified approximately by genetic instructions, perhaps concerning the frequency range and general pattern of sounds to be learned, the physical appearance of organisms from which information is to be learned, and so on, but the precise characteristics of cultural traits are acquired only through learning. Thus, genetics acts as the disciplinarian for the learning of culture and subdues eclectic and dilettante tastes in the pupil.

Cultural evolution differs from genetic evolution in several important respects. A cultural trait can spread more rapidly through a population because it may be learned by siblings, parents, and unrelated individuals as well as by progeny; a genetic trait can be passed only from parent to offspring. The spread of a new cultural characteristic need not rely on differential survival and reproduction through many generations. Cultural traits are subject to selection inasmuch as the adoption of a behavior pattern by an individual affects the fitness of the individual. Cultural evolution may, however, occur independently of genetic evolution. For example, while the early development of human language was accompanied by genetically inherited modifications of the brain and vocal apparatus, the subsequent elaboration and diversification of languages have been entirely cultural.

Man's astounding biological success may be attributed in part to the extremely rapid rate of cultural evolution. Man's culture evolves too fast for most organisms to keep pace with it genetically. Only bacteria and other small organisms, with short generations and rapid turnover of individuals in populations, can evolve genetically at a fast enough rate to keep abreast of such cultural innovations as antibiotics and pesticides.

The modes by which cultural information is transmitted more frequently produce mistakes than occur in genetic inheritance, thereby increasing the variability available for cultural evolution. Cultural transmission by learning cannot match the precision of molecular replication involved in the inheritance of genetic information. But the most significant aspect of cultural evolution for man, and possibly for some other higher vertebrates, is that of inventiveness, or directed change in culture. Whereas mutation and natural selection are basically random and opportunistic, man can purposefully change his culture to solve specific problems. In this way, cultural evolution is greatly speeded, giving man a significant edge over his evolutionarily more conservative cohabitants on this planet.

The genetic relationship between individuals and agents of selection influences the pattern of adaptation.

Because closely related organisms have genetic traits in common, they share a common evolutionary future and may evolve cooperative or closely coordinated adaptations to serve their mutual evolutionary interests. Unrelated organisms interact only through competition or exploitation, and their adaptations are usually mutually antagonistic and can lead to extinction of populations. Agents of selection in the environment of an organism may be usefully distinguished by their genetic relationship to the organism: the *physical-chemical* environment (abiotic); the *biotic* environment (organisms belonging to other species); the *social* environment (unrelated or distantly related individuals of the same species); the *sexual* environment (adult individuals of opposite sexes, related through their progeny); and the *parent-offspring* environment.

The physical-chemical environment is evolutionarily passive and does not respond adaptively to evolutionary changes in living organisms. The physical-chemical environment provides more or less stable constraints for biological evolution. The biotic environment (including predators, parasites, prey, and competitors) is the vital world of adaptation and counteradaptation. In the biotic environment, the individual interacts with individuals of other species whose adaptations are mostly antagonistic: swift, sharp-eyed predators, efficient competitors, and parasites. Unrelated species occasionally enter into cooperative relationships that involve co-evolution of their adaptations for mutual benefit, but the opportunities for such mutualism are restricted.

The social environment, consisting of unrelated individuals of the same species, has a functional and genetic relationship to the organism different from that of the biotic environment. The relationship is usually competitive but results in evolutionary change by the population rather than extinction and replacement by another population. The sexual environment derives from the interaction between two individuals of opposite sex for the purpose of mating and rearing young. This is a truly cooperative venture because both parents contribute to the genotypic makeup of their offspring. Adaptations for sexual interaction exhibit a high degree of mutual accommodation, but an equally close relationship exists between parent and offspring. It is here that we see the greatest integration of adaptations, to the point where, as with the mammalian embryo in the uterus of its mother, the two organisms are essentially inseparable.

The next series of chapters takes up the relationship between the organism and the various components of its environment, organized according to the degree of genetic relationship just outlined.

Part 4

Organisms in Their Physical Environments

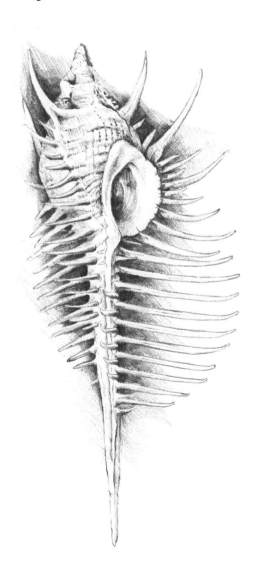

10

Basic Properties and
Requirements of Life

Life is an extension of the physical world. Biological systems have unique properties, but they nonetheless must obey the constraints imposed by physical and chemical properties of the environment and of organisms themselves. Life is intimately bound to the physical world for three essential needs: water, as a medium for life processes; energy, to fire the engines of living systems; and chemical nutrients (carbon, nitrogen, phosphorus, and others), which are the building blocks of life.

But in spite of the utter dependence of life on the physical environment, the essence of living systems is that they maintain themselves out of equilibrium with the physical world. This standoff posture is expensive. Its cost may be measured by the energy transformations of living systems, just as one measures the cost of transportation by fuel used per mile or the cost of electric heating by kilowatt-hours.

This and the following four chapters are about life in its physical and chemical milieu. Here we contrast the physical properties of organisms and their surroundings, examine the suitability of the environment for life processes, and consider the interface between life and its environment. To do so, we shall examine in detail some biological solutions to the problems of gas exchange and salt balance. There is a perpetual tension between the living and nonliving, for although living systems need much of what surrounds them, they must also maintain their distinctiveness. Terrestrial animals obtain oxygen from the atmosphere, but they must guard against the loss of water from their bodies at the same time. Marine fish drink water and pump salts out of their bodies through their gills and kidneys to maintain the proper salt concentration of their blood, which is only half that of seawater. Animals and plants are open, dynamic systems, continually exchanging with the environment to maintain their activity, continually using the fruits of their activities to maintain their integrity.

Living systems have always had to come to terms with their surroundings. These terms are the subject of this discussion. The physiologist Lawrence J. Henderson remarked in his book, *The Fitness of the Environment* (1913), that "Darwinian fitness is compounded of a mutual relationship be-

117

tween the organism and the environment. Of this, fitness of environment is quite as essential a component as the fitness which arises in the process of organic evolution; and in fundamental characteristics the actual environment is the fittest possible abode of life." Although we understand that the physical world does not adapt to the presence of life, it is no accident that life, as we know it, is both constrained by the physical world and takes advantages of its opportunities.

Life processes take place in an aqueous medium. All organisms are composed mostly of water, whether they dwell in the oceans, lakes, and rivers, or on the land. Because the physical and chemical properties of water are well suited to the requirements of life, it is no accident that life is a water-based phenomenon. Water is generally abundant on the Earth's surface, and within the temperature range normally encountered, it is present in a liquid state. This property is fundamental to the occurrence of life, for liquids are dense and, at the same time, fluid. If life processes are to work, various compounds must be brought into close proximity to react with each other. Gases are too diffuse and have too little substance to support such processes; solids are too rigid and constraining. Water also has many thermal and solvent properties that are favorable to life. A large amount of heat energy must be added to or removed from water to change its temperature, and water, therefore, does not change temperature rapidly. Because of this property, and the fact that heat travels rapidly through water, the temperatures of organisms and aquatic environments tend to be relatively constant and homogeneous. In addition, the formidable capacity of water to dissolve inorganic compounds makes these materials accessible to living systems.

Heat and temperature are often used interchangeably to refer to some property we sense in bath water, but the terms have quite distinct meanings. *Heat* is a measure of the energy content of a substance, the total kinetic energy of its molecules. *Temperature* is a measure of the rate of motion of molecules in a substance. At a given temperature, individual molecules in different substances have similar kinetic energies, but the substances may contain different amounts of heat energy depending on their density and the relative weights of their molecules. For example, a cubic meter of water at 30 C contains about 500 times as much heat as the same volume of air at 30 C, because the water contains so many more molecules.

Changes in the amount of heat energy in a substance are related to temperature change by a quantity called *specific heat*, the amount of heat energy that must be added to or removed from a substance to change its temperature by a specific amount. A small amount of heat applied to a given volume of air changes its temperature rapidly, but the same amount of heat added to an equal volume of water would change its temperature little. This is why it takes longer to bring a large pot of water to a boil than to heat up an oven, and why air temperatures fluctuate between greater extremes than the temperatures of oceans and large lakes.

Physicists measure the specific heat of a substance by the amount of heat energy that must be added or removed to cause a one degree Celsius increase or decrease in its temperature. By definition, one calorie (cal) of heat energy is the amount required to raise the temperature of one gram of water by one degree Celsius. In contrast, only one-tenth as much energy is needed to change the temperature of iron. The specific heats of all substances are compared in terms of calories per gram per degree Celsius ($cal/g \cdot C$ or $cal \cdot g^{-1} \cdot C^{-1}$). Whereas the specific heat of water is $1 \, cal \cdot g^{-1} \cdot C^{-1}$, that of iron is $0.1 \, cal \cdot g^{-1} \cdot C^{-1}$. By way of comparison, the specific heats of two common liquids, ethanol and carbon tetrachloride, are 0.59 and 0.21 $cal \cdot g^{-1} \cdot C^{-1}$. Not only does water resist change in temperature, it also resists change of state between solid (ice), liquid, and gaseous (steam) phases. Evaporation requires the addition of 536 calories per gram of water (*heat of vaporization*); freezing requires the removal of 80 calories per gram (*heat of fusion*).

Another curious thermal property of water is that whereas most substances become more dense when they are cooled, water becomes less dense as it is cooled below 4 C. One consequence of this property is that ice floats. This fortuitous circumstance not only makes ice-skating possible but also prevents the bottoms of lakes and oceans from freezing, thereby permitting the persistence of plants and animals at the bottoms of large water bodies.

The immense solvent properties of water are based on the strong attraction of water molecules for other compounds. Molecules are composed of electrically charged atoms. Common table salt, sodium chloride (NaCl), is made up of a positively charged sodium atom (Na^+) and a negatively charged chlorine atom (Cl^-). When salt is placed in water, the attraction of the water molecules for the charged sodium and chlorine atoms is so great, compared with the bonds that hold the molecule together, that the salt molecule readily dissociates into its component atoms. (Electrically charged atoms, or parts of molecules, are referred to as *ions*.) Thus, as the salt molecules dissolve in water, they dissociate and become closely held by hydrogen bonds to water molecules. The dissociation of sodium chloride into its component ions may be written

$$NaCl \rightleftharpoons Na^+ + Cl^-$$

or as

$$NaCl + H_2O \rightleftharpoons H_2ONa^+ + Cl^-$$

if the role of the water molecules as a solvent is to be portrayed. The arrows indicate that even in solution, ions continually rejoin as well as dissociate. It is the equilibrium between these processes that determines the solubility of the substance. The ability of a substance to dissolve in a liquid depends, in part, on the strength of attraction of the solvent molecules compared with the intramolecular bonds of the substance to be dissolved. For biological processes to occur, it is important that substances be dissolved in water because the dissociated ions of the substance are available to react with other ions in the solution (Miller 1969, Williams 1970, De Vries 1971, Hochachka and Somero 1973).

Life processes occur optimally within a narrow range of temperature.

Life processes, as we know them, are restricted to the temperatures at which water is liquid: 0 C to 100 C. Relatively few organisms can survive above 45 C. A few kinds of bacteria occur in hot springs close to the boiling point of water, and blue-green algae are found in water as hot as 75 C (Brock 1970, Brock and Darland 1970). The properties that permit existence at high temperatures are not well known, but those organisms with greatest heat tolerance are prokaryotes, that is, belonging to groups that lack nuclear membranes and a complicated spindle apparatus for separating the genetic material of daughter cells during division. These structures may be so heat sensitive as to limit the upper temperature distribution of all eukaryotes, organisms that do contain them, to less than 45 C.

Temperatures greater than 50 C are relatively rare—being found only in hot springs and at the soil surface in hot deserts—but temperatures below the freezing point of water occur commonly over large portions of the Earth's surface. If living cells are frozen, the crystal structure of ice prevents the occurrence of most life processes and damages delicate cell structures, rapidly leading to death. A large number of species successfully cope with freezing temperatures in their environments either by maintaining their body temperatures above the freezing point of water or by activating one of a number of mechanisms to resist freezing or to tolerate its effects.

Many organisms reduce the freezing points of their body fluids by adding large quantities of glycerol or other similar organic compound to their blood. These substances act like antifreeze and allow, for example, Antarctic fish to remain active in sea water that is colder than the normal freezing point of blood for fish in temperate or tropical seas. Other species are known to be able to become supercooled. That is, their body fluids can fall below the freezing point without ice crystals being formed. Ice crystals generally form around some object, called a seed, which can be a small ice crystal or some other particulate matter. In the absence of crystal seeds, water or blood may be cooled more than 30 C below its melting point.

Within the range of temperatures from the freezing point of water to the highest temperatures normally encountered on the Earth's surface, temperature has two opposing effects on the life processes. First, heat increases the kinetic energy of molecules and thereby accelerates chemical reactions; the rate of biological processes commonly increases between two and four times for each 10 C rise in temperature (Giese 1968, Schmidt-Nielsen 1975). Second, the specific biological compounds that catalyze biological reactions (enzymes) become unstable and do not function properly at high temperatures. The combination of these two factors results in an optimum temperature range for the occurrence of biological systems. Enzymes are usually adapted to function best within the particular temperature range that corresponds to the normal body temperature range of the organism. No enzyme is fully active over a broad range of temperatures. Enzymes in organisms that inhabit cold environments usually function better in cold temperatures than the enzymes of organisms from warmer climates. Even within an organism the temperature characteristics of certain enzymes can vary. In many arctic mammals and birds, the temperatures of the extremities and surface of the body are lower than they are deep inside the body, and the temperature optima of enzymes from these different parts of the body vary accordingly.

Life depends on the availability of inorganic nutrients.

Organisms require a wide variety of chemical elements to form their structure and maintain their proper function. The elements required in greatest amount, after hydrogen, oxygen, and carbon (which are assimilated by plants during photosynthesis), are nitrogen, phosphorus, sulfur, potassium, calcium, magnesium, and iron. Their primary functions are summarized in Table 10-1. Many other nutrients are known to be required in smaller quantity (Treshow 1970).

Mineral nutrients are acquired by plants in the form of dissolved ions, electrically charged parts of compounds that have dissociated in water. Plants obtain nitrogen in the form of ammonia ion (NH_4^+) or nitrate ion (NH_3^-), phosphorus in the form of phosphate ion (PO_4^-), calcium and potassium in the form of their simple (elemental) ions (Ca^{++}, K^+), and so on. The solubility of these substances, which determines their availability, varies with temperature, acidity, and the presence of other ions.

All natural waters contain some dissolved substances. Although rain water is nearly pure, it invariably acquires some dissolved minerals from dust particles and droplets of ocean spray in the atmosphere (Ingham 1950). Most lakes and rivers contain 0.01 to 0.02 per cent dissolved minerals, roughly 1/20 to 1/40 the average salt-concentration of the oceans (3.5 per cent), where salts and other minerals have accumulated over the millennia. In hot climates, where dissolved substances are concentrated by the evaporation of

TABLE 10-1

Major nutrients required by living organisms, and their functions.

Element*	Function
Nitrogen (N)	Structural component of proteins and nucleic acids
Phosphorus (P)	Structural component of nucleic acids, phospholipids, and bone
Potassium (K)	Major solute in animal cells
Sulfur (S)	Structural component of many proteins
Calcium (Ca)	Regulator of cell permeability; structural component of bone and material between woody plant cells
Magnesium (Mg)	Structural component of chlorophyll; involved in function of many enzymes
Iron (Fe)	Structural component of hemoglobin and many enzymes
Sodium (Na)	Major solute in extracellular fluids of animals

* Chemical symbol in parentheses

water more rapidly than rainfall dilutes them, lakes without natural drainage outlets—the Dead Sea and the Great Salt Lake of Utah are examples—may contain up to ten per cent dissolved substances.

Dissolved minerals in fresh and salt water differ in composition as well as in quantity (Table 10-2). Sea water abounds in sodium and chlorine, with respectable amounts of magnesium and sulfate. Fresh water contains a more even distribution of diverse ions, but calcium is usually the most abundant cation (positively charged ion) and carbonate and sulfate the most abundant anions (negatively charged ions). The difference in composition between fresh and salt water is due to the different solubilities of different substances. Calcium carbonate is relatively insoluble (0.0014 per cent in pure water) and precipitates to form limestone sediments before attaining a high concentration in the oceans. At the other extreme, the solubility of sodium chloride (35.7 g/100 g water, or 26.3 per cent) far exceeds its concentration in sea water; nearly all the sodium chloride that has been washed into ocean basins remains dissolved. The solubility of magnesium sulfate (26.0 g/100 g water) is intermediate.

The pervasive tension between biological systems and the physical world extends to minerals in the environment. On one hand, organisms must obtain minerals from the soil, water, or their food. On the other, they must maintain concentrations of some minerals in their bodies at different levels from those that occur in the environment. For terrestrial and fresh-water organisms this means conserving ions within the body fluids and preventing their being washed out of the body either through the skin or in urine. For many salt-water organisms, abundant ions in the surrounding milieu must be kept out. The blood plasma of the frog, and most vertebrates whether terrestrial or aquatic, resembles sea water in its ionic composition, although it is only one-third as concentrated (Table 10-2). The intracellular medium contains abundant potassium, but plasma ions are kept out to a large extent.

TABLE 10-2

Percentage composition of dissolved minerals in rivers (fresh water), in sea water, and in the blood plasma and cells of frogs (from Reid 1961 and Gordon 1968).

Mineral ion	Delaware River*	Rio Grande River*	Sea water	Frog plasma	Frog cells
Sodium	6.7	14.8	30.4	35.4	1.3
Potassium	1.5	0.9	1.1	1.3	77.7
Calcium	17.5	13.7	1.2	1.2	3.1
Magnesium	4.8	3.0	3.7	0.4	5.3
Chlorine	4.2	21.7	55.2	39.0	0.8
Sulfate	17.5	30.1	7.7	—	—
Carbonate	33.0	11.6	0.4	22.7	11.7

* The percentages of the negatively charged ions (Cl^-, $SO_4^=$, and $CO_3^=$) exceed those of the positively charged ions because, ion for ion, anions are much the heavier. The numbers of positive and negative ions are approximately equal.

Kidneys and other salt-concentrating organs maintain the ionic balance of an organism.

Left to their own devices, ions diffuse across cell membranes from regions of high to low concentration, thereby tending to equalize their concentrations. Water also moves across membranes (*osmosis*) toward regions of high ion concentrations, tending to dilute concentrations of dissolved minerals. Maintaining an ionic imbalance between the internal medium of the organism and the surrounding environment (*osmoregulation*) against the physical forces of diffusion and osmosis requires the expenditure of energy by organs specialized for salt retention or excretion (Potts and Parry 1964).

In terrestrial and fresh-water organisms, ion retention is critical. In fresh-water fish, for example, water continually enters the body by osmosis through the mouth and gills, which are the most permeable tissues exposed to the surroundings. (The skin is relatively impermeable.) To counter this influx, water is continually eliminated in the urine, but if dissolved ions were not selectively retained the fish would soon be a lifeless bag of water. Retention is effected by the kidneys, where salts are actively removed from the urine and fed back into the bloodstream. In addition, the gills are capable of selectively absorbing dissolved ions from the surrounding water and secreting them into the bloodstream. Terrestrial animals acquire minerals in the water they drink and the food they eat; plants absorb ions dissolved in soil water.

It would be quite futile for a salt-water fish, with blood containing half the concentration of mineral ions as sea water, to try to dilute the oceans with water from its body. Keeping ions out of the body is as big a problem for salt-water fish as the retention of ions is for fresh-water species. In contrast to their fresh-water relatives, the gills and kidneys of salt-water fish actively *excrete* ions to counter the tendency of ions to diffuse into the body from the surrounding water. Marine fish also drink sea water to replenish water lost in urine and by osmosis across the surfaces of the gills and, to a lesser extent, the skin. The sharks and rays have achieved a rather elegant solution to the problem of osmotic (water) balance. Urea—$CO(NH_2)_2$—a normal nitrogenous waste product of metabolism in vertebrates, is retained in the bloodstream, instead of being excreted in the urine, to raise the ionic concentration of the blood to the level of sea water without having to increase the concentration of sodium chloride. Although sharks and rays must regulate the diffusion of specific ions in and out of the body, the high level of urea in the blood effectively cancels the tendency of water to leave the body by osmosis. And inasmuch as sharks do not have to drink water to replace osmotic losses of water, they also do not ingest large quantities of salt. It is interesting to note that the few fresh-water sharks and rays do not accumulate urea in their blood.

Salt excretion is not limited to marine fishes. Oceanic birds and reptiles ingest more salt in their food than their kidneys can excrete, and many species have additional regulatory organs known as salt glands, often located in the socket of their eye. Salt glands excrete salt at extremely high concentrations to help rid the body of excesses while conserving water

(Peaker and Linzell 1975). Some species of mangrove trees, whose roots are immersed in sea water, actively secrete salt from their leaves for the same purpose.

Light energy is the driving force of life processes.

Light is the primary source of energy for the ecosystem. Green plants absorb light and assimilate its energy into manufactured organic compounds by *photosynthesis* (literally, a putting together with light). The resulting chemical energy in turn is used by plants, and by the animals that consume plants, as a source of energy for other biological processes. The basic reaction in photosynthesis is as follows:

carbon dioxide + water + light energy →
glucose (high in chemical energy) + oxygen

Light energy is absorbed by organic pigments, such as chlorophyll and various types of carotenes. This energy is then used, by way of a long series of steps, to combine carbon dioxide and hydrogen obtained from water molecules into glucose ($C_6H_{12}O_6$). The excess oxygen produced by photosynthesis is released into the atmosphere or into the water surrounding the plant. Glucose may be split up and its atoms rearranged and combined to form other organic compounds. Other elements (nitrogen, sulphur, phosphorus, and so on) may also be added along the way to form proteins, complex sugars, fats, nucleic acids, and pigments.

The absorption of light by plant pigments is the first step in the production of all organic compounds. But not all the light striking the Earth's surface is useful in photosynthesis. Rainbows and prisms show that light is made up of a spectrum of different wavelengths that we perceive as different colors. Wavelengths of light are expressed interchangeably in terms of a variety of units. The micron (μ) is one-millionth of a meter or one-thousandth of a millimeter; the visible spectrum extends between about 0.4μ (violet) and 0.7μ (red). The millimicron ($m\mu$), used in this book, is one thousandth of a micron, hence 400 to 700 $m\mu$ define the visual spectrum. The Angstrom unit (Å) is one-tenth of a millimicron, or 10^{-10} meters. The visual spectrum lies between 4000 and 7000 Å. The energy content of light varies with wavelength and hence with color; short-wavelength blue light represents a higher energy level than longer wavelength red light.

The light that reaches the Earth from the Sun actually extends in quality far beyond the visible range: through the ultraviolet region toward the short-wavelength, high-energy X rays at one end of the spectrum, and through the infrared region to such extremely long wavelength, low energy radiation as radio waves at the other end of the spectrum (Figure 10-1). As light passes through the upper atmosphere of the Earth, most of its ultraviolet components are absorbed, primarily by a form of oxygen known as ozone, which occurs in the upper atmosphere. Because of its high energy level, ultraviolet light can damage exposed cells and tissues. (Sunburn is a symptom of overexposure to ultraviolet radiation.)

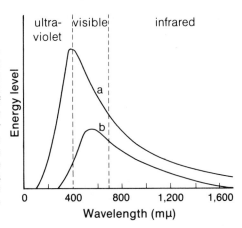

Figure 10-1   The spectral distribution of the energy in sunlight above the Earth's atmosphere (a) and at the Earth's surface (b). Most energy in the ultraviolet portion of the spectrum, specifically below 290 mμ, is absorbed in the upper atmosphere by ozone (O_3). The range of wavelengths to which the human eye is sensitive, which is the visible part of the spectrum, occurs within the region of highest energy level (400 to 700 mμ) (after Hutchinson 1957).

Vision and the photochemical conversion of light energy to chemical energy by plants occur primarily within that portion of the solar spectrum containing the greatest amount of energy. Absorption of radiant energy depends on the nature of the absorbing substance. Water has relatively little capacity to absorb light whose characteristic wavelengths fall in the visible region of the spectrum of energies and, as a result, its appearance is "colorless." Dyes and pigments are strong light absorbers of some wavelengths in the visible region and reflect or transmit light of definite colors that become identifying characteristics. Plant leaves contain several kinds of pigments, particularly chlorophylls (green) and carotenoids (yellow), that absorb light and harness its energy. Carotenoids, which give carrots their orange color, absorb primarily blue and green light (Figure 10-2) and reflect light in the yellow and orange portions. Chlorophyll absorbs light in the red and violet portions of the spectrum, and reflects green and blue. When chlorophylls and carotenoids occur together in a leaf, green light is absorbed least, hence the color of the leaf.

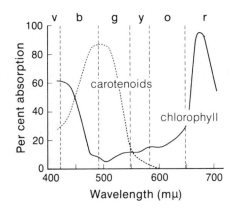

Figure 10-2   Absorption of light of different wavelengths (the absorption spectrum) by two groups of plant pigments—chlorophyll and carotenoids—which capture light energy for photosynthesis. The colors of the spectrum are violet (v), blue (b), green (g), yellow (y), orange (o), and red (r) (after Emerson and Lewis 1942).

The availability of oxygen limits biological activity.

The chemical energy in organic compounds is made available to living systems by the converse process to photosynthesis, called *respiration.* In the respiratory metabolism of a simple sugar like glucose, the sugar molecule is oxidized to produce carbon dioxide and water (the raw materials for photosynthesis) and release energy to drive other biological processes. The overall equation for respiration is

$$\text{glucose} + \text{oxygen} \rightarrow \text{carbon dioxide} + \text{water} + \text{energy}$$

Because oxygen plays such an important role in making energy available, its occurrence in the environment can limit the level of metabolic activity of the organism. The manner in which different kinds of organisms procure oxygen demonstrates the importance of the physical environment to the design of organisms (Table 10-3).

Small aquatic organisms usually obtain oxygen by diffusion from the surrounding environment into their tissues. Carbon dioxide produced by respiration also escapes the body by the same route. If the concentration of oxygen in an organism's tissues is lower than that of the surrounding

TABLE 10-3
Summary of some problems in the evolution of high rates of oxygen delivery to the tissues of large organisms.

Problem	Solution	Biological occurrence
Not critical in small or inactive organisms	Oxygen is obtained by simple diffusion through cells	Protozoa, sponges, coelenterates
Diffusion distance from surface to body core is too great in large organisms	Circulatory system to pump fluids from surface to core	Widespread: pumping by body muscles in roundworms; open system without capillaries in arthropods and many molluscs; closed capillary systems in vertebrates
The solubility of oxygen in water limits oxygen transport by circulating fluids	Incorporation of oxygen-binding proteins (*e.g.,* hemoglobin) in blood	Hemoglobin is widespread in vertebrates, but sporadic in lower groups where other pigments may be found; insects notably lack blood pigments because air is carried directly to cells by a tracheal system
High concentrations of protein increase osmotic level of blood	Respiratory proteins are tightly packed in red blood cells	All vertebrates, some molluscs and echinoderms

medium, oxygen will diffuse into the body. And because the oxygen is used in respiratory metabolism, its concentration in the body is kept low.

Diffusion can satisfy the oxygen needs of very small aquatic organisms (and the leaves of terrestrial plants), but in large organisms the distance between the external environment and the center of the body is too great for diffusion to ensure a rapid supply of oxygen. In fact, diffusion is ineffective at distances greater than about one millimeter (Schmidt-Nielsen 1975:15). One solution to this problem has been the evolution of circulatory systems. Oxygen that diffuses across the surface of the organism, or across the surfaces of specialized structures with large areas in direct contact with the environment (lungs and gills) is carried to other parts of the organism by circulating body fluids.

The movement of aqueous fluids through a system of vessels greatly aids the distribution of oxygen and other materials throughout the body, but water itself often cannot carry enough dissolved oxygen to support a high rate of activity. The solubility of oxygen in water (up to 1 per cent by volume or about fourteen parts per million by weight) just cannot supply sufficient oxygen to active tissues.

To increase the oxygen-carrying capacity of their blood, most groups of animals have complex protein molecules, such as hemoglobin, to which oxygen molecules readily attach and thereby are taken out of solution. When oxygen becomes attached to hemoglobin—four molecules of oxygen for each one of hemoglobin—a complex known as oxyhemoglobin is formed. The binding process must be reversible so that oxygen can be released to the tissues. While blood plasma itself carries only limited oxygen in solution, up to fifty times more is transported in the bloodstream, bound to oxygen-carrying molecules. Hemoglobin is most effective in binding oxygen when its molecules are present in very high concentration, close to the point of crystallization. If such high concentrations of hemoglobin occurred in the plasma, the osmotic level of the blood would become too great for proper physiological function, particularly for proper salt balance, so hemoglobin is concentrated inside red blood cells (erythrocytes), each of which may contain upward of a quarter billion hemoglobin molecules (Schmidt-Neilsen and Taylor 1968). A further advantage to packaging hemoglobin in red blood cells may be that the close association of hemoglobin molecules may alter their interaction among each other and with the surrounding blood environment in such a way as to facilitate oxygen binding and release. An analogous arrangement is found in several worms and snails, which have hemoglobin but lack special blood cells. In these species, hemoglobin is aggregated into large groups of, perhaps, twenty to fifty molecules (Schmidt-Nielsen 1975:83).

Another recurrent adaptation to enhance the uptake of dissolved oxygen from water is countercurrent circulation, a particular arrangement of the structure of gills whereby water and blood flow in opposite directions (Figure 10-3). In a countercurrent system, as blood picks up oxygen from the water flowing past, it comes into contact with water having progressively greater oxygen concentration. This is possible because the water has flowed past a progressively shorter distance of the gill lamella (Figure 10-4). With this

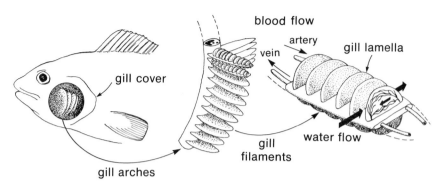

Figure 10-3   A fish's gill consists of several gill arches, each of which carries two rows of filaments. The filaments bear thin lamellae oriented in the direction of the flow of water through the gill. Within the lamellae, blood flows in a countercurrent manner in opposite direction to the movement of water past the surface (from Randall 1968).

arrangement, the oxygen concentration of the blood plasma can approach very nearly the concentration in the surrounding water. If blood and water were to flow together through the gill, an equilibrium oxygen concentration would be established with equal and intermediate levels in the blood and water. The countercurrent system keeps the blood and water out of equilibrium and maintains a constant gradient across which oxygen can flow.

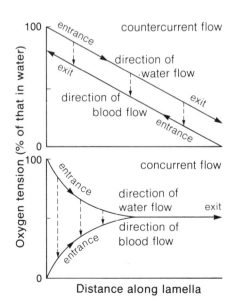

Figure 10-4   A schematic comparison of changes in oxygen tension of blood and water in concurrent and countercurrent systems. In the latter, a constant gradient is maintained across which oxygen can diffuse, resulting in a higher oxygen tension in the blood leaving the system, approaching that of the incoming water (adapted from Schmidt-Nielsen 1975:22).

The rate of oxygen uptake depends, over and above the oxygen gradient between the gill and the surrounding water, on the area of the gills, or other respiratory surface. Active fish with high oxygen requirements have large gills compared with their more sedentary relatives (Gray 1954). Pound for pound of body weight, the gills of the predatory mackerel have fifty times the surface area of the gills of the sedentary, bottom-dwelling goosefish.

The oxygen-binding capacity of hemoglobin is a compromise between oxygen availability and requirement.

Active organisms require an abundant supply of oxygen for cell respiration. Hemoglobin, and other oxygen-carrying pigments such as hemocyanin, act as go-betweens for the uptake of oxygen from the surrounding environment and its release to the cells. The two functions are conflicting, however, for the hemoglobin molecule that readily binds oxygen that has diffused from the environment into the blood plasma tends to be tenacious when the oxygen is to be released to active tissues.

The compromise between the oxygen-uptake and -release functions of hemoglobin, or other blood pigments, can be seen in its oxygen dissociation curve (Figure 10-5). The dissociation curve portrays the amount of oxygen bound to hemoglobin, expressed as a per cent of the total possible, as a function of the concentration of oxygen in the blood plasma. The binding of

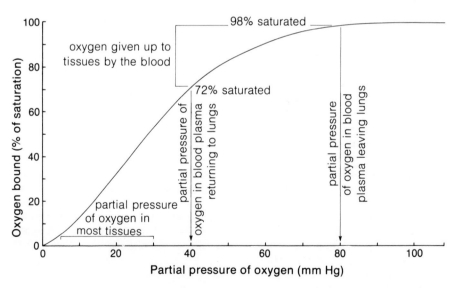

Figure 10-5   The oxygen dissociation curve for human hemoglobin. About 25 to 30 per cent of the oxygen bound to hemoglobin is released to the tissues, which have oxygen tensions of 5 to 30 millimeters of mercury.

oxygen to hemoglobin is reversible, the per cent of bound and unbound oxygen reaching an equilibrium that is described by the expression

$$Hb + O_2 \rightleftharpoons HbO_2$$

As oxygen is added to the blood plasma in the gills or lungs, the equilibrium is shifted to the right and additional oxygen is bound to hemoglobin. As oxygen is removed from the blood as it passes through active tissues, the equilibrium shifts toward the left, and oxygen is released from the oxyhemoglobin complex. In general, the proportion of hemoglobin with bound oxygen increases with greater concentration of dissolved oxygen in the blood plasma until a saturation level is reached and no more oxygen can be bound. Each hemoglobin molecule can bind only four molecules of oxygen.

The concentration of oxygen, or any other gas, dissolved in a liquid is commonly expressed in terms of oxygen tension or partial pressure, whose units correspond to the familiar scale of a barometer: millimeters of mercury (mm Hg). At every water-air interface, gas molecules are continually entering into solution and escaping back into the air. For a given gas under a given pressure, these comings and goings achieve an equilibrium in which there is no *net* movement of the gas molecules either into or out of solution. Under these conditions, the concentration of the gas dissolved in water is expressed in terms of the pressure of the gas in air. Physiologists have adopted this convention because the movement of a gas between air and water (or blood plasma) depends on the difference between oxygen tensions, not on absolute concentrations.

The sea-level atmospheric pressure is 760 millimeters of mercury (mm Hg), or about 30 inches of mercury. Because approximately one-fifth of the atmosphere is oxygen, the partial pressure of oxygen is about one-fifth of 760, or 159 mm Hg. Thus, when oxygen dissolved in water reaches an equilibrium with a sea-level atmosphere, it has a partial pressure of 159 mm Hg. By comparison, the absolute concentration of oxygen in the atmosphere is 209 ml per liter, but dissolved in water it is only 5 to 10 ml per liter at equilibrium (partial pressure = 159 mm Hg), depending on the temperature and salinity.

In humans, oxygen enters the lungs at a partial pressure of 159 mm Hg and diffuses rapidly into the bloodstream of capillaries surrounding the air sacs. When the blood leaves the lungs, oxygen dissolved in the plasma has a partial pressure of 80 mm Hg, and the hemoglobin is about 98 per cent saturated with oxygen. (Saturated blood carries more than 200 ml of oxygen gas per liter of blood, almost fifty times the amount of oxygen that can be dissolved in water at the normal temperature of the human body and very nearly the absolute concentration of oxygen in the atmosphere.) The partial pressure of oxygen in the various tissues of the body ranges from 5 to 30 mm Hg, and oxygen therefore diffuses readily from the blood plasma into the tissues.

The rate at which tissues receive oxygen is proportional to the difference between the partial pressures of the oxygen in the tissues and in the blood: the higher the gradient, the higher the rate of diffusion. When dissolved oxygen leaves the bloodstream to enter the tissues, the partial pressure of oxygen in the bloodstream is reduced. As the difference between oxygen concentrations in the blood and in the tissues decreases, the diffusion of oxygen slows. But as the partial pressure of oxygen in the blood plasma decreases, the oxyhemoglobin releases some bound oxygen, thereby tending to restore the oxygen concentration. Thus oxygen drawn from the blood plasma into the tissues is partially replenished by that bound to hemoglobin.

As blood travels through the human body, hemoglobin loses 25 to 30 per cent of its bound oxygen. The blood returns to the lungs carrying about 144 cc of oxygen per liter. The partial pressure of oxygen in the blood plasma (40 mm Hg) is only a little higher than its concentration in many tissues. Such depleted blood can, however, continue to release oxygen to tissues because of a phenomenon known as the Bohr effect (Riggs 1960, Schmidt-Nielsen 1975:87). As carbon dioxide is released into the blood by active tissues, where it is produced during cellular respiration, the blood plasma becomes increasingly acid owing to the formation of carbonic acid out of carbon dioxide and water. For the hemoglobin of most species, increasing acidity shifts the oxygen dissociation curve to the right, facilitating the release of bound oxygen from hemoglobin and raising the partial pressure of oxygen in the blood plasma.

Small animals generally have oxygen dissociation curves that are shifted to the right for easier unloading of oxygen to satisfy the demands of their high metabolic rates. The dissociation curve of a mammalian embryo is shifted to the left of that of its mother because its oxygen supply derives from the mother's blood, which has a lower partial pressure of oxygen than the air the mother breathes. As the blood supplies of the mother and fetus come into contact in the placenta, the partial pressure of oxygen in the fetal blood nearly doubles (from about 20 to 40 mm Hg in sheep), at which point it is nearly saturated with oxygen (Barron and Meschia 1954). Adult hemoglobin is only about half saturated at the same concentration of oxygen in the blood plasma. Dissociation curves of the aquatic tadpole larvae of many frogs lie to the left of the air-breathing adult stage because of lower concentrations of oxygen in water compared with that in air. Similarly, curves for fish inhabiting stagnant waters that are low in oxygen are shifted to the left compared with those that inhabit well-oxygenated streams and lakes (carp compared with trout, for example). The design of the hemoglobin molecule must allow it to cope with properties of the blood other than oxygen tension. For example, oxygen binding in most vertebrate hemoglobin is inhibited by moderate concentrations of urea in the blood. In sharks and rays, which use high concentrations of urea to adjust the osmotic level of their blood, the hemoglobin is insensitive to urea (Bonaventura et al. 1974).

Whereas the hemoglobin molecule is adapted to both the availability of oxygen in the environment and the oxygen requirement of the organism, the total amount of hemoglobin in the bloodstream is also subject to adjustment. For example, the goosefish has a total oxygen capacity in its blood of 5 per cent by volume; the mackerel, 16 per cent (Lagler, Bardach, and Miller 1962). This difference reflects the relative hemoglobin concentration in the blood and parallels the relative sizes of the gills in the two species, as seen above.

The gas-exchange characteristics of the blood of an individual can also change in response to the environment. When humans from sea-level elevations were transported to an elevation of 17,600 feet for several weeks, the oxygen-carrying capacity of their blood increased from 21 to 25 per cent by volume (210 to 250 ml per liter), but did not reach the 30 per cent capacity of the local high-altitude residents (Prosser and Brown 1961). Again, these differences are related largely to the concentration of hemoglobin in the blood. Increased lung size, breathing rate and volume, heart size, rate and stroke volume of the heart beat, and capillary density are also important adaptations that influence the overall rate of activity that can be sustained at high altitude (Frisancho 1975).

Adaptations for procuring oxygen illustrate a set of solutions to the problems organisms must confront more generally at the interface between themselves and their environments. The design of the hemoglobin molecule—influencing simultaneously its oxygen-binding and -releasing properties—further emphasizes the constraint of compromise in evolution and some of the mechanisms, like the Bohr effect, that enable organisms to eat at least a little bit of the cake. The interface between organism and environment is developed further in the next chapter by considering the rather different problems of terrestrial and aquatic environments.

11

Aquatic and Terrestrial Environments

Life arose in the sea. Conditions in shallow coastal waters were ideal for the development and diversification of the first plants and animals. Temperature and salinity varied little; sunlight, dissolved gases, and minerals were abundant. Water itself is buoyant and supports both delicate structures and massive bodies with equal ease.

The difficulty of the first step in colonizing the land can be measured by the gap of several hundred million years between the time life began to flourish in the sea and the appearance of life on land. Yet in spite of the harshness of terrestrial environments, life has generally attained a higher degree of organic diversity and productivity on the land.

Perhaps we should not distinguish environments as being primarily aquatic or terrestrial, for the sea is as surely underlain by land as the terrestrial environment is drenched in an ocean of air. To appreciate fully the distinction between aquatic and terrestrial environments, we should contrast the properties of water and air rather than those of water and earth. The qualities of water that overwhelmingly determine the form and functioning of aquatic organisms are its density (about 800 times that of air) and its ability to dissolve gases and minerals. Water provides a complete medium for life; most marine organisms are independent of the land beneath them except for those that use it as a site for attachment in shallow water or a place to burrow. In contrast, terrestrial life is narrowly confined to the interface between the atmosphere and the land, each of which makes essential contributions to the environment of life. Air provides oxygen for respiration and carbon dioxide for photosynthesis, while soil is the source of water and minerals. Air also offers less resistance to motion than water, and thus constrains movement less.

Water and air differ in buoyancy and viscosity.

Because water is dense, it provides considerable support for organisms that after all are themselves mostly water. But organisms also contain bone, proteins, dissolved salts, and other materials that are more dense than salt or fresh water, and thus generally tend to sink. To counter this, aquatic plants and animals have a variety of adaptations either to reduce their density or to retard their rate of sinking. Such adaptations are crucial to the tiny plants and animals of the plankton that cannot move actively. Many fish have swim bladders, small enclosures within the body that are filled with gas to equalize the density of their bodies and the surrounding water (Denton 1960). Many large kelps, a type of seaweed found in shallow waters, have analogous gas-filled organs. The kelps are attached to the bottom by holdfasts, and gas-filled bulbs float their leaves to the sunlit surface waters. Many microscopic unicellular plants (*phytoplankton*) that float in great numbers in the surface waters of lakes and oceans contain droplets of oil that are less dense than water and compensate for the natural tendency of cells to sink (Gross and Zeuthen 1948, Steen 1970). Fish and other large marine organisms also make use of lipids to provide buoyancy (Hochachka and Somero 1973:304). Most fats and oils have a density of about 0.90 to 0.93 g · cc^{-1} (90 to 93 per cent of the density of water) and thus tend to float. Aquatic organisms also lighten their bodies by reducing skeleton, musculature, and perhaps even the salt concentration of the body fluids. It has been argued that the low osmotic concentration of the blood plasma of aquatic vertebrates (about one-third to one-half that of sea water) is an adaptation to reduce density. Accumulation of less dense oils and fats and reduction of more dense body components are particularly important to organisms that inhabit very deep water where, under high pressure, the density of gases in the swim bladder increases to nearly the density of water and hence provides little buoyancy.

The high viscosity of water lends a hand to some organisms that would otherwise sink more rapidly, but hampers the movement of others. Tiny

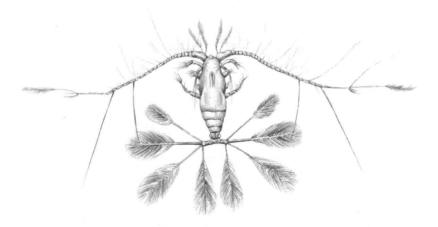

Figure 11-1   Filamentous and feathery projections from the body of a tropical marine planktonic crustacean, *Calocalanus pavo.* Overall length is about 1.2 millimeters (after Wimpenny 1966).

marine animals often have long, filamentous appendages that retard sinking, just as a parachute slows the fall of a body through air (Figure 11-1). The wings of maple seeds, the spider's silk thread, and the tufts on dandelion and milkweed seeds provide a similar function and increase the dispersal range of land species. But to reduce the drag encountered in moving through a medium as viscous as water, fast-moving animals must assume streamlined shapes. Mackerel and other schooling fish of the open ocean closely approach the hydrodynamicist's body of ideal proportions (Figure 11-2). Of course, air offers far less resistance to movement, having less than 1/50 the viscosity of water.

Because water is more buoyant than air, gravity does not limit the maximum size of organisms to the extent that it does on land. Blue whales may attain more than 100 feet in length and can weigh more than 100 tons, dwarfing the largest land animals. (Large elephants weigh only seven tons.) That water provides excellent support against gravity is illustrated by the skeletons of sharks, which are composed of flexible cartilage, and therefore would offer little support on land. Even the air-breathing whales suffocate quickly when they are accidently stranded on a beach because their great weight deflates their lungs. By contrast, more terrestrial organisms have strong supporting structures to keep their bodies upright against the pull of gravity. The bony internal skeletons of vertebrates, the chitinous exoskeleton of insects, the rigid cellulose walls of plant cells all provide the same function: support. Rigid structures occur in aquatic organisms more for protec-

Figure 11-2   The streamlined shapes of young mackerel reduce the drag of water on the body and allow the fish to swim rapidly with minimum energy expenditure (courtesy of the U.S. Bureau of Commercial Fisheries).

tion (the shells of molluscs) or to provide rigid attachment sites for muscles (the shells of crabs and bony skeletons of fish) than to support the weight of the body.

The attenuation of light in water limits photosynthesis in aquatic environments.

Light strikes the surface of the oceans and the land with equal intensity. On land, most light is absorbed or reflected by the leaves of plants. But unlike air, water absorbs or scatters light strongly enough to severely limit the depth of the sunlit zone of the sea. The transparency of a glass of water is deceptive. In pure sea water, the energy content in the visible part of the spectrum diminishes to 50 per cent of its surface value within a depth of 10 meters, and to less than 7 per cent within a depth of 100 meters (Sverdrup 1945). Furthermore, different wavelengths are affected differently by water (Weisskopf 1968). Longer wavelengths are absorbed more rapidly. In fact, virtually all infrared radiation is absorbed within the topmost meter of water. Short light waves (violet and blue) tend to be scattered by water molecules and thus do not penetrate deeply. A consequence of the absorption and scattering of light by water is that with greater depth, green light tends to predominate (Figure 11-3). The photosynthetic pigments of plants are adapted to this spectral shift. Plants near the surface of the oceans, such as the green alga *Ulva* (sea lettuce), have pigments resembling those of terrestrial plants and mostly absorb light of blue and red color (compare Figures 11-3 and 10-2). But the deepwater red alga *Porphyra* has additional pigments that enable it to utilize green light more effectively in photosynthesis.

Because photosynthesis requires light, the depth to which plants are found in the oceans is limited by the penetration of light to a fairly narrow zone close to the surface, in which photosynthesis exceeds plant respiration, called the *euphotic* zone. The lower limit of the euphotic zone, where photosynthesis just balances the rate of respiration, is called the *compensation point.* If algae in the phytoplankton sink below the compensation point or are carried below it by currents, and are not soon returned to the surface by upwelling, they will die.

In some exceptionally clear marine and lake waters, particularly in tropical seas, the compensation point may be a hundred meters below the surface, but this is a rare condition. In productive waters with dense phytoplankton, or in turbid waters with suspended silt particles, the euphotic zone may be as shallow as one meter. In some polluted rivers, little light penetrates beyond a few centimeters.

Because plants require light, large benthic algae (forms attached to the bottom) occur only near the edges of continents where the depth of the water does not exceed a hundred meters. In the vast open reaches of the ocean, as well as in shallower coastal waters, one-celled floating plants compose the phytoplankton of the euphotic zone. The small floating animals (*zooplankton*) that prey upon the phytoplankton are also found primarily in this region where their food is most abundant. But animal life is not restricted to the upper layers of water. Even the deepest parts of the ocean, under several

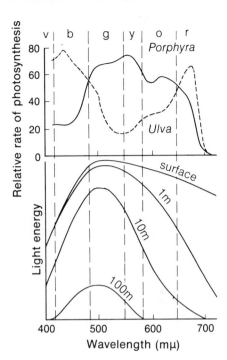

Figure 11-3 Above. Relative rates of photosynthesis by the green alga *Ulva* and the red alga *Porphyra* as a function of the color of light. Note that the red alga is photosynthetically more active in the middle regions of the visible spectrum (after Haxo and Blinks 1950). Below. Relative levels of energy in light of different wavelengths as a function of depth in pure sea water (after Sverdrup 1945).

miles of water, harbor a diverse fauna, supported by the constant rain of dead organisms that sink from the sunlit regions above.

Oxygen is scarce in water.

Nearly all organisms, including green plants, require oxygen for respiration. Although oxygen abounds in the atmosphere, comprising about one-fifth its volume, oxygen does not dissolve readily in water. The solubility of oxygen is affected by both temperature and salinity, but even at its maximum possible solubility, at 0 C in fresh water, the concentration of oxygen is about ten milliliters per liter (1 per cent by volume), or one-twentieth that of air (Krogh 1947). Such concentrations of oxygen are never reached in natural bodies of water, where values probably range from a maximum of about 6 $ml \cdot l^{-1}$ to zero, that is, completely *anaerobic* (no-oxygen) conditions.

To remove oxygen from the environment, aquatic organisms must bring water into contact with their respiratory structures. Inasmuch as the high viscosity of water makes it difficult to move water about, this property also restricts the availability of oxygen to aquatic organisms. Terrestrial animals can move air in and out of their lungs rapidly, compared to the rate at which a fish or clam can pump water past its gills. Schmidt-Nielsen (1975:27) has estimated that with a favorable concentration of oxygen (7 $ml \cdot l^{-1}$, an aquatic organism must pump 100,000 grams of water (about 26 gallons) past its gills to obtain one gram of oxygen. Air breathers would have to inhale only five grams of air (about one gallon) to obtain the same amount of oxygen. Clearly, oxygen uptake in the aquatic environment requires a considerable expenditure of energy to move water, if it is to be done rapidly. It is not surprising that in most water breathers water flows in one direction through their respiratory organs. This arrangement eliminates the need to use energy to periodically stop the flow of a large mass of water and accelerate it in the opposite direction. Such breathing in and out presents a negligible cost to terrestrial organisms because of the small quantities of air involved.

Although oxygen is evenly distributed in the atmosphere, the concentration of dissolved oxygen varies considerably. Oxygen is generally most abundant near the air-water interface, and its concentration decreases with depth. Still water usually contains less oxygen than rapidly flowing water, where mixing of water and air occurs in stream riffles, waterfalls, and waves. Photosynthesis by aquatic plants can also be an important source of dissolved oxygen. But the consumption of oxygen by animals and microorganisms in poorly mixed waters has a more dramatic effect on oxygen tension. Whereas photosynthesis occurs mostly in sunlit surface waters that are normally well aerated, animal and microbial respiration occurs at all depths and is often most intense in sediments at the bottom. In a stratified lake, respiration in the *hypolimnion,* the poorly mixed layer of water below the thermocline (see page 41), tends to deplete the oxygen, leading to anaerobic conditions and a slowing or stopping of life processes. It was such anaerobic conditions in stagnant swamps and deep ocean basins that allowed organic sediments to escape microbial decomposition and become what are now oil and coal deposits.

In addition to oxygen for respiration, plants require carbon dioxide for photosynthesis. Carbon dioxide is one of the lesser gases of the atmosphere, accounting for only 0.03 per cent of the volume of air ($0.3 \, ml \cdot l^{-1}$). But the solubility of carbon dioxide in water is roughly thirty times that of oxygen, and under ideal conditions water contains about 0.3 ml of dissolved carbon dioxide per liter, the same concentration as in the atmosphere. In addition, carbon dioxide and water readily form bicarbonate ion (HCO_3^-), which is extremely soluble in water (69 g of $NaHCO_3$ per liter of water, for example). Sea water normally contains the equivalent of 34 to 56 ml carbon dioxide per liter in the form of bicarbonate (Nicol 1967). Because bicarbonate and dissolved carbon dioxide exist in equilibrium

$$H_2O + CO_2 \rightleftharpoons HCO_3^- + H^+$$

dissolved carbon dioxide utilized by plants is readily replenished from bicarbonate ion. And for this reason, shortages of carbon dioxide rarely occur.

Water loss is a critical problem for terrestrial organisms.

Terrestrial environments pose a severe problem to life: the conservation of water within the body. This problem has had an overriding influence on adaptations of form and function (Hadley 1972, Schmidt-Nielsen 1964). The outer coverings of most truly terrestrial organisms—the chitin of arthropods (insects, spiders, and others), the skin of reptiles, birds, and mammals, the bark and cuticle of flowering plants and conifers—are nearly impermeable to water. Moreover, the respiratory organs, whose surfaces must be kept moist to effect gas exchange, have been relocated from external positions (as in the gills of fish and amphibians) to more protected internal positions: the lungs of vertebrates and the tracheae (air passages) of insects. Gas exchange in terrestrial plants is limited to small openings (*stomata*) distributed over the surface of the leaf (Figure 11-4). These morphological adaptations considerably reduce the loss of water through evaporation. Terrestrial organisms lacking well-developed water-conserving adaptations—earthworms, for example—are restricted to moist soil within which air is saturated with water vapor.

Where fresh water is scarce, animals cannot drink to replenish water lost by evaporation from the lungs and in the excrement and urine. To achieve water balance under these conditions, organisms must restrict avenues of water loss if they are to survive. Some desert mammals conserve water so efficiently that their need for water may be satisfied by the water produced by respiratory metabolism. Among mammals, the kangaroo rat is well adapted for life in a nearly waterless environment (Schmidt-Nielsen and Schmidt-Nielsen 1952, 1953). In the large intestine of the kangaroo rat's digestive tract, water is resorbed from waste material to such an extent that the feces are nearly dry. Much of the water that evaporates from the lungs is recovered by condensation in enlarged nasal passages (Schmidt-Nielsen *et al.* 1970). The tissues within the nose are cooler than the lung surfaces, and water exhaled in the warm air from the lungs is condensed there. This simple mechanism takes advantage of the high heat of vaporization of water and works as follows: When the kangaroo rat inhales dry air, moisture in its nasal passages evaporates, cooling the nose and saturating the inhaled air with water. When moist air is exhaled from the lungs, much of its water is condensed and retained in the nasal passages. By alternating condensation with evaporation during breathing, the kangaroo rat minimizes its respiratory water loss. Recovery of water in the nose is greatly enhanced by the enlarged surface area of the nasal passages (Figure 11-5).

Water loss and salt balance are as intimately linked as are the physical processes of diffusion and osmosis. Animals that eat meat or other animal food obtain salts in their food in excess of their requirements. Where water is abundant, organisms merely have to drink large quantities of water to flush out salts that would otherwise tend to accumulate in the body. Where water is scarce, however, organisms must produce a concentrated urine to conserve water. Animals in moist and arid habitats excrete similar amounts of salt, but desert animals cannot afford the copious loss of water that would result from a dilute urine. And so, as one would expect, desert animals have champion kidneys (Schmidt-Nielsen 1964). For example, whereas humans can concentrate salt ions in their urine to about four times the level in blood plasma, the kangaroo rat's kidney produces urine with a salt concentration as high as eighteen times that of the blood (Prosser and Brown 1961).

Carnivores consume excess nitrogen, as well as excess salts, in their food. This nitrogen, ingested in the form of proteins, must be eliminated from

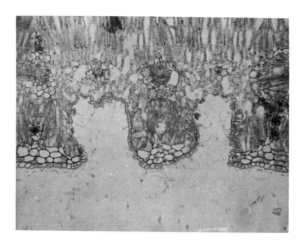

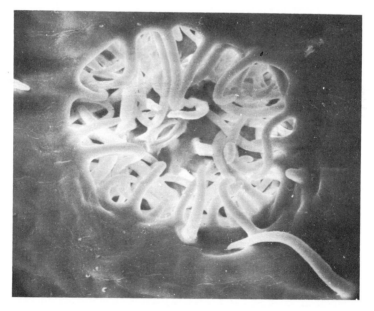

Figure 11-4   The leaf of oleander, a drought-resistant plant. At top, a cross section shows the location of the stomata—openings to the leaf interior through which gas exchange takes place. The stomata lie deep within hair-filled pits on the leaf's undersurface. The hairs reduce air movement and trap moisture, thereby reducing water loss from the leaf. At bottom, a scanning electron micrograph shows a hair-filled pit from underneath the leaf surface magnified about 500 times (courtesy of P. Green and M. V. Parthasarathy).

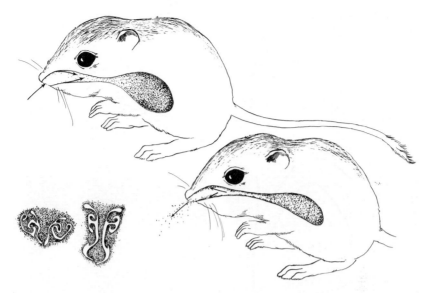

Figure 11-5   Water conservation in the respiratory passages of the kangaroo rat. Inhaled air is warmed and humidified, and at the same time the nasal passages are cooled by the evaporation of water from the surface membranes (upper left). Water-saturated air is cooled as it is exhaled through the nose, thereby leaving behind condensed water (right), which is then evaporated again by inhaled air. Cross-sections of the nasal passages at a depth of three and nine millimeters, respectively, from the external opening of the nose (lower left) exhibit convolutions to increase the surface area of the nasal membranes (after Schmidt-Nielsen, Hainsworth, and Murrish 1970).

the body when proteins are metabolized (Campbell 1970, Schmidt-Nielsen 1972). Animals lack the biochemical mechanisms possessed by some micro-organisms for producing nitrogen gas and, consequently, they cannot dispose of nitrogen as a gas that would escape from the blood through the lungs. Many inorganic forms of nitrogen, nitrate for one, are highly poisonous and cannot be produced in quantity without toxic effects. To solve the problem of nitrogen excretion, most aquatic organisms produce the simple metabolic by-product, ammonia (NH_4^+). Although ammonia is mildly poisonous to tissues, aquatic organisms can rapidly eliminate ammonia in a copious, dilute urine before it reaches a dangerous concentration. Terrestrial animals cannot afford to lose so much water for the sake of nitrogen excretion, and they produce a less toxic metabolic by-product of protein metabolism, which can be concentrated in the blood and urine without dangerous effects (Hochachka and Somero 1973:144). In mammals, this waste product is urea—$CO(NH_2)_2$—which, because it dissolves in water, requires some urinary water loss, the amount depending on the concentrating power of the kidneys. Birds and reptiles have carried adaptation to terrestrial life one step further by producing uric acid—$C_5H_4N_4O_3$—as a nitrogenous

Figure 11-6   Stem of the aborescent prickly pear cactus (*Opuntia echios*) from the Galápagos Islands (left) and the jumping cholla (*Opuntia fulgida*) of the Sonoran Desert (right).

waste product. Uric acid has the distinct advantage in desert environments of being crystallized out of solution and thereby greatly concentrated in the urine. This water-conserving adaptation allows birds and reptiles to be active during the heat of the day in the desert when small mammals are forced to retreat to their underground burrows.

Water loss is a greater problem for terrestrial plants than for terrestrial animals for, although most plants do not use water to excrete salts and nitrogenous wastes, they lose much more water through normal gas exchange. The reason for this difference is quite simple. Animals breathe to obtain oxygen, which constitutes 20 per cent of air. Plants require carbon dioxide, which constitutes only 0.03 per cent of air. Therefore, to obtain one milliliter of carbon dioxide, plants must expose themselves to almost 700 times as much air, and to 700 times the opportunity to lose water vapor to the atmosphere, as an animal does in procuring the same volume of oxygen. No wonder plants have roots! Their demand for water requires a continuous supply from the soil.

Drought-adapted plants do have numerous adaptations that reduce water loss (Hadley 1972). These involve modifications to reduce transpiration across plant surfaces, to reduce heat loads (rate of evaporation increases with temperature), and to tolerate higher temperatures and thereby avoid having to evaporate water to reduce temperature during hot spells. When plants absorb sunlight, they heat up. Overheating can be avoided by increasing the surface area for heat dissipation and protecting the plant surface from direct sunlight with dense hairs and spines (Figure 11-6). Spines also produce a still, boundary layer of air that traps moisture and reduces evaporation from the plant surface. Transpiration is further reduced by covering the plant's surface with a thick, waxy cuticle that is impervious to water and by recessing the stomata (openings in the leaf for gas exchange) in deep pits (Figure 11-4). In many species, biochemical pathways of photosynthesis are adapted either for rapid uptake of carbon dioxide, whereby the period required for a given amount of gas exchange is reduced, or for storage of carbon dioxide in an altered chemical form so that gas exchange can occur at night when temperatures are lower and moisture stress is reduced (Black 1973, Hochachka and Somero 1973:89, Laetsch 1974, Medina 1974, Neales *et al.* 1968).

And so we see some of the major differences between aquatic and terrestrial habitats and the unique problems each poses for plants and animals. The characteristics of aquatic and terrestrial environments are determined primarily by the density and viscosity of water and air and the relative availability of water, oxygen, and carbon dioxide. Adaptations to these conditions emphasize once again the tension between living systems and their surrounding environments. This dynamic relationship between organism and environment is made all the more tenuous and difficult by the inconstancy of the physical environment. Organisms must continually adjust their relationship to the outside world in order to maintain the constancy of their internal environments. How plants and animals regulate their function in response to environmental change is the subject of the next chapter.

12

Regulation and Homeostasis

Change pervades an organism's surroundings—the annual cycle of the seasons, daily periods of light and dark, and frequent unpredictable turns of climate. The survival of each organism depends on its ability to cope with change in the environment. As we look around us, we are constantly aware of organisms, including ourselves, responding to change. Responses include both the use of the physiological apparatus and changes in the apparatus itself. When we step from a warm room into the outdoors on a cold day, we shiver to generate heat. A few weeks on the beach and our skin darkens to block damaging radiation from the sun. When we shiver, we make use of a muscle response that is always present and available on short notice. Morphological changes like the production of pigment granules in the skin in response to sunlight involve altering the physiological apparatus itself. Such changes require more time. Responses involving the structure and function of an organism are usually reversible, as they must be to follow the ups and downs of the environment. Plants and animals also exhibit more or less permanent and irreversible developmental responses to the particular environments in which they grow up.

All these responses are directed toward a single goal, that of maintaining the internal conditions of the organism at some optimum level for its proper functioning (*homeostasis*). But what determines the best internal condition? What is the proper rate of functioning for an organism? And what is the most effective mechanism of response to environmental change? The ultimate measure of "effectiveness" is the net influence of a response on the fitness of the individual—the number of descendants it leaves. But the path of influence between shivering on a cold day and number of great-grandchildren is difficult to follow. Homeostatic responses must be treated, like a problem in economics, in terms of costs and benefits. In the case of shivering, the benefit is clearly measured in terms of survival during the next few minutes or hours by an organism that must maintain a high body temperature to function properly. The cost is measured in terms of energy released to produce body heat, which in turn may deplete the fat reserves of the animal and render its life precarious in the face of a sudden food shortage. One way to reduce the cost of temperature regulation is to lower the regulated temperature, just as we turn down the thermostat in our houses and office buildings to save fuel. But turning down the fire of an organism's life also reduces its rate of activity and hence its food-gathering capacity. In the context of the many factors affecting costs and benefits of a particular response, the optimum becomes a subtle concept. In this chapter, we shall explore some facets of homeostasis to understand why different organisms regulate their internal conditions at different levels, or not at all, and why different means of response to environmental change are chosen.

Biological processes occur optimally within a narrow range of cellular conditions.

The molecules responsible for biological function are extremely sensitive to changes in conditions of temperature, pH, and salt concentration (Hochachka and Somero 1973). Under high temperatures, the structure of enzyme molecules is altered and their function inhibited; exposed to intense light, photosynthetic pigments in plants may break down; with variation in salt concentration, the configuration of protein molecules may change. All of these effects have serious consequences for the functioning of the organism.

We do not need to look far to find examples of the influence of environment on the activity and well-being of the organism. In many cases, the effects of varied conditions not only are important to the animal or plant in question but also have economic consequences for man. It is not surprising, therefore, that many of our best-studied neighbors in the natural world are either part of our diet or pests on our crops and livestock. The oyster is such a neighbor.

Oysters pass their larval stages in the brackish waters of small bays and estuaries. The oyster larvae grow most rapidly when the concentration of salt in the water remains between 1.5 and 1.8 per cent, about halfway between fresh and salt water. High salinity depresses growth slightly. Lower salinity slows growth markedly and causes death: at 1 per cent salinity, 90 to 95 per cent of the larvae die within two weeks; at 0.25 per cent salinity, growth ceases and all larvae die within one week (Davis 1958).

In the example of the oyster, our measure of environment was that of the surrounding water, not the conditions within the cells of the larval oyster. It is not immediately apparent whether salinity directly determines the salt concentrations within the oyster larva or secondarily influences the cost to the oyster of maintaining its cells at an optimum salt concentration. In either case, salinity weighs heavily on survival.

For some properties, the external environment provides a reasonable measure of internal conditions. Temperature is such a property for small cold-blooded organisms. Almost any measure of activity, such as the swimming speed of goldfish portrayed in Figure 12-1, will exhibit a marked temperature dependence, often with a peak rate within a narrow range of optimum temperature (Brett 1956, 1971).

One general class of effects in which a property of the environment could not directly influence conditions within the cell is the effect of atmospheric

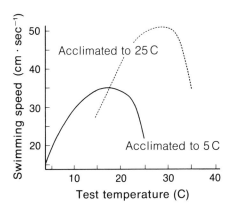

Figure 12-1   Swimming speed of the goldfish (*Carassius auratus*) as a function of temperature. Separate curves are graphed for individuals normally maintained at 5 C and 25 C (after Fry and Hart 1948).

humidity on the function of terrestrial organisms, particularly invertebrates. Migratory locusts are most fecund at about 70 per cent relative humidity (Hamilton 1950). Starved adult tsetse flies live longest in the laboratory at 40 to 50 per cent relative humidity (Buxton and Lewis 1934). The chain of events linking relative humidity to biological activity in the locust or tsetse fly is not understood in detail. Inasmuch as biochemical reactions take place in an aqueous solution within the cells, the relative humidity of air cannot describe the conditions within which cellular processes occur. But humidity nonetheless influences water balance, and, in turn, the regulation of salt concentration in the blood plasma. Considerations of water and salt balance may indirectly constrain other processes that eventually influence fecundity and survival. Both rate of breathing, which contributes to evaporative water loss from respiratory surfaces, and nitrogen excretion, which inevitably involves loss of water in urine, are intimately associated with rate of activity and protein metabolism. The point to be made is this: When the conditions in the surrounding environment differ from the optimum for cellular processes, organisms face a choice between impairment and reduction of cellular function on one hand and paying the metabolic price to maintain the proper internal conditions on the other. The price of maintenance may be exacted either in terms of an energetic cost, as in heat production to maintain body temperature, or in terms of constraints on organism function, as in the measures required to conserve body water in a dry environment.

Homeostatic responses are a type of negative feedback.

Homeostasis refers to the ability of the organism to maintain constant internal conditions in the face of a different and usually varying external environment. All organisms exhibit homeostasis to some degree, although the occurrence and effectiveness of homeostatic mechanisms varies.

Regulation of body temperature is one of the most restricted forms of homeostasis, being fully developed only in the so-called warm-blooded animals: the birds and mammals. Most mammals, including man, closely regulate the temperature of their bodies around 37 C, even though the temperature of the surrounding air may vary from -50 to $+50$ C. Such close regulation guarantees that biochemical processes within cells can occur under constant temperature (homeothermic) conditions, the cost of homeothermy to the organism notwithstanding. Over the same temperature range within which mammals and birds maintain constant body temperatures, the internal environment of cold-blooded (poikilothermic) organisms, such as frogs and grasshoppers, would conform to external temperature. Of course, frogs cannot possibly function at either high or low temperature extremes, so they are active within a narrow part of the range of the environmental conditions over which mammals and birds are active.

How is body temperature regulated? Most homeotherms have a sensitive thermostat in their brain. This thermostat responds to changes in the temperature of the blood by secreting hormones into the bloodstream to slow down or accelerate the generation of heat in body tissues. In addition, most homeotherms partly regulate body temperature by altering gains and losses of heat from the environment. For example, humans put on heavy clothes in cold weather and avoid standing in the sun when it is hot; birds fluff up their feathers to provide greater insulation against the cold.

Because the so-called cold-blooded organisms cannot generate heat without increasing their rate of movement, many adjust their heat balance behaviorally by altering their exposure to the sun. This may involve such simple behavior as moving into or out of the shade, or orienting the body with respect to the sun (Bogert 1949, McGinnis and Dickson 1967). Horned lizards can increase the profile of their bodies exposed to the sun by lying flat against the ground or decrease their exposure by standing erect upon their legs (Heath 1965). By lying flat against the ground, horned lizards are also able to gain heat from the sun-warmed surface. Such behavior, widespread among reptiles, effectively regulates body temperature within a narrow range, with that range elevated considerably above the temperature of the surrounding air (Cowles and Bogert 1944, Hammel et al. 1967, Huey 1974). But this strategy relies on the availability of direct sunlight and, of course, cannot be practiced at night or by inhabitants of continually shaded habitats.

Regardless of the particular mechanism of regulation, all homeostasis exhibits properties of a negative feedback system. If we walk from a dark room into bright sunlight, the pupils of our eyes rapidly contract, which restricts the amount of light entering the eye. A sudden exposure to heat brings on sweating, which increases evaporative heat loss from the skin and helps to maintain body temperature at its normal level. Behavior, too, can be homeostatic because it serves either to modify the environment or to change the individual's relationship to its environment. Putting on an overcoat serves a regulatory purpose, the maintenance of body temperature. Social behavior may be viewed in the same context. A man confronted by a threatening person may try to appease his antagonist. If handing over a wallet satisfies the mugger, he, and the threat he poses, will go away.

Homeostatic responses act to maintain the internal constancy of the organism. To achieve this goal, responses are controlled in the manner of a thermostat. If a room becomes too hot, a temperature-sensitive switch turns off the heater; if the temperature drops too low, the temperature-sensitive switch turns on the heater. This pattern of response is called *negative feedback,* meaning that if external influences alter a system from its norm, or desired state, internal response mechanisms act to restore that state. A man driving a car down a straight road embodies a negative feedback system. If a sudden gust of wind forces the car to veer to the right, the driver immediately responds by turning the steering wheel to the left, and the car returns to its normal path.

Homeostasis requires energy.

To maintain internal conditions significantly different from the external environment requires work and the expenditure of energy. This fact is entirely compatible with what we know about the laws of diffusion and heat transfer: substances and energy flow more rapidly across large gradients than they do across small ones. The greater the difference between an organism and its environment, the greater the energy cost of homeostasis.

The metabolic cost of homeostasis may be demonstrated by examining temperature regulation by homeotherms in cold environments. Birds and mammals maintain constant high body temperatures, generally between 35 and 45 C, depending on the species. At progressively lower air temperatures, the gradient between the internal and external environments increases, and the rate at which heat is lost across the body surface increases proportionately. An animal that maintains its body temperature at 40 C loses heat twice as fast at an ambient (that is, surrounding) temperature of 20 C (a gradient of 20 C) as at an ambient temperature of 30 C (a gradient of 10 C). To maintain a constant body temperature, an organism must replace heat that is lost by releasing heat energy metabolically. Thus, the rate of metabolism required to maintain body temperature increases in direct proportion to the difference between body and ambient temperature (Figure 12-2).

If the only function of metabolism were temperature regulation, metabolic rate would be zero when body temperature equaled ambient temperature and there would be no net movement of heat between the organism and its environment. But organisms release energy to maintain functions not related to temperature regulation—such as heartbeat, breathing, muscle tone, and kidney function—regardless of the ambient temperature. The metabolic rate of an organism that is resting quietly, not having recently eaten, called *basal metabolism* and is the lowest level of energy release under normal conditions. Even at this basal level, the rate of energy release is sufficient to maintain body temperature when ambient temperatures exceed a certain level, the *lower critical temperature* (T_C). At lower temperatures, metabolism must increase to maintain a constant body temperature.

An organism's ability to maintain a high body temperature while exposed to very low ambient temperatures is limited over the short term by its physiological capacity to generate heat, and over the long term by its ability to gather food to supply the energy for metabolism. The maximum rate at which an organism can perform work is generally no more than ten to fifteen times its basal metabolism (Kleiber 1961, King 1974). When the environment

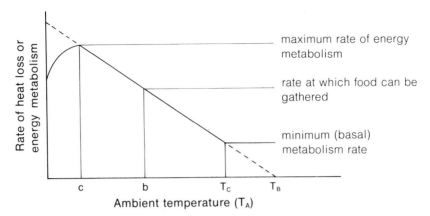

Figure 12-2   Relationship between energy metabolism and ambient temperature for a homeothermic bird or mammal whose body temperature is maintained at T_B. T_C is the lower critical temperature, below which metabolism must increase to maintain body temperature. Points c and b are the lower lethal temperature and the lowest temperature at which the organism can maintain itself indefinitely.

becomes so cold that heat loss exceeds the organism's ability to generate heat (point c of Figure 12-2), body temperature begins to drop, a condition that is fatal to most homeotherms. The lowest temperatures that homeotherms can survive are usually determined by their ability to gather food (point b) rather than their ability to metabolize the energy in food. Animals can quite literally starve to death at low temperatures because they release energy for heat production more rapidly than they can find food to supply their metabolic requirement. At warmer temperatures, surplus energy is available for such functions as reproduction. Points c and b in Figure 12-2 might appropriately be called the critical physiological temperature and the critical ecological temperature. Below c, organisms die. Between c and b, organisms survive but only for relatively brief periods and with a negative energy balance. Above b, energy balance is positive and not only can the individual survive indefinitely, but also can engage in other energy-consuming activities besides foraging. This picture is, of course, simplified. It would be greatly complicated if we properly accounted for the energy expended to gather food, or if we acknowledged that the availability of food changes as ambient temperature and other climate conditions change.

When homeostasis costs more than the organism can afford to spend, certain economy measures are available. For example, the temperature of portions of the body may be lowered, thereby reducing the difference between air and body. Because the legs and feet of birds are usually not insulated by feathers, they would be a major avenue of heat loss in cold regions if they were not maintained at a lower temperature than the rest of the body (Figure 12-3). This is accomplished in gulls by a countercurrent heat-exchange arrangement in which the arteries carrying warm blood to the feet are cooled by passage close to the veins that carry cold blood back from the feet. In this way, heat is transferred to venous blood and transported back

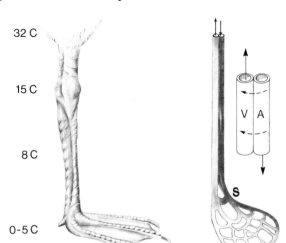

Figure 12-3   Skin temperatures of the leg and foot of a gull standing on ice. Countercurrent heat exchange between arterial blood (*A*) and venous blood (*V*) is diagramed at right. Arrows indicate direction of blood flow and heat transfer (dashed arrows). A shunt at point *S* allows the gull to constrict the blood vessels in its feet, thereby reducing blood flow and heat loss further, without having to increase its blood pressure (after Irving 1966).

into the body rather than lost to the environment. In addition, the flow of blood to the cold feet can be reduced by constricting the blood vessels. When the vessels in the foot are constricted, blood flow largely bypasses the foot area by way of shunts in the legs. With the flow of blood through the feet restricted, heat loss is reduced—remember, a gull may be standing on ice whose temperature is thirty degrees below freezing.

Because of their small size, hummingbirds have a large surface area relative to their weight and consequently lose heat rapidly compared with their ability to produce heat. As a result, hummingbirds need very high metabolic rates to maintain their at-rest body temperature near 40 C. Species that inhabit cool climates would risk starving to death overnight if they did not become *torpid,* that is, lower their body temperatures and enter into an inactive state resembling hibernation. The West Indian hummingbird, *Eulampis jugularis,* lowers its temperature to 18 to 20 C when resting at night. It does not cease to regulate body temperature, but merely changes the setting on its thermostat to reduce the difference between ambient and body temperature (Hainsworth and Wolf 1970).

What determines the level of regulation?

Animals that maintain constant internal environments are usually referred to as *regulators*; those which allow their internal environments to follow external changes are called *conformers* (Figure 12-4). Few organisms

Figure 12-4   Relationship between internal and external environments in idealized regulating
and conforming organisms. Regulators maintain constant internal
environments with homeostatic
mechanisms, whereas conformers
allow their internal environments to
follow changes in the external environment. The difference between
the curves represents the gradient
that regulators maintain between
internal and external environments.

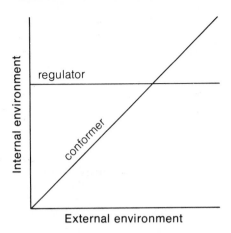

are ideal conformers or regulators. Frogs regulate the salt concentration of
their blood but conform to external temperature. Even warm-blooded animals are partial temperature conformers: In cold weather, our hands, feet,
nose, and ears (in other words, our exposed extremities) become noticeably
cool.

Organisms sometimes regulate their internal environments over
moderate ranges of external conditions, but conform under extremes. Small
aquatic amphipods of the genus *Gammarus* regulate the salt concentrations
of their body fluids when they are placed in water with less concentrated salt
than their blood, but not when they are placed in water with more concentrated salt (Figure 12-5). The fresh-water species *G. fasciatus* regulates the
salt concentration of its blood at a lower level than the salt-water species
G. oceanicus, and thus begins to conform to concentrated salt solutions at a
lower level. In their natural habitat, however, neither the fresh-water species
nor the salt-water species encounters salt more concentrated than that in

Figure 12-5   Salt concentration
in the blood of three gammarid
crustaceans from different habitats
as a function of the salt concentration of their external environment.
The normal salt concentration of
sea water is 3.5 per cent. The inset
shows *G. fasciatus* (from data of H.
Werntz in Prosser and Brown,
1961).

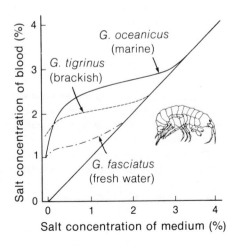

their blood. Among animals that inhabit salt lakes and brine pools, however, the salt concentrations in the blood are actively kept below that of the surrounding water. The brine shrimp *Artemia,* for example, maintains the salt concentration of its blood below 3 per cent even when placed in a 30 per cent salt solution (Croghan 1958).

Whereas most fully aquatic invertebrates cannot regulate the salt concentration of their blood below that of the surrounding water, land crabs and many intertidal invertebrates, which are periodically exposed to air, possess this capability (Prosser 1973). This seems puzzling until one recalls that land animals must sometimes tolerate the loss of body water. If blood volume decreases, salt concentration would increase dangerously unless the organism could excrete salt. In addition to the excretory function of salt-secreting organs, impermeable outer skins or shells reduce the movement of salt into the body while restricting water loss. Thus biological solutions to problems of water and salt balance imposed by terrestrial environments enable organisms to regulate the salt concentrations of their body fluids to levels below those in their environment.

The extent of temperature regulation among homeotherms exhibits a level of variability similar to that of the effectiveness of osmoregulating mechanisms. Most birds maintain body temperatures between 40 and 43 C, and mammals keep theirs between 36 and 39 C depending on the species (Prosser 1973), although some forms are poorer regulators. The body temperature of the opossum, a marsupial, varies widely between 29 and 39 C, and the temperatures of sloths and armadillos may drop below 30 C. Although temperature regulation by these primitive mammals is irregular (heterothermic), temperature may be precisely regulated at a high level by pregnant females to provide the embryo the constant environment it requires for proper and rapid development.

Metabolic generation of heat for temperature regulation is not restricted to birds and mammals. Pythons maintain high body temperatures while incubating eggs (Hutchison *et al.* 1966, Vinegar *et al.* 1970). Some large fish, such as the tuna, maintain temperatures up to 40 C in the center of their muscle masses while they are swimming (Carey *et al.* 1971). Countercurrent heat exchange between vessels carrying blood from the muscles to the skin and back keeps the outer layers of the tuna's body from warming up and thereby reduces heat loss. A preflight warm-up period is frequently encountered in large moths (Heinrich and Bartholomew 1971). Even among plants, temperature regulation has been discovered in the floral structures of philodendron and skunk cabbage (Nagy *et al.* 1972, Knutson 1974).

Clearly, organisms other than birds and mammals are physiologically capable of generating heat to maintain elevated body temperatures, and many do so under certain conditions. Why, then, is the distribution of homeothermy throughout the animal and plant kingdom so limited? Part of the answer certainly lies in a consideration of body size. Birds and mammals are relatively large compared to representatives of other groups. As body size increases, volume increases relatively more rapidly than surface area, across which heat is lost. In general, the lower the ratio of surface to volume, the more comprehensive and precise regulation can be made.

Although body size may explain why mammals are homeotherms and insects generally are not—large moths that exhibit preflight warm-up approach the size of small mammals—large fish and reptiles also have not made the adaptive shift to homeothermy, at least not in a big way. For most fish, the high rate of heat loss in the aquatic environment, owing to the thermal conductance of water, and the low availability of oxygen must preclude the high metabolic rates necessary for temperature regulation in all but the most active species. The metabolic rate of resting birds and mammals may be ten times that of fish, amphibians, and reptiles of similar size (Schmidt-Nielsen 1972). Marine mammals can maintain high body temperatures partly because they breathe air, which provides a rich source of oxygen, partly because they have either fur or blubber for insulation, and partly because they can keep their body surface cold by countercurrent heat exchange and thereby reduce heat loss (Irving 1969).

Why have reptiles not evolved temperature regulation? After all, the body sizes of contemporary reptiles are comparable to small mammals, and they are air breathing. The major innovation for temperature regulation appears to have been the insulation provided by fur and feathers. Insulation functions to reduce the exchange of heat between the body and its surroundings. This aids temperature regulation when the ambient temperature fluctuates, but impedes regulation when rates of activity, and thus of heat production, vary. Clearly, homeotherms must assume precise control over the dissipation of heat as well as its generation and conservation.

Homeothermy implies a constant body temperature; but, furthermore, most homeotherms maintain their body temperatures considerably above that of the surrounding air. Logic tells us that if the energetic cost of temperature regulation is to be minimized, the regulated temperature should be close to the average ambient temperature. Why are the body temperatures of birds and mammals so high and so completely independent of the environment (McNab 1966)? The advantages of high body temperature are poorly understood, but could be several. First, increased temperature raises the level of sustained activity that is possible and may increase alertness, both contributing to capture of food and avoidance of predators. Second, elevated body temperature reduces the need for frequent use of evaporative cooling to dissipate body heat. For terrestrial organisms, potential savings in water may justify the energetic cost of elevated temperature. If this were a major consideration, however, one might expect a stronger relationship between body temperature and survival. Third, the large gradient in temperature generally maintained between body and environment may allow more rapid control over rate of heat loss in response to changes in rate of activity than is possible when a small gradient is maintained. This in turn would allow body temperature to be regulated more closely. These considerations illuminate only one facet of the optimization of physiological functions, a general problem in the relationship between organism and environment that has barely been touched.

What is clear, however, is that organisms have developed a variety of physiological mechanisms, morphological devices, and behavioral ploys to lessen the tension in their relationship to the physical environment. This versatility may be seen particularly well in the adaptations of desert birds and mammals to their stressful environment.

Temperature regulation and water balance in hot deserts require diverse homeostatic adaptations.

Heat stress is one of the most critical factors to an organism's survival. Warm-blooded animals maintain their body temperatures only 6 C, or thereabouts, below the upper lethal maximum. In cool environments, animals can quickly dissipate excess heat, generated by activity or absorbed from the Sun, by conduction and radiation to their surroundings. But when air temperature exceeds body temperature, evaporative water loss must become the primary route of heat dissipation, and desert animals cannot afford to use scarce water to regulate their temperatures. Inactivity, use of cool microclimates, and seasonal migrations provide escape from heat stress for desert animals, but also limit the animal's ability to exploit the desert environment (Figure 12-6).

Temperature regulation and water balance are closely linked. Where fresh water is scarce, organisms have a wide variety of behavioral, morphological, and physiological adaptations for conserving water and using it efficiently to dissipate heat (Schmidt-Nielsen 1964). The daily activity patterns of desert animals are closely tied to problems of temperature regulation. Many small mammals, such as kangaroo rats, which eat only dry seeds, appear aboveground only at night when the desert is cool. Because they avoid hot temperatures, kangaroo rats can survive in the desert without having to drink. Their only source of water is the small amount in the seeds they consume and that produced during the metabolism of foodstuffs. (Remember that both carbon dioxide and water are by-products of oxidative respiration.) Ground squirrels, in sharp contrast, remain active during the day, but they conserve water by allowing their body temperatures to rise when they are aboveground. Before their body temperatures become dangerously high, ground squirrels return to their cool burrows where they dissipate their heat load by convection rather than evaporative heat loss. By alternately appearing aboveground and retreating to their burrows, ground squirrels extend their activity into the heat of the day and pay a relatively small price in water loss (Hudson 1962).

Among vertebrates, birds are perhaps the most successful inhabitants of the desert. They remain active in the heat long after other animals have sought refuge. The success of birds derives from their low excretory water loss (nitrogen is excreted as crystallized uric acid rather than urea) and from feeding on insects, from which they obtain some free water. Even some seed-eating birds can persist without water in the desert, provided they avoid both full sun and shade temperatures above 35 C.

The behavior of the cactus wren, a desert insectivore (Figure 12-7), shows that it too must respect the physiological demands of the hot desert climate. In cool air, wrens lose two to three milliliters (ml) of water each day in the air they exhale. Water loss increases rapidly above 30 to 35 C, to over 20

Figure 12-6   A jackrabbit seeking refuge from the hot sun of southern Arizona in the shade of a mesquite tree. The large ears and long legs of desert jackrabbits effectively radiate heat from the body when the temperature of the surroundings is lower than body temperature (courtesy of the U.S. Fish and Wildlife Service).

ml per day at 45 C; active birds might use five times that much water to dissipate their heat load. (The wren's body contains about 25 ml of water.) In the cool temperatures of the early morning, wrens forage throughout most of the environment, actively searching for food among foliage and on the ground (Ricklefs and Hainsworth 1968). As the day brings warmer temperatures, wrens select cooler parts of their habitat, particularly the shade of small trees and large shrubs, always managing to avoid feeding where the temperature of the microhabitat exceeds 35 C. When the minimum temperature in the environment rises above 35 C, the wrens become less active. They even feed their young less frequently during hot periods.

The cactus wren apparently does not obtain enough water in its diet to allow it to dissipate excess heat through evaporation of water; one almost never observes cactus wrens panting. The behavior of several seed-eating birds, such as the English sparrow and the house finch, contrasts sharply with that of the cactus wren. These species occur only where there is water, and because they can drink freely they can remain active even during the hottest parts of the day, when temperatures may reach 60 C on the ground. They pant at a furious rate under such conditions.

Figure 12-7   The cactus wren (*Campylorhynchus brunneicapillus*), a conspicuous resident of deserts in the southwestern United States and northern Mexico.

Birds lack sweat glands. Dissipation of heat by evaporative cooling occurs primarily from the mouth and respiratory surfaces. At high temperatures, the rapid ventilation needed for heat dissipation draws carbon dioxide from the blood stream passing through the lungs rapidly enough to upset the delicate chemistry of the blood. Panting also requires energy expenditure, and thus itself increases the heat load on the organism. Some birds avoid these problems by gular flutter: passing air rapidly in and out of the mouth and throat, without increasing the ventilation of the lungs (Bartholomew *et al.* 1968, Lasiewski and Snyder 1969). Gular flutter localizes evaporative cooling to the oral surfaces. Muscles expand and contract the mouth and throat cavity in a regular rhythm like a vibrating rubber band. Because the frequency of gular flutter is adapted to coincide with the natural resonating frequency of muscles lining the mouth cavity, gular flutter requires little energy. It occurs, however, only in such birds as doves and nighthawks, which have large distensible throats for ingesting or storing large quantities of food.

Many desert birds build enclosed nests or place their nests in holes in the stems of large cacti, where the young are protected from the Sun and from extremes of temperature. The cactus wren builds an untidy nest, resembling a bulky ball of grass, with a side entrance. Once a pair of wrens have built their nest, they cannot change its position or orientation. For a month and a half, from the beginning of egg laying until the young fledge, the nest must provide a suitable environment day and night, in hot and cool weather. Cactus wrens usually nest several times during the period of March through September. Early nests are oriented with their entrance facing away from the direction of the cold winds of early spring; during the hot summer months, nests are oriented to face prevailing afternoon breezes, which circulate air through the nest and facilitate heat loss (Ricklefs and Hainsworth 1969; Figure 12-8). Nest orientation is an important component of nesting success, particularly during the hot summer months. Nests oriented properly for the season are consistently more successful, 82 per cent, than nests facing the wrong direction, 45 per cent (Austin 1974).

Cactus wrens are so conservative about their use of water that they do not even let the water in the feces of their young go to waste. The fecal sacs, which adults remove from the nest during cool parts of the year, are left in the nest during the hot weather. The evaporation of water from the fecal sacs presumably helps to cool the air in the nest, and the increased humidity reduces respiratory water loss. The re-use of fecal and excretory water is fairly common in desert organisms. In the desert iguana, secretions from salt glands, located in the orbits of the eyes, run down to small pits at the entrance of the nasal passages, where the water evaporates into the inhaled air and reduces respiratory water loss. Some storks have been observed to defecate on their legs during periods of hot weather, thus benefiting from evaporative cooling of their fecal water. Adult roadrunners eat the fecal sacs

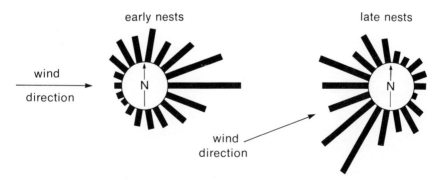

Figure 12-8   Orientation of nest entrances of the verdin (*Auriparus flaviceps*) during the early (cool) and late (hot) part of the breeding season, based on 182 and 206 nests, respectively, from near Las Vegas, Nevada and Tucson, Arizona (data courtesy of George T. Austin).

of their young; the kidneys of adults apparently have greater urine-concentrating abilities than the kidneys of the young, so the adults can extract water from the ingested fecal sacs (Calder 1968). These examples merely emphasize the diversity of adaptations—morphological, physiological, and behavioral—that are focused on the solution of a single problem: heat dissipation and water balance in desert organisms.

The secret of life is the distinctiveness of living systems compared to the physical environment and their ability to utilize energy to maintain this difference in an open dynamic state. The energetic cost of maintaining organisms depends in part on how different they are from the surrounding physical environment and how their behavior differs from the passive existence of inanimate objects. The success of adaptations depends on minimizing this cost relative to the rate of energy intake. We have seen the many ways in which organisms can adjust their adaptations to the physical environment. Foremost among these are reducing the surface area of the body across which exchange with the environment occurs, seeking out those portions of the habitat in which the costs of regulation are minimum, utilizing physical mechanisms of regulation, such as absorption of solar radiation and, conversely, convective cooling, where possible, and employing conservation and recycling, particularly of water.

Homeostasis is made all the more difficult and costly by the inconstancy of the physical world both in time and space. Homeostatic mechanisms that suffice under one set of conditions, for example those experienced in summer, may not be adequate under other conditions. Rapid changes in the environment are countered by rapidly adjusting the rate of physiological processes, but slower changes allow time for altering both the homeostatic mechanism itself and the interface between the organism and its environment. In the next chapter, we shall examine the mechanisms by which animals and plants adjust to spatial and temporal variation in their environments.

13

Organisms in Heterogeneous Environments

Variation in the environment is a fact of life for all plants and animals, except perhaps for inhabitants of the abyssal depths of the oceans. Inasmuch as the adaptations of an organism are best suited for particular conditions, variation poses a critical problem to most forms of life. Those changes that occur rapidly—from within a few seconds to within a few hours or days at most—are accommodated by the homeostatic response mechanisms described in the last chapter. Such responses involve primarily change in the rate of physiological processes, or they may be behavioral reactions. They hold in common the property that they are rapidly reversible; responses to short-term variation must be as labile as the environment. But other changes in an organism's surroundings happen more slowly and are more persistent. Seasonal changes, in particular, permit a long response time for the organism, which is thereby freed to undergo substantial morphological and physiological change.

The type of response adopted by the organism is dictated by the rate and amount by which the environment changes. In this chapter, we shall explore some of the means by which plants and animals adjust to long-term, persistent changes in the environment. Furthermore, for organisms that are immobile—plants and many sessile and slow-moving animals—spatial variation has many of the implications of long-term temporal variation.

The environment is inconstant in time and space.

Daily, seasonal, and tidal cycles produce highly predictable changes in the environments of most organisms. Furthermore, the celestial cycles that produce these changes also provide cues about the future. Lengthening shadows herald the end of the day, which signifies changes in light intensity, temperature, relative humidity, predator activity, and so on, to whichever plants and animals can make use of this information. Summer follows a period of increasing daylength; extremely high and low tides occur with the new and full moons.

Beyond these regular cycles are the unpredictable changes that cause climate to vary from seasonal norms and the changes that result from biological cycles, such as fluctuations in the abundance of prey populations and developmental phases of hosts or parasites. These changes frequently happen on short notice or without warning, and the organism must therefore include their expectation in its bag of response tricks.

Spatial variation in the environment poses different problems for animals and plants. Where animals are faced with different types of habitats or places within habitats, they can choose among them (Hilden 1965, Wecker 1964). And, of course, animals must express habitat preferences, for one cannot be everywhere at once. In contrast to animals, plants are, with few exceptions, stuck where they land as seeds. They must make do with the particular conditions of their surroundings, or they die.

British plant ecologist John Harper and his co-workers have shown that for proper germination, seeds require quite specific combinations of light, temperature, and moisture, which vary even among closely related species (Harper et al. 1965). Irregularities in the surface of natural soils provide the variety of conditions needed to allow the germination of many species, but Harper dramatized the differences between species by creating an artificially heterogeneous soil environment. Three species of plantains, common lawn and roadside weeds in the genus *Plantago,* sown in seed beds responded differently to the variations in environment produced by slight depressions, by squares of glass placed on the soil surface, and by vertical walls of glass or wood (Figure 13-1). In fact, relatively few seeds of any species germinated on the smooth areas of soil surface that had not been disturbed experimentally.

Temporal fluctuations sometimes interact within the three-dimensional structure of the habitat to produce more complex patterns of variation. The edge of the sea is alternately covered by water and exposed to the air by a twice-daily cycle of tides (Figure 13-2). Over the region between the highest and lowest tides, called the *intertidal zone,* the proportion of time during which the surface is covered by seawater changes continuously from 0 to 100 per cent. Above the intertidal zone, conditions are essentially terrestrial; below, they are marine. Within the zone, animals and plants are exposed to the influences of both realms to a varying degree. Organisms that live high in the intertidal zone must tolerate occasional submergence, but must also be able to withstand the desiccating influence of air and the wide variations in temperature so typical of terrestrial environments. Near the high-tide mark, one finds only barnacles and periwinkles, which have doorlike coverings to the openings of their shells and can close themselves off completely from the air when the tide is out. The highest seaweeds, *Ulva* and *Fucus,* begin to appear a foot or so lower. Along the west coast of the United States, mussel beds with associated species of gooseneck barnacles, limpets, chitons, snails, starfish, and worms occupy a narrow band in the middle intertidal region (Ricketts and Calvin 1952, Stephenson and Stephenson 1972). The seaweed *Laminaria* forms a dark brown mantle over even richer and more diverse communities at the lower edge of the intertidal zone, where forms less tolerant of terrestrial conditions, such as sea anemones, sea urchins, nudibranches, tunicates, sponges, and myriad crabs and amphipods, abound.

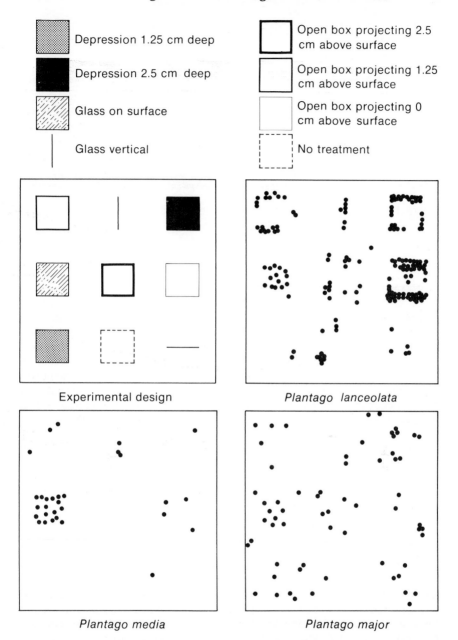

Figure 13-1   Germination of seedlings of three species of plantains (genus *Plantago*) with respect to artificially produced variation in the soil surface (after Harper *et al.* 1965).

Figure 13-2   A portion of the coast of Kent Island, New Brunswick, at the mouth of the Bay of Fundy during high tide (left) and during low tide (right). The daily tidal range in this area is about 25 feet.

The graininess of the environment.

Fluctuations that are of long duration require a different response from those which are of short duration. The choices presented to organisms by spatial variation depend on the distance between patches of different habitat type. But what is long and brief, or near and far, depends on the life span and response time of the organism and on its mobility. The upper and lower surfaces of a leaf do not matter much to us as we push aside the branch of a tree, but they are different worlds to the aphid that sits on the undersurface of a leaf and sucks plant juices. A tropical storm or passing cold front means little more to us than perhaps spoiling plans for a picnic, but may completely encompass the adult life of a mayfly during which it must mate and lay eggs.

These different perspectives on variation in the environment are summarized by the concept of environmental *grain* (Levins 1968). Imagine for a moment that variation consists of patches of uniform conditions that are distributed like a mosaic in time and space, resembling a patchwork quilt or the pattern made by alternating fields of different crops. The patches have different sizes, and patterns of different kinds of patches can be superimposed. For example, patches of air temperature tend to be much larger than patches of soil moisture because topographic influences subdivide the soil moisture landscape more finely. The concept of grain relates the size of the patch to the activity space of the organism. We define a *coarse-grained* environment as one in which the patches are relatively so large that the individual can choose among them. In a *fine-grained* environment, the patches are so small that the individual cannot usefully distinguish among them, and the environment appears essentially uniform. To anyone other than a trained botanist, a field appears to be a uniform carpet of plants—a patch that we distinguish from patches of forest, marsh, beach, and so on. But to the caterpillar, a single plant within the field may be its home for the duration of larval life. Even grasshoppers, which can fly from plant to plant, choose carefully among the various species in the field, for some provide better feeding than others. Therefore, the patches that are individual plants in a field are fine-grained to us, but coarse-grained to small insects. Needless to say, grain also depends on activity. If we set out in the field to pick flowers of a particular kind, we would perceive individual plants as coarse-grained patches.

Patches may be thought of as occurring in time as well as in space. At one extreme, conditions that fluctuate through a daily cycle or over a shorter period may be thought of as fine-grained to most organisms, inasmuch as animals and plants do not have time to undergo major adjustments in response. At the other extreme, seasonal changes and longer cycles are decidedly coarse-grained for most organisms. Given persistent changes of several months' duration, plants and animals can complete morphological changes, such as dropping their leaves and increasing the thickness of their fur, and corresponding changes in physiological mechanisms, and in the case of animals, undertake migrations to areas with more favorable conditions.

An organism's choice of environment patches defines its activity space.

Animals actively move among patches of conditions in a coarse-grained environment as the environment changes temporally. Conditions within each spatial patch change over diurnal and seasonal cycles, and by moving among them, animals can remain within ranges of conditions close to their optimum. Although plants cannot pick themselves up and move off to another patch of Earth, most regulate their activity levels according to the suitability of conditions at a particular time. Simply by closing the stomata on their leaves, plants are able to shut themselves off from some unfavorable conditions.

As the Sun changes its position each day and throughout the year, the thermal environment of organisms in exposed habitats changes drastically (Porter and Gates 1969). The surface of the desert, cool in the early morning, becomes a furnace at noon, and drives animals to shaded sites or to their burrows (Figure 13-3). As the conditions within different patches change, the activity space also changes.

The diurnal behavior cycle of lizards is geared to the varying temperatures of habitat patches (Heatwole 1970). Although lizards do not generate heat metabolically for temperature regulation, they do take advantage of solar radiation and warm surfaces to maintain their body temperatures within the optimum range. At night, these sources of heat are not available, and the lizard's body temperature gradually drops to that of the surrounding air. The mallee dragon *(Amphibolurus fordi)*, an agamid lizard of Australia, is most active when its body temperature is between 33 and 39 C (Cogger 1974). In the early morning, before its body temperature has risen above 25 C and when its movement is still sluggish, the mallee dragon basks within large

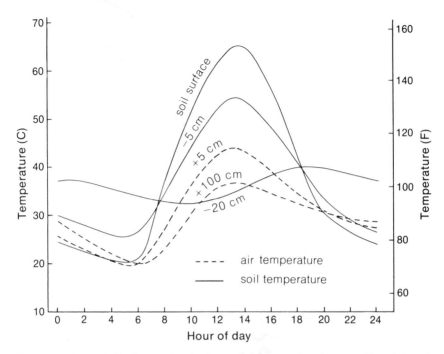

Figure 13-3   Daily fluctuation in the temperature of a desert soil and the air above it. Measurements indicate the depth below the surface, solid lines, and height above it, dashed lines (after Hadley 1970).

protective clumps of grass of the genus *Triodia* (Figure 13-4). In fact, *Amphibolurus* is so dependent upon these grass clumps that it occurs only where *Triodia* is established. When the temperature of an individual rises above 25 C, it moves out of the *Triodia* clump and basks in the sunshine nearby, with its head and body in direct contact with the ground surface from which additional heat is absorbed. When body temperature enters the normal activity range (33–39 C), *Amphibolurus* ventures farther from *Triodia* clumps to forage, its head and body normally raised above the ground as it moves. When body temperature exceeds 39 C, the lizards move less rapidly and seek the shade of small *Triodia* clumps; above 41 C, they re-enter large *Triodia* clumps, at whose centers they find cooler temperatures and deeper shade. The lizards may also pant to dissipate heat by evaporative cooling. If heat stress is not avoided, *Amphibolurus* loses locomotor ability above 44 C, and will die if body temperature exceeds 46 C.

On a typical summer day, during which air temperature varies from about 23 C at dawn to 34 C at midday, the mallee dragon does not begin to forage until about 8:30 A.M. By 11:30 A.M., it has become too hot for normal activity and most individuals seek shade and inactivity. By 2:30 P.M., the air has cooled off enough so that foraging is resumed until 6 P.M., after which time the lizards retreat back into *Triodia* clumps and their bodies rapidly cool. If

Figure 13-4   The mallee dragon (*Amphibolurus fordi*) at different points of its activity cycle: top, early morning basking in *Triodia* grass clump; middle, mid-morning basking on ground (note body flattened against surface to increase exposed profile and contact with warm soil); bottom, normal foraging attitude (photographs courtesy of H. G. Cogger, from Cogger 1974).

they remained in the open after this time of day, they would easily be caught by warm-blooded predators.

The desert iguana *(Dipsosaurus dorsalis)* of the southwestern United States faces a more severe environment, with greater annual fluctuation, than the mallee dragon (Beckman *et al.* 1973, Norris 1953, Porter *et al* 1973, DeWitt 1967). In summer, air temperatures can reach 45 C; in winter, temperatures frequently drop below freezing. During mid-July, the thermal environment changes so rapidly between extremes that the desert iguana can be normally active within its preferred body temperature range of 38 to 43 C for about 45 minutes in the middle of the morning and a similar period in the early evening (Figure 13-5). During the remainder of the day, lizards seek the shade of plants or the coolness of their burrows, where the temperature rarely rises above the preferred range (for example, see Figure 13-3). At night, the lizards enter their burrows, partly to escape predation and partly because at dawn the burrow is warmer than the surface, hence the early morning warm-up period for the lizard is reduced.

Whereas the activity period of the desert iguana in summer is restricted to two brief bouts separated by a long period of inactivity to avoid midday heat stress, spring is more favorable. In May, the thermal environment does not exceed the preferred range of *Dipsosaurus,* and individuals remain active

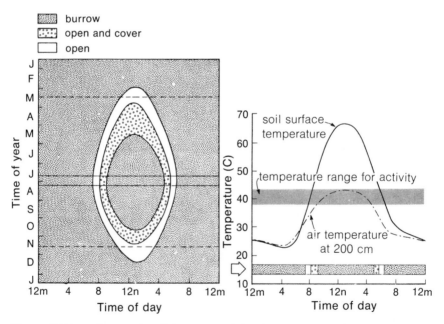

Figure 13-5   Seasonal activity space of the desert iguana (*Dipsosaurus dorsalis*) in southern California. At left, the daily activity budget is portrayed for an entire seasonal cycle. At right, the activity budget for July 15 is shown with the time course of environmental temperature (after Beckman *et al.* 1971).

aboveground from 9 A.M. to 5 P.M., occasionally seeking the shade of plants to cool off. In winter, cold temperatures restrict the activity of *Dipsosaurus* to brief periods in the middle of the day when body temperature warms up enough so that individuals can come aboveground and forage. Between early December and the end of February, most days are so cold that the desert iguana remains inactive in its burrow.

The temporal characteristics of homeostatic responses vary.

The time course of a response to changing conditions must be shorter than the period of environmental change. Otherwise today's response may be effective only for yesterday's conditions. For each type of response mechanism, some types of environmental fluctuations are coarse-grained, others are fine-grained. Response mechanisms may be placed in three general categories, which are, in order of decreasing response rate, regulatory, acclimatory, and developmental responses. *Regulatory responses* involve changes in the rate of physiological processes or changes under behavioral control. Stimulated metabolism in response to cold and shade-seeking behavior of the mallee dragon are examples. Regulatory responses do not include extensive change in morphology or the biochemical apparatus in the cell. Such changes, including thickening of fur in winter, increase in number of red cells in the blood at high altitude, and production of enzymes with different temperature optima are more properly called *acclimatory responses.* These may be thought of as a shift in the range of the regulatory responses of an individual. Both regulatory and acclimatory responses are reversible, as they must be to follow the ups and downs of the environment.

When environmental changes are persistent and a given set of conditions may prevail during the adult life span of an individual, responses may become fixed during the development period and remain stable until the organism dies. Such *developmental responses* are slow and generally irreversible.

The type of response adopted depends upon the rate of environmental change relative to the life span of the organism. Changes that occur over minutes or hours are usually too rapid for all but regulatory responses, discussed in detail in Chapter 12. Changes occurring over a period of days to weeks allow time for major adjustment of morphology: acclimation. For small organisms with short life spans, developmental responses may be appropriate. Seasonal cycles seem to divide plants and animals into two major groups: annuals (those living less than one year) and perennials (those with a reasonable expectation of living well beyond their first birthday). For perennial organisms, seasonal changes must be accommodated by acclimation because the response must be reversible. For annuals whose adult life spans may be a few months or less, developmental responses may be called for. In addition to finding their place in response to coarse-grained temporal change, developmental adjustments are widely employed by plants finding themselves as seeds in a spatially coarse-grained environment without the ability to choose among or move between patches.

Acclimation is a major response mechanism of perennial plants and animals.

Acclimation is the modification of an organism's morphology or physiology in response to long-term environmental change. Many birds have a heavier plumage, with greater insulating properties, during the cold winter months than during hot summer months. These species replace their body feathers only twice each year, and each plumage must be suited to the average conditions of the environment between each molt. The willow ptarmigan, a ground-feeding arctic bird, sheds its lightweight brown summer plumage in the fall for a thick white winter plumage, which provides both insulation and camouflage against a background of snow. With increased insulation, ptarmigans require less energy expenditure to maintain their body temperatures during the winter (Figure 13-6). Seasonal change in plumage thickness effectively shifts the regulatory response range to match the prevalent temperature range of the season. In winter, maintenance of body temperature at −40 C requires the same expenditure of energy as temperature regulation at −10 C in summer (a conceivable temperature in the ptarmigan's arctic home). The metabolic response curves of the ptarmigan and most other species of homeotherms would seem to suggest that winter-acclimated individuals are energetically most efficient over all temperatures within both winter and summer ranges. But there is no free lunch. Heat-conserving devices of winter-acclimated individuals retard the dissipation of heat, and

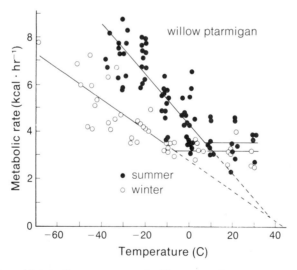

Figure 13-6   Metabolic responses of willow ptarmigan acclimatized to summer and winter temperatures. Winter-acclimated birds have thicker plumages providing better insulation than summer birds. Hence their metabolic rates are lower at any given temperature, and their lower critical temperature is also lower (after West 1972).

would lead to heat prostration with normal activity in summer. Adjusting insulation to enhance heat conservation in winter and to facilitate heat dissipation in summer maintains constant body temperature at the least possible cost. Acclimatory responses in body insulation occur only in species that are active and maintain high body temperatures during cold weather (Hart 1957). Red squirrels, which hibernate during the winter in Alaska, exhibit no winter increase in the insulation of their fur.

Temperature-conforming animals and plants also acclimate to seasonal changes in their environment (Schmidt-Nielsen 1975). By switching between enzymes and other biochemical systems with different temperature optima, cold-blooded animals adjust their tolerance ranges in response to prevalent environmental conditions (Hochachka and Somero 1973). Experiments with lobsters have shown that the upper lethal temperature increases as the temperature to which the lobsters are acclimated increases (McLeese 1956). Lobsters kept at 5 C die when exposed to 26 C, but tolerate 28 C when acclimated to 15 C, and tolerate 31 C when acclimated to 25 C. Acclimation does not, however, allow an organism to respond infinitely to environmental change. Regardless of their previous temperature experience, lobsters cannot be physiologically acclimated to withstand temperatures above 31 C, because their physiological systems are adapted to much colder environments and the capacity of lobsters to acclimate to extreme temperature is limited.

The upper and lower lethal temperatures of fish vary with acclimation temperature, producing a pattern of temperature tolerance that also varies among species (Figure 13-7). One feature of this relationship is that the upper lethal temperature never increases in parallel with acclimation temperature. The maximum temperature that a species can tolerate regardless of its previous temperature regime is that point at which the lines for upper lethal temperature and acclimation temperature meet; an organism cannot become acclimated at a temperature that is beyond its tolerance range. Another feature of temperature tolerance common to most species is that they can withstand cold temperatures to a greater degree than hot temperatures. As

Figure 13-7   Upper and lower lethal temperatures of the goldfish and chum salmon as a function of acclimation temperature. The vertical range between upper and lower lethal temperatures at a given acclimation temperature is the activity space for that acclimation temperature and indicates tolerance of short-term temperature fluctuations. Long-term acclimatory response adjusts this range (after Fry et al. 1942, Brett 1956).

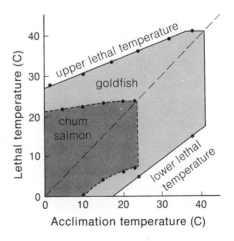

long as it does not freeze, cooling merely slows processes, whereas heating can change the structure of molecules and disrupt processes.

Although acclimatory responses can produce remarkable adjustment of a species to a wide range of conditions, they nonetheless operate within somewhat limited ranges. If we consider swimming speed in fish as an index to activity level, its relationship to water temperature shows at once the capabilities and limitations of acclimation (Fry and Hart 1948). Goldfish swim most rapidly when they are acclimated to 25 C and placed in water between 25 and 30 C, conditions that closely resemble their natural habitat (see Figure 12-1). Lowering the acclimation temperature to 15 C increases the swimming speed at 15 C but reduces it at 25 C. (Increased tolerance of one extreme often brings reduced tolerance of the other.) Reducing acclimation temperature further, to 5 C, well beyond the normal lower range of temperatures experienced by goldfish in nature, does not increase swimming speed at 5 C, but does reduce swimming speed at moderate temperature.

Acclimatory changes in physiology enable a basically warm-water fish, like the goldfish, to better tolerate temperatures near the upper and lower ends of the normal temperature range in its environment. When brook trout, a cold-water species, are tested over a similar range of temperatures, they perform best when acclimated to 15 C, approximately the summer temperature of their environment. The difference between goldfish and trout represents the accumulation of evolutionary modifications to the different temperature ranges of their respective habitats.

Developmental responses are conspicuous in plants and animals with several generations per year.

Developmental responses indicate a flexibility of growth and development processes that are sensitive to environmental variation. These responses are generally not reversible. Once fixed during development, they remain unchanged for the remainder of the organism's life. Because developmental responses have distinct limitations for short-term environmental changes (long response time and irreversibility), plants and animals are likely to exhibit developmental flexibility only in environments with persistent variation for the individual. For plant species whose seeds may settle in many different kinds of habitats, the strategy of developmental flexibility makes good sense. Also, when environmental changes occur slowly compared to the life span of an organism, developmental responses may be the most appropriate type of adjustment. For example, an organism that lives much less than a year can adjust itself to seasonal conditions by developmental response.

A striking example of developmental flexibility occurs in several species of aquatic plants whose morphology varies according to whether the plant grows on the land, partly submerged in water, or completely underwater (Sinnott 1960). Leaves of arrowleaf plants that grow underwater are not structurally rigid, and because they lack a waterproof waxy cuticle, they may absorb nutrients directly from the water into the leaves (Figure 13-8). If these

Figure 13-8   Variation in the morphology of the arrowleaf plant when it grows on land (left), partially submerged in water (center), and fully submerged (right) (after Wallace and Srb 1964).

underwater leaves are removed from the water they collapse. The aerial leaves of the plant are broad, more rigid than underwater leaves, and are covered with a thick cuticle to decrease transpiration. The stiff aerial leaf would probably snap off if it were exposed to water currents. An arrowleaf that grows partly submerged has both kinds of leaves. Terrestrial individuals have large root systems compared with those of aquatic individuals. The latter can absorb nutrients through their leaves, whereas terrestrial plants must obtain all their nutrients from the soil by way of their root system. These differences in growth form result from developmental responses of the arrowleaf to the environmental conditions of each habitat. In the aquatic environment, the development of thick cell walls in leaf tissue is suppressed; in well-drained soil, the growth of roots is more pronounced; and so on.

Light intensity is another factor that can influence the course of development in plants. Loblolly pine seedlings grown in shade have smaller root systems and more foliage than seedlings grown in full sunlight (Bormann 1958). Because the shaded environment has lower water stress, shade-grown seedlings can allocate more of their production to stem and needles. Sun-grown seedlings must grow more extensive root systems to obtain water. The larger proportion of foliage of the shade-grown seedlings results in a higher rate of photosynthesis per plant under given light conditions, particularly when the light intensity is low (Table 13-1). The different developmental responses of shade-grown and sun-grown seedlings adapt

TABLE 13-1

Proportions and rates of photosynthesis of loblolly pine (*Pinus taeda*) seedlings grown under shade and full sunlight (after Bormann 1958).

	Shade-grown	Sun-grown
Per cent of dry weight in*		
roots	35	52
needles	47	37
stem	18	11
Photosynthetic rate (mgCO$_2$ · hr^{-1} · gm^{-1})†		
low light intensity	1.9	1.0
moderate light intensity	4.6	4.0
high light intensity	7.2	6.6

* Six-month-old seedlings.
† Four-month-old seedlings. Light intensities were 500, 1,500, and 4,500 foot candles.

each to the different conditions of light intensity and moisture stress in its environment.

A fascinating and complicated case of developmental response involves the wing development of water striders, fresh-water bugs of the genus *Gerris* (Brinkhurst 1959, Vepsalainen 1973, 1974a, 1974b, 1974c). European species fall into four categories of wing length depending on the type of habitat in which they live (Table 13-2). At one extreme, species inhabiting large permanent lakes have short wings or are wingless and do not disperse between lakes. At the other extreme, species inhabiting temporary ponds are usually long-winged and disperse to find suitable habitats for breeding each year. Between these extremes, species that inhabit small ponds, that are more or less persistent from year to year but tend to dry up during the summer, frequently exhibit both long- and short-winged forms.

The life cycle of most *Gerris* species in central Europe, England, and the southern parts of Scandinavia includes two generations per year. The first, or

TABLE 13-2

Wing lengths of water striders (*Gerris*) found in habitats with different levels of permanence and predictability (after Vepsalainen 1974a, 1974c).

Characteristic of habitat	Characteristic wing length*	Mechanism of determination
Permanent	short	genetic
Fairly persistent but unpredictable	both short and long	genetic dimorphism
Seasonally temporary	seasonally dimorphic (summer)	developmental switch
Very unpredictable	long	genetic

* Short wings are not functional and prevent dispersal.

summer, generation hatches during the spring, reproduces during the summer, and then dies. The second, hatched from eggs laid by females of the summer generation, develops to the adult stage during late summer, then overwinters before breeding the following year in early spring. In species that inhabit seasonal ponds, the summer generation is dimorphic, having both long- and short-winged forms (Figure 13-9). All the individuals in the winter generation are long-winged and able to fly. They leave the pond in late summer and move into nearby woodlands where they overwinter. In the spring, they return to small bodies of water to lay eggs.

Dimorphism in the summer generation reflects two extreme strategies. The long-winged forms are capable of flying to other habitats if the pond in which they developed happens to dry up, especially if this happens early in the season. The short-winged forms bank on the persistence of the habitat into late summer, and they convert nutrients that would have become wing and flight muscles in long-winged forms into eggs. Thus short-winged individuals tend to be more fecund, but their progeny are sometimes destroyed when a pond dries up earlier than usual. No true compromise between the long- and short-winged strategies is possible, because long wings and fully developed flight muscles are needed to fly at all. A water strider cannot make it out of the pond on half-sized wings.

Because all the winter generation have long wings, seasonal dimorphism must be controlled during development. Vepsalainen (1971) has found that wing length is determined primarily by daylength. If the daylength increases continually during larval development and exceeds eighteen hours (in southern Finland) during the last larval stage prior to change into the adult form, the individual will have short wings. If the daylength begins to decrease before the end of larval development (as it would if the development period

Figure 13-9   Alary polymorphism in the water strider *Gerrisodontogaster*. The macropterous (winged) form is on the left and the micropterous (short-winged) form is on the right (courtesy of K. Vepsäläinen).

extended beyond June 21), a long-winged adult is produced. Thus summer generation individuals become short- or long-winged depending upon when they hatch and upon their rate of larval development. The switch between wing length determination is also influenced by temperature (Vepsalainen 1974b), with high temperatures, which could lead to early drying of ponds, favoring the development of long-winged forms.

The kind of mechanism described by Vepsalainen for developmental determination of wing length in water striders is only representative of a large number of similar developmental responses in various insects with short generations (*e.g.* Lees 1966, Young 1965, Lamb and Pointing 1972, Steffan 1973).

A striking example of developmental response to environmental variation is to be found in the coloration of several species of locusts and grasshoppers that inhabit arid regions of the tropics. Although temperature is relatively constant there, rainfall varies seasonally. In the wet season, vegetation is lush and the habitat is essentially green. During the early part of the dry season, the vegetation turns brown and dies, often exposing red-brown earth. Toward the end of the dry season, natural and man-set fires sweep through some dry areas, leaving vast expanses blackened. As a result, the course of a year brings a regular seasonal progression of color in the dry tropics from green to brown to black, and back to green again. Grasshoppers develop so rapidly that the life span of each individual may coincide largely with one background color. And grasshoppers do, in fact, match the background coloration of their environment very closely, regardless of season. In one species of grasshopper, *Gastrimargus africanus,* coloration is controlled by environmental conditions, particularly the quality and intensity of the light to which the grasshopper is exposed (Fraser Rowell 1970). Details of the color polymorphism are complicated, but the basic scheme is as follows: The epidermis (outer skin) of the grasshopper has a pigment system that permits any given area of the skin to be either green or brown; both colors may occur on a single animal but not in the same area. The green and brown colors are known to result from small biochemical variations on a single pigment molecule. Where brown pigment occurs in the epidermis, additional pigments may produce colors ranging from yellow, through orange and red, to black. Furthermore, black pigment (melanin) may be deposited in the cuticle which covers the epidermis.

The color response in *Gastrimargus* is most pronounced in the early development stages; adults have a more characteristic and unvarying pattern of color distribution. Between each developmental stage (instar) the skin of the grasshopper is shed, carrying with it the color pattern. A new skin develops underneath, and thus the grasshopper may change its color with each molt. The environmental factors that affect epidermal pigment systems in the grasshopper are probably more complicated than the systems themselves. High-intensity light, a characteristic of the dry season, leads to the predominance of the brown and black pigment systems, whereas low-intensity light combined with high humidity tends to increase the proportion of green coloration. The latter conditions prevail during the rainy season, a time of heavy cloud cover and lush green vegetation. When the surroundings of the grasshopper *reflect* little of the incident light, for example when an area is burned over and covered with black ash, the black pigment system is stimulated. Thus *Gastrimargus* has evolved a pigment system that automatically adjusts its color by developmental response to match that of its background. The color pattern is fixed in the epidermis of the grasshopper and cannot be changed, but because the grasshopper nymph molts its skin periodically during development, the color response may be repeated several times.

Many animals migrate between seasonally suitable locations.

Under conditions of extreme drought or cold, physical conditions may become sufficiently stressful, and food sufficiently difficult to find, that plants and animals can no longer maintain normal activity. Faced with these conditions, some organisms leave the environment, seeking more favorable conditions elsewhere, while others enter a dormant state, sealing themselves off from the rigors of the environment.

Migrations are widespread in nature, particularly among flying animals: birds, bats, and some insects (Kramer 1961). Many of them perform impressive feats of long-distance navigation each year (Figure 13-10). Shorebirds, like the golden plover, breed in the Canadian Arctic and spend their winters as far south as Patagonia at the southern tip of South America. They make yearly round trips of up to 25,000 miles. The arctic tern probably holds the record for long-distance migration with a yearly round trip of 36,000 miles between its North Atlantic breeding grounds and its Antarctic wintering grounds (Welty 1963). A few insects, like the monarch butterfly, perform impressive migratory movements each year, but most species of insects overwinter in a dormant state as eggs or pupae (Dingle 1972).

Each fall hundreds of species of land birds move out of temperate and arctic North America in anticipation of cold winter weather and dwindling supplies of their invertebrate food. Montane birds similarly make altitudinal migrations of several thousand feet, the mountain quail making its annual trek on foot. Because the Southern Hemisphere winter is less harsh than the Northern Hemisphere winter (see page 38), South American species have less need to escape the southern winter to take advantage of the temperate summer occurring to the north.

Mammals, other than bats, do not have the migratory abilities of birds and some insects, but some mammals do exhibit impressive seasonal movements. The barren-ground caribou of northern Canada migrates from its summer home on the tundra into the spruce forest for the winter because its food (lichens and mosses) is covered by snow on the tundra, but remains accessible in the protected spruce forest habitat (Kelsall 1968).

Some marine organisms also undertake large-scale migrations to reach spawning grounds, to follow a food supply, or to keep within suitable temperature ranges for development. The migration of salmon from the ocean to their spawning grounds at the headwaters of rivers and the reverse migration of adult fresh-water eels to their breeding grounds in the Sargasso Sea are

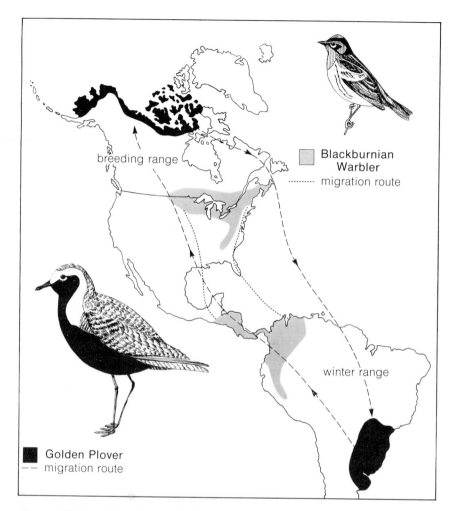

Figure 13-10   Breeding and wintering ranges of the golden plover (left) and Blackburnian warbler (right). Migration routes are indicated by dashed lines.

striking examples (Jones 1968). Lobsters undertake less conspicuous annual migrations up to several hundred kilometers off Long Island and Massachusetts, from deep waters in summer to shallow in winter, always staying within zones of cool temperature (Cooper and Uzmann 1971). Predatory gastropods and some sea urchins are known to undertake seasonal movements into shallow waters for feeding.

Some populations exhibit irregular or sporadic movements that are tied to food scarcity during particular years rather than to seasonal conditions (Svärdson 1957). The occasional failure of cone crops in coniferous forests of

Canada and the mountains of the western United States forces large numbers of birds that rely on seeds to move to lower elevations or latitudes. Birds of prey that normally feed on rodents disperse widely when their prey populations decline sharply. Snowy owls move southward from their arctic hunting grounds into the northern United States and adjacent Canada every four years on the average, corresponding to a periodic scarcity of lemmings (Gross 1947, Snyder 1947). In desert areas, irregular rainfall forces animals such as the budgereegah (an Australian parakeet) into a nomadic existence in a continual search for areas where rain has recently fallen (Keast 1966, Frith 1967). Even insects are subject to irregular movements. Outbreaks of migratory locusts, from areas of high local density where food has been depleted, can reach immense proportions and cause extensive crop damage over wide areas (Gunn 1960, Waloff 1966) (Figure 13-11). The behavioral traits that lead to population irruption are actually a developmental response to population density (Uvarov 1961, Krebs 1972:289). When the locusts grow up in sparse populations, they develop into a solitary phase as adults. In dense populations, frequent contact with other individuals stimulates a developmental switch to the gregarious phase, which can often lead to population outbreaks following local depletion of food resources.

Figure 13-11   A dense swarm of migratory locusts in Somalia, Africa, in 1962 (courtesy of the U.S. Dept. of Agriculture).

Many organisms become dormant during extremely unfavorable periods.

When the environment becomes so extreme that normal life processes can no longer function, or that the maintenance of normal activity would quickly lead to starvation or desiccation, plants and animals incapable of migration must enter into a physiologically dormant state. For many small invertebrates and cold-blooded vertebrates, freezing temperatures directly curtail activity and lead to dormancy. Many mammals enter a dormant state because of a lack of food, rather than because of physiological inability to cope with the physical environment.

Lack of water is the key factor in the adoption of the deciduous habit by plants (Vegis 1964). Many tropical and subtropical trees shed their leaves during seasonal periods of drought (see Figure 4-10). Temperate and arctic broad-leaved trees shed their leaves in the fall to avoid desiccation. Moisture frozen in the soil is unavailable to plants; if these trees kept their leaves through the winter, transpiration of water from the leaves would wilt them as quickly as if the tree were cut down.

Many physiological changes accompany dormancy (Prosser 1973). The onset of hibernation in mammals is anticipated by the accumulation of a specific type of fat, with a low melting point, that will not harden and cause stiffness at low temperature. Blood chemistry changes to prevent clotting in capillaries at reduced rate of blood flow. Hibernating ground squirrels have heartbeat rates of 7 to 10 per minute compared with 200 to 400 for active individuals (Svihla et al. 1951). Body temperature falls 30 C, to about 6 C, as the ground squirrel enters hibernation, and its metabolism drops to 1 to 5 per cent of normal (Landau and Dawe 1958).

Insects enter into a resting state known as *diapause,* in which water is chemically bound or reduced to prevent freezing and metabolism drops to near nought. In summer diapause, drought-resistant insects either allow their bodies to dry out and tolerate desiccation or secrete an impermeable outer covering to prevent drying. Plant seeds and the spores of bacteria and fungi have similar dormancy mechanisms (e.g., Koller 1969, Wareing 1966). Regardless of the mechanism, dormancy serves the single purpose of shutting off the organism from the environment and reducing exchange between the two. In this way animals and plants ride out unfavorable conditions and await better ones before reassuming an active and interactive state.

Organisms store resources to moderate the effects of environmental fluctuations.

Although homeostasis helps maintain function in the face of a changing physical environment, environmental changes often plunge organisms from feast into famine. When the environment becomes barely tolerable and small fluctuations in food or water supply can mean disaster, many plants and animals store food and water reserves. Desert cacti absorb water during rainy periods and store it in their succulent stems. Cacti use these reserves during the long dry intervals between desert rains. Many temperate and arctic animals store fat during periods of mild weather in winter as a reserve of energy for periods when heavy snow covers food sources. Tropical animals sometimes store fat prior to the onset of seasonal dry periods (Ward 1969a, 1969b). Instead of accumulating body fat, many winter-active mammals (beavers and squirrels) and birds (acorn woodpeckers and jays) cache food underground or under the bark of trees (Ritter 1938, Swanberg 1951). During winter months, piñon jays of the western United States normally feed on insect grubs in the soil, but they cannot do so when snow covers the ground. In the fall, the jays harvest the vast crops of piñon pine nuts and bury them in caches near the base of trees. The nut stores are usually placed at the south side of a tree's trunk, and after a snowfall, the snow first melts on the south side, thus exposing the cache.

Fat is often stored in anticipation of greatly increased energy demands—as before long migrations or periods of reproduction. Many species of migrant birds store large quantities of fat, often half their normal weight, particularly if they undertake long flights over water (Odum 1960, Odum *et al.* 1961). The side-blotched lizard (*Uta stansburiana*) stores fat during the fall and winter to provide energy for egg formation in the spring (Hahn and Tinkle 1965). The fat bodies of lizards may also serve as an insurance policy against poor conditions, for some species accumulate larger fat stores where winters are harsh than they do in milder regions (Pianka 1970a). Unlike most vertebrates and insects, molluscs lack specific energy storage sites, such as the liver, fat bodies, and subcutaneous regions of fat deposition (Giese 1969). Energy storage in the intertidal snail *Thais lamellosa* is accomplished by the seasonal growth of viscera tissues. Hence storage is primarily in the form of protein rather than fat. The tissues are broken down and energy released during the spring period of population aggregation and production of young, during which the snails do not feed (Figure 13-12).

All deciduous plants store materials during the summer and early fall to provide energy and nutrients needed for flowering and the early growth of

Figure 13-12   Seasonal cycle of protein content in the visceral mass of male and female *Thais lamellosa*. Periods of aggregation (non-feeding periods) and production of offspring are indicated by horizontal bars (after Stickle 1975).

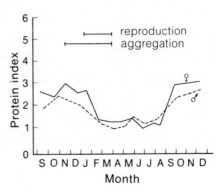

Figure 13-13   Root-crown sprouting by chamise following a fire in the chaparral habitat of southern California. The upper photograph was taken on May 4, 1939, six months after the burn. The lower photograph, showing extensive regeneration, was taken on July 16, 1940 (courtesy of the U.S. Forest Service).

leaves in the spring. Just before trees begin to leaf out, sap rises in the trunk to the tips of the branches, carrying sugars and other nutrients.

Many plants store nutritive materials in their roots to allow recovery after their shoots have been destroyed by fire or by defoliating insects. Every few years in the northeastern United States, tent caterpillars defoliate black cherry trees in the early spring. The cherry trees respond by putting out new sets of leaves, drawing on untapped reserves of nutrients in their roots. Where fires frequently sweep through habitats—as in the chaparral of southern California—many plants store food reserves in fire-resistant root crowns, which sprout and send up new shoots shortly after a fire has passed (Figure 13-13). Root sprouting promotes the recovery of vegetation and stabilization of the soil more rapidly than would be possible if the vegetation could grow back only from seed. The seeds of many annual plants are also fire-resistant, and so get an early foothold in burned-over areas, growing up and producing seed before shrubby vegetation crowds them out.

Organisms must anticipate environmental changes.

To undertake homeostatic responses, particularly those requiring a long period to complete, organisms frequently must estimate environmental conditions removed either in time or in space from their present surroundings. What stimulus indicates to birds wintering in the tropics that spring is approaching in the northern forests or urges salmon to leave the seas and migrate upstream to spawning grounds? How do aquatic invertebrates in the arctic sense that if diapause is delayed a few days, a quick freeze may catch them unprepared for winter and kill them? Most acclimatory and developmental responses require considerable time, and therefore must be initiated before changes in the environment occur. If hawks and owls are to feed their young during spring and summer when food is plentiful, reproductive activities must be started in late winter because nest building and incubation of the eggs take several months.

J. R. Baker (1938) was the first to distinguish *"proximate factors"*—cues by which organisms can assess the state of the environment—and *"ultimate factors"*—features of the environment that are directly important to the well-being of the organism. We are concerned here with the cues that organisms use to time the initiation of homeostatic responses when the changes to be accommodated, the ultimate factors of food supply, ice thaw, and so on, must be anticipated.

Daylength is the most precise indicator of seasonal change.

Days are long in summer and short in winter. At the spring and fall equinoxes in March and September, day and night are equal in length. Seasonal change in daylength varies with latitude. At the Equator there is essentially none, whereas poleward of 67° north and south latitude, daylength ranges from 24 hours in summer to zero hours in winter. Across the middle latitudes of the United States (40°N), daylength varies from about nine to fifteen hours. Almost all plants and animals are sensitive to the length of the day *(photoperiod)* as an indicator of season (Lofts 1970, Bunning 1967).

The occurrence of diapause in insects is controlled by many factors, the most important of which is the photoperiod. Between eight and thirteen hours of daylight are most effective for the inception of diapause in the oriental fruit moth in southern California (Figure 13-14). These light periods correspond to the daylength in early fall and winter when occasional frosts would quickly kill individuals not in diapause. Temperature and availability of food also influence entrance into diapause. A twelve-hour photoperiod stimulates diapause in more than 90 per cent of larvae at 21 to 26 C, but the growth and development process is not arrested by diapause at the same photoperiod when the temperature is as low as 12 C or as high as 30 C (Dickson 1949).

Similar organisms may have strikingly different responses to photoperiod in different areas. The photoperiodic responses of populations of side oats gramma grass vary according to latitude. Under controlled cycles of light and dark, southern populations of the grass (30°N) flowered when daylength was thirteen hours, whereas more northerly populations (47°N) flowered only if the light period exceeded sixteen hours each day (Olmsted 1944). The longer period of light suppressed flowering in the southern populations.

At 45°N in Michigan, populations of small fresh-water crustaceans, known as water fleas *(Daphnia),* form diapausing broods at photoperiods of twelve hours of light (mid-September) or less (Stross and Hill 1965). In Alaska, at 71°N, diapause occurs when there are fewer than twenty hours of daylight, which corresponds to mid-August (Stross 1969). Warm temperature and low population densities tend to shorten the daylength necessary for diapause (and hence delay the inception of diapause in the fall), indicating that these are more favorable environmental conditions for *Daphnia.*

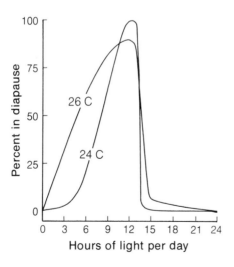

Figure 13-14   The influence of daylength upon entrance into diapause by larvae of the oriental fruit moth (*Grapholitha molesta*). Note that the sharpest break in the response occurs between thirteen and fourteen hours of daylength, corresponding to early fall, which is biologically the most critical point for the inception of diapause (after Dickson 1949).

How does an organism know whether a daylength of twelve or thirteen hours indicates fall or spring? In many temperate-zone birds, a daylength of about thirteen hours in the spring initiates breeding activity. After the end of the breeding season, however, there is a refractory period (Bissonnette and Wadlund 1932), during which the same daylength cannot elicit reproductive behavior (Farner 1964). When some species are maintained artificially on photoperiods that mimic midsummer days, they may continue to breed indefinitely. But once breeding activity has ceased and the reproductive organs have regressed, reproduction cannot be resumed until reproductive potential is restored at the end of the refractory period, no matter what the daylength. The refractory period is frequently terminated in response to short daylength in the late fall, well after the photoperiod becomes too short to stimulate reproductive activities (Lofts and Murton 1968). After the end of the refractory period, reproductive behavior can be initiated by the appropriate artificial photoperiod, even in winter, but unless other conditions, such as temperature, are also satisfactory, birds usually will not go so far as to lay eggs.

Daylength is not always an accurate predictor of the environment. Photoperiod is useless right at the Equator where daylength does not vary during the year; it is also useless where environmental conditions are extremely irregular and thus not predictable, particularly in deserts where rainfall is sporadic. The activity of equatorial organisms, nevertheless, often has a pronounced annual cycle. Other cues, such as climate, or the effects of climate on vegetation, are used to mark the seasons. In such highly unpredictable environments as deserts, organisms are forced to adopt a conservative strategy of readiness through the entire period when the sporadic rains are likely to occur. Photoperiod stimulates the development of reproductive organs in some desert birds to a point just short of breeding. The gonads are maintained in this state of readiness throughout the period in which rains are likely to occur, but the stimulus that finally kicks off the completion of their physiological development and the initiation of breeding may be more closely linked to rainfall itself (Marshall and Disney 1957). Because only two to three weeks separate the first rains from the abundance of food necessary to feed young, all the activities required to initiate breeding and have young in the nest (territory establishment, courtship and mate selection, nest building, egg laying, and incubation) could not possibly be crammed into the period between the first (and sometimes only) rains of the season and the period of abundant food during which adults can feed their young. Hence desert birds accomplish most of the preliminaries to reproduction, up to the point of nest building, before the sporadic rains are likely to occur, and they maintain this state of readiness for a long period if necessary.

In this chapter, we have seen how animals and plants respond to variation in their environments to maintain their activity at as high a level as is possible. But in spite of the battery of homeostatic defenses available to the organism, its activity space within the entire range of conditions possible is greatly limited. This restriction takes on an additional geographical perspective when we consider the distribution of organisms and the factors that delimit geographical ranges in the next chapter.

14

Environment, Adaptation, and the
Distribution of Organisms

No single type of plant or animal can tolerate all the conditions found on Earth. Each thrives within relatively narrow ranges of temperature, precipitation, soil conditions, and other environmental factors. The geographical range of any population cannot exceed the geographical distribution of suitable environmental conditions. Moreover, the preferences and tolerances of each species differ, and so, although the distributions of species broadly overlap one another, no two are found under exactly the same range of conditions.

This chapter is concerned with the factors that determine the distributions of species' populations. On a local scale, environment plays an overwhelming role in distribution, confining populations to regions and habitats within which the species thrives and perpetuates itself. On a global scale, geographical barriers and historical accidents of distribution assume prominent roles in determining the presence or absence of a particular species. One could easily find localities in Asia, North America, and Europe with closely matched climate and soils; but although the vegetation of each locality would superficially resemble the vegetation of the other two, most of the species would differ. The plants of each region are adapted to tolerate similar environments, but barriers to dispersal—oceans, deserts, and mountain ranges—restrict their distributions. That species often can flourish beyond their natural geographical ranges is demonstrated by the success of dandelions, Norway maples, starlings, and honeybees, all of which were introduced from Europe to the United States, sometimes intentionally and sometimes accidentally, by man.

The geographical distributions of plants are related to climate patterns.

The range of the sugar maple, a common forest tree in the northeastern United States and southern Canada, is limited by cold winter temperatures to the north, hot summer temperatures to the south (Figure 14-1). Sugar maples cannot tolerate average monthly temperatures above 75 to 80 F (24 to 27 C) or below about 0 F (−18 C) (Fowells 1965). The western limit of the sugar maple,

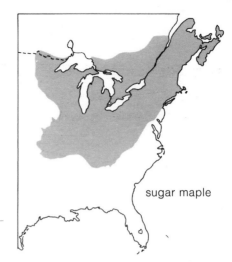

sugar maple

Figure 14-1   The range of the sugar maple in eastern North America (after Fowells 1965).

determined by dryness, coincides closely with the western limit of forest-type vegetation in general. Because temperature and rainfall interact to determine availability of moisture, sugar maples tolerate lower annual rainfall at the northern edge of their range (about 20 inches, or 50 cm) than at the southern edge (about 40 inches, or 100 cm). To the east, the distribution of the sugar maple is limited abruptly by the Atlantic Ocean.

Within its geographical range, sugar maple is more abundant in northern forests, where it sometimes forms single-species stands, than in the more diverse forests in the south. Sugar maples occur most frequently on moist, podsolized soils that are slightly acid.

The range of the sugar maple overlaps three other tree-sized species of maples: black, red, and silver (Figure 14-2). The range of the black maple falls almost entirely within the distributional limits of the sugar maple, indicating similar but narrower tolerance of temperature and rainfall extremes. In fact, the two maples are so similar that foresters only recently have recognized them as different species, by slight differences in the shape of the fruits and by the presence or absence of hairs on the underside of the leaves. The northern limit of red maple nearly coincides with that of the sugar maple, but red maple extends south to the Gulf Coast, and appears to be less tolerant of drought in the midwestern United States. Silver maple also extends beyond the southern limit of sugar maple, but unlike the red maple, it is found farther to the west, extending along stream valleys where soils are moist.

Where their ranges overlap, maples exhibit distinct preferences for local environmental conditions created by differences in soil and topography. Black maple frequently occurs together with sugar maple, but prefers drier, better-drained soils with higher calcium content (therefore less acidic). Silver maple is widely distributed, but prefers the moist, well-drained soils of the Ohio and Mississippi river basins. Red maple is peculiar in preferring either

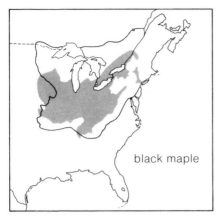

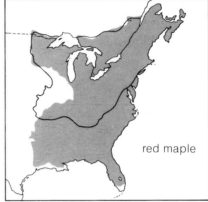

Figure 14-2 The ranges of black, red, and silver maples in eastern North America. The range of the sugar maple is outlined on each map to show the area of overlap (after Fowells 1965).

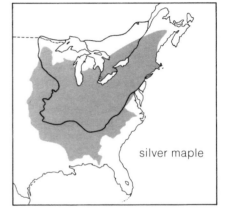

very wet, swampy conditions (it is often called swamp maple) or dry, poorly developed soils. Whether these extremes have some other soil factor in common that the red maple likes or whether red maple consists of two distinct physiological types, each with difference preferences, is not known.

The local distribution of plants is determined largely by topography and soils.

The effects of different factors on the distribution of plants are manifested on different scales of distance. Climate, topography, soil chemistry, and soil texture exert progressively finer influence on geographical distribution. For example, the perennial shrub *Clematis fremontii,* variety *riehlii,* exhibits a hierarchy of distribution patterns revealed by examining its distribution on different geographical scales (Figure 14-3). Climate and perhaps related species in other areas restrict this species of *Clematis* to a small part of the midwestern United States. Variety *riehlii* is found only in Jefferson

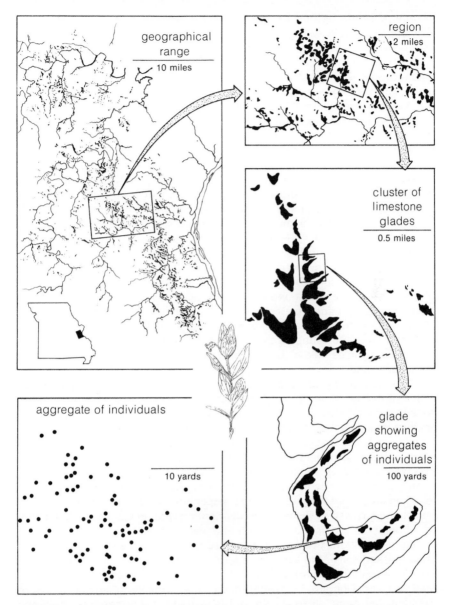

Figure 14-3   Hierarchy of distribution of *Clematis fremontii*, variety *riehlii* (after Erickson 1945).

County, Missouri. Within its geographical range, *Clematis fremontii* is restricted to dry rocky soils on outcroppings of dolomite, which are distributed with respect to mountain and stream systems. Small variations in relief and soil quality further restrict the distribution of *Clematis* within each dolomite glade to sites with suitable conditions of moisture, nutrients, and soil structure. Local aggregations occurring on each of these sites consist of many, more or less evenly distributed individuals (Erickson 1945).

Elevation, slope, exposure, and underlying bedrock—factors that greatly influence the plant environment—vary most in mountainous regions. The heterogeneity of such regions breaks the geographical ranges of species into isolated areas with suitable combinations of moisture, soil structure, light, temperature, and available nutrients. Ecologists frequently turn to the varied habitats of mountains to study plant distribution. Along the coast of northern California, mountains create conditions for a variety of plant communities ranging from dry coastal chaparral to tall forests of Dougas fir and redwood (Waring and Major 1964). Moisture exerts the strongest influence of all environmental factors on the distribution of forest trees. When localities are ranked on scales of available moisture and exchangeable calcium, the distribution of each species among the localities exhibits a distinct optimum (Figure 14-4). The coast redwood dominates the central portion of the moisture gradient and frequently forms pure stands. Cedar and Douglas fir, and two broad-leaved evergreen species with small thick leaves—manzanita and madrone—are found at the drier end of the moisture gradient. Three deciduous species—alder, big-leaf maple, and black cottonwood—occupy the moister end.

The distribution of species along the moisture gradient may coincide with distribution along an available nutrient gradient. For example, the dry soils in which cedar thrives are also poor in nutrients. Low exchangeable calcium indicates that the soil has few cation-exchange sites or that these are largely occupied by hydrogen H^+ ions; in either case, the soil is relatively infertile. Because moisture and nutrient availability are so highly correlated, however, we cannot determine whether water, nutrients, or a third factor related to both, exerts the greatest influence on the distribution of cedar. In fact, cedars are virtually restricted to serpentine barrens in northern California. There may also be instances of distribution with no apparent relationship to nutrients and moisture. Soil nutrients vary widely over the range of soil moisture conditions that favor the redwood, which occurs in all but the most impoverished soils—the serpentine barrens.

The ecological distributions of some species apparently affect others. Madrone and Douglas fir overlap each other broadly along the moisture gradient, but occur apart on the nutrient gradient. Furthermore, madrone occupies the central part of the soil nutrient gradient, with Douglas fir forming two peaks of abundance, one on poorer soils and the other on richer soils. The separate peaks of abundance may represent distinct subpopulations of Douglas fir, each with different tolerance ranges.

Environmental variables do not change independently of each other. Increasing soil moisture usually alters the status of available nutrients in the soil. Variation in the amount and source of organic matter in the soil creates

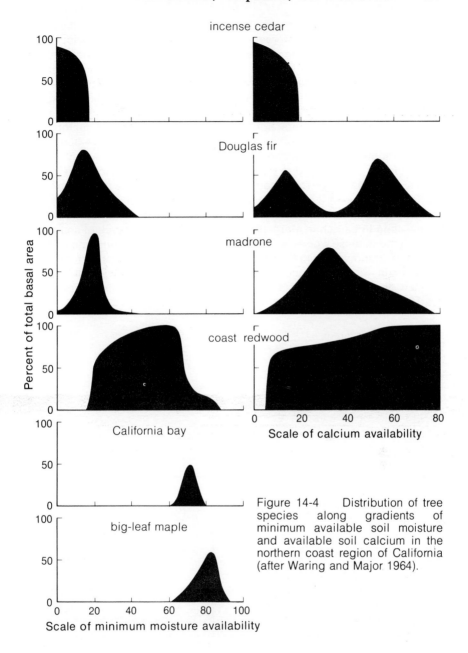

Figure 14-4   Distribution of tree species along gradients of minimum available soil moisture and available soil calcium in the northern coast region of California (after Waring and Major 1964).

parallel gradients of acidity, soil moisture, available nitrogen, and so on. Because these variables interact, one should examine the distribution of plants with respect to all the variables at the same time. We cannot visualize more than three axes on a graph owing to the three-dimensionality of our concepts, and so analyses of plant distributions with regard to more than three dimensions must be left to the computer (see, for example, Marchand 1973, Wali and Krajina 1973, Loucks 1962, Barkam and Norris 1970). We can, however, visualize the interaction between two environmental factors on a simple graph, which will suffice to make the point. In Figure 14-5, distributions of some forest-floor shrubs, seedlings, and herbs in woodlands of eastern Indiana are related to levels of organic matter and calcium in the soil (Beals and Cope 1964). These soils contain between 2 and 8 per cent organic matter and between 2 and 6 per cent exchangeable calcium. Furthermore, levels of calcium and organic matter are interrelated: soils rich in organic humus tend to be rich in inorganic nutrients. Within the range of soil conditions found in these woods, each species shows different preferences. Black cherry is found only within a narrow range of calcium but is tolerant of variation in the percentage of organic matter. Bloodroot is narrowly restricted by the per cent of organic matter in the soil but is insensitive to

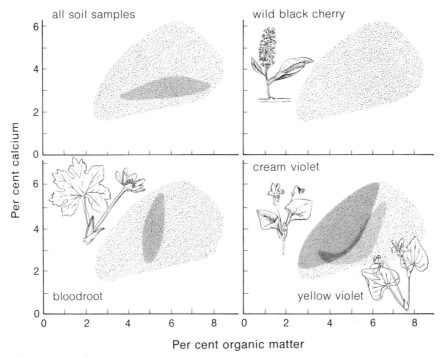

Figure 14-5   The occurrence of four forest-floor plants with respect to the calcium and organic-matter contents of the soil in woodlands of eastern Indiana (after Beals and Cope 1964).

variation in calcium. The distributions of yellow violets and cream violets extend more broadly over levels of organic matter and calcium in the soil, but they do not overlap. Cream violets prefer relatively higher calcium and lower organic-matter content than yellow violets; where one is found, the other is usually absent. In nature, the distributions of species on all scales of geographical distance are not only related to physical and chemical properties of the environment but are influenced by the presence of other species as well.

Populations are self-maintaining within a narrow range of environments.

For each organism there exists some combination of environmental conditions that is optimum for its growth, maintenance, and reproduction. To either side of the optimum, biological activity falls off until the organism ceases altogether to be supported by the environment. We see this pattern over and over again, whether we examine the dependence of photosynthetic rate on leaf temperature, the distribution of plants along moisture and nutrient gradients, the vertical range of seaweeds and marine snails within the intertidal zone, or the size of prey captured by predatory animals. Within the range of environmental conditions that will adequately support an individual, the perpetuation of a population is confined more narrowly. The successful population requires resources for growth and reproduction in addition to those required for individual maintenance.

We may visualize the general relationship of an organism to its environment on a graph on which we relate rate of biological activity (by whatever measure we choose) to a gradient of environmental conditions. This relationship is portrayed in Figure 14-6 as a bell-shaped curve showing a distinct optimum over the middle of the gradient. In nature, these curves can be quite asymmetrical. Oyster development, for example, proceeds best in moderate salinities but is depressed more by low salinity than high salinity (at least up to the maximum for seawater). Optimum oxygen concentration for brook trout occurs toward the upper end of the oxygen gradient in lakes and streams (Shepard 1955).

Regardless of the shape of the activity curve, the ecological distribution of a species along the gradient is governed by three levels of tolerance. First, extreme conditions may totally disrupt critical biological functions, resulting in rapid death. Such lethal conditions occur beyond points c and c' in Figure 14-6. Along a temperature gradient, for example, temperatures below freezing and above 50 C are lethal for most organisms.

Second, organisms must sustain a certain level of activity to maintain themselves in a steady state for long periods. Within points b and b' on the environmental gradient, the organism can exist indefinitely. Outside these limits, the organism's activity level is too low to be self-maintaining, and the organism can venture beyond these limits only briefly.

Third, populations can maintain their size only if reproduction balances death. Reproduction requires resources, hence biological activity, over and above the level needed for self-maintenance. Populations, therefore, persist

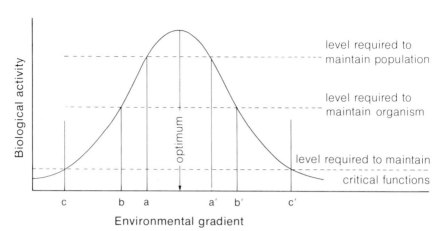

Figure 14-6   The general relationship of biological activity to a gradient of environmental conditions. Levels of activity required to maintain critical biological function, the organism, and the population, determine the lethal extremes (c and c'), the limits of persistence for organisms (b and b'), and populations (a and a').

within a narrower range of conditions than the individual can tolerate. Individuals may live in environments that are inadequate for population maintenance, but their numbers can be maintained only by immigration from populations in more suitable habitats where reproduction exceeds death.

The suitability of the environment determines the success of a population.

Because organisms can live only under characteristic combinations of favorable environmental conditions, one should be able to determine the suitability of a particular place for any species. For example, optimum climates for the Mediterranean fruit fly lie between 16 and 32 C, and 65 to 75 per cent relative humidity. Populations thrive under these conditions provided, of course, that food is available. Outside this optimum, fruit flies can maintain populations under conditions between about 10 and 35 C, and 60 to 90 per cent relative humidity; individuals can persist at temperatures as low as 2 C, and relative humidity as low as 40 per cent; more extreme conditions are usually lethal (Figure 14-7). The Mediterranean fruit fly is a major agricultural pest, but populations reach outbreak proportions only where conditions are within the biological optimum most of the year and rarely exceed the limits of tolerance. Thus Tel Aviv, Israel, where mean monthly temperature varies between 7 and 31 C, and humidity ranges between 59 and 73 per cent, is frequently plagued by the fly and requires extensive control measures, although conditions favor outbreaks more in some years than others. The climate of Paris, France, is generally too cold for fly populations to reach damaging levels, and Phoenix, Arizona, is too dry. The climates of Honolulu, Hawaii, and Miami, Florida, where the pest has been accidentally introduced, are quite suitable for rapid population growth.

As we saw in the last chapter, seasonal changes in climate usually exceed the optimum range of conditions of most species. Consequently, population growth is favored during restricted periods, and populations exhibit seasonal cycles of abundance and activity. Temperature governs the seasons in most temperate and arctic regions, restricting the growing season to the warm summer months.

In the tropics, rainfall sets the seasonal pace of activity. Arboreal mosquitoes are abundant in Panama only during the rainy months of May through December. Most mosquito populations persist through the dry season as desiccation-resistant eggs. Species of arboreal mosquitoes whose eggs cannot withstand drying must find permanent sources of water in which to lay their eggs; adults do not live long enough to span the entire dry season (Galindo *et al.* 1956).

Seasonal changes in the food value of plants often determine shifts in diet and the occurrence of reproductive cycles in herbivores (Dasmann 1964). In the chaparral regions of California, winter rainfall and warming temperatures in the early spring stimulate plant growth, greatly increasing the abun-

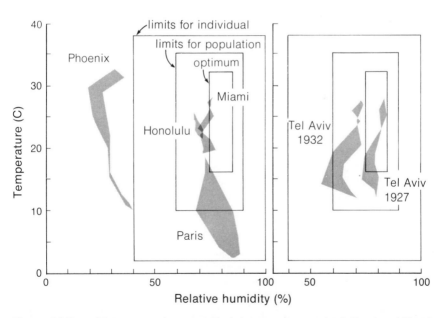

Figure 14-7   The seasonal course of air temperature and relative humidity at selected locations in relation to conditions favorable for the Mediterranean fruit fly. The inner rectangle encloses conditions optimum for growth, the middle rectangle encloses conditions suitable for development, and the outer rectangle delimits the extreme tolerance range (after Bodenheimer, in Allee *et al.* 1949).

dance and protein content of food plants for deer (Figure 14-8). Range quality declines during the dry summer months. Deer require 13 per cent protein in their diet for optimum growth and reproduction; 7 per cent protein barely provides a maintenance diet. Chamise is a common chaparral shrub used extensively by deer, but it does not provide a maintenance diet during most of the year and cannot adequately support a deer population without supplemental foods. Mature chaparral vegetation, including all species of plants used by deer, provides an adequate maintenance level of protein during most of the year and supports moderate reproduction and growth in early spring.

Heterogeneity of soils in the coast ranges of northern California has resulted in striking cases of limited plant distribution. In particular, several species of pines (*Pinus*) and cypresses (*Cupressus*) are restricted to serpentine soils (e.g., Sargent cypress), while others are found only on extremely acid soils (e.g., Bishop pine, lodgepole pine, pygmy cypress) (Kruckeberg 1954, McMillan 1956). When grown on soils from different localities, seedlings of these endemics often do best when planted in soil resembling that on which the population is normally found. Thus, lodgepole pine grows only in acid soils and Sargent cypress, the serpentine endemic, grows somewhat better on serpentine soil than on "normal" soil, and not at all on acid soil (Figure 14-9). Not all endemics perform best on their home soil. When given the chance in an experimental garden, pygmy cypress, normally restricted to acid soils, grows much better on "normal" and serpentine soil. Wherefrom it derives its tolerance for serpentine soils is not known. Nor is it known why pygmy cypress is excluded from soils on which its seedlings grow vigorously. Clearly, edaphic conditions are not the only ones limiting the distributions of these species.

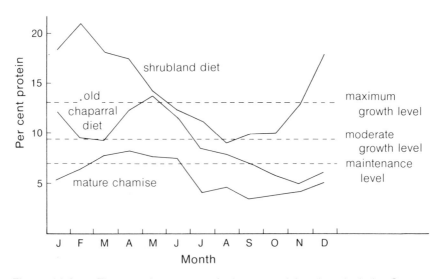

Figure 14-8   The protein content of plants used by deer in Lake County, California (after Dasmann 1964).

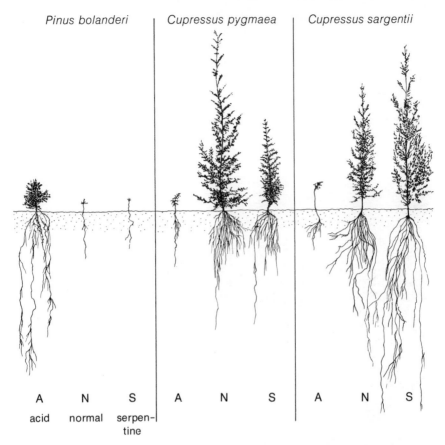

Pinus bolanderi   *Cupressus pygmaea*   *Cupressus sargentii*

A   N   S   A   N   S   A   N   S

acid   normal   serpen-
tine

Figure 14-9   Seedling growth of lodgepole pine, pygmy cypress, and Sargent cypress in acid (left), normal (center) and serpentine (right) soils. Acidity (pH) of the three soils was 4.5, 6.8, and 7.1; exchangeable calcium (millequivalents per 100g) was 0.67, 12.90, and 3.40; exchangeable magnesium (me per 100g) was 0.05, 8.9, and 17.0 (after McMillan 1956).

Adaptations of plants and animals match the conditions within their environments.

It is no accident that different species find different portions of the environmental spectrum suitable for maintaining their populations. The adaptations of an organism—its form, physiology, and behavior—cannot easily be separated from the environment in which it lives. Organism and environment go hand in hand. Insect larvae from stagnant aquatic environments in ditches and sloughs can survive longer without oxygen than related species from well-aerated streams and rivers; species of marine snails that occur high in the intertidal zone, where they are frequently exposed to air, can withstand a greater degree of desiccation than species from lower levels.

In general, the forms of organisms are closely linked to their ways of life and to the environmental conditions in which they live. Compare the leaves of deciduous forest trees with those of desert species. The former are typically broad and thin, providing a large surface area for light absorption and for water loss. Desert trees have small, finely divided leaves—or sometimes none at all (Figure 14-10). Leaves heat up in the desert sun. Structures lose heat by convection most rapidly at their edges, where wind currents disrupt insulating boundary layers of still air. The more edges, the cooler the leaf and the lower the water loss (Vogel 1970). Moreover, small size means a large portion

Figure 14-10   Leaves of some desert plants from Arizona. Mesquite leaves (top) are subdivided into numerous small leaflets, which facilitate the dissipation of heat when the leaves are exposed to sunlight. The paloverde carries this adaptation even farther (center); its leaves are tiny and the thick stems, which contain chlorophyll, are responsible for much of the plant's photosynthesis (hence the name paloverde, which is Spanish for green stick). Cacti rely entirely on their stems for photosynthesis; their leaves are modified into thorns for protection. Unlike most desert plants, limberbush (bottom) has broad succulent leaves, but limberbush plants leaf out only for a few weeks during the summer rainy season in the Sonoran Desert. Photographs are about one-half size.

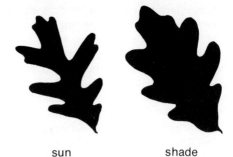

sun   shade

Figure 14-11   Silhouettes of sun and shade leaves of white oak.

of each leaf is given over to its edge. One may find many exceptions to the relationship between leaf size and moisture availability, but the pattern holds for most species. Even on a single plant, leaves exposed to full sun are likely to show adaptations for dissipating heat and conserving water that shade leaves lack. In the white oak, sun leaves are more deeply lobed than shade leaves and hence have more edge per unit of surface area (Figure 14-11).

The vertical distribution of species of algae within the intertidal range is closely paralleled by their rates of photosynthesis in water and when exposed to air (Table 14-1). Along the coast of central California, several common seaweeds have vertical ranges of 0.5 to 2.5 feet within the approximately

TABLE 14-1

Characteristics of intertidal species of algae from various height zones (after Johnson *et al.* 1974).

Characteristic	Species				
	Prionitis	*Ulva*	*Iridaea*	*Porphyra*	*Fucus*
Mean height above sea level (ft)	−1.0	+0.5	+1.0	+3.0	+3.0
Time exposed to air (per cent)	5	15	25	40	40
Leaf area per unit weight ($dm^2 \cdot g^{-1}$ dry weight)*	0.4	3.3	1.3	1.9	0.5
Photosynthesis rate ($mgCO_2 \cdot g^{-1} \cdot hr^{-1}$)					
in air	1.1	12.2	5.2	17.7	5.8
in water	1.2	16.7	1.7	6.3	0.9
air to water ratio	0.9	0.7	2.9	2.8	6.6
Photosynthesis drops to one-half submerged rate at					
water loss (per cent)	10	24	32	47	60
hours	0.4	0.6	1.3	3.6	5.3
Rate of water loss ($\% \cdot hr^{-1}$)	25	40	25	13	11

* dm = decimeter; 1 dm^2 = 100 cm^2

8-foot intertidal range. Height above sea level determines the proportion of time exposed to air on an annual basis. Exposure is about 10 per cent at 0 feet, 40 per cent at 3 feet and 90 per cent at 6 feet. Algae from the lower part of the tide range (*Prionitus* and *Ulva*) have a reduced photosynthetic rate while exposed; algae from the middle and upper parts of the tide range have elevated rates of photosynthesis in air (Johnson *et al.* 1974). Maximum rate of photosynthesis, whether obtained while exposed or submerged, is directly related to the surface area of the alga per unit of dry weight. The relative rates of photosynthesis while submerged and exposed depend on tolerance of desiccation. In algae from the lower portion of the intertidal, photosynthesis after exposure to air drops to one-half the value while submerged in 0.4 to 0.6 hour after a 10 to 24 per cent loss of water (at a rate of 25 to 40 per cent per hour). In algae from the middle and upper intertidal zone, photosynthesis first increases upon exposure. It does not drop to one-half the value while submerged for 1.3 to 5.3 hours after a 32 to 60 per cent loss of water (at a rate of 11 to 25 per cent per hour). Thus, the adaptations of these algae—low rates of water loss and relative insensitivity to desiccation—suit them to the long periods of exposure to air experienced in the upper parts of the tidal range.

But not only do species in different environments have different adaptations, species living together are also adapted to utilize different parts of the environment. That is, they have different activity spaces within the same habitat.

The water relations of coastal sage and chaparral plants in southern California demonstrate divergent courses of adaptation (Harrison *et al.* 1971, Mooney and Dunn 1970). Chaparral habitats generally occur at higher elevation than the coastal sage habitats, and thus are cooler and moister. Both vegetation types are exposed to prolonged summer drought, but water deficiency is greater in the sage habitat. Plants of the coastal sage habitat are typically shallow-rooted with small, delicate deciduous leaves (Figure 14-12). Chaparral species have deep roots, often extending through tiny cracks and

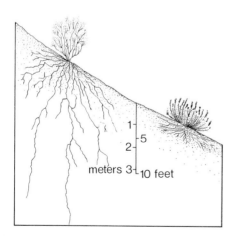

Figure 14-12   Profiles of the root systems of chamise, a chaparral species (left), and black sage, a member of the coastal sage community (after Hellmers *et al.* 1955).

TABLE 14-2

Characteristics of chaparral (left) and coastal sage (right) vegetation in southern California (after Harrison *et al.*1971, Mooney and Dunn 1970).

	Vegetation Type	
Characteristics	Chaparral	Coastal sage
Roots	Deep	Shallow
Leaves	Evergreen	Summer deciduous
Average leaf duration (months)	12	6
Average leaf size (cm^2)	12.6	4.5
Leaf weight (g dry wt $\cdot$ dm^{-2})*	1.8	1.0
Maximum transpiration (g H$_2$O $\cdot$ dm^{-2} $\cdot$ hr^{-1})	0.34	0.94
Maximum photosynthetic rate (mg C $\cdot$ dm^{-2} $\cdot$ hr^{-1})	3.9	8.3
Relative annual CO$_2$ fixation	49.8	46.8

* dm = decimeter; 1 dm^2 = 100 cm^2

fissures far into the bedrock. Their leaves, typically thick and with a waxy outer covering (cuticle) that reduces water loss, are persistent. The more delicate leaves of coastal sage species are often dropped during the summer drought period (see page 178). Leaf morphology influences photosynthetic rate in conjunction with its influence on transpiration (Table 14-2). The thin leaves of coastal sage species lose water rapidly, but also carry on photosynthesis rapidly when water is available to replace transpiration losses. When leaves are clipped from plants and placed in a chamber where transpiration and photosynthesis can be measured, both functions decline as the leaves dry out and their stomata close to prevent further water loss. Coastal sage species, such as the black sage, have high photosynthetic and transpiration rates at the beginning of the transpiration experiment, but shut down quickly owing to rapid water loss (Figure 14-13). Chaparral species such as the toyon (rose family) have maximum photosynthetic rates which are only one-fourth to one-third those of coastal species, but they resist desiccation and continue to be active under drying conditions for longer periods. In this property, the chaparral species resemble the desiccation-resistant algae of the mid-intertidal zone mentioned earlier. For the latter, however, drought comes twice each day.

When chaparral and coastal sage species grow together near the overlapping edges of each other's range, they exploit different parts of the environment: deep perennial sources of water and shallow ephemeral sources of water. In spite of these differences and the corresponding adaptations of leaf morphology and drought response, the annual productivity of both types of species is about the same where there are intermediate levels of water availability. In drier habitats, the prolonged seasonal absence of deep water tips the balance in favor of the adaptations found in deciduous coastal sage vegetation. Increasing availability of deep water at higher elevations favors evergreen chaparral vegetation.

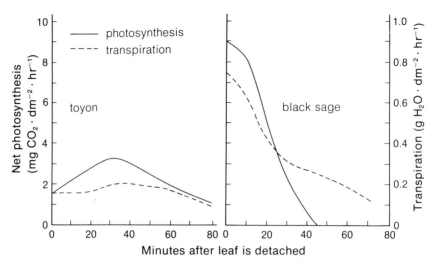

Figure 14-13   Time-course curves for photosynthesis and transpiration under standard drying conditions for a chaparral species (left) and a coastal sage species (right). Note that transpiration continues well after photosynthesis has been shut off; hence leaf dormancy is an ineffective long-term solution to drought (after Harrison *et al.* 1970).

Ecotypic differentiation results in an expanded ecological range.

Botanists have long recognized that a species, when grown in different habitats, may exhibit various forms corresponding to the conditions under which it is grown. In the hawk-weed *Hieracium umbellatum,* for example, woodland plants generally have an erect habit, those from sandy fields are prostrate, and those from sand dunes are intermediate in form. Leaves of the woodland ecotype are broadest, those of the dunes ecotype are narrowest, and those of sandy fields, intermediate. Plants from sandy fields are covered with fine hairs, a trait the others lack. When individual plants of the same species are taken from different habitats and grown in a garden under identical conditions, differences in growth form often persist generation after generation. About fifty years ago, the Swedish botanist Göte Turesson collected seeds of several species of plants that occurred in a wide variety of habitats and grew them in his garden. He found that even when grown under identical conditions, many of the plants exhibited different forms depending upon the habitats from which they were originally taken. Turesson (1922) called these forms *ecotypes,* a name that persists to the present, and suggested that ecotypes represented genetically differentiated strains of a population that are restricted to specific habitats. Because Turesson grew these plants under identical conditions, he realized that the differences between the ecotypes must have a genetic basis, and they must have resulted from evolutionary differentiation within the species according to habitat.

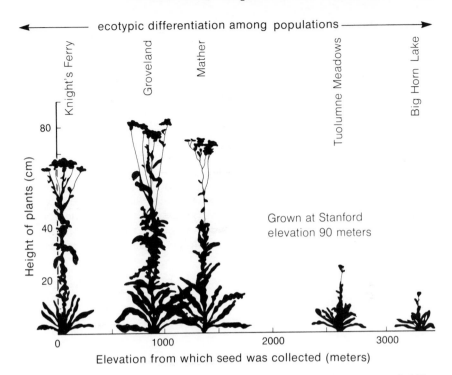

←——— ecotypic differentiation among populations ———→

Grown at Stanford
elevation 90 meters

Figure 14-14   Ecotypic differentiation in populations of the yarrow, *Achillea millefolium,* demonstrated by raising plants derived from different elevations under identical conditions in the same garden (after Clausen *et al.* 1948).

Jens Clausen and co-workers David Keck and William Hiesey (1948) conducted similar experiments on an introduced species of yarrow, *Achillea millefolium,* in California. *Achillea,* a member of the sunflower family, grows in a wide variety of habitats ranging from sea level to more than 10,000 feet elevation. Clausen collected seed from plants at various points along the altitude gradient and planted them at Stanford, California, near sea level. Although the plants were grown under identical conditions for several generations, individuals from montane populations retained their distinctively small size and low seed production (Figure 14-14), thereby demonstrating ecotypic differentiation within the population. Such regional and habitat differences in adaptations undoubtedly broaden the ecological tolerance ranges of many species by dividing the species into smaller subpopulations, each differently adapted to local environmental conditions (Eickmeier *et al.* 1975, Kruckeberg 1951, Hiesey and Milner 1965, McMillan 1959, Antonovics 1971).

The physical environment clearly plays an immense role in shaping the adaptations of plants and animals, and in determining their distributions. In the last two chapters, we have seen how the activity space of an organism is shaped by the interaction of adaptation and environment and how the concept of activity space can be extended on a geographical scale to encompass the distribution of the species. Not only do these spaces differ from one species to the next, but they may also vary within a population in response to the different conditions encountered in different parts of the species' range.

No single species can thrive under all conditions found on Earth. The compromises dictated by their adaptations restrict them to a narrow portion of the ecological spectrum. Until now, we have dealt only with the physical and chemical dimensions of the environment. In the next section, we shall begin to examine biological influences on the organism. Predation and competition restrict the activity space of organisms even within their range of physiological tolerance, while some mutually beneficial interactions between species may expand activity space. Organisms must also respond to other individuals of the same species, including competitors, who vie for resources, and parents and offspring, through which the thread of inheritance runs. All these interactions make up the biological realm of the organism.

Part 5

Organisms in Their Biological Environments

15

Predators and Their Prey

"Struggle for existence" and "survival of the fittest" conjure up images of ferocious predators and their hapless prey. Although sometimes fanciful, these images depict two basic truths: Animals must capture food to survive and reproduce, and they themselves must avoid capture. In this chapter, we shall consider some of the problems facing organisms that search for, and are sought as, food. The diverse array of predator-prey interactions is interwoven so intricately that it is difficult to generalize the adaptations that serve as solutions to these problems.

Regardless of the particular interaction between predator and prey, predation nearly always limits the distribution of a species to a narrower range of conditions than is permitted by physiological adaptations. This is so because, by removing individuals from a population, the predator increases the level of biological activity required to maintain the population. More young must be produced per capita to replace losses to predators. Consider a graph on which biological activity is related to some gradient of environment conditions (Figure 15-1). The level of activity required to maintain the population delimits the range of conditions over which the population will persist. If the maintenance level of activity were raised, as it is when predators remove individuals from a population, the range of conditions that can support the population would become more narrow. And so we see that predation not only shapes the adaptations of the organisms, but also influences the ecological distribution of the population.

The effects of key predators in a biological community may extend far beyond the populations of their prey. When an efficient predator greatly reduces the numbers of a particular prey species that would otherwise be abundant, it makes room for other species whose ecological requirements are similar to the prey species but are not themselves eaten (Paine 1966). The crown-of-thorns starfish (*Acanthaster*) is such a predator (Chesher 1969, Porter 1972a). *Acanthaster* eats corals primarily, and it may become so abundant as to devastate large areas of reef (Figure 15-2). In the Pacific Ocean off the coast of Panama, corals grow as isolated structures on the rocky bottom, rather than forming large reefs as they do elsewhere. Where *Acanthaster* populations were sparse (about two individuals per hectare), corals of two to

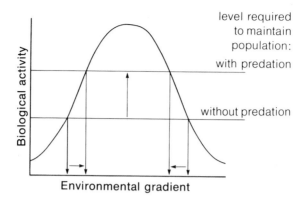

Figure 15-1   The influence of predation on the level of activity needed to maintain a population. Note that predation tends to restrict the gradient of environmental conditions over which a prey population can persist.

four species covered 71 per cent of the bottom. Where *Acanthaster* reached densities of 2,000 individuals per hectare, coverage of the bottom by corals decreased to 21 per cent, but the number of species increased by three to seven (Porter 1972a). Thus, *Acanthaster* not only reduces the abundance of corals in general, but also preys most intensively on certain dominant species, thereby allowing other corals, once only rarely encountered, to become more common.

That *Acanthaster* prefers certain species of corals over others suggests that their quality as prey differs. The quality of prey is, in part, a function of adaptations to avoid being eaten, each species having different adaptations aimed at thwarting their most persistent predators, but all having chinks in their armor. Predators and their prey are engaged in a continual evolutionary struggle to gain the upper hand in their relationship. We shall discuss the various outcomes of this struggle below.

The relationship between predator and prey is structured according to their relative sizes.

When we think of predator and prey, we usually think of lynx and rabbit, of bird and beetle, and so on—of predators that pursue, capture, and eat individual prey (Errington 1963). The prey are generally smaller than the predator, but large enough to be worth pursuing. But some predators, the filter feeders, differ in that they consume minute organisms in vast numbers. The blue whale weighs many tons, but eats small shrimp, fish fry, and the like. This size discrepancy can also apply on a smaller scale: Clams and mussels pump water through fine filtering devices that trap minute plankton; many protozoa and rotifers filter bacteria and other microorganisms from the water. Most filter feeders either live in the sea or gather their food from aquatic environments. Prey are not abundant enough in the atmosphere to make filter feeding worthwhile.

Figure 15-2   The crown-of-thorns starfish *Acanthaster* attacking the coral *Pavona* on the Pacific coast of Panama. White areas in the upper middle and lower left of the photograph are exposed skeletons of incompletely digested *Pavona* (courtesy of J. W. Porter, from Porter 1972).

As the size ratio of prey to predator increases, prey become more difficult to capture, and predators become more specialized for pursuing and subduing their prey. Eventually a limit is reached beyond which predators can no longer capture prey, simply because they lack sufficient strength and swiftness. Lions will attack prey that are their own size or a little larger, but they leave full-grown elephants strictly alone. Some species have extended the upper end of the size range of their prey by the evolution of social cooperation in predation. Pack hunting has been adopted by wolves, hyenas, and army ants (Mech 1970, Rettenmeyer 1963), but by few others.

At the other end of the relative size spectrum are the "live-in" predators, or parasites—the myriad viruses, bacteria, protozoa, worms, and others that invade the body of the host and feed on its tissues or blood or, more directly, on the digested food in its intestine. The strategy of parasitism differs utterly from predation or filter feeding, because the survival of the parasite depends entirely upon the survival of its host, rather than on its death; the parasite normally develops and reproduces within the body of its host. Parasites could not be large, for if they were, they would literally drain the life from their host. External parasites, such as fleas, ticks, and vampire bats, which can move from host to host, are not constrained by this restriction, although they still must remain relatively small to escape detection and removal by their hosts.

Herbivores feeding on plants are analogous to either predator or parasite, depending upon how they do it. Plants tolerate being eaten more than animals do. If we apply the criterion that parasites live off the productivity of a host organism without killing it, we must classify deer, which browse on trees and shrubs, as parasites. If a cow consumes an entire plant, pulling it up by the roots and macerating it into lifeless shreds, the cow is functionally a predator. The beetle larva that develops within a seed, thereby destroying the embryonic plant organism it contains, is also a predator. While we often view herbivores as distinct from carnivores, they have the same functional relationship to their prey; they are distinguished more by the specialized adaptations each needs to pursue and consume its prey.

Predator adaptations demonstrate the importance of the biotic environment as an agent of natural selection.

The morphological modifications of predators for feeding are too obvious to warrant detailed discussion. Filter feeders remove small food particles from the aquatic environment, usually by means of a comblike filter through which water, containing the minute prey, is forced. Clams continually pump water in through their siphons to a filter where food particles are carried to the mouth by the rhythmic beating of small hairlike cilia. The baleen whales, on the other hand, gulp a vast quantity of water into their mouths, which they then force with their tongues out through the comblike plates (the baleen, or whalebone) that protrude from the upper jaw and trap small organisms.

The nature of the diet often has a profound effect on the evolution of teeth. Herbivores, especially those that eat grass, have teeth with large grinding surfaces to break down tough, fibrous plant materials (Figure 15-3). The

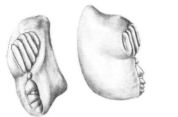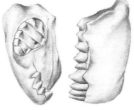

Figure 15-3   Relationship between diet and the structure of the mandible in grasshoppers. Left, granivore (front and side views); center, herbivore (front and side views); right, carnivore (side view only) (after Iseley 1944).

teeth of predators have cutting and biting surfaces that both immobilize the prey in the mouth and cut it into pieces small enough to be swallowed. Seemingly simple differences in dentition, for example between horses and other ungulates such as cows, sheep, and deer, can have important ecological consequences. Horses have strongly opposed upper and lower incisors that can cut the fibrous stems of grasses. Other ungulates lack upper incisors; their lower teeth press against the upper jaw at an angle for gripping and pulling plant material. This difference partly explains the success of zebras, which are closely related to horses, in dry, grassy areas of Africa; gazelles and other ungulates require lush, leafy vegetation for browse (Gwynne and Bell 1968).

Many predators use their forelegs to manipulate their food. Fish and snakes obviously cannot do this, but some species have very large mouths with distensible jaws that enable them to swallow large prey whole. Such adaptations emphasize the important role the appendages play in the ingestion of food. The only predators that regularly tear their food into small morsels are those which can use their limbs for this function. Among birds, for example, only the hawks, eagles, and owls, with their powerful sharp-clawed feet and hooked beaks, can do so. Diving birds often eat large fish, but must swallow them whole because their hind legs are specialized for swimming and diving rather than for grasping and dismantling prey.

The quality of the diet influences adaptations of the predator's digestive and excretory systems as well. Because plant foods have a high cellulose content and are, therefore, more difficult to digest than the high-protein diet of carnivores, the digestive tracts of herbivorous animals are often greatly elongated. With the larger volume of the intestines, the plant food remains in the digestive tract longer and can be more completely digested. The other side of this coin is that herbivores have to carry a lot of undigested food with them, adding weight and reducing mobility.

Food quality may also impose demands on other body functions. The high salt content of the diets of sea birds and mammals taxes their kidneys and salt glands as they strive to regulate the osmotic level of their body fluids. The blood diet of a vampire bat poses a more unusual problem. Because its food is mostly water, a vampire bat must consume a great quantity to obtain adequate nourishment, so much that it becomes too heavy to fly. After a meal, the bat crawls into a crevice or under a rock or log while its kidneys produce a copious dilute urine to rid the body of excess water. Later, it must rid its body of nitrogenous waste products because most of what the bat consumes in the blood is protein, and so the kidneys must reverse their earlier function to produce a high concentrated urine that prevents excessive water loss.

To locate and capture food, predators are endowed with senses commensurate with their habitat, feeding tactics, and their prey's ability to avoid detection (Gordon 1968, Prosser 1973). We ourselves locate food primarily by vision. Yet our sight is paltry compared to that of hawks and falcons (Fox *et al.* 1976). Most insects are sensitive to ultraviolet light—wavelengths shorter than those we see as violet—and their world is all the more different from ours (Figure 15-4). Insects are also capable of detecting rapid movement, such as that of wings beating 300 times per second, that is only a blur to us.

Among the more unusual sensory organs of predators are the pit organs of pit vipers, a group that includes the rattlesnakes. The pit organs, which are located on each side of the head in front of the eyes (Figure 15-5), are designed to detect the infrared (heat) radiation given off by the warm bodies of potential prey, a sort of "seeing in the dark" (Grinnell 1968). Infrared radiation has a very low energy level, and so it is usually not included within

Figure 15-4   The appearance of flowers to the human eye (left) and to eyes that are sensitive to ultraviolet light (right). Above: marsh marigolds; below: five species of yellow-petaled Compositae from central Florida (from Eisner *et al.* 1969, courtesy of T. Eisner).

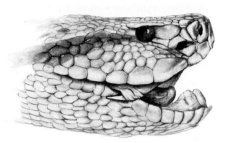

Figure 15-5   Head of the western rattlesnake (*Crotalis viridis*), showing the location of the infrared-sensitive pit between, and slightly lower than, the eye and the nostril (after a photograph in Burkhardt, Schleidt, and Altner 1967).

the range of most visual perception. But nocturnal predators are forced to rely on much weaker stimuli. Nonetheless, pit vipers are so sensitive to infrared radiation that they can detect a small rodent several feet distant within less than a second. Moreover, because the pits are directionally sensitive, the snakes are able to locate warm objects precisely enough to strike at them.

A sensory capability that has been developed only rarely is the detection of electrical fields. Some fish, like the electric ray *Torpedo,* can release powerful electrical currents, up to 50 volts at several amperes, which are used in defense or to kill prey. The shocks are produced by specialized muscle organs that are layered like the plates of a battery to increase the voltage. Other species of electric fish use weak continuous discharges to create electrical fields around themselves. Nearby objects distort the electric field because their properties of electrical conduction are different from those of water, and these changes in the electrical field are picked up by electrically sensitive receptors on the surface of the fish. As one might expect, this sensory mechanism is most highly developed in fish that inhabit murky waters where visibility is poor. In similar habitats, such bottom-dwelling species as the catfish have elongated fins, or barbels, around the mouth that are sensitive touch and taste receptors.

Web-weaving spiders seemingly have little need for elaborate prey-recognition senses. They simply take ensnared food items from their webs, like canned soup from a grocer's shelf. But a closer look reveals a much more complex picture. It is true that web-building spiders have poor eyesight, but their sense of touch is delicate and the slightest vibration of the web sends the spider to the prey, lest it escape. If a prey struggles in the web, it is easily located by the vibrations sent along the silken fabric. If the prey stops moving, the garden spider, *Argiope,* goes to the center of its web and plucks the lines radiating out from the hub in a 360-degree circle until it detects the prey by the dampened vibrations of the plucked line. Removing prey from the web is further complicated for the spider because moths and butterflies often escape from webs, leaving only their detached scales sticking to the threads. The garden spider quickly identifies the type of insect by touch; moths and butterflies are immediately wrapped with silk to prevent their escape, whereas other insects are bitten and injected with a poison to immobilize them (Robinson 1969a).

Many predators have poorly developed senses and rely on bumping into prey items by chance. Of course, their prey must be equally oblivious of what is going on around them if this strategy is to be successful. But even in such seemingly random interactions, the searching patterns of predators are often adapted to maximize the chances of encountering prey. For example, the predatory larvae of the ladybird beetle feed on mites and aphids that infest the leaves of certain plants, and they must physically contact their prey to recognize them. Their movements on the leaves are not oriented toward the prey, but neither are they random. The veins and rims of leaves make up less than 15 per cent of the leaf surface, yet the larvae spend most of their time searching on these areas, where almost 90 per cent of the aphids are distributed (Dixon 1969).

Prey can hide, flee, or fight to escape predation.

The adaptations of prey organisms to avoid predators are as diverse as the foraging adaptations of the predators. Organisms that are extremely small compared to their predators—those captured by filter feeders—have little recourse from predation and exhibit few obvious adaptations to avoid being caught. But a larger prey may try (a) to flee its predator, (b) to defend itself by fighting, tough armor, or spines, or (c) to disguise itself as either a distasteful species or an enemy of the predator. Many circumstances of the predator-prey relationship determine which strategy, or combination of strategies, will minimize chances of being caught. For example, there are no hiding places for large ungulates on a short-grass prairie, so they must rely on swiftness for escape. Because plants cannot move, they cannot flee predators and must rely on thorns and toxic chemicals to keep herbivores away.

Protective defenses rarely involve physical combat because few prey are a match for their predators, and predators carefully avoid those that are. Instead, foul-smelling or stinging chemical secretions are often used to dissuade predators (Eisner and Meinwald 1966). Whip scorpions and bombardier beetles direct sprays of noxious liquids at threatening animals. A tenebrionid beetle, commonly known as the stinkbug, discharges a foul-smelling secretion from glands at the tip of its raised abdomen. (Grasshopper mice can circumvent this problem by picking up the beetles and jamming them butt-end into the earth.) Many plants and animals are inedible because they contain substances that are toxic to predators (Whittaker and Feeney 1971, Feeney 1975). Slow-moving animals, such as the porcupine and armadillo, have evolved spines or armored body coverings; this avenue of defense is not open to flying organisms, however, because such defensive structures weigh too much. All defense strategies are costly to the prey organism because they require the allocation of time, energy, and materials, which are limited in supply. The fact that organisms are willing to accept this burden of antipredator adaptations suggests that predation has a major influence on fitness and that predators are important agents of natural selection.

Cryptic animals demonstrate the complexity of antipredator adaptations.

Our fascination with the natural world is perhaps most strongly stimulated by the defensive adaptations of prey, particularly insects, against visually hunting predators. This is not so surprising as we ourselves are visually oriented. But the perfection of cryptic appearances and attitudes to avoid detection by predators will always remain an impressive testament to the force and pervasiveness of natural selection (Cott 1940, Wickler 1968).

One means of achieving *crypsis* is for an organism to match the color and pattern of the background upon which it rests, as in the peppered moth (Figure 8-2). Another is for the organism to be marked with disruptive patterns to obliterate the telltale outlines of its body against the background (Figure 15-6). A third is countershading, which eliminates the shadowed areas of the body by which prey are often located. In countershaded organisms, the surface of the body that faces the sun (usually the dorsal surface) is dark, and the shaded side is light colored. Normal lighting makes the dark dorsal surface appear the same brightness as the lighter-colored but more poorly illuminated ventral surface.

Many prey organisms have evolved special resemblances to inedible objects. Among the most obvious of these are the small invertebrates that very closely resemble sticks, leaves, and flower parts—even bird droppings. The great pains that evolution has taken to conceal the head, antennae, and legs of organisms underscores the importance of these cues to predators for recognizing prey. In the stick-mimicking phasmids (stick insects) and leaf-

Figure 15-6   Examples of crypsis achieved by disruptive coloration in the jackknife-fish (left) and the angelfish (right). If the contrast is strong enough, one tends to see the pattern rather than the fish itself (after Laglar, Bardach, and Miller 1962).

Figure 15-7   Some examples of protective coloration and form in forest-dwelling animals in Panama. The three praying mantids (A, B, and C) avoid detection by blending with different backgrounds. The mantid in A is green and blends with leafy vegetation; the mantid in B (genus *Acanthops*) resembles a dead, curled-up leaf; the mantid in C has a long, thin body that can easily be confused with a twig. The spider in D has a greatly enlarged orange abdomen, whose appearance is similar to flower parts that frequently drop into its web. The swallow-tailed butterfly in E is strikingly colored with a bold, iridescent green and black pattern as a warning to predators that it is distasteful. The frog in F escapes notice among the leaf litter of the forest floor. When viewed from above, the caterpillar in G blends perfectly with the twig on which it is resting; from the side, it is betrayed by its silhouette. The moth *Hyperchiria* (H) folds its wings over its back to break up its symmetry. The body of the caterpillar in I is covered with numerous spiny projections with toxic hairs to dissuade predators.

mimicking katydids, legs are often concealed in the resting position either by being folded back upon themselves or the body, or by being protruded in a stiff, unleglike fashion (Robinson 1969b). The dead-leaf–mimicking mantis *Acanthops* partially conceals its head under its folded front legs (Figure 15-7B). Symmetry is also a sure giveaway for animals, but it is difficult to conceal. The leaf-mimicking moth *Hyperchiria nausica* produces the appearance of an asymmetrical midvein by folding one forewing over the other (Figure 15-7H). Moths will sometimes rest with a leg protruding to one side, but not to the other, or with the abdomen twisted to one side, both of which achieve the same effect.

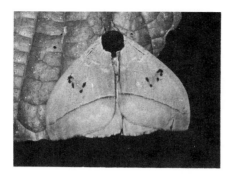 

Figure 15-8   The eyespot display of an automerid moth from Panama: normal resting attitude (left); reaction when touched (right).

The behavior of cryptic organisms must correspond to their appearance. A leaf-mimicking insect resting on bark, or a stick insect moving rapidly along a branch, will not likely fool many predators. Mimicry involves behavior as well as form. The behavior patterns of cryptic insects also suggest which characteristics of prey their predators are most sensitive to. For example, many katydids and mantids apparently avoid detection by predators either by moving very slowly or by rocking from side to side while they move, mimicking natural leaf movements (Robinson 1969b). Another prey strategy is to display occasional flashes of bright coloration while moving. It is believed that predators fixate upon the place where these flashes occur and thereby lose track of the prey. Some cryptically colored frogs show bright colors on the inside of their hind legs as they jump. The frogs quickly assume a concealed position when they land, leaving predators to search where they saw the frog jump from.

When cryptic organisms are discovered by predators, the prey may respond with a variety of second-line defenses, including startle-and-bluff displays, and various attack-and-escape mechanisms (e.g., Blest 1964). The caterpillar of the hawkmoth Leucorampha omatus is green and countershaded, and normally assumes a cryptic position. When disturbed, however, the caterpillar puffs up its head and thorax to look like the head of a small poisonous snake, complete with a false pair of large shiny eyes; then it begins to weave back and forth, hissing in a serpentine manner (Robinson 1969b). The eyespots displayed by many moths and other insects when disturbed (Figure 15-8) frighten predators because they resemble the eyes of large birds of prey (Blest 1957). The use of such displays is sometimes constrained by other features of the environment. Large-bodied moths at low elevations in the tropics rarely exhibit such protective displays as eyespots or flash coloration. When they are resting during the day, the air temperature is only 6 C below the optimum working temperature of the flight muscles, and escape by flight is often possible. At higher elevations (4,000–4,500 ft.), moths require a preflight warm-up period because air temperature may be 15 C lower than the working temperature of the flight muscles and escape is not possible. At these elevations, a large proportion of moth species have special protective displays (Blest 1963). This example should remind us that adaptations must be interpreted in the broadest possible context.

Distasteful or noxious organisms are often conspicuously marked to warn predators.

Crypsis is a strategy of palatable animals. When an individual is high on a predator's shopping list, it is best to be active only at night when motion is not an immediate giveaway; by day the individual hides in some protected cranny or fades away into the habitat, posing as bark, a leaf, a twig, or some equally commonplace object. Other organisms have rejected crypsis for a bolder approach to predator defense. Instead they produce noxious chemicals and advertise the fact with conspicuous color patterns. Predators learn quickly to avoid such conspicuous markings as the black and orange stripes of the monarch butterfly. One attempt to eat such a butterfly is a memorable enough experience to make a predator avoid monarchs for some time to come (Brower 1969). It is interesting to note that many *aposematic* forms adopt similar patterns of warning coloration: black and either red or yellow stripes characterize such diverse animals as yellow jackets and coral snakes.

Batesian mimics fool predators by resembling distasteful forms.

Distasteful animals and plants that display warning coloration often serve as models for the evolution of mimicking color patterns by palatable forms (Wickler 1968, Rettenmeyer 1970). This relationship is known as *Batesian mimicry,* named after is discoverer, the nineteenth century English naturalist Henry Bates. In his journeys to the Amazon region of South America, Bates found numerous cases of palatable insects that had forsaken the cryptic patterns of close relatives and had come to resemble brightly colored distasteful species. Bates rightly guessed that these insects were successfully tricking predators into avoiding them.

Mimicry may involve more extensive changes in morphology than altered color pattern. The buprestid beetle *Acmaeodera* mimics bees and wasps with yellow and black bands on its abdomen (Figure 15-9). Because the mimicry pattern must appear both in flight and at rest, the elytra (the hard forewings of beetles that are not used in flight, but only for protection) became fused, and remain fixed in position over the back even in flight (Silberglied and Eisner 1969). Most beetles spread their elytra in flight and do not resemble bees and wasps in the least. Some beetles have taken to mimicking flies, which are perfectly edible, but fly so fast that predators cannot catch them (Hespenheide 1973). Because of their speed, flies are effectively inedible and certainly not worth pursuing to some predators.

Experimental studies have demonstrated that mimicry does confer advantage to the mimic. Brower and Brower (1962) showed that toads fed live bees thereafter avoided the palatable drone fly that mimics bees; if the toads

Figure 15-9   Mimicry of wasps by a buprestid beetle, *Acmaeodera pulchella* (left). The mimic bears yellow bands on a black abdomen, which is typical of wasps. Note that the elytra (forewings) of *Acmaeodera* are fused, whereas most beetles spread their elytra during flight (right) (photographs courtesy of Thomas Eisner, from Silberglied and Eisner [1969]).

were fed only dead bees from which the stings had been removed, they ate the drone fly mimics quite readily. Similar results were obtained with blue jays as predators; the distasteful monarch butterflies were the models and their viceroy butterfly mimics were the experimental subjects (Brower 1958).

Attempts to demonstrate advantage in mimicry in the natural environment have been less successful. The Browers' and their associates performed a series of experiments in natural environments in Trinidad, involving the palatable prometheus moth from the United States and the distasteful native bufferflies *Parides* and *Helaconius* (Brower *et al.* 1964, Brower *et al.* 1967). Wild birds were the most important predators. In these studies, survival rates were determined by releasing marked male prometheus moths and subsequently recapturing them by attracting them to caged females. In the first experiment, prometheus moths were altered to mimic *Parides,* which is red, black, and green; in the second, they were painted to resemble another distasteful butterfly, the black and red *Helaconius.* In both experiments, the altered prometheus moths were recaptured with the same frequency as the controls that were painted black, the natural background color of the prometheus moth (Table 15-1), suggesting that the artificial mimicry conferred no advantage. Puzzled by these results, the Browers designed a third experiment in which they painted prometheus moths to be conspicuous and unlike any other bufferfly in Trinidad. Significantly fewer of these were recaptured than controls, and so the conspicuous experimental moth was apparently found more readily by predators.

Their failure to find an advantage to mimicry led the Browers to consider the possibility that the experimental techniques were improper. In the first three experiments, large numbers of moths were released simultaneously in

TABLE 15-1

A summary of the results of experiments with artificial mimicry in a natural environment (from Brower *et al.* 1964 and Brower *et al.* 1967).

EXPERIMENT I. *Hylaphora* (edible) altered to mimic *Parides* (inedible).

	Control	Experimental
Release	51	52
Recapture	18 (35%)	16 (31%)

Conclusion: No advantage detected.

EXPERIMENT II. *Hylaphora* altered to mimic *Helaconius* (inedible).

	Control	Experimental
Release	162	156
Recapture	44 (27%)	45 (29%)

Conclusion: No advantage detected.

EXPERIMENT III. *Hylaphora* altered to be conspicuous and unique.

	Control	Experimental
Release	42	43
Recapture	16 (38%)	8 (19%)

Conclusion: Unique conspicuous markings are disadvantageous to a palatable moth.

EXPERIMENT IVa. *Hylaphora* altered to mimic *Parides* and released in high numbers at one locality.

	Control	Experimental
Release	414	414
Recapture	129 (31%)	95 (23%)

EXPERIMENT IVb. *Hylaphora* altered to mimic *Parides* and released at different times in many localities.

	Control	Experimental
Release	194	199
Recapture	37 (19%)	53 (27%)

Conclusion: Results are consistent with the hypothesis that advantage is conferred only when mimics are less common than models.

one place. The artificial mimics may have outnumbered natural unpalatable models so greatly that predators quickly learned that they were, in fact, edible. A second series of experiments was designed in which prometheus moths were again painted to mimic *Parides*. In one experiment, they were released simultaneously in large numbers at one location. In the second experiment, they were released at different points over a longer period to reduce their numbers relative to the models. In the first experiment, altered moths returned less frequently than controls, but in the second group the experimentally produced mimicry conferred a substantial advantage. The results indicate that mimicry works only when the mimics are less common than the models (Brower 1960).

Predators must learn to distinguish acceptable and unacceptable prey items. Naïve animals will often eat unpalatable aposematic prey—once anyway. Predators probably do not "remember" their experiences with prey indefinitely, but continually resample potential prey items. If the noxious taste or sting associated with a particular color pattern is not reinforced by eating an occasional model, a predator will continue to feed on the palatable mimics. It has been shown that a predator will remember an unpleasant experience with a prey item longer if the experience was strong. Thus, the greatest advantage is conferred to mimics by models that are both common and extremely distasteful.

Extensive studies by Brower (1969, 1970) and his colleagues (Brower et al. 1968) on the monarch-viceroy butterfly mimicry complex have revealed a system, more complicated than was at first perceived, that involves mimicry within the monarch population (automimicry). Monarch caterpillars feed on milkweeds of the genus Asclepias. Of the many species of milkweeds, only the tropical and subtropical species, A. humistrata and A. curassavica, produce cardiac glycosides that make the adult butterflies noxious to birds. Larvae raised on the more northern species of milkweed, A. tuberosa and A. incarnata, produce perfectly palatable adults. All four species of milkweed occur on the Florida peninsula, so there are both palatable and unpalatable individuals in the same population. In effect, the palatable monarchs mimic their noxious relatives.

Müllerian mimicry complexes involve many noxious species.

Müllerian mimicry, named for the nineteenth century German zoologist Fritz Müller, involves the resemblance of two or more species of unpalatable organisms. Müllerian mimicry complexes often involve dozens of species, both closely and distantly related. The advantage of Müllerian mimicry to unpalatable species is that predators must learn to avoid fewer different types of noxious organisms. By resembling one another, diverse species can share the burden of the predator learning experience.

Although there are many thousands of Batesian and Müllerian mimicry complexes, a further variation on these basic themes seems important enough to mention, namely that predators sometimes use mimicry to avoid being detected by unsuspecting prey—the proverbial wolf in sheep's clothing. This is often called aggressive mimicry. The mimetic egg coloration and adult behavior of the giant cowbird, a brood parasite of the chestnut-headed oropendola, is a familiar example (see page 80). The zone-tailed hawk of Mexico closely resembles the turkey vulture both in appearance and behavior (Willis 1963), and it frequently soars in the midst of vultures that, because they eat only carrion, are ignored. By virtue of their resemblance to vultures, zone-tailed hawks largely go unnoticed by rabbits and other small mammals, many of which have unwittingly fallen prey to "turkey vultures" that suddenly folded their wings and plummeted down out of the sky at them.

The firefly *Photuris versicolor* (actually a beetle) is the femme fatale of the insect world. Females mimic the flashing patterns of the females of other species of fireflies to attract unsuspecting, or otherwise-suspecting males to their dinner plate. Some versatile females of *P. versicolor* have the flashing patterns of two or three other species in their repertory of aggressive mimicry (Lloyd 1975).

The relationship between predator and prey is characterized by adaptations of each to foil the other. We have considered many such adaptations in the preceding examples. Parasitism and herbivory are also forms of predation that pose unique problems solved by specialized adaptations. Some of these will be described below.

Host specificity and complex life cycles that facilitate dispersal are characteristic of most parasites.

Parasites are usually much smaller than their prey, or hosts, living either on their surfaces (*ectoparasites*) or inside their bodies (*endoparasites*). Internal parasites and, to a lesser extent, ectoparasites demonstrate characteristic adaptations to their way of life. First, because parasites generally live inside of, or in close association with a larger organism, they expend little effort to maintain a constant internal environment. This function is largely accomplished by the homeostatic mechanisms of the host. Thus, endoparasites are often stripped-down organisms that contain little more than food-processing and egg-producing machinery. Second, parasites are faced with the problem of dispersal through the hostile environment between hosts. This is commonly accomplished through complicated life cycles, one or more stages of which can cope with the external environment.

Ascaris, an intestinal roundworm that parasitizes man, has a relatively simple life cycle. A female *Ascaris* may lay tens of thousands of eggs per day, which pass out of the host body in the feces. In regions where sanitation is poor, or where human excrement is applied to farm lands as fertilizer, the eggs may be inadvertently reingested and start a new life cycle. The only stage during the life cycle of *Ascaris* that occurs outside the host is the egg, which is well protected by a sturdy, impermeable outer covering.

Schistosoma, a trematode worm that commonly infects man and other mammals in tropical regions, has a more complicated life cycle that involves a fresh-water snail as an intermediate host. Male-female pairs of adult worms live in the blood vessels that line the human intestine or the bladder, depending on the species of worm. The eggs pass out of the host body in the feces or urine. If the eggs end up in water, they develop into a free-swimming larval form (the miracidium), which burrows into a snail within 24 hours. In the snail, the miracidium produces cells that develop into free-swimming cercariae, which can penetrate skin. Once inside the body of a human or other host, the cercariae travel a circuitous route through blood vessels until they become lodged in the appropriate location where they metamorphose into the adult worm (Jordan and Webbe 1969).

A snail infected with one miracidium may liberate from 500 to 2,000 cercariae per day over a period of a month (McClelland 1965). It is not known precisely how many eggs are produced by the adults during their lifetime, which averages about four and a half years in man, but the reproductive rate is very high, perhaps several hundred eggs per day. Relatively little is known about the ecology and population dynamics of *Schistosoma* (cf. Hairston 1965), although the parasite, which causes the incurable disease schistosomiasis, probably affects 200 million humans (Maldonado 1967) and countless domestic animals (Soulsby 1968).

The life cycle of the protozoan parasite that causes malaria is similar to that of *Schistosoma* in that it involves two hosts, a mosquito and man or some other mammal or bird (Mattingly 1969). A major difference is that whereas the sexual reproductive phase of the schistosome life cycle occurs in man, that phase of the malaria parasite's life cycle takes place within the mosquito. When an infected mosquito bites a human, cells called sporozoites are injected into the bloodstream with the mosquito's saliva. The sporozoites enter the red blood cells and feed upon hemoglobin. When the malaria cell becomes large enough, it undergoes a series of divisions (asexual reproduction), and the daughter cells break out of the red blood cell. Each daughter cell enters a new red blood cell, grows, and repeats the cycle, which takes about 48 hours. (When the infection has built up to a high level, the emergence of daughter cells corresponds to periods of high fever.) After several of these cycles, some of the cells that enter red blood corpuscles change into sexual forms. If these are swallowed by a mosquito along with a meal of blood, the sexual cells are transformed into eggs and sperm, and fertilization (sexual reproduction) takes place. The fertilized egg then divides and produces sporozoites, which work their way into the salivary glands of the mosquito and may be transmitted to a new intermediate host, thereby completing the life cycle.

In spite of impressive successes in combating some parasite-caused diseases of man, little is known about parasites in natural populations of animals and plants, and even less is understood about the control of disease epidemics and the regulation of parasite populations. The population dynamics of parasites must be tied closely to populations of principle hosts, availability of alternate hosts, immunity and susceptibility of hosts, and environmental conditions for free-living stages of the life cycle.

A few generalizations about parasite-host relationships can be made. First, parasites are generally host-specific, each species being restricted to one host organism or to a few closely related ones in each stage of its life cycle. Second, most parasite-host relationships have evolved to the point of being finely balanced; parasites rarely impair the health of their host. The benign character of most parasites may be explained as follows. Any strain of parasitic organisms that is so virulent as to kill its host also kills itself. Parasites that have only benign levels of infection, therefore, are strongly selected. The delicate balance achieved depends in part upon the adaptation of the parasite to its host environment. Disease or other stresses can change the internal environment of the host and upset the host-parasite balance, often leading to the death of the host (Latham 1975). The famous black plagues of medieval Europe are thought to have followed periods of widespread famine, which generally increased susceptibility to all disease (Cartwright and Biddis 1972). Moreover, parasites introduced to populations that have not developed the proper mechanisms of immunity can increase rapidly with disastrous results. Time and time again, foreigners have introduced parasites to susceptible native populations with the result that normally benign disease organisms become virulent.

The mutual accommodation of parasites and their hosts is poorly understood (Jackson, Herman, and Singer 1969, 1970). How, for example, does the intestinal worm avoid being digested? How do organisms that invade the bloodstream of vertebrates avoid being destroyed by immunity mechanisms? Vertebrate organisms do not tolerate foreign proteins (antigens), the introduction of which into the bloodstream usually stimulates the formation of antibodies that specifically attack the introduced protein. Successful parasites have found several ways to circumvent the immune mechanism. Some microscopic disease organisms produce chemical factors that suppress the immune system of the host (Schwab 1975). Others have surface proteins that mimic host antigens and thus are overlooked by the immune system (Damien 1964). Some schistosomes are known to excite an immune response when they enter the host but do not succumb to antibody attack because they coat themselves with proteins of the host before antibodies become numerous (Smithers *et al.* 1969). A consequence of this tactic is that parasites that are latecomers to a host are met by a barrage of antibodies stimulated by the earlier entrance of now-entrenched parasite individuals. Schad (1966) has suggested that parasites could use the immune response of their hosts to exclude other parasites. This response is also apparently effective between closely related species, a phenomenon known as *cross-resistance* (Cohen 1973, Kazacos and Thorson 1975). For example, most predominantly human forms of schistosomiasis are extremely virulent in man, but if a man has previously been infected by other schistosome organisms—some of which have little effect on man—the effects of the human parasite will be moderated considerably (Lewart 1970).

Plants use structural and chemical defenses against herbivores.

The relationship between herbivores and their plant food resembles that between parasite and host in that both rely primarily on biochemical weapons. Plant defenses against herbivores are based partly on the inherently low nutritional value of most plant tissues and partly on the toxic properties of so-called *secondary substances* produced and sequestered for defense (Whittaker and Feeny 1971, Robbins and Moen 1975). Structural defenses such as spines, hairs, and tough seed coats are also important (Figure 15-10).

Figure 15-10   Spines protect the stems and leaves of many plants: a lowland forest tropical tree from Panama (left), a branch of *Parkinsonia* (bean family) from the Galápagos Islands (center), and an agave (century plant) from Baja California (right).

The nutritional quality and digestibility of plant foods is critical to herbivores. Because young animals have a high protein requirement for growth, the reproductive success of grazing and browsing mammals depends upon the protein content of their food. Herbivores usually select plant food according to its nutrient content (Gwynne and Bell 1968). Young leaves and flowers are frequently chosen because of their low cellulose content, and fruits and seeds are particularly nutritious (Short 1971).

Many plants are capable of making their proteins unavailable to herbivores. For example, the tannin deposited in oak leaves combines with leaf proteins in such a way that the proteins cannot be digested by caterpillars and other herbivores, thereby slowing their growth considerably (Feeny 1969). With the buildup of tannin in oak leaves during the summer, fewer and fewer leaves are attacked by herbivores (Feeny 1968, 1970, Feeny and Bostock 1968). Tannins are usually deposited in vacuoles near the surface of the leaf where they will not interfere with the functioning of the plant's proteins. When herbivores eat the leaves, the vacuoles are broken and the tannins combine with the proteins. Leaf-mining beetle larvae have gotten around this problem by burrowing through and consuming the leaf's inner tissues, completely avoiding the tannin-filled vacuoles.

The seeds of leguminous plants (the pea family) are frequently infested with larvae of bruchid weevils, a kind of beetle. The adult weevil lays its eggs on developing seed pods. The larvae then hatch and burrow into the seeds, which they consume as they grow. To counter this attack, legumes have evolved a variety of defensive adaptations (Janzen 1969, Center and Johnson 1974). Each larva feeds on only one seed. To pupate successfully and metamorphose into an adult, the larva must attain a certain size, which is ultimately limited by the amount of food in the seed. As a result, many legumes have evolved extremely tiny seeds that do not contain enough food to support the growth of a single bruchid larva and thus they are not suitable as host plants (Janzen 1969a). Most legumes contain substances that reduce the digestibility of proteins by inhibiting the effectiveness of the enzymes produced in the herbivore's digestive organs to break down proteins. While these inhibitors probably evolved as a general biochemical defense against insects, they are futile against bruchid weevils, which have metabolic pathways that either bypass or are insensitive to the presence of the inhibitors (Applebaum 1964, Applebaum, Gestetner, and Birk 1965). Among legume species, soybeans are resistant to attack even from most bruchid species. If bruchid eggs are laid on soybeans, the first instar larvae die soon after burrowing beneath the seed coat; chemicals have been isolated from soybeans that completely inhibit the development of bruchid weevil larvae in experimental situations.

The production of toxic compounds forms a common defense by plants (Fraenkel 1959, 1969, Beck 1965). Evidence is beginning to accumulate showing that where predation is most intense, plants have more varied and more concentrated toxins (*e.g.,* Dolinger *et al.* 1973). This response only stimulates adaptations that enable some herbivores to detoxify some poisonous substances, and generally leads to biochemical specialization of herbivores to certain restricted groups of plants with similar toxins, or to single species. The tobacco hornworm (larval stage of the sphynx moth *Manduca sexta*) can tolerate nicotine concentrations in its food far in excess of levels that kill other insects. Nicotine disrupts normal functioning of the nervous system by preventing the transmission of impulses from nerve to nerve. The hornworm's solution to this problem is to keep nicotine from crossing the membrane of the nerve; in other species of moths, nicotine readily diffuses into nerve cells (Yang and Guthrie 1969).

Resistance to nicotine enables *M. sexta* to feed on tobacco (*Nicotiana tabacum*), a member of the tomato family (Solanaceae), but some other species of *Nicotiana* have other alkaloid toxins that the tobacco hornworm cannot handle. Presumably, the mechanism that prevents nicotine from entering the nerve cells fails to recognize other related toxins, or these toxins have a different mode of action. Regardless of the physiological effect, when tobacco hornworms were grown on 44 species of *Nicotiana* in greenhouse experiments, growth was normal on 25 species, but was retarded or stopped completely on the others. In addition, fifteen of the species caused moderate to severe mortality (Parr and Thurston 1968). Specialization on tobacco has some favorable consequences for *M. sexta* over and above access to vast crops of food grown for it by man. In addition to being toxic to most herbivores, tobacco also dissuades many would-be parasites. Tiny wasps that might parasitize hornworm eggs become trapped in the gummy fluid produced by trichomes (fine hairs) on the tobacco plants. Eggs laid on other solanaceous plants that lack trichomes, like Jimson weed (*Datura*) and tomato, are readily parasitized (Rabb and Bradley 1968). In another plant-herbivore system, terpenoid resins produced by pines as a general herbivore defense are in turn used by sawflies (*Neodiprion sertifer*) as a chemical defense against parasitic wasps and flies (Eisner *et al.* 1974).

The potential defenses of plants against herbivores extend beyond direct confrontation with structural and chemical weapons. Daniel Janzen (1968, 1970) has suggested that many plants avoid intense herbivore pressure by being widely scattered in space and by breeding at irregular intervals. Populations of herbivores and seed predators cannot reach high levels when their food supply is erratic and difficult to locate. Many trees produce large crops of seeds at intervals of several years. During the interval between crops, populations of seed predators die out, or are maintained at low levels on alternate foods. During mast years, the seed crop floods the market, quickly exceeding the ability of herbivores to consume it. Most of the seeds, therefore, come through unscathed.

Many species interact to their mutual benefit.

Not all interactions between species are antagonistic. You may recall that under certain environmental conditions the relationship between oropendolas and cowbirds is beneficial to both; the oropendola provides food for the cowbird young, and the cowbird young protect the oropendola nestlings from fly parasites (see page 82). In a sense, the oropendola buys its anti-predator protection from the cowbird. Interactions that benefit both parties in this way are described as mutualistic. *Mutualism* is not as common in the natural world as predation and parasitism, but for many species mutualistic interactions are helpful in obtaining food or avoiding predation: the pollination of flowering plants by animals involves a mutualistic interaction; lichens are a mutualistic association of algae and fungi; ruminants and termites both have the ability to digest cellulose because of their mutualistic association with specialized microorganisms in their intestinal tracts; the ability of legumes to fix nitrogen from the air depends on mutualistic bacteria in their root nodes; and so on. We shall consider these and other mutualistic interactions in the paragraphs that follow.

Most mutualistic relationships probably evolved by way of host-parasite, predator-prey, or plant-herbivore interactions. This origin is particularly evident in the association between cowbirds and oropendolas, in which the cowbird is parasitic on the oropendola in certain colonies. The presence of insects flying from flower to flower to eat the nutritious pollen may have provided the opportunity for plants to adapt their flower structure to enhance and control insect pollination. Plants have evolved showy flowers to attract pollinators; nectar, which is little more than a weak sugar solution, provides an inexpensive food incentive for insects (and also for many birds and bats) to visit the flowers; the structure of flowers has also been modified to ensure that pollinators will efficiently transfer pollen from one flower to another (Faegri and Pijl 1971, Grant and Grant 1965).

Plant-pollinator relationships are highly developed in the orchid family, with its variety of flower shapes, colors, and smells. The intricate tie between flower and pollinator is exemplified by the orchid *Stanhopea grandiflora* and the euglossine bee *Eulaema meriana.* The attraction of euglossine bees to the particular orchids they visit is unusual in that no nectar is produced by the flowers, and only male bees visit them. The flowers are extremely fragrant, and each species of *Stanhopea* orchid has a unique combination of fragrances (Dodson *et al.* 1969) that attracts only one species of bee pollinator (Dressler 1968).

When a male euglossine bee visits an orchid, it brushes parts of the flowers with specially modified forelegs, and then appears to transfer some substance to the tibia of the hind leg, which is enlarged and has a storage cavity. The function of the bee's behavior is not understood, but the orchids may provide the bee with a sort of perfume used for mate attraction. For *Eulaema meriana* to pollinate *Stanhopea grandiflora,* the bee enters the flower from the side and brushes at a saclike modification on the lip of the orchid (Figure 15-11). The surface of the lip is very smooth, and the bee often slips when it withdraws from the flower. (The orchid fragrances may also intoxicate the bees and cause them to lose their footing on the lip of the orchid (Dodson and Frymire 1961). If the bee slips, it may brush against the

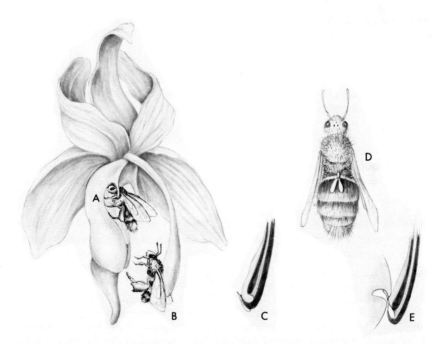

Figure 15-11   Pollination of the orchid *Stanhopea grandiflora* by *Eulaema meriana.* The bee enters from the side and brushes at the base of the orchid lip (A). If it slips (B) the bee may fall against the pollinarium, which is placed on the end of a column (C), and the pollinarium becomes stuck to the hind end of the thorax (D). If a bee with an attached pollinarium falls out of a flower, the pollinarium may catch in the stigma (E), which is so placed on the column that the flower cannot be self-fertilized (after Dressler 1968).

Figure 15-12   The mohave yucca (*Yucca schidigera*) and the yucca moth (*Tegeticula maculata*) that pollinates it (after Powell and Mackie 1966).

column of the orchid flower, where the pollinaria are precisely placed to stick to the hindmost part of the thorax of the bee. If a bee with an attached pollinarium slips and falls out of another flower, the pollinarium will catch on the stigma and pollinate the flower (Figure 15-11).

An even more unusual plant-pollinator relationship occurs between species of yucca plants and moths of the genus *Tegeticula* (Figure 15-12). Their curious interrelationship was first described by C. V. Riley a century ago, but has been considerably elaborated since (*e.g.,* Powell and Mackie 1966). The moth enters the yucca flower and deposits one to five eggs on the ovary. Later, when the eggs hatch, the larvae burrow into the ovary, where they feed on the developing seeds. But after the moth has laid her eggs, she scrapes pollen off the anthers in the flower and rolls it into a small ball, which she grasps with specially modified mouthparts. She then flies to another plant, enters a flower, and proceeds to place the pollen ball onto the stigma of the flower before laying another batch of eggs.

The relationship between the moth and the yucca is *obligatory.* The moth can grow nowhere else, the yucca has no other pollinator. In return for pollinating its flowers, the yucca seemingly tolerates the moth larvae feeding on its seeds, but the extent of this loss of potential reproduction is small, rarely exceeding 30 per cent, and more nearly half that value, on average, in *Yucca whipplei.*

Daniel Janzen (1966, 1967a) has described a fascinating case of complete mutualistic interdependence between certain kinds of ants and swollen-thorn acacias in Central America. The acacia plant provides food and nesting sites for ants in return for protection that the ants provide from insect pests. The bull's-horn acacia (*Acacia cornigera*) has large hornlike thorns with a tough woody covering and a soft pithy interior (Figure 15-13). To start a colony in the acacia, a queen ant of the species *Pseudomyrmex ferruginea* bores a hole in the base of one of the enlarged thorns and clears out some of the soft material inside to make room for her brood. In addition to housing the ants, the acacias provide food for the ants in nectaries at the base of their leaves, and in the form of nodules, called beltian bodies, at the tips of some leaves (Figure 15-13). As the colony grows, more and more of the thorns on the plant are filled and, in return, the ants protect the plant from insect pests. A colony may grow to more than a thousand workers within a year, and eventually may have tens of thousands of workers. At any one time, about a quarter of the ants are actively gathering food and defending the plant against herbivorous insects. The relationship between *Pseudomyrmex* and *Acacia* is obligatory: Neither the ant nor the acacia can survive without the other. Other ant-acacia associations are *facultative*. That is, the ant and the acacia can co-occur to mutual benefit, but they can both exist independently as well. Species of acacia that lack the protection of ants altogether frequently produce toxic compounds in their leaves (Rehr *et al.* 1973).

The mutualism between ants and acacias has been accompanied by adaptations of both parties to increase the effectiveness of the association. For example, *Pseudomyrmex* is active 24 hours a day, an unusual trait for ants, and thereby provides continuous protection for the acacia. In a similar adaptive gesture, the acacia has leaves throughout the year, and thereby provides a continuous source of food for the ants. Most related species lose

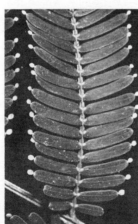

Figure 15-13   Modifications of swollen-thorn acacias for mutualistic interactions with ants. Many of the thorns are enlarged and have a soft pith that the ants excavate for nests (left, *Acacia hindsii*). The acacias provide food for the ants from nectaries at the base of leaves (center) and in the form of nutritious nodules at the tips of modified leaves (right, *A. collinsii*) (photographs courtesy of D. Janzen, from Janzen 1966).

their leaves during the dry season. Both adaptations are necessary to preserve the mutualistic relationship.

To test the influence of ants on the growth and survival of acacia plants, Janzen kept ants off new acacia shoots and measured their growth compared with shoots that had ants (Table 15-2). Differences between experimental and control plants demonstrate strikingly the advantages to the plant of having ants. The growth and survival of shoots without *Pseudomyrmex* was reduced by herbivorous insects; when such a shoot was transferred to a plant inhabited by an ant colony, the offending insects were immediately removed.

Some animals are not to be outwitted evolutionarily by the ant-acacia mutualism. Several species, including a weevil, a fly, and a predatory spider, mimic the ant so closely in their appearance and behavior that the ants do not recognize them as aliens, and they are left alone to go about their business on the acacia (Janzen 1967a). A close relative of the acacia ant, *Pseudomyrmex nigropilosa,* parasitizes the ant-acacia mutualism by eating the food provided by the plant but not providing protection against herbivores. *P. nigropilosa* is not even active at night. Unlike some of the ant mimics noted above, however, *P. nigropilosa* is recognized by the acacia ant *P. ferruginea* and ousted when it is discovered. The generally short-lived colonies of *P. nigropilosa* have a correspondingly high reproductive rate and produce broods of reproductive individuals quickly (Janzen 1975).

The ant *Pseudomyrmex* grooms the acacia plant in return for food and protection. A similar function, cleaning parasites from the skin, is performed for many marine fish by other fish or by shrimp that benefit from the food value of the parasites they remove. Such relationships, often referred to as *cleaning symbiosis* (Feder 1966, Limbaugh 1961), are most highly developed in the clear warm waters of the tropics, where many cleaners display their striking colors at particular locations, called cleaning stations, to which other fish come to be groomed. As you might expect, a few species of predatory fish mimic the cleaners; when other fish come and expose their gills to be groomed, they get a bite taken out of the gills instead.

The importance of cleaning and grooming mutualism in the overall functioning of natural communities is overshadowed by the many mutualistic relationships between organisms that fulfill specialized nutritional requirements. Among animals, both ruminant mammals (cows and sheep, for example) and termites have specialized microorganisms (bacteria and protozoa, respectively) in their digestive tracts that are biochemically equipped to break down the cellulose in plant food. Cellulose would normally be unusable to most organisms without the help of specialized symbionts (Koch 1966, Howard 1966).

Many plants, particularly the legumes, which include peas, peanuts, and alfalfa, have symbiotic nitrogen-fixing bacteria in nodules in their roots. These bacteria can utilize nitrogen directly from air spaces in the soil and make it available to their hosts (Lange 1966, Quispel 1974). Most higher plants require nitrogen in the form of nitrates that diffuse into the nodules or of ammonia; where soils are deficient in these compounds, leguminous plants may thrive while other plants cannot. Alfalfa crops are often planted in rotation with other crops, such as corn, which deplete the soil of nitrogen,

TABLE 15-2

Effect of the presence of the ant *Pseudomyrmex* on the growth and survival of bull's-horn acacia in Oaxaca, Mexico (from Janzen 1966).

	With ants	Without ants
Growth of suckers from stumps (cm)		
May 25–June 16	31	6
June 16–August 3	73	10
Size of suckers after ten months' growth		
Number of stumps sampled	72	66
Total wet weight of suckers (gm)	41,750	2,900
Total number of leaves	7,785	3,460
Total number of swollen thorns	7,483	2,596
Survival of stumps (per cent) from October 18 until:		
March 13	87	62
June 10	72	51
August 6	72	44

because the alfalfa can enrich the soils with nitrogen obtained from the atmosphere. The biochemical relationship between the nitrogen-fixing bacterium *Rhizobium* and its host plant is extremely complex. Nitrogenase, the enzyme needed to fix nitrogen, is found only in bacteria and blue-green algae, but the enzyme cannot function efficiently in the presence of dissolved oxygen. The nodules in the roots of legumes provide an anaerobic environment because although oxygen diffuses into the nodule along with nitrogen, it is utilized in respiration by root tissues before it reaches the center of the nodule where the bacteria are found (Hardy and Havelka 1975). Nitrogen-fixing bacteria are also found in mutualistic association with green algae, which provide specialized, nonphotosynthetic cells for their guests. (The oxygen liberated by photosynthesis in normal algae cells would seriously reduce the rate of nitrogen fixation.)

Although free oxygen reduces nitrogen fixation, *Rhizobium* must have a source of oxygen for its own respiration. In the nodules of legumes, this is transported into the nodule bound to a special type of hemoglobin, called leghemoglobin. The mutualism between bacteria and plant has proceeded so far that *Rhizobium* carries the gene for the heme part of the molecule and the legume has the gene for the globin component (Hardy and Havelka 1975). No one knows how this arrangement got started, but the early stages in the development of the mutualism may have been like the loose association that currently exists between another nitrogen-fixing bacterium, *Azotobacter,* and the roots of certain grass species.

Lichens are a close association of algae and fungi (Ahmadjian 1967, Hale 1967). The lichens are frequently found on such surfaces as tree trunks and bare rock, substrates from which most plants cannot obtain the water or nutrients necessary for growth. In the lichen symbiosis, algae provide the photosynthetic mechanism for producing organic compounds, while the fungi are thought to secrete substances that dissolve nutrients out of bare rock surface, making them available for synthetic processes. It has also been shown that the algae can photosynthesize under lower water potential when associated with the fungus than when cultured separately. This may be due to carbohydrate molecules stored in the fungal cells that raise the osmotic potential enough to absorb water from air (Brock 1975).

Soil fungi grow together with the roots of trees in what are called *mycorrhizal* associations. These fungi sometimes penetrate the cells of the roots, and they fulfill the important role of dissolving mineral nutrients that would otherwise be unavailable to the trees. In return, the fungi obtain organic nutrients from the roots (Rovira 1965, Wilde 1968, Harley 1969, Marks and Kozlowski 1973). We shall consider the function of mycorrhyzae in more detail later in connection with nutrient cycles in the ecosystem.

In this chapter, we have covered many aspects of the predator-prey relationship and how it affects the adaptations of the species involved in the interaction. We have also seen that many predator-prey associations, particularly those involving herbivory and parasitism, have become modified to the mutual benefit of both parties, often to the extent that each is totally dependent on the other. But the basic character of the predator-prey relationship, which is the exploitation of one population by another, is sometimes lost sight of among the diverse array of adaptations that occur in these relationships. Predators and prey are major components of the biotic environment of every organism. In the next chapter, we shall consider another important aspect of the environment: competition between organisms for resources. Like predation, competitive relationships between populations and among individuals in a single species strongly influence adaptations.

16

Competitors

Competition is perhaps the most elusive and controversial of all ecological phenomena. We may define *competition* as the use of a resource (food, water, light, space) by an organism that thereby reduces the availability of the resource to others. If competing individuals belong to the same species, their interaction is called *intraspecific* (or within-species) competition; if the individuals belong to different species, their interaction is called *interspecific* (or between-species) competition. In either case, a resource consumed by one individual can no longer be used by another. When a fox captures a rabbit, there is one less rabbit for other foxes, or for bobcats, hawks, and other predators to prey upon.

When competitors reduce the abundance of resources, the level of biological activity that can be achieved by an individual is also reduced. When placed on a graph showing the level of activity over an environmental gradient, we see that the effect of competition is to restrict the activity space of a population. Furthermore, if one species is most active toward one end of the environmental gradient, the activity space of its competitor is shifted toward the other end of the gradient, as well as being restricted (Figure 16-1).

The interaction between competitors is usually more subtle than that between predator and prey. Individuals need not ever encounter each other and yet may feed off the same prey population and drink from the same pool. Even in their serenity, trees strive to absorb light, water, and nutrients at the expense of their neighbors.

Competition is a pervasive force in the environment of the individual. And, for the most part, individuals have little control over their competitors, in the sense that prey can hide from or flee their predators. But we shall see in this chapter that competition may be managed through the diverging adaptations of competitors to exploit different resources or through direct interference in each other's activities.

Similar species coexist by utilizing different resources.

We frequently see similar species coexisting in nature, clearly using many of the same resources. Closely related species of trees grow in the same forest, all needing sunlight, water, and soil nutrients. Coastal estuaries

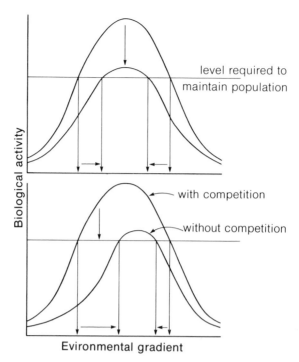

Figure 16-1   Influence of competition on the activity space of a population through reducing resources, hence level of biological activity. Above: restriction of activity space owing to evenly distributed effects of competition. Below: shift in activity space owing to strong competition at one side of resource continuum.

and inland marshes harbor a variety of fish-eating birds, including egrets, herons, terns, kingfishers, and grebes. As many as six kinds of warblers (small insectivorous birds), all with similar morphology and feeding habits, may be found in one locality in forest habitats in the northeastern United States (MacArthur 1958) (Figure 16-2). The shallow waters of Florida's Gulf Coast are frequented by up to eight species of large predatory snails (Paine 1963).

These examples are not remarkable. Several lakes have witnessed the diversification of some lineages into many hundreds of closely related species. The fauna of Lake Baikal, an ancient lake in southwestern Siberia, includes 239 species of gammarid crustacea, tiny aquatic organisms related to crabs and brine shrimp (Kozhov 1963). Lake Malawi in Africa has more than 200 species of cichlid fish, many of which appear to have similar ecological characteristics (Fryer 1959).

Whenever ecologists have examined groups of similar species in the same habitat, they have found small but significant differences in size or foraging behavior that enable the species to use slightly different resources and avoid intense competition (Schoener 1974).

Imagine that we are standing on a rocky coast, watching two kinds of cormorants—large dark-colored diving birds—feed in the same area and apparently in the same manner: both swim on the surface of the water and dive for food. The cormorants seem to be competing for food, but, in fact, the appearance is deceptive. Had we followed the cormorants beneath the water and observed their feeding habits directly, we would have found that one feeds primarily at the bottom and the other at intermediate depths. Because the species feed in different parts of the habitat, their diets do not overlap greatly, and they do not compete intensely for food. This fact was not appreciated until studies of stomach contents revealed major differences in the diets of the two species (Lack 1945). More than 80 per cent of the stomach contents of the great cormorant consisted of sand eels and herring-like fish, which swim at various depths in the water but well above the bottom. These are almost completely lacking from the diet of the shag cormorant, which specializes on several bottom-living flatfish and shrimp.

Four species of bumblebees *(Bombus)* common in fields in England avoid competition through morphological and behavioral specializations to slightly different food resources (Brian 1957). All four species gather nectar and pollen from flowers, but each has a different tongue length and visits flowers of correspondingly different size. The species with the longest tongue, *B. hortorum,* feeds only at flowers with long corollas. The other species feed at a wide variety of flowers with short corollas. Of the three short-tongued honeybees, only *B. lucorum* feeds primarily in fields. *B. pratorum* and *B. agrorum* are more restricted to shrubby habitats and rarely venture more than twenty yards from such cover into open fields. These two species compete

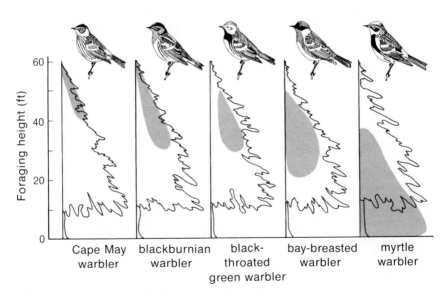

Figure 16-2   Feeding locations of five species of warblers (genus *Dendroica*) in spruce forests in Maine (after MacArthur 1958).

less intensively among themselves because B. pratorum is active at an earlier season than B. agrorum.

The morphology and behavior of predatory snails on the Gulf Coast of Florida emphasize the variety of ways ecological overlap can be avoided (Paine 1963). The small Murex snail, for example, is not strong enough to pry or chip open the shell of a large clam or oyster. Instead, it uses the filelike teeth (radula) at the tip of its proboscis to drill a neat hole in the shell of its prey, through which it scrapes away the flesh. Because Murex has a poorly developed foot, it can neither move fast enough to pursue other predatory gastropods or dig clams out of the sand or mud. Its diet is therefore restricted to such prey as the rock cockle Chione, which rests exposed on the sea floor. The large conches of the genus Busycon use the heavy edges of their shells to chip at the edges of large clam shells or to wedge them open. The large Busycon contrarium opens the thick-shelled Chione by grasping the prey and aligning the shell margin at a 45- to 90-degree angle to the lip of its own shell. Using its shell as a hammer, the conch then chips away enough of the clam's shell to insert its proboscis or to wedge the clam open. Conches pry open thin-shelled clams and species that do not close completely simply by using the edge of their shell as a wedge and pressing hard on the margin between the two valves of the clam's shell. The smaller Busycon spiratum has a thinner shell than B. contrarium and primarily attacks incompletely closing or thin-shelled clams. The large horse conch (Pleuroploca gigantea) and tulip shell (Fasciolaria tulipa) prey upon predatory gastropods smaller than themselves. The smaller Fasciolaria hunteria has a particularly long proboscis that is adapted to attack tube-building worms in their burrows beneath the mud. These differences in size and feeding technique enable the snails to avoid intense competition.

The foregoing examples clearly show that similar species potentially avoid intense competition by being adapted to exploit somewhat different resources, but this fact raises two important questions. First, does competition influence the activity levels of individuals in natural environments? Second, does competition play a role in the evolution of differences between species in their use of resources?

Competition is a potent ecological force.

The most striking demonstrations of the effects of competition come from studies of plants, whose growth and development are so flexible and sensitive to the environment as to provide an index to resource levels. The depressing effect of intraspecific competition on growth of trees has been demonstrated in forest-thinning experiments. The acceleration of growth of young longleaf pine trees in response to selective thinning of trees more than fifteen inches in diameter is shown in Figure 16-3. Each core of wood was obtained by boring into a tree trunk, from the bark to the center, with a long, tubular device called an increment borer. The core of wood removed by the borer tube provides a record of annual ring growth without cutting down the entire tree. These cores show an increased growth rate, particularly in sum-

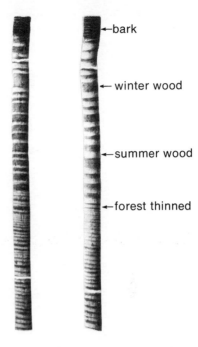

←bark

← winter wood

←summer wood

←forest thinned

Figure 16-3 Cores of two longleaf pine trees obtained near Birmingham, Alabama, showing the effect of removing large trees on subsequent growth (courtesy of the U.S. Forest Service).

mer (light wood), during the eighteen years between the time the forest was logged and the time the cores were taken.

An experimental demonstration of the influence of competition for light and water on plant growth comes from the tropical forests of Surinam, where forest ecologists set out to determine whether the growth of commercially valuable trees could be improved by removing species of little economic importance (Schulz 1960). The foresters poisoned about 70 per cent of undesirable trees with girths greater than 30 centimeters in one area and greater than 15 centimeters in another, leaving the desirable species untouched. The increase in girth of the desirable trees was then measured over a year's time in experimental plots and in control plots that had not been selectively thinned (Figure 16-4). Removing trees greater than 30 centimeters in girth increased the penetration of light to the forest floor by six times. The additional light greatly stimulated the growth of trees remaining on the experimental plot; the improvement was greatest among small individuals, which are normally most shaded by competing species. Trees whose girth exceeded 100 centimeters did not grow appreciably faster after forest thinning. Although removal of small trees (15 to 30 cm girth) further increased light penetration by only one-third, it led to a striking response in growth rate, particularly among the remaining large trees. The improved growth could not have been caused by increased light because many of the trees that responded were much taller than the trees that were poisoned. The added growth stimulus, therefore, probably resulted from reduced competition for either water or mineral nutrients in the soil.

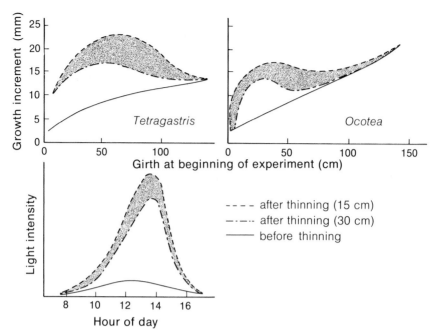

Figure 16-4   Effect on the increase in girth of two species of tropical forest trees, *Ocotea* and *Tetragastris,* achieved by removing competing trees greater than 15 centimeters or 30 centimeters in girth. Increase in light intensity that resulted from thinning is shown at left. Shaded area represents the difference between plots with 15 cm and 30 cm girth trees removed (after Schulz 1960).

Because most animals achieve a relatively fixed body size, the effects of competition are more usually expressed in terms of survival or reproductive rate, both of which are often more difficult to measure in natural populations than is plant growth.

Competition between animals can manifest itself by reducing the numbers of one or both competing species. Such situations have been discovered when one species is removed from an area and the range or population size of another species with similar resource requirements increases. For example, two species of voles (small mouselike rodents of the genus *Microtus*) co-occur in some areas of the Rocky Mountain states. In western Montana the meadow vole inhabits both dry habitats and the wetter habitats surrounding ponds and water courses, whereas the montane vole is restricted to dry habitats. When meadow voles were trapped and removed from an area of wet habitat, montane voles began to move in from surrounding dry habitats (Koplin and Hoffman 1968). Stoecker (1972) obtained the complementary result at another site in Montana. After trapping for nine days an area of dry habitat, which was apparently occupied solely by the montane vole, meadow voles began to be caught, presumably after having moved in from surrounding moister habitats in response to the removal of the montane voles. These results suggest that each of the species excludes the other from its preferred habitat by aggressive behavior or else avoids nonpreferred habitats where the other species of vole is common. Both types of behavior appear to be common among rodents and tend to reinforce habitat preferences of different species (Grant 1972).

Competition between two species in different genera of barnacles within the intertidal zone of the rocky coasts of Scotland was demonstrated by Connel (1961). Normally, adult *Chthamalus* occur higher in the intertidal zone than *Balanus,* and the line of demarcation between the two species is sharp, although the vertical distributions of newly settled larvae of the two species overlap broadly within the tide zone. Connell demonstrated that adult *Chthamalus* are not restricted to the portion of the intertidal zone above *Balanus* due to physiological tolerance limits. When *Balanus* are removed from rock surfaces, *Chthamalus* thrive in the lower portions of the intertidal zone where they are normally absent. Thus, the two species appear to compete for space (see Figure 16-5). The heavier shelled *Balanus* grow more rapidly than *Chthamalus,* and as individuals expand, the shells of *Balanus* edge underneath the shells of *Chthamalus* and literally pry them off the rock. *Chthamalus* can occur in the upper parts of the intertidal zone because they are more resistant to desiccation than *Balanus,* which cannot grow there.

The removal experiments described in the preceding paragraphs demonstrate that in spite of the specializations that reduce resource overlap, competition can be a potent ecological force, and it greatly influences populations in natural communities.

Figure 16-5   Competition for space among barnacles on the Maine coast. Above their optimum range in the intertidal zone the barnacles are sparse, and there are bare patches for the young to settle on (left). Lower in the intertidal (right) the barnacles are so densely crowded that there is no room for population increase; the young barnacles are forced to settle on older individuals (courtesy of the American Museum of Natural History).

Is competition a potent evolutionary force?

The evidence for competition between individuals or between populations in nature is overwhelming. It is also evident that differences between species in structure, physiology, and behavior result in different resource requirements and enable species to avoid intense competition. Whether the differences between species that reduce competition are fortuitous or are the direct result of the influence of competitive interactions on the evolution of these differences is a matter of considerable controversy (Cody 1974). One could argue that differences between species represent the accumulations of adaptations to the different environments in which they occur, and where the species happen to overlap these differences act to reduce competition between them. Or one could argue that the differences in habitat or resource preference between two species have evolved as a mechanism to reduce interspecific competition. Ecological divergence is seen as evolving in the following manner. Consider two hypothetical species that use many of the same resources. Suppose that the size range of prey eaten by each species overlaps that eaten by the other, as shown in the upper diagram of Figure 16-6. Where the species overlap most, they compete most intensely. Individuals of species A that eat prey smaller than the average have a greater evolu-

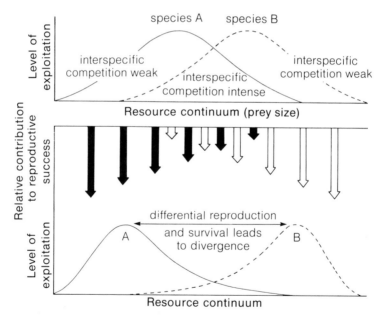

Figure 16-6   Diagram showing evolutionary divergence between populations caused by the influence of interspecific competition on the survival and reproduction of individuals that are adapted to utilize different parts of the resource continuum.

tionary fitness than the population as a whole because their productivity is influenced less by interspecific competition. The same is true of individuals of species B that consume prey larger than the average. If variation in resource utilization within a population is determined, at least in part, genetically, competition will influence evolution because it fosters the gradual replacement of individuals that compete intensively with other species by individuals that avoid interspecific competition.

It is not difficult to accept that divergence can occur as the result of competition, but to demonstrate the process is challenging because it must occur too slowly to be observed. We may, however, infer that divergence has occurred if two ecologically similar species differ more where they are found together than they would in comparisons of individuals from nonoverlapping parts of their ranges. This phenomenon, referred to as *character displacement* (Brown and Wilson 1956), would provide convincing evidence that ecological isolation is the result of competition between species, but evidence for character displacement is sparse and often equivocal. The classic example of character displacement occurs between two species of rock nuthatches (genus *Sitta*), which are small birds that probe for insects on rocks and on the ground. The ranges of the two species overlap broadly in the Middle East, though their distributions individually extend far beyond the area of overlap. Where the western and the eastern rock nuthatches occur separately, the lengths of their beaks are nearly identical. Beak length presumably measures some important aspect of resource exploitation, such as the range of prey-size (Hespenheide 1971). Where the two species occur together, their beak lengths differ by about 15 per cent, which suggests that competition for prey within the same size range has resulted in evolutionary divergence. One of the difficulties with this example is that within the range of overlap, the two species of *Sitta* occupy distinct habitats and thus do not compete. Although the habitat differences may be the result of character displacement, bill-dimension differences probably are not and likely reflect the different habitats of each species.

A more convincing case of character displacement has been described in the ground finches of the genus *Geospiza* on the Galapagos Islands (Lack 1947). On islands with more than one species, the finches usually have beaks of different size. For example, on Bindloe and Abingdon Islands, the ranges in beak size of the three resident species of ground finches do not overlap (Figure 16-7). On Charles and Chatham islands, the two species *G. fuliginosa* and *fortis* have different-sized beaks. On Daphne Island, however, where *fortis* occurs in the absence of *fuliginosa,* its beak is intermediate in size between the two species on Charles and Chatham islands. On Crossman Island, *fuliginosa* occurs in the absence of *fortis,* and its beak is intermediate in size. Since habitats on the islands are similar in most respects, the simplest explanation for these patterns is that where species co-occur, competition has caused character divergence. Lack (1971) has collected from the literature on birds many other examples of character displacement in such characteristics as morphology, nest site, host specificity of brood parasites, habitat, and feeding behavior.

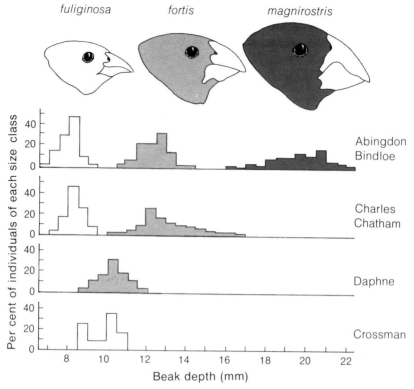

Figure 16-7   Proportions of individuals with beaks of different sizes in populations of ground finches (*Geospiza*) on several of the Galapagos Islands (after Lack 1947).

How does competition occur?

Competition may be expressed either indirectly, through *exploitation* of the same resource by different individuals, or directly by *interference* (Miller 1967, 1969). Competition through interference usually occurs for space. The disputed space is not often the ultimate object of competition, but it usually contains the resource that is. The territories of birds and other animals are defended areas from which other individuals are excluded. The prey included within the territory would be impossible to defend individually, so competition is transferred from the ultimate resource, food, to the defense of an area. When food resources are concentrated and conspicuous, they can be defended directly. For example, hummingbirds often defend flowering bushes that have high nectar production (Wolf 1969). Direct interference between the barnacles *Chthamalus* and *Balanus* (see page 239) is a case of competition for a place to live.

Competition is probably manifested more often as the effect organisms produce on resource levels than as direct interference. Indirect competiton through exploitation of resources differs from direct interference because effects of indirect competition are slowly expressed through differential survival and reproduction, leading either to extinction or to slow evolutionary divergence. Direct interference, on the other hand, can cause the immediate exclusion of a competing individual or population from a resource. Because direct interference is the stronger form of competition, one would expect it to occur primarily where the overlap between individuals is greatest, that is, between individuals of the same species. Territorial defense is usually practiced against members of the same species. Occasionally, direct interference is adopted as a mode of competition between similar species, as in the voles mentioned on page 238 (also see Orians and Wilson 1964, and Cody 1969).

Direct interference is not limited to active animals. Barnacles pry each other off rocks, although they do not go out of their way to do so. Evidence is beginning to accumulate that parasites can exclude competitors from a host by stimulating the host's immune defenses against them (Schad 1966). Even plants are capable of hostile interaction, although they must resort to shading and chemical warfare rather than behavior (Whittaker and Feeny 1971). It has been suggested that the abundant oils in the leaves of eucalyptus trees of Australia promote frequent fires in the leaf litter on the forest floor and burn out potential competitors once the eucalyptus have become established (Mutch 1970).

The decaying leaves of walnut trees release toxic substances into the soil that inhibit seedling growth in many other species of trees (Massey 1925). Such toxic restraints are frequently referred to as *allelopathy*. Perhaps the best known case of chemical interference involves several species of sage of the genus *Salvia* (Muller 1966, Muller *et al.* 1968). Clumps of *Salvia* are usually surrounded by bare areas separating the sage from neighboring grassy areas (Figure 16-8). Sage roots extend to the edge of the bare strip but not into the bare area beyond. Thus, it seems unlikely that a toxic substance is exuded into the soil by the roots. The leaves of *Salvia* produce volatile terpenes (a class of organic compounds that includes camphor) that apparently affect nearby plants directly through the atmosphere. Heavy rainfall washes the toxic compounds out of the atmosphere, reducing the prohibitory effect of the sage on other species. This may explain the general absence of direct chemical competition in wet climates.

Although terpenes produced by *Salvia* have been shown to suppress the growth of other species, *Salvia* may have evolved the ability to produce volatile chemicals as a means of attracting insects to pollinate its flowers or of dissuading herbivores from eating its leaves. Many of you have tasted sage honey. The close association of the honeybee *(Apis mellifera)* and sage plants is underscored by the Latin names of two species of sage. *S. apiana* and *S. mellifera.* Furthermore, one of the terpene compounds produced by sages (cineole) acts as a powerful attractant to many kinds of bee, including the attraction of euglossine bees to orchids (page 227). *Salvia* may have only secondarily adapted attractants to function as toxins to other plants.

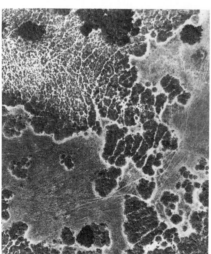

Figure 16-8   Top: bare patch at edge of a sage clump includes a two-meter wide strip with no plants (A-B) and a wider area of inhibited grassland (B-C) lacking wild oat and bromegrass, which are found with other species to the right of (C) in unaffected grassland. Left: aerial view of sage and California sagebrush invading annual grassland in the Santa Inez Valley of California (courtesy of C. H. Muller from Muller 1966).

Who competes?

Competition is generally thought to be most intense between similar species, usually close relatives. Differences in structure, physiology, and behavior increase the likelihood that species will utilize different resources and thereby avoid intense competition. But structural difference need not imply difference in resource. Animals may adopt very different forms as the result of alternate ways of getting at the same resources. Rodents, sparrows, and ants are worlds apart in form and function, yet all three compete for seeds in many areas. Similar cases of diverse groups of organisms in direct competition are squids and fish (Packard 1972), lizards and birds (Pianka 1971), and moths and hummingbirds (Snow and Snow 1968).

Although studies of ecological diversification usually center upon closely related species, most species are subject to the effects of *diffuse competition,* which is the sum of many weak competitive interactions among less closely allied individuals. As yet, ecologists have not been able to measure completely the competitive environment for any one species. Unlike acts of predation, competition is generally more subtle and can be determined unequivocably only by observing the response of one species when another is added or removed. For this reason, competition remains an area of active research and much controversy, one that we shall return to later.

Intraspecific competition can promote the diversification of individuals within a population.

If the intensity of competition were to increase in direct relation to the ecological similarity of two individuals, competition would be most strongly felt among members of the same population. Although different populations may avoid intense competition by diverging ecologically, such a response is not normally available to individuals within a population because there is no simple genetic basis for divergence. Different populations have separate gene pools and genetic differences between populations can, therefore, accumulate rapidly. Within a population, individuals at least potentially share most of their genes. Differentiation may nonetheless occur within a population based upon sexual dimorphism, size differences related to age, learned differences in behavior, and, rarely, genetic polymorphism.

Divergence within a population usually occurs only where the pressure of interspecific competition is reduced. This condition is often met on islands where biological diversity is low compared with mainland areas. For example, male and female woodpeckers of species in the genus *Centurus* usually do not vary in size in mainland populations, where other species of different sizes occur. On many Caribbean islands where one species of *Centurus* may be the only woodpecker present, sexual dimorphism occurs frequently (Selander 1966). Equally impressive size differences occur between sexes of falcons and accipiters (hawks) (Storer 1966), but more subtle behavioral differences between the sexes with respect to foraging location and technique have also been discovered (*e.g.,* Robbins 1971). *Anolis* lizards on Caribbean islands exhibit within-population variation in the utilization of habitat and food items, based upon sexual dimorphism and upon size differences related to age (Schoener 1967).

Like most reptiles and many fish, lizards continue to grow throughout their lives. As they increase in size, individuals shift their activity space to different parts of the habitat and capture different-sized prey. A single population of lizards may, therefore, contain individuals of a variety of sizes that compete less among each other than if they were of a uniform size.

An unusual manifestation of within-population variation has been observed in the oyster catcher, a shorebird with a long blade-shaped beak. Oyster catchers use two techniques to open mussel shells—hammering and stabbing—but each individual employs only one of these behaviors (Norton-Griffiths 1967). Hammerers search for smaller mussels on firm sand well above the tidemark. These are flipped over, exposing their weaker valve, and hammered at until the shell cracks. Stabbers look for larger shells that are just submerged and, therefore, have their valves slightly apart. The oyster catcher quickly jabs its beak through the opening and severs the muscle that holds the valves of the shell together. Eating a mussel incapacitated thusly is a simple exercise left to the imagination of the reader. Each behavior is learned from the parents, which feed the young at the mussel beds from an early age. Hence the trait is passed culturally from generation to generation.

Genetic polymorphisms not related to sexual differences are not commonly employed to reduce competition within a population. We have encountered one case of dimorphism in the wing lengths of water striders (page 172), but this trait is an adaptation for dispersal and does not result in diversification of resource use within the population. Polymorphism is more commonly encountered within populations of organisms that mimic several different models (Ford 1971). If the models are considered a resource, the mimics reduce competition among themselves by resembling a variety of different models.

Striking as such examples are, most organisms are subjected to intense competition from other individuals of the same species. In this chapter we have seen that competition is a pervasive and powerful ecological force that can exert strong selective pressure on populations to evolve differences that enable them to avoid intense ecological overlap. But there are relatively few instances in which intraspecific competition is reduced through intrapopulation divergence. In the next chapter, we shall examine some of the ways in which competition between individuals of a species is organized.

17

Individuals in Groups

Competition is most intense between individuals of the same species because the adaptations that define their activity spaces are most similar. Whereas different species are capable of diverging over evolutionary time to reduce the intensity of competition, individuals frequently use aggressive behavior—such as the maintenance of individual distance or the defense of spatially defined territories—to reduce competition from others in the same population. Under certain conditions, territoriality and other behavioral mechanisms that space individuals are not tenable. The resulting breakdown of spatial organization can lead to the formation of social groups, within which the outcome of competition is, in many species, determined by dominance hierarchies. In exceptional circumstances, group existence has led to the evolution of cooperation and specialization of function—what we normally think of as attributes of societies. In this chapter, we shall examine the kinds of social structures that occur in populations, and how these are affected by the environment.

Members of a population maintain an individual distance from others.

The spatial distribution of individuals within a population is the outcome of two opposing social tendencies. On one hand, individuals may be attracted to others of the same (or sometimes different) species and, therefore, assume a clumped distribution. On the other hand, they may avoid each other and tend to form an evenly spaced distribution. The simplest spacing mechanism is the maintenance of individual distance (Conder 1949), either by not tolerating individuals that approach too closely, or by avoiding others. The threshold distances between individuals that evoke aggressive behavior vary widely among and within species and under different conditions. For example, when chickadees and kinglets are foraging in flocks, the intrusion of an individual to within two to four feet of another generally produces an antagonistic response on the part of one of the birds (Morse 1970). Mated pairs will usually tolerate each other more than they will other individuals in the population and, of course, the individual distance between young animals and their parents is negligible when the young are being fed. In cold weather, many animals reduce their individual distance to zero as they huddle to conserve heat.

Individual distances are difficult to measure in mobile populations. Behavioral ecologists have instead frequently resorted to quantifying aggressive interactions to determine the tolerance of individuals for others. Basic characteristics of aggressive interactions between individuals have been described by Recher and Recher (1969) for migrant shorebirds that stopped to feed along mudflats and sandy beaches. The Rechers scored interactions between individuals according to their intensity, attaching the lowest value (1.0) to threat displays and the avoidance of other individuals, intermediate values to the displacement of one individual by another (2.0) and to active pursuit (3.0), and the highest value (4.0) to active fighting between individuals. Most of the scored aggressive encounters occurred between individuals of the same species, 96 per cent in the case of the semipalmated sandpiper. Among western sandpipers, intraspecific encounters had an average score of 2.5. Interspecific interactions were both less frequent and less intense. Encounters between western sandpipers and least sandpipers had an average score of just over 1.0. Aggressive interactions appear to be directed toward the closest competitors. Most encounters occurred between individuals of the same species, and of 138 interspecific interactions recorded, 88 involved species so closely related that they were placed in the same genus.

In the Rechers' study, the frequency and intensity of aggressive interactions increased as population density, and thus competition, increased. The quantity of the food resource was another factor in the relationship between aggression and population density. As the receding tide exposed new feeding areas, aggressive interactions decreased, regardless of the density of foraging individuals. On the other hand, rising tides restricted suitable foraging areas at the water's edge, and the frequency of aggressive interaction between birds increased.

The defense of territory represents the extension of individual distance to space and objects.

That birds defend territories against other individuals of the same species has been recognized for more than fifty years, since the publication of H. E. Howard's classic book *Territory in Bird Life* (1920). The behavior of most temperate-zone birds changes dramatically in spring. In winter, many species congregate in loose flocks, but these break up in spring and individuals disperse widely over habitats suitable for breeding. On any fine spring morning, one may find male white-throated sparrows vehemently proclaiming their territories from singing perches. Throughout the day they restrict their activities to well-defined boundaries and will not permit the intrusion of other males into their defended area, within which they will soon mate and rear their young. At the end of the breeding season, after the young have been reared and are no longer dependent on their parents, the behavior of the birds changes; the territory is abandoned and individuals join flocks that gradually move south ahead of the oncoming winter.

Although the white-throated sparrow may be regarded as typical, territorial behavior is so diverse that it is difficult to generalize our concept of the phenomenon. A territory is often defined as "any defended area," but because one rarely observes territorial defense, "any exclusive area" is a more practical definition. Although territoriality is most conspicuous in birds, territorial defense occurs in other groups of organisms as well: ants (Pickles 1944), field crickets (Alexander 1961), owl limpets (Stimson 1970), fish (Assem 1967), and crayfish (Bovbjerg 1970), to name a few.

Territories may serve a variety of purposes, and they persist for varying periods (Nice 1941). Individual birds often set up territories on their wintering grounds, and we have seen that migrating species, such as the shorebirds mentioned above, sometimes establish transient feeding territories around them. These last for only minutes or hours. A territory below the high-water line can last only as long as the area is exposed by low tides. A bird may abandon a territory for a few minutes to forage elsewhere, and then return to re-establish its claim.

In one instance, Recher and Recher (1969) observed a group of sanderlings, small shorebirds that feed in large flocks while on migration and during winter, that had established individual territories over a particularly rich food source—the eggs of spawning horseshoe crabs—and kept other birds out of the area. Territorial sanderlings spent an average of 32 per cent of their time defending their feeding areas, and nonterritorial birds spent 39 per cent of their time engaged in aggressive interactions trying to enter these areas. Territorial birds could feed between aggressive encounters, but nonterritorial birds could not, and they soon left the area. The time required for territorial defense in this case was extremely high. As territories become more firmly established and persist longer, the effort expended in territorial defense diminishes. Once territorial limits are well defined by aggressive encounters with neighboring individuals, intrusions become relatively infrequent and territorial challenges are rarely observed.

The nature of breeding territories varies considerably. During the breeding season, all the nesting and foraging activities of the white-throated sparrow occur on the territory, but this arrangement is not practical for many kinds of organisms. For example, seabirds defend only small territories in the immediate vicinity of their nests, often about as far as they can reach while on the nest. The adults disperse over such vast areas of the ocean in search of food that it would be both impractical and unnecessary to defend a foraging territory. For other species, large territories may be effectively secured by defending some smaller area or object that is in short supply. If nest holes are rare, defense of the hole itself will exclude other breeding birds from the vicinity. For many species of flycatchers and hawks that inhabit open grassy areas and feed from perches, defense of the perches has the effect of excluding birds having similar feeding requirements from the area. Also, territory does not have to be defended continually. Once birds have constructed their nests and breeding activities are well under way, foraging areas become tied to the nest site, and the defense of territory can become superfluous.

Several cases of group territoriality, the cooperative defense of a common area by several individuals, are known. These vary from the transient and loose defense of group territories of migrating shorebirds to the highly structured group defense efforts exhibited by some more complexly social organisms. A most interesting territorial system occurs in the Australian magpie, a crowlike bird that inhabits savanna habitats in Australia (Carrick 1963). The magpies require trees for roosting and nesting, and pastures for feeding; the best habitats for magpies include some of both vegetation types. The magpie populations are divided into small territorial groups of two to ten birds, which breed regularly, and large nonterritorial flocks in which no breeding occurs. Among the territorial groups, reproductive success parallels the quality of the territory occupied. The so-called permanent groups inhabit the optimum territory and do most of the breeding. "Marginal" groups occupy less desirable areas and are less prolific. The least desirable habitats are large pastures with few trees; these areas are held by "mobile" groups, which often defend feeding and roosting territories separated by more than a mile. Birds in mobile groups rarely attempt to breed, so although their life style classes them as territorial rather than nonterritorial, they do not reproduce—probably because their resources are too meager.

The territories of permanent groups in optimum habitats are stable from year to year. Between 1955 and 1963, for example, sixteen of 38 permanent groups under observation occupied the same territory, and about 20 per cent of the adults banded at the beginning of the study remained to the end.

What are the functions of territorial defense?

The result of territorial defense is to exclude other individuals from a particular area. The *purpose* of territorial defense is more difficult to pinpoint (Hinde 1956). When territories include concentrations of food or most of an individual's foraging grounds, the purpose of territorial defense must be to secure resources against their use by other individuals.

Defense of territory may aid pair formation and reproductive behavior by eliminating the behavioral confusion that might arise from intrusions by other individuals. Particularly in species that gather food outside their territories, defense of territory may occur for this purpose. Seabirds that nest in crowded colonies on rocks and small islands may defend territories to prevent other adults from killing their chicks, which is a major mortality factor in seabird colonies (Nelson 1966). Territoriality may also slow the spread of disease by increasing the distance between susceptible individuals. For example, during one particularly wet winter many birds in the large nonterritorial flocks of the Australian magpie died from infection by *Pasturella pseudotuberculosis,* while none of the territorial birds succumbed. Little is known of the occurrence or influence of disease in natural populations. It is difficult to imagine, however, that the primary purpose of territorial behavior is to reduce this factor.

The most controversial of the postulated functions of territory is the regulation of population size (Wynne-Edwards 1962, Lack 1966, Brown 1969). There can be little doubt that in many cases territoriality effectively regulates population density. Territorial behavior tends to set an upper limit to the number of individuals that can occupy an area, and others are forced to go elsewhere or not breed. This was first demonstrated inadvertently by a series of experiments designed to examine the effects of bird predation upon an outbreak of the spruce budworm. An attempt was made to remove all birds in a forty-acre plot of coniferous forest in northern Maine, but almost as quickly as birds were shot, others appeared and established territories (Hensley and Cope 1951, Stewart and Aldrich 1951). In 1949, 148 pairs (296 individuals) were counted at the beginning of the study, but before the end of the breeding season 420 adults, mostly males, were shot in the area. In an identical experiment the following year, 154 pairs (308 individuals) were counted, and 528 individuals, again mostly males, were removed. Clearly, a substantial portion of the populations of some of the species did not hold territories in the area at the beginning of the experiment and would not have attempted to breed under normal conditions.

One of the most intensely studied of all territorial birds is the great tit, *Parus major,* a European relative of chickadees and titmice. A series of observations and experiments in Wytham Wood, near Oxford, England, has revealed the nature of territorial behavior in this species (Krebs 1971). The tits nest in natural holes or in boxes. Their preferred habitat is woodland, although individuals will nest in hedgerows or similarly broken patches of woods. Birds nesting in hedgerows are usually one-year-olds and are generally unsuccessful in breeding. The older individuals that secure territories in woodlands successfully raise 90 per cent of their broods. When territorial males are removed from woodland habitat, they are rapidly replaced by newcomers from hedgerows, mostly first-year birds, and by the expansion of established territories (Figure 17-1). The experiments suggested that continual song advertisement is an important component of territorial behavior. Hedgerow birds apparently could recognize the individual songs of territorial birds in neighboring woodlands, and quickly moved into the woodland when one of these was no longer heard.

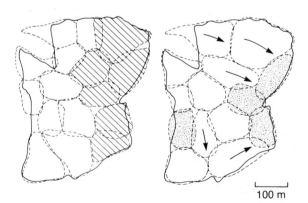

100 m

Figure 17-1   An example of the replacement of birds removed from a population of great tits. Six pairs were shot in late March 1969 (striped areas). Within three days, four new pairs had taken up residence in the wood (shaded areas), and the remaining area was filled by expansion of existing territories (arrows) (after Krebs 1971).

There can be no question that territory limits the size of the breeding population for the great tit, but territory size was not related to the availability of food. Rather, Krebs felt that territorial behavior evolved in the tit as a spacing-out mechanism that secondarily influenced population density. The primary function of spacing for the individual may be to reduce predation: nests closer than 45 meters to their nearest neighbors suffered twice the predation of nests at greater distances. The actual size of the territory defended, hence the degree of spacing, probably depends on the number of territory seekers, at least defended areas are small when the tits are numerous in winter prior to the establishment of territories. Furthermore, the level of aggression maintained during the period of territory establishment varies in direct proportion to the number of neighboring territories. Hence a large territory requires more time to defend against intrusion and the expansionist desires of neighbors.

A study of territoriality in the great tit in the Netherlands showed that although territory size is flexible, during years of population abundance many individuals are forced to breed in suboptimal habitats. Kluijver and Tinbergen (1960) sampled populations using nest boxes in two habitats: a uniform Scots pine plantation and a narrow strip of mixed woods along a brook. The latter is the favored habitat of the species, and the number of individuals nesting in the mixed woods was higher and varied less proportionately (26 to 43 pairs over ten years) than in the neighboring Scots pine plantation (0 to 23 pairs). The territorial behavior of great tits in the mixed forest apparently set a more or less rigid upper limit on the number of pairs that could occupy the habitat, and excess individuals were forced to nest in the less desirable pine woods.

Within limits, territory size expands to fill unoccupied areas or can be compressed by the pressure of high population density. Weeden (1965) found that the density of tree sparrows in one study area in Alaska ranged between 21 and 35 pairs per hundred acres over a three-year period. When territories were large, a relatively small proportion of the total territory (about 15 per cent) was used intensively, but when territorial area was reduced owing to high population density, a larger portion (about 40 per cent) of the territory was frequently visited. Most of the territories contained patches of habitat that were unsuitable for foraging and were not visited.

There must be a certain size below which a territory cannot provide sufficient food for normal nesting activities. On the other hand, there must be an upper limit to territory size: The advantages of a large territory for food gathering can be offset by the time and effort required to patrol the boundaries of the area and repel intruders. Furthermore, at high population densities, optimum territory size may be smaller than when intraspecific competition for space is less stringent (Krebs 1971).

Territory size is related to the role of the territory in providing food resources.

Territory size varies in a fairly regular manner among species that gather all their food within defended areas. The relationship of territory size to body

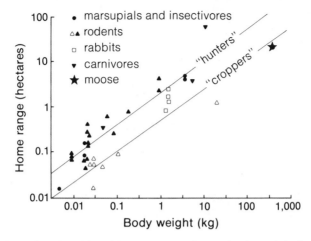

Figure 17-2   Relationship of home range size to body weight in mammals. Home ranges of hunters (solid symbols) are about four times larger than the home ranges of croppers (open symbols) (after McNab 1963).

size, feeding behavior, and habitat shows the importance of food resources to territoriality.

Home ranges in mammals do not correspond exactly to defended territories because the home ranges of individuals frequently overlap, but the size of the home range generally increases with the adult body weight of the species (Figure 17-2), presumably in accordance with the increased metabolic requirements of large mammals (McNab 1963). The relationship is complicated by the fact that large and small mammals do not feed on prey of the same size range, and certain assumptions must be made about the relative abundances of their prey. The size of the home range also depends on the dispersion of the food resource. Home ranges of carnivores and of herbivores that search for fruits and seeds ("hunters") are on average about four times the size of home ranges of species that graze and browse on leafy vegetation ("croppers").

If the function of territory is to secure food resources, one would expect territory size to be large in habitats that are relatively unproductive (Table 17-1). In desert habitats annual plant production, which ultimately determines the level of food resources for all species, is less than 70 grams of dry organic matter per square meter (Odum 1971). One typical desert area supported only 22 pairs of birds per 100 acres. The productivity of prairie, shrub, and mountain forest habitats is intermediate, varying from 300 to 800 grams, and that of typical moist lowland forests is 1,500 grams. The number of breeding birds in each habitat corresponds to its productivity.

The territory size of some wide-ranging species varies with habitat. The song sparrow has been studied in many localities. In shrubby habitats in Ohio, the average size of song sparrow territories is about 2,700 square meters, with extremes of about 2,000 and 6,000 square meters (0.5 and 1.5

TABLE 17-1
Censuses of birds in a series of six habitats with increasing plant productivity
(data from Audubon Field Notes; production estimates from Whittaker and
Likens 1973).

Habitat	Locality	Net primary production $(g \cdot m^{-2} \cdot y^{-1})$	Total breeding pairs per 100 acres	Abundance of the three commonest species (pairs per 100 acres)
Desert	Mexico	70	22	4,3,3
Prairie	Saskatchewan	300	92	28,18,11
Chaparral	California	500	190	48,38,21
Pine forest	Colorado	800	290	45,35,35
Hardwood forest	West Virginia	1,000	320	63,40,30
Flood-plain forest	Maryland	1,500	581	83,69,59

acres; Nice 1937). Variation in territory size within a habitat may result from differences in the quality of territories and the resources they contain (Stenger 1958). Along Minnesota lake shores, song sparrows defend territories averaging about 1,900 square meters, somewhat smaller than in shrubby habitats in Ohio (Suthers 1960). In the salt marshes of San Francisco Bay, territory size shrinks to about 400 square meters (Johnston 1956), less than one-sixth the size of those in the Ohio habitats. The high productivity of the salt marsh presumably increases food availability, so birds do not need to defend large territories.

On Mandarte Island off the coast of British Columbia, song sparrows nest in a strip of shrubby habitat in the center of the island, but feed partly in the grassy habitat along the fringe of the island. Territories are defended only in the shrubby habitat, and their size (200 to 300 square meters) is much smaller than that of the mainland territories, in which the sparrows do all their foraging as well as nesting (Tompa 1964). Territories of unmated males on Mandarte Island were much smaller than territories in which reproduction occurred.

The utilization of a territory varies with the stage of the nesting cycle. When the young are being fed, food requirements are greatest, and the territory receives its most extensive use by the foraging adults (Figure 17-3). The increased utilization of territory by ovenbirds during the mating period presumably corresponds to increased territorial defense at that time (Stenger and Falls 1959).

Territorial behavior can break down under high density or with unpredictable resources.

Breeding territories are generally stable if only because the young cannot be moved during much of their development period. Nests, dens, and other breeding sites tie the activites to a small area. Even if a feeding area is

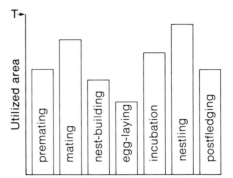

Figure 17-3   The utilization of territory by ovenbirds (*Seiurus aurocapillus*) at different stages of the nesting cycle. Utilized area is expressed as a fraction of the total area (T) used by the pair during the entire nesting cycle (after Stenger and Falls 1959).

not defended, space around the nest itself is usually vigorously defended against trespassers. Territories maintained solely for the purpose of feeding can be more transient. Such territories are often established and abandoned on the spur of the moment as the resources change or when a formidable challenger appears.

Territorial defense is often abandoned and group behavior taken up when a population becomes so dense that competition for space precludes the formation of territories. This type of switch in behavior with varying density is illustrated by the behavior of the dragonfly *Leucorrhinia rubicunda* in Finland (Pajunen 1966). In one dense population, dragonflies were spaced less than one-half meter apart, on the average, along the edges of pools. In another area, where the population was less dense, individuals were spaced between three and seven meters apart. Pajunen scored the intensity of interactions between the males (only the males are territorial) from low values of 1 for no interaction, through intermediate values, corresponding to varying degrees of threat and pursuit, to the highest score (5) for threats followed by fighting. Where the dragonflies were sparse, the frequency and intensity of aggressive interactions were high (average score 4.2) compared with densely inhabited areas (3.1) because of frequent territorial defense in the first area. The level of interaction in the sparse population was more intense between neighboring territorial males than between territorial males and intruders from a distance (3.2). In dense populations, individual dragonflies rarely returned to resting sites, which are usually stems of emergent vegetation. Instead, they flew over larger areas of the ponds and alighted to rest less frequently than did dragonflies in less dense populations (Figure 17-4). Both the level of territorial defense and the level of site tenacity decreased as the density of dragonflies in an area increased.

Unpredictability in the temporal and spatial distribution of food resources may also result in the breakdown of territoriality. There is little point in defending an area if it is unlikely to provide adequate food for a reasonable period. Habitats with low rainfall tend to have less predictable food resources than habitats with abundant rainfall. African weaverbirds inhabiting grassland, savanna, and deep forest reveal differences in territorial systems that are consistent with greater uncertainty of food resources in dry areas (Crook

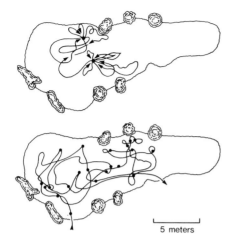

Figure 17-4   Examples of male flight activity in a pool population of the dragonfly *Leucorrhinia rubicunda* showing well-developed site attachment (above) and weak site attachment (below). The periods of observation are one and thirteen minutes, respectively (after Pajunen 1966).

5 meters

1964). Most of the forest species are insectivorous, feeding on a predictable and well-dispersed food supply. Most of the savanna and grassland species are granivorous and must rely for food on the irregular occurrence of seeds in small patches of habitat that happened to receive adequate rainfall during a particular season. Almost without exception, the insectivorous species of the forests nest as solitary pairs and defend large territories. None of the granivorous species defend feeding territories. Most of the species that inhabit savannas nest in large colonies in the few trees that are available, reminiscent of large colonies of seabirds on oceanic islands. Within these immense colonies, individuals defend small areas immediately around their nests. Grassland birds also do not defend areas for the purpose of feeding, although they do maintain small nesting territories in "neighborhoods" with other pairs. Owing to the absence of trees, grassland species are forced to nest on the ground where a small dense colony could easily be found by predators. In this case, the spatial dispersion of nest sites as a result of territorial behavior may serve an antipredator function primarily.

The adaptive value of close association in groups may lead to the evolution of mutual attraction between individuals.

The breakdown of territorial defense does not necessarily mean that groups based on the mutual attraction of individuals will necessarily form. For that matter, the close association of individuals in nature does not necessarily imply that mutual attraction exists. Many more or less fortuitous associations of individuals result from the limited extent of suitable habitat. The aggregation of seabirds on small oceanic islands, or nesting weaver finches on trees in the savanna, occurs because suitable places for reproduction are limited. In species whose food is abundant, territories may be very small and closely packed, giving the appearance that the population is a large social group. Prairie dog colonies, for example, occur in extremely dense populations but actually are strongly territorial.

Whether organisms are brought together at first by fortuitous associations or by some accident of behavior, association with other individuals of the same species may enhance survival and reproduction. The advantages of group living are varied. Most biologists feel that groups increase the efficiency of foraging or of detecting and dissuading predators. The "more eyes" hypothesis suggests that the more individuals in a group, the sooner predators are detected and defensive action can be taken. An alarm call from one individual will alert all the others in the group. The defensive responses of individuals to predators may augment one another and confer advantage to living in close proximity. For example, nests of Brewer's blackbirds in dense colonies were preyed upon less frequently than nests in more diffuse colonies (Horn 1968). Horn suggested that if the defended area around a nest is sufficiently large, other nests built within the defended area will benefit from the additional protection. Hamilton (1970) has argued that an individual is more likely to escape predation when in a group than when alone because predators tend to pick off isolated and marginally situated individuals. In this case, purely selfish motivations should draw organisms to the center of a group like a social magnet. Group formation may also lead to greater feeding efficiency (Cody 1971, Pulliam 1973) because localized, abundant patches of food are more likely to be found if many individuals search, searching patterns are more highly organized and efficient, and insects and other animals flushed by one individual might be caught by a second individual if the first fails.

Membership in a group for reason of gaining protection in numbers or keying in on food resources found by others is basically a self-serving phenomenon. On the whole, all individuals are better off in a group than alone. Being alert to the reaction of another individual at the approach of a predator does not require any degree of social organization within the group. Nonetheless, once groups have formed, the social environment may begin to mold the behavior and morphology of individuals in the direction of mutualistic cooperation.

The group environment can lead to the evolution of social adaptations.

Even in the most loosely organized groups of animals, those held together by what appears to be a minimum social structure, one may find adaptations specifically directed toward group living. Many kinds of fish gather in schools because of the mutual attraction of individuals for each other. Yet as loosely organized as fish schools seem to be, several adaptations to encourage schooling behavior have evolved. In the pilchard, a small fish whose schooling behavior has been studied by Cullen and his associates (1965), the distance and orientation to the nearest fish in the school are well defined. The fish maintain distances of three to six centimeters, but rarely follow directly behind one another so as to avoid the tail swing and water turbulence from the individual ahead. Generally, fish of the same size school together because they swim at approximately the same rate, about three lengths per second for the pilchard.

In addition to behavioral orientation and spacing, several morphological adaptations facilitate schooling. First, species that form large schools are often sleek in appearance and silver colored, which permits orientation and association even in low light intensities. (During the night schools disband; they reorganize each morning.) The pectoral fins of most schooling species are reduced in size and are relatively immobile, thereby preventing individuals from swimming backward and helping to maintain the oriented movement of the school. Why do fish school? No one is certain, but increased feeding efficiency, predator detection and defense, efficiency of movement, and ability to find mates all may be important (Breder 1967, Shaw 1970).

One frequently finds more complex social arrangements, particularly among mammals and birds. To describe one such set of social adaptations at this stage of our discussion, we shall consider the rutting behavior of goats and sheep. Males of most species vie for access to females just before the mating season by highly ritualized and frequently spectacular combat (Geist 1966, 1971, Schaffer 1968). The closing speed of mountain sheep rams during a charge has been estimated at fifty to seventy miles per hour (Figure 17-5). But the rams are careful to strike only at the base of the back-curving horns where the skull is specially reinforced to withstand the impact (Schaffer and Reed 1972). Moreover, a ram will not take advantage of another that is not prepared for the combat, a level of chivalry rivaling that of medieval jousting matches.

Rutting behavior generally matches the development of the horns and skull. Sheep with smaller or less massive horns (Barbary sheep, tahr, and most goats and ibex) do not engage in so athletic a rut as mountain sheep. The more distantly related chamois and Rocky Mountain goat have short sharply-pointed horns and limit combat to body butting, or have none at all. The noncombative arrangement of these species is sensible because their horns are so like daggers that serious injury could result from intensive combat.

Dominance hierarchies reduce aggression within the social group.

The abandonment of territoriality by no means eliminates behavioral competition between individuals; rather it allows competition to be organized differently, and at the same time permits some degree of mutual benefit or cooperation. In most groups, aggressive interactions are ordered according to a *dominance hierarchy*. In the simplest manifestation of a hierarchy, individuals are ranked in a linear series according to the outcomes of aggressive encounters: the top individual dominates all others, the second-ranking individual dominates all except for the top individual, and so down the line, to the

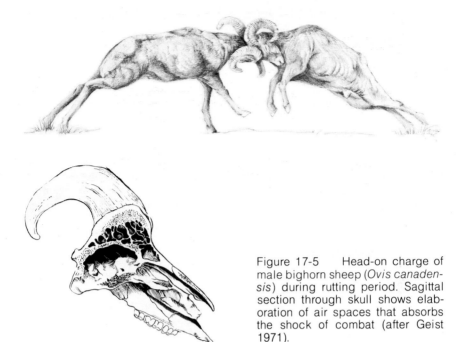

Figure 17-5   Head-on charge of male bighorn sheep (*Ovis canadensis*) during rutting period. Sagittal section through skull shows elaboration of air spaces that absorbs the shock of combat (after Geist 1971).

lowest individual who is subordinate to all others. An individual's rank is determined, usually at an early age, during brief periods of fighting with other individuals in the group. After rank has been established, it is not often contested, except by newcomers to the group or when increasing age and size contribute to rank.

One function of the dominance hierarchy is to reduce aggression within the social group. If the outcome of aggressive interactions with other individuals can be predicted beforehand on the basis of previous experience, the need for aggression disappears. For this reason, dominance-subordinance encounters rarely involve prolonged physical contact. They are more often expressed as subtle avoidance, supplanting, occasional pursuit, and a considerable amount of bluff.

The structure of dominance hierarchies is not rigidly fixed, and there are many variations on the basic theme. The dominance relationship between two individuals is sometimes influenced by a third party. In permanently established troops of monkeys, rank may be inherited from the parents. In some rhesus macaques, young take the rank of their mother (Koford 1963). Females of many species assume a rank commensurate to that of their mate, but in others, females and males have separate dominance orders (see, for example, Sabine 1959) or females and males enter independently into a single dominance hierarchy (Kikkawa 1961). The expression of the hierarchy and the relative positions of the sexes may change seasonally with the cycle of reproductive activities. Size and age also may contribute to the rank of an individual. Newcomers to a flock usually assume a low rank at first, regardless of their status elsewhere, but may work their way up the hierarchy later.

Observations by Greenberg (1947) on the behavior of green sunfish in small laboratory tanks is typical of experimental studies on dominance hierarchy. The individuals used in these experiments—immature fish that were 1 to 1.5 inches long—were collected in ponds in Illinois. The experiments were conducted in bare freshwater aquaria of various sizes. Most aggressive behavior consisted of "driving"—quick dashes by dominants at lower ranking individuals. Dominant fish occasionally threatened others by spreading their gill covers and swimming toward them. The threatened fish usually responded by assuming a submissive posture, turning broadside to the approaching dominant and moving back and forth in a rigid vibrating fashion. Such behavior dampened the aggressive tendency of the threatening fish.

In an initial experiment, Greenberg kept eight groups of four fish in one-gallon aquaria for two to three weeks. Dominance hierarchies were quickly formed. The highest-ranking fish was also the most aggressive (Table 17-2), and its position was relatively more stable than that of lower-ranking fish. Top-ranking fish maintained their position in the hierarchy for 97 per cent of the duration of the experiment, compared with 70 to 74 per cent for second- and third-ranking fish, respectively.

Greenberg also found that territoriality could be interchanged with the dominance hierarchy. Territories were established in the corners of the aquaria at various times, usually according to the rank of the fish; the top-ranking fish was most likely to set up a territory, followed by the second-ranking fish. When aquaria were fitted with a partition and a small gate, thereby separating the tank into two halves with limited access between them, the tendency to establish territories greatly increased. Three-quarters of the experimental groups exhibited territory formation by at least two individuals, whereas only one-quarter of the control series without partitions had similar territorial arrangements. In almost every case, the first- and second-ranking fish occupied different compartments in the partitioned aquaria.

TABLE 17-2

Rate of aggression (drives/hour) of green sunfish maintained in groups of four in one-gallon aquaria. Reverses in dominance are given in the lower-left portion of the table (from Greenberg 1947).

	Rank of driven fish			
Rank of driving fish	1	2	3	4
1	—	45	44	44
2	1	—	27	32
3	0	3	—	15
4	0	2	3	—

TABLE 17-3

Relationship between degree of crowding, territoriality, and level of aggression in the green sunfish. Four fish were placed in each aquarium (from Greenberg 1947).

Size of aquarium (gallons)	Number of groups with		Total	Aggression (drives/hour)
	Territories	Hierarchies		
0.4	2	8	10	32
6.9	6	4	10	40
39.9	3	5	8	9

In another series of experiments, the size of the aquarium was varied, but the number of fish was kept at four. Crowding was found to influence the alternative formation of territories or dominance hierarchies (Table 17-3). In crowded conditions, territoriality generally broke down and aggressive interactions between individuals decreased somewhat, as Pajunen observed for dragonflies. At the other extreme, in large tanks territories were infrequently held and the level of aggression dropped. Aggression was most frequent at intermediate densities at which territories were most actively disputed.

Observations such as those on the green sunfish suggest that territoriality and dominance hierarchies represent the expression of aggressive interaction over different objectives—in one case, space, and in the other case, rank in a hierarchy. One might consider territoriality as a complex spatial organization of dominance hierarchies in which an individual holds the highest rank in the hierarchy in its own territory but is lower ranking beyond the limits of the territory. Niko Tinbergen (1953) devised a simple but elegant experiment to test this hypothesis with stickleback fish. When more than one male was placed in an aquarium, individuals usually defended territories. Tinbergen put territory-holding sticklebacks in glass tubes, so he could move them where he wished in the aquarium. He found that a male stickleback always dominated an intruding male on its own territory, but the same individual became subordinate to the other male when moved to any other part of the aquarium. Similar spatial patterns of dominance have been found in birds by Brown (1963) and Willis (1967).

The position of an individual in the dominance hierarchy is sometimes expressed by its spatial position within the group. We have seen earlier (page 86) that in baboon troops, dominant males occupy the center of the troop and the younger, subordinate males travel at the periphery. The position of individual weaver-birds (*Quelea*) on perches is related to their position in the dominance hierarchy (Dunbar and Crook 1975). Large foraging flocks of wood pigeons are organized similarly (Murton 1967). Individuals that are low in the dominance hierarchy tend to occupy positions at the periphery of the flock. Murton supposed that these birds were more exposed to predators. They appeared to be nervous, and because they spent a large portion of time looking up from their feeding, they were often undernourished. The dominant birds in the center of the flock were generally calmer and fed more because they were protected from the surprise attack of a predator by the individuals at the periphery of the group. In later papers, Murton ascribed the primary advantage of flocking to feeding efficiency rather than to predator escape (Murton 1970, Murton, Isaacson, and Westwood 1971). Pigeons apparently watch each other feed, and thereby continually monitor which are the best components of the food supply. Dominant birds coordinate their feeding behavior, and thus adjust their foraging according to the general experience of many individuals in the flock. Subordinate individuals, mostly juveniles, concentrate at the front of the flock where they can learn how dominant birds are feeding. But as they are constantly pushed ahead by dominant birds, they cannot feed so efficiently.

Why do submissive individuals remain with the group if they are more susceptible to predation, starvation, and reduced mating success? To answer this question, we must weigh the advantages of group living against the disadvantages of low status. While a low-ranking individual may have a reduced prospect of survival and reproduction, its success may be much higher than if it were on its own. As social groups become larger, their size may become disadvantageous to some or all members of the group. Competition within a group is organized by dominance hierarchies that are based, to a large extent, on personal knowledge of all individuals. If the group becomes so large that such familiarity is no longer possible, the dominance hierarchy could break down and aggression increase. Also, large groups may develop a problem faced by large organisms, namely that of supply and distribution of resources. The rate of movement of winter feeding flocks or birds increases in proportion to the size of the flock (Morse 1970). Flocks could conceivably become so large that the time spent moving would reduce foraging time and cause starvation. As yet, however, the organization and purpose of such social groups is poorly understood.

Behavioral facilitation, cooperation, and specialization can result in elaborate social organization.

A firmly established social environment can lead to adaptations that promote the interdependence of individuals in the group. The simplest of these adaptations involves the coordination of activities. Dependence on the group environment can proceed to the point that behavior will not occur unless stimulated by interaction with other members in the group. This type of stimulation is often referred to as *social facilitation.* In colonial birds, large colony size leads to early breeding, increased synchrony of breeding within the colony, and higher reproductive success of the individuals (Coulson 1968). Such social behavior represents an intermediate situation between one in which group life is merely advantageous to the individual and one in which it is absolutely necessary. The purpose of social facilitation probably varies for different behaviors and populations. It may be advantageous for the individual to time its activities to coincide with those of its neighbors: In a synchronous population, aggressive interactions would be limited to the briefest period possible.

An intriguing possibility, for which, unfortunately, there is little evidence, is that the synchrony of nesting in seabird colonies, and perhaps in other groups, results from a kind of voting procedure for the initiation of breeding. Suppose birds spread out over vast areas of ocean in search of food and then return to the colony, having experienced either good or poor feeding conditions. If an individual experienced conditions good enough to support reproduction, it might behave as if interested in beginning to breed. If a bird returned to the colony highly motivated to breed, the uninterested behavior of neighbors could dampen its enthusiasm if their feeding conditions were generally poorer than it had experienced that day. Thus, an individual could assess the availability of food in the surrounding ocean from the overall level of interest expressed in breeding. As the food resources improve, more birds will exhibit breeding intentions, and finally reproductive activities in the entire colony will be initiated.

Ward and Zahavi (1973) have suggested that breeding colonies, roosting congregations, and other group formations may serve as "information centers" for food finding where the food supply is unpredictable in time and space. The authors envision individuals spreading out over vast areas in search of food, conveying information about feeding conditions when they return to the assemblage, and as they leave to feed, following birds that have located good foraging areas.

Cooperation involves the coordination of individual behavior toward a common goal: the defense of the herd by male musk oxen, the pack-hunting behavior of wolves and other large predators, the coordinated behavior of many aquatic birds that corral fish into a small area where they can be fed upon easily (Bartholomew 1942, Emlen and Ambrose 1970). Such instances of cooperation are not common in nature, but where they do occur, they represent one of the highest levels of social behavior in which interaction serves not to organize competition between individuals but is mutualistic.

One of the most remarkable examples of social organization is that of the piñon jay (*Gymnorhinus cyanocephalus*) (Figure 17-6), native to the western United States, described by Balda and Bateman (1971, 1973). Flock size varies up to a maximum of about 250 individuals in the Flagstaff, Arizona, region. The flocks have a complicated organization: year-old birds are excluded from breeding, some cooperative feeding of young exists, sentinels watch for predators while the rest of the flock feeds, individuals show strong group cohesion behavior and tolerance of others, and piñon nuts are stored in large communal caches.

So long as the nut crop of the piñon pine is good, the flock remains within a small area, typically about eight square miles, which is not intruded upon by other flocks. During years of nut crop failure, the flock may join others in wandering over hundreds of miles in search of food, but always returning eventually to its home base. The flock habitat includes ponderosa pine forest for nesting and piñon-juniper woodland for gathering pine nuts. General foraging for insect larvae occurs widely.

Figure 17-6   The piñon jay *Gymnorhinus cyanocephalus*) (after Balda and Bateman 1973).

Because this species is so unusual, and has been studied so intensely, its behavior will be described in detail. In the fall, roughly October to January, loosely organized flocks of 200 to 250 individuals of all age groups are spread over about two acres while feeding and six acres during inactive periods. There is virtually no aggression within the flock, even though individuals feeding on snow-free patches of ground or at a salt lick may be crowded shoulder to shoulder. The flock usually moves one to two miles per hour while feeding, but long-distance group movements also occur after considerable mutual stimulation within the group. At first, a few birds fly off a short distance and return to the flock. This encourages other birds to do the same. Eventually this leaving behavior overwhelms the flock as a whole, and all the birds fly off together.

Feeding in the fall is concentrated on insect grubs in the soil and on the few piñon nuts left in cones from the summer crop. Seeds that are not eaten are buried in the ground on the south side of a tree trunk, where the winter sun first melts the snow and thaws the ground. The jays also choose their mates during the fall, sealing the bond with a ceremony in which the male feeds seeds to his prospective mate. These ceremonies become more intense as the year wears on, and by winter, mated pairs stay together within the flock.

An unusual feature of the jay flock is the sentries, numbering four to a dozen, which remain at the periphery of the flock perched on high vantage points. Individuals apparently take turns being sentries. At the appearance of hawks, foxes, and other predators, the sentries give warning calls that send the flock into trees and to protective cover, and occasionally the sentries mob the intruder while the rest of the flock moves off to a new area.

Breeding activities begin in earnest during February when courtship becomes more frequent, often involving courtship chases during which the pair flies rapidly through and over trees, performing sharp turns and steep dives, and mated pairs begin to leave the flock to feed. By this time, the flock consists mostly of first year birds during the day, but is joined by mated pairs in the early evening after a calling ceremony. The young birds begin the ceremony by uttering soft calls, which then become louder and culminate in a long flight, during which the courting birds rejoin the flock.

Nest building commences in March over an area of 120 acres, within which nests are spaced 50 to 500 feet apart. First-year birds stay in a small flock that remains within one-half mile of the nesting area. While the eggs are being incubated by the females, the males form a separate feeding flock, from which individuals occasionally go to feed their mates on the nest.

The nests are extremely synchronized, with the first eggs of the entire flock being laid within a three- or four-day period. As a result, most of the young are of the same age and leave their nests at the same time. During the greatest part of the nest period the young are fed by their own parents, but during the last four to five days before fledging birds share the responsibility of feeding, with up to seven adults providing food at any one nest.

Foster feeding continues for three weeks after fledging, during which time the adults and young from a group of nearby nests form an aggregation. At any one time, a few of the adults act as sentinels for the aggregation, others remain hidden with the young in bushes, while the rest fly off as a group to gather food. Within the whole flock, the individual aggregations remain distinct, even after the young begin to forage for themselves, and do not mix with others.

During August, when the seed crop has ripened on the piñon trees, all the aggregations converge in the piñon woodland to gather the nuts and store them in the breeding area for future use. The harvest lasts for two to three weeks and marks the reformation of the large group flock and the completion of the annual cycle.

A final note to this story. In addition to being seed predators of the piñon pine, the jays act as dispersal agents. The seed crop in August is too large to be consumed on the spot, and so the jays obligingly store the seeds. For those seeds that are forgotten or otherwise overlooked later, this behavior amounts to planting. In return, the piñon makes it easy and profitable for the jays to gather the nuts. The cones open wide and do not have protective spines. The cones are also placed upright on branches so the seeds do not fall out. Furthermore, the seeds are extremely large compared with other pines and do not have a heavy seed coat (Figure 17-7).

It is not known how such a complex social organization as the piñon jay's evolved, or what particular circumstances of habitat or food promoted it (Brown 1974). The example does illustrate the level of organization and interdependence possible and should caution us about too-simple interpretations of social behavior that is usually more difficult to observe than it is in the

Figure 17-7   The cones and seeds of the piñon pine (*Pinus edulis,* left) and the ponderosa pine (*P. ponderosa,* right). Cones are life size (after Sudworth 1917).

conspicuous jays. We shall return again and again to various aspects of social behavior as an organizing principle in population biology. In the next two chapters, we shall consider special cases of social interaction to which cooperation is the key: The sexual behavior of males and females, and the relationship between parent and offspring.

18

The Sexes

To reproduce, the sexes must get together in a cooperative effort that requires varying levels of commitment, depending on the species. Because of its cooperative nature, reproductive behavior contrasts sharply with competition between individuals in territorial and dominance systems, and creates an entirely different environment for evolution. In this chapter, we shall consider some of the behavioral adaptations that coordinate the sexes. As we shall see, these behavior patterns are influenced by the ecological background and provide a basis for the evolution of differences in the ecological roles of the sexes.

Courtship is any behavior between individuals of opposite sex that facilitates mating.

In his book *Social Behavior in Animals,* the ethologist Niko Tinbergen ascribed four functions to courtship behavior: timing, persuasion, orientation, and reproductive isolation between species. In each species, courtship may serve any number of these functions.

The basic physiological changes that occur during the reproductive cycle are determined by daylength or some other quality of the environment that stimulates breeding. Reproductive behavior such as copulation, egg laying, care of the young, and so on, may occur throughout a long breeding season for the population as a whole, but for each pair of individuals these activities must be well synchronized to ensure successful reproduction. A female fish that spawns her eggs when no male is present to fertilize them wastes her effort. By the same token, a male motivated only to copulate with his mate would be of little help when young had to be fed. One purpose of courtship behavior is to facilitate the synchronization of behavioral activities within a pair.

Daniel Lehrman (1961, 1964a, 1964b) studied the influence of courtship and environmental stimuli on the timing of reproductive behavior in birds and mammals, particularly in the ring dove. Lehrman found that courtship initiates a series of physiological changes in both sexes that synchronizes their

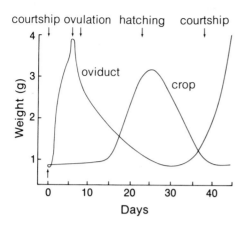

Figure 18-1   Change in weight of the female oviduct and the crop of the ring dove after the initiation of courtship (after Lehrman 1964).

behavior with the stage of the breeding cycle (Figure 18-1). The female's reproductive organs rapidly increase in size during the first week; ovulation occurs on the sixth and eighth days after courtship begins. After the eggs are laid, the size of the oviduct decreases; and the crop, a pouchlike enlargement of the gullet from which the young are to be fed, grows greatly by the time hatching occurs on the twenty-third day. In budgereegahs (small Australian parakeets), certain vocalizations of males stimulate ovarian development and egg laying in females—even when they are maintained alone in constant darkness (Brockway 1965). Hormones act as the intermediaries between external stimuli and physiological function, but their mode of action and specific timing are not well understood.

Persuasion, another function of courtship, occurs when one sex, usually the male, advertises its sexual readiness. Bird song and the chorusing of frogs and crickets have this function. In part, the same function is served by the bright coloration sported by the males of some species during the breeding season. But persuasion requires more than attraction. The individual must also convince potential mates that it is the appropriate sex and is relatively desirable compared with others, either due to its personality or the territory it possesses. In species with pronounced sexual dimorphism this is no problem. When male and female look alike, however, recognition of sex depends on subtle behavioral differences. This is certainly the case for many species of flycatching birds (Tyrannidae) in which the sexes resemble each other and adults are aggressive toward one another regardless of sex throughout most of the year. At the beginning of the breeding season, when males become involved in territorial defense, a female can enter a male's territory for courtship only if her behavior is different from that of neighboring territorial males (Smith 1969a). Some females complicate the situation by being as aggressive as the males, and considerable time may pass before the male and female realize that reproduction is the name of the game and stop attacking each other.

A male spider is faced with the particularly delicate task of persuading the voracious and much larger female that he is a suitable mate, and not a tasty morsel. The male's courtship behavior plays up to the female's best-developed senses. In the web-spinning spiders, whose vision is poor but whose tactile sense is well developed, courtship may involve tapping a specific rhythm on the web, or a bout of fencing with the forelegs. In the jumping spiders, whose vision is excellent, courtship involves a stereotyped series of movements of the body and the front pair of legs (Figure 18-2).

Courtship often serves to orient reproductive behavior in space as well as in time. This function is well demonstrated by the studies of Tinbergen (1951, 1952) on the courtship behavior of the three-spined stickleback. These small fresh-water fish normally form schools, but during the breeding season males isolate themselves from the school in territories. At the same time, the male takes on a new appearance: His eyes become a shiny blue and his back changes from dull brown to greenish, while his underparts turn a bright red. The male builds a nest in a depression that he scoops out by mouthfuls from the sand at the bottom of a pool in his territory. The dome-shaped nest is constructed of filaments of algae glued together by a substance secreted from the male's kidneys. A tunnel is opened through the side of the dome, where females will eventually lay their eggs.

After preparations have been completed, the appearance of a female in the territory of the male stimulates a sequence of ritualized courtship behavior (Figure 18-3). First, the male performs what is called a zigzag dance, a series of short charges at the passing female. If the female is not ready to spawn her eggs, she flees. But if the female is ready to breed, she courts the male by adopting a more or less upright posture (an attitude that is nearly the opposite of the male threat position shown in Figure 18-4). The male then leads the female down to the nest at the bottom of the pool, and points to the nest entrance with his head. The female follows and enters the nest. When the female is fully inside, the male commences to tremble, prodding the tail of

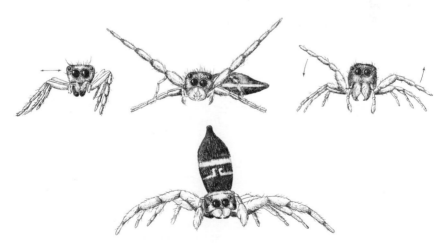

Figure 18-2   Courtship postures in salticid spiders, illustrating differences between four species (after Crane 1949a, 1949b).

MALE   FEMALE

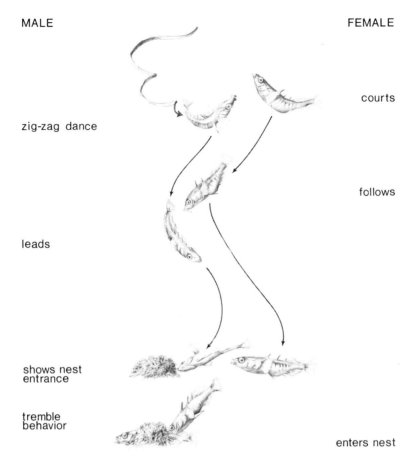

courts

zig-zag dance

follows

leads

shows nest
entrance

tremble
behavior

enters nest

Figure 18-3   Sequence of behavior during the courtship of the three-spined
stickleback (after Tinbergen 1951).

the female with his snout. If he does not stimulate the female in this manner, she will not deposit her eggs. After spawning, the female leaves the nest and the male enters to fertilize the eggs. The male may court two, three, or more females in a few days before his sexual drive wanes. By this time, the fertilized eggs are beginning to develop and require intensive parental care, which the male provides alone.

Courtship may culminate in mating, after which the behavioral bond between the male and female is broken. In the stickleback, as well as most other animals in which only one sex, or neither, cares for the young, the pair bond dissolves shortly after mating. Courtship is also brief. Mating may occur after a few seconds of courtship in many insects, or minutes in the stickleback and other organisms with similarly complex behavior patterns. Days or weeks of courtship may be required when courtship behavior is also

Figure 18-4   Aggressive postures of two male three-spined sticklebacks during a boundary encounter. The individual on the left has assumed an intense threat posture (after Tinbergen 1951).

responsible for initiating and synchronizing physiological changes prerequisite to mating. In birds and mammals, individual temperaments are varied and longer courtship is needed to solidify the pair bond.

When both sexes contribute to the care of the offspring over a long period, the pair bond is typically long-lived. Persistent pair bonds also form in stable territorial or social systems, in which pairs may remain mated even when no young are being cared for. Evidently, the efficiency in reproduction that is gained by forming a permanent pair bond more than balances any degree of competition between mates during nonbreeding periods.

Differences in courtship behavior among species enhance reproductive isolation.

Reproductive isolation prevents the appearance of unfit hybrids between individuals of different species, which are a waste of both time and effort on the part of the parents. Behavioral isolating mechanisms vary widely. The nuptial coloration and pattern of male sticklebacks differs among species, and females respond only to those males that are properly attired. In fireflies, variation in the length and timing of the light pulses among different species provides an effective isolating mechanism (Buck 1937). In anoline lizards, isolation is effected by differences in color and pattern on the dewlap, a flap of skin beneath the throat that is extended during both courtship and aggressive encounters (Figure 18-5).

When behavioral differences between species are not sufficient to totally prevent interbreeding, a second line of defenses may evolve. For example, the male sexual organs of some closely related species of insects are often so different that males are mechanically prevented from mating with females of other species by a sort of lock-and-key arrangement (Alexander and Otte 1967, Figure 18-6). Other groups have complicated biochemical incompatibility systems that prevent the fertilization of ova by male gametes of the wrong species; such systems are particularly common among plants whose pollen is dispersed at random by wind or by unspecialized pollinators that visit many kinds of flowers.

Figure 18-5　　A male *Anolis auratus* with his dewlap retracted (left) and extended (right). The color and markings on the dewlap (blue flecked with white) are distinctive of the species.

Mating systems and territorial systems are closely related.

If mating and pair-bond formation always occurred between one male and one female who cooperated in the rearing of their mutual offspring, there would be little more to say about mating systems beyond matters of descriptive interest. But monogamy may be more the exception in a world of polygamous and promiscuous mating systems (Orians 1969c).

The ecology of diverse mating systems is best illustrated by birds. The normal situation here is a monogamous one in which the sexes cooperate in reproductive activities on a more or less equal basis. A few species have *polygamous* mating systems in which an individual forms pair bonds with two or more individuals of opposite sex. (If more than one female is involved,

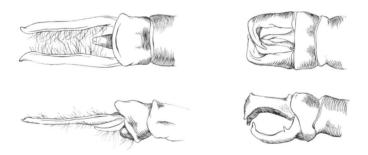

Figure 18-6　　The male genital organs of two species of dragonflies (left, *Plathycantha acuta*, from New Guinea; right, *Onychogomphus forcipatus*, from Europe), illustrating morphological differences that prevent interspecific matings. Above, top view; below, side view (after Portmann 1961).

which is the common case, the relationship is called *polygynous;* when more than one male is involved, it is *polyandrous.*)

Why does polygamy occur? The advantage of polygyny to a male is straightforward: He can fertilize more females, thereby fathering more offspring than a monogamous male. The advantage to the female, which has to share one male with other females, is less obvious. When does a female make a wiser choice by pairing with a male that already has a mate than with a male that is unmated? Verner and Willson (1966) have suggested that polygyny is selected in habitats in which the relative quality of territories differs considerably. To illustrate their argument, consider two species—the relative values of territories for gathering food or for suitable nest sites are about equal in one species and uneven in the other (Figure 18-7). A female must choose between a mated male and a bachelor. If two females occupied a territory, each would reduce its value for the other because they would compete for nesting sites, food, and the parental investment of the male in their offspring. If territories had about equal intrinsic value, a female should always choose the unmated male. If the values of the territories vary greatly, and if males with the best territories mate first, a female may still do better to pair with a mated male who holds a superior territory than to pair with an unmated male that holds a poor territory. This leads directly to polygyny.

Polygyny is frequently associated with dry, open habitats, where the distribution of resources is more heterogeneous than in moister habitats. Marshes are also heterogeneous, and the quality of territories among marsh-nesting birds can vary widely. Verner (1964) found that the mating success of male marsh wrens depended largely on the amount of emergent vegetation in their territories. In one marsh, the territories of unmated males averaged 9,787 square feet; monogamous males, 14,215 square feet; and bigamous males, 17,394 square feet. Emergent vegetation (in this case, cattails) covered an average of 54, 80, and 95 per cent of the territories, respectively.

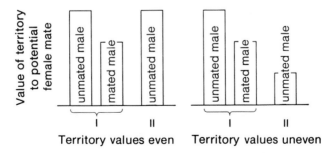

Figure 18-7   Diagram of the influence of territory value on the mating system of a species. Where territories I and II have about the same intrinsic value, monogamy is favored (left), and where their values differ greatly, polygamy is favored (right).

If Verner and Willson's hypothesis were correct, one would expect the number of young produced per male to increase greatly with level of polygyny and the number produced per female to remain constant with increasing harem size, at the very least. In other words, if females in large harems fared poorly, one would expect their better choice to be with unmated males or with males having few females in their territories. Data on the reproductive success of red-winged blackbirds bear out the hypothesis fully (Holm 1973). In marshes in eastern Washington, harem size varies from one to six, with two and three being most common. The productivity of males increases in direct proportion to the number of females in the harem (Table 18-1). The productivity of females also increases slightly, suggesting that territories established by males with large harems are of vastly superior quality to those of less attractive males. In the indigo bunting, a small finch common in the eastern United States, bigamous males produce twice as many offspring as monogamous males, hence the respective females are equally productive (Carey and Nolan 1975).

Where relative values of territories vary from area to area, and from year to year, polygyny may occur only locally. Where environmental heterogeneity always results in large discrepancies in the value of territories, polygyny may become so firmly established that it influences the evolution of the species. One frequently find strong sexual dimorphism in the behavior and the appearance of species that are always polygynous (Selander 1965), often to the extent that males do not participate at all in the care of the young, but are solely responsible for the defense of territory (Verner and Willson 1969).

Polygyny also occurs in some species that, unlike the marsh wren, gather most of their food outside the nesting territory. In such species, the value of the territory must be related to its suitability for nest building. For example, the value of a nesting site for many polygamous marsh-nesting blackbirds and grackles varies with the density of the vegetation and depth of the water, which determine the degree of protection from predators (Goddard and Board 1967).

TABLE 18-1

Breeding productivity of red-winged blackbirds (*Agelaius phoeniceus*) in relation to harem size in eastern Washington (from Holm 1973).

	Harem Size					
	1	2	3	4	5	6
Number of males	15	29	25	14	14	4
Number of nests	17	77	90	75	87	28
Number of eggs	66	222	276	225	292	93
Rate of fledging						
per egg	0.15	0.19	0.30	0.24	0.30	0.43
per male	0.67	1.48	3.32	3.86	6.29	10.00
per female	0.67	0.74	1.11	0.96	1.26	1.67

Polygyny is also common in deer, sheep, and other ungulates. Reproduction is not tied to a nest site in these species, nor are males territorial, and yet males commonly have large harems. The number of females that a stag can gather in his harem is determined by combat with other males during a rutting period. The attractiveness of a stag is not related to a particular parcel of land but rather to his general health and often to his ability to defend females and their young from predators. Because the young are nourished exclusively by the female, each female gives up relatively little by sharing a male with others.

Polyandry, the situation in which a female mates with several males, is extremely rare. There are few cases in birds (Jenni 1974, Hays 1974), and parental care of the young is relegated to the males of some monogamous species, particularly the phalaropes (Hohn 1969). The strain of laying many sets of eggs and the danger of defending territory when an egg is in the oviduct seem to have ruled out polyandrous mating systems in most birds, and similar restrictions undoubtedly act in other groups that form permanent pair bonds. Polyandry does occur with some regularity in the shorebirds (sandpipers and other waders). Although most species hold territories and are strictly monogamous, the breeding systems of a few opportunistic species open the door to a weakened pair bond and polyandry (Pitelka et al. 1974). The females of these species lay two sets of eggs in rapid succession, the first incubated by the male and the second by the female. With each sex tending to a different clutch, the pair bond is weakened and many females will mate with a second male to produce the second clutch.

In some species, courtship occurs on specific areas, called *leks* or *arenas* depending on the species, where males join together in communal displays. Lek behavior is fairly rare and occurs sporadically throughout the animal kingdom, but it is worth considering inasmuch as it may help us to understand more common mating systems better.

Lek and arena behavior involve the complete separation of male and female breeding territories. Considering how intimately the female's reproductive activities are related to the male's territory in most species, the evolution of lek behavior is difficult to envision. For nomadic species, whose reproductive activities are not fixed spatially, this does not raise a serious problem. The localization of aggressive encounters between males on a traditional arena seems a fairly simple evolutionary step from a nomadic existence, and it has occurred in several African ungulates, the Uganda kob and the lechwe (Buechner 1961, Leuthold 1966). The origin of lek behavior in birds poses a more difficult problem because the reproductive activities of females are confined to a nest site. The intermediate steps are not clear between a monogamous situation, in which reproduction takes place within a territory defended by the male, and typical lek behavior, in which males hold small courtship territories on communal display grounds and attract females to them solely for mating. In birds, lek behavior has been discovered in two situations: grouse and some shorebirds that breed in prairie or tundra habitats and in which males normally do not care for the young (ruff, Hogan-Warburg 1966; grouse, Kruijt et al. 1972, Wiley 1974); and several fruit-eating species inhabiting tropical forests (cock-of-the-rock, Gilliard 1962; manakins, Snow 1962, Lill 1974).

The evolution of lek behavior in grouse and other precocial species in which the young feed themselves is usually explained by the abundance and dispersion of food resources. Kruijt *et al.* (1972) have suggested that populations of black grouse are reduced so much in winter that male territories could not cover all the suitable habitat. This would free the females, which alone care for the brood, to nest outside male territories and to pick and choose among territorial males for mates. With breeding occurring elsewhere, the male territory could be greatly compressed to small areas within an arena. Pitelka *et al.* (1974) have suggested that in shorebirds that breed opportunistically in local areas of resource abundance, all the incubation duties can be undertaken by one sex, the female, which can gather sufficient food for itself in brief periods away from the nest. Thus emancipated, males may actively seek to mate with many females (promiscuity). If several males courting together attract females more strongly than lone males, a lek may easily be formed. Such a system is reinforced by the advantage to males of promiscuous mating and the advantage to females of dispersing widely to breed so as to avoid competition and interference from other females (Downhower and Armitage 1971).

Lek behavior is more difficult to explain in tropical forest species in which the young are fed by their parents and the male can, therefore, increase its own fecundity beyond contributing its gametes.

I can imagine four routes that the evolution of lek behavior from monogamy might have taken, but there may be other alternatives (for example, see Moynihan 1963a). First, lek behavior may have evolved through the breakdown of territoriality and adoption of colonial nesting behavior. This would bring the males into close proximity to facilitate the evolution of communal displays and the weakening of pair bonds. But one would also have to postulate that the male's role in caring for the young would decrease and that females would subsequently revert to territorial behavior. A second route could have involved extreme polygyny by territorial males, which then led to the defense of territory by the females within the male's territory. The diminishing role of the male in territorial defense could possibly have led to promiscuous mating and, in turn, to communal display. A third alternative for placing males in close proximity is the retention of male offspring in a family group of prolonged duration. A family courtship display might have the advantage of more strongly attracting females than individual displays do. To go from a family display to an enlarged communal display by incorporating unrelated males does not seem to be a difficult evolutionary step. The high density of males on courtship grounds could then have forced females to separate their reproductive activities from the males and to establish more widely dispersed territories. A fourth possibility might resemble the situation in shorebirds in which abundant food, in this case fruit, might have made the male's contribution to feeding young superfluous. Thus freed of parental duties, males could adopt a promiscuous, then lek type of behavior. These possibilities are only speculation, and the origin of lek behavior remains a major unsolved problem for behavioral ecologists.

Are the sexes always separate?

The occurrence of two sexes, usually referrable to male or female gender by the relative size of the gametes, is the general, but by no means universal, rule in nature. In fact, the occurrence of two separate sexes is the exception in many groups. Bisexual individuals (hermaphrodites) are the rule in plants, and in many of the so-called lower forms of animal life. Most bisexual organisms have mechanisms that prevent self-fertilization, so more than one individual is required for mating (see orchid pollination on page 227). A hermaphroditic organism's male and female gametes may mature at different times, or the sexual organs may be located in such a way that self-fertilization is impossible. In some molluscs and fish, the male and female roles of an individual are separated by age; such organisms are often *protandrous,* that is, male early in life, female later (for example, the slipper shell *Crepidula*: Coe 1944, 1953, Gould 1952; other molluscs: Fretter and Graham 1962, Hyman 1967). The protandrous system occurs where the number of eggs produced by the female depends on her size, but fertilization is internal and males need not produce large quantities of sperm to do the job. Male reproduction requires little energy and does not slow growth rate, so the individual may reach large size, at which he (soon to be she) can produce large batches of eggs. For species in which males determine mating privilege by aggression and territorial defense, as in the case of many fish, large size is at a premium for males and the sex sequence is usually *protogynous,* that is, female then male (Warner 1974, Warner *et al.* 1975).

In many single-celled protozoa, sexes do not exist as we know them, but a population can have several different "mating types" (Baker and Parker 1973). Exchange of genetic material always occurs between individuals of different mating type, never within the same type. Many plants that have normal bisexual individuals also have mating types superimposed upon the primary sexual pattern to prevent self-fertilization.

Sexual dimorphism reflects the relative roles of the male and female in reproductive activities.

Males and females are distinguished primarily by the relative sizes of their gametes (the sperm and egg, respectively) and by mechanical aspects of fertilization. Superimposed upon the basic male-female dimorphism are many secondary sexual characteristics that are indirectly related to the physiology of reproduction, and which reflect differences in the roles of the male and female in reproduction.

I mentioned earlier that size dimorphism between the male and female may separate the sexes ecologically and reduce intraspecific competition (page 245). In this particular context, dimorphism is not necessarily related to sexual roles, and either the male or the female may be the larger sex, all else being equal. In other situations, the role of the male in reproduction determines its relative size. The male whose only role is to provide sperm for fertilization is often small. The relative energetic cost of producing sperm is

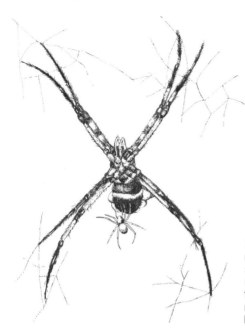

Figure 18-8   Extreme sexual dimorphism in size in the garden spider, *Argiope argentata*. The male is much smaller than the female, which is portrayed in her normal resting position at the hub of her web.

less than that of producing the large egg that provides nutrition for the developing embryo. (When males contribute more than their sperm to the reproductive effort—for example, by feeding young or defending territory—their size closely approaches or exceeds that of the female.) Gross sexual dimorphism is often found in spiders (Figure 18-8) and in some fish. Dimorphism in the deep-sea angler fish is extreme to the point that the male is attached to, and fully parasitic upon, the female (Pietsch 1975). In species in which males are small compared with the female, fertilization is always internal and relatively few sperm are needed. In species such as sea urchins, which shed their gametes into the water, fertilization occurs at random and there is a great evolutionary advantage for males to produce as many sperm as possible. Males of such species are usually the same size as females.

Many birds exhibit sexual dimorphism in plumage coloration, in which case the males are almost always the more conspicuous sex. The tendency toward dimorphism is greatest among species that live in forest habitats and in migratory species that have little time for courtship and pair formation after they reach their breeding grounds (Hamilton and Barth 1962). Male plumage is also frequently conspicuous in polygynous and promiscuous species (Verner and Willson 1969).

The pattern of sexual dimorphism in which the male assumes the brighter coloration cannot be attributed to any particular characteristic of maleness, but must be tied to the role of the male in reproduction. This point is demonstrated by a group of water birds, the phalaropes, in which the roles of the sexes are nearly reversed; the female defends the territory and the male incubates the eggs. In the phalaropes, females are the more brightly colored sex (Hohn 1969).

We may ask whether the bright coloration of males has evolved through competition among the males for territory or through competition to attract mates. In other words, has bright coloring evolved because of selection upon the male by other males, or by females? Experiments have shown that bright coloration plays a major role in aggressive interaction among males. In the stickleback, for example, males attack other fish, or even crude models, with male coloration, but ignore others lacking the coloration (Tinbergen 1951). David Lack (1943a) experimented with the role of the English robin's bright red breast in aggressive interactions. Male robins attack other males but ignore females and juveniles. Even a tuft of red cotton may be sufficient to stimulate the male's ire.

However important the role of bright coloration in male interactions, the role of sexual selection is often evident. For example, in the African swallow-tailed butterfly *Papilio dardanus,* the females mimic distasteful species in some areas, but the males always retain the original coloration of the species, presumably because female conservatism in mate choice selects against any change in coloration. The force of sexual selection is also expressed by the curious displays of bowerbirds, in which the male does not bear bright plumage; instead, he constructs a bower, a structure of grass decorated with bright-colored stones and other objects (Gilliard 1956, 1963). Since bowers are stationary, they cannot possibly play a role in territorial defense or male-dominance interactions, and must have evolved strictly by sexual selection.

The diverse ecological and social roles of the sexes illustrate how interactions between individuals within a species apply selective pressure that molds the adaptations of each sex differently and in accordance with other environmental constraints. Relationships between the sexes are mostly cooperative, as we have seen, so long as both contribute to the number and well-being of offspring. Where parental care is minimal, even if only for the male, sex goes little beyond the act of mating, and promiscuous breeding systems become the rule. Where the pair cooperates to raise the young, the pair bond is necessarily stronger. But nowhere can the bond between individuals be closer than between parent and offspring, which are mutually dependent upon each other for continuing the genetic line. As we shall see in the next chapter, adaptations of parent and offspring are complementary, at least until the young are sufficiently independent that they begin to compete with their parents. We shall also see that the development of the individual is dictated by both the physical and biological environments, and by those environments as they are modified by parental care.

19

Life History Patterns

Every animal and plant has a life story, beginning with its conception and ending in death. In between, organisms grow, often passing through several distinct developmental stages, and reproduce. The principle of natural selection tells us that the life pattern of each species is adapted, within the constraints of the environment and its own structure and function, to provide the greatest productivity of offspring. How this is done varies widely. In this chapter we shall explore the tremendous variety of adaptations of growth and development, reproductive life span, parental care, and family life.

Parental care and fecundity are inversely related.

The function of parental care is to increase the survival of the offspring to reproductive age. Parental care may be expressed in a variety of ways, from provisioning the egg with a yolk supply to protecting and feeding the young for extended periods. With more intense parental care, fewer offspring can be cared for. We can see this most easily where the adults must allocate a limited amount of food among their brood. If more food is given to each young, fewer young can be supported. The principle of allocation applies generally to the time and effort parents spend caring for their offspring.

Species that provide a high degree of protection to the developing offspring have fewer offspring. In a survey of marine-bottom invertebrates (including sea urchins, starfish, worms, molluscs, and crabs), Thorsen (1950) found that species that are viviparous (live-bearing) or otherwise provide a high degree of brood protection have between twenty and a hundred offspring; species with large eggs or primitive brood protection have clutches of 100 to 1,000 eggs; species that provide no protection generally lay at least 1,000 eggs, and as many as 500 million eggs, as in the sea hare *Aplysia*. Variation in the number of eggs among species with poorly developed brood protection partly reflects the size of the egg—that is, the amount of yolk provided for the embryo. Large, well-provisioned eggs mean fewer young.

Similar patterns occur among frogs, depending on the environment of the egg and developing larvae (Salthe 1969, Salthe and Duellman 1973). In general, the size of the egg mass laid varies in direct proportion to the size of the frog, and number of eggs per mass is inversely related to egg size. The largest eggs are laid by terrestrial species that carry their tadpoles on their backs or have direct development and bypass the tadpole stage altogether (actually, the metamorphosis from tadpole to frog occurs in the egg before hatching). The smallest eggs belong to species whose larvae develop as tadpoles in ponds, as do most species in temperate regions.

Brood protection can take many forms. Various fish lay their eggs at random (cod), in large masses (perch), or even inside another animal (the bitterling lays its eggs inside a mussel) (Balon 1975). Adults of some species provide additional egg protection by constructing and defending nests and fanning the eggs to increase the flow of dissolved oxygen through the egg mass. In a few groups of fishes, the parent carries its eggs externally. But a more important step toward increased parental care is the retention of the eggs within the female's body, where the embryos develop, followed by their live birth (vivipary) as in the surf perch, mosquitofish, and the sea horse (Laglar, Bardach, and Miller 1962). In one species of cichlid fish, *Symphysodon,* the small fry feed on a mucus that the adult secretes through its skin (Hildeman 1959). A species of frog retains its eggs in the oviduct and gives birth to swimming tadpoles. Another broods its eggs and developing young in its stomach (Corben *et al.* 1974).

Among birds, too, fecundity and parental care are inversely related. Temperate zone species that feed their young through most of the growth period lay two to ten eggs, with an average of about four. Species in which the young are self-feeding, usually lay between eight and twenty eggs, with an average of about twelve. Although fecundity is linked to the adults' role of providing food for the young, such other functions as brood protection and providing parental care to older offspring can influence reproductive rate. For example, lengthening the period of care for one brood could delay a second brood. Although extended parental care may reduce the number of eggs laid, it may actually increase the number of young that survive to reproduce.

The optimum investment of parental care depends partly upon how much the adults can improve the survival of their young. As the young grow, they can take better care of themselves and the value of parental care is diminished. The degree to which adults should provide parental care also depends partly on how easily the young can be replaced. If the average survival from the egg until the end of the first month of life is 25 per cent, a one-month-old fledgling is worth at least four newly laid eggs, not considering the time invested in upbringing. The intensity of parental care increases directly with time during the nesting cycle and with mortality rate of the brood as the young become progressively more difficult to replace, hence more valuable to their parents (Ricklefs 1969c). This is nicely demonstrated by the reaction of adult black-billed magpies, which are close relatives of crows and jays, to the presence of a human intruder at the nest (Figure 19-1). Adult

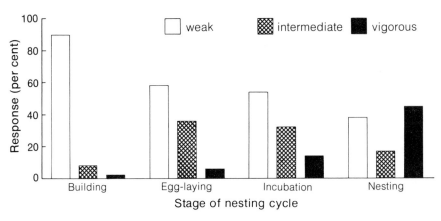

Figure 19-1   Responses of black-billed magpies to human intruders at the nest at different stages of the nesting cycle (after Erpino 1968).

responses were scored on three levels: weak, in which the adults left the nest area until after the intruder had disappeared; intermediate, in which the adults generally remained in the vicinity of the nest; and vigorous, in which the adults usually tried to mob the intruder. As the nesting cycle progressed from nest building to egg laying, incubation, and feeding the nestlings, weak responses were less frequent and vigorous calling and mobbing became more prominent.

Parental care removes the burden of self-maintenance from the young and permits rapid development.

If young are fed, brooded, and protected by their parents, they have no need to develop such capabilities as temperature regulation, food gathering, and predator escape as rapidly as if they were on their own. The young can thus use more of their energy and nutrients for growth, and development occurs more rapidly (Ricklefs 1969a, 1973b). The relationship between growth rate and the attainment of mature functioning is apparent when we compare the development of *altricial* and *precocial* birds. Altricial young of such species as the robin and starling hatch after 12 to 15 days of incubation without feathers, with their eyes closed, and completely helpless (Figure 19-2). For the first three weeks of life they are fed and brooded in the nest by the parents. The Japanese quail is typical of precocial species. Quail eggs hatch after 17 to 23 days of incubation, and the young are open-eyed, covered with down, and able to run about gathering their own food. Parental care after hatching extends only so far as to provide protection for the brood, show food items to the young, and warm them during cold weather. Compared with the altricial nestling, the precocial chick hatches with a well-developed complement of mature functions and parental care is all the less

Figure 19-2   Newly hatched young of an altricial species, the starling (*Sturnus vulgaris,* left), and a precocial species, the lapwing (*Vanellus vanellus,* right) (after Schiess 1963 and Portmann 1967).

important. Because the starling can allocate more of its energy and tissues to embryonic functions, the young grow more rapidly than the quail. In fact, the newly hatched altricial young is little more than a food-processing and growing machine. Twice as much of its body weight is allocated to digestive organs—the stomach, intestines, and liver—as the quail's. Little tissue is allocated either to muscle for movement or to feathers for insulation. Through the course of development, the altricial young undergoes a rapid transition from a primarily embryonic organism to a self-sufficient adult (Figure 19-3). This acquisition of mature function is reflected in a steadily slowing relative growth rate of the young during the development period (Ricklefs 1968).

The altricial and precocial development types differ primarily in the timing of certain maturation processes. In the precocial species, the transition occurs mostly in the egg. Prior to hatching, the embryo passes through all the developmental stages of the altricial chick. In fact, the precocial chick at hatching is often equated developmentally with the altricial young on the verge of fledging.

Why have some species adopted the altricial mode of development, and others the precocial mode? The distinction appears to be related to the type of food eaten. Precocial species are mostly ground feeders or aquatic birds that eat immobile or slowly moving prey: plant material, insect larva, and the like. All these prey can be caught by small, inexperienced, and physically somewhat inept chicks. In contrast, most altricial species feed in parts of the habitat (aerial or foliage) or on highly mobile prey, or both, requiring flight and considerable experience and agility to capture the prey. Such prey would

Figure 19-3   Nestling starlings at the age of about two days (left) and seventeen days (right). Note the ear openings, closed eyes, and wide flange on the mouth of the younger bird. The flange and some of the down are still visible two weeks later, but otherwise the transformation to adult characteristics is nearing completion.

clearly be beyond the reach of any newly hatched bird, no matter how precocial, and hence must be caught by the parents and fed to the young. This arrangement established, the rapid growth made possible by altricial development provides the greatest energetic efficiency for raising young.

Parent-offspring conflict can arise over the timing of independence.

Although parent and offspring are engaged in a generally cooperative venture, disagreement can arise over when to cut the ties (Trivers 1974). After a point, when the young are fairly self-sufficient, the parent might better spend its effort to start a new brood of young, but the young may be better off waiting a bit longer before setting off on their own. Hence the conflict of interest. This conflict often erupts in open aggression toward the young and increased postures of submission, appeasement, and begging toward the parent as the time of weaning approaches. The parent-offspring conflict is usually resolved in the parent's favor, if only because the parent is larger and more experienced than its progeny. Trivers suggests, however, that offspring may apply psychological ruses to trick the parent into providing more than he or she should give optimally. By crying, withholding gestures that communicate satisfaction, and reversion to infantile behavior, the young may be able to use normal communication signals between offspring and parent to solicit more than its due. Biologists are just beginning to focus their attention on such problems of behavior and social interaction that have traditionally been left to psychologists and anthropologists (Wilson 1975). Although fraught with the difficulty of anthropomorphism, we can look forward to much exciting discovery and reinterpretation in the future.

The relationship of the organism to its environment changes with development.

Most organisms are sent into life with a minimum of parental investment, usually no more than the nutrients packed into the egg or seed. For these, the primary concern is nutrition and survival to reproductive age. The transition from fertilized egg (*zygote*) to adult exposes the individual to a variety of environments, each of which selects different adaptations of form and function.

With increased body size, the probability of death generally decreases because the individual outgrows many of its smaller predators. Furthermore, the probability of death from exposure decreases because the reduced ratio of surface to weight enhances homeostasis and buffers the individual against environmental change.

Few large animals feed on plants when they are small because they need a rich source of protein in their diet to support rapid growth. Iguanas (large tropical lizards) feed on leafy vegetation as adults, but their young are insectivores. Similarly, most species of birds that eat seeds and other plant materials as adults feed insects to their young (Morton 1973). Mammals that graze and browse on plant material nurse their young with protein-rich milk. The fastest-growing mammals need the most protein relative to their total energy requirements, and the protein content of their milk is correspondingly higher. Rabbits grow about ten times faster than horses, pound for pound; their milk contains 10.4 per cent protein, whereas that of horses contains only 2.3 per cent protein (Cambell 1966).

One of the great virtues of parents feeding their young is that the young can be nourished by the same food supply as the adult. With nursing and other forms of parental feeding, growing organisms need not maintain adaptations for foraging and food processing that they may not need as adults. Furthermore, the inexperienced, incompletely matured, and perhaps even awkward young of large animals are not forced to compete for food with experienced adults of smaller species. Pond (1977) has suggested that primitive mammals' parental nourishment of the young, once perfected, may have been a crucial factor in their being able to replace the dinosaurs that had reigned on Earth for the previous 100 million years.

Metamorphosis separates life-history functions in different developmental stages.

In many organisms, such functions as food gathering, reproduction, and dispersal are separated in time during the life history by one or more complete changes in the body form (*metamorphosis,* see Figure 19-4). The larval forms of butterflies and moths (caterpillars) accomplish the entire growth

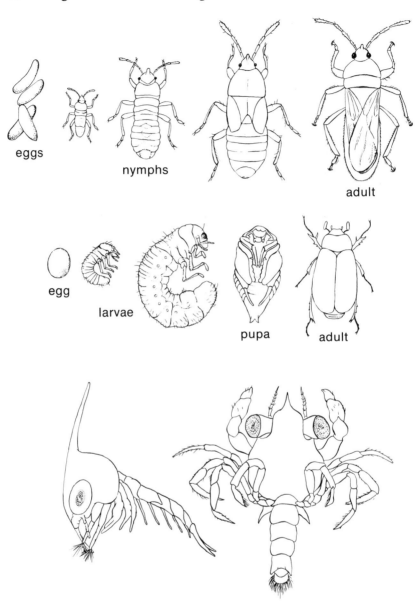

Figure 19-4   Examples of metamorphosis in several invertebrates. Top row: gradual metamorphosis in the chinch bug (Hemiptera). Center: complete metamorphosis in the beetle (Coleoptera). Below left: larval (zoea) stage of the crab, *Atele cyclus;* right: postlarval (megalops) stage of the blue crab, *Callinectes* (after Ross in Barnes 1968).

phase of the life history. The winged adult which emerges after the pupal stage (cocoon or chrysalis) is responsible for reproduction and dispersal. In many moths, the long, tubular mouth parts, used by other species to obtain nectar from flowers, are not functional. The adults do not feed; they live only long enough to mate and lay their eggs. The roles of the larvae and adults are the reverse of butterflies and moths in some groups. For example, the pelagic larval form of barnacles is a dispersal stage. Most of the feeding, growth, and reproduction is accomplished by adults that are firmly attached to rocks.

Metamorphosis accomplishes the complete separation and specialization of function in different stages of the life history, which presumably enhances the efficiency in performing each function. Because metamorphosis involves a complete reorganization of the life strategy, it must be relatively difficult to evolve. The presence or absence of metamorphosis in insects is characteristic of whole taxonomic orders, suggesting that metamorphosis arose early in the diversification of some insect groups and is not readily available as an evolutionary strategy. A complete metamorphosis enables flies, beetles, butterflies, and others to utilize the environment in an entirely different manner from grasshoppers and true bugs, which undergo only minor changes in structure and function during their life history (Figure 19-4, top).

Metamorphosis characterizes species belonging to a variety of ecological types. We have already seen that many parasites have complex life-cycles, including several complete changes in form to get them from one host to another (Chapter 15).

Most benthic species in rivers, lakes, and oceans have pelagic larval stages that may last anywhere from a few hours to several months. The purpose of the larval stage is not well understood and may vary from species to species. One function of the pelagic larva is to disperse from its place of hatching, but it is not clear that dispersal *per se* is adaptive. On one hand, dispersal may enable individuals to escape local areas of high population density, but on the other, it may carry them to unknown regions from an area where the species has proved successful. Another important function of larval life for many species is feeding. The shores and bottoms of the oceans are crowded with organisms of all types, hardly leaving room for their myriad offspring, which themselves could become food for someone else. Cast into the open waters, where planktonic algae abound, pelagic larvae of worms, molluscs, and arthropods escape a teeming world of predators and competitors and can feed on a bountiful food supply until they are large enough to settle to the bottom, undergo metamorphosis, and take up their adult life style.

In the amphibians, many of which undergo a substantial metamorphosis, this change represents the transition from aquatic to terrestrial life. Metamorphosis involves change in locomotory appendages and the breathing apparatus primarily, although some species additionally overhaul their feeding and digestive systems to accommodate change from an algae and detritus diet to one rich in insects and other animal prey.

Metamorphosis may be an unalterable event in the life cycle of the species, but the length of the larval and adult stages may be adjusted within wide limits (Istock 1967). In many insects, the adult life stage is reduced to a few hours or days, its sole purpose being to mate and lay eggs (May flies, many moths and butterflies). At the other extreme, tsetse flies and some cave beetles produce larvae that immediately undergo metamorphosis without having fed at all (Deleurance-Glaçon 1963). It is not altogether clear why in some species one stage or another of the life cycle is greatly reduced, although detailed studies of their ecology would provide tentative answers at least. The amphibians exhibit a particularly wide variety of life-cycle patterns, varying from some salamanders that never undergo metamorphosis but reach sexual maturity in their larval form to other salamanders and a few frogs with direct development, metamorphosis being accomplished during the embryonic period within the egg. Wilbur and Collins (1973) have suggested that aquatic amphibian larvae are subjected to opposing pressures to metamorphose and enter the terrestrial realm. On one hand, competition from other larvae as they grow and deplete the limited resources of a pond, or the uncertainty of ephemeral ponds as habitats as the summer wears on and the ponds dry up, favors early metamorphosis (Low 1976). On the other hand, persistent aquatic habitats with uninhabitable surroundings (deserts, alpine tundra and bare rock, caves) favor complete suppression of metamorphosis. In intermediate situations, the timing of metamorphosis may be quite flexible, allowing the individual to respond to the conditions prevalent during the particular year of its larval life.

Organisms with indeterminate development must seek an optimum balance between growth and reproduction.

Many invertebrates, fish, reptiles, amphibians, and plants do not have a characteristic adult size. They grow, at a continually decreasing rate, throughout most of their adult lives. This pattern of development is often called indeterminate growth. Fecundity is directly related to body size in most species with indeterminate growth (see Figure 19-5). Egg production and

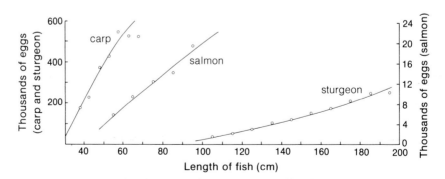

Figure 19-5   Relationship between annual fecundity and size in carp, salmon, and sturgeon. The left-hand ordinate pertains to the carp and the sturgeon; the right-hand ordinate pertains to the salmon (after Nikolsky 1963).

TABLE 19-1

Numerical comparisons of the strategies of slow growth–high fecundity (A) and rapid growth–low fecundity (B) in a hypothetical fish.

			Year			
	1	2	3	4	5	6
Strategy A						
Body weight*	10	12	14.4	17.3	20.8	25.0
Growth increment†	2	2.4	2.9	3.5	4.2	5.0
Weight of eggs	8	9.6	11.5	13.8	16.6	20.0
Cumulative weight of eggs*	8	17.6	29.1	42.9	59.5	79.5
Strategy B						
Body weight	10	15	22.5	33.8	50.7	76.1
Growth increment	5	7.5	11.3	16.9	25.4	38.1
Weight of eggs	5	7.5	11.3	16.9	25.4	38.1
Cumulative weight of eggs	5	12.5	23.8	40.7	66.1	104.2

* Body weight + growth increment = next year's body weight; cumulative weight of eggs to last year + weight of eggs = cumulative weight of eggs to this year.
† Growth increment and weight of eggs in each year are equal to the body weight.

growth draw upon the same resources of energy and nutrients available to the organism. Hence, increased egg production during any one year reduces growth and must be weighed against reduced expectation of fecundity in future years (Williams 1966a). In general, if an organism has a long life expectancy, growth should be favored over fecundity during each year. This strategy will increase total fecundity over the organism's life span. But if an organism has little chance of living to reproduce in future years, resources allocated to growth instead of to eggs are largely wasted and we expect fecundity to be all the higher.

Consider two fish, each weighing ten grams at sexual maturity, but having different growth strategies. Both gather enough food each year to reproduce their weight in new tissue or eggs. Species A allocates two-tenths of its production to growth and eight-tenths to eggs, whereas species B allocates one-half each to growth and eggs. Calculated growth, fecundity, and accumulated fecundity (Table 19-1) show that if the fish live four or fewer years, on the average, the strategy of high fecundity and slow growth gives the greater overall productivity, whereas if the fish live longer than four years, the strategy of rapid growth and low fecundity is superior. The adult mortality of the fish, therefore, will determine the optimum allocation of its resources between growth and reproduction.

The allocation of resources between growth and reproduction has been measured in few species, partly owing to technical difficulties and partly because interest in this aspect of adaptation is so recent, dating from the mid-1960s at the earliest. Annual plants, including most of our crop plants, have been the subject of several recent studies (Harper 1967). The proportional distribution of dry weight among the various tissues of the composite plant groundsel (*Senecio vulgaris*) during the growing season (Figure 19-6) shows a progressive shift in allocation from roots to leaves, stems (flowering

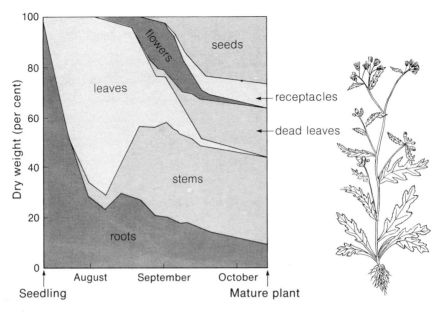

Figure 19-6   The proportional distribution of the dry weight among different plant parts by groundsel, *Senecio vulgaris* (Compositae), during its life cycle (based on J. Ogden in Harper 1967).

stalks) and flowers, and finally seeds. By concentrating production in roots initially, the seedling can become firmly established and obtain soil water and nutrients. Leaves then provide the photosynthetic area needed for rapid production. Reproductive organs do not develop until the end of the growing season so as not to reduce the earlier growth potential of the plant. Subsequent studies by Gadgil and Solbrig (1972) on dandelions and by Abrahamson and Gadgil (1973) on goldenrods have shown that the allocation of production favors growth in well-shaded sites with intense competition from other species and favors reproduction under disturbed conditions where few plants have become established.

Reproduction may begin very soon after germination under favorable but ephemeral conditions, such as follow rainstorms in the desert, or where herbivory and predation are so intense as to wipe out any benefits that might accrue from a longer growth phase. The timing of reproduction in plants may also be determined to some extent by the availability of pollinators. Beyond the timing of reproduction and the allocation of resources between reproduction and growth, plants face an equally significant choice between annual and perennial life cycles. This choice is partly made for the plant by the persistence of its habitat. Old fields undergoing succession back to forests change so rapidly that initial colonists probably could not survive in the habitat a year later. In deserts, sufficient rainfall for growth and reproduction of herbaceous plants may come at intervals of several years, separated by periods of drought. Under such conditions, most herbaceous plants are annuals and wait out the poor years as seeds. But the choice between annual and perennial life cycles also involves allocation of resources. To survive the winter or dry period of inactivity, plants must store large quantities of nutrients in roots, bulbs, tubers, or whatever, to give them a good start the following spring. This allocation of resources to storage organs necessarily reduces seed output and is evolutionarily justified only if the plant's prospects for survival and reproduction in the future are good.

In long-lived plants and animals, the onset of reproduction may be delayed for several years. In plants, fish, and reptiles, delayed maturity allows the individual to attain larger size, hence greater fecundity, at sexual maturity. Resource allocation cannot, however, fully explain delayed maturity. In species with internal fertilization, males produce relatively few sperm and reproduction does not greatly reduce their growth potential. Yet the onset of sexual reproduction in males is often delayed beyond that of the females. In birds and mammals, the independence of growth and reproduction is all the more evident because most species grow rapidly and attain adult size long before they become sexually mature. Sea gulls do not breed until their third year. Other seabirds, large eagles and condors, and large mammals do not breed until six to twelve years of age, and man usually not until fifteen or more years of age.

The development of sexual function is not delayed by physiological limitations; mature reproductive appearance and behavior can be induced by administering sex hormones early in life (Boss 1943). In these species, reproductive development is delayed by constraints of intraspecific competition until individuals acquire the status and experience necessary to win mates and successfully raise young (for example, Lack 1966). Furthermore, species with delayed maturation often use specialized methods of feeding that may require considerable practice to perfect (Ashmole and Tovar 1968, Orians 1969a).

If rank in a dominance hierarchy or success in establishing territory is largely, or even partly a function of age, it may be advantageous to delay reproductive maturity until the young have gained enough experience to compete with older adults. This is particularly true if adults are long-lived and few places in the breeding population fall vacant each year. Delayed maturation of sexual function generally includes delayed development of secondary sexual characteristics, which give an external appearance that communicates reproductive maturity to others in the population. In gulls, an immature plumage distinct from the bold black and white pattern of the adult is retained for two years (Figure 19-7). Why should an immature individual suffer in aggressive encounters when no reproduction is at stake and retention of an immature appearance can appease adults?

Extensive preparation for breeding and uncertain or ephemeral environmental conditions may favor a single, all-consuming reproductive effort.

Some species of salmon have adopted a course of rapid growth for several years, culminating in a single immense reproductive effort, in which a

Figure 19-7   Immature and adult herring gulls, illustrating the differences in their plumage. Note the hunched begging posture of the young birds (courtesy of Susan White).

large portion of the body is converted to eggs, followed shortly after spawning by death. Gadgil and Bossert (1970) reason that the effort salmon expend migrating upriver just to reach their spawning grounds is so great that the best strategy is to make the trip just once during their lifetime and to produce as many eggs as possible, even if the supreme reproductive effort requires the conversion of muscle and digestive tissue to eggs and guarantees death from starvation soon after reproduction.

The salmon life-history pattern, sometimes referred to as big-bang reproduction but more properly as *semelparity,* is rarely encountered among animals and plants that live for more than one or two years. Usually, the effort and allocation of resources required to survive between growing seasons are so much greater than those used in preparation for breeding that, once a perennial life form has been adopted, reproduction every year seems the most productive pattern.

The best known cases of big-bang reproduction in plants occur in the agaves (century plants) and the bamboos (Janzen 1976), two of the most distinctly different groups one could imagine. Most bamboos are tropical or warm-temperate zone plants that form dense stands in disturbed habitats. Reproduction does not appear to require substantial preparation, as in the growth of a heavy flowering stalk. But opportunities for successful seed germination are probably rare. Once established, a bamboo plant increases by asexual reproduction, continually sending up new stalks, until the habitat in which it germinated is fairly packed with bamboo. Only at this point, when vegetative growth is becoming severely limited, is there any benefit to be gained from seed production. In a sense, bamboos are like the annual plants in early field succession (they are members of the grass family), but because they grow rapidly and attain large size they can stave off the intrusion of other forest species for years. Furthermore, the tropical environment does not place heavy demands on plants for living more than one year.

The environment of the agave is at the opposite end of the spectrum from that of the bamboo. Most species of agave inhabit arid climates with sparse and erratic rainfall. Plants grow vegetatively for several years, the number varying from species to species, then send up a gigantic flowering stalk. After the seeds have been produced, the agave dies (Figure 19-8). One curious

Figure 19-8   Stages in the life cycle of the Kaibab agave (*Agave kaibabensis*) in the Grand Canyon of Arizona. The plant grows as a rosette of thick, fleshy leaves (above) for up to fifteen years. Then it rapidly sends up its flowering stalk and sets fruit (right), after which the entire plant dies (above, left-hand rosette).

fact about agaves is that they frequently live side by side with yuccas, a group of plants with a growth form similar to that of the agave (see Figure 15-12) but which flower year after year. Yuccas and agaves differ primarily in their root systems. Yuccas are deep-rooted and can tap persistent sources of ground water. Agaves have shallow, fibrous root systems that catch much of the water percolating through the surface layers of the soil after a shower, but which are left high and dry during drought periods. The erratic water supply of the agave makes it impossible for it to flower and produce seed every year, and the period between suitable years may be very long. Under these conditions, the best flowering strategy for the agave may be to grow and store nutrients until an extremely wet year comes—perhaps one in ten or even one in a hundred—and then to put all resources into reproduction.

Our interpretation of big-bang reproduction and other patterns of life-history phenomena are largely intuitive speculations that seem reasonable within the context of the principles of natural selection and adaptation. In the section of this book concerned with the genetics of populations, we shall explore some of these phenomena more rigorously with the aid of mathematical models.

Interactions between siblings may influence the reproductive potential of their parents.

Reproduction brings about some curious conflicts of evolutionary interest, such as that between parent and offspring over the timing of independence. Whereas all the offspring in a brood are part of the parent's reproductive strategy, the offspring are themselves individuals and potentially compete with each other and with their parents. To the degree that the progeny of a single mating differ genetically, competition between them can have important evolutionary implications. And because competition within a brood can disrupt reproduction, parents often exercise some form of control over interactions among siblings. In one simple situation, adults can manipulate their offspring to reduce the impact of competition between them. Where the food supply is highly variable and unpredictable, birds are not able to nourish their young properly in some years. Many species have overcome this problem by adopting the strategy of asynchronous hatching and selective starvation of young (Lack 1954, Ricklefs 1965). By starting to incubate the eggs before the entire clutch has been laid, the adult gives the first-laid eggs a head start on development. Because the first-hatched chicks are larger on a given date than their younger sibs, they compete more effectively for food delivered by the parents, and the smallest of the brood starves when feeding conditions are poor. Selective starvation brings the energy requirements of the brood into line with the food-gathering rate of the adults and ensures that all of the young raised are adequately nourished. Inasmuch as hatching order has no genetic component, the development rate of the chick cannot be adapted to counter this form of parental manipulation.

Some tropical saturniid moths exhibit adaptations that are most readily interpreted as having evolved in response to selection from within the brood (Blest 1963). Some of the species are distasteful to bird and mammal predators and have warning displays: either a black-and-yellow-banded abdomen or foul odor (Figure 19-9). Other moths are palatable and have adopted cryptic coloration, usually by mimicking leaves. The adult moths do not feed, and they die shortly after laying their eggs. The adults of unpalatable aposematic species also do not feed, but survive for a longer period after reproduction than the cryptic species (Figure 19-10). Even after reproduction, distasteful moths may confer protection on their brothers and sisters by

Figure 19-9   Warning display of the unpalatable saturniid moth *Dirphia*. The curled abdomen is banded with black and yellow (after Blest 1960).

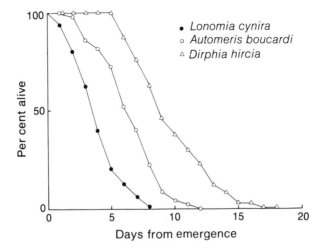

Figure 19-10   Survival, after emergence as adults, of three saturniid moths. *Lonomia,* a leaf mimic, is very palatable; *Automeris* is also a leaf mimic, but it has eyespots on the hind wings, and is somewhat palatable; *Dirphia,* which has a banded abdomen and discharges a foul odor, is rejected by predators (after Blest 1963).

serving as a predator's learning experience, thereby allowing a newly emerged moth to reproduce. Postreproductive survival parallels the degree of aposematism exhibited by the moths.

If large numbers of cryptic moths were to survive after reproduction, their abundance could key predators in on prereproductive individuals. Palatable species exhibit adaptations that facilitate the removal of postreproductive individuals from the population. Their scales are soft and rub off easily, whereupon their appearance changes. As a consequence, predators do not associate these worn postreproductives with newly emerged moths in full garb and do not form strong search images for prereproductive individuals. Furthermore, the moths become intensely active after egg laying. This activity uses up food reserves rapidly, leading to early starvation, and also helps to beat the soft scales off the wings.

If these behavior patterns were adaptations to increase the survival of brothers and sisters in the manner outlined by Blest, aposematic moths should lay their eggs in masses and siblings should remain close together; cryptic forms should disperse their eggs widely (Table 19-2). We might also predict that the development period of the larvae of cryptic moths will vary widely so adults will emerge over a long period. These expectations do not hold true for saturniid moths because we have failed to consider several important factors (Blest, personal communication). First, many species have very brief mating seasons that require a high degree of synchrony in emergence from the pupae. Any selection of variable emergence times to reduce predator pressure is countered by selection of synchronous emergence to facilitate mating (and perhaps to overwhelm predators by

TABLE 19-2

Contrasting strategies of predator avoidance evolved by palatable and unpalatable saturniid moths (adapted from Blest 1963).

	Strategy I	Strategy II
Taste*	Palatable	Unpalatable
Appearance	Cryptic	Aposematic
Length of adult life	Brief	Longer
Scales	Soft, to obliterate pattern	Durable, to preserve pattern
Level of post-reproductive activity	Increased	Decreased
Expected pattern of egg placement	Dispersed	Clumped
Expected development span and hatching	Variable and asynchronous	Uniform and synchronous
Expected dispersal of adults from hatching locality	High	Low

* Evolved in response to selection on the individual exerted by predators.

quickly satiating them. Second, larvae of all species of the particular group of saturniids that Blest studied are covered with poisonous spines (see Figure 15-7) and tend to remain together in small groups, as we might expect of distasteful forms. Little is known about the dispersal of adult saturniids in Panama, but Benson (1971) has presented evidence that among heliconiid butterflies of Trinidad the distasteful species tend to be sedentary and congregate in large roosts, whereas closely related edible species disperse widely.

One particularly interesting case of cooperation among progeny involves a form of specialization within broods of tent caterpillars (Wellington 1960, 1965). Caterpillars in each colony can assume two extreme behavior types (sluggish or active) or any number of intermediates, depending on the amount of nutrition provided in the egg. Thus the female parent determines the variety of behavior morphs among her brood by the amount of yolk she provides each egg—a blatant expression of parental manipulation. Active caterpillars forage farther from the protection of the colony's tent and gather more food than their more sedentary siblings. In fact, sluggish caterpillars rarely eat enough to pupate and become adults. Why are they present at all if they do not survive to lay eggs as adults? Because their behavior indirectly enhances the survival of their active brothers and sisters. The tent is spun by the caterpillars as they forage over the branches (Figure 19-11). The active forms move widely over the tree, spinning a large open web that does not

Figure 19-11   Tent caterpillars congregated on their tent in a black cherry tree. The tree was almost completely defoliated by the time the caterpillars were ready to pupate.

provide good protection from weather and parasitic insects. The sluggish forms spin a much more compact tent that provides protection not only for themselves but for the active caterpillars in the brood. Hence labor within the colony is divided among active caterpillars, which feed well and generally survive to reproduce, and sluggish caterpillars, which spin the protective portions of the tent but rarely survive to reproduce (Figure 19-12). The specialization of the brood into individuals that eventually breed and those

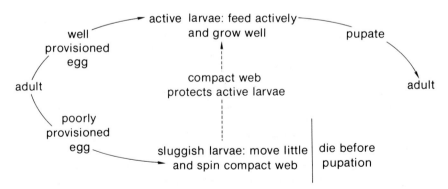

Figure 19-12   Relationships between active and sluggish larvae in broods of the tent caterpillar, *Malacasoma pluviale.*

that do not, and the control of this specialization by the level of nutrition the adult supplies to the egg—a form of parental manipulation—is suggestive of early stages in the evolution of insect societies. These societies are elaborately extended family groups whose basic organization is not unlike that of the tent caterpillar colony.

Offspring may cooperate with their parents to rear brothers and sisters.

In the case of the tent caterpillar some offspring are coerced into helping their sibs. But a more cooperative and voluntary effort may arise when an individual can further its own genotype more by helping to rear brothers and sisters than by attempting to raise its own offspring. Among birds, particularly in the tropics, one frequently finds broods of young being raised not only by their parents but also by brothers and sisters produced by previous nestings (Skutch 1961, Fry 1972). Eventually these helpers will leave the family group and attempt to establish their own territories. Why do the young stick around at all? Why don't they direct their effort toward raising their own young? These questions are not fully resolved, but the answers probably lie in one of two directions. First, the disappearance rate of adults from the population may be so low (as it often is in tropical birds) that young cannot hope to secure breeding territories in their first year. If breeding on their own is delayed, helping their parents may provide experience that will help them later or may be token payment for being allowed to remain in the territory until they are old enough to set off on their own. Second, genetic similarity of individuals is the same between siblings (50 per cent of their genes are identical by descent, on average) as it is between one parent and its offspring. Each progeny receives one-half its genotype from each parent. All other things being equal, an individual's fitness would be served equally by contributions to the upbringing of siblings and to its own progeny. This principle, applied to the case of helpers among birds, may greatly strengthen social bonds within the family group (Brown 1974, Ricklefs 1975).

When helping at the nest was first described, it was reported as a form of *altruism,* referring to a broad class of behavior in which an individual benefits another at his own risk or expense. Any helpful behavior provided to an individual other than a son or daughter was considered altruistic in an evolutionary sense because it could not further the donor's direct genetic line.

W. D. Hamilton (1964) revolutionized our thinking about social behavior and altruism by pointing out that an individual's genes are also carried by siblings, uncles, nieces, cousins, and so on, in direct proportion to their degree of relationship. The more distantly related, the fewer genes two individuals are likely to share. By aiding one's brother or sister, the altruist aids many of the same genes that he carries. If the loss in fitness that the altruist suffers is less than half the gain in fitness that he provides his brother, his genotype benefits. Altruistic behavior may ultimately promote an individual's genotype if the cost or risk assumed is less than one-quarter the benefit provided to nephews or one-eighth the benefit provided to cousins. In general, the ratio of cost to benefit must be less than the degree of genetic relationship. This relationship is the basis of Hamilton's principle of *inclusive fitness,* which includes mutual influence of related organisms on each other's fitness. Selection of behaviors toward relatives that increase inclusive fitness is called *kin selection.* Although kin selection undoubtedly exists, the importance of its role in the evolution of social behavior, particularly among the social insects, is poorly understood and much disputed, as we shall see below.

The insect society is an extended family.

The social insects, properly called *eusocial* (Wilson 1971), are set apart from all others by a complex social organization dominated by an egg-laying queen. Nonreproductive progeny of the queen gather food and care for developing brothers and sisters, some of which become reproductively mature, leave the colony to mate, and establish new colonies. All insect societies are huge families, usually the progeny of a single queen, providing a relatively uniform genetic environment within which cooperative and altruistic behavior has its highest expression.

Eusociality has evolved only in two orders of insects: Isoptera (termites) and Hymenoptera (ants, bees, and wasps). The termites are probably little more than somewhat modernized and socialized cockroaches, a group with which they share many characteristics, including symbiotic protozoa in their guts that help digest the wood and other sources of cellulose they eat. The termite colony consists of an egg-laying queen, a king that keeps the queen fertile, and several forms of sterile workers and soldiers. Development is direct (without metamorphosis), and the growing nymphs do their share of the work in the colony. The most primitive forms of termites, known from fossils in rock deposited 50 million years ago, form small colonies, feed mostly on wood, and are symbiotic with ciliated protozoa primarily. More advanced forms, appearing 20 million years ago, feed on a wider variety of plant material, often rely on bacteria to digest the cellulose they eat, and form large colonies that are housed in specially built nests.

Eusociality apparently arose many times in the Hymenoptera: at least twice in the aculeate wasps, eight times in bees, and once in the ants, all of which are social. In fact, bees and ants are relatively recent derivatives of the aculeate wasp line. Hymenopteran societies differ from termite societies in that colonies have no kings; a single mating is sufficient to fertilize a queen for life. The Hymenoptera feed primarily on pollen and nectar, or on animal diets, and lack the symbiotic relationship with a gut fauna and flora characteristic of the termites. Beyond these differences between termite and wasp-type sociality there are numerous additional distinctions of behavior, physiology, and genetics, some of which are mentioned below, and almost endless variations on the basic theme of insect society.

Insect societies: Happy brothers and sisters or despotic mothers?

Both the origins of insect societies and the evolutionary interpretation of their origins are unsettled problems. The most widely accepted sequence of evolutionary steps in the development of insect societies is that envisioned by Wheeler (1928). The steps must have included a lengthened period of parental care of the developing brood, either guarding the nest or continual provisioning of the larvae in a manner similar to birds feeding their young. If the parent lived and continued to produce eggs after its first progeny emerged as adults, it would be in a position to have her first progeny help to raise the subsequent brood. This overlapping of generations, which is not common among insects, and extensive parental care are necessary ingredients in the recipe for eusociality. Once progeny remain with their mother after they attain adulthood, the way is open to relinquishing their own reproductive function solely to support their mother's. Only Michener (1958, 1969) has proposed a serious alternative to Wheeler's scenario. He suggested that insect societies originated in the aggregation of unrelated adult individuals, most of which eventually became subordinated to a single reproductive female. Although some of the early stages in this sequence are known from bees, the jump from primitive aggregation to the eusocial insect colony is difficult to envision, and most biologists subscribe to Wheeler's extended-family hypothesis.

If the exact sequence of events leading to eusociality were known, one would still not understand *why* insect societies developed. Speculation about the evolution of brood care by effectively sterile individuals has created a major controversy. One side (Hamilton 1972, Eberhard 1975) believes that kin selection of cooperative, altruistic behavior underlies sociality. The inclusive fitness of an individual can be greater when siblings rather than offspring are raised. An opposite viewpoint, stated primarily by Alexander (1974), is that the queen manipulates the development and behavior of her progeny through nutrition and chemical signals (*pheromones*), retaining many in the colony to build its numbers and allowing a few to achieve sexual maturity.

The kin-selection hypothesis is made attractive by the peculiar sex-determining mechanism of the Hymenoptera. It is variously called haplodiploidy and arrhenotoky and works as follows. Fertilized eggs become females having one complement of genes from each parent, and unfertilized eggs become males having only one complement of genes, that of the mother. By controlling fertilization a female can control the sex of her offspring. By this curious arrangement, the degree of genetic relationship between sisters is .75, provided they have the same father, but between mother and daughter it is only .5. The principle of inclusive fitness suggests that under the haplodiploid system it is better to raise sisters than daughters. It may be no accident that workers in hymenopteran societies are female. The degree of genetic relationship of brother to sister is .25, hence females should definitely prefer raising sisters to raising brothers.

Alexander (1974) has voiced some serious objections to the haplodiploidy theory for social evolution. First, termites have a more usual sex-determining mechanism; both males and females develop from fertilized eggs, and siblings share one-half their genes regardless of sex. Second, workers in bee and ant colonies raise both male and female *reproductives* in about equal numbers when the sexual brood is produced. So although the haplodiploid system may result in some asymmetrical relationship between sexes within the colony, it is unlikely that this sex-determining mechanism confers sufficient strength to kin selection to account for eusociality. Much undoubtedly remains to be discovered about the social insects, including additional clues to the origin of their sociality.

For a detailed and readable account of social insects, you should read E. O. Wilson's book, *The Insect Societies* (1971). These animals are of sufficient general interest to ecologists and evolutionary biologists that we shall conclude this chapter on life-history strategies with a brief discussion of castes in social insects and a detailed description of one social insect colony, that of the army ant.

A caste system accompanies division of labor.

Bee societies are simply organized; females are divided among a sterile worker caste and a reproductive caste that is produced seasonally. Whether an individual will become a sterile worker or a fertile reproductive is controlled by the quality of nutrition given the developing larvae (Light 1942–43). The development of sexual forms can be inhibited by substances produced by the queen and fed to the larvae. In bees, the worker caste represents an arrested stage in the development of the reproductive female, stopped short of sexual maturity.

Ant and termite colonies often have a continuous gradation of worker castes, ranging from very small individuals that are primarily responsible for the nutrition of the colony to larger individuals that are specialized morphologically to defend the colony (Figure 19-13). The so-called soldiers are often equipped with formidable mandibles and stings (worker bees have stings and defend the colony as well as gather food). The soldiers of some species of termites can produce noxious gases (*Rhinotermes*) or direct a stream of noxious fluid at an intruder (*Nasutotermes*). Nonreproductive castes of termites include both males and females, in contrast to the all-female castes of the Hymenoptera. In many ants, a caste of individuals called repletes, or honeybarrel ants, are specialized to digest and store food for the rest of the colony. Workers feed the repletes, which process the food and store it in their extensible abdomens. The workers later recover the processed food from the repletes in the form of a liquid exudate that resembles honey.

Although most bee colonies have only one sterile worker caste, the worker's role in the colony varies with age. In honeybee colonies, the workers function primarily as nurses for the first two weeks of adult life, caring for developing larvae. Between the ages of 10 and 25 days, the workers are called house bees and perform other tasks within the hive itself, cleaning out wastes and ventilating the colony by beating their wings at entrance ways. Between 20 and about 55 days, the end of their normal lifespan, the workers are field bees and gather nectar and pollen to feed the developing brood. The ant

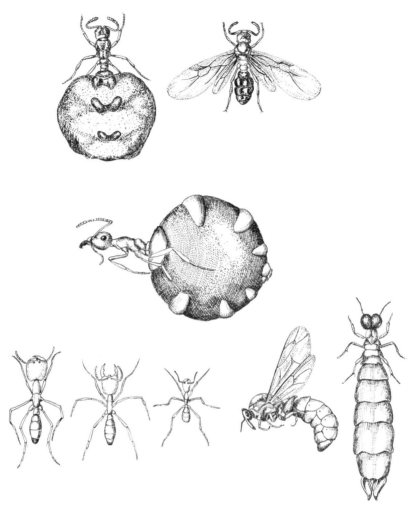

Figure 19-13   Castes of several species of ants. Top right, a virgin queen; top left, an old egg-laying queen of the workerless social parasite, *Anergates atralulus,* of Europe. Center, a replete of the honey ant, *Myrmecocystus melliger,* from Mexico. Bottom, from left: three sizes of blind workers; a winged male; and a queen, blind and wingless, of the visiting ant *Dorylus nigricans,* an African relative of the New World army ants (after Wheeler 1923, and Grassé 1951).

Myrmica exhibits similar behavioral changes with age (Weir 1958). During the first season of its life, the worker provides a nurse function, caring for the larvae. During the second season, the worker is primarily responsible for construction and repairs to the nest, and only in the third year does the worker forage extensively outside the nest.

The queen in most insect societies is highly specialized as an egg-laying machine. The queens of some termites may lay as many as 6,000 to 7,000 eggs per day for many years. Termite queens cannot store sperm, so sexually active males must accompany them continuously. On the other hand, most bee and wasp queens can store millions of sperm for many years and produce an entire colony from a single mating. Some queen ants have been known to be sexually active for as long as fifteen years, laying more than a million eggs.

The extreme specialization of the queen is a problem at colony-founding time because there are no progeny to take care of the queen's first brood. The female must initially provide food for her young. Honeybee and army ant queens have overcome this problem by taking part of the parental colony in which they were born to help found their own colonies and to help care for the first broods (Schnierla 1956).

In most other species of ants and in termites, new queens feed their first larvae with secretions from their body, and then wait until their brood is large enough to gather food before continuing to reproduce. In some of the more primitive ponerine ants of Australia, the queen herself, when founding a new colony, builds the first shelter and gathers food for her brood (Wheeler 1933). Colony growth in these ants is slow at first because the queen must first perform most of the tasks eventually provided by workers before she can develop into an efficient egg-laying machine.

Army ants alternate nomadic and sedentary periods.

The army ants are among the most feared of all animals, and an impressive array of folk literature, much of it myth, has developed around them. It is true that army ants will bite like mad if one stands in the middle of their swarm, but stories of people being stripped to the bones in a matter of minutes are overstated, and army ants sometimes help people by swarming through their houses and cleaning out roaches and garbage.

The army ants are interesting over and above their voracious habits because they are one of the few social insects that have abandoned permanent nests to range over wide areas. To accommodate their migratory habit, the army ants have adopted unique patterns of social and reproductive behavior.

The army ants belong to the subfamily Dorylinae, within which most of the species are grouped in the genus *Eciton*. In the New World tropics, where army ants have been studied intensively by the late T. C. Schnierla (1956) and his colleagues, two common species, *E. hamatum* and *E. burchelli*, occur. Rettenmeyer (1963) has summarized their behavior in detail. Army ants alternate between two phases of activity: a sedentary or "statary" phase, during which raiding occurs from a single locality for several days, and a nomadic phase, during which the entire colony moves each night (Table 19-3).

The statary phase lasts 17 to 22 days, during which eggs are laid and larvae that developed from the eggs laid during the previous statary phase pupate. Egg-laying rates of 180 per hour, or about 48,000 eggs in ten days, have been recorded for queens of *Eciton hamatum*. The hatching of the eggs and the emergence of workers from the cocoons mark the end of the statary phase and the beginning of the 16- to 18-day nomadic phase. During this

TABLE 19-3

Summary of the activity cycle of army ants of the genus *Eciton* (from Rettenmeyer 1963).

Stage	Duration	Current brood	Previous brood
Statary	17–22 days	Eggs are laid	Cocoons are spun and the larvae pupate
Nomadic	16–18 days; 8–13 days for sexual broods	Larvae develop	Ants are first active as adults

Arrows follow the life cycle of the army ant.

period, the larvae grow, and the bivouac (temporary nest) of the army ants is moved each night to a new locality. When a sexual brood is raised, the nomadic phase lasts only eight to thirteen days because sexual larvae develop much faster than larvae that are destined to be workers.

The army ant colony has only one queen, and the workers are divided into several castes, including major workers (soldiers) and a wide range of various-sized, but smaller minors. A colony produces one sexual brood each year, and the winged sexual males swarm in great numbers through tropical forests in search of colonies with unmated females. New queens take a part of their parent colony with them to found new colonies.

Colony size in army ants varies considerably, but a typical colony of *Eciton hamatum* contains from 100,000 to 500,000 adult workers, an egg brood of 50,000 to 200,000, and as many as 10,000 to 60,000 full-grown larvae. Colonies of *E. burchelli* are usually larger, containing perhaps 300,000 to 1,500,000 adult workers.

The sexual brood of *E. hamatum* consists of about 15,000 eggs, which are usually somewhat larger than eggs destined to be workers. Of the 15,000 or so eggs that are laid, about 1,000 to 2,000 males, and perhaps only six females, or new queens, emerge. The females emerge first and are immediately receptive to mating. Males cannot mate for several days after emergence and in some species they have to perform a dispersal flight before they are physiologically able to mate. Thus the males virtually never mate with females from their own colony. After one or more of the females in a colony has been inseminated, the colony divides and the new queen or queens each take a part with them.

Eciton hamatum and *E. burchelli* can coexist in forested areas of Panama largely because their foraging behavior differs. *E. hamatum* forms narrow columns during raids and feeds mostly on the larvae of wasps and ants captured in raided nests (Figure 19-14). *E. burchelli* deploys a swarm raid, which may be as wide as twenty meters at the front. Large numbers of insects, millipedes, and spiders flushed from the forest floor are captured by the ants—and also by the numerous parasitic flies and ant-following birds that accompany the raiding swarms (Willis 1967).

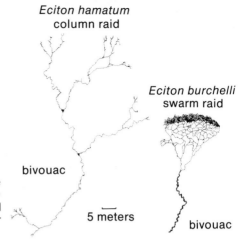

Eciton hamatum
column raid

Eciton burchelli
swarm raid

bivouac

Figure 19-14   The shapes of the
column raid of the army ant *Eciton
hamatum* (left) and the swarm raid
of *E. burchelli* (after Rettenmeyer
1963).

5 meters

bivouac

Army ants raid close to the ground, although *E. burchelli* may foray into high trees as well. Raids of *hamatum* extend as far as 300 meters from the bivouac, whereas the swarming raids of *burchelli,* being much broader, are confined to within 75 to 150 meters. The bivouac itself is a cluster of ants hanging under logs or in hollow trees for protection from the rain. Booty is stored, the larvae cared for, and the queen attended, all within this hanging, clinging, basketball-sized mass of ants.

E. hamatum begins to forage shortly after dawn, although raiding activity is generally low during midday. Raids are largest during the latter part of the nomadic phase when the food requirements of the large developing larvae are greatest. During statary phases, the traffic along column raids is about one-quarter to one-half that observed during periods of intense feeding. During the nomadic phase, traffic picks up in the early afternoon, and culminates in an emigration to a new bivouac site. The emigration lasts late into the evening, and extends in a 150- to 200-meter-long column of ants carrying larvae and booty from the old bivouac to the new one. When the colony emigrates, guard workers scattered along the column attack anything that tries to cross the emigration path. The queen always leaves the old bivouac after dark, when about three-quarters of the ants have left.

By taking up a nomadic existence, army ants have greatly extended the area over which they can forage. The ants can maintain very large colonies with high productivity and yet do not risk depleting their food resources.

Army ants are the culmination of a long evolutionary line within which interaction among individuals has been made progressively more elaborate, until the family has achieved a level of complexity and integration more commonly associated with the organism. The complete subservience of an individual to another's reproductive fitness emphasizes the subtlety and, perhaps, unpredictability, of the results of evolutionary change. In fact, many ecological phenomena are difficult to interpret without a better understanding of the evolutionary process than we have needed until now.

In the first part of this book we have examined adaptations of organisms to their physical, chemical, and biological environments, always maintaining the organism in a central ecological role. In the following sections, we shall explore ecological phenomena on the population and community levels where patterns are formed by the interactions of the adaptations of many different kinds of individuals. But first, we shall turn our attention to the mechanism of evolution, to learn more precisely how selection operates and the consequences that it has for the ecology and evolution of organisms. It is fitting that the following chapters are introduced by a brief historical sketch of Charles Darwin and his role in the development of modern evolutionary thought.

Part 6

Ecological Genetics and Evolution

20

Charles Darwin and the
Theory of Evolution

Over a hundred years ago Thomas Huxley "... defended Darwinian evolution because it seemed to constitute, for terrestrial life, a scientific truth as significant and far-reaching as Newton's for the stellar universe—more particularly, because it seemed to promise that human life itself, by learning the laws of its being, might one day become scientifically rational and controlled" (Irvine 1955). Let us go back a century and more, to the world of Darwin, Huxley, and their predecessors, to examine the roots of modern evolutionary thought. Before the publication of *On the Origin of Species* by Darwin, people widely believed that species arose independently and uniquely as the divine creations of God, and that their forms, being perfect, could not undergo change.

Explorations into the nature of the biological world intensified during the late eighteenth century as man sought to classify and categorize the diversity of plants and animals. Naturalists realized that some species were closely related in form and others were only distantly allied, but this posed no difficulty for those who believed in special creation. That animals and plants could be changed under domestication into a wide array of varieties began to raise doubts about the immutability of species. New species were never formed, however; the varieties derived from a single stock always could be bred together.

Some thinkers did begin to entertain the notion that species were not fixed—that they could be, and had been, transformed. Charles Darwin's grandfather, Erasmus Darwin, was one of them, but he offered no proof that species had changed or even could change. A little later, the French biologist Jean Baptiste de Lamarck introduced the idea of transformism more formally. Lamarck's arguments had several critical weaknesses and he assumed the truth of his idea without bothering to prove it. Europe at that time was so beset by speculation, and sufficiently tempered by the popularization of the physical sciences, that logical minds could not accept his arguments.

A part of Lamarck's theory was concerned with the inheritance of acquired characteristics. Lamarck stated that the effects of use and disuse on organs were transmitted from parent to offspring. The long neck of the giraffe could thus be explained by postulating generations of giraffes that had to reach progressively higher vegetation and had stretched their necks in doing so. Lamarck and others paved the way for a theory of evolution by raising the possibility that species change. It remained for Charles Darwin to present convincing evidence for evolution and to discover the mechanism of evolutionary change.

When Darwin left on his famous voyage round the world on the H.M.S. *Beagle,* he had no reason to doubt the special creation of the varieties of life by divine power. But during his explorations and throughout his lifetime, Darwin's observations of nature slowly and painstakingly led him to the theory of evolution through natural selection.

Darwin was born in 1809, the son of a well-to-do medical doctor. As a youth, Darwin's interest lay primarily with beetle collecting and snipe shooting, but his father wanted more of Charles and sent him to study medicine at Edinburgh. Darwin turned out to be too squeamish for the study of anatomy and nineteenth century surgery, and soon returned home. It was then decided that he study for the ministry. Charles agreed with the decision and went to the university at Cambridge, where his interest in beetle collecting led him to study biology and geology while pursuing his theological training. In 1831, John Henslow, a professor of botany, recommended him for the post of naturalist on the H.M.S. *Beagle,* which was to sail on a five-year voyage to survey the coast of South America. After family trepidations and a variety of circumstances that nearly caused him to miss the sailing, Darwin finally set forth with the *Beagle* on December 27, 1831.

Never having been on a major ocean voyage before, Darwin immediately became violently seasick, a malady that periodically plagued him for most of the journey. In the calm weather that lay between the Cape Verde Islands and the coast of Brazil, Darwin read Charles Lyell's *The Principles of Geology.* Lyell explained geology as the result of very slow processes, acting over very long periods of time. Previous theories about the geology of the Earth's crust involved cataclysmic explanations for the appearance of mountain ranges and other geological features, and thus were like the theory of special creation: The sudden appearance of new forms by forces that were not apparent in nature. Darwin began to apply Lyell's new concept of changes in the Earth's crust to changes in living things.

Darwin's genius was of a very slow sort. He was not given to flashes of inspiration, and the theory of evolution grew slowly in his mind. His journals of the voyage of the *Beagle* give little hint of the idea that would rock the intellectual foundations of the civilized world. What is not often recognized is that Darwin was one of the greatest naturalists of all times. Had he never discovered the theory of evolution, we would still remember him as an eminent scientist whose ability to observe nature has rarely been surpassed. But Darwin's studies in South America and in the Pacific, particularly on the Galápagos Islands, had sown the seeds of the concept of evolution by natural selection.

Darwin returned to England after his global voyage with an idea that he knew was heretical to the scientific dogma of the day and blasphemous to religious orthodoxy and Victorian ethics. Not wanting to become a public figure, Darwin confided his ideas only to a few of his close associates, including Charles Lyell. Darwin began a notebook on evolution in 1837, and had produced a handwritten draft of his ideas by 1844. But fifteen more years would elapse before Darwin published his work as *On the Origin of Species.*

Darwin kept very much to himself and his family during these intervening years, and was in poor health most of the time. He slowly accumulated evidence about evolution, desiring to make his case invulnerable to criticism. But others were also playing with the ideas of evolution. Lyell, fearing that Darwin might be pre-empted, urged him to publish his ideas. But Darwin deferred. He had not accumulated enough evidence yet, he said.

Darwin devoted much of his time to publishing the results of his world expedition, and spent seven years on an intensive study of barnacles, which remains to this day the classic work in the field. His book on coral atolls was widely acclaimed, as were his other works, and Darwin rose to scientific eminence.

In 1852, Herbert Spencer, a philosopher and editor of the *Economist* in London, published an article on population in which he outlined an emerging theory of natural selection: "... from the beginning, pressure of population has been the proximate cause of progress. For those prematurely carried off must, in the average of cases, be those in whom the power of self-preservation is the least, it unavoidably follows that those left behind to continue the race, are those in whom the power of self-preservation is greatest—are the select of their generation."

Finally, in 1858, Darwin received a manuscript from a young naturalist, Alfred Russell Wallace, who was interested in receiving Darwin's criticism on the paper. The manuscript destroyed the quiet seclusion of Darwin's life, for it contained the basic points of Darwin's theory of evolution. Wallace had arrived at the theory of natural selection independently of Darwin while collecting specimens in Indonesia. Urged on by Wallace's manuscript and close friends, Darwin wrote a brief summary of his ideas, and presented his paper jointly with Wallace's paper to the Linnean Society of London in 1858. Only thirty of the more than four hundred members of the society were present at the meeting, and the paper was received with a grave silence. The president of the society remarked in his annual report that nothing of importance had happened that year.

The following year, Darwin published *On the Origin of Species,* and although only an abstract of an intended larger work, which would comprise all his evidence, the *Origin* was bought out by booksellers the first day and immediately created an uproar. Still wishing to be left alone and to avoid publicity, Darwin left the case for evolution to be defended by others, primarily by the botanist Asa Gray in the United States, Ernst Haeckel in Germany, and Thomas Huxley in England.

Darwin's thinking was molded by the nineteenth century intellectual environment and profoundly influenced by other writers. The most important was the English economist Thomas Malthus, whose essay on population, published in 1798, was one of the earliest papers to contest the Utopian concept that the progress and perfectibility of man are unlimited. Malthus began: "I think I may fairly make two postulata. First, that food is necessary to the existence of man. Secondly, that the passion between the sexes is necessary and will remain nearly in its present state." He then extended these observations to the natural world and suggested that in populations of organisms, including man, the tendency to increase is checked by the limited availability of food:

> Population, when unchecked, increases in a geometrical ratio: Subsistence increases only in an arithmetical ratio. A slight acquaintance with numbers will shew the immensity of the first power in comparison of the second.
>
> By that law of our nature which makes food necessary to the life of man, the effects of these two unequal powers must be kept equal.
>
> This implies a strong and constantly operating check on population from the difficulty of subsistence. This difficulty must fall some where and must necessarily be severely felt by a large portion of mankind.
>
> Through the animal and vegetable kingdoms, nature has scattered the seeds of life abroad with the most profuse and liberal hand. She has been comparatively sparing in the room and the nourishment necessary to rear them. The germs of existence contained in this spot of earth, with ample food, and ample room to expand in, would fill millions of worlds in the course of a few thousand years. Necessity, that emperious all pervading law of nature, restrains them within the prescribed bounds. The race of plants, and the race of animals shrink under this great restrictive law. And the race of man cannot, by any efforts of reason, escape from it.

Darwin echoes these ideas in *On the Origin of Species:*

> As more individuals are produced than can possibly survive, there must in every case be a struggle for existence, either one individual with another of the same species, or with those of distinct species, or with the physical conditions of life. It is the doctrine of Malthus applied with manifold force to the whole animal and vegetable kingdoms; for in this case there can be no artificial increase of food, and no prudential restraint from marriage.

Let us return to South America with Darwin to trace the development of evolutionary theory in his mind. It is impossible to pinpoint the exact place or time of Darwin's inspiration about evolution. His notebooks are filled with the observations of a perceptive and imaginative naturalist. He saw the bones of giant animals in the dry arroyos of the pampas and identified their owners as extinct predecessors of modern forms, although the local inhabitants insisted that these were bones belonging to living forms that grew in size after their death through the magical powers of the waters of certain rivers.

The one locality that most influenced Darwin's thinking was perhaps the Galápagos archipelago, a group of volcanic islands located on the Equator about 600 miles off the western coast of South America and described in this passage from *The Voyage of the Beagle:*

> The natural history of these islands is eminently curious, and well deserves attention. Most of the organic productions are aboriginal creations, found nowhere else; there is even a difference between the inhabitants of the different islands; yet all show a marked relationship with those of America, though separated from that continent by an open space of ocean, between 500 and 600 miles in width. The archipelago is a little world within itself, or rather a satellite attached

to America, whence it has derived a few stray colonists, and has received the general character of its indigenous productions. Considering the small size of these islands, we feel the more astonished at the number of their aboriginal beings, and at their confined range. Seeing every height crowned with its crater, and the boundaries of most of the lava-streams still distinct, we are led to believe that within a period, geologically recent, the unbroken ocean was here spread out. Hence, both in space and time, we seem to be brought somewhat near to that great fact—that mystery of mysteries—the first appearance of new beings on this earth.

A fertile seed planted in Darwin's imagination on the Galápagos Islands had grown to maturity by the time *On the Origin of Species* was published:

The Galapagos Archipelago lies 500 and 600 miles from the shores of South America. The naturalist, looking at the inhabitants of these volcanic islands feels that he is standing on American land. Why should the species which are supposed to have been created in the Galapagos Archipelago, and nowhere else, bear so plainly the stamp of affinity to those created in America? In the conditions of life, in the geological nature of the islands, in their height and climate there is a considerable dissimilarity from the South American coast. On the other hand, there is considerable resemblance between the Galapagos and Cape Verde Archipelagoes; but the inhabitants of the Cape Verde Islands are related to those of Africa. Such facts admit of no explanation on the view of independent creation.

Darwin's observations on the Galápagos Islands, which had brought him "... somewhat near to that great fact—that mystery of mysteries—the first appearance of new beings on this earth," had gradually convinced him that "... such facts admit of no explanation on the view of independent creation."

Darwin's evidence for evolution came primarily from the geographical distribution of species, the kind of evidence that led Wallace to formulate the same theory. Darwin reasoned that if the creation of each species was independent of the creation of all others, it would be difficult to explain the existence of a small, closely related group of finches on the Galápagos Islands, or that the faunal and floral affinities of the Galápagos Islands were shared more closely with South America, the nearest mainland, than with other regions of the Earth. Darwin was also impressed by other evidence. For example, in dry, savanna-like habitats in South America there exists a group of rodents that resembles rabbits. Darwin pondered why there should not have been real rabbits in the pampas. Why a rabbit-like creature derived from a different group? Darwin also noted that the extinct mammals, whose bones he had found in the clays of South America, would have resembled some of the living mammals that occurred only in that region, and nowhere else in the world. Darwin began to form a theory based upon the descent of modern species from ancestral forms that seemed to bring together all of the accumulating evidence.

Darwin gathered additional material for his work on evolution after he returned to England. If man could achieve marked changes in domesticated animals in a short period of artificial selection, could not a very new species be formed by similar processes acting in nature over much longer periods? Why should the wings of birds and the arms of man, both modified for different purposes, have homologous bones if they did not have a common ancestor (Figure 20-1)? Why should snakes have the bony remains of a pelvic girdle if they had not evolved from forms with legs?

While the evidence was overwhelmingly in favor of evolution through natural selection, there were also many problems with the theory. Darwin admitted to some of these. For example, few transition forms are found between carnivores and whales, reptiles and birds, and so on, where the theory of evolution would have predicted the existence of intermediate types. Where were these missing links? How could an organ so perfect as the vertebrate eye have evolved when any state of development less than its present perfection probably would result in the death of its possessor? As Darwin phrased the problem in *On the Origin of Species,*

> To suppose that the eye with all its inimitable contrivances for adjusting the focus to different distances, for admitting different amounts of light, and for the correction of spherical and chromatic aberration, could have been formed by natural selection, seems, I freely confess, absurd in the highest degree.

Darwin dealt effectively with this problem by proposing the gradual evolution of such organs through many primitive stages and intermediate links, many of which may still be extant:

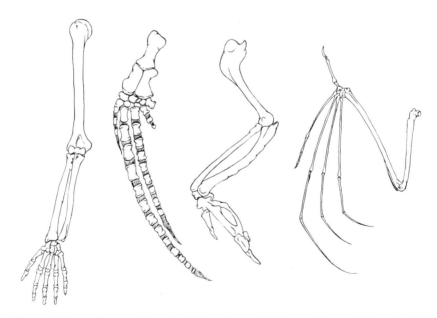

Figure 20-1   Bones of the forelimbs of man (left), a whale, a bird, and a bat (right), illustrating how a structure has been modified to serve different purposes without completely giving up its basic form. Darwin used such evidence to infer common ancestry for diverse groups.

When we reflect on the graduated range of structure in the eyes of lower animals the difficulty ceases to be great in believing that natural selection may have converted an optic nerve, coated with pigment and invested by transparent membrane, into an optical instrument as perfect as is possessed by any of the articulate class. He who will go thus far ought to admit that a structure as perfect as an eagle's eye might thus be formed, although he does not know the transitional states.

Darwin's theory of natural selection was plagued by other difficulties, the greatest of which was that heredity was poorly understood. Darwin's contemporaries did accept the fact of heredity—it was manifest all around them, even in their own families—but the mechanism of heredity, including the nature of variability that was so central to Darwin's theory, remained obscure. Darwin was also unable to explain how a single species could give rise to two or more new species.

General acceptance of the theory of evolution was slow. Since the Copernican revolution, scientific ideas had upset man's view of his own place in nature. Many were unwilling, finally, to give up center stage. How could Darwin dare to suggest that man, formed in the image of God, was only a modified monkey? The problem of heredity became such a difficulty that evolution had almost been forgotten by the end of the last century. Alfred Russell Wallace (1889) remained one of the theory's few supporters. But during the twentieth century, Darwin's theory was revitalized by the development of genetics, which provided an explanation for the mechanism of heredity. Population geneticists conferred mathematical rigor on the theory of natural selection, and the recent emergence of evolutionary ecology has provided convincing demonstrations of evolutionary processes in nature, to the point that evolution is now largely taken for granted.

After the publication of *On the Origin of Species,* Darwin, in his quiet way, actively continued his study of evolution and natural history, and wrote many influential books and articles on such topics as human evolution, carnivorous plants, and earthworms. Charles Darwin died of a heart attack in 1882, and was buried in Westminster Abbey, close to the grave of Sir Isaac Newton.

21

The Nature of Inheritance

One of the difficulties Darwin faced while grappling with his theory of evolution was the mechanism of inheritance. He expressed this difficulty in *On the Origin of Species:*

> The laws governing inheritance are for the most part unknown. No one can say why the same peculiarity in different individuals of the same species, or in different species, is sometimes inherited and sometimes not so; why the child often reverts in certain characters to its grandfather or grandmother or more remote ancestor; why a peculiarity is often transmitted from one sex to both sexes, or to one sex alone, more commonly but not exclusively to the like sex.

Darwin believed, along with others at the time, that inherited traits were transmitted by a fluid composed of minute secretions from every cell in the body. This concept was called *pangenesis* (Ghiselin 1975). One of the tenets of the theory of pangenesis was that the characteristics contained in the fluids of the male and female parent blended. Many traits do, in fact, appear to blend: A tall father and short mother tend to produce children of intermediate height; progeny of black and white parents tend to have intermediate skin color, and so on. But other characteristics do not exhibit blending inheritance. Eye color is transmitted faithfully from generation to generation and does not exhibit mixing. Moreover, two brown-eyed parents can have a blue-eyed child if some of their ancestors had blue eyes.

The theory of blending inheritance posed an additional problem for Darwin's theory of evolution in that it made no provision for maintaining the genetic variation that was both observed in nature and necessary for the process of natural selection. Blending inheritance would tend to even out differences between individuals in a population until all had approximately the same genetic composition.

The later discovery of the principles of genetic inheritance removed the last obstacles to the development of a complete evolutionary theory.

Mendel demonstrated that traits are inherited by the transmission of particulate factors.

Although Gregor Mendel's work on the inheritance of traits in garden peas was to lay the foundation for modern genetics, his life was not surrounded by the publicity and aura of intellectual revolution that accompanied Darwin after the publication of *On the Origin of Species.* Mendel was born in 1822 and attended the Augustinian monastery in Brunn, Austria (now Brno, Czechoslovakia), when he was 21. Four years later he was ordained a priest. In 1851, the Augustinian order sent him to the University at Vienna to study natural science so he could teach that subject upon his return to Brunn. Mendel's studies awakened a deep-rooted curiosity about the nature of inheritance that, in 1857, led him to begin his studies on the garden pea. These studies culminated nine years later with the publication of his results in the annual proceedings of the Natural History Society of Brunn. Mendel had demonstrated the particulate nature of inheritance and defined an entirely new field of science, but his paper went unnoticed and lay forgotten for 34 years. It was rediscovered independently in 1900 by three biologists who had begun to perform similar experiments. It is profoundly ironic that the answer to Darwin's dilemma about inheritance lay buried in an obscure journal even while he was writing several revisions of *On the Origin of Species.*

Mendel's experiments with the garden pea were both simple and elegant. Many high school students have repeated the experiments, but in 1857 the possibility of understanding inheritance by experimentation seemed no more real than space travel, and the subject was neglected. Mendel's ability to recognize the problem and to design suitable experiments was part of his genius, but discovery and understanding do not lie only in experiments, they lie also in the analysis of experimental results and their interpretation within a new conceptual framework.

Mendel chose the garden pea because numerous strains had been developed with distinguishing characteristics of flower color and seed color that always bred true. That is, the strains never produced deviant types. Mendel experimented with seven characteristics, each with distinct alternative forms, among which were the form of the ripe seed (round versus wrinkled), the color of the cotyledons of the seed (yellow versus green), and the color of the flower (white versus violet-red).

When Mendel crossed two plants with the same characters, their progeny always exhibited the parental form. If, however, plants will one trait were fertilized with pollen from plants with the alternative trait, only one form of the trait appeared in the offspring. For example, when plants grown from round seeds were crossed with plants grown from wrinkled seeds, all of the progeny had round seeds. Similarly, yellow seeds and violet-red flowers always appeared in the progeny of crosses between parents with alternative seed and flower coloration. One of the traits was always "lost" in the progeny of hybrid crosses. But if the hybrid plants were in turn mated with each other, those traits reappeared in about one-quarter of their progeny. Thus, a particular trait was not lost, rather it became latent, and it certainly did not blend with the trait expressed in the hybrid offspring. In Mendel's words, ". . . those characters which are transmitted entire, were almost unchanged in the hybridization, and therefore in themselves constitute the characters of the hybrid, are termed *dominant,* and those become latent in the process, *recessive*."

The results of Mendel's experiment had a remarkable quantitative consistency. For example, the hybrids of strains with round and wrinkled seeds, when crossed and allowed to fruit, produced a total of 7,324 seeds: 5,474 were round and 1,850 were wrinkled, a ratio of 2.96 round seeds to each wrinkled seed. Over 8,000 seeds from hybrids of strains with yellow and green seeds gave a ratio of 3.01 yellow seeds to each green seed. To Mendel, the simplest explanation for these results was to postulate that each trait was determined by two factors that were transmitted from parent to offspring as discrete units (Figure 21-1). He called these units of inheritance *genes.* Even though its expression was suppressed in the hybrid plants, the presence of the genetic factor for wrinkled seeds was demonstrated by its reappearance in the offspring of hybrids mated with each other.

Because of the dominance of one genetic trait over another, the *genotype* of an organism does not always correspond to the outward expression of the character, or *phenotype.* Through a series of crosses of hybrid offspring with the parental strains (backcrosses), Mendel demonstrated that the progeny of hybrid crosses contained all three expected genotypes: *RR* (round phenotype), *Rr* (round phenotype), *rr* (wrinkled phenotype), even though only two phenotypes appeared. A genotype that contains two of the same *allele,* or form of the gene (*RR* or *rr*) is called a *homozygote.* A genotype that contains different alleles of a gene is called a *heterozygote.*

Because the heterozygous genotype can be produced in two ways (by the round gene coming from the male parent or coming from the female parent) there are three equally probable genotypes that give a round phenotype and one genotype for the wrinkled phenotype. Mendel consistently obtained the three to one ratio in his experiments.

Not all traits exhibit complete dominance; hybrid offspring can have an intermediate phenotype for some traits. In crosses between snapdragons

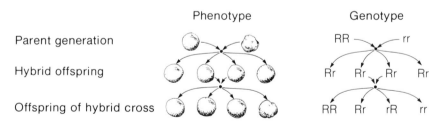

Phenotype   Genotype

Parent generation   RR ———— rr

Hybrid offspring   Rr   Rr   Rr   Rr

Offspring of hybrid cross   RR   Rr   rR   rr

Figure 21-1   Diagrammatic representation of the phenotypes Mendel observed in his preliminary experiments with hybrid strains (left), and his genetic interpretation of the results (right). *R* = round seeds, the dominant trait; *r* = wrinkled seeds, the recessive trait. If an individual with *RR* genotype among offspring of two hybrids is backcrossed to an *rr* parent, all the progeny have the *R* phenotype; if an *Rr* or *rR* individual is backcrossed to the *rr* parent, half the progeny have the *R* phenotype and half the *r* phenotype.

with red and white flowers, the hybrid generation all have pink flowers. In their progeny, the traits segregate in the ratio of one red to two pink to one white, as one would expect on the basis of Mendel's law. Hence the apparent blending of traits in hybrid snapdragons results from the phenotypic expression of an interaction between genetic traits, rather than the blending of the genetic traits themselves.

Mendel made another important discovery, that genetic traits for different characters were inherited independently. That is, the appearance of a particular genetic trait is not influenced by the segregation of genes that determined other phenotypic characters. Mendel crossed pure strains of plants in which one parental type always produced round yellow seeds and the other always produced wrinkled green seeds; the progeny of the cross all produced round yellow seeds. But the progeny of crosses between these hybrid offspring were of four types, representing all possible combinations of seed shape and color: round yellow, wrinkled yellow, round green, and wrinkled green. Although the genes for each trait were unaltered, combinations of genes were broken up during reproduction and redistributed at random in the progeny of the hybrids.

Genes are arranged into linkage groups that correspond to chromosomes.

At the same time Mendel was studying the genetics of the pea, cytologists were discovering structures in cells that later would be correlated with genetic results. We now know that during cell division strands of material called *chromosomes* appear in the nucleus of the cell. The cells of any one species always contain the same number of chromosomes (White 1973), 6 in the mosquito, 14 in the pea, 46 in man, 254 in the shrimp, and so on (for example, see Figure 21-2). In the *gametes,* the egg and sperm, there are exactly half as many chromosomes as in other cells of the body. But the

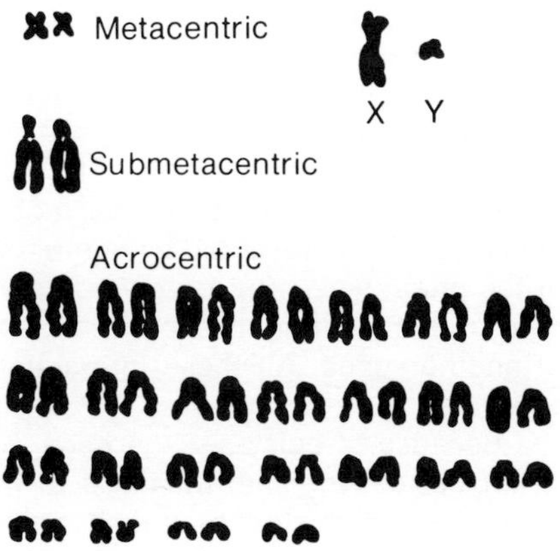

Figure 21-2   The chromosome complement, or karyotype, of the pocket mouse, *Perognathus gold-mani,* from Sinaloa, Mexico. The chromosomes have been arranged to show the homologous pairs. Metacentric refers to paired chromosomes that attach to each other at their centers; acrocentrics attach at their ends. The sex-determining chromosomes (*X* and *Y*) are not homologous (after Patton 1969).

chromosomes of the male and female gametes match each other in form, with the exception of the sex-determining chromosomes in many species of animals. The *zygote,* the cell formed by the fertilization of the egg by the sperm, and most of the cells of the organism, produced from the zygote by cell division, contain two sets of chromosomes, one contributed by the male gamete and one by the female gamete. This condition is referred to as *diploidy;* the term *haploidy* is applied to the single set of chromosomes in a gamete. Thus, the 46 chromosomes in human cells represent 23 pairs of homologous chromosomes. (Actually, the sex-determining *X*- and *Y*-chromosomes do not form a strictly homologous pair because they differ greatly in form and function.)

Chromosomes and genes exhibit parallel behavior during the life cycle. Both are individual units that maintain their integrity throughout reproduction; each gamete contains a full complement of chromosomes just as each gamete has a full complement of genetic traits. Furthermore, Mendel's experiments demonstrated that genes exist in pairs, one gene contributed by each parent; chromosomes also exist in pairs. So far so good. But there are not nearly enough chromosomes in the cell to account for all of the traits exhibited by an organism. The pea has only seven pairs of chromosomes. Mendel had experimented with seven traits, which could hardly have made a dent in the number of genetic factors that must be needed to make a pea, let alone an entire pea plant. The problem was quickly resolved when genes were discovered in peas for two characteristics that did not assort independently. The traits remained linked, and the offspring of crosses between hybrids consisted almost entirely of the original parental genotypes. Some traits thus appeared to be inherited as a single genetic unit, called a *linkage group,* which later proved to correspond to the chromosome. As we shall see in the next chapter, genetic material is often exchanged between homologous chromosomes during reproduction (recombination), thereby producing new combinations of genetic traits for natural selection to work on.

Interactions within the genotype influence the expression of genes.

Mendel's experiments were somewhat misleading in that he carefully selected traits whose expression was unaffected by other genes. In early experiments this restriction was necessary so as not to obscure the simplest and most basic features of heredity. Such simple inheritance may be the exception to the rule, however. The most conspicuous gene interactions are: (a) two genes complementing each other in the expression of a single trait (if either gene is represented by nonfunctioning alleles, the trait will not appear in the phenotype); (b) a gene completely blocking the function of another gene and preventing its expression; and (c) a gene modifying the expression of another gene. All these interactions are collectively known as genetic *epistasis.*

Because of epistasis, the gene pool of a population, out of which the genotypes of every individual are drawn, is carefully balanced by selection to contain only those genes whose interactions promote the functioning of the

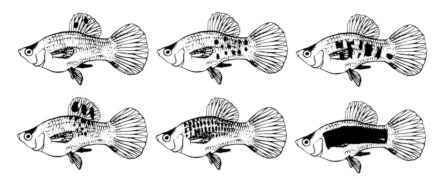

Figure 21-3   Expression of several genes affecting the pattern of melano-phores in the platyfish *Xiphophorus maculatus*. Left, two phenotypes of the "dor-sal-spotted" gene with different modifiers; top center, expression of "spot-sided" gene; bottom center, "stripe-sided" gene; right, two phenotypes of the "black-sided" gene (from Gordon and Gordon 1957).

organism. The expression of a gene and its influence on fitness are affected not only by the environment of the organism, but also by the genetic envi-ronment of the gene itself. The mutual adjustment of genes to each other is frequently termed the *coadaptation* of the genotype (Wallace 1968: Ch. 18).

Even though coadaptation is a fundamental feature of the gene pool of a population, its influence is difficult to detect in simple genetic experiments. But when crosses are made between different populations of a species that have been sufficiently isolated from one another to accumulate numerous genetic differences, the integrating effect of coadaptation frequently be-comes evident. In a series of crosses between populations of the platyfish *Xiphophorus (Platypoecilus) maculatus* from isolated river basins, Gordon and Gordon (1957) showed that the proper expression of genes that con-trolled color pattern depended on other modifying genes unique to each population. The fish normally have distinct patterns of small dark spots along their sides. These markings act as species-specific signals for mate recogni-tion. When a male and female from distant populations are mated, their offspring no longer develop the normal patterns. Instead, the black spots become very large and they often fuse, obliterating all pattern (Figure 21-3). Evidently the genes from different populations do not interact properly, and when combined they produce a phenotype with reduced fitness. We shall encounter similar examples in subsequent chapters frequently enough to demonstrate the pervasiveness of coadaptation.

Continuously varying traits have a genetic basis made up of the additive effects of many genes.

Even after the rediscovery of Mendel's law in 1900, the nature of inheri-tance and the role of genes in evolution was disputed for another decade and more. The reluctance of many well-known scientists to accept Mendelian inheritance as fundamental to evolution was based partly on the existence of continuously varying metrical traits such as height, weight, body proportions, rates of physiological function, timing of seasonal change—in fact, most of the variation in nature that seemed important to the ecology of organisms. The differences that we observe between species belong predominantly to categories of size, shape, amount of pigment, and the like. Surely these factors were not under the same kind of genetic control as seed color in peas?

The reconciliation of continuous variation in traits with the particulate inheritance of Mendelian genetics may be attributed primarily to the American geneticist E. M. East (1910). He interpreted continuous variation as the result of many genes, each contributing a small modifying influence above or below the average phenotypic expression of a trait in a population. Different combinations of genes having small positive and negative effects produced the range of phenotypes observed within the population. East demonstrated his point by crossing strains of corn that bred true for short ears and long ears. The hybrids had ears of intermediate length as one might expect, but their progeny had more variable ear lengths, including some that approached the extremes of length of the two parental strains (Figure 21-4).

East interpreted the results of the cross between the hybrid corn plants as showing the independent segregation of many genes influencing ear

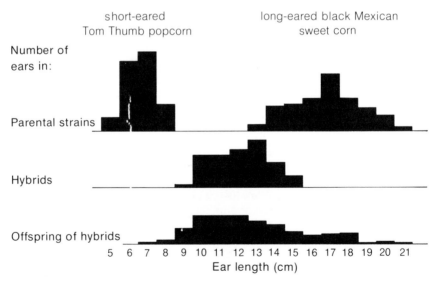

Figure 21-4   A demonstration of multiple factor inheritance of ear length in corn. The parental strains (short-eared Tom Thumb popcorn and long-eared black Mexican sweet corn) were highly inbred and hence genetically uniform. Hybrids between the strains were intermediate. Progeny of hybrids exhibited increased phenotypic variation with extremes similar to the parental strains (after East 1910).

length. We may demonstrate this effect by considering a simple hypothetical case in which two nonlinked genes influence the length of a certain organ. At each of the genetic loci, gene x increases length one-half unit above the mean and gene o decreases length one-half unit below the mean when present on one chromosome. In a diploid organism, two genes can produce five possible genotypes ranging from (oo/oo), with length two units less than the population mean, to (xx/xx), with length two units above the mean. If one were to cross pure strains having minimum and maximum length, all the hybrids would possess two x genes and two o genes (xo/xo) and would have organs of intermediate length. If two of these hybrid individuals were crossed, the progeny would exhibit all possible genotypes, including the original parental extremes. The proportion of each genotype among the progeny can be predicted from probability theory. When lumped according to number of x and o alleles in the genotype, regardless of their parentage or genetic locus, the proportions of genotypes correspond to the coefficients of the terms of the binomial expansion, $(a + b)^n$, where n is the number of segregating alleles and a and b are the proportions of the two alleles in the population, 0.5 and 0.5 in the case of a hybrid cross. For the one-locus, two-allele case (e.g., flower color in snapdragons), we have

$$(a + b)^2 = a^2 + 2ab + b^2$$

and the coefficients of the terms of the expanded binomial expression are 1, 2, and 1. If we let a represent the x (or red) allele and b the o (or white) allele, the various genotypes are to be expected in the ratio 1(xx): 2(xo): 1(oo), or one red: two pink: one white. In the two-locus model described above, there are four alleles and the binomial expansion becomes

$$(a + b)^4 = a^4 + 4a^3b + 6a^2b^2 + 4ab^3 + b^4$$

giving coefficients of 1:4:6:4:1. Thus, in our cross between (xo/xo) hybrids, we would expect $\frac{1}{16}$ of the progeny to have genotype (xx/xx), $\frac{4}{16}$ to bear genotypes (ox/xx), (xo/xx), (xx/ox), or (xx/xo), all having identical phenotypic expression, and so on. As the number of additive genes influencing the expression of a metrical character increases, the number of genotypes increases in proportion, and genetic variation approaches more closely the continuous variation that we observe in populations (Figure 21-5).

The genetic analysis of metrical traits is based on statistical interpretations of phenotypic variance.

Quantitative inheritance cannot be analyzed in the manner that Mendel used to discover the rules of heredity in traits controlled by one gene locus. Too many genes with similar function are involved. Worse yet, geneticists cannot accurately estimate the number of genes involved in the inheritance of continuously varying traits, nor do they know whether dominance and epistasis (interactions between genes) are important. All these properties must be estimated indirectly by analysis of phenotypic *variance*.

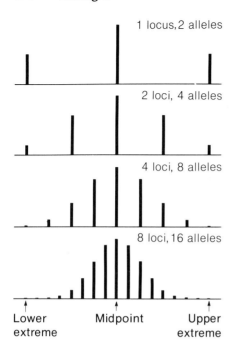

Figure 21-5   Proportions of individuals having phenotypes distributed between extreme values when the expression of a metrical trait is influenced by 1, 2, 4, and 8 genetic loci. Proportions were calculated from coefficients of the binomial expansion. The alleles were assumed to exhibit no dominance.

Variance is a statistical measure of the deviations of individual values from the mean of the population. It is calculated as the average of the squared deviations of values from the mean, or

$$V = \frac{1}{n} \sum_{i=1}^{n} (x_i - \bar{x})^2$$

where V is the variance in x in a population of n individuals; x_i is the value for each individual i ($i = 1$ to n) and $\bar{x}$ is the mean value of x in the population. The calculation of variance is shown by example in Table 21-1.*

Continuously variable traits usually exhibit a bell-shaped distribution of values in natural populations, with most individuals clustered near the mean and with the frequency diminishing rapidly at extreme values (Figure 21-6). The distribution of values resembles the proportions derived from the binomial expansion with a large number of terms (see Figure 21-5).

Each value of a particular trait (the phenotypic value) is determined by deviations from the mean caused by genetic factors and by environmental influences (Falconer 1960, Brewbaker 1964). Both sources of deviation enter into the calculation of variance for all values in the population. Hence we may

---

* If one cannot measure the entire population, and a small sample of individuals is used to estimate the variance of a large population, the sum of the squared deviations is divided by $n - 1$ rather than by n. Also, statisticians have devised other shortcut formulas for calculating variances of large samples (Sokal and Rohlf 1969, Steel and Torrie 1960).

TABLE 21-1

Calculation of variance in height in a small population ($n = 10$).

Individual (i)	Height (inches) (x_i)	Deviation from mean $(x_i - \bar{x})$	Squared deviation $(x_i - \bar{x})^2$
1	75	5	25
2	73	3	9
3	72	2	4
4	72	2	4
5	71	1	1
6	70	0	0
7	68	-2	4
8	67	-3	9
9	67	-3	9
10	65	-5	25
	Total = 700		Sum of squared deviations = 90
	Mean $(\bar{x})$ = 70		Variance = 9.5

speak of phenotypic variance (V_P) as having two components, one attributable to genetic constitution (V_G) and one to environmental influences (V_E). The two components added together equal the total phenotypic variance, or

$$V_P = V_G + V_E$$

The genotypic variance can be broken down further into additional components:

V_A = additive variance determined by the expression of the alleles in homozygous form

V_D = dominance variance determined by the interaction of the alleles in heterozygous form

V_I = interaction variance, comprising the influences of different genes on the expression of alleles at a particular locus.

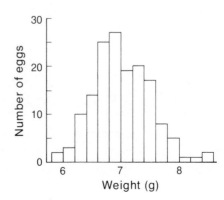

Figure 21-6 Frequency distribution of the weights of eggs of the European starling Sturnus vulgaris near Philadelphia, Pennsylvania.

In general, $V_G = V_A + V_D + V_I$, although variance may be incremented by the correlation between particular genotypes and particular environments and other, usually minor factors. The relationship between additive genetic variation and dominance deviations is shown in Figure 21-7. The additive genetic deviation attributed to each allele (a) is one-half the difference between the phenotypic expressions of each homozygote, (oo) and (xx). The dominance deviation (d) is the difference between the phenotype expression of the heterozygote (xo) and the mid-point between the homozygotes, $\frac{1}{2}(oo + xx)$. The degree of dominance is expressed by the ratio of d to a, which varies from 0 (no dominance, heterozygote intermediate) to 1 (complete dominance, heterozygote indistinguishable from one of the homozygotes). Because the interaction component of variance is quite complicated, its value is usually the variance left over when other components have been accounted for.

The greatest challenge of quantitative genetics is to estimate the magnitude of the several components of phenotypic variance. This is made necessary by the fact that selection can act only on the additive genetic component of variance, because only it reflects the genetic diversity of the population—that is, the different alleles that replace others, and are replaced, during evolutionary change. The remaining components of phenotypic variance are due to various interactions within the genome and between the organism and its environment, which are not directly amenable to change by natural selection, unless there is additive genetic variation in the interaction itself.

The methods for partitioning phenotypic variance into its several components are complicated and generally of a statistical nature (Falconer 1960: Chapters 8 and 9), and will not be dealt with in detail here. Let it suffice that the environmental and genetic components of variance may be separated by comparing natural populations and highly inbred (genetically uniform) lines reared under identical—usually laboratory—conditions. The results of such an experiment involving the length of the thorax in the fruit fly *Drosophila melanogaster* is shown in Table 21-2. The additive genetic variance can be partitioned from the total phenotypic variance by correlating phenotypic values among close relatives, usually parents and offspring or brothers and sisters. The observation that the height of children is correlated with the height of their parents is an expression of the additive genetic component of variance in height.

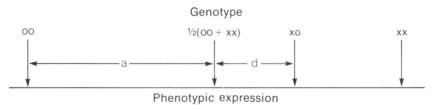

Figure 21-7   Schematic representation of the additive (a) and dominance (d) deviations in the phenotype resulting from genetic variation.

TABLE 21-2

Calculation of the components of variance in the thorax length of *Drosophila melanogaster* (after Robertson 1957).

Population	Components of variance	Observed variance (10^{-4} mm^2)
Genetically mixed	$V_G + V_E$	0.366
Genetically uniform (inbred)	V_E	0.186
Difference	V_G	0.180

By combining studies on inbred strains with phenotypic correlations among relatives, the population geneticist can provide a complete breakdown of genetic variance into its components. The proportion of the variance in each component varies with the trait, as we see in Table 21-3. One usually finds the highest proportions of additive genetic variance in characters that seemingly have little bearing on the overall fitness of the organism—like bristle number—and low proportions in traits with clear adaptive value—like number of eggs laid.

The proportion of the phenotypic variance due to additive genetic factors, hence subject to selection, is often expressed as their ratio, called the *heritability* (h^2) of a trait,

$$h^2 = V_A/V_P.$$

Hence the heritabilities of characters in *Drosophila melanogaster* for which the partitioning of variance is set out in Table 21-3, are bristle number 0.52, thorax length 0.43, ovary size 0.30, and number of eggs 0.18. As one might expect, most studies of heritability involve traits of commercial value in livestock, poultry, and crops. Some representative values of h^2 are given in Table 21-4, among which there are several trends. Sizes have higher heritabilities (0.50–0.70), hence less environmental influence, than weights (0.20–0.35). Among traits involved in production and fecundity, those creating the greater drain on energy and nutrients have the lower heritabilities.

TABLE 21-3

Partitioning of variance in four characters in *Drosophila melanogaster* (based on studies by F. W. Robertson and his co-workers, summarized by Falconer 1960: 140).

Variance component (per cent)		Bristle number	Thorax length	Ovary size	Number of eggs
Additive genetic	V_A	52	43	30	18
Nonadditive genetic	$V_D + V_I$	9	6	40	44
Environmental	V_E	39	51	30	38
Total phenotypic	V_P	100	100	100	100

TABLE 21-4

Heritabilities of several traits in domesticated and laboratory animals and plants (after Falconer 1960: 167 and Brewbaker 1964: 12).

Character	Organism	Heritability
Size or length		
plant height	corn	0.70
root length	radish	0.65
tail length	mice	0.60
length of wool	sheep	0.55
body length	pigs	0.50
Weight		
body weight	sheep	0.35
body weight	pigs	0.30
body weight	chickens	0.20
Production		
butterfat content of milk	cattle	0.60
egg weight	chickens	0.60
thickness of back fat	pigs	0.55
weight of fleece	sheep	0.40
milk yield	cattle	0.30
Fecundity		
egg production	chickens	0.30
yield	corn	0.25
litter size	pigs	0.15
litter size	mice	0.15
conception rate	cattle	0.05
Life history		
age at onset of laying	chickens	0.50
age at puberty	rats	0.15
viability	chickens	0.10

Thus, butterfat content is under strong genetic control ($h^2 = 0.60$) while total milk production has a lower heritability (0.30); variation in egg size in chickens has a large additive genetic component ($h^2 = 0.60$) while rate of egg production has a lower heritability (0.30). Any character that requires a large commitment of resources must be sensitive to environmental variations in those resources. The heritabilities of fecundity and life history characteristics are generally very low (0.05–0.50).

Virtually all estimates of heritability pertain to domesticated or laboratory populations in which the environmental component of variance is minimized. In spite of the fundamental importance of genetic variation to evolutionary change and the key roles of metrical characters, such as size, life span, and fecundity, in determining the fitness of the organism, we know next to nothing of genetic variation in natural populations, at least for traits having clearly interpretable ecological consequence. As we shall see in the next chapter, our knowledge of genetic variation in natural populations derives mostly from genes with simple and clear-cut effects in the phenotype. It has, unfortunately, proved too difficult to obtain the measurements on related individuals necessary to calculate heritability in natural populations. Furthermore, quantitative genetics has traditionally been practiced by animal and plant breeders and ignored by population geneticists and ecologists. As a result, the discussion in subsequent chapters of genetic variation and its role in the population ecology and evolution of natural populations will be shamefully narrowed by this provincialism.

In this chapter, we have done no more than examine the rules of heredity. In most organisms, the phenotype is specified by a set of genetic units, called genes, that are passed from parent to offspring during the act of reproduction. In most higher organisms, each individual carries two sets of genes (genomes), one provided by each parent. Although many genes have unique expression in the phenotype, most probably interact with others to provide subtle quantitative variation in continuously varying traits.

The mechanism of heredity and the action and interaction of the genes are elucidated by the genetic variation within populations. Without variation, Mendel could not have performed the experiments that revealed to him the particulate nature of genes. But how does genetic variation arise? How and why is variation maintained in the population? We shall answer these questions in the next chapter. The broader issue of the significance of genetic variation to ecology and evolution will be taken up later.

22

Genetic Variation in Natural Populations

Genetic variation is a controversial topic. Because of its underlying importance to evolutionary change, biologists have long viewed variability as beneficial to a population insofar as it provides the genetic material necessary for evolutionary response to changing environments. For the individual, however, variability can reduce evolutionary fitness when the individual carries the less fit variants in the population's gene pool. Even for the population, there must be some upper limit to the amount of genetic variation that is desirable, depending both on the rate of environmental change and on the intensity of competition with other populations having fewer unfit genes.

Although genetic variation provides the material basis for evolution, we do not understand the evolutionary significance of variability itself. We do not know whether the level of genetic variation in a population is optimized by natural selection or is an unavoidable consequence of intrinsic properties of biological systems. In this chapter, and in several to follow, we shall examine genetic variability: its cause, its occurrence in populations, and its importance to ecological and evolutionary processes.

Mutations are the primary source of variation.

During cell division, the genetic instructions in the nucleus of the cell are duplicated, and copies of these instructions are passed on to the progeny cells, a process, called *mitosis*, that has occurred continually since the first appearance of living beings with a genetic mechanism of inheritance. The replication of the genetic material is not perfect, however, and occasional copies are not faithful to the original. The term *mutation* covers a variety of such genetic mistakes, ranging from changes in the fine structure of the genetic material itself to gross aberrations of whole chromosomes.

To understand genetic variation, it will be helpful to review the molecular basis of genetics and mutation. Genetic information resides in the structure of a long organic molecule called *deoxyribonucleic acid* (*DNA*, for short), which has four kinds of molecular subunits: thymine (T), cytosine (C), adenine (A), and guanine (G). DNA molecules are paired in complementary strands in which thymine always occurs in conjunction with adenine on the other strand and cytosine is always found opposite guanine. The genetic information encoded in the DNA molecules is specified by the sequence of base pairs. During replication, each strand acts as a template for the construction of a new complementary strand, and the original detailed structure of the molecule is thus preserved.

The information contained in the sequence of subunits in the DNA molecule is transferred to structural components of the cell and the organism (the phenotype) in two steps (Figure 22-1). First, a molecule of *ribonucleic acid (RNA)*, which is a close chemical relative of DNA, is transcribed from a segment of the DNA molecule just as one makes a copy of a tape recording. The segment of DNA transcribed includes all of the base subunits required to specify a complete protein molecule; hence the unit of transcription is usually thought of as a gene. The subunits of the RNA molecule are complementary to those on the DNA molecule from which it was transcribed, and thus they retain the same sequential information. The RNA molecule is then used as a template on which structural proteins and the enzymes that control cell metabolism are produced, a process referred to as translation. Proteins, in turn, are directly responsible for the structural and functional details of the cell, hence the phenotype of the organism.

Proteins consist of strands of units called *amino acids.* The twenty or so common amino acids can be arranged into a nearly infinite variety of diverse proteins. Each amino acid is specifically encoded in the RNA molecule by one or more combinations of three consecutive bases (subunits). The three bases taken together are called a *codon.* The RNA base sequence UUU is specific for phenylalanine, CUC for leucine, GAA for glutamic acid, and so on (in the

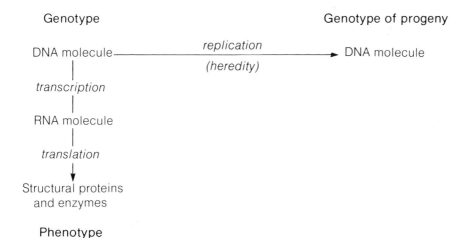

Figure 22-1   A summary of processes involved in transforming the genotype into the phenotype. Environmental influences on the developmental processes are not shown.

RNA molecule, the base uracil, U, replaces the thymine of DNA). If the codon consisted of two bases only, RNA could specify only 16 (4^2) different amino acids, fewer than actually occur in nature. On the other hand, three bases per codon results in 64 (4^3) possible combinations, many more than the number of amino acids. Consequently, the genetic code is redundant: Several codons can specify the same amino acid. For example, leucine is encoded by six different base sequences: UUA, UUG, CUU, CUC, CUA, and CUG.

A change in one of the bases in a DNA codon will alter the RNA codon and, possibly, the amino acid that it specifies. For example, consider the following sequence of base pairs in an RNA molecule and the corresponding amino acids:

RNA:	CUU	UCC	GCU	CUU	UAU	CCC	A . . .
Amino acid:	leucine	serine	alanine	leucine	tyrosine	proline	

If the cytosine base occupying the eighth position were changed to adenine, the third codon would be altered from GCU to CAU and would encode the amino acid aspartine instead of alanine. Because of redundancy in the genetic code, changes in some bases would not alter the amino acid specified by a codon and, therefore, will not be expressed in the phenotype.

Since the base sequence is read by groups of three, the deletion of a single base pair will offset the triplets by one place from their original position and will change the sequence of bases in all the ensuing codons. For example, if adenine were deleted from the eighth base pair of the following RNA molecule,

RNA:	CUU	UCC	GAU	CUU	UAU	CCC	A . . .
Amino acid:	leucine	serine	aspartine	leucine	tyrosine	proline	

the third triplet of the code would be read as the seventh, ninth, and tenth bases rather than the seventh, eighth, and ninth, as in the original molecule. The sequence of amino acids in the protein would be altered as follows:

RNA:	CUU	UCC	GUC	UUU	AUC	CCA . . .
Amino acid:	leucine	serine	valine	phenyl-alanine	isoleucine	proline

In general, deletions and additions have a far more pervasive effect on the gene than do substitutions, because they change the base sequence in all the codons that follow the point of mutation.

A few proteins have been characterized so well genetically and biochemically that it is possible to show all the links between genotype, phenotype, and fitness. The hemoglobin molecule is one of these. Hemoglobin is composed of four protein subunits; the predominant type of hemoglobin in adult humans has two types of subunits, referred to as alpha chains and beta

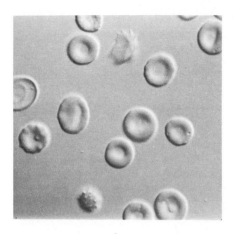 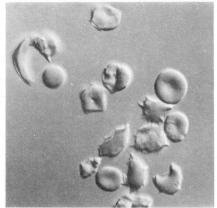

Figure 22-2   The appearance of sickle-cell erythrocytes (red blood cells) under well-oxygenated conditions, when they resemble normal cells (left) and in any oxygen-deficient medium, which causes the cells to assume contorted shapes (right) (photographs by J. Jamieson, courtesy of A. Cerami, from Cerami and Manning 1971).

chains. Each subunit surrounds a so-called heme group that contains an atom of iron. Separate genes are responsible for each type of subunit and for the heme group, and the entire hemoglobin molecule is assembled from these smaller components with the aid of specific enzymes. The disease sickle cell anemia results from the mutation of a single base in the gene controlling the production of the beta chain of hemoglobin (Harris 1970). If the seventeenth base, which is normally adenine, is changed to uracil, the amino acid glutamic acid (whose codon is GAA) is replaced by valine (GUA). This genetic defect alters the structure of hemoglobin so that when the molecules release oxygen, they become stacked close together in long helices, impairing their function. The aberrant hemoglobin molecules give red blood cells a peculiar sickle-like shape, hence the name of the disease (Figure 22-2).

Specific mutations are not stimulated by the environment of the phenotype.

Because so many steps separate the genetic code and its expression as the phenotype, it seems unreasonable that the environment could influence mutation directly as a response to the evolutionary needs of the population. It seems impossible to predict, for example, whether a specific change in the amino acid sequence of a protein should make it more or less sensitive to high temperature. How could one link specific amino acid changes to the pattern of a butterfly's wing or the predisposition to learn a specific song pattern?

Before genetics was understood, it was widely believed that the environment could induce inherited changes in plants and animals. This doctrine was first espoused by the French zoologist Lamarck and became known as Lamarckianism, or the inheritance of acquired characteristics. Darwin's theory of pangenesis led naturally to a belief in Lamarckianism. Inasmuch as structure and function were influenced by the environment or by the use and disuse of organs, these changes would be reflected in the pangens produced by each cell. As Darwin wrote in *On the Origin of Species:*

> Changed habits produce an inherited effect, as in the period of the flowering of plants when transported from one climate to another. With animals, the increased use or disuse of parts has had a more marked influence; thus I find in the domestic duck that the bones of the wing weigh less and the bones of the leg more, in proportion to the whole skeleton, than do the same bones in the wild duck; and this change may be safely attributed to the domestic duck flying much less, and walking more, than its wild parents.

The idea that the environment could influence the genetic material was not finally put to rest until Lederberg and Lederberg (1952) demonstrated that mutations in the bacterium *Escherishia coli* were produced spontaneously and independently of the environment. This conclusion was based on the results of experiments designed to test whether resistance to a bacterial virus (phage) was present in some bacteria prior to exposure to the phage or was stimulated by the presence of phage in the culture medium of the bacteria. The Lederbergs grew bacteria colonies in Petri dishes on normal culture media and then transferred bacteria from the original plate to other Petri dishes in which the culture medium was coated with phage. To maintain the spatial relationship of the bacteria colonies, the original Petri dish was first pressed against a piece of velvet, which picked up an imprint of bacteria from each of the colonies. The velvet containing the bacterial samples was then successively pressed against several Petri dishes containing fresh culture medium coated with virus. This technique is known as replica plating. If the mutation for resistance to the phage appeared in response to contact with the phage-coated medium, one would expect the appearance of resistant colonies on the replicated plates to be independent of each other and thus randomly distributed among the plates. Conversely, if the mutation had occurred in the original plate, prior to exposure to the phage, the progeny of the resistant bacterium would have been transferred to the same position on all of the replicate plates. As this is what the Lederbergs observed, they concluded that phage resistance was previously present in a few individuals in the *E. coli* population and was not induced by contact with the phage itself.

Rates of point mutation vary with the generation time of the organism.

The rate of mutation of a gene can be measured only within a limited range of circumstances. Some mutations do not produce visible phenotypic effects, and their presence can be detected only by direct biochemical assay of the DNA molecule or the specific proteins that it encodes. But even with these techniques we can detect only the extent of variation in a population,

TABLE 22-1

Comparison of mutation rates, expressed per generation and per unit time (rate per generation data from Lerner 1968).

Species	Range of mutation rates per generation	Midpoint	Generation time (days)	Modal mutation rate per day
Bacteria	10^{-10} to 10^{-6}	10^{-8}	10 hours ($10^{-0.4}$)	$10^{-7.6}$
Fly	10^{-7} to 10^{-5}	10^{-6}	3 weeks ($10^{1.3}$)	$10^{-7.3}$
Mouse	10^{-6} to 10^{-5}	$10^{-5.5}$	2 months ($10^{1.8}$)	$10^{-7.3}$
Corn	10^{-6} to 10^{-4}	10^{-5}	1 year ($10^{2.6}$)	$10^{-7.6}$
Man	10^{-6} to 10^{-3}	$10^{-4.5}$	20 years ($10^{3.9}$)	$10^{-8.4}$

not the actual rate of mutation. Mutation rate can be determined most easily in such organisms as bacteria which have only one chromosome; mutations, therefore, cannot be masked by dominant alleles. In species with two sets of chromosomes, mutation rate can be measured directly only for dominant alleles with clear expression in the phenotype. Such dominant human traits as retinoblastoma (a type of eye tumor), achondroplastic dwarfism, and aniridia (the absence of an iris in the eye) appear spontaneously in about five to fifty individuals per million born. These dominant mutations are not usually inherited because they either cause death or drastically curtail reproduction. Therefore, their expression must, in every case, represent a mutation in the germ cells of the parents. Mutation rates can also be calculated for some lethal recessive genes from their known frequency of occurrence in populations.

The rate of mutation for different genes varies considerably, by as much as several orders of magnitude in a single species. But mutation rate also appears to vary among species in a consistent pattern related to the length of the generation (Table 22-1). Mutation rates are usually measured per generation because altered genotypes are exposed with each generation of newborn young. Furthermore, most mutations are thought to occur during *meiosis,* the special cell division that directly precedes gamete production (Lindgren 1975). The generation times of different species vary considerably, however, particularly when one compares such diverse organisms as bacteria and man. Mutation rates may be compared on an absolute time scale by dividing the rate per generation by the length of the generation, giving mutation rate per unit of time. When this is done mutation rates among different kinds of organisms fall more closely into line with each other, although the reason for this uniformity is not understood.

Mutations usually reduce evolutionary fitness.

Mutations usually reduce the evolutionary fitness of an organism. This is partly due to the way in which we define mutation: The "normal" genotype is that possessed by most of the population, and deviants from the prevalent genotype are called mutations. As we ascribe the term mutant to the rare phenotype, the normal, or prevalent genetic trait is almost by definition more fit because it has been incorporated into the population in competition with alternative alleles. In a well-adapted population, one in which the gene pool and the environment are close to evolutionary equilibrium, rare alleles are maintained only by mutation of normal genes to their alternate forms.

One may compare the fitness of an organism by analogy to the focus of a microscope or the tuning of a string on a piano or guitar. If the instrument is perfectly tuned, any minor change, no matter how small, will make the instrument out of tune. If the organism is nearly, but not quite perfectly adapted to its environment—a string just slightly out of tune—a small adjustment in a random direction could improve the system half of the time. But a large adjustment, much greater than the amount by which the string is out of tune, can only make the situation worse. And so it is with mutations. Mutations with gross effects on the phenotype generally have a deleterious effect on fitness, whereas small phenotypic changes are more likely to confer some advantage to the organism, depending upon how perfectly the organism is adapted to its environment.

Recombination can produce new gene arrangements.

Phenotypic characters express the interaction of many different genes. By the same token, a novel gene mutation, which by itself may not be advantageous to its bearer, might increase the evolutionary fitness of an

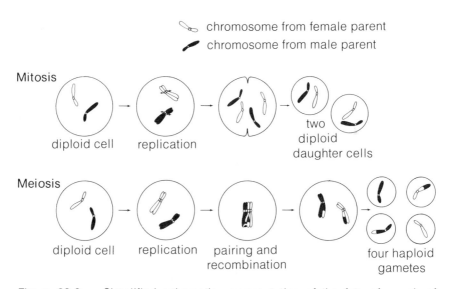

Figure 22-3   Simplified schematic representation of the fate of a pair of homologous chromosomes during somatic cell division (mitosis) and gametogenesis (meiosis).

organism in combination with another gene mutation. If such a fortunate double mutation involved genes located on the same chromosome, the new mutations could appear together in the progeny of an individual. But if the mutations occurred on different chromosomes of a complementary pair, the genes would be separated during reproduction: Paired chromosomes are apportioned to different gametes, and the advantageous combination of mutations would be lost. Through *recombination,* genes occurring on different chromosomes of a homologous pair can be brought together within the same chromosome, and therein be transmitted to the offspring as a linked unit. If the genes were located close together on the chromosome, they could persist in the population long enough to ensure their establishment before being separated again. The probability of recombination between two genetic traits, either to unite or separate them, increases in direct proportion to the distance by which they are separated on the chromosome.

Just as mutation does not bring permanent change to the genotype because of the possibility of further mutation or reversion to the original form (back mutation), recombination produces genetic changes that are even more transient and unstable. Recombination between two gene loci is a potent evolutionary force only because it occurs much more frequently than mutation.

Recombination occurs during a special cell division, called meiosis, that immediately precedes gametogenesis.

To understand how recombination occurs, a brief outline of the chromosomal events during the life cycle of the organism may be helpful. Reproduction effects the transmission of genetic material through a successive lineage of cells. During the life cycle of the organism, there are two types of cell division, *mitosis* and *meiosis,* which differ in several important respects. During mitosis, cell division produces two daughter cells whose chromosomes are like the mother cell's chromosomes in every respect (Figure 22-3). But during meiosis, when a cell in the germ line divides to produce male or female gametes, the chromosome number of each daughter cell is halved—only one chromosome of each pair is given to each gamete. Thus through meiosis, the diploid (2N) chromosome number of the gamete-precursor cell is reduced to the haploid (1N) number in each gamete. The diploid chromosome number is restored at fertilization, when the male and female gametes contribute a complement of chromosomes to the zygote.

The basic features of the meiotic division, diagrammed in Figure 22-3, involve (a) replication of each chromosome, (b) pairing of complementary homologous chromosomes to form a tetrad (a group of four chromosome strands), and (c) two successive divisions of the nucleus that produce four haploid gamete nuclei from the tetrad. Cytological crossing over between chromosomes, which results in genetic recombination, takes place within the tetrad. Because chromosomes from the male and female parent segregate and assort independently during the meiotic divisions, the genetic derivation of the four gametes can differ completely as a result of recombination.

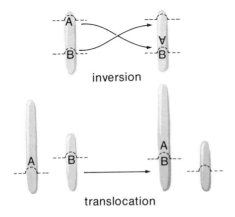

inversion

translocation

Figure 22-4   Rearrangement of gene loci on chromosomes through inversion of a segment within a chromosome and translocation (exchange) of segments between chromosomes. Inversions and translocations occasionally bring genes with related functions very close together on a chromosome to form a supergene.

Chromosome rearrangement is a form of mutation.

Chromosomes occasionally break. The freed segments can reattach in their original positions, but sometimes they become inverted on the chromosome, attach to other chromosomes, or fail to reattach anywhere and simply float off during meiosis. Most chromosome aberrations are sufficiently drastic that organisms cannot survive their genetic effects. But chromosome rearrangements occasionally bring interacting genes into close proximity on a chromosome, thereby preventing beneficial combinations from being broken up during some future meiosis by the recombination process.

The inversion of a piece of a chromosome in its original position may bring two genes with mutual interactions into close proximity (Figure 22-4). This type of inversion could maintain a gene combination as a unit by reducing recombination between gene loci. Genes with interacting functions are sometimes so close together on a chromosome that little, if any, recombination occurs between them. Such genes are often referred to as *supergenes* because they appear to act as a single unit of recombination. Genes on nonhomologous chromosomes can also be brought together into linkage groups by reciprocal translocation between the chromosomes (Figure 22-4).

Chromosomal aberrations probably occur infrequently and rarely produce viable, much less, improved chromosomes. Nonetheless, differences in chromosome structure among species, even within closely related groups, demonstrate that cytological rearrangement of genetic material on chromosomes plays an important role in evolution (White 1973, Stebbins 1971).

The evolution of the hemoglobin molecule demonstrates the course of genetic change.

Molecular evolution provides us with a model of evolutionary change. And as it is possible to determine the exact sequence of amino acids in a simple protein, such as hemoglobin, we can also perceive the structure of the gene itself. Although several important proteins have been studied in detail, hemoglobin is perhaps the best known and medically one of the most important. Among the first genetic diseases discovered and characterized were several caused by mutation of the hemoglobin gene.

In man and most other vertebrates, hemoglobin is composed of four subunits, two each of two types, out of four possible types of subunit (Ingram 1963, Manwell and Baker 1970). The four subunits are designated by the Greek letters α (alpha), β (beta), δ (delta), and γ (gamma). The α subunit consists of a string of 141 amino acids, the other three have 146. These subunits combine in pairs to form three types of hemoglobin in humans: fetal ($α_2γ_2$), adult A_2 ($α_2δ_2$), and the commonest type, adult A ($α_2β_2$). Genetic analyses have shown that the gene for the α subunit is located on one chromosome and that the genes for the other three subunits are closely linked on a different chromosome. The amino acid sequences also demonstrate that there have been many substitutions of one amino acid for another in the evolution of the hemoglobin subunits. If we were to line up the various subunits side by side, we could count 78 differences between the amino acids of the α and β chains (a little more than half the total), 83 differences between α and γ, 36 between β and γ, and only 6 between β and δ. According to Harris (1970), the primitive hemoglobin subunit that eventually gave rise to the various chains that we know at present probably had 148 amino acids. The β chain lacks amino acids at positions 21 and 22, the α chain lacks positions 2, 41, and 56 through 60. These gaps indicate deletions in the genes for each of the chains.

Ingram (1963) has suggested that the genes for all four types of hemoglobin subunits evolved from a single gene (Figure 22-5). At first hemoglobin was a much simpler molecule consisting of only one subunit and encoded by only one gene. The next step in hemoglobin evolution may have involved the duplication of the entire gene caused by an error in recombination, and

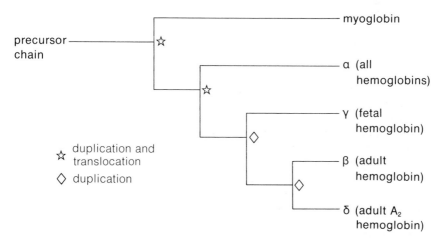

Figure 22-5   Schematic diagram of the evolutionary diversification of hemoglobin subunits (after Ingram 1963).

subsequent translocation placing the duplicate genes on different chromosomes. One of these remained as a single subunit that is currently found only in muscle and is known as *myoglobin*. After this split, the blood form of hemoglobin underwent gradual change and eventually became a *dimer,* that is, made up of two of the subunits. Not only did dimerization improve the oxygen-binding characteristics of hemoglobin, it also provided further opportunity for differentiation of the two subunits in the hemoglobin molecule.

The next step in hemoglobin evolution involved duplication and translocation of the subunit gene into lines that lead to the α chain on one hand, and the ß, δ, and γ chains on the other. The last three became distinguished by gene duplications not involving translocation. Hence the genes for these chains are located next to each other on one chromosome.

The hemoglobin of all vertebrates consists of *tetramers* (four subunits) of α- and ß-like chains. Because some of the vertebrate evolutionary lines have been separate for several hundred million years, we might expect to find considerable differences between species in the amino acid sequences of their homologous chains. Comparisons among the α and ß chains of human, horse, mouse, and rabbit hemoglobin reveal an average of about twenty substitutional differences (range 14 to 25) between the species. Because their evolutionary lines diverged about 75 million years ago, the differences suggest a rate of substitution on the order of 27 amino acids per 100 million years. Extrapolating this figure to the number of differences between different types of chains, we can estimate that the α- and ß-type chains diverged perhaps 300 million years ago. This figure must be slightly conservative because the fish and amphibian lines split about 400 million years ago, and yet both have α- and ß-type hemoglobin chains. At a rate of 27 substitutions per 100 million years, the ß and γ (fetal) lines would have diverged about 130 million years ago, when mammals were actively evolving from reptile lines.

In addition to the prevalent form of each type of chain, there are also many mutant forms of each chain in every population. For example, by 1970, 59 mutants of hemoglobin had been identified in humans. Of these, eleven cause anemia owing to the instability of the hemoglobin molecule, five affect the binding of the subunit to the heme group, and 43 have no apparent physiological effect and were identified by extensive biochemical screening procedures. Only a small fraction of the human population has been checked for hemoglobin mutations, and it has been estimated that all possible substitutions of amino acids probably exist somewhere in the human population at any one time. The tremendous variety of mutants in a population, when contrasted with the slow pace of hemoglobin evolution, emphasizes the overwhelming conservativeness of natural selection. It would not appear that genetic variation could limit the rate of evolution, yet if mutations were beneficial only in certain combinations, the likelihood of finding that particular combination could seem infinitesimal.

How much genetic variation occurs in natural populations?

This question is not easy to answer because variability cannot be measured readily, particularly for genes that are maintained at very low frequencies in a population, and for genes that have small additive effects on phenotypic expression. Also, how much variation is a lot? We have virtually no way of assessing desirable levels of variability for a population faced with a changing environment, nor can we easily measure the detrimental effects of excessive genetic variation on the population or on the individual.

Sampling limitations make it almost impossible to measure variation at the extremely low gene frequencies maintained by mutation in most natural populations. The frequencies of some extremely rare genes in human populations are known accurately because of their obvious phenotypic effects and the large number of individuals that can be surveyed. For example, the genetic trait of albinism (lack of pigment in the skin, hair, and iris of the eyes) is expressed in about one in ten thousand individuals. Because albinism is due to a recessive gene, the probability of its being expressed in the phenotype by occurring on both chromosomes of a homologous pair (0.0001) must be approximately equal to the square of probability of the allele occurring on only one of the chromosomes, which is therefore about one in a hundred. (The probability of occurrence of two independent events together is equal to the product of the probabilities of each event: $0.01 \times 0.01 = 0.0001$). Recessive genes that have more drastic effects on the phenotype usually occur with lower frequencies in the population because they are removed more vigorously by natural selection.

Marked genetic polymorphisms are easier to detect than the presence of rare mutations. Geneticists usually refer to a gene as polymorphic if a second allele has a frequency exceeding 1 per cent. Many polymorphisms are known in humans, among which are eye color, color blindness (about 10 per cent of males exhibit this trait), and blood type polymorphisms such as the ABO, MN and Rh groups (Race and Sanger 1962). The frequencies of the alleles of a polymorphic gene often vary greatly among populations (Table 22-2).

A recently developed biochemical tool, protein electrophoresis, has become an important means of detecting genetic polymorphism (Hubby and Lewontin 1966, Lewontin and Hubby 1966). Electrophoresis can distinguish protein molecules with slightly different electrical charges caused by the substitution of one amino acid by another amino acid with different electrical properties. Proteins extracted from the bodies of flies or the blood of mice, for example, are placed on a specially prepared gel through which an electric

TABLE 22-2

Proportions of ABO blood types in representative human populations (from Dunn and Dobzhansky 1946).

	Blood group (per cent)			
	O	A	B	AB
Caucasian (Scottish)	54	32	12	3
Asian (Chinese)	31	25	34	10
American Indian (Navaho)	75	25	0	0
African (Pygmy)	31	30	29	10
Australian (Aborigine)	48	52	0	0

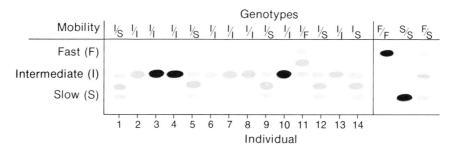

Figure 22-6   Differential migration patterns of a specific enzyme (malate de-hydrogenase) taken from fourteen horseshoe crabs (*Limulus polyphemus*). Three additional genotypes not present in the sample are shown at the right (from Selander *et al.* 1970).

current is passed. The proteins separate according to their rate of movement through the gel. By staining the gel specifically for certain proteins, one can detect protein polymorphisms as a series of bands at different distances from the starting point on the gel. For example, in Figure 22-6, three distinct genotypes may be recognized on the basis of slow, intermediate, and rapid mobility. In this particular sample, neither the slow nor fast alleles occurred as homozygotes. The presence of intermediate bands in heterozygotes shows that the enzyme is composed of two subunits, each of which may be fast, intermediate, or slow.

Electrophoresis cannot detect all variability because many amino acid substitutions do not change the electrical properties of proteins. Neverthe-less, the technique has been used to obtain some preliminary indications of the occurrence of enzyme polymorphisms in several species (Table 22-3). Because electrophoresis is time consuming, only the frequencies of common alleles can be measured. Electrophoresis can tell us little about the frequency of relatively rare mutations. A more critical difficulty with electrophoretic technique is that the biological significance of enzyme polymorphisms is not clearly understood (Lewontin 1974). Several geneticists believe that most enzyme variants have neutral evolutionary fitness (that is, the fitnesses of alleles are identical) and they are transient in populations, their frequencies being under the influence of random change (Kimura 1968a, 1968b; Kimura and Ohta 1971a, b; King and Jukes 1969). Others are equally adamant about the selective value of these alleles (e.g., Ayala 1975, Ayala and Anderson 1973). The controversy is far from settled, and ultimately may be resolved by accepting the presence of both neutral and selected alleles.

The detectable level of enzyme polymorphism is high and rather similar among widely diverse species. The level of polymorphism appears to be unrelated to rates of evolution: It is just as high in the horseshoe crab (*Limulus*), whose appearance has changed little over the past 200 million years, as it is in the evolutionarily plastic rodents. Furthermore, some groups (for example, fruit flies) appear to be more polymorphic than average while

TABLE 22-3

Levels of polymorphism in natural populations, observed by electrophoretic methods.

Species	Number of proteins tested	Per cent polymor- phic	Per cent hetero- zygous per indi- vidual	Reference
Human (red blood cell antigens)	33	36	16	Lewontin 1967
Human	71	28	7	Harris and Hopkinson 1966
House mouse (*Mus*)	41	22	9	Selander, Hunt, and Yang 1969
Field mouse (*Peromyscus*)	32	23	6	Selander et al. 1970
Various passerine birds and frogs (*Hyla*)	—	15	—	Selander et al. 1970
Cricket frog (*Acris*)	20	14–23	—	Dessauer and Nevo 1969
Lizard (*Uta*)	—	15–20	6–8	Selander et al. 1970
Horseshoe crab (*Limulus*)	25	25	6	Selander et al. 1970
Fruit flies (*Drosophila*)	14–28	25–86	11–18	Lewontin and Hubby 1966, Prakash et al. 1969, Prakash 1969, Kojima, Gillespie, and Tobari 1970, Ayala et al. 1972.

others (frogs, lizards, and passerine birds) are less polymorphic, although it is not clear why.

The evolutionary significance of genetic variation is poorly understood. Clearly, variation is needed for evolutionary change. But how much is optimum for an environment that is changing at a given rate? To what extent is variation detrimental to the fitness of the individual? How is the level of variability in a population regulated? Or is the level of variability beyond the control of evolved mechanisms? Some of these questions will be explored in the next chapter.

23

The Organization of Variability

Most higher plants and animals are diploid and engage in sexual reproduction. These characteristics are themselves adaptations that help to organize genetic variability in populations. In fact, if there were no genetic variation, both diploidy and sexual reproduction would be superfluous. In this chapter, we shall consider some of the evolutionary consequences of sexuality, diploidy, and other mechanisms for managing genetic variation, and we shall discuss some of the ecological situations in which the advantages of these adaptations are outweighed to the point that they are abandoned as evolutionary strategies.

While it may be argued that genetic variation provides populations with evolutionary flexibility, most adaptations discussed in this chapter clearly have evolved to reduce the detrimental effects of mutations on the fitness of individuals. This strongly implies that observed levels of genetic variation are not adaptations of the population to optimize evolutionary flexibility, rather they are determined by molecular and biochemical properties of the genetic material that are beyond the reach of evolutionary modification.

Diploidy allows such allelic interactions as dominance and heterosis.

There appears to have been a persistent trend toward diploidy in the evolutionary history of life forms. Most primitive unicellular plants and animals are haploid for the greater part of their life cycles, and are diploid only during a brief period of sexual reproduction when two haploid cells fuse and exchange genetic material (Figure 23-1). The haploid stage is reduced to a lesser position in the life cycle as one proceeds to more complex levels of organization. Most multicellular animals are diploid throughout their life cycles, except for the gamete stage. Haploidy is more persistent in plants. In green algae the diploid stage has brief reign, ending when the zygote undergoes meiosis to form haploid cells, called spores. The spores divide and grow into the mature algal plant, which is haploid. In mosses and liverworts, the diploid generation is extended beyond the zygote stage by mitosis, but the diploid part of the plant is small and parasitic upon the haploid stage of the

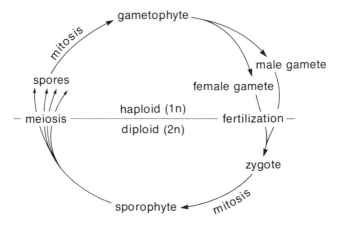

Figure 23-1   The alternation of haploid and diploid generations as a result of sexual reproduction. Chromosome number is halved to a single chromosome complement (1N) by meiosis. The diploid chromosome number (2N) is reconstituted by the fusion of gametes at fertilization. Cell proliferation by mitosis may follow either meiosis, fertilization, or both, but in higher animals it is restricted to the diploid stage of the life cycle.

life cycle (Figure 23-2). In ferns, the diploid plant (*sporophyte*) has assumed the dominant part of the life cycle, and the haploid *gametophyte* is reduced to a small prostrate leaflike structure that is easily overlooked in nature. In the conifers and flowering plants, the haploid generation is reduced to the point that the few cells of the haploid plant are either enclosed in the female parts of the flower or form the pollen grains. Why should the extent of the gametophyte generation vary so widely in plants? An examination of the evolutionary consequences of diploidy provides some clues.

Diploidy can be advantageous to the individual in several ways, the most obvious of which is by masking the expression of deleterious recessive

Figure 23-2   Moss plants with spore-producing capsules. The sporophyte plant is diploid, and because it lacks chlorophyll it is completely dependent upon the leafy haploid stage of the life cycle.

alleles. Although the step from haploidy to diploidy confers immediate advantage to the individual, that advantage is soon lost (Muller 1932). When recessive alleles are masked, the strength of selection to remove them is reduced. As a result, the frequency of deleterious recessive alleles is increased by recurring mutations until the effects of their expression in the homozygous form reach the same level as in the haploid stage. (This will be shown mathematically in Chapter 28). Once diploidy has become established, it cannot be lost without exposing a substantial amount of deleterious genetic variation. The evolution of diploidy, therefore, has the short-term advantage of masking recessive alleles, but when this advantage is subsequently lost, it is virtually impossible to revert to haploidy. Diploidy may ultimately lead to evolutionary disadvantage because if a deleterious trait is expressed to any degree in heterozygous individuals, the efficiency of selection to remove it is reduced. Selective death of heterozygous individuals with partially recessive deleterious genes also removes beneficial alleles.

Potential long-term disadvantage to diploidy accrued through incomplete dominance may be countered, in part, by interactions between alleles that result in the heterozygote having a greater fitness than either homozygote. This property is referred to variously as *heterosis, heterozygote, superiority,* and *overdominance.* The widespread occurrence of polymorphism in nature (see Table 22-3) has led some geneticists to suggest that heterosis may be fairly common; as we shall show more rigorously in a later chapter, heterozygote superiority maintains alleles in stable polymorphism.

Diploidy may also confer advantage by masking somatic mutations (Crow and Kimura 1965). Mutation is by no means limited to the germ line. All the cells in an organism should have the same genes because they are derived by mitosis from a single cell, the zygote. But, instead, the cells of the body undergo slow genetic differentiation through somatic mutation. If the organism were so complex that it depended on the interaction of numerous cells to function properly, mutations in somatic cells could be detrimental to the organism. Diploidy could well be an important mechanism to suppress the expression of deleterious somatic mutations. At mutation rates of one per million cells per generation, or even higher, for each gene locus, a large proportion of the cells in an organism could accumulate defects in several of the many hundreds or thousands of genes responsible for its functioning. If such defects were expressed, they would seriously reduce the fitness of the organism. Perhaps it is more than a coincidence that haploidy is found either among single-celled organisms, in which the cells in a clone are independent and defective individuals do not influence the fitness of others, or among organisms, particularly the lower plants, in which the level of tissue differentiation is primitive and defective cells are readily replaced by cell proliferation.

Is dominance a feature of the gene or an evolved property of the genotype?

An overwhelming proportion of mutants is recessive to the normal or prevalent allele, and these mutants are expressed only when homozygous. Is it fair to assume, then, that most mutations are recessive? How do we account for the fact that common alleles must, at one time, have been rare mutations? There are two hypotheses: first, that the degree of dominance is an inherent property of a gene and evolution proceeds primarily by the incorporation of dominant mutations (Wright 1934, Haldane 1939); second, that the expression of a mutant is altered by modifying genes (Fisher 1931).

Little is known about the biochemistry of gene expression. Mutants that fail to produce a protein molecule with normal function seemingly would be recessive because the complementary normal allele would continue to produce the enzyme or structural protein product, thereby masking the defective mutant. If a mutation were to change the function of a protein, its expression would depend largely on how that protein interacted with others. English statistician and population geneticist R. A. Fisher (1928, 1931) suggested that if a mutant allele resulted in an intermediate heterozygote—if its expression were neither completely dominant nor completely recessive—and if the mutation were superior to the normal allele, modifying genes would tend to increase the dominance of the mutant allele. That is, any heterozygous individual that also had modifiers for increasing the dominance of the advantageous allele would be selected in place of a heterozygote that lacked such modifiers. Such modifiers, which derive their advantage by increasing the fitness of another gene, are rapidly incorporated into the gene pool by natural selection (*e.g.,* Kettlewell 1965).

Recombination is associated with sexual reproduction.

Through recombination, new combinations of genes may arise in loosely linked groups. Most geneticists feel that this is the primary evolutionary advantage to sexual reproduction. How this advantage is actually conferred, however, is a point of considerable controversy. Recombination can occur if homologous chromosomes are present in the same cell at some time during the life cycle of the organism, as in all species with sexual reproduction. The evolutionary advantages of recombination pertain to populations that are haploid through most of their life cycle as well as to populations of organisms that are predominantly diploid. Thus the advantages of sexual reproduction must not be misconstrued as being related only to diploidy.

Recombination does not directly improve the fitness of an individual; beneficial alleles of two genes are a part of the genotype, whether both are located on the same chromosome or on different chromosomes of a homologous pair. But by bringing two beneficial alleles together on the same chromosome and thereby preserving the combination through linkage, recombination may greatly enhance the fitness of an individual's progeny.

Several theories have been proposed to account for the apparent advantage conferred by recombination: first, that it permits the simultaneous incorporation of favorable mutations that occur spontaneously within a population (Crow and Kimura 1965); second, that recombination preserves favorable gene combinations appearing as the result of gene flow between populations with different genetic composition (Maynard-Smith 1968a, 1971); and third, that recombination produces occasional offspring in a brood with exceptionally high fitness (Williams 1966b, 1975, Williams and Mitton 1973). The difference between these theories is subtle, but it is fundamental to understanding the evolutionary advantage of sexual reproduction and recombination. Suppose genes A and B both have mutant forms A' and B' that are advantageous over the wild type. Crow and Kimura suggested that the advantageous combination A'B' will occur more rapidly in sexual populations because an individual with one of the mutant alleles can mate with an individual carrying the other mutant allele. In asexual populations this is not possible, and the combination A'B' can arise only through the successive mutation of A to A' and B to B' in a single lineage, or clone. Unless both mutations arose in the same lineage, one would outcompete the other, and only the fitter gene would be incorporated into the population.

Maynard-Smith (1968a) questioned Crow and Kimura's assumption regarding mutation as a unique event. He argued that mutation at a given gene locus occurs continually and mutant alleles build up to such a level that combinations of two or more mutants are about as likely to occur in asexual clones as they are through sexual reproduction. The advantage to recombination thus cannot be related to the preservation of new combinations formed out of the genetic variation present in a population, but it must be related to the occurrence of more unique combinations, perhaps of alleles that occasionally enter a population by migration from outside. The genetic interaction of two populations that are subjected to slightly different environments may produce combinations of genes that are superior to either parental population in a third habitat that differs slightly from the environments of either parental population.

Because mutants are generally rare, very small populations can hold relatively little genetic variability and mutations are more nearly unique events in the sense of Crow and Kimura's model. Hence the relative importance of the two theories for the advantage of recombination must depend, in part, on population size.

Williams has argued that sex must confer evolutionary advantage to the individual, rather than to the population, because the individual is the fundamental unit of adaptation. Recombination continually produces variations in the genotypes of offspring. The greater the genetic variability among the progeny of an individual, the greater the chance that at least one of these progeny will have high fitness, and outcompete others in the population in the particular conditions into which they are born. But all these arguments are speculation, and none would seem to account for the prevalence of sex among such a wide array of animals and plants. The evolutionary advantage of sex remains one of the most elusive and fundamental problems remaining for biologists to solve.

Can recombination be prevented from breaking up favorable gene combinations?

Although recombination can aid in bringing together favorable combinations of genes, it is equally effective in breaking them apart. The disruptive effects of recombination are not a major problem where a gene combination is so prevalent that there are no alternative alleles. But when several alleles are maintained at relatively high frequencies in a stable polymorphism, or are present in a population while one is replacing another, recombination must frequently break up favorable gene combinations.

Gene combinations can be protected in several ways, the most drastic of which—giving up sexual reproduction altogether—may seriously reduce the evolutionary plasticity of the population. Gene combinations are also preserved when gene loci are brought close together on the same chromosome by suitable inversions. Crossing over between gene loci is reduced in this way, thereby causing the loci to behave as a single unit of recombination. Such *supergenes* are common and have been carefully studied in reference to background color and banding patterns in some land snails (discussed below). In addition, rates of recombination within the entire genome or within specific regions on chromosomes are under the direct influence of selection (Valentin 1973, Chinnici 1971, Kidwell 1972). Recombination can be suppressed completely during meiosis by cytological mechanisms involving specific types of inversions on chromosomes. As we shall see below, these inversions allow crossing over to occur, but the chromosome products of recombination are not viable, and thus are never incorporated into gametes. As might be expected, this unusual way of suppressing recombination is fairly rare. It has been best documented in that genetic workhorse, the fruit fly *Drosophila,* to which we shall return after first considering the supergenes of land snails in further detail.

Genes for background coloration and banding pattern are tightly linked in the land snail *Cepaea nemoralis.*

A lengthy research effort in England and continental Europe has centered upon the ecological genetics of the common land snail *Cepaea nemoralis* and its close relatives (Cain and Sheppard 1954, Jones 1973). *Cepaea* exhibit marked polymorphism both for color, brown being dominant over yellow, and for pattern, unbanded being dominant over banded (Figure 23-3). It is believed that the polymorphisms are maintained by heterosis based on pleiotropic physiological characteristics. But the importance of coloration in camouflaging the snail against visual predators is demonstrated by the patterns of predation on the snails by the song thrush (Sheppard 1951, Carter 1968). The thrushes capture snails on the forest floor and carry them to certain rocks, known as anvil stones, where they break them open and eat their flesh. In early spring, 40 per cent of the snail shells found at these stones have yellow background color even though yellow individuals constitute only 25 per cent of the total population in the woods: The predators apparently remove yellow snails selectively. As the season progresses and more green vegetation appears on the forest floor, the proportion of yellow shells at the anvil stones decreases, to between 22 and 27 per cent in early May and between 9 and 15 per cent by early June. These observations suggest that yellow coloration confers a greater degree of protection on a background of

Figure 23-3   Basic color and banding patterns of the polymorphic land snail *Cepaea nemoralis*. Left to right: yellow-unbanded, yellow-banded, and brown-unbanded (after Ford 1971).

green vegetation than of brown leaf litter. As one might expect, the proportion of yellow shells is greater in field habitats than in forest habitats. Furthermore, most yellow shells are also banded, whereas most brown shells lack banding. A solid yellow snail undoubtedly would be conspicuous, even against a green background, whereas banding patterns break up the outline of the snail and increase its apparent camouflage. The general scarcity of yellow, unbanded recombinants is due to the linkage of genes for background color and banding pattern into a single supergene, within which little recombination occurs (Ford 1971).

The polymorphic land snail *Bradybaena similaris* of Japan has more rapid development and simpler genetics than *Cepaea,* and so it is a better subject for genetic experiments to determine the presence of supergenes. Like *Cepaea, Bradybaena* is polymorphic both for color (brown dominant over yellow) and pattern (banded dominant over unbanded). Several supergenes—yellow-banded, brown-unbanded, and yellow-unbanded—occur in the Japanese population; few recombinant types appear in the progeny of matings between these types (Komai and Emura 1955). The basis for the polymorphism is problematical, but it may be caused by heterosis for some physiological function, as suspected in *Cepaea.*

Chromosome inversions preserve favorable gene combinations.

Recombination rates in some species of *Drosophila* flies are reduced by the presence of polymorphisms for inversions on their chromosomes. Chromosome inversions may be identified easily when homologous chromosomes are paired during meiosis, because heterozygous combinations of inverted and normal chromosomes produce unusual, but highly characteristic pairing configurations. When two identical chromosomes pair during meiosis, homologous regions on the two chromosomes must match for pairing to take place. When one of the chromosomes has an inversion, pairing can occur only if the chromosomes assume a looplike configuration over the inverted region, with one of the chromosomes twisting inside the loop (Figure 23-4).

The fruit fly *Drosophila pseudoobscura* of western North America has sixteen different inversion forms (arrangements, as they are called) of chromosome III (there are five pairs of chromosomes altogether), each named after the locality in which it was first described—Olympic, Estes Park, Chiricahua, and so on (Dobzhansky and Epling 1944). *D. pseudoobscura* shares one-third chromosome arrangement (Standard) with the closely related species *D. persimilis,* which itself exhibits eleven other arrangements. Any one population of *pseudoobscura* may have up to seven different forms of chromosome III; in some tropical species of *Drosophila,* more than a dozen arrangements have been reported from one locality. With such high numbers of chromosome arrangements, most of the chromosome pairs in cells that are undergoing meiosis are heterozygous for inversions. The most highly polymorphic species known is another fly, the midge *Simulium,* with 134 different inversions recorded (White 1973).

Pairing of two chromosomes that are heterozygous for an inverted region can cause cytological complications during meiosis if crossing over occurs within the inverted region. Crossing over within a paracentric inver-

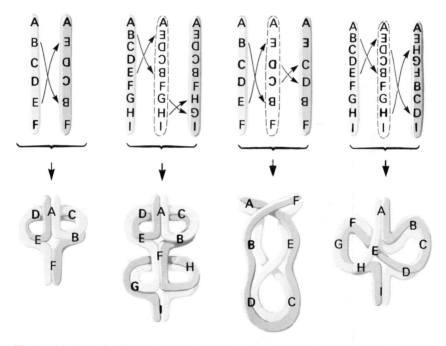

Figure 23-4   Configurations of paired homologous chromosomes that are heterozygous for one or more inverted regions. From left to right: a single inversion; two independently placed inversions; one inversion completely including another inversion; and one inversion partly overlapping another inversion (after Dobzhansky 1951).

sion, one that does not include the centromere of the chromosome (the point on the chromosome from which the pair is pulled apart during meiosis), produces nonviable chromosome fragments when the pair segregates. Although crossing over between the chromosomes still occurs, its effects are not passed on to the progeny. Demonstrating how heterozygous inversions prevent the expression of recombination is like explaining a can of worms, but some satisfaction might be gained from following the process as it is outlined in Figure 23-5. If the centromere is included in the inverted region (a pericentric inversion), the recombined chromosomes are viable in that they behave normally during meiosis, but because one lacks part of the chromosome and the other has a duplicated region, the gametes formed with such chromosomes usually are not viable. It is not surprising that such pericentric inversions are virtually unknown in flies. Although few animals other than flies exhibit chromosome polymorphism, many plants do (for example, see Walters 1942 and Swanson 1940). Plants have other cytogenetic mechanisms to suppress crossing over, including translocation heterozygotes that make the entire chromosome complement behave as a single genetic unit within which no recombination takes place (Stebbins 1950, 1970).

The occurrence of chromosome polymorphism in populations of *Drosophila* and other flies, almost as much as its absence from most other groups, raises many questions about its evolutionary significance (Van Valen and Levins 1968, Dobzhansky 1970). Perhaps inversions prevent locally advantageous combinations of genes from being broken up by recombination with genes introduced from other populations. That chromosome arrangements are adapted to local conditions in *Drosophila* populations was shown by laboratory experiments on competition between flies with different chromosome arrangements (Dobzhansky 1947). The success of a particular chromosome type depended upon the locality from which the flies were obtained. The results of these experiments suggest that the genotype can vary independently of the structure of the chromosome. Although unique

Figure 23-5   The fate of crossing over in heterozygous inversions. Left: the inversion loop in the chromosome pair after replication, showing the location of the crossing over. Center: appearance of the chromosomes after the first meiotic division. Right: the two inviable chromosomes have either two centromeres or none, and thus do not segregate properly during meiosis. They lack whole segments of the chromosome and duplicate others (after Dobzhansky 1951).

local inversions could protect a population from the detrimental effects of gene flow, local adaptation and the formation of supergenes probably are not the primary purpose of inversion polymorphism.

Inversion polymorphism could enhance the degree of heterozygosity within a population by maintaining different alleles of many genes on chromosomes with different inversions. Experiments on competition between individuals bearing different chromosome arrangements show that heterozygous individuals are usually the most fit (Dobzhansky 1947). This result explains the persistence of the polymorphism through heterosis. Furthermore, heterozygotes of chromosomes from different areas usually are *not* more fit than homozygous pairs, hence chromosome arrangements are locally coadapted to produce heterozygote superiority and balanced polymorphism. It would seem that one function of balanced polymorphism is to maintain genetic variability in a population for responsiveness to changing environments without the risk of losing variability altogether. If certain chromosome arrangements were clearly inferior to others under certain conditions, they could easily be eliminated from local populations.

Beyond functioning to protect local gene combinations against recombination and to preserve genetic variation, inversion may also speed the evolutionary substitution of some alleles by separating them from heterotic loci. If a gene is closely linked to a polymorphic gene maintained by heterosis, it would be difficult to fully incorporate the new mutant into the population without recombination between the two loci. The probability of recombination is increased by separating the loci on the chromosome and inversion could handle this quite easily. We should note here that flies usually carry only four to six chromosomes, fewer than any other group, hence their genes are arranged into a few large linkage groups. Inversions in the chromosomes effectively increase the number of chromosome equivalents in flies by breaking up close linkages into a variety of evolutionarily independent arrangements.

In sexual, diploid populations, inbreeding increases homozygosity.

The advantage of diploidy lies primarily in the interaction between alleles at each gene locus. The phenotypic advantages of dominance and heterozygote superiority can be expressed only in the heterozygous state. *Inbreeding* (mating with close relatives or with oneself—*selfing*—which occurs in many plants) promotes homozygosity, thereby permitting the expression of deleterious recessive alleles and reducing heterosis. The effects of homozygosis are thought to have led to the genetic decline of many of the European royal families, within which inbreeding between brother and sister and first cousins was a common practice.

If an individual carries a deleterious recessive allele in a heterozygous combination, chances are one in two that the allele will be passed on to any one offspring, and the chances are, therefore, one in four that the gene will be present in any two offspring. (We assume that the deleterious trait is rare and that our heterozygous individual mates with an individual homozygous for the normal gene.) If the progeny of this union happen to be male and female, and if they mate with each other, the deleterious gene will be homozygous in their progeny one time in four. Thus, when brother and sister mate (sib mating), a recessive trait inherited from one of their parents has one chance in sixteen of being homozygous in their offspring. The probability is one in four when selfing occurs because, in effect, two heterozygote genotypes mate with each other.

Inbreeding and homozygosis usually reduce the fitness of an inbred line. For example, yield in selfing lines of corn decreased to less than one-half after ten generations and to less than one-quarter after twenty generations. Similar effects of inbreeding on fitness in humans, which include increased physical abnormalities and mental retardation, are summarized by Lerner (1968).

Dobzhansky (1967) has described diploid populations in nature as "... saturated with injurious mutant genes." To show the extent to which chromosomes contain deleterious recessive alleles, Dobzhansky and Spaasky (1963) devised an experiment by which they could produce flies that were homozygous for any particular chromosome. To do this, they first mated male flies from natural populations of *Drosophila pseudoobscura* with females from laboratory strains carrying dominant markers on the second, third, and fourth chromosomes (Figure 23-6, step 1). The male progeny were then backcrossed to the laboratory strains to obtain males and females heterozygous for the same wild chromosomes (step 2). When these flies were sib-mated (step 3), the relative fitness of individuals with homozygous wild-type chromosomes (those lacking the markers in the laboratory strain) could be determined directly. The effects of homozygosis on viability and fertility were drastic. One-quarter to one-third of the chromosomes contained lethal or semilethal genes, and most had genes that reduced fitness to a lesser extent; 4 to 18 per cent of the chromosomes produced male or female sterility when homozygous.

The degree to which inbreeding reduces heterozygosis in a population depends on how closely mates are related. The most extreme case of inbreeding is selfing. In a two-allele system, with selfing, the alleles at a single gene locus may be combined in three possible matings: *aa* with *aa*, *ab* with *ab*, and *bb* with *bb*. The first and third produce only homozygous progeny. Matings between heterozygous individuals produce heterozygous and homozygous offspring in the ratio of one to one. Each generation of selfing, therefore, reduces the proportion of heterozygotes in a population by one-half. Notice that the frequencies of the alleles in the population do not change through inbreeding; only the proportion of genotypes changes. The reduction of heterozygosis in a population through inbreeding proceeds less rapidly when mates are less closely related.

Population geneticists calculate a coefficient of inbreeding (F), which varies between 0 and 1, on the basis of deviations from random mating within a population (see Li 1955, 1967). Different levels of inbreeding have been established through controlled matings in laboratory and domestic populations to quantify the relation between inbreeding and loss of fitness. For example, egg-to-adult survival in several species of *Drosophila* decreases by

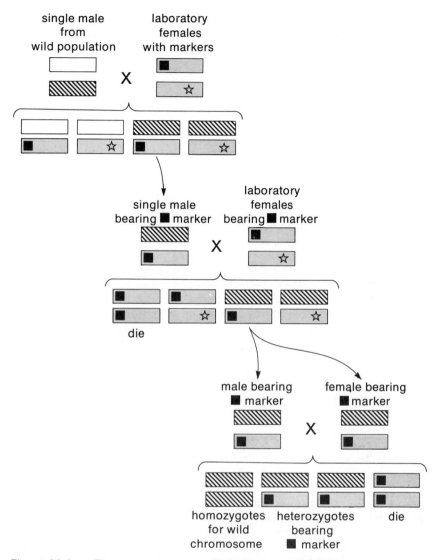

Figure 23-6   The method used by Dobzhansky and Spaasky (1963) to produce homozygosity in chromosomes from natural populations. According to the laws of probability, homozygous wild-type individuals should be one-third as common as the individuals carrying the laboratory marker. Deviations from this ratio may be used as a relative measure of the fitness of the homozygous genotype. The wild chromosomes W are all individually distinct genetically (W_1, W_2, W_3, . . .); the laboratory chromosomes (L) are uniform and have a dominant marker. In all of these experiments, crossing over was suppressed by inversions.

about 6 per cent for each increase of 0.1 in F (Mettler and Gregg 1969). Similar data have been compiled by Falconer (1960), who found that phenotype values decreased an average of 4.9 per cent (range, 0.8 to 12.5 per cent) for each 0.1 increase in F. Survival of Japanese quail embryos to hatching decreases by about 10 per cent for each 0.1 increase in F (Lucotte 1975). In an extreme case, inbreeding in laboratory populations of mites led to the complete breakdown of reproductive function and extinction of the populations after one year (Poe and Enns 1970).

How do populations prevent inbreeding?

Inbreeding is a serious problem in natural populations in which random mating is limited. The sedentary habits of many species, particularly of plants, could lead to a high degree of consanguineous mating were it not for the evolution of mechanisms to reduce or altogether prevent inbreeding.

Selfing does not arise as a problem when sexes are separate. In bisexual species, inbreeding among progeny is often reduced by the dispersal of young prior to reproductive maturity, although prevention of inbreeding may not be the primary evolutionary purpose of dispersal. When progeny do not disperse, other mechanisms may prevent the occurrence of inbreeding. That armadillos always give birth to identical quadruplets may be such a mechanism: All four young are the same sex, and thus they cannot mate with each other.

When both sexual organs occur on the same individual, selfing is a potential problem, but the physical arrangement of the male and female organs frequently makes self-fertilization impossible, even for a hermaphroditic Houdini. As we have already seen (page 277), some hermaphrodites are first male and then female, and avoid the problem of selfing by separating their sexes in time.

Most plants have both male and female sexual organs on the same individual, often in the same flower. Because plants rely on wind or animals to effect pollination, their sex organs are precariously exposed to the possibility of self-fertilization through the haphazard distribution of pollen. Frequently this problem is solved by the differential maturation of the male and female organs; the stigma is usually receptive to pollen before pollen is produced by the same flower or plant. In the insect-pollinated passionflowers (Passifloraceae), the flowers open each day, first exposing the female flower parts and later the male flower parts (Figure 23-7).

The heterostyle system reduces inbreeding in the primrose.

Flowers of the primrose (*Primulus*) contain both male and female sexual organs, and they are pollinated by long-tongued insects, mostly bees and moths. Self-pollination would be a major problem if it were not for the occurrence of two types of flowers, the "pin" and the "thrum," in most natural populations (Figure 23-8). In pin flowers, the stigma is elevated on a

Figure 23-7   A mechanism to reduce self-fertilization in the passionflower, *Passiflora foetida*. In the newly opened flower (left), the stigmas are raised above the horizontal anthers; they cannot be contacted by a bee in this position. About twenty minutes later (right), the stigmas are below the anthers, where they can be pollinated by bees taking nectar from the base of the flower. Photographs courtesy of D. Janzen (from Janzen 1968).

long style to the opening of the corolla tube, and the anthers are located deep within the tube. The arrangement of sexual organs is reversed in the thrum flower. The thrum style is short, and so the stigma resides deep within the corolla tube; the anthers are above, arranged in a ring around its mouth. (*Thrum* is an old English word for an unraveled end of yarn, which the thrum flower resembles.) That the spatial separation of the male and female sexual organs in the primrose flower prevented self-pollination was first appreciated by Charles Darwin (1877). Because the stigma and anthers are located in complementary positions in pin and thrum flowers, pollen from a pin flower

Figure 23-8   Exposed view of flowers of the primrose, *Primulus vulgaris*, showing the position of the stigma and anthers: left, a pin flower; center, a thrum flower; and right, a long homostyle flower (after Sheppard 1958).

that sticks to the tongue of a bee will contact the stigma of the thrum flower when the tongue is inserted into it, and vice versa.

Inbreeding still occurs occasionally in primroses, when pollen falls by gravity onto the stigma of a thrum flower, or when an insect's tongue brushes pollen against the stigma as it is withdrawn from a pin flower. The primrose has, however, additional mechanisms to reduce inbreeding. The stigma can receive pollen several days before the anthers are ready to shed their pollen. Delayed maturation of the male gametes has the great advantage that for the first few days after the flower opens, outcrossing is absolutely guaranteed, but if there are few insect pollinators or if primroses are uncommon in the area, inbreeding is possible after the anthers mature.

Inbreeding is also reduced by several so-called legitimacy mechanisms that control the growth of pollen tubes through the style on their way to fertilize the ova. Legitimacy is usually determined by a series of self-sterility alleles at one or more gene loci. When the same allele occurs in both the pollen and the stigma (which would follow self-pollination), the pollen tube does not penetrate the style or grows slowly and gives pollen of a different genotype the advantage in fertilization (Bateman 1956).

The differences between pin and thrum flowers are controlled by a supergene that incorporates anther height, style length, legitimacy mechanisms, pollen size, and the structure of the stigma (Ford 1971). The thrum superallele (*S*) is dominant to the pin superallele (*s*). Thrums are normally heterozygous (*Ss*) in nature because most matings are of the pin-thrum (*ss* × *Ss*) type. Note the similarity between the pin-thrum mechanism and male-female determination in most organisms. The primrose, with bisexual flowers, has adopted a second level of "sexuality" to ensure outcrossing.

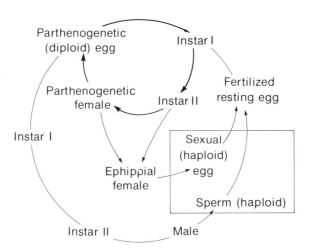

Figure 23-9   The life cycle of the chydorid cladoceran *Pleuroxus denticulatus*. The parthenogenetic part of the life cycle is outlined by a heavy line and the sexual cycle by thin lines. Females can change between the sexual and parthenogenetic forms, depending on environmental conditions (after Shan 1969).

The characteristics of the flower parts of the primrose are controlled by separate gene loci within the supergene: short style (*G*) is dominant to long style (*g*), and high anther (*A*) is dominant to low anther (*a*). Thus the thrum genotype (*S*) is *GA* and the pin genotype (*s*) is *ga*. However closely the genes within the supergene are linked, recombination occasionally breaks up the supergene. Recombinant flowers have low stigmas and low anthers (short homostyles) or long stigmas and high anthers (long homostyles). Small numbers of homostyles, especially the long type, are found in some natural populations of the primrose. Although the fitness of the recombinant genotype is eventually reduced considerably through inbreeding (selfing) and the homozygosity that follows, homostyle plants can have an advantage over heterostyle plants where animal pollinators are rare, and selfing leads to high fertility and efficient reproduction. Recombinant homostyle genotypes have been known to spread rapidly through a population as a highly inbred line until the deleterious effects of inbreeding caused the population to die out (Crosby 1949, 1959). Inbreeding is maintained in some environments, such as windy seacoasts, from which flying insect pollinators are absent. Coastal populations of *Primulus* are frequently predominantly homostyle individuals.

Can asexuality and inbreeding be advantageous?

Sexual reproduction and outcrossing stave off the deleterious consequences of homozygosis, but a price is often exacted through inefficiency of reproduction. In most populations, fecundity is ultimately limited by the number of females; males contribute only genetic information to the zygote and do not influence the number of eggs produced by the females. Where parental care is well developed, males can contribute to the survival of those eggs, but if males do not serve this function they are a costly genetic necessity.

When rapid population growth is a major factor in the evolution of the species, the efficiency of asexual reproduction may outweigh its long-term genetic disadvantages. Many species of small animals whose populations are built up newly each year from a small number of overwinter survivors have adopted *parthenogenesis* (the formation of diploid eggs that develop without fertilization) to increase their reproductive rates. Many plants that live in highly disturbed habitats and must continually colonize newly disturbed areas have also abandoned sexual reproduction. Other species that readily colonize disturbed or otherwise marginal habitats have a high degree of inbreeding built into their sexual systems.

The life cycle of many water fleas, or cladocera (tiny crustaceans found in lakes and ponds), includes, during the favorable spring and summer months, a phase of rapid expansion of an all-female population that reproduces by means of parthenogenesis. As environmental conditions worsen toward the end of the growing season, a sexual generation may be produced (Banta and Brown 1939, Figure 23-9). Parthenogenetic eggs then produce diploid males, instead of females, and females form haploid eggs that are fertilized by males. The zygote overwinters in the egg at the bottom of the pond and hatches out as a parthenogenetically reproducing female in the spring. Seasonally parthenogenetic cladocerans thus have achieved the high reproductive efficiency of asexual reproduction without completely giving up the genetic advantages of sexual reproduction. They share this pattern with many protozoa, rotifers, and aphids (e.g., Kennedy and Stroyan 1959) that undergo similar seasonal phases of rapid population growth.

All-female populations without interspersed sexual generations have been found in some fish, reptiles, and amphibians. Most of these populations are derived from hybrids between parents of different species (Schultz 1973). A parthenogenetic diploid egg is usually produced by a modified meiosis. This would lead immediately to homozygosis at all gene loci if the second meiotic division were suppressed to maintain the diploid chromosome number (see Figure 22-3). But, in fact, heterozygosity is probably fixed by employing a premeiotic mitosis to increase the chromosome number in the cell to tetraploid (4N), and then proceeding with the regular meiosis (Uzzell 1970).

Several all-female species are known among fresh-water cyprinodont fish of the genera *Poecilia* and *Poeciliopsis* (Schultz 1969). The reproductive systems of the fish differ from species to species, but fall into two general classes: gynogenesis and matriclinus sexual inheritance (Schultz 1961, 1967). In the first case, diploid eggs are formed and the female mates with a male of a *different* species in the same genus, but the male's sperm function only to initiate the development of the diploid eggs; the sperm do not contribute genetically to the offspring. In matriclinus inheritance, also called hybridogenesis, haploid eggs are fertilized by males of another species and the paternal characteristics are expressed in the offspring, as in any sexual union. But when meiosis occurs in the offspring, maternal and paternal chromosomes do not pair and only the chromosomes from the female parent reach the haploid eggs. Thus, male characteristics are not transmitted from generation to generation, but rather are supplied newly with each fertilization.

All-female species of *Poeciliopsis* are as interesting ecologically as they are genetically because they must coexist with ecologically similar bisexual species. On one hand, abandonment of the male sex increases the reproductive efficiency of the all-female species, and therefore increases their ability to compete with respect to bisexual species. On the other hand, the unisexual species are at a disadvantage in mating because males prefer to mate with females of the bisexual species (McKay 1971). As a result, the unisexual and bisexual species achieve stable coexistence (Moore and McKay 1971). When bisexuals are relatively rare, bisexual females are fertilized selectively, and the proportion of the bisexual population in the mixed-species association increases because there are few males to fertilize the unisexual females. If the bisexuals become abundant, however, many of the males are kept from breeding with bisexuals by dominant fish, and they fertilize the unisexual females, allowing that population to increase. The position of the equilibrium is influenced by the relative ecological fitnesses of unisexual and bisexual females.

Mating systems in plants have received comparatively little attention. Most species of flowering plants are bisexual (*monoecious*), having both male and female sexual organs on the same plant, often in the same flower (hermaphroditic). The advantages of monoecy are not well understood; many monoecious plants are self-incompatible and cannot realize increased reproductive efficiency by having the reproductive organs of both sexes. For species in which individuals are self-compatible, increased reproductive efficiency may be a major evolutionary advantage. Regardless of degree of self-compatibility, the ability to produce both male and female gametes may be a way to achieve a degree of reproductive flexibility analogous to the extreme degree of developmental flexibility found in most plants.

Dioecious species (separate sexes) are uncommon in temperate floras, comprising only 2 per cent of the flowering plants of the British Isles (Lewis 1942), but are more common on oceanic islands (Carlquist 1966) and in tropical forests, particularly in the understory (Ashton 1969). Carlquist suggested that dioecy increases outcrossing and thereby maintains a high degree of genetic variability in small island populations. Janzen (1970) and Levin (1975) have applied the same genetic argument to tropical forest species, suggesting that tropical plants must maintain a high evolutionary potential to cope successfully with the adaptations of the herbivores that exploit them. In temperate forests, Janzen argued, the environment is dominated by physical rather than biological characteristics, and so much evolutionary potential is not needed.

Most discussions of mating systems in plants treat level of outcrossing as a more general expression of adaptation than are the specific and varied morphological and physiological mechanisms that promote or retard outcrossing (Lewis 1954, Fryxell 1957, Bowler and Rundel 1975). Inbreeding is most commonly associated with annual life cycles (Stebbins 1957, Grant 1975), long-distance dispersal (Baker 1965), marginal habitats (Lloyd 1965, Solbrig 1972), and cold environments (Savile 1972); outcrossing is conspicuous among species having long life cycles and living in stable habitats, particularly the tropics (Ashton 1969, Bawa 1974).

Correlations between ecology and mating system do not necessarily reveal the adaptive significance of level of outcrossing, which may, after all, be influenced by many factors. Inbreeding is thought to promote reproductive efficiency and guarantee seed set either in the absence of pollinators or in sparse populations. Inbreeding also leads to population uniformity, close genetic relationship between parent and offspring, and reduced gene flow, all of which promote local adaptation and ecotypic differentiation. Outcrossing is seen as maintaining genetic variability through recombination and the preservation of genetic polymorphism (Levin 1975). So far, however, the role of outcrossing in adaptation and evolution remains an important unsolved problem in evolutionary ecology.

Can natural selection favor genetic polymorphism?

Many of the mechanisms discussed in this chapter are evolutionary strategies to reduce the deleterious effects of genetic variability on the fitness of individuals. We have also seen that the superior fitness of a heterozygote genotype can maintain genetic polymorphism in a population because two

Figure 23-10   Mimicry polymorphism in some East African butterflies. The three butterflies in the left-hand row are unpalatable species that serve as models for several mimetic species: center row, *Papilio dardanus*, with non-mimetic male (top) and three mimetic forms of females; upper right, *Hypolimnas misippus,* with nonmimetic male (top) and mimetic female; lower right, *H. dubius*, in which both sexes are dimorphic mimics (after Wynne-Edwards 1962).

alleles constitute the most fit genotype. The unavoidable homozygous genotypes that appear under a heterotic system are less fit than the heterozygote. Hence polymorphism resulting from heterosis does not benefit all individuals in a population and is not selected in and of itself. Under some circumstances, however, genetic variation may actually increase the fitness of all individuals. Mimicry polymorphisms, in which a species that is palatable to predators mimics more than one model, provides a good example. When mimics are common compared with models, the advantages conferred by mimicry are reduced because the negative reinforcement of the predator's learning experience occurs less frequently (see page 217). By mimicking many different models, therefore, the absolute abundance of mimics relative to any one model is reduced, thereby permitting increase in the population of the palatable species. The advantage that mimicry polymorphism confers on the individual lies in the individual's resemblance to a model species that is mimicked by relatively few palatable individuals. That is, there is a great advantage to being atypical, and this leads directly to the diversification of mimetic forms within a population (Figure 23-10).

In the evolution of mimicry polymorphism, predators select for diversity based on natural variation in the environment (in this case, a variety of models). This general mechanism may be extended to select variability *per se,* irrespective of the environmental background, if predators selectively remove the most abundant prey items. This mechanism is usually referred to as *apostatic selection* (Clark 1969) and was first proposed by Popham (1941, 1942) and by Cain and Sheppard (1950, 1954) to explain polymorphism in the appearance of bugs and the snail *Cepaea nemoralis* (see page 349). Selection for diversity of appearance between species is evident from the bewildering array of adaptations that have been evolved by insects and other forms to confer cryptic properties (Blest 1963, Rand 1967). Could such diversity of appearance evolve within a single population? Owen (1963, 1966) suggested that it has for the African land snail *Limicolaria martensiana* (Figure 23-11). *Limicolaria* is highly polymorphic in areas where its population is dense, but tends to be monomorphic for the most cryptic form in sparse populations (Table 23-1). Owen suggested that where predators encounter the snails frequently, they form a search image for the snail, and any individual that *differs* from the common morph, whether it is more conspicuous or less, may gain some selective advantage from being the odd man out. Payne (1967) suggested that apostatic selection also could be responsible for plumage

Figure 23-11   Color forms of the land snail, *Limicolaria martensiana.* The background color is pale buff and the streaking is either dark brown or pale brown. Adult snails are about three centimeters long. From left to right, the forms are referred to as streaked, pallid 1, pallid 2, and pallid 3 (after Owen 1966).

TABLE 23-1

Relative frequency of color forms as a function of density in populations of the land snail *Limicolaria martensiana* around Kampala, Uganda (see Figure 23-11; from Owen 1963, 1966).

Population	Density (number per square meter)	Streaked	Color form (per cent)		
			Pallid 1	Pallid 2	Pallid 3
1	> 100	44.4	13.8	36.2	5.6
2	> 100	55.0	21.0	20.0	3.9
3	> 100	61.4	16.2	19.7	2.7
4	26	68.4	15.2	12.9	3.5
5	21	75.7	10.2	12.3	1.8
6	8	77.8	10.7	11.5	—
7	5	92.6	7.4	—	—
8	< 5	97.7	2.3	—	—
9	< 1	100.0	—	—	—
10	< 1	100.0	—	—	—

polymorphism in some species of parasitic cuckoos, and Manwell and Baker (1970) proposed the same mechanism to explain the diversity of alleles for disease resistance in some populations.

The hypothesis of apostatic selection has encountered some difficulties (Carter 1967), among which is the observation that shell color polymorphism occurs in several species of snails for which predation is probably a minor factor. Additionally, the few experiments that have been designed to demonstrate apostatic selection (Allen and Clark 1968) have met with questionable success. Nonetheless, apostatic selection remains an attractive and testable hypothesis (Clarke 1969).

Mechanisms for generating genetic polymorphism are not limited to heterosis and apostatic selection. A mosaic of patches of habitat, each of which selects different alleles, can result in a polymorphic population, if patch size and quality and movement of individuals between patches are of the right magnitude (Levene 1953, Levins and MacArthur 1966, Gillespie 1974). Selection by one sex, usually the female, may favor polymorphism in traits in the other sex (Rhijn 1969, Cooke and McNally 1975, O'Donald et al. 1974). The nature of these mechanisms will become more apparent in later chapters when we discuss population genetics and natural selection.

We have seen how the occurrence of genetic variability in natural populations is influenced by adaptations of the life cycle of organisms and the mating structure of populations. Most of these adaptations clearly enhance the fitness of the individual and its progeny. But there is also a geographical component of genetic variation within populations. Individuals that inhabit slightly different environments are exposed to different selective pressures, and these differences are reflected in the genetic constitution of a population over its geographical range. In the next chapter, we shall discuss geographical variation in populations and how patterns of variation are influenced by the movements of individuals. Because the population is the unit of evolution, it is important that we understand how the geographical extent of a population is delimited. We shall find that populations do not always have precise boundaries and that the existence of populations as evolutionary units is sometimes best regarded as an abstract concept.

24

The Spatial Occurrence of Variation in Populations

The study of geographical variation within populations began with the early attempts of taxonomists to classify organisms into types or species. Organisms collected in a particular locality usually could be assigned to one or another well-differentiated type without ambiguity. Distant areas may have been inhabited by closely related organisms, but the organisms frequently differed enough to be distinguished by a museum taxonomist as separate species. As collections of organisms became more nearly complete, some of the geographical gaps were filled in with intermediate forms, and the distinction between some species became less certain.

When is a species a species? When do two populations differ sufficiently to be categorized as distinct species? Historically, accumulating evidence of phenotypic variation within populations made it ever more difficult to establish other than nonarbitrary criteria for species distinctions based on morphology. As a result, the species concept has gradually been transformed during this century from the typological notion of the museum specialist to its present biological definition that derives from viewing species as genetically integrated evolutionary units (Mayr 1963).

The *species* may be defined as a collection of populations that actually or potentially interbreed under natural conditions. The species includes all individuals whose genes potentially may be shared by their descendants, and, therefore, embodies the genetic continuity of life. Hence, the species is the basic unit within which evolution takes place. Or is it? As we shall see, the biological concept of species, built so painstakingly over many decades, is being questioned by modern genetic studies of populations, just as the typological concept was challenged earlier by phenotypic studies.

In this chapter, we shall first examine geographical variation in populations, and then try to define the evolutionary unit. We must remember, however, that the geographical patterns we observe are thin slices in time through populations, like snapshots of continuing population processes. Whereas geographical variation represents divergence within a population over time, genetic monitoring of populations over periods of a decade or so

366

has shown that the time scale for some change can be extremely short and that populations can be genetically dynamic.

Gene flow and introgression introduce genetic variation from other populations.

The primary unit of genetic variability, the mutation, can spread geographically by the dispersal of individuals in the population (*gene flow*). The importance of gene flow to the evolution and ecology of populations is a controversial matter. If individuals move freely between areas with diverse environments in which different genes or gene combinations are selected, local populations will be subjected to a continual influx of genes that are better suited for other environments. In this way, gene flow has a deleterious effect on the fitness of local populations. But, by increasing the genetic variability of a population, gene flow can also increase the number of gene combinations for natural selection to work on, thereby enhancing the adaptability of the population. Which is gene flow, culprit or benefactor? The question is difficult to answer because data are lacking on rates of gene flow in natural populations and on the influence of gene flow on genetic variability and adaptation to local conditions.

Gene flow sometimes occurs between species that are not completely isolated from each other reproductively. Genetic exchange between species by hybridization is usually called *introgression,* and it may introduce enough genetic variability into a population to alter its evolutionary prospects. The role of introgression in the evolution of a natural population was revealed unexpectedly in studies on the expansion of the Queensland fruit fly (*Dacus tryoni*) into cooler regions of Australia during the past hundred years. Extension of the fly's range occurred largely because of the evolution of increased physiological tolerance of low temperatures that may have been facilitated by introgression of genes from another species. *Dacus tryoni* depended upon the fruit of tropical rain-forest trees in Queensland before Europeans established fruit orchards in Queensland, in the northern part of Australia. The flies quickly spread into the new habitat where they encountered and occasionally mated with the closely related species *Dacus neohumeralis. D. neohumeralis* is also restricted to the northern tropical areas of Australia, and hybrid flies with characteristics intermediate between *tryoni* and *neohumeralis* are rare outside of the zone of *sympatry* or geographical overlap (Gibbs 1967, 1968).

To test whether introgression could have played a role in the evolution of increased tolerance of extreme temperatures by *tryoni,* Lewontin and Birch (1966) maintained populations of pure *tryoni* and of *tryoni-neohumeralis* hybrids in the laboratory at extreme temperatures for two years. Natural selection was allowed to work on the populations in the different temperature environments. Initially the hybrids did not fare as well as the pure *tryoni* populations, but their reproductive performance improved more rapidly than that of the pure strain of *tryoni,* and had exceeded that of the pure strain in extreme environmental conditions by the end of the experiment. Lewontin and Birch's experiment suggests that the hybrid population, although suffering from an initial disadvantage, was able to evolve tolerance of new conditions more rapidly than the parental population because of its increased genetic variability.

Balanced against the effects of gene flow and introgression in any population is the selection of adaptations for the local conditions of the environment. With unchanging conditions, these unifying and diversifying forces will tend toward an equilibrium that is expressed both in local variability and geographic differentiation within a population.

How extensive is geographical variation within the gene pool?

When steplike changes in phenotypic characters occur abruptly within a population, such changes are often correlated with physical barriers to dispersal. For example, in the San Francisco Bay area, three morphologically distinct races of the song sparrow occur in salt-marsh habitats. Although this situation superficially appears to be one of differentiation within a single population, the salt-marsh races are well isolated from each other and from an upland race by habitats that are unsuitable for the sparrows (Marshall 1948). These isolated populations have undergone evolutionary changes independently of one another, and thus have behaved as separate evolutionary units, at least for a short period. In this chapter, we shall concentrate more on variation within uniformly distributed populations. Such variation is usually *clinal*. That is, characteristics change gradually over the geographical range of the population, sometimes in association with apparent environmental gradients and sometimes not.

Typical patterns of geographical variation are well illustrated by three Japanese invertebrates: the Emma field cricket *Teleogryllus,* the ladybird beetle *Harmonia,* and the land snail *Bradybaena*. In the Emma field cricket, body size and the duration of nymphal development both increase clinally from north to south. Masaki (1967) demonstrated that the clines have a genetic basis by raising individuals from various localities throughout Japan at the same temperature and photoperiod. Under these conditions, the development period of the larvae and the size of the adults retained their regional differences (Figure 24-1). Masaki suggested that the adaptive significance of the clinal trends is that the suitable season for insect development in northern latitudes is relatively brief, and therefore the larval development period and final adult size are reduced compared with southern populations.

The ladybird beetle *Harmonia axyridis* exhibits marked polymorphism for color pattern that is controlled by a series of supergenes (Figure 24-2). The relative frequency of two of the morphs changes in a clinal pattern: *succinea* decreases from north to south, while the proportion of *conspicua* increases (Figure 24-3). The frequency of *axyridis* and *spectabilis,* which are intermediate in the degree of melanism and in their relative genetic dominance, does not change significantly. The geographical distribution of color patterns exhibited by *Harmonia* is often referred to as a morph-ratio cline.

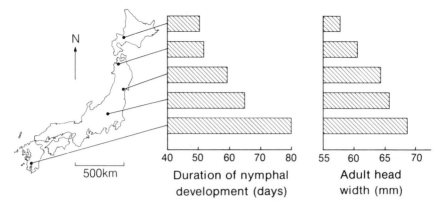

Figure 24-1   Clinal geographic variation in duration of nymphal development and the width of the adult head in males from six local strains of the Emma field cricket (*Teleogryllus emma*) in Japan. The nymphs were raised at 27–28 C on a cycle of a sixteen-hour day and an eight-hour night (data from Masaki 1967).

Morph frequencies in *Harmonia* have varied little from year to year, except near Sapporo on the island of Hokkaido, where *succinea* was common (84 per cent) in 1923, but had decreased to 43 per cent by 1944. In the earlier collections from Sapporo, the morph frequencies closely resembled continental populations from Korea and China, suggesting that the Hokkaido population may have resulted from a recent invasion from the mainland (Komai *et al.* 1950). The morph frequencies in the population quickly changed to resemble northern Honshu populations, but the Hokkaido beetles still retain a remnant of their continental character in the rarity of the *axyridis* morph and the presence of a ridge on the elytra (forewings) of most individuals.

The morph frequencies in local populations of the polymorphic land snail *Bradybaena similaris* do not exhibit clinal trends in geographical variation (Figure 24-4), nor is there any apparent correlation between morph frequency and habitat. The absence of morph-ratio clines in *Bradybaena* may be related to the fact that the snail was introduced into Japan, probably

Figure 24-2   Morphs of the ladybird beetle *Harmonia axyridis* in Japan. From left to right: *succinea, axyridis, spectabilis,* and *conspicua* (after Dobzhansky 1951).

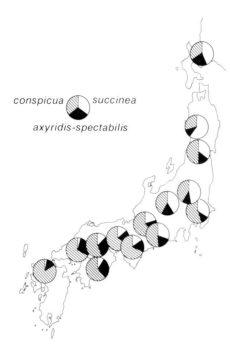

Figure 24-3   Geographical variation in the frequencies of morphs of the ladybird beetle *Harmonia axyridis* in Japan: hatched *conspicua*; black *spectabilis* and *axyridis*; white *succinea* (from data of Komai *et al.* 1950).

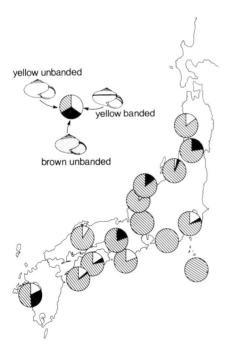

Figure 24-4   Geographical variation in the frequencies of morphs of the land snail *Bradybaena similaris* (from data of Komai and Emura 1955).

numerous times, all within the past 200 years, with the widespread cultivation of sugar cane (Komai and Emura 1955). Variation in morph frequencies among populations could have occurred if each local population was established by a small group of colonists that contained a random, but not necessarily representative, sample of the genetic variation of the parent population. The change in gene frequency as a result of colonization by small populations has been referred to by Ernst Mayr (1942) as the *founder principle*.

What is the significance of chromosomal polymorphism in *Drosophila*?

Polymorphism for inversions in the third chromosome of *Drosophila pseudoobscura* (see page 350) exhibits widespread geographical variation in the western United States and northern Mexico (Dobzhansky 1970). The frequencies of chromosome arrangements and the particular genes they carry vary seasonally in some areas (Dobzhansky 1948, Dobzhansky and Ayala 1973). The chromosome arrangements also exhibit ecological morph-ratio clines (Dobzhansky and Epling 1944). For example, in the Sierra Nevada Mountains of California, the proportion of the Standard chromosome arrangement decreases from about 50 per cent at sea level to less than 10 per cent near timberline (3,000 meters), a horizontal distance of about 100 km. At the same time, the frequency of the Arrowhead chromosome type increases from about 25 to 50 per cent, and the Chiricahua arrangement from about 15 to 25 per cent. Furthermore, over the same altitude gradient, *Drosophila pseudoobscura* is gradually replaced by the closely related species, *D. persimilis*.

The geographic distribution of the chromosome inversions is still more striking. Standard is found primarily throughout the western part of the range at low elevations, but nowhere exceeds 60 per cent of the total; Arrowhead is more widespread, but reaches its greatest abundance in the Rocky Mountains and the Basin and Range province of the western United States, where it may comprise an entire local population. The Pike's Peak chromosome arrangement is localized in the Rocky Mountain area, which is the extreme eastern part of *pseudoobscura's* range. The treeline arrangement is widespread outside of arid regions but does not comprise a large percentage of any population except on the Mexican plateau.

Before we can determine the significance of geographical variation, we must understand why chromosome polymorphism should exist at all. Several hypotheses are plausible: Polymorphism may be stabilized through heterosis; polymorphism may result from gene flow between populations in which different chromosome arrangements are most fit; chromosome variability may enable a population to exploit a broad range of local ecological opportunities, that is, it may be an expression of intraspecific divergence; inversions may facilitate the incorporation of mutant genes by separating them from stable polymorphic loci.

Heterozygote superiority was demonstrated by Dobzhansky (1947), who found that Standard and Chiricahua chromosome arrangements coexisted in laboratory populations, with Standard comprising 60 to 70 per cent of the chromosomes at equilibrium. But balanced polymorphism between two chromosome arrangements occurred only if they were derived from the same area; when chromosomes were derived from different populations, one arrangement usually eliminated the other by competition. These experiments indicated that the specific genetic characteristics of each chromosome arrangement vary geographically, a point that was demonstrated for electrophoretically detectable proteins of *D. willistoni* by Ayala et al. (1972). Although each chromosome arrangement does not necessarily contain identical alleles throughout its geographical range, the ecological specificity of chromosome types suggests that they have some degree of genetic uniformity.

Carson (1958, 1959) suggested that chromosome polymorphism should be greatest in the center of a species' range, where population density is greatest and where the species is relatively free from interspecific competition. In this situation, variation in chromosome arrangements may enable individuals to exploit different parts of the environment, and thereby reduce intraspecific competition. Carson further suggested that at the periphery of a species range, where environmental conditions are more "marginal" for the species and where interspecific competition may be more important, increased chromosome homozygosity would allow the population to adapt more precisely to local environmental conditions, thereby enhancing its competitive ability compared to other species. Chromosome homozygosity also permits increased levels of recombination that provides new combinations of genes for selection. In *Drosophila robusta,* which is distributed throughout most of the eastern United States, the highest degree of chromosome polymorphism occurs near the center of the range, in Missouri and Tennessee, where the flies are also most abundant. In such peripheral areas as Vermont and northwestern Nebraska populations tend to be homozygous (Carson 1955, 1956).

Several lines of evidence do not support Carson's hypothesis, however. First, chromosome polymorphism in *Drosophila pseudoobscura* is as great at high altitudes in California, where *D. persimilis* occurs in greater abundance, as it is at low elevations, where *D. pseudoobscura* is the more abundant species. Second, chromosome homozygosity in a population does not necessarily imply reduced genetic variation. For example, Townsend (1952) has shown that deleterious recessive alleles occur as frequently in peripheral populations of *Drosophila willistoni* that are vitually homozygous for chromosome arrangement as in highly polymorphic populations from the center of the range in Brazil. Prakash, Lewontin, and Hubby (1969) have further shown that the degree of genetic polymorphism in *D. pseudoobscura,* measured by gel electrophoresis, is independent of the level of chromosome polymorphism in a population (Table 24-1). (The low genetic variation in the isolated Bogotá, Colombia, population is probably due to recent colonization in 1960 by a small number of individuals.) The same has been demonstrated for *D. robusta* by Prakash (1973). Third, if Carson's hypothesis is correct, one would expect a high degree of polymorphism on islands, where there are

TABLE 24-1

Chromosome and enzyme polymorphism in four populations of *Drosophila pseudoobscura* (from Prakash *et al.* 1969).

	Per cent of third chromosome arrangements*						Per cent of loci polymor- phic‡	Per cent of genome heterozy- gous per individual
	ST	AR	PP	CH	TL	Other		
Strawberry Canyon, California	47	9	3	19	14	8	46	14
Mesa Verde, Colorado	1	98	1				42	11
Austin, Texas	6	8	78	1	6	1	38	12
Bogotá, Colombia					37	63†	25	4

* ST = Standard, AR = Arrowhead, PP = Pike's Peak, CH = Chiricahua, TL = Tree Line
† Santa Cruz arrangement
‡ Twenty-four loci were checked

relatively few species of *Drosophila* and where the population of each species tends to be dense. Yet the converse seems true: Populations on West Indian islands are practically homozygous for chromosome arrangement, but have levels of genic polymorphism similar to mainland populations (Ayala 1972). The geographical occurrence of a chromosome arrangement seems to be determined partly by ecological conditions, as we have seen for *Drosophila pseudoobscura,* and partly by the relative accessibility of a place. Peripheral localities and islands exhibit relatively little polymorphism.

It also seems unlikely that polymorphism results from gene flow between areas in which different chromosome arrangements are adapted to local conditions, or that polymorphism has evolved to protect local gene combinations from gene flow. The distributions of chromosome arrangements seem too broadly overlapping to be accounted for entirely by gene flow, and the occurrence of heterozygote superiority indicates complicated evolutionary interactions between chromosome arrangements.

An intriguing observation was made by the Russian scientists Dubinin and Tiniakov (1946), who found that populations of *Drosophila funebris* were highly polymorphic for chromosome arrangement in Moscow (up to 19 per cent of the individuals contained inversion heterozygotes), whereas in rural areas, 60 to 200 kilometers from Moscow, less than 2 per cent of the individuals contained inversion heterozygotes. What ecological characteristics do urban populations of *Drosophila funebris* share with populations of *D. willistoni* in central Brazil? It is unfortunate that we know so little about the ecology of a genus whose genetics are so familiar.

Genetic variation within populations can occur on a local geographical scale.

We have been looking at genetic variation on a continental scale, over which changes in gene frequencies are usually related to changes in important ecological factors. Considerable genetic variation can also occur locally, however, and this variation often is not clearly associated with particular ecological conditions. For example, Danish populations of the house mouse differ by as much as 50 per cent in the frequencies of alleles of polymorphic genes within a few miles (Selander, Hunt, and Yang 1969). A similar situation occurs in populations of the small annual plant *Linanthus parryae*. In one small area in southern California, blue flowers—the normal flower color is white—are haphazardly distributed without apparent relation to environmental factors (Epling and Dobzhansky 1942, Wright 1943a, Figure 24-5; cf. Epling, Lewis, and Ball 1960).

Local geographical variation sometimes occurs across small gaps in the distribution of a population. *Clarkia biloba,* a colonial, insect-pollinated annual plant, which grows in the foothills of the Sierra Nevada Mountains, is separated into three subspecies with morphologically distinct flowers:

	brandegeae	biloba	australis
Petal width	broad	broad	narrow
Petal lobe	shallow	deep	shallow
Petal color	lavender	reddish	magenta

Colonies of *Clarkia biloba,* which may have up to 10,000 individuals, are separated by distances of a few yards to several miles. In *C. biloba,* flower color changes abruptly between the subspecies *biloba* and *australis* at the Merced River, which carves a mile-wide canyon through the foothills, disrupting suitable habitat (Roberts and Lewis 1955). Other subspecific characteristics do not change precisely at the river itself, but nearby. In other species of *Clarkia,* gaps between populations as narrow as 50 feet have been found to block gene flow completely (Lewis 1953). The two northern subspecies of *C. biloba, brandegeae* and *biloba,* are not separated by any obvious ecological barrier, and they remain distinct through reproductive isolating mechanisms inherent in the plants themselves. Artificial crosses between the subspecies have very low fertility. In *Clarkia,* as in *Linanthus,* genetic changes over small distances seem inexplicable on ecological grounds.

Isolation is even more restrictive in the evening primrose *Oenothera organensis,* an extremely rare species known only from the Organ Mountains of southern New Mexico. The plants are perennial and colonial, and they usually grow near springs and small water courses in canyons. The entire species numbered only about 500 plants in the 1930s (Emerson 1939). As one might expect of a species with so few individuals, there are many self-sterility alleles to ensure outcrossing. Emerson identified these alleles by crosses between clones grown from plant clippings taken from the wild populations. In two areas, one along the north fork and the other along the east fork of a

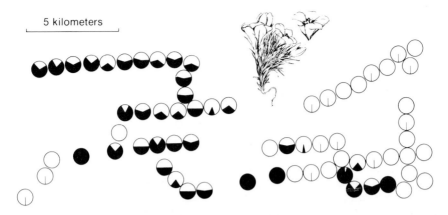

Figure 24-5   Distribution of blue flowers (frequency indicated by black portion of circle) in populations of *Linanthus parryae* from a small region of southern California. Sampling localities were located along roads, which accounts for their distribution pattern (after Epling and Dobzhansky 1942, and Abrams 1951).

small stream, separated by about 600 feet, only thirteen of the 22 self-sterility alleles of both populations were held in common.

It seems remarkable that the genetic composition of two populations separated by the length of two football fields could differ so greatly. Because the 500 individuals of *Oenothera organensis* are divided into perhaps fifty local groups, averaging ten individuals each, one might expect differences to arise between the populations by chance alone. Assuming that new self-sterility mutants arose at the rate of one in 10,000 individuals per generation, Wright (1939) calculated that the observed population differences in self-sterility alleles would be produced if about 2 per cent of the pollen that fertilized flowers in a local population came from outside.

Wright's analysis suggested that random processes were responsible for the genetic differentiation of small geographical isolates. As the movement of individuals or gametes between subpopulations decreases, the genetic independence of local populations increases. In slowly dispersing organisms, such as the land snails *Cepaea* and *Bradybaena,* gene frequencies may vary by 30 to 40 per cent between populations separated by less than a mile in the same habitat (Cain and Sheppard 1954; Komai and Emura 1955); populations of the more readily dispersing fruit flies *Drosophila* exhibit genetic uniformity over relatively large areas (Prakash, Lewontin, and Hubby 1969, Ayala *et al.* 1972).

Gene frequencies of small semi-isolated populations can diverge as a result of a random process called *genetic drift* (Wright 1951). Imagine that a large diploid population contains two alleles at one gene locus, each with a frequency of 50 per cent. If we remove several subpopulations of two individuals each from the larger population, there is one chance in sixteen that these subunits will be homozygous for a particular gene (each individual has one chance in four of being homozygous for a particular gene), and only half the subpopulations will initially have the same gene frequency as the parent population. Genetic drift can also pertain to changes that occur over time within a population. At any given time, the individuals in a population are descended from a relatively small proportion of the individuals in ancestral populations. To the extent that births and deaths occur at random with respect to genotype, genetic drift owing to the statistical properties of small samples can cause changes in allele frequency.

There has been a considerable disagreement over the role of genetic drift in evolution (see Mayr 1963). The importance of genetic drift increases in progressively smaller populations. But the effect of drift, which eventually can lead to the fixation of one allele or another, is countered both by natural selection and by mutation and back mutation pressure. Although the existence of genetic drift cannot be argued, its potency as an evolutionary force must be assessed in relation to population structure and selective forces. For many, perhaps most, genes, selective forces are so great, or populations are so large, that drift is not an important factor.

Marked genetic differentiation is known to have occurred between populations with little or no spatial isolation but that have experienced selection for different genes. For example, species of grass growing on mine tailings are tolerant of heavy metals, whereas neighboring populations on normal soils a few feet away are not (Antonovics and Bradshaw 1970); populations of *Drosophila melanogaster* living in wine cellars in Australia have a different gene for an enzyme involved in metabolizing alcohol than do those living outside the cellars (McKenzie 1975); many species of grasses grown on adjacent experimental plots with different fertilizer treatments accumulated substantial genetic differences between subpopulations over distances of thirty meters and within fifty years (Snaydon and Davies 1972); populations of speedwell (*Veronica peregrina*) growing not more than five meters apart at the periphery and center of temporary ponds in central California have evolved differences in plant form and seed output (Linhart 1974). In many of these cases, particularly among plants, populations have evolved mechanisms to reduce gene flow as a means of reinforcing genetic differences between subpopulations (Allard 1975). For example, subpopulations of the grass *Agrostris* growing on mine tailings have evolved a high level of self-fertility to prevent pollination by plants growing on normal soils and lacking the genetic basis of heavy metal tolerance (Antonovics 1968a).

How is the genetic uniformity of the species maintained?

The definition of the species as "a group of actually or potentially inter-breeding populations" has led to the widely held belief that species are integrated by gene flow, "genetic communication," within the population, in spite of evidence, presented above, that considerable local differentiation may appear in the face of gene flow.

Islands provide the most clear-cut examples of completely isolated populations. Populations are generally differentiated between islands rather than within an island, even though the habitat distribution of a species on each island usually varies more than the habitats of the species differ between islands. Such patterns seem explicable only by the hypothesis that the lack of significant gene flow between islands has enabled populations on different islands to evolve independently. Divergence is caused primarily by differential selection pressures acting on local populations. A logical corollary to this hypothesis is that the genetic uniformity of a species is maintained by rapid gene flow between subpopulations that overwhelms the force of local selective pressures. Ehrlich and Raven (1969) have taken a point of view directly opposite to the "isolation-differentiation" hypothesis, claiming that uniform selection pressure over the range of a population, not gene flow, is responsible for observed genetic uniformity in populations. They point out that completely isolated populations often show very little divergence. The butterfly *Euphydras editha* is extremely sedentary; individuals rarely fly more than a few hundred feet from their place of birth (Ehrlich 1961, 1965, Brussard *et al.* 1974). Yet this species exhibits little morphological variation over a range of several hundred kilometers (Ehrlich *et al.* 1975).

Estimates of gene flow have relatively little value unless one knows how effectively gene flow prevents divergence. The dispersal of individuals varies considerably among different types of organisms. Some organisms disperse widely, such as large mobile animals, plants that are pollinated by wide-ranging insects or by the wind or whose fruits are eaten by wide-ranging animals, and aquatic organisms that are carried by ocean currents. Other organisms are extremely sedentary (see Wolfenbarger 1946).

Regardless of the rate of gene flow within a population, distant portions of the population will usually be sufficiently isolated genetically to exhibit the effects of divergent selection. Wright (1943b) termed this property of populations "isolation by distance" (Rohlf and Schnell 1971). Slatkin (1973) and Endler (1973) have further shown by mathematical model and by experiment that marked clinal changes in gene frequency can be produced by selection in the face of strong gene flow.

To discover whether gene flow or selection is the primary integrative factor of gene pools, we must look for situations that could falsify each of the hypotheses. It is not sufficient to point to a lack of divergence between populations with no gene flow and say that such evidence refutes the role of gene flow in population integration. The populations may, in fact, occupy habitats with similar selective pressures, so that one would not expect divergence under any circumstances, gene flow or not. Can we find cases in which gene flow prevents differentiation among subpopulations that occur in different habitats, or alternatively, cases in which divergence has occurred within a population in the face of strong gene flow?

Both kinds of evidence are at hand in the literature. The occurrence of seemingly uniform populations over vast areas and in a wide variety of habitats would appear to discount the role of selection compared to gene flow in population integration, but we do not know what aspects of the environment are pertinent for selection. Keast (1961) presented some relevant evidence for birds of the Australian subcontinent. He divided species into two groups: (a) nomadic species that inhabit the arid interior of Australia

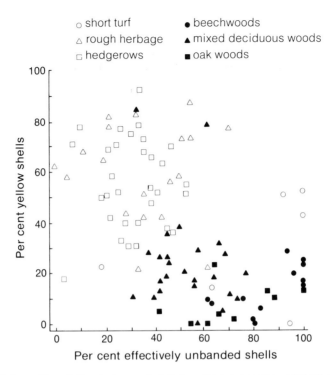

Figure 24-6   Correlation between habitat and the proportion of shell color and pattern morphs in the land snail *Cepaea nemoralis* near Oxford, England (after Cain and Sheppard 1954).

and range over vast areas in search of temporary food resources; (b) species that inhabit moister regions along the coasts of Australia and are frequently subdivided into isolated subpopulations in patches of wet forest habitat. Birds of moist habitats are more sedentary than desert birds because they can rely more surely upon finding persistently adequate conditions in local environments. Keast showed that interior species rarely exhibit taxonomic differentiation throughout their range, whereas in the moist habitats scattered along the coasts of Australia, subspecific differentiation is prevalent. We cannot measure environmental variation over the range of each species, but we may reasonably assume that selection for the plumage characteristics taxonomists often use to distinguish subspecies, which are important to birds mostly in intraspecific behavioral interactions, is largely independent of the abiotic environment. The degree of differentiation in such characteristics, therefore, is a reasonable index of the degree to which a population is broken up into isolated evolutionary subunits.

Variation in the environment is critical for evolutionary divergence. In species that are divided into small, reproductively isolated or nearly isolated subunits, differences in selective pressures along environmental gradients have undoubtedly been responsible for the establishment of genetic differences between populations. For example, in the land snail *Cepaea,* morph frequencies differ between habitats over and above the considerable local variation among populations in similar habitats (Figure 24-6). Snails from forest habitats tend to have unbanded, brown shells, whereas individuals from open fields and hedgerows tend to have predominantly banded, yellow shells. Such data re-emphasize the importance of natural selection for divergence, but they say relatively little about the contribution of gene flow to population integration. In snail populations where gene flow appears to be extremely limited differentiation has, as we might expect, occurred. But as we do not know *how much* differentiation to expect, perhaps even more divergence should have occurred.

The unifying effect of natural selection has been amply demonstrated by the convergence of different species to a similar form under similar environmental selection pressures (Figure 24-7). An extreme example that comes to mind is Christiansen and Culver's (1968) study of collembola (a group of primitive insects) in caves of the southeastern United States. Through an analysis of fifteen morphological characteristics, the authors concluded that the recognized species *Pseudosinella hirsuta* is actually not one species at all, but arose independently in at least four evolutionary lines that came to resemble each other through convergent specialization for cave life.

I am more impressed by the fact that many species are virtually unvarying over broad geographical ranges, which often include a wide variety of habitats. Furthermore, gene flow in some of these species is demonstrably low. How can we hope to explain such uniformity? On one hand, if a population were relatively flexible genetically, its uniformity could be derived from the rapid spread of a small uniform population. This would not be an equilibrium situation, rather it would represent the early stages of a widely distributed population that eventually would become differentiated geographically (for example, see Johnston and Selander 1964). On the other

Figure 24-7 Convergence in appearance of unrelated grassland birds: left, the eastern meadowlark (*Sturnella magna*), an American bird related to blackbirds and orioles (*Icteridae*); and right, the yellow-throated longclaw (*Macronix croceus*), an African bird related to larks and pipits (*Motacillidae*) (after Fisher and Peterson 1964).

non-mimetic male   mimetic female

hybrid offspring

Figure 24-8   Cross between a mimetic female of *Papilio dardanus* (form *cenea*) from South Africa (right) and a male from Madagascar, where *Papilio* is nonmimetic (left). Mimicry breaks down in the offspring (center) because the male parent lacks the modifiers required to perfect the pattern (after Ford 1971).

hand, a genome may be relatively stable because of a high degree of integration. In this case, divergent selection pressure may be effectively balanced by conservative selection from within the genotype for coadapted gene combinations (page 321). Some threshold of selection would have to be exceeded before divergent evolution could occur. But once a new gene complex were formed, it would spread rapidly through the population.

The integration of the gene pool is strikingly demonstrated by mating experiments with the African swallowtail butterfly *Papilio dardanus* (Clarke and Sheppard 1959, 1960a-d). The females have evolved a complex polymorphic mimicry system (see page 362) determined by a series of allelic supergenes that exhibit simple dominance-recessiveness relationships. If a particular morph from a South African population is crossed with individuals of the same species from Madagascar (none of the Madagascar population is mimetic), the mimicry breaks down almost completely in the offspring because the Madagascar race lacks the particular modifier genes needed to perfect the expression of the major mimicry supergenes (Figure 24-8). Similarly, if South African butterflies are mated with individuals from Uganda that exhibit the same mimicry pattern (for example, the morph *hippicoonides* from South Africa with *hippocoon* from Uganda, both of which mimic *Amauris niavius;* see Figure 23-10), the offspring exhibit partial breakdown of the mimetic resemblance because the gene pools in one locality lack the particular genes necessary to modify the expression of a gene from another locality. Because the gene pool is highly coadapted it is resistant to the introduction of foreign genes and can exhibit considerable stability in the face of gene flow (Selander, Hunt, and Yang 1969).

We certainly will not be able to understand the genetic integration of populations until geographical variation within species is more systematically studied in relation to environmental change and gene flow. We know so little about gene flow and selection for divergence that it is impossible to draw any firm conclusions from the assortment of facts at hand. It has been demonstrated that gene flow, selection, and coadaptation all play important roles in determining the level of genetic uniformity within species, and we should expect that the relative importance of these factors varies both among species and among traits within species.

25

Phyletic Evolution and Speciation

The irrevocable separation of populations into independent evolutionary units is called *speciation*. The reproductively isolated evolutionary unit is straightforward in concept but, as we saw in the last chapter, difficult to pin down in nature. As we try to enlarge our definition of a species in space and time, its boundaries become less distinct, and the rigid typological definition of the species becomes strained by its extended dimensions. So long as we recognize that a geographical definition of the species is only an attempt to summarize the complex interactions between populations, practical difficulties of the biological species concept will not detract from our understanding the local ecological interactions that lead to species formation.

The genetic organization of populations transcends the simple Mendelian view of the individual gene locus or a small number of interacting genes. Major evolutionary change in a population cannot be equated with gene substitutions at a few loci, because such change often represents a major genetic rearrangement of the gene pool. It is difficult to measure the genetic differences between species because individuals of different species cannot be mated. The gene pool of each species exhibits such complex integration that the introduction of a foreign allele into a local genome can greatly upset its expression in the phenotype. We need only to extrapolate such findings to the genetic differences between species to realize that interspecific crosses may disrupt the genetic system so much that the development and functioning of the organism break down. Transpecific evolution should be viewed as an extension of Mendelian genetics from the scale of the gene locus to that of the genotype.

Speciation, of course, means more than evolutionary change in a population. Speciation requires divergent evolutionary change in two or more independently evolving populations to the point that accumulated genetic differences between them are so great that they cannot normally interbreed. Speciation thus has two components: phyletic evolution and reproductive isolation of populations. (These processes may be bypassed by certain plants in which major chromosome "mutations" can lead to instant speciation. We will return to this problem a little later.)

Phyletic evolution is necessary but not sufficient for speciation.

That extensive evolutionary change has occurred needs little documentation here. We are living proof of this fact. If the genetic structure of populations had fossilized like bone, we could trace our lineage back through time, past gene substitution after gene substitution through an impressive phylogeny of ancestors, to that first point (or one of them) at which the phenotype was encoded in a DNA molecule. Some 50 or 60 million years ago the soma borne by our genetic line resembled that of the present-day tree shrew; 300 million years before that, the soma was embodied in the fins and scales of some primitive fish.

Evolution proceeds slowly by everyday standards. Time, which seems so critical to such impatient creatures as we, has hardly constrained evolution. One hundred fifty million years ago the world was crawling with animals so formidable that even their fossilized remains stagger our imagination. One-half billion years ago, when living organisms first were preserved in abundance in the sediments of ancient swamps and seas, animals and plants would have appeared different compared to present-day forms, but not unfamiliar. By then, the basic cellular and genetic properties of life, upon which all subsequent evolution has been based, were firmly established.

Rates of phyletic evolution have been uneven within evolutionary lines, and indeed within large groups of organisms, during the evolutionary history of the earth. Most phylogenetic lines appear to have gone through bursts of evolutionary activity, followed by quiescent periods leading either to further bursts of activity or to the termination of the line. The lungfishes, for example, first occur in the fossil record about 325 million years ago during the Devonian Period of the Earth's history (named after rock formations, dating from that period, near Devon, England; see Figure 26-1 for a chronological table of the Earth's history). Lungfishes underwent a period of rapid evolution lasting about 75 million years in which about 80 per cent of the features that characterize modern forms appeared (Figure 25-1). During the past 250 million years, the lungfishes have exhibited relatively little evolutionary activity (Westoll 1949, Schaeffer 1952).

The detailed fossil record of horses and their primitive ancestors reveals the evolution of a single line over a long period. The paleontologist George Gaylord Simpson (1953) demonstrated that various characteristics of the modern horse evolved during different periods over the past 60 million years (Figure 25-2). Early changes during the Eocene Epoch involved an increase in the size of the molar teeth, presumably associated with an increase in the amount of plant roughage in the diet. The Oligocene Epoch brought additional important changes: body size began to increase, a trend that has continued through the ancestral line of the modern horse virtually to the present; the crown of the lower teeth evolved a pattern of crests of hard enamel interspersed with areas of soft enamel to grind up plant food more easily; the structure of the forefoot was modified by the gradual reduction and loss of one of the toes. During the Miocene Epoch changes in tooth and foot structure reflected a gradual shift from browsing on leafy vegetation to grazing on much tougher grasses. This shift probably involved a change in

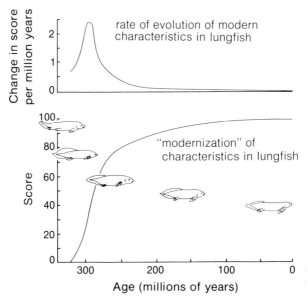

Figure 25-1   Evolution of "modern" characteristics in the lungfishes (Dipnoi). Scores range from wholly primitive characters (0 per cent modern) to 100 per cent modern characters (from Simpson 1953, after data in Westoll 1949; fish outlines from Schaeffer 1952).

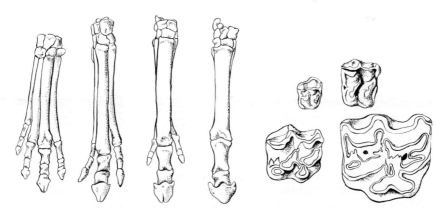

Figure 25-2   Evolution in the lineage that culminated in the modern horse, showing the progressive simplification of the foreleg and the increase in size and complexity of the upper molars. From left to right: *Eohippus*, Lower Eocene; *Miohippus* (leg) and *Mesohippus* (molar), both Oligocene; *Merychippus*, late Miocene; *Equus*, Pleistocene (from Romer 1966).

habitat from forest and forest edges to savannas and plains and, perhaps, a general drying trend in climate. The molar teeth began to increase in height, presumably to balance the increased wear suffered by the teeth. Also, horses began to stand on their toes instead of on the pads of their feet, which gave them the spring needed for moving rapidly in open habitats. During the Pliocene Epoch, about 10 million years ago, the last important evolutionary modification of the horse occurred, the reduction of the foot structure to one toe.

The evolution of the modern horse can be traced back through a single phylogenetic line, but the line also had many offshoots in the past. The divergent equine groups met with varying evolutionary success. A major adaptive radiation of new species occurred during the Miocene Epoch, apparently in conjunction with the shift of part of the equine line from browsing to grazing. The history of each phylogenetic line is marked time and time again by offshoots representing new evolutionary starts, isolated from the phylogeny of the modern horse. But the phyletic changes revealed by the fossil records of horses occurred over so vast a time scale that we cannot resolve the events that take place during speciation. The details of phyletic change have to be inferred from direct observation of present-day evolutionary processes. Speciation normally occurs too slowly to be experienced directly, but too rapidly to be preserved in detail in the fossil record.

The morphological characters preserved in the Earth's sediments do not tell the whole story of evolution. In Chapter 22, we traced the evolution of the protein, hemoglobin, as it can be inferred from the present occurrence of hemoglobin variation both within and among species. The sequence of amino acids in a protein brings us close to the sequence of bases in the genetic material itself. But the genome evolves on the chromosome level as well. In plants and some groups of animals, chromosomal rearrangement may be an important mechanism of evolution and speciation (Stebbins 1971, White 1973). Regardless of their role in the diversification of phylogenetic lines, the number, size, and shape of chromosomes have undergone considerable evolutionary change as revealed by the variety of karyotypes among present-day forms.

Chromosome complements are altered in several ways. One is by the translocation of segments between chromosomes. A second is by the duplication of small portions of any one chromosome. And a third is by the duplication of the entire genome.

Translocations between nonhomologous chromosomes can both increase and decrease the number of chromosomes in the genome (Figure 25-3). For example, by exchanging a short arm for a long arm from another chromosome, one chromosome may be eliminated completely. Similarly, a chromosome may be added when part of one chromosome arm is added to another chromosome. In either case, success is limited to those exchanges whose products do not have overlapping chromosome sections in the gametes (Figure 25-3, top). If they did, gamete formation in the offspring would be disrupted by the pairing of nonhomologous chromosomes. Although translocation has been a major factor in the evolution of many groups of plants and animals, the purpose of exchanges between chromosomes and

original chromosome complement

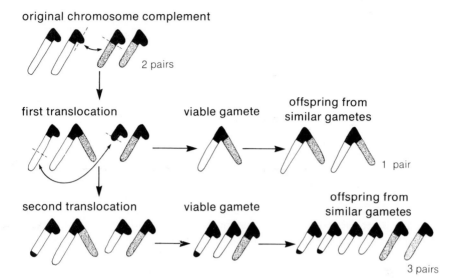

Figure 25-3   Chromosome translocations leading to the addition and deletion of a chromosome from the genome. Following translocation, few of the gametes formed are viable owing to deletion or duplication of genetic material (after Stebbins 1971).

their consequence for genome size is not well understood. Chromosome number and size could affect evolution directly by altering levels of recombination and independent assortment, but translocations may also be selected when they separate or bring together combinations of gene loci.

Ohno (1970) has argued that gene duplication has been a major creative force in evolution because duplicated genes are freed of the constraining effects of conservative selection. With one of the duplicated loci maintaining the original function of the gene, the other site is able to undergo selection more readily. It is widely believed that the various subunits of hemoglobin (see page 339) are the products of duplicated genes. Similar duplications have been documented in trypsin and chymotrypsin, the subunits of immunoglobin, and other molecules made up of homologous subunits. The role of gene duplication in evolution also can be appreciated through the general increase in the amount of DNA in cells with increasing biological complexity and evolutionary recency (Mirsky and Ris 1950).

Small portions of chromosomes are thought to be duplicated occasionally by errors in crossing over (genetic recombination) during meiosis. If the homologous chromosomes do not break in exactly the same place, crossing over will result in one of the chromosomes having a small duplicated region, the other a small deleted region (Figure 25-4). Of course, the chromosome with the deletion would probably not produce a viable zygote because it lacked a portion of its genetic instructions, and such nonreciprocal recombination events usually would be selected against. Furthermore, the successful

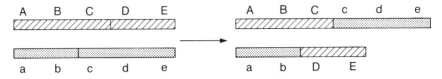

Figure 25-4   Nonreciprocal crossing over leading to gene duplication in one of the homologous chromosomes.

chromosome with a duplicated region could have difficulty pairing properly during meiosis with a normal homologous chromosome lacking the duplicated region. But so long as the region were small, this would only result in a small kink in the chromosome pair, and if the chromosome with the duplicated region were to become common in the population, whether through selection or random drift, chromosome pairing would again be normal. Through repeated gene duplications of this sort, the size of the genome can increase slowly. More rapid increase may come about through the duplication of the entire complement of chromosomes owing to the failure of chromosome complements to segregate during meiosis. We shall discuss this mechanism in more detail a bit further on.

The fine structure of the phyletic line can affect evolution.

The course an entire population takes through time does not follow parallel and interdependent lineages through parent and offspring. Subgroups in a population occasionally intermingle, occasionally separate, much as water courses over the gravel bed of a braided stream. Population subunits with varying degree and duration of reproductive isolation are continually segregating and merging.

Fragmentation of the phylogenetic line has important implications for the rate of evolution, particularly because the influence of random events and inbreeding is felt more strongly in small populations. A population divided into small evolutionary units will exhibit more genetic variability overall than the same number of individuals in larger evolutionary units. This has been demonstrated for laboratory populations of *Drosophila pseudoobscura* by Dobzhansky and Pavlovsky (1957). They established a series of large and small populations of flies with 50 per cent Pike's Peak chromosomes, and then followed changes in the chromosome frequencies. During the first four months of the experiment, the proportion of Pike's Peak chromosomes decreased to between 35 and 45 per cent of both the large and the small populations. By the end of the following year, however, the chromosome frequencies in the small populations were more variable (15 to 50 per cent Pike's Peak) than in the large populations (20 to 35 per cent). In several of the small populations, the frequency of the Pike's Peak chromosomes even increased after the uniform decrease experienced during the early part of the experiment. This could not be explained solely by sampling errors, and perhaps reflected substantial genetic changes in the chromosomes of small populations.

A similar situation involving color polymorphism in spittlebugs on small islands off the coast of Finland shows the effects of random factors in small colonizing populations (Halkka *et al.* 1974). Recently exposed islands (50–300 years) exhibited widely divergent frequencies of the color morphs in sharp contrast to the strikingly uniform distribution of frequencies on older islands and on the mainland.

New mutants and gene combinations can be incorporated more rapidly in small populations because each individual carrying the change is a larger proportion of the total. Although mutation and recombination are stronger forces in large populations by sheer weight of numbers, small population size accelerates the incorporation of variants when they do occur. By extrapolation, we can see how fragmentation of populations into small subunits can increase evolutionary rate. We do not know to what extent the genetic constitution of an isolated fragment contributes to the mosaic of populations that comprise the species. This point is crucial, however, because the gene pool of a small population is greatly diluted when it is incorporated into a large population.

The temporary isolation of a small evolutionary unit can accelerate the evolution of a larger population in two ways. First, Maynard-Smith (1968a) suggested that if population fragments evolved under slightly different environmental conditions, recombination between the populations could produce new gene combinations more rapidly than recombination within a local population (page 348). Second, the occurrence of a recessive mutation that is potentially beneficial will be more readily expressed in the homozygous form in a small inbred group than in a large outcrossed population, and the mutant, therefore can be incorporated into the gene pool much more rapidly in a small population than in a large population. Mechanisms that promote inbreeding and, in plants, selfing effectively fragment the population into numerous, partly independent evolutionary lines (Allard 1975).

Small population size in and of itself may create a genetic bottleneck, reducing the genetic variation and, in turn, the evolutionary flexibility of a population. In spittlebug populations on islands off the coast of Finland, polymorphism at gene loci for electrophoretically detectable enzymes is considerably reduced on small isolated islands (Table 25-1). A more striking example of genetic uniformity produced by a population bottleneck has been revealed by electrophoretic analysis of enzymes in the elephant seal of coastal California (Bonnell and Selander 1974). The seals were hunted to near extinction during the late nineteenth century, when they were reduced to perhaps twenty to a hundred animals concentrated at one locality. Although the population has increased under protection to about 30,000 individuals, all 21 proteins surveyed were monomorphic. Apparently, severe inbreeding associated with small population size and random sampling errors eliminated most of the genetic variability from the population. Without other subpopulations of elephant seals to interbreed with, the genetic line has become greatly restricted.

TABLE 25-1

Relationship between population size, age, and isolation and genetic variation in populations of spittlebugs (*Philaenus spumarius*) on islands near the coast of Finland (from Saura *et al.* 1973).

Island	Age (years)	Distance index	Area of habitat (m²)	Population (ind)	Polymorphic loci (per cent)	Average heterozygosity (per cent)
Segelskär	200	17.7	500	1,100	17	2.3
Flatgrund	100	0.7	50	90	39	7.1
Storsundsharun	300	2.4	400	500	43	6.3
Fyrholmen	200	0.6	500	1,000	42	9.9
Skyffelskär	400	6.5	600	3,000	50	9.6
Mellankobben	200	1.0	100	200	61	11.2
Tvärminne*	1,300	0.0	80,000	30,000	64	15.3

* Mainland locality

Speciation usually requires the extended isolation of gene pools.

Speciation requires the accumulation of genetic differences between populations to the point that individuals from the different populations can no longer interbreed. For most groups of organisms, genetic differences can accumulate only if two gene pools are completely isolated (Mayr 1942, 1963). This model is usually called *geographical,* or *allopatric, speciation.* Populations may be isolated geographically when individuals cross some dispersal barrier, such as a mountain range, river, or ocean, and establish a population in a suitable habitat effectively isolated from the parent population, or when major climatic changes contract and fragment the habitat of a population as, for example, in periods of glaciation (Keast 1961, Haffer 1967, 1969).

Subspecies (subpopulations that are distinguishable by morphological characteristics but nonetheless are reproductively compatible) are often regarded as evidence of incipient speciation. The song sparrow, for example, has been subdivided into 34 subspecific populations (Miller 1956), most of which occur along the western coast of North America, where mountain ranges, islands, and sharp climate and habitat changes create barriers to dispersal.

The taxonomic characteristics used by the museum specialist to distinguish subspecies do not always represent substantial genetic differences between populations. Furthermore, so-called subspecific differences may not actually stem from geographical disjunction; isolation by distance may create the same apparent result. Distinctive population differences are described frequently after examining arbitrary isolated points along a continuous cline. For example, the striping pattern of Burchell's zebra (*Equus burchelli*) in South and East Africa changes over a continuous gradation. The black stripes, which cover the entire body in northern populations, gradually disappear from the back and flanks in southern populations (Figure 25-5). Although Burchell's zebra has been given different names in different areas, these names do not correspond to well-differentiated populations. Only Grevy's zebra (*Equus grevyi*) to the north and the mountain zebras (*Equus zebra*) in the south of Africa are distinct, reproductively isolated species (*e.g.,* Keast 1965).

Where populations are separated by some physical barrier to dispersal, we can feel confident that they are, at least temporarily, independent evolutionary units. In Australia, moist forest habitats were formerly much more widespread over the continent than at present. Drying trends in the Australian climate during the past several thousand years have restricted forest habitats to the eastern coast of Australia and to other isolated areas around the periphery of the island continent. Most of the interior of Australia is now extremely arid, and savanna- or forest-dwelling birds do not venture there. As the moist forest contracted to peripheral areas, subpopulations of many widespread animals and plants became isolated and morphologically differentiated. This pattern is illustrated by the *Climacteris picumnus* group of savanna-dwelling tree creepers (Figure 25-6). The eastern coastal populations are assigned to the species *C. picumnus,* in which a northern and a southern subspecies are recognized. Along the northern and northwestern

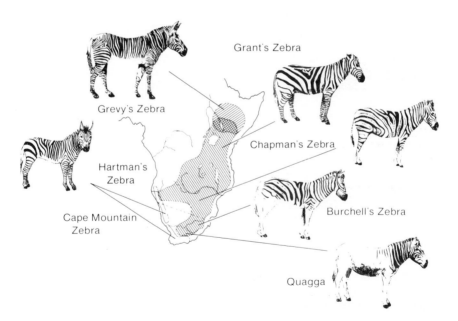

Figure 25-5   Distribution and striping patterns in zebras. Grant's zebra, Chapman's zebra, Burchell's zebra, and possibly the quagga, represent poorly defined races, or subspecies, of the wide-ranging *Equus burchelli*. Grevy's zebra (*E. grevyi*) to the north and the mountain zebras (*E. zebra*) to the south represent distinct, reproductively isolated species (after Mettler and Gregg 1969).

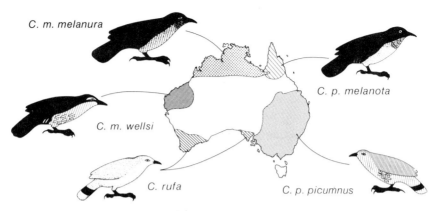

C. m. melanura

C. p. melanota

C. m. wellsi

C. rufa

C. p. picumnus

Figure 25-6   Distribution of the savanna-dwelling tree creepers of the *Climacteris picumnus* group. Each of the peripheral areas of savanna habitats has a distinctive form. The northwestern species *C. melanura* has extended its range into that of *C. picumnus,* to the east, and the two do not interbreed. The southwestern *C. rufa* is so distinctive that it is recognized as a different species (after Keast 1969).

coasts, another distinct species, *C. melanura,* is also divided into two subspecies. The southwestern corner of Australia is inhabited by *C. rufa,* which has secondarily invaded a small area adjacent to the range of *C. picumnus.* That the taxonomically recognized species are actually reproductively isolated is shown in one instance where the range of *C. melanura* extends into the range of *C. picumnus* at its eastern border, and the two populations do not interbreed.

Perhaps the most convincing demonstration of the role of isolating barriers in speciation may be found in island groups, where barriers are clearly defined. Differentiated populations of birds, for example, are rarely found on an isolated island, even of moderate size. In contrast to the failure of bird populations to speciate on individual islands, such archipelagos as the Galápagos and Hawaiian islands present a very different pattern. Each island of an archipelago can accumulate many species from a single ancestral type through the following series of events: (a) initial invasion of the island group; (b) isolation of island populations due to water barriers; (c) differentiation of populations on separate islands; (d) reinvasion of populations onto other islands in the group. A single type of finch that colonized the Galápagos Islands at some unknown time in the past, but probably within the last million years, has radiated to form at least thirteen species, up to ten of which can be found on some islands in the group (Lack 1947a).

Although it may seem inconsistent to state, in one breath, that birds can fly between islands and establish new populations and, in the next, that populations on different islands are reproductively isolated, this apparent dilemma may be resolved by comparing the relative rate of immigration needed to colonize an island with that required to provide effective levels of gene flow. The infrequent dispersal of a few individuals to an island may be sufficient to establish a population. Once the population has grown, however, the occasional arrival of a few immigrants from a distant population would have a greatly reduced effect on the gene pool.

Incomplete speciation emphasizes the weakness of a geographical notion of the biological species.

Reproductive isolation between two species occurring together can be determined with certainty. But what if the "species" are reproductively isolated in one locality and not in another? Two species of garter snakes, *Thamnophis hydrophila* and *T. elegans,* are sympatric over a broad area in northern California and southern Oregon without interbreeding. But in the Klamath River basin of northern California the two species form hybrid populations (Fox 1951). The garter snakes do not strain our biological concept of the species; the two forms are perfectly good species in most areas, but they behave as two races of a single species in the Klamath area. Populations are not uniform over their entire distribution, and there is no reason to believe that the same degree of reproductive isolation must be achieved wherever two populations meet.

The peculiar situation of so-called ring species deserves mention here. Sometimes a species is distributed in suitable habitats that form a ring around an unfavorable area. When a species spreads round such a ring, and the two ends meet on the far side, they may have become different enough that they cannot interbreed. Even though gene flow can occur between the two ends of the population by traveling the full circuit of the ring, the two recently joined ends of the population are effectively isolated from each other by distance.

Salamanders of the genus *Ensatina* have formed a ringlike distribution through foothill habitats surrounding the dry Central Valley of California. The salamanders extend southward from British Columbia through the Cascade Mountains of Washington and Oregon, and then along two separated areas of suitable habitat in the Pacific Coast Ranges and the Sierra Nevada of California (Figure 25-7). Near the southern end of the range, the coastal and inland populations meet again near the border of California with Mexico, but they are so distinct that they do not interbreed. Also, a small enclave of the Coast Ranges population appears to have crossed the Central Valley of California and established a foothold in the Sierras where it is almost completely reproductively isolated from the Sierran population. The disjunct distribution of the Sierran form of *Ensatina* in southern California suggests that it may have arrived first and become established over a large area of southern California, only to be pushed out by the expanding Coast Ranges population. The coastal form is undoubtedly better adapted to the hot, dry habitats of the southern coastal mountains.

Can isolation be achieved by factors other than allopatry?

Speciation does not necessarily require geographical separation of populations, although this seems to be the easiest way to achieve reproductive isolation in most groups of animals and plants. Reproductive isolation

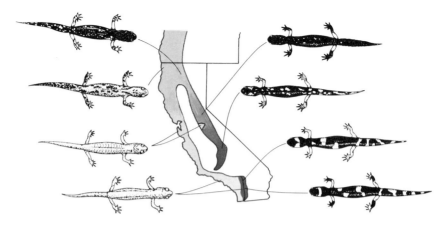

Figure 25-7   Distribution and morphological variation in the salamander *Ensatina* in western North America. The Sierra Nevada and Coast Ranges subpopulations have achieved secondary contact in southern California, where they do not interbreed (after Stebbins 1949).

between sympatric species is often based on differences in the timing of reproductive activities and on specific ecological requirements of the species. Several theoretical models demonstrate that such differences can arise within a single population without geographical isolation, that is to say, by *sympatric speciation* (Alexander and Bigelow 1960, Brues 1924, Maynard-Smith 1966, Smith 1941). Still, however valid these hypotheses may prove to be, they will probably continue to be overshadowed by allopatric speciation.

Bush (1969) outlines a mechanism for sympatric speciation in frugivorous tephritid flies (*Rhagoletis*) based upon the formation of local host races. Courtship and mating in *Rhagoletis* occur on the larval host plant; mate and host selection are, therefore, highly correlated. As a result, minor changes in the genotype associated with host plant selection could lead to local ecological isolation. Developmental induction of host preference by the specific chemical qualities of the food plant itself (for example, Jermy *et al.* 1968, Manning 1967) would greatly facilitate isolation. Bush further suggested that infestation of host plants having different fruiting periods can lead to rapid allochronic isolation, as apparently has happened in the apple and hawthorn races of *Rhagoletus pomonella* in eastern North America.

If habitat or host selection were strong, that is, if the entire life cycle of the individual, from egg to egg, took place in the same habitat or host, then a population could become subdivided into a mosaic of ecologically defined subunits that could have sufficiently low genetic interchange to lead to speciation. Such cases, in which the geographical scale in geographical isolation is drastically reduced, may be more prevalent than we are led to believe by our biased concept of geographical scale and the size of ecological patches.

One further model of speciation through *disruptive selection,* proposed by Levene (1953), Mather (1953, 1955), and Thoday (1958), can be made to work in the laboratory, but is difficult to relate to natural conditions. Disruptive selection occurs when extreme phenotypes have greater evolutionary fitness than intermediate phenotypes. Disruptive selection discourages matings between extreme types, which produce offspring with less fit intermediate phenotypes. Polymorphic populations can be produced in the laboratory by selectively removing intermediate phenotypes, as Thoday (1959) and Thoday and Boam (1959) have done for abdominal bristle number in the fruitfly *Drosophila melanogaster.* Although the high- and low-selected subpopulations were continually crossed with each other, the experiment resulted in a bimodal distribution of chaetae number within the population. Thus, even when flies in one subpopulation were forced to mate with individuals in the other, maximizing gene flow between the subpopulations, polymorphism could be evolved in a single culture of flies, the first step to sympatric speciation. Complete reproductive isolation between subpopulations was eventually achieved (Thoday and Gibson 1962) but more recent attempts to repeat the experiment have failed (Scharloo 1971). Still, regardless of the success of laboratory experiments with disruptive selection, the conditions necessary for its existence in natural situations are difficult to imagine.

How much genetic difference is needed for reproductive isolation?

Although the steps in allopatric speciation have been outlined in detail by evolutionary biologists, next to nothing is known about the genetics of speciation (Lewontin 1974). Where genetic crosses cannot be made, except between closely related species in the laboratory, genetic differences cannot be determined with certainty. Recently, electrophoresis has enabled geneticists to identify genes indirectly, through their protein products, with reasonable certainty and without having to make genetic crosses. Allele frequencies may then be used to calculate an index of genetic distance (*D*) between two species, which is an estimate of the number of electrophoretically detectable amino acid (codon) differences between homologous proteins (genes) in different species.

Genetic surveys by electrophoresis reveal a consistent pattern of genetic difference between species that varies in accordance with degree of taxonomic differentiation. Different populations of the same species and different species that nevertheless mate readily and successfully in the laboratory (semispecies) have genetic distances less than 0.5 substitutions per protein. Sibling species, which are closely related and often difficult to distinguish but are reproductively isolated, have values of *D* between 0.2 and 1.5. For well-characterized species in the same genus, *D* usually exceeds 1.0, sometimes by a large amount. Unfortunately, as more evidence accumulates, the distinctions become less clear-cut. Most data have been obtained from *Drosophila* and rodents, with a few odd groups thrown in for flavor. When King and Wilson (1975) applied the same measure of distance to man and chimpanzee, each belonging to quite distinct genera, *D* was found to be 0.62, in the middle of the range of sibling species, which *Homo* and *Pan* clearly are not. King and Wilson underscored these controversial data by pointing out that most electrophoretically detectable enzymes are concerned with cell function and not with the differences in development and gene expression that distinguish species and genera. They argued that most evolution proceeds by change in genes that regulate the expression of genes responsible for cell function. One could argue that most tissue functions are similar even in widely divergent species. Their differences may reside in the amounts of various gene products produced in different tissues and at different times during development.

Problems encountered in attempting to measure genetic differences between species are illustrated by analyses of relict species of pupfish (*Cyprinodon*) in the Death Valley region of California (Turner 1974). The time of separation of the phylogenetic lines varies from about 20,000 to more than a million years. All five species are morphologically and behaviorally distinctive and inhabit quite different habitats, ranging from saline lakes with fluctuating temperatures to freshwater springs with constant temperatures. In spite of the clear morphological separation of the species, they are remarkably similar with respect to electrophoretically detectable proteins. Similarities of alleles between the species vary from 80 to 97 per cent, with an average of about 90 per cent. Furthermore, the genetic similarities do not agree with taxonomic and geographic evidence on degree of relationship. To make matters more confusing, even though specific characters are maintained under uniform laboratory conditions and thus are genetic, four of the species readily mate with each other and produce fertile offspring. Their ranges do not overlap, and so hybridization does not occur in the wild. Whatever their genetic differences, they are not sufficient to prevent interbreeding (the chromosomes of the five species are indistinguishable), but they have achieved levels of morphological divergence comparable to what are known to be good species.

Plants may achieve reproductive isolation in one generation by hybridization or by chromosomal mutation followed by asexual reproduction.

Speciation is complicated in plants by the fact that asexual reproduction can turn cytological mistakes into reproductively isolated populations that are incipient species (Stebbins 1950, Briggs and Walters 1969). The low degree of specificity of many pollination mechanisms frequently results in hybridization between plant species. Through subsequent reproductive and cytogenetic events, new species containing the chromosomes of both parental species (*allopolyploids*) may arise.

Let us consider in more detail how allopolyploids are formed, and why they are reproductively isolated from the parental species. We will represent the paired chromosomes of the parental species by *AA* and *BB*, respectively. The gametes produced by the two species are represented by *A* and *B*, respectively, and the hybrid zygote is *AB* (Figure 25-8). If the chromosomes of *A* and *B* are very different, they will pair improperly or not at all during meiosis, and the hybrid will not produce viable gametes. During meiosis, unpaired chromosomes are parceled out at random among the gametes, and the probability that a gamete will receive all the chromosomes present in the hybrid is small. The chance that a chromosome will go to a given side of the cell during the first meiotic division is one in two, the chance that any two will go to that pole is one in four ($\frac{1}{2} \times \frac{1}{2}$), and so on. For ten chromosomes, the chances are about one in 1,000, and for twenty chromosomes, about one in a million. Depending on the chromosome number, the probability that meiosis will lead to the formation of gametes containing all of the chromosomes of one parent or the other is remote. Occasionally, however, chromosomes fail to segregate during meiosis and diploid gametes (*AB*) are formed. If the hybrid were to self and if two of these diploid gametes should happen to fuse during fertilization, the zygote would be an allotetraploid (*AABB*), which contains a full complement of chromosomes from each parent and normally is fully fertile.

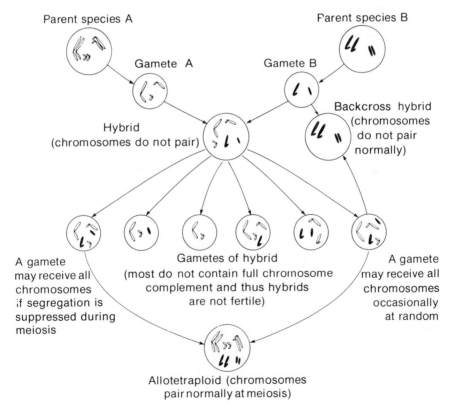

Figure 25-8   Diagram of events that occur in the formation of allopolyploid species of hybrid origin.

The formation of fertile gametes by chance occurs with such low probability in hybrids that allopolyploids with thirty and forty chromosomes seem highly unlikely. Too few viable gametes would be formed, and even if they were, the chances of their forming a zygote would be minuscule. Yet, there are hybrid species with more than seventy chromosomes. Clearly, such allopolyploids could be formed only if gametogenesis occurred through mitosis, or a modified meiosis in which chromosome pairing were not necessary and full complements of both parental chromosomes were distributed to each of the daughter cells.

The new hybrid species is genetically isolated from both parental species because backcrosses, for example between A and AB gametes, create the same difficulties in pairing at meiosis as the original hybrid, and thus do not generally produce fertile offspring. Note, however, that the backcross hybrid itself could give rise to a new, reproductively isolated, polyploid population (AAAABB). Because allopolyploids can mate only with other similar allopolyploids, new species of hybrid origin can occur only in plant groups that either self or include asexual reproduction as a normal part of their life cycle.

Hybridization is not the only means by which instant isolation can occur in plants. Any chromosomal rearrangement that upsets normal pairing during meiosis can lead to reproductively isolated lines. For example, Clarkia rhomboidea (haploid chromosome number, n = 12) is an allopolyploid derivative of two species, C. virgata (n = 5) and C. mildrediae (n = 7), that are locally distributed in the Sierra Nevada Mountains of California. C. rhomboidea itself is spread widely over the western United States, and 26 populations examined by Mosquin (1964) exhibited six different chromosome rearrangements, none of which overlap geographically. Two of the chromosome arrangements are widespread, while four are local. Apparently, chromosome rearrangement in Clarkia results in almost complete reproductive isolation, and the arrangements confer different selective advantages depending upon the environment.

The reproductive isolation of populations through chromosome rearrangements occurs in animals, although much less frequently than in plants (White 1968, Patton 1969). Allopolyploid hybrid animal species are also known, but they are not common.

The significance of polyploidy in plants is not well understood. The phenomenon is closely associated with self-compatibility and inbreeding and occurs most frequently in high latitudes (Table 25-2). It does not seem reasonable to me that polyploidy is a fortuitous result of sloppy pollen transfer and self-compatibility or asexual reproduction, and without ecological and evolutionary significance of its own. A potential advantage of polyploidy to self-compatible species is the preservation of heterozygosity. Selfing normally reduces the frequency of heterozygotes in a population and erodes genetic variability. When the genome size is multiplied by allopolyploidy, the genetic diversity of two species is combined into one. A special kind of polyploidy involves the multiplication of chromosome number within a single line without hybridization. Autopolyploidy, as this process is called, fixes the heterozygosity of the parent in the offspring, even in selfing populations.

TABLE 25-2

The occurrence of polyploid species in the floras of several regions (from Morton 1966).

Region	Percentage of polyploid species
West Africa	26
Northern Sahara	38
Great Britain	53
Iceland	66
Greenland	71
North Greenland	86

These considerations lend credibility to the hypothesis that polyploidy serves to maintain the genetic diversity of the genome in situations that strongly favor selfing and vegetative reproduction.

Reproductive isolation between sympatric populations may be reinforced by selection against hybrids.

When two isolated populations re-establish contact, they may be completely reproductively isolated, completely interfertile, or any of the intermediate spectrum of possibilities. If individuals of the two populations interbreed successfully, population characteristics will tend to merge in the area of sympatry, resulting in a recognizable *hybrid zone* that fills both the geographical and phenotypic gaps between differentiated populations (Sibley 1965, Sibley and Short 1959).

When hybrids between populations are less fit than either parent, individuals that make reproductive mistakes are strongly selected against. Any characteristic that emphasizes differences between populations (for example, in courtship and mating) will be selected because the individuals that carry those characteristics are less likely to be involved in illegitimate matings (Dobzhansky 1937, Grant 1963). Just as character divergence reduces competition between sympatric populations (page 240), the evolution of behavioral isolating mechanisms reduces the production of unfit hybrids.

Artificially produced hybrids between the red and black forms of the stickleback *Gasterosteus* have a lower fitness than either of the parental forms (McPhail 1969). The fertility rate of stickleback eggs of hybrid origin is nearly equal to that of control eggs (exhibiting only a 5 per cent loss of viability), and, similarly, the fertilization success of hybrid males is only about 5 per cent less than controls. But the hatching success of eggs produced by hybrids was nearly 50 per cent lower than controls, and that of their progeny, nearly 70 per cent lower. The low reproductive success of hybrids is caused by a breakdown in the behavior of males, which do not care properly for their eggs.

The inferiority of hybrids in the stickleback has promoted the evolution of secondary isolating mechanisms (based on mate preference by females in this case) where the black and red forms are sympatric. When females from allopatric populations of the two forms are offered both black and red males in courtship-choice experiments, both strongly prefer males that are red, the normal color of the courting male stickleback. But in the zone of contact between the two populations, black females show a weak preference for black males, in contrast to the behavior of allopatric black females (McPhail 1969).

Whereas premating isolation mechanisms have not been strongly developed in the stickleback, two Australian tree frogs, *Hyla ewingi* and *Hyla verreauxi,* have evolved extremely strong behavioral isolation in areas of sympatry. *H. ewingi* is distributed throughout much of Victoria and Tasmania; *H. verreauxi* occurs along the east coast of Australia, and overlaps the distribution of *H. ewingi* in eastern Victoria (Figure 25-9). Mating calls of the frogs exhibit character displacement: those of allopatric populations of the two species are similar, and those of sympatric populations are quite distinct (Figure 25-9).

The importance of mating call differences in reproductive isolation was demonstrated experimentally by Littlejohn and Loftus-Hills (1968). They placed female frogs in the center of a long enclosure; when frog calls were played through speakers at either end of the enclosure, females readily distinguished calls of the two species from the area of sympatry, and always hopped toward the call of the appropriate species. Females often could not distinguish calls of the two species from allopatric populations.

The outcome of experimental crosses between *H. verreauxi* and *H. ewingi* varied, depending on whether crosses were between sympatric or

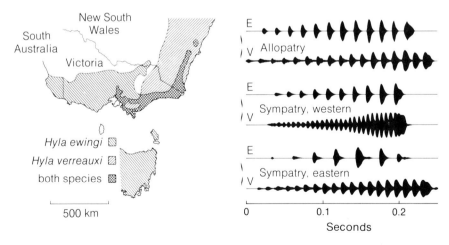

Figure 25-9   Distribution of the tree frogs, *Hyla ewingi* and *H. verreauxi,* in southeastern Australia, and oscillograms of mating call notes from areas of allopatry and sympatry (after Littlejohn 1965).

TABLE 25-3

Summary of artificially produced hybrids between populations of *Hyla ewingi* and *H. verreauxi*. Sympatric crosses are enclosed in the upper left-hand quadrant of the table (based on data in Watson and Martin 1968).

Female parent	Male parent			
	Sympatric *H. verreauxi*	Sympatric *H. ewingi*	Allopatric *H. verreauxi*	Allopatric *H. ewingi*
Sympatric *H. verreauxi*	normal	reduced hatching success, abnormal embryos*	normal	not done
Sympatric *H. ewingi*	abnormal embryos, failure to metamorphose	normal	abnormal embryos	not done
Allopatric *H. verreauxi*	not done	not done	abnormal	not done
Allopatric *H. ewingi*	normal	normal	normal	normal

* Embryo abnormalities include oedema, exogastrulation, brachycephaly, and spina bifida.

allopatric populations (Watson and Martin 1968). Allopatric crosses, both within and between species, usually produced offspring with normal viability compared with controls—crosses within the same species from one locality (Table 25-3). Sympatric crosses between the two species were less fertile because of abnormalities during embryonic development or failure to metamorphose from the tadpole to the adult stage. Therefore, not only has premating isolation been strengthened in the area of sympatry, but postmating reproductive isolation has increased through reduced viability of hybrids. Postmating isolation confers no advantage on the individual because gametes, and perhaps parental effort, are completely wasted. The loss of viability in sympatric hybrids must be a secondary consequence of rapid evolutionary divergence in areas of sympatry. Furthermore, sympatric hybrids must have had at least slightly reduced fitness initially, if premating isolation mechanisms were to be selected.

It is becoming increasingly evident that subtle behavioral and morphological differences may effectively isolate populations reproductively. For example, four species of gulls breed in close proximity in a small area of Baffin Island without interbreeding (Smith 1967). The dark-backed Thayer gull and the herring gull are obviously distinct from the Kumlein gull and the glaucous gull, which have light backs. But beyond back coloration, the most conspicuous species differences are the dark iris and dark eye ring of the Thayer and Kumlein gulls, and the light iris surrounded by a light-colored eye ring of the herring and glaucous gulls. In a series of difficult experiments,

TABLE 25-4

Effects of experimental reversal of eye ring color on the formation and maintenance of the pair bond in gulls (from Smith 1967).

Subjects	Control	Results	Experimental
Mated females light eye ring			
changed to dark	6	per cent of	74
dark to light	6	pairs broken	57
Unmated males			
light to dark	78	per cent	12
dark to light	15	rejected	73
Mated males			
light to dark	—	per cent of	2
dark to light	—	pairs broken*	6

* Males do not copulate after their eye rings are altered because of inappropriate female responses to their premating invitations. Some females understandably pack up and leave.

Smith (1967) painted light eye rings dark and dark eye rings light in several hundred birds, and demonstrated that eye coloration is probably the most crucial factor in species recognition. Altering the eye ring color of mated females led to the dissolution of the pair bond in the majority of cases, and unmated males with altered eye ring color were rejected by the majority of females that they courted (Table 25-4). That reproductive behavior can be totally upset by a change in such a minor part of the animal's anatomy as its eye ring demonstrates at once the simplicity of pair bond formation and the extreme power of some social signals over behavior.

We may interpret speciation as an ecological and genetical process resulting from evolutionary change within isolated subpopulations. But speciation may, in turn, play an important role in ecological interactions through its influence on the number of interacting species in a community. Species diversity on islands is thought to reflect the balance between immigration and extinction rates, but it is clear that the diversity of finches on the Galápagos Islands has been overwhelmingly influenced by speciation within the archipelago rather than by any ecological limit to the number of species that can coexist on an island. The relationship between speciation rate and species diversity can be extended to mainland faunas, at least in principle, but the relative roles of speciation and extinction in producing geographical variation in species number is a controversial topic, which we shall defer until Chapter 37.

26

The History of Evolution

Phyletic change and speciation can be followed through a sometimes adequate, sometimes fragmentary fossil record. Perhaps the most conspicuous feature of the history of evolution is its heterogeneity. Phylogenetic groups, and even whole faunas, have gone through periods of rapid evolution and diversification, followed by quiescent periods, and even periods of major extinctions, conspicuous only over tens of millions of years. Yet, we should be able to interpret evolutionary trends through our understanding of the relation between the evolutionary process and the ecology of populations. We might also better understand present-day ecological relationships by gaining an evolutionary perspective on their origin.

In this chapter, we shall review some of the major features of evolution, particularly *adaptive radiation,* and their ecological implications. To provide a timetable reference for this discussion, the major geological ages and their biological events are outlined in Figure 26-1.

The basic design of organisms was accomplished more than 500 million years ago.

When the first major assemblages of fossils were preserved in the Earth's crust, organisms had body plans similar to many of the present-day forms of marine plants and invertebrates. A lobster would not have looked out of place among trilobites on the floor of a Cambrian sea. Some genera of bivalves fossilized in early Paleozoic deposits have living representatives. Of course, many of today's important groups had not yet appeared. Fishes do not occur in the fossil record until the Ordovician Period, nearly 100 million years after the beginning of the Paleozoic Era. A similar stretch of time would pass before plants and animals first invaded the land. But the basic morphological and physiological features of organisms had been fully developed by the beginning of the Cambrian Period. Evolution since that time has merely brought the progressive modification of successful themes.

Prior to the opening of the Cambrian Period, fossils are exceedingly rare; they consist mostly of primitive aquatic plants, some as old as 3 billion years

ERA	PERIOD	EPOCH	DISTINCTIVE FEATURES	YEARS BEFORE PRESENT
CENOZOIC	Quarternary	Recent	modern man	11,000
CENOZOIC	Quarternary	Pleistocene	early man	3,000,000
CENOZOIC	Tertiary	Pliocene	large carnivores	13,000,000
CENOZOIC	Tertiary	Miocene	first abundant grazing animals	25,000,000
CENOZOIC	Tertiary	Oligocene	large running mammals	36,000,000
CENOZOIC	Tertiary	Eocene	many modern types of mammals	58,000,000
CENOZOIC	Tertiary	Paleocene	first placental mammals	63,000,000
MESOZOIC	Cretaceous		first flowering plants; extinction of dinosaurs and ammonites at end of period	135,000,000
MESOZOIC	Jurassic		first birds and mammals; dinosaurs and ammonites abundant	180,000,000
MESOZOIC	Triassic		first dinosaurs; abundant cycads and conifers	230,000,000
PALEOZOIC	Permian		extinction of many kinds of marine animals, including trilobites	280,000,000
PALEOZOIC	Carboniferous	Pennsylvanian	great coal-forming forests, conifers; first reptiles	310,000,000
PALEOZOIC	Carboniferous	Mississippian	sharks and amphibians abundant; large primitive trees and ferns	345,000,000
PALEOZOIC	Devonian		first amphibians and ammonites; fishes abundant	405,000,000
PALEOZOIC	Silurian		first terrestrial plants and animals	425,000,000
PALEOZOIC	Ordovician		first fishes; invertebrates dominant	500,000,000
PALEOZOIC	Cambrian		first abundant record of marine life; trilobites dominant, followed by massive extinction at end of period	600,000,000
	Pre-cambrian		fossils extremely rare, consisting of primitive aquatic plants	

Figure 26-1   The geological ages, their approximate span of time, and the major biological features of each age. The fossil record is relatively brief compared with the four-billion-year age of the Earth (from Newell 1963).

(Schopf 1970). The age of the Earth is thought to be about 4 billion years. Clearly, life has not always existed as we see it now. The basic development of cell structure and function and the development of hereditary material must have preceded the evolutionary elaboration and diversification of what we know as organisms. The origin of life can only be surmised, as all evidence has been obliterated (Oparin 1953, Bernal 1967, Miller and Orgel 1974, Cloud 1974). Even the conditions under which life arose no longer exist, except in laboratory reconstructions.

Life originated in an environment different from any now found on Earth.

The primitive environment of the Earth differed greatly from its present condition. As the Earth condensed and cooled, its surface had little water or atmosphere. Gradually, volcanic outgassing (the venting of gases produced and trapped within the interior of the Earth, still going on in some areas) began to contribute to a primitive atmosphere. There was no oxygen and little nitrogen, only methane (CH_4), ammonia (NH_3), water vapor (H_2O), and hydrogen (H_2). Because hydrogen is light, most of it rose to the top of the thin atmosphere and escaped into space. As the Earth cooled, water vapor condensed and began to rain down upon the Earth's surface, creating the first primitive seas. These may have been the conditions under which life originated.

How could life have developed in such an environment? Living organisms, even the simplest cells, are such complex combinations of organic chemicals that the chances of their coming together in the right amounts and combinations at one time are infinitely small. Consider the probability of producing just a simple word, like *cat*, by random combinations of the letters *c, a,* and *t* out of the 26 letters of the alphabet. If the chance of drawing any one of the letters is the same, one in 26, then the probability of drawing the word *cat* is one in $(26)^3$, or 17,576. To produce a living cell at random would be equivalent to randomly drawing a good short story—five thousand words, say, in length. But such an improbable event as the origin of life was favored in two ways: It did not all have to occur in one step, and there was plenty of time—a billion years or more.

By way of analogy, consider again the chance origin of the word *cat*. Either of the two-letter combinations, *ca* or *at*, would be drawn at random with a probability of one in 676 (26^2), which is not too unlikely if one kept at it for awhile. If these two-letter combinations could be preserved, perhaps owing to an inherent stability, the completion of the word by drawing a *c* or a *t* becomes highly probable (one in 26). And so it must have been with the origin of life: Certain of the simplest building blocks appeared first, probably amino acids, simple sugars, and other organic compounds. These combined in various ways, some of them stable (and thus successful) and others unstable, until the first properties of living beings appeared.

The early environment of the Earth was, in many ways, ideal for the formation of life. In fact, it is exceedingly unlikely that the same events could take place in the present environment of the Earth. In order to build organic molecules there must be the proper raw materials: carbon, nitrogen, oxygen, and hydrogen. Energy is needed too, because the energy content of organic molecules is higher than the raw materials from which they are made. The raw materials were present in abundance in the atmosphere of the primitive Earth, and energy could have been supplied either by electrical discharges (lightning) in the atmosphere, or by ultraviolet radiation that penetrated the primitive Earth atmosphere. The present protective mantle of ozone (O_3) in the upper atmosphere, which absorbs ultraviolet light, had not yet developed.

Conditions just like these were first reproduced in the laboratory by Stanley L. Miller (1953). Miller was able to make several amino acids by circulating methane, ammonia, water vapor, and hydrogen past an electrical discharge. Others have since obtained amino acids and other organic compounds using ultraviolet light as a source of energy.

Ultraviolet light may provide the energy to form organic molecules, but it will break them down just as quickly. The primitive life molecules were probably protected from the radiation when they were washed into the sea by heavy rains; water efficiently absorbs ultraviolet radiation. Over a long period, the accumulation of inorganic and organic molecules gradually transformed the primitive sea into a thin "soup."

Life probably arose slowly by many steps.

We are less sure of the course that the early life-building process took after the first appearance of organic molecules, but we can surmise some of the important steps (*e.g.,* Miller and Orgel 1974). For any biosynthesis to have occurred in the aquatic environment, there would necessarily have been a source of energy other than ultraviolet radiation or electrical discharges. Organic molecules must have been broken down to make available their chemical energy for biosynthesis. The first metabolic process may have been a form of hydrolysis, in which a carbon-carbon bond (as in carbohydrates and fats) or a carbon-nitrogen bond (as in proteins) is broken by the insertion of a water molecule. Perhaps the first "living" molecules capable of hydrolyzing other molecules were self-replicating enzymes or nucleic acids that used the energy released by breaking down organic compounds to synthesize other molecules like themselves. Imagine that several of these molecules may have clumped together, as many organic molecules do of their own accord, and formed aggregates. If different sorts of molecules came together in aggregates, greater functional efficiency or new functions could possibly have been attained.

Somewhere along the line, control over the function of the aggregate must have been localized onto one kind of self-replicating molecule, the ancestor of DNA, which then became specialized for heredity and did not directly participate in metabolic function. This step, which required most of the early period of the development of life, gave life forms a continuity through time and made evolution possible. From there, the development of life must have proceeded rapidly to the stage at which the organization of life forms began to take on the appearance of what we know as cells.

At first, life probably relied for energy on inefficient fermentation processes (which still occur in most cells under anaerobic conditions), because there was no oxygen in the atmosphere or seas to support oxidative respiration. As life became abundant, the availability of freely floating organic molecules declined simply because they were consumed by life forms faster than they were produced in the atmosphere. This may have been the point at which life forms were first limited by ecological constraints.

Probably the first major innovation after the advent of the cellular stage was the development of autotrophic mechanisms, such as photosynthesis, for the production of food. Light energy and carbon dioxide were abundant in the surface waters of the seas, and autotrophy would have freed cells from their reliance on organic molecules for a source of energy. This step heralded the evolution of plants from the primitive bacteria-like scavengers that preceded them.

Photosynthesis was not only a major step for life, it was also a major step for the environment. One important result of photosynthesis was the production of oxygen, which diffused into the water and the atmosphere, where it had two major ecological consequences. First, oxygen permitted the development of efficient oxidative metabolism that provided energy to cells at a higher rate than had been possible before. Second, some of the oxygen was transformed into a thin layer of ozone at the top of the atmosphere, which blocked the penetration of ultraviolet radiation to the surface of the Earth. With these damaging rays blocked, life could take its first step onto the land.

Sexual reproduction created the first populations.

Probably long before the development of autotrophy, life had taken another major step, that of sexual reproduction (Maynard-Smith 1970, Baker and Parker 1973). With the exchange of genetic material between cells, and the recombination and independent assortment of genes in progeny, populations were formed, and the rate of evolution was greatly enhanced. Populations probably soon began to exhibit patterns of organization in the form of primitive colonies or groups of progeny that remained together. Such organization in turn led to specialization of structure and function, and hence to many-celled (metazoan or metaphyton) organisms. The organism stage in the development of life must have been attained more than 2 billion years ago, judging from the occurrence of primitive multicellular algae fossilized in rocks of that age.

During the next billion and a half years prior to the beginning of the Paleozoic Era, the basic plans of body symmetry evolved and with them, the foundations for all later life forms. In the pelagic environment of the early seas, organisms probably floated about haphazardly and could not metabolize energy rapidly enough to sustain locomotion. Under such conditions, spherical symmetry—facing the environment equally in all directions—was the most appropriate body architecture. Many modern planktonic organisms have spherical symmetry. When organisms settled on the ocean bottom, they faced an environment that was uniform in all directions around them except up and down. Under such conditions, symmetry is usually specialized by the adoption of a top and a bottom—the radial symmetry exhibited by starfish, sea urchins, and most erect plants.

As the habit of predation developed, organisms began to move under their own power toward prey or away from predators. Now the environment had a temporal sequence; it was encountered head first, which undoubtedly stimulated the evolution of a head end (and a rear end as a result), with its concentration of sensing organs and feeding apparatus. Of course, with a head and tail superimposed on a top and a bottom (or back and front), one is automatically endowed with a right and a left (bilateral symmetry). This constitutes the basic body plan of virtually all organisms that are capable of rapid movement.

Other innovations, such as the development of circulatory and nervous systems, opened the way for larger and more complex organisms that could cope with the increasingly more demanding environments to be found on land. These innovations, in turn, created more efficient predators that further stimulated the evolution of prey defenses. By the beginning of the Cambrian Period, most of the major evolutionary steps in the development of life had been taken, and the seas flourished with sponges, bivalves, snails, crustacea, and other forms, a few of which have descended virtually unchanged to the present.

By the Paleozoic Era, basic life patterns had become established in a physical environment not much different from environments now found on the Earth. The subsequent history of life can be interpreted in the light of our understanding of ecological and evolutionary processes as they occur at present. We will first consider some of the major features of evolution—emergence into new adaptive zones and adaptive radiation—and then we will pass on to geographical aspects of evolution, the rate of production and extinction of forms, and to the question of whether there is progress in evolution.

The penetration of new adaptive zones has led to extensive adaptive radiation.

Since the beginning of the Paleozoic Era, animals and plants periodically have entered new *adaptive zones* (Simpson 1953) not previously inhabited by other forms of life. In a sense, these pioneers were misfits because they diverged in the extreme from the typical mode of adaptation of the time. The fish that first began to foray out of the water for brief periods were, indeed, weird fish. The first reptiles that developed feathers and began gliding into a new mode of locomotion would not have been given very high marks today by bird or reptile enthusiasts, except perhaps for their oddity. Yet each of these innovators accidently entered a vast ecological region of unexploited possibilities. Once they had made the step, their lineage began to spread and diversify with incredible speed, producing virtual explosions of evolutionary activity amidst the quiet of evolution's normal pace.

The penetration of a new adaptive zone, for example the step between reptiles and birds, must be a substantial evolutionary undertaking that leaves behind few intermediate forms—none in the case of the reptile-bird sequence. The general absence of intermediates among fossil or living forms gives the superficial appearance that the transition into a new zone is made in a single giant step, a view that is inconsistent with current evolutionary theory. The lack of intermediate forms suggests that they were somehow unfit and that the bridge between two adaptive zones was fraught with environmental and evolutionary obstacles. But there is no reason to doubt that intermediate forms were as well adapted to their environment as any other extant population was to its particular environment. The intermediate way of life may have been less productive than the typical life styles at the time, but there were also fewer competitors. Some evolutionary lines must have been forced into adaptive extremes by intense competition from other groups. Once one of the extreme lines had emerged into a new and potentially unlimited adaptive zone, the adaptive radiation that soon followed, together with pressure from the original stock, possibly drove the now intermediate forms out of existence.

Many new adaptive zones were inherent in the physical environment; others have been created by the evolution and diversification of life forms. Evolution is self-accelerating in that environmental complexity produced by life forms creates additional opportunity for the evolution of new forms. Selection for aquatic animals to venture onto the land could not have been strong—except, perhaps, to escape predators—until there were terrestrial plants to eat. Similarly, there must not have been strong selection for flight in birds and insects, other than to escape predators, until flowers, fruits, and flying insects had evolved. For whatever reason insects and birds took to the air, the evolution of flight and the adaptive radiation of flying forms was considerably accelerated by the development of flowering plants during the Cretaceous Period. Until the end of the Mesozoic Era, wind-pollinated conifers dominated the terrestrial environment; the development of terrestrial flowering plants led to the coevolution of plant-animal pollination and fruit dispersal complexes. The variety of flower types in the phlox family, Polemoniaceae, demonstrates the adaptive radiation of flower morphology in conjunction with diverse dispersal agents (Figure 26-2).

Adaptive radiations tend to fill gaps left by anomalies of geographical distribution.

Spectacular adaptive radiations have occurred within several remote oceanic island groups that received relatively few colonists from mainland areas. Immigrants that do reach isolated archipelagos enter an environment that is relatively free of competing species and thus affords excellent opportunities for speciation and evolutionary diversification. Isolated island groups are like new adaptive zones, separated by a geographical barrier rather than an evolutionarily intermediate zone, and colonizing populations have sometimes responded to these conditions with spectacular adaptive radiations. Two well-known examples among birds are Darwin's finches (Geospizinae) on the Galápagos archipelago (Lack 1947a) and the honeycreepers (Dre-

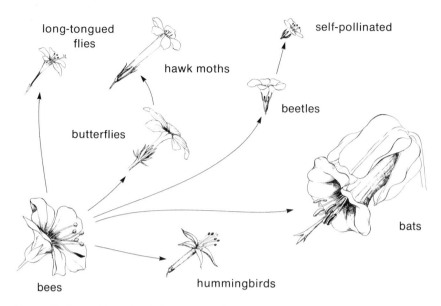

long-tongued
flies

self-pollinated

hawk moths

beetles

butterflies

bats

bees

hummingbirds

Figure 26-2   Diversity of flower types in the phlox family, Polemoniaceae, with their major pollinators indicated. The arrows suggest possible pathways in the evolution of the various pollination systems (after Ehrlich and Holm 1963, based on Verne Grant).

panididae) on the Hawaiian Islands (Amadon 1947, Figure 26-3). Each group probably evolved from a single colonizing population that arrived on the Galápagos as recently as one million years ago and on the Hawaiian Islands about five million years ago. The precise origins of the groups are not known, although both were clearly derived from finchlike ancestors of American origin. Through speciation and competitive displacement, the groups radiated to exploit a variety of features of the environment.

Evolutionary history has a heterogeneous geographical distribution.

Although most large groups of plants and animals have become widely distributed over the globe, the dispersal powers of many taxa are limited, and their adaptive radiation has not extended to all continents or ocean basins. Other groups, once enjoying widespread occurrence, are now restricted to relict refuges or have disappeared completely from vast areas of their former range. Competitive exclusion appears to have occurred on the family or even the ordinal taxonomic level. Radiating phylogenetic lines never seem to fully escape their bonds with the parental stock from which they came—such is the conservatism of evolution.

Patterns of geographical distribution among taxonomic orders higher than species may be illustrated by the distributions of several families of fresh-water fish (Figure 26-4). In most cases, dispersal is limited by salt water. New fresh-water drainage systems can be invaded only when they are confluent with others. Dispersal is often a chance phenomenon determined by irregularities of geography and the history of geographical change. The fresh-water minnows of the family Cyprinidae are widespread throughout the world, from subarctic regions, through the tropics, and into south temperate regions in South Africa (Figure 26-4A). Minnows disperse easily and they are an ecologically versatile group. They originated in the Old World, probably in central or southern Asia, and spread outward from there crossing into the Western Hemisphere at the Bering land bridge (the floor of the Bering Straits between the North Pacific Ocean and the Arctic Ocean has risen above the surface of the water several times in the last geological era). The distributional capability of the minnows is limited, however, as they have failed to reach Australia, South America, and even Madagascar, which has not been connected to mainland Africa for more than 100 million years.

The absence of a group from a large geographical region does not necessarily indicate limited dispersal ability: It may be the result of competitive exclusion. Sometimes the only clue to the former presence of a group in a region is a relict population whose occurrence can be explained only by postulating previous widespread distribution of the group. For example, the iguanid lizards comprise a large and successful family in the New World, but they are almost completely absent from the Old World, where they are replaced by agamid lizards. The presence of iguanids on Madagascar is a lone reminder of an extensive former distribution throughout the world.

Cichlid fish (Cichlidae), which are ecologically similar to the American sunfish family (Centrarchidae), are limited primarily to tropical and subtropical fresh waters (Figure 26-4B), but evidently they can disperse across salt water because they are found both in tropical Africa (including Madagascar)

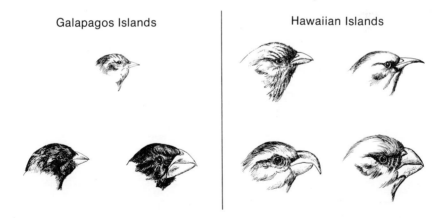

Galapagos Islands   Hawaiian Islands

Figure 26-3   Heads of representative species of Darwin's finches from the Galápagos Islands (left) and honeycreepers from the Hawaiian Islands (right) to demonstrate the respective ranges in beak depth and beak length in the two groups (after Lack 1947 and Amadon 1947).

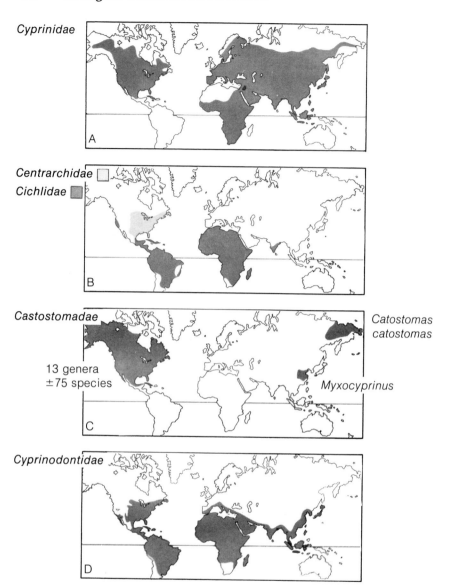

Figure 26-4   Geographical distributions of several major families of fresh-water fish: (A) minnows (Cyprinidae); (B) cichlids (Cichlidae) and their ecological counterparts in North America, the sunfish (Centrarchidae); (C) suckers (Catostomidae); and (D) killifish (Cyprinodontidae), whose occurrence in fresh water is secondary (after Laglar, Bardach, and Miller 1962).

and South America, and they have recently invaded Central America and India.

The mud minnows (Umbridae), a family with only four extant species, are clinging tenaciously to their restricted ranges in Eastern Europe, the Atlantic coastal plain and Great Lakes region of the eastern United States, and the Olympic Peninsula of Washington State. The suckers (Catostomidae) probably arose in the Old World, but they are now represented there only by an ancient representative of the family (*Myxocyprinus*) with primitive characteristics, although there are about 75 species in North America (Figure 26-4C). One of these (*Catostomus catostomus*) has recently emigrated back across the Bering Straits to invade eastern Siberia, and perhaps to begin a new radiation of the family in the Old World.

Dispersal barriers differ for each group, and they greatly influence geographical distribution. The killifishes (Cyprinodontidae) are secondary fresh-water fishes that can tolerate salt water and can cross sea barriers. Although the killifishes are limited to tropical and warm temperate regions, their distribution is nearly world wide and includes many island groups which often lack fresh-water forms (West Indies, Madagascar, Bermuda, Celebes; Figure 26-4D).

Convergent evolution obscures diverse evolutionary backgrounds.

Because of irregularities of geographical distribution, diverse areas are inhabited by organisms with diverse origins and evolutionary backgrounds. But organisms that inhabit similar environments in different geographical localities often resemble each other even though their evolutionary backgrounds differ. Adaptive radiation and convergent evolution tend to compensate for irregularities in distribution. Form and function converge under the mantle of similar selective forces in the environment; the contemporaneous environment acts as a unifying principle.

Trees are found virtually everywhere, and with them occur various insects that burrow through the wood of the trunks and branches. Among birds, the woodpeckers have met with singular success in exploiting this food resource, which has led to their widespread distribution throughout forest habitats. But as most woodpeckers inhabit deep forests, they lack strong flight and so do not disperse well across water barriers. As a result, woodpeckers do not occur on such remote islands as the Galápagos, the Hawaiian, and New Zealand. Yet on each of these island groups, birds unrelated to woodpeckers have evolved to exploit the food resource normally utilized by them. Moreover, each approaches this specialized task in a different way. The woodpeckers themselves combine a chisel-like bill for hacking away bark and wood with a long tongue that can be extended into crevices and burrows to extract insects (Figure 26-5). On the Hawaiian Islands, honeycreepers of the genus *Heterorhynchus* have separated the excavating and probing functions onto their lower and upper mandibles, respectively. They tap at wood with their short lower mandibles and probe with their elongated upper mandibles. The Galápagos woodpecker-finch digs trenches in soft wood with

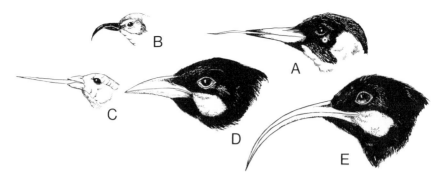

Figure 26-5   Birds that have become adapted to extract insects from wood in different ways: (A) the European green woodpecker (*Picus viridis*) excavates with its beak and probes with its long tongue; (B) the Hawaiian honeycreeper (*Heterorhynchus*) taps with its short lower mandible, probes with its long upper mandible; (C) the Galápagos woodpecker-finch (*Camarhynchus pallidus*) trenches with its beak and probes with a cactus spine; and the New Zealand huia (*Heterolocha acutirostris*) divided foraging roles on the basis of sex—the male (D) excavated with his short beak, and the female (E) probed with her long beak (after Lack 1947).

its stout beak and probes with a cactus spine that it holds in its beak. The huia of New Zealand separated these functions between the sexes. The male excavated with his short stout beak, and the female probed with her long beak. The birds were supposed to have fed in pairs, but as the species is now extinct, this arrangement evidently did not work out satisfactorily. Other trunk-feeding birds either excavate (nuthatches) or probe (creepers), but do not do both (Lack 1947a).

Evolutionary change is not uniform.

Because evolutionary change depends on the replacement of individuals within a population, we might expect evolution to proceed most rapidly in populations in which the turnover rate of individuals is greatest. For example, insects have high reproductive potential, and they have undergone a tremendous adaptive radiation involving extensive divergent evolution. But the phylogenetic changes that have occurred in most insect groups were no more rapid than the evolutionary changes that led to the development of the modern horse (see page 382). In fact, cockroaches and dragonflies have remained essentially unchanged since their first appearances in the latter part of the Paleozoic Era. Wilson and Taylor (1964) described segments of an ant colony, fossilized in amber more than 30 million years ago, that closely resemble present-day members of the same genus. Even more remarkable examples of stable evolutionary lines are found in the marine bivalves; some living genera have fossil representatives 500 million years old.

Turnover rates of populations apparently have little to do with rate of evolution. Some large mammals, even whales, which have low reproductive potential, have evolved as rapidly during the Tertiary Period as any other group, and yet others, like the opossum, have remained essentially unchanged for more than 60 million years. Evolutionary potential and evolutionary rate cannot be equated. The following argument may explain why some large animals have evolved rapidly, while smaller animals have changed little. If we consider the environment as a mosaic of environmental patches with different conditions, small organisms may be so specialized that their environments include only one small patch. Changes in the environment would affect a small organism only if its own piece of the mosaic were to disappear. But the environments of large organisms encompass larger portions of the whole mosaic, and so almost any change in the environment forces them to readjust their adaptations. The more specialized the organism, the less likely it is to change, but the greater its risk of extinction.

Genetic variation, hence variation in the fitnesses of genotypes, should enhance the rate of evolution. But the fossil evidence again points to independence of evolutionary rate and the degree of phenotypic variation in evolving characteristics. (Unfortunately, we have no way of assessing genetic variation in the past.) Simpson (1953) showed that the rate of evolutionary change in two tooth characteristics of horses varied greatly even though the degree of variation in the tooth measurements at any one time remained unchanged (Table 26-1). In addition, the two measurements evolved at different rates, although they had similar levels of morphological variation. Admittedly, Simpson's measures were phenotypic and not genetic, but there is no reason not to assume that they provide a reasonable index of genotypic variation. Since rate of evolution is a direct function of variation in fitness, the data in Table 26-1 suggest that morphological variation influenced fitness differently during the evolution of the horse, probably in proportion to the rate of change in the environment of the horse.

The rate of phyletic evolution in dipnoid fish and horses (page 382) varied considerably, and diverse characteristics often evolved at different

TABLE 26-1

Variation in tooth measurements of horses and their rate change during evolution (from Simpson 1953).

Segment of phylogenetic line	Coefficient of variation (per cent) in earlier form		Mean change per million years (per cent)	
	Paracone height	Ectoloph length	Paracone height	Ectoloph length
Hyracotherium-Mesohippus	6.2	5.7	2.4	1.5
Mesohippus-Merychippus	4.8	4.6	7.9	2.9
Merychippus-Neohipparion	5.9	5.3	5.9	0.6

times within a phylogenetic line. Fits and starts of evolutionary progress have not been limited to individual evolutionary lines. Bursts of radiation and diversification followed by quiescent periods have characterized the history of most major groups of organisms. The diversification in the bony fishes since their appearance during the Devonian Period exemplifies evolutionary activity within a large group. The first proliferation of fish during the Devonian involved the appearance of many basic lines—orders—but relatively little diversification within these lines—genera (Figure 26-6). During the Mississippian and Pennsylvanian periods, many lines of fishes underwent extensive adaptive radiation, bringing about the appearance of many new families and genera, but no new orders. Although the rate of adaptive radiation slowed considerably in the Permian, several new orders arose. A new round of diversification and adaptive radiation occurred in the Triassic Period, involving all levels of taxonomic hierarchies. The Jurassic Period again witnessed a slowing in the rate of diversification of the bony fishes, while the Chondrichthyes (cartilaginous fishes: sharks and rays) underwent their last major radiation. During the Cretaceous and Tertiary periods, the bony fishes engaged in an immense diversification that made them the dominant vertebrates in the seas.

Periods of intense evolutionary activity do not necessarily correspond among different groups of organisms, and may be completely discordant. Among the terrestrial vertebrates, bursts of diversification have seemingly progressed from the amphibians to the reptiles and then to the mammals, leading to a greater diversification of new forms each time in a pattern similar to the succession of dominance among aquatic vertebrates from the agnathes to the placoderms, the sharks and rays, and finally to the bony fishes.

The origination of new forms is not a complete measure of the evolutionary success of a group, because the number of forms alive at any one time also reflects the persistence of forms that arose in the past. A genus sometimes may "disappear" from the fossil record by changing enough to warrant the naming of a new genus, but most genera disappeared without leaving descendants. As with adaptive radiation and the appearance of new forms, phylogenetic lines have also witnessed irregular bursts of extinctions interspersed by more favorable times (Newell 1963). Nearly half of the animal families on the Earth disappeared at the end of the Permian Period. Of course, the extinctions did not occur at once; the Permian Period was 50 million years long, and many forms disappeared before its end. Nevertheless, the Permian was a period of major extinctions, particularly of marine animals. The ends of the Cambrian, Devonian, Triassic, and Cretaceous periods have also marked peaks in the rate of disappearance of lineages.

The persistence of evolutionary lines has differed considerably among groups. Simpson (1953) compared the longevity of genera of bivalve molluscs and mammalian carnivores (Figure 26-7) and found that bivalve genera living today are an average of 78 million years old, whereas the extant carnivore genera are only about 8 million years old. Of course, compared to bivalves, which are known from the beginning of the Paleozoic, carnivores are comparatively recent arrivals. But even when the life spans of extinct genera are considered, bivalves are still more persistent than carnivores.

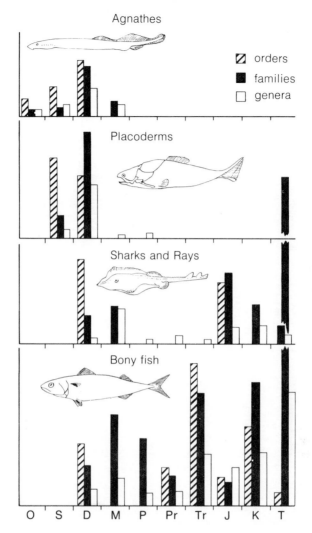

Figure 26-6 Rates of origination (expressed as groups per million years) of orders, families, and genera of the four classes of fish: agnathes, placoderms, sharks and rays, and bony fishes. The scale for genera is ten times smaller than for families and fifty times smaller than for orders. The time scale is designated by geological periods from the Ordovician to the Tertiary (after Simpson 1953).

Moreover, such a large proportion of known bivalve genera are still living (139 of 423 genera compared to 33 of 262 carnivore genera) that the life span of extinct genera greatly underestimates the actual mean life span of bivalve genera.

Rates of origination and extinction of genera in the bivalve and carnivore lines are compared in Figure 26-8. In both groups, origination and extinction

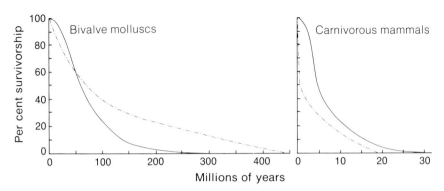

Figure 26-7   Survivorship curves for genera of bivalves and terrestrial carnivores that are now extinct (solid line). Mean age of living bivalve genera
(dashed line) is 78 million years, and of living carnivore genera 8.1 million years
(after Simpson 1953).

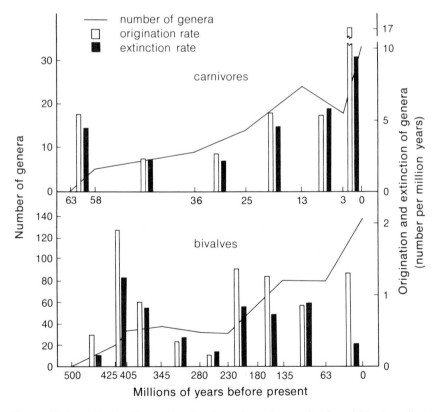

Figure 26-8   Production and extinction of carnivores (top) and bivalves (bottom), and the net change in total number of genera in each group (based on
data in Simpson 1953).

rates have followed a more or less parallel course: Changes in the total number of genera have been small compared to the rate of turnover, or replacement, of genera. The number of genera of bivalves underwent major periods of proliferation during the Silurian and Triassic-Jurassic periods. The carnivores, being a more recent group, are still engaged in their first major radiation. As in bivalves, the origination of carnivore genera has consistently exceeded extinction, hence the number of carnivore genera has grown steadily.

The correspondence between extinction and origination rates opens the question of cause and effect. Is radiation initiated when major phylogenetic lines drop out, or does the rapid radiation of a group lead to extinction of older groups, or both? Newell (1963) suggested that increases in the rate of radiation and evolutionary activity usually followed periods of massive extinctions. On the other hand, Simpson (1950) has pointed out that the immigration (which cannot necessarily be equated with origination) of many new forms of mammals to South America, when a land bridge to North America was formed during the Pliocene Epoch, caused widespread extinctions among the endemic South American fauna. A similar phenomenon occurred in North America during the late Pleistocene Epoch when invaders across the Bering land bridge from Asia caused the extinction of many local North American forms (Webb 1969). But it may not be entirely valid to compare immigration from outside a region to the origination of new taxa within a region. We shall consider the relationship between the proliferation and extinction of taxa in more detail later.

Is there progress in evolution?

Evolution progresses in the sense that evolutionary change charts a forward course through time. New forms arise, old forms disappear. But the question whether the more recent products of evolution are any better than those whose heyday is past cannot be easily resolved. Because the success of a species can be evaluated only in the context of its own particular environment, it is difficult to compare the relative competitive abilities of old and modern forms. In an absolute sense, all forms are successful so long as they live; extinction is the mark of utter failure. Modern forms apparently fail just as often as their predecessors did.

Can we judge success by numerical abundance or geographical distribution? Is not the Kirtland's warbler, restricted to a small area of secondary jack pine forests of Michigan, just as successful *at what it does* as the European starling? Is it important to an individual warbler whether there are 10,000 or 100 million others that share its plumage characteristics? It is certain that the starling will be around longer than the Kirtland's warbler, but the starling will just as surely become extinct in its time.

Success, by whatever measure, waxes and wanes as species replace each other in an endless continuum. But we might also ask whether evolution is, on the whole, building better organisms than it used to. Has the basic design of organisms been improved in the course of the countless minor adjustments that evolution has made? Surely the diversity of form and function in the living world is greater now than it ever has been. But it would be preposterous to argue, for example, that birds and mammals are an improvement over their fish progenitors—most birds and mammals are hopelessly inept in water. Whether evolution has produced better fish remains an unanswered question.

Quite apart from any judgment of evolutionary success, several evolutionary trends appear repeatedly within diverse phylogenetic groups. One example is "Williston's Law," which states that modern forms tend to have fewer, relatively more specialized segments and appendages than older forms. The comparison between early arthropods, such as trilobites, and modern crustaceans, such as crabs, illustrates Williston's Law (Figure 26-9). The meaning of Williston's Law, and others like it, is open to question. Such "laws" no doubt represent one-directional evolutionary trends. The reduction in number and increase in specialization of appendages has rarely been reversed in the fossil record. But the mere existence of a trend is not an impressive argument for progress in evolution. All we know is that forms with reduced numbers of specialized legs are living today, whereas the trilobites have long since disappeared.

Another persistent phyletic trend has been one of increasing body size, "Cope's Law" (Cope 1896, Haldane and Huxley 1927, Newell 1949, Rensch

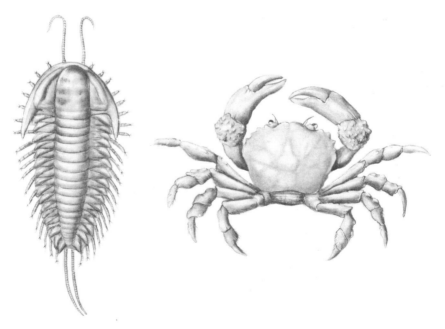

Figure 26-9   An example of Williston's Law: The ancient trilobite (left) has many similar segments and appendages; the modern crab (right) has relatively few specialized segments and legs (after Simpson 1949).

1948). This trend is apparently widespread, although there have been some notable exceptions (for example, Kurtén 1965). Cope's Law applies only within phyletic groups; the average size of organisms on Earth is not actually increasing. Present-day mammals are no larger on the average than the reptiles that lived 100 million years ago, and perhaps they are smaller; insects are probably smaller on the average now than they were during the Paleozoic Era; modern molluscs and crustaceans are no bigger than their ancient relatives.

Increased size can bring many advantages: increased buffering against environmental fluctuations; decreased requirement, pound for pound, for energy and nutrients; and increased success in direct competition with other individuals. But the cause of phyletic size increase has never been adequately explained. If phyletic increase in size is a general trend, and if the average size of organisms does not increase as a whole with time, then most adaptive radiations must be initiated by relatively small species. Does this mean that large size is an evolutionary dead end? And if so, why? Perhaps, because the ecological relationships of large species encompass a relatively large portion of their environment, they cannot become specialized to exploit many of the microhabitat opportunities that are available to smaller species. To large species, the habitat is relatively more homogeneous and presents less opportunity for adaptive radiation than it presents to small species. Following this argument, it is no surprise that there are many more kinds of insects than birds and mammals.

If large species are less likely to produce diversified lines than small species, we might consider large size to be an evolutionary liability. The lineage that produces few species is likely to be completely eliminated by extinctions. The same may be said of any specialization that limits the potential for adaptive radiation; specialized forms almost always arise from more generalized species rather than from other specialized forms. A high degree of specialization within an entire phyletic group, perhaps caused by intense interspecific competition with other members of the lineage, or with newly

Figure 26-10   Representative species of ammonites, illustrating the usual, simply coiled type (left), and several atypical forms (right), which indicate unusual specializations (after Simpson 1949).

radiating competitors, could decrease the rate of new adaptive radiations within the group and lead to the extinction of the entire lineage. Several paleontologists have considered the appearance of many unusual specialized forms within a phyletic group just prior to its extinction as a sign of racial "old age" (Figure 26-10).

In this chapter, we have briefly reviewed some of the major historical patterns evident in the fossil record. Clearly, the historical scale of evolution elucidates many phenomena and processes that cannot be observed or inferred from contemporary nature. Historical patterns can also help us to understand directly observable phenomena. But we must also recognize that these patterns can be understood only in terms of population processes that occur within our own period of existence. In the chapters that follow, we will explore these processes in detail, starting with an exposition of demography, followed by further discussion of the mechanism of natural selection.

Part 7

Genetics of Populations

27

Demography

Demography is the study of the birth and death processes that determine the age structure of populations and the rate of change in population size. The basic statistics of demography are the probability of death and the fecundity of individuals of each age in the population. These factors reflect the adaptations of the individual: its ability to procure resources, survive extreme climate conditions, and avoid predators. From them we can calculate the intrinsic growth potential of a population.

Demography is important to understanding natural selection, competition, predation, and even ecosystem stability, because populations, or genetically distinguishable subpopulations, having different birth and death statistics by reason of their various adaptations also have different inherent population growth potentials. This growth potential under a given set of conditions is an expression of evolutionary fitness, competitive ability, and predator efficiency. The response of the intrinsic growth potential to varying conditions is a measure of the inherent stability of a population. Demography also is an important concept for ecologists because it links the structure and functioning of organisms directly to population processes and thereby bridges two levels of biological organization. Processes on each level of biological organization summarize and integrate processes on the levels below.

Populations grow by multiplication.

Although we may measure population growth by the number of individuals added to a population, the size of this increment depends on the original size of the population. Two populations, one of ten individuals and one of a hundred individuals, both exposed to the same environment, would not be expected to increase by the same absolute amount. It stands to reason that increase is proportional to initial size. Populations grow by multiplication (proportional or geometric increase) rather than by addition (absolute or arithmetic increase), because all individuals potentially can contribute to population growth by reproducing.

Consider a simple model of population growth: Reproduction occurs only during a restricted period; each female gives birth to four (two male and two female) offspring, and then dies. All newborn individuals live for one year, at the end of which they reproduce and die. In this population, the generations do not overlap—that is, individuals mate only with others of the same age. According to the fecundity and mortality specified, the population will double in size each year (each mother has two daughters), so if the population initially has 100 individuals, there will be 200 after one year, 400 after two years, 800 after three years, and so on. Population growth rate is most conveniently measured as the annual increase per individual. In the example above, this rate is 2.0. The size of the population in any one year is always twice that of the previous year, even though the number *added* each year increases as the population grows. The growth of the population could be predicted from the survival and reproductive rates of individuals without having to measure the size of the population. Population growth rate is determined by its demographic statistics.

Now suppose we consider another population, one in which individuals live for several years but with different probabilities of survival and fecundity at each age; mother, daughter, and granddaughter might all live to give birth in the same year (see Figure 27-1). If we can measure the size of the population each year, we can determine the growth rate of the population, just as we could for the simpler model above. Suppose we count 100 individuals in the first year, 160 in the second, 256 in the third, 410 in the fourth. The per capita population growth rate may be calculated by dividing the number of individuals in one year by the number in the previous year. Thus $160/100 = 1.6$, $256/160 = 1.6$, $410/256 = 1.6$, and so on. Our population clearly is growing by a factor of 1.6 each year. We shall see later that the growth rate of this population could have been calculated from the survival and fecundity of individuals of different ages without having to measure population size directly. But before illustrating these calculations, we shall characterize the growth rate of a population a little more precisely.

Population growth rate may be described by geometric and exponential constants (λ and r).

Population biologists denote population size by the symbol N and time by t. The size of a population at time t is designated $N(t)$. This mathematical shorthand greatly simplifies communication about population processes and allows us to treat population growth as a set of equations amenable to mathematical manipulations. For example, the statement, "the per capita population growth rate may be calculated as the number of individuals in one year divided by the number in the previous year," may be expressed

$$\lambda = \frac{N(t)}{N(t - I)} \qquad (27\text{-}1)$$

where λ (the lower case Greek letter *lambda*) denotes the *geometric growth rate* of the population.

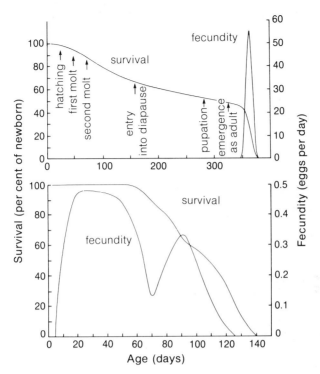

Figure 27-1   Age-specific survival (l_x) and fecundity (b_x) in two invertebrates. Top: the garden chafer (*Phyllopertha horticula*), a rutelid beetle with annual, nonoverlapping generations and highly seasonal reproduction (from data in Laughlin 1965). Bottom: a laboratory population of the chydorid cladoceran (*Chydorus sphaericus*) a species with broadly overlapping generations; the population was maintained at 15 C (from data in Keen 1967).

Equation 27-1 can be rearranged algebraically to give an expression for population growth from one year to the next

$$N(t) = N(t - 1)\lambda \tag{27-2}$$

which is equivalent to

$$N(t + 1) = N(t)\lambda \tag{27-3}$$

Suppose we wished to follow the growth of a population whose initial size (time zero) is $N(0)$. After one year (or any other time unit), the population $N(1)$ increases by factor λ over its initial size $N(0)$. Thus, $N(1) = N(0)\lambda$. After two years, $N(2) = N(1)\lambda$, but because $N(1) = N(0)\lambda$, $N(2) = N(0)\lambda\lambda = N(0)\lambda^2$. It follows that after t units of time

$$N(t) = N(0)\lambda^t \tag{27-4}$$

In the two examples of population growth described earlier, λ in equation 27-4 would be 2.0 and 1.6, respectively. We may use equation 27-4 to predict the size of any population after a period during which its growth rate is constant. Thus, a population with $N(0) = 100$ individuals initially and a constant rate of growth of $\lambda = 1.6$ for $t = 15$ years will attain a size $N(15) = N(0) \cdot 1.6^{15}$, or approximately 115,000 individuals.

The geometric constant for rate of population increase (λ) is difficult to use when one wishes to compare the growth of two populations. For example, suppose two populations with initial sizes of 100 individuals had geometric growth rates of 2.0 and 1.6, and we wished to know how much larger than the second population the first would be after t years. To evaluate the ratio $N(t)_1/N(t)_2$, we must first calculate the population sizes by equation 27-4. We could avoid this mathematical step and calculate the ratio of population sizes directly, however, if we expressed population growth rate as an *exponential* constant (r) rather than the geometric constant (λ). This may be done simply by substituting e^r for λ. The constant e is the base of the natural logarithms, whose numerical value is approximately 2.72. Although e may seem an awkward number at first glance, its mathematical properties greatly simplify algebraic manipulations of exponential and logarithmic expressions. (If you are unfamiliar with geometric series, exponential equations, and logarithms, you may wish to read the note at the end of this chapter (page 444).

With e^r substituted for λ, the equation for population growth becomes

$$N(t) = N(0)e^{rt} \qquad (27\text{-}5)$$

where r is a measure of the exponential growth rate of the population. The constants r and λ are interconvertible through the expression $\lambda = e^r$ and $r = log_e \lambda$. Thus for $\lambda = 2$, $r = log_e (2) = 0.69$; for $\lambda = 1.6$, $r = log_e (1.6) = 0.47$. Important guideposts to growth rate constants are their values for a population whose size is constant. Inasmuch as the size of such a population does not change year after year, the ratio $N(t + 1)/N(t)$ is 1, which is the value λ. The corresponding value of r is the natural logarithm of 1, or 0. The relationship between r and λ is expanded upon in Table 27-1.

Using the exponential expression for population growth greatly simplifies certain calculations because exponents of the same base (e in this case) are added when two exponential functions are multiplied and subtracted when one exponential function is divided by another. Accordingly, the ratio of the size of one population to that of another may be simplified by the following steps:

$$\frac{N(t)_1}{N(t)_2} = \frac{N(0)_1 e^{r_1 t}}{N(0)_2 e^{r_2 t}}$$

$$= \frac{N(0)_1}{N(0)_2} e^{(r_1 - r_2)t} \qquad (27\text{-}5)$$

For the two populations described above, both having 100 individuals initially and values of r of 0.69 and 0.47, respectively, equation 27-5 becomes $N(t)_1/N(t)_2 = (100/100)e^{(0.69-0.47)t} = e^{0.22t}$. After three years, the ratio would be $e^{0.66} = 1.94$. (As we saw earlier, the populations had 800 and 410 individuals after three years of geometric growth; 800/410 = 1.95—the discrepancy is due to a rounding error.) After fifteen years, the ratio would be $e^{3.30} = 27$.

TABLE 27-1

The relationship between the exponential (r) and geometric (λ) growth rates and change in population size.

Change in population size	Growth constant	
	Exponential (r)	Geometric (λ)
Decreasing	less than 0	between 0 and 1
Constant	0	1
Increasing	greater than 0	greater than 1

The value of r may be calculated directly from population growth data by dividing the difference between the logarithms of the population size at the beginning and the end of a period of growth by the length of the period. To derive this relation, we first take the logarithm of both sides of equation 27-4, obtaining

$$\log_e N(t) = \log_e N(0) + \log_e (e^{rt}) \qquad (27\text{-}6)$$

(when numbers are multiplied, their logarithms are added). Because the natural logarithm of an exponential number with base e is the exponent itself, i.e., $\log_e (e^{rt}) = rt$,

$$\log_e N(t) = \log_e N(0) + rt. \qquad (27\text{-}7)$$

This equation may be rearranged to give

$$r = \frac{\log_e N(t) - \log_e N(0)}{t} \qquad (27\text{-}8)$$

In the culture of yeast cells described in Table 27-2, the population density increases from an average density, for example, of 71.1 to 174.6 individuals during the interval between four and six hours. Substituting these values into equation 27-8, $r = (\log_e 174.6 - \log_e 71.1)/2 = (5.16 - 4.26)/2 = 0.45$ per hour—an average rate of increase of 45 per cent per hour during the period.

r is an instantaneous, relative measure of population growth rate.

This is an appropriate point to provide further clarification of the difference between exponential and geometric measures of population growth. You will remember that the geometric growth constant λ was calculated as the ratio between the sizes of a population at the beginning and end of a specified period. The value of λ varies, however, with the period of growth specified. For an interval of one year, $\lambda = 160/100 = 1.6$, but for two years $\lambda = 256/100 = 2.56$, and so on. In contrast, r is an instantaneous measure of population growth rate, expressing the percentage rate of increase in the population at every instant in time. The mathematical difference between λ and r has a biological counterpart. In most populations, reproduction is

TABLE 27-2

Analysis of the growth of yeast cells in laboratory culture (from data in Pearl 1925).

Time in hours (t)	Density of individuals (N)	Absolute increase in density	Relative rate of increase (r)*
0	9.6		
		19.4	0.55
2	29.0		
		42.1	0.45
4	71.1		
		103.5	0.45
6	174.6		
		176.1	0.35
8	350.7		
		162.6	0.19
10	513.3		
		81.1	0.08
12	594.4		
		46.4	0.04
14	640.8		
		15.1	0.02
16	655.9		
		5.9	0.01
18	661.8		

$$* \, r = \frac{\log_e N(t + 2) - \log_e N(t)}{2}$$

seasonal; population size, whether it increases or decreases in the long run, increases each year during the season of reproduction and decreases during the rest of the year as individuals die. To obtain a useful measure of long-term population growth, individuals must be counted at intervals of one year so that successive counts are obtained at the same season each year. In other populations, such as that of man, reproduction continues at a relatively constant level throughout the year, and all other things being equal, the growth rate of the population does not vary seasonally. Growth curves for these two types of populations are shown in Figure 27-2 for equivalent values of $\lambda = 1.6$ and $r = 0.47$. We note that over a period of one year the population increases 60 per cent (by a factor of 1.6), yet the equivalent exponential growth constant is only 47 per cent per year. The discrepancy is resolved when one remembers that exponential growth is like compounded interest on a bank account—the interest is continually added to the principal. The geometric growth constant is like simple interest, which is posted only once each year. The factor of 1.6 is applied only to the initial population size (1.6 × 100 = 160). But the exponential growth rate is applied continuously as the population grows exponentially. At time zero the growth rate is 0.47 × 100 = 47 individuals per year; at six months it is 0.47 × 127 = 60 individuals per year; and after one year of growth it is 0.47 × 160 = 75 individuals per year. (The value of 127 individuals after six months was obtained by the expression, $N(0.5 \text{ years}) = N(0)e^{r(0.5)} = 100e^{0.47 \times 0.5} = 100e^{0.235} = 127$.)

It should be clear that although e^r and λ are mathematically equivalent, they are not so biologically. We may say that over a period of many years a population grows at an instantaneous rate of 47 per cent per year, and this value may allow an accurate prediction of population size in the future, yet the population may never actually exhibit that particular rate of growth. In

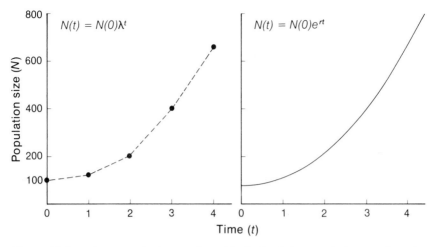

Figure 27-2   Increase in the number of individuals in populations undergoing geometric (left) and exponential growth (right) at equivalent rates ($\lambda = 1.6$, $r = 0.47$).

this sense, r is an abstraction of demography applicable only to the hypothetical population growing under constant conditions. But r is nonetheless useful both as a description of population growth rate in nature and as a device for conceptualizing the relationships between demographic variables and population growth.

Mathematicians use different symbols to distinguish instantaneous increase and increase over an interval. The first is the slope of the exponential growth curve at any particular instant dN/dt (d may be read "a differential" so dN/dt is a differential in number with respect to a differential of time). The second is the difference between two measurements separated by an interval, or difference in time, $\Delta N/\Delta t$ (Δ may be read "a difference in"). Now the slope of a curve or the difference between two measurements is an absolute value, whereas exponential and geometric growth rates are expressed on a "per individual" basis and are often called relative growth rates, in contrast to absolute growth rates. The exponential growth rate r is dN/dt divided by N, hence

$$r = \frac{1}{N}\frac{dN}{dt} \tag{27-9}$$

The geometric growth rate is the population size after an interval of growth—$N(0) + \Delta N/\Delta t$—divided by the initial size, hence

$$\lambda = \frac{N(0) + \Delta N/\Delta t}{N(0)}$$

$$= 1 + \frac{1}{N(0)}\frac{\Delta N}{\Delta t} \tag{27-10}$$

The analogy between $(1/N) \cdot dN/dt$ and $(1/N(0)) \cdot \Delta N/\Delta t$ is obvious. In fact, as the interval Δt is made smaller and smaller, $\Delta N/\Delta t$ approaches the instantaneous growth rate dN/dt and λ approaches $1 + r$.

The value of r may be calculated directly from age-specific data on longevity and fecundity.

Population growth rate is a function of the reproductive rate of individuals and the length of a generation, which in turn are related to age-specific schedules of fecundity and survival for the population. We can estimate the growth rate of a population from these data without having to measure its size or growth rate.

Population biologists define the number of progeny that an individual may be expected to produce during its lifetime as the net reproductive rate (R_0), which is

the average fecundity at age one times the probability of survival to age one, plus the average fecundity at age two times the probability of survival to age two, plus the average fecundity at age three times the probability of survival to age three, and so on.

For convenience, we may express the terms for net reproductive rate in the following mathematical notation:

$$R_0 = l_1 b_1 + l_2 b_2 + l_3 b_3 + \cdots \qquad (27\text{-}11)$$

where

$x = $ age,

$l_x = $ survival to age x; the probability that a newborn individual or freshly laid egg will be alive after x years (or other units of age),

$b_x = $ age-specific fecundity; the expected number of female eggs or offspring per female alive at age x; referred to as m_x in some texts.

Demographic variables are normally tabulated only for females because the fecundity of males is usually impossible to ascertain. Hence the net reproductive rate is the number of female progeny born to a female, on average, during her lifetime.

The expression for R_0 is a sum of similar terms ($l_x b_x$) in which only the age (x) changes (from $x = 1$ to the age at the end of the longest life span of an individual). The series of terms may be represented in condensed mathematical notation as the sum

$$R_0 = \sum_{x=1}^{\infty} l_x b_x \qquad (27\text{-}12)$$

where Σ (the Greek letter *sigma*) indicates that the terms $l_x b_x$ from age (x) one to infinity are added together.

R_0 is the total expected reproduction by a female during her lifetime. Fecundity is only half of the equation for population growth rate. The other half is the mean generation time (T) of the population. The relationship between R_0, T, and the geometric growth rate λ may be derived from the basic equation for population growth

$$N(t) = N(0)\lambda^t$$

rearranged to give

$$\frac{N(t)}{N(0)} = \lambda^t$$

Because the net reproductive rate (R_0) is equivalent to the expected size of a generation (all of the female progeny born to a female), and the generation time (T) is the period required for a population to increase by the factor R_0, the ratio $N(T)/N(0)$ should be equal to the net reproductive rate (R_0). Therefore,

$$\frac{N(T)}{N(0)} = \lambda^T = R_0$$

which can be rearranged to give

$$\lambda = R_0^{1/T} \tag{27-13}$$

The equivalent expression for r is derived from the equation

$$\frac{N(T)}{N(0)} = e^{rT} = R_0$$

which, after taking the logarithm of both sides,

$$rT = \log_e (R_0)$$

can be rearranged to give

$$r = \frac{\log_e (R_0)}{T} \tag{27-14}$$

According to this equation, the exponential rate of population growth is directly related to the logarithm of the net reproductive rate and inversely related to the mean generation time.

We may find an approximate value of the mean generation time (T_c) by calculating the average age at which a female lays her eggs or gives birth to her offspring. Because $l_1 b_1$ eggs are produced at age 1, $l_2 b_2$ eggs are produced at age 2, and $l_x b_x$ eggs are produced at age x, the average age at reproduction is the sum of $x l_x b_x$ terms for all ages divided by the total number of progeny (R_0),

$$T_c = \frac{\Sigma x l_x b_x}{\Sigma l_x b_x} \tag{27-15}$$

T_c is most useful as an estimate of the mean generation time T when generations do not overlap (reproduction is highly seasonal and individuals do not live for more than one season) or a population is composed primarily of individuals of similar age, which might occur, for example, during the initial stages of population growth of a colony founded by a few immigrants (Laughlin 1965). If these conditions do not prevail, T_c may provide a poor estimate, usually an underestimate, of r. For this reason, population biologists usually distinguish values of r based on T_c by the subscript c, hence r_c.

A life table summarizes the survival and reproductive performance of a population under specified conditions.

Life tables provide the vital statistics for calculating growth performance of a population (Dublin and Lotka 1935, Deevey 1947, Birch 1948). In a life table, survival and fecundity are laid out for females in each age class in the population. These data are not, however, always readily obtained.

If a population's size remains constant over a long period, the age structure of the population parallels the survival of individuals because the number of individuals alive at a given age decreases according to the mortality rate for that age. A serious restriction to using age structure as a basis for survival curves is that one must assume that individuals alive at each different age are the descendants of an equal number of newborn, in other words, that the number of newborn does not change from year to year. This condition is rarely met in natural populations. The influence of birth rate on population age structure can be visualized in various human populations (Figure 27-3). In many nonindustrialized countries (for example, Costa Rica), where birth rates are high and populations are expanding rapidly, the number of newborn increases each year. This tends to make the age distribution bottom-heavy and suggests lower survival than actually occurs. In some industrialized countries of Europe (Sweden, for example), where populations are growing slowly, the age structure more closely parallels the survival curve of the population.

Age-specific survival is normally estimated for natural populations from the survival of individuals marked at birth or from the deaths of individuals of known age (Farner 1955), but both methods have serious limitations. The first requires that individuals do not emigrate from the area in which they were marked; otherwise they would be erroneously counted as dead. The second method, commonly used in studies of bird populations (Farner 1948), requires that large numbers of individuals be marked to obtain sufficient recoveries of dead individuals. For example, of 180,718 robins banded as adults in the United States between 1946 and 1965, only 2,444 were subsequently retrapped or found dead; of 7,604 brown pelicans banded as nestlings, only 375 were ever seen again (Henny 1972).

When the ages of individuals can be determined by their appearance, without knowing their dates of birth, the construction of a survival table becomes a relatively simple task. Among plants, for example, the age of forest trees in temperate zones can be determined from growth rings in their wood. Among animals, the age of fish can be determined from growth rings

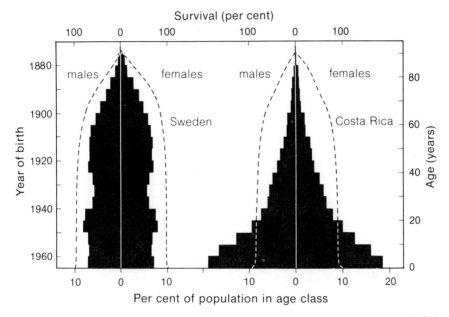

Figure 27-3   Population age-structure and survival, calculated separately for males and females, in Sweden (1965) and Costa Rica (1963). Because Sweden's population has grown slowly, its age structure resembles the survival curve. Declining birth rates during the Depression and the baby boom that followed World War II are responsible for irregularities in the age-structure of Sweden's population. Costa Rica's rapid population growth has resulted in a bottom-heavy age structure (after data in Keyfitz and Flieger 1968).

in their scales or ear bones; the age of mountain sheep can be estimated from the size of their horns.

Estimated age of death, judged by the size of horns on skulls, was used by Deevey (1947) to calculate survival from Murie's (1944) data for the Dall mountain sheep (Figure 27-4) in Mount McKinley National Park, Alaska. In all, 608 skeletal remains of sheep were found: 121 of the sheep were judged to have been less than one year old at death; seven between one and two years, eight between two and three years; and so on, as shown in Table 27-3. We may construct a survival table using the following reasoning: all 608 dead sheep must have been alive at birth; all but 121 that died during the first year must have been alive at the age of one year (608 − 121 = 487); all but 128 (121 dying during the first year and seven dying during the second) must have been alive at the end of the second year (607 − 128 = 480); and so on, until the oldest sheep died during their fourteenth year. Survival (right-hand column in Table 27-3) was calculated by converting the number of sheep alive at the beginning of each age interval to a decimal fraction of those alive at birth. Thus, the 390 sheep alive at the beginning of the seventh year represented 64.0 per cent (decimal fraction 0.640) of the 608 newborn in the sample.

TABLE 27-3

Life table for the Dall mountain sheep (*Ovis dalli*) constructed from the age at death of 608 sheep in Mount McKinley National Park (based on data in Murie 1944, quoted by Deevey 1947).

Age interval (years)	Number dying during age interval	Number surviving at beginning of age interval	Number surviving as a fraction of newborn (l_x)
0–1	121	608	1.000
1–2	7	487	0.801
2–3	8	480	0.789
3–4	7	472	0.776
4–5	18	465	0.764
5–6	28	447	0.734
6–7	29	419	0.688
7–8	42	390	0.640
8–9	80	348	0.571
9–10	114	268	0.439
10–11	95	154	0.252
11–12	55	59	0.096
12–13	2	4	0.006
13–14	2	2	0.003
14–15	0	0	0.000

Figure 27-4   A group of Dall mountain sheep in Alaska. The size of the horns increases with age (courtesy of the American Museum of Natural History).

TABLE 27-4

Life table of a hypothetical population, demonstrating the calculation of the net reproductive rate and mean generation time.

Age (x)	Survival (l_x)	Fecundity (b_x)	Expected offspring ($l_x b_x$)	Product of age and expected offspring ($x l_x b_x$)
0	1.00	0.0	0.0	0.0
1	0.50	1.0	0.5	0.5
2	0.40	4.0	1.6	3.2
3	0.20	4.0	0.8	2.4
4	0.10	2.0	0.2	0.8
5	0.00	0.0	0.0	0.0
		Net reproductive rate = 3.1		
		Total weighted age =		6.9

Survival provides only half the life table. Field biologists have devised numerous techniques for estimating fecundity from embryo counts in mammals, nest checks in birds, ratios of juveniles to adults in many kinds of animals, and direct counts of eggs in amphibians, insects, marine invertebrates, and others.

The life table provides the basis for estimating population growth rate, as shown for the hypothetical population whose demographic statistics are given in Table 27-4. The first step in calculating λ or r is to calculate the products of l_x and b_x, and of x, l_x and b_x, for each age. The net reproductive rate R_0 is the sum of the $l_x b_x$ terms, 3.1 in this example. The estimated generation time T_c is the sum of the $x l_x b_x$ column divided by R_0 (equation 27-15), or 6.9/3.1 = 2.22 years in this example. From equation 27-13 we can now calculate λ by the equation $\lambda = 3.1^{1/2.22} = 3.1^{0.45} = 1.66$. Similarly, r is calculated from equation 27-14 as $r_c = \log_e (3.1)/2.22 = 1.13/2.22 = 0.51$.

Populations growing at a constant rate assume a stable age distribution.

A basic principle of the theory of population growth states that if a population remains under constant conditions and, therefore, has constant age-specific survival and fecundity for sufficient time, the growth rate of the population will become constant and the population will assume a *stable age distribution* (Lotka 1922). Under these conditions, all age groups grow or decline at the same rate as the entire population, so their relative proportions remain constant. That populations do, in fact, attain a stable age distribution under constant conditions can easily be demonstrated by following the course of a hypothetical population through several generations.

Consider a population with the following life table:

Age (x)	Survival (l_x)	Age-specific survival (l_{x+1}/l_x)	Fecundity (b_x)	Expected offspring ($l_x b_x$)
0	1.0	0.5	0	0
1	0.5	0.5	2	1
2	0.25	0.0	4	1
3	0.0	—	0	0

We shall assume that mating takes place once each year during a restricted season. Survival is 50 per cent each year during the first two years and zero from the second to third years, so one-half of the newborn survive to breed during the first year, and one-half of those (one-quarter of the newborn) survive to breed during the second year. None survives past the second reproductive season. Surviving individuals have two young during the first year and four during the second. Because the net reproductive rate of the population is two, the population should increase.

Now suppose that just prior to the reproductive season there are ten individuals, age one, and five individuals, age two, in the population. During the first reproductive season each of the ten one-year-old individuals gives birth to two offspring, and each of the five two-year-old individuals gives birth to four, making a total of forty newborn, as follows:

Number of newborn (n_0)   40
Number of one-year-olds (n_1)   10   (10 × 2 = 20 births)
Number of two-year-olds (n_2)   5   (5 × 4 = 20 births)
Number of three-year-olds (n_3)   0

Between the first and second reproductive seasons, mortality occurs in the population:

	First Year	Second Year
n_0	40	
		½ die → 20
n_1	10	
		½ die → 5
n_2	5	
		all die → 0
n_3	0	0

When the forty newborn individuals make up the one-year-old age class in the next reproductive season, their numbers have dwindled by half to twenty. Similarly, when the one-year-old individuals make up the two-year-old age class, their numbers have been reduced by half to five. All the two-year-olds, which would have been three years old at the next breeding season, die before reaching that age.

In the second reproductive season, the number of offspring produced is (20 × 2) plus (5 × 4), a total of 60. In turn, half these 60 individuals will form the one-year-old age class during the next productive season, and so on:

Year (t):	1	2	3	4	5	6	7
$n_0(t)$	40	60	100	160	260	420	680 (71.2%)
$n_1(t)$	10	20	30	50	80	130	210 (22.0%)
$n_2(t)$	5	5	10	15	25	40	65 (6.8%)
$n_3(t)$	0	0	0	0	0	0	0
Total $N(t)$	55	85	140	225	365	590	955
λ		1.55	1.65	1.61	1.62	1.62	1.62

Each age cohort behaves as prescribed by the life table. For example, the 100 newborn offspring in the third year dwindle to 50 by their first year (at which time they give birth to 100 young), to 25 by their second breeding season (at which time they give birth to 100 young), then die out completely before their third year. After a few generations of growth, the age distribution of the population quickly settles down to a steady state (71.2 per cent newborn, 22.0 per cent age one, and 6.8 per cent age two), showing that each of the age classes, hence the entire population, is growing at the same rate and that a stable age distribution has been achieved (Figure 27-5). As the population attains a stable age distribution, its growth rate also reaches a steady state. The geometric growth rate (λ) is the ratio of population sizes in successive generations ($N(t)/N(t-1)$). For example, between year one and year two, λ is 1.55 (85/55). After some small fluctuations, λ assumes a constant value of 1.62 at about the time a stable age distribution is achieved. As an exercise, try calculating λ for this population, using equation 27-13.

In the previous example, the initial age structure of the population was fairly close to the stable age distribution. If we now set up another pencil-and-paper population with the same life table, but with an initial age distribution of $n_1 = 5$ and $n_2 = 10$, the population will take longer to attain a stable age distribution and assume a constant growth rate:

Figure 27-5   Growth of a population and of its component age classes, demonstrating the existence of a stable age distribution. Data are from the numerical example presented on page 438. The growth of a population with the same life table but different initial age structure (page 438) is shown for comparison (dashed line). When population size is plotted on a logarithmic scale, exponential growth appears as a linear function of time.

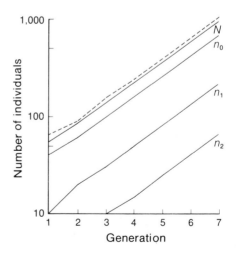

Year (t):	1	2	3	4	5	6	7
$n_0(t)$	50	58	106	162	266	426	670 (70.6%)
$n_1(t)$	5	25	29	53	81	133	213 (22.4%)
$n_2(t)$	10	2*	12	14	26	40	66 (7.0%)
$n_3(t)$	0	0	0	0	0	0	0
Total $N(t)$	65	85	147	229	373	599	949
λ		1.31	1.73	1.56	1.63	1.61	1.59

* We will not count half-individuals here.

Although the age distribution of the population after seven years is nearly identical to that of the previous example, a constant population growth rate still has not been attained.

To demonstrate how the stable age distribution and rate of population growth reflect changes in the life table, we shall change fecundity in our hypothetical population to one offspring at one and two years of age:

Age (x)	Survival (l_x)	Age-specific survival (l_{x+1}/l_x)	Fecundity (b_x)	Expected offspring ($l_x b_x$)
0	1.0	0.5	0	0
1	0.5	0.5	1	0.5
2	0.25	0.0	1	0.25
3	0.0	—	0	0

Because the net reproductive rate is now less than one, $R_0 = (0.50 + 0.25 = 0.75)$, the number of individuals in the population will decrease. If we allot to this new population the same number of individuals as the first population contained after seven generations of growth, it gradually decreases and its age distribution shifts to favor older age classes:

Year (t):	1	2	3	4	5	6	7
$n_0(t)$	680	445	392	307	251	201	162 (50.0%)
$n_1(t)$	210	340	222	196	153	125	100 (30.9%)
$n_2(t)$	65	105	170	111	98	76	62 (19.1%)
$n_3(t)$	0	0	0	0	0	0	0
Total $N(t)$	955	890	784	614	502	402	324
λ		0.93	0.88	0.78	0.82	0.80	0.81

In this example, a stable age distribution is approached within seven or eight generations, at which time the population assumes a steady rate of decrease of about 19 per cent per year.

The foregoing examples show that the growth rate of a population can be estimated from the life table only when the population has a stable age distribution. For this reason, we must distinguish two measures of growth rate: one is an empirically determined rate calculated from change in population size, regardless of the age structure; the other is an *intrinsic rate of increase* determined by survival and fecundity schedules and obtained only under stable age conditions. When speaking of exponential increase, population biologists have reserved the symbol r_m for the intrinsic rate of increase (e.g., Andrewartha and Birch 1954); r_m is often called the *Malthusian parameter*, hence the subscript *m*.

The intrinsic rate of increase of a population (r_m) is determined by its stable age characteristics.

The intrinsic rate of increase is an important index of potential population performance because it depends only on the present ecological conditions of the population and is independent of past environmental conditions that are reflected in the contemporary age structure and growth rate of the population.

Knowing that in a stable age population each age class grows at the same exponential rate, we can derive a relationship for calculating precisely the intrinsic exponential growth rate of the population. First, we express the number of newborn individuals (n_0) at a given time (t) as the sum of offspring born to individuals of each age class in the population:

$$n_0(t) = \Sigma n_x(t) b_x \tag{27-16}$$

Because the number of individuals of age *x* alive at time *t* ($n_x(t)$) is equal to the number of newborn *x* years ago times their survival to the present (l_x), we may substitute the expression

$$n_x(t) = n_0(t - x) l_x$$

in equation (27-16), giving

$$n_0(t) = \Sigma n_0(t - x) l_x b_x \tag{27-17}$$

Because the size of each age class has been growing by a factor e^{r_m} each year, the number of newborn in the population *x* years ago is equal to the current number of newborn times $e^{-r_m x}$. We now substitute

$$n_0(t - x) = n_0(t) e^{-r_m x}$$

in equation (27-17), obtaining

$$n_0(t) = \Sigma n_0(t) e^{-r_m x} l_x b_x \tag{27-18}$$

Dividing both sides of equation (27-18) by $n_0(t)$ we obtain the sum

$$1 = \Sigma e^{-r_m x} l_x b_x \tag{27-19}$$

Equation 27-19 and the equivalent expression for geometric growth

$$1 = \Sigma \lambda^{-x} l_x b_x \tag{27-20}$$

contain only the intrinsic rate of growth and the known life table parameters: survival (l_x) and fecundity (b_x).

Unfortunately both equations are difficult to solve directly, and r_m and λ must be evaluated by trial and error substitution to determine which value makes the sum closest to 1. The calculations for determining the exact value of r_m from the life table of the hypothetical population presented above are outlined in the following table:

Age (x)	l_x	b_x	$l_x b_x$	$x l_x b_x$	Values of $e^{-r_m x}$ and $e^{-r_m x}l_x b_x$ for r_m equal to							
					−0.25		−0.22		−0.21		−0.18	
0	1.0	0	0	0	1.00	0.00	1.00	0.00	1.00	0.00	1.00	0.00
1	0.5	1	0.5	0.5	1.28	0.64	1.25	0.62	1.23	0.61	1.20	0.60
2	0.25	1	0.25	0.5	1.65	0.41	1.55	0.39	1.52	0.38	1.43	0.36
3	0.0	0	0	0								
					$\Sigma e^{-r_m x}l_x b_x = 1.05$		1.01		0.99		0.96	

We obtain an initial estimate of r_m from the equation

$$r_c = \frac{\log_e R_0}{T_c} \tag{27-21}$$

In this example, $R_0 = \Sigma l_x b_x$ (0.75) and $T_c = \Sigma x l_x b_x / R_0$ (1.0/0.75 = 1.33). Therefore, $r_c = (\log_e 0.75)/1.33 = -0.29/1.33 = -0.22$. Taking −0.22 as our first estimate of r_m, we calculate $e^{-(0.22)x}$ for x equal to 0, 1, and 2 ($e^0 = 1$, $e^{+0.22} = 1.25$, and $e^{+0.44} = 1.55$). Multiplying $e^{-r_m x}$ times $l_x b_x$ for each age gives 0, 0.62, and 0.39, respectively, and a total of 1.01. Therefore, −0.22 is only slightly less than the true value of r_m for the population. Trying $r_m = -0.21$ gives $\Sigma e^{-r_m x}l_x b_x = 0.99$, so r_m actually lies between −0.21 and 0.22. Note that $e^{-0.21} = 0.81$, which is the value of λ after a stable age distribution had been attained in the example on page 439.

More detailed calculations of the exact value of r for laboratory populations of the cladoceran *Chydorus sphaericus* are presented in Table 27-5. Initial estimates of r, calculated from equation 27-21, were 0.0698 at a temperature of 15 C and 0.138 at 25 C. The sum $\Sigma e^{-r_m x}l_x b_x$ was calculated for a range of values between $r_m = 0.05$, and $r_m = 0.30$. By plotting the sums as a function of r_m and drawing a line through the points, a close estimate of r_m can be obtained graphically as the point at which $\Sigma e^{-r_m x}l_x b_x = 1$. By this method, second approximations to r_m were 0.146 and 0.213 for the 15 C and 25 C populations, respectively. These values give sums of 1.005 and 1.006, which are reasonably accurate.

Knowing r_m and the net reproductive rate, we can calculate the mean generation time from the equation

$$T = \frac{\log_e R_0}{r} \tag{27-22}$$

In this case,

$$T = (\log_e 34.96)/0.146 = 24.34 \text{ days (15 C)},$$
$$T = (\log_e 19.12)/0.213 = 13.85 \text{ days (25 C)}.$$

Because the generations of *Chydorus* overlap greatly, the mean generation time in a rapidly growing population is about half the average age at which a female gives birth to her offspring (T_c). Thus when population growth is rapid and generations overlap, r_c can be a highly biased estimate of r_m.

TABLE 27-5

Life table data and calculation of r_c and r for the cladoceran *Chydorus sphaericus* raised under laboratory conditions at temperatures of 15 C and 25 C (life table from Keen 1967).

Age interval (days)	Midpoint of interval	15 C				25 C			
		l_x	b_x	l_xb_x	xl_xb_x	l_x	b_x	l_xb_x	xl_xb_x
0–4	2	1.00	0.00	0.00	0.0	1.00	0.00	0.00	0.00
5–14	10	1.00	2.88	2.88	28.8	0.948	8.27	7.84	78.40
15–24	20	1.00	4.80	4.80	96.0	0.692	6.94	4.80	96.05
25–34	30	1.00	4.80	4.80	144.0	0.440	8.18	3.60	107.98
35–44	40	1.00	4.64	4.64	185.6	0.340	6.05	1.84	73.57
45–54	50	1.00	4.56	4.56	228.0	0.184	5.00	0.92	46.00
55–64	60	0.988	3.28	3.24	194.4	0.036	3.33	0.12	7.19
65–74	70	0.888	1.53	1.36	95.1	0.000	0.00	0.00	0.00
75–84	80	0.804	3.03	2.44	194.9				
85–94	90	0.656	4.39	2.88	259.2				
95–104	100	0.584	3.28	1.92	191.6				
105–114	110	0.488	2.30	1.12	123.5				
115–124	120	0.328	0.98	0.32	38.6				
125–134	130	0.136	0.00	0.00	0.0				
Sums			40.47	34.96	1,779.7		38.57	19.12	409.19
R_0				34.96				19.12	
T_c			1,779.7/34.96 = 50.90				409.19/19.12 = 21.40		
r_c			$(\log_e 34.96)/50.90 \doteq 0.0698$				$(\log_e 19.12)/21.40 = 0.138$		

$\Sigma e^{-r_m x}l_xb_x$		15 C	25 C
for $r =$	0.05	5.93	7.65
	0.10	2.07 ⎫ *	3.75
	0.15	0.95 ⎭	2.03
	0.20	0.49	1.16 ⎫ *
	0.25	0.27	0.68 ⎭
	0.30	—	0.40
	0.35	—	0.24
	r	0.146	0.213
	T	24.34	13.85

* Actual value of r_m lies between bracketed values.

r varies with habitat, genotype, and population density.

The intrinsic rate of population increase varies considerably among species, genotypes, and environmental conditions. By evaluating growth rate only under optimum (usually laboratory) conditions, one can eliminate the unique environmental factors that influence r in each species. Under such conditions, the exponential growth rate is often referred to as the population's *innate capacity for increase* (r_0), hence it is the maximum realization of its biological potential for reproduction and growth.

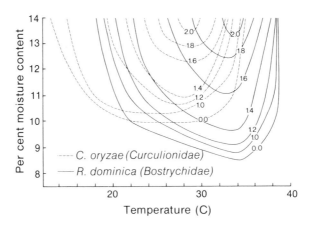

Figure 27-6   The geometric rate of increase (λ) as a function of temperature and moisture for the grain beetles *Calandra oryzae* and *Rhizopertha dominica* living in wheat. Rates of increase are indicated by contour lines that describe conditions with identical values of λ (after Birch 1953a).

Chydorus sphaericus has a much greater intrinsic growth rate at 25 C than it does at 15 C, even though the total number of eggs laid per female at the higher temperature is little more than half that at the lower temperature. The difference lies primarily in the fact that, at the higher temperature, females lay more of their eggs sooner, thereby reducing generation time and increasing r_0. In an extensive study on the conditions for population growth in grain beetles, Birch (1953a, 1953b) determined the intrinsic rate of increase (λ) of *Calandra oryzae* and *Rhizopertha dominica* over a wide range of temperature and moisture (Figure 27-6). The optimum conditions for growth differ between the species. In particular, *Rhizopertha* is more tolerant of desiccating conditions than *Calandra* (Perttunen 1972). When temperature and humidity are maintained within grain storage sheds so that λ is less than one for both species, neither can become a serious pest.

Landahl and Root (1969) compared the life tables of a temperate and tropical species of milkweed bug (*Oncopeltus*) raised in the laboratory under conditions simulating the summer climate of their natural environments. Individuals of *O. fasciatus* were obtained from Ithaca, New York; *O. unifasciatellus* was obtained from Cali, Colombia. Individuals of the two species developed at similar rates. Survival curves were similar up to about 150 days of age, after which the temperate species died off at a faster rate (Figure 27-7). The two species differed most in the timing and rate of fecundity. Females of the temperate *O. fasciatus* began to lay eggs at 68 days of age while the onset of reproduction in *O. unifasciatellus* occurred on day 76. In addition, although both species laid an average of 26 eggs per clutch, temperate females produced clutches more frequently during their shorter life span (Figure 27-7) and had the higher egg production (556 versus 324 eggs per adult during its life span). These life table data were utilized to calculate

r_0, which was slightly higher in the temperate species (0.044 per day) than in the tropical species (0.034). In another study, values of r_0 ranged between 0.037 and 0.086 per day for *O. fasciatus* reared at different combinations of temperature, photoperiod, and density (Dingle 1968).

A final example of the influence of environment on the life table comes from extensive studies summarized by Myers (1970) on populations of the European rabbit introduced to Australia. Survival and fecundity vary tremendously, being greatest in mesic, Mediterranean climates and least in arid, alpine, and wet coastal climates (see Table 27-6). Even though female rabbits survive longer in arid climates than in Mediterranean climates, their fecundity is much lower and young are born later in life, on average. Hence the intrinsic growth rate of Mediterranean-climate populations ($r_c = 0.109$) is almost three times that of arid-climate populations ($r_c = 0.037$).

Variations in genotype also influence survival and fecundity. The relationship between the genotype and the expression of population parameters in the phenotype underlies all of natural selection theory, because *r* measures the competitive ability of a genotype. The relationship between the exponential growth rate of a population and its density lies at the heart of our understanding of population regulation and the interaction between different populations, but we will defer that topic until we have discussed the theory of population genetics.

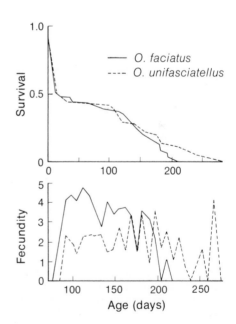

Figure 27-7   Survival (l_x) and fecundity (b_x) curves for *Oncopeltus fasciatus* (solid line, Temperate Zone) and *O. unifasciatellus* (dashed line, Tropical Zone) (after Landahl and Root 1969).

TABLE 27-6

Some life-table parameters of European rabbits in arid and Mediterranean climates of Australia (after Myers 1970).

Age (months)	Pivotal age* (months)	Survival (l_x)	Proportion of female population (per cent)	Females pregnant (per cent)	Litter size	Production per female	Production per age group (per cent)
Mediterranean
3–6	4.5	0.220	17.7	23	4.4	1.8	7.4
6–12	9.0	0.160	35.9	43	5.6	8.7	35.4
12–18	15.0	0.076	30.3	53	5.9	9.4	38.4
18–24	21.0	0.028	10.7	45	5.9	2.8	11.5
> 24	37.5	0.006	5.3	55	6.2	1.8	7.3
Arid
3–6	4.5	0.570	14.0	5	3.8	0.2	2.2
6–12	9.0	0.457	26.5	15	4.7	1.8	16.1
12–18	15.0	0.302	21.7	32	4.6	3.1	27.9
18–24	21.0	0.186	19.2	37	4.3	3.1	27.8
> 24	37.5	0.061	18.7	34	4.5	2.9	26.1

* Midpoint of the age interval, frequently used to calculate generation time when reproduction is more or less continuous, as it is in rabbits.

Note

In mathematics, there are two basic progressions of numbers: the arithmetic progression, in which each element of a series of numbers increases by a constant difference,

2   4   6   8   10 . . . (constant difference of 2)

and the geometric progression, in which each element progresses by a constant factor,

2   4   8   16   32 . . . (constant factor of 2)

In the case of population growth, the number of generations increases with time in an arithmetic fashion, whereas the size of the population increases by a geometric progression, each element of the progression being some constant factor times its previous size.

When each element in a geometric series progresses by a factor of two, we define the *base* of the progression as two, because each element is separated from the next by that factor. We can now write each of the elements in a geometric progression as a complex number, including the base (in this case 2) and a superscript (called an exponent) which indicates how many times the number 1 is multiplied by the base to obtain the element of the progression. For example, the number 2 in the geometric series above is the first power of the series because it is 1 multiplied once by the base. We write this number as 2^1, where 2 is the base of the scale and the superscript 1 is the exponent and designates what is known as the power of the element. The number 8 equals $1 \times 2 \times 2 \times 2$. We write $8 = 2^3$. By the same token, $32 = 1 \times 2 \times 2 \times 2 \times 2 \times 2 = 2^5$, and so on.

We now convert the geometrical series into a power series as follows:

Geometrical series	2	4	8	16	32	...
Exponential series	2^1	2^2	2^3	2^4	2^5	...

Note that whereas the elements of the geometric progression are a constant factor apart, the superscripts of the power progression are a constant difference apart. Taken by themselves, they form an arithmetic series called the logarithmic progression:

Logarithmic series (base 2)	1	2	3	4	5	...

The power scale can be extended indefinitely in either direction by progressively adding or subtracting 1 from the exponent for each factoral increase or decrease in the progression. The power that corresponds to 1 in the geometrical progression is 2^0 (1 multiplied by the base zero times), and that which corresponds to $\frac{1}{2}$ is 2^{-1} (1 divided by the base one time), and so on.

To describe a logarithmic scale, the base of the power progression must be defined. The logarithm to the base 2 of 8 is 3, or in mathematical notation, $\log_2 8 = 3$. To reconvert the logarithm to a power we say that the antilogarithm to the base 2 of 3 (2^3) is $1 \times 2 \times 2 \times 2 = 8$ (formally, $\text{antilog}_2 3 = 2^3 = 8$). By changing the base of the logarithm, the logarithm takes on a different value. For example, $\text{antilog}_{10} 3 = 10^3 = 1 \times 10 \times 10 \times 10 = 1{,}000$ (that is, $\log_{10} 1{,}000 = 3$).

Logarithms belong to a peculiar class of mathematical relationships known as transcendental functions, whose values cannot be determined directly and must be approximated indirectly. For example, it is impossible to evaluate $2^{2.5}$ directly because it is meaningless to multiply 2 by itself two and one-half times. But you should be able to satisfy yourself that it is possible to evaluate $2^{2.5}$ numerically by sketching the graph of the geometrical progression with base 2 as a function of its logarithms. By drawing a smooth curve through the points, you can read the value (5.66) directly. Fortunately, mathematicians have provided us with convenient tables of logarithms and powers and engineers have provided us with electronic calculators, both of which greatly facilitate working with the mathematics of population growth.

28

Natural Selection

A genetic change that influences fecundity and survival alters the fitness of individuals that bear the mutation. By applying the methods of population genetics theory, which is firmly grounded in demography, we can predict the success of a new gene and assess its effect on the evolution of a population. In this chapter we shall derive mathematical relationships for changes in allele frequency in a population as they are influenced by fitness, mutation, and interactions between alleles.

Putting relationships that often seem intuitively obvious into mathematical terms is much more than an exercise in rigor and quantification. Mathematics provides a concise and precise form of expression. We often obtain more information, and are surer of our conclusions, by testing quantitative hypotheses rather than qualitative hypotheses. Furthermore, our intuition is often wrong or naïve, and its shortcomings can be revealed by quantifying relationships. Mathematical treatment allows us to use deductive and analytical tools to solve problems that simply cannot be handled in any other way.

We should be careful, however, not to treat mathematical expression as an end in itself. It is only a powerful tool that can be used to describe and analyze biological problems—it is the biologist who must provide solutions.

Natural selection tends to increase the average fitness of a population.

Let us consider a population in which individuals breed during a brief reproductive season, and then die immediately after reproduction. This population is similar to the one we used in the previous chapter for our model of geometric growth. The generation time is one year, and generations do not overlap. Further, let us assume that the population is haploid, so all genes are expressed in the phenotype. Now suppose that the population has two alleles, A and B, at a given gene locus and that the net reproductive rates (R_0) of individuals that carry the alleles are 1.2 and 0.5, respectively.* Each succes-sive generation will have 1.2 times as many A alleles and 0.5 times as many B alleles as the last. If the population initially contained 100 individuals of each genotype, there would be 120 of genotype A and 50 of genotype B after one generation. As the population continued to change through time, the number of A individuals would gradually increase while the number of B individuals gradually dwindled, as follows:

Generation	1	2	3	4	5	6
Number of As	100	120	144	173	207	248
Number of Bs	100	50	25	12	6	3
Total population	200	170	169	185	213	251
Per cent A	50	71	85	94	97	99
Average fitness of the population	0.85	1.00	1.10	1.16	1.18	1.19

This example illustrates two fundamental results of natural selection. First, although the total population initially decreased before it began to grow, the frequency of the A allele increased throughout the course of selection. Second, the average fitness of the population continuously in-creased. The average fitness of the population is calculated by weighting the fitness of each genotype according to its frequency in the population. Thus the average fitness of the initial population is $(0.5 \times 1.2) + (0.5 \times 0.5) = 0.60 + 0.25 = 0.85$. The average fitness of the population after one genera-tion of selection is $(0.71 \times 1.2) + (0.29 \times 0.5) = 0.852 + 0.145 = 0.999$ (rounded off to 1.00 in the table above). The increase in the total population from one generation to the next is determined by the average fitness of the population. For example, from generation three to generation four the pop-ulation increases by a factor of 1.09 (185/169) which, within the limits of rounding errors in the table, corresponds to the average fitness of the pop-ulation at the beginning of the third generation (1.10).

We can construct a general quantitative expression for natural selection based on the example above of a haploid population with nonoverlapping generations and seasonally restricted reproduction. Suppose the alleles A and B are solely responsible for differences in fitness between individuals. Proportion p of the individuals in the population have the A allele, which confers fitness (net reproductive rate) R_A to them. Proportion q of the indi-viduals have allele B and fitness R_B. The values p and q are decimal fractions (for example, 0.75 and 0.25), and together they must add to 1.00. In any generation, the population has the following genetic characteristics:

---

* Because the generation time in this population is one year, the geometric growth rate λ is equal to R_0 and the exponential rate r is equal to $\log_e (R_0)$ (equations 27-13 and 27-14). Population geneticists usually use the symbol W for fitness, but for consistency we shall retain the symbols of demography.

	Individuals with		All
	allele A	allele B	individuals
Proportion of each allele in the population (allele frequency)	p	q	1
Number of each allele in a population with N individuals	Np	Nq	N
Fitness (net reproductive rate) of each allele	R_A	R_B	$\bar{R}$
Number of alleles in the next generation	NpR_A	NqR_B	$N\bar{R}$

$\bar{R}$ (the average fitness of the population) is the sum of the fitness of each allele weighted by its frequency in the population,

$$\bar{R} = pR_A + qR_B \tag{28-1}$$

After one generation of selection, the number of each allele will be its initial number in the population times its fitness. The number of both A and B alleles in the next generation is NpR_A plus NqR_B, which equals $N\bar{R}$. The new frequency of the A allele (p') will be the number of A alleles after one generation of selection divided by the total number of alleles, hence

$$p' = \frac{NpR_A}{NpR_A + NqR_B} \tag{28-2}$$

By dividing the numerator and denominator by N, equation (28-2) becomes

$$p' = \frac{pR_A}{pR_A + qR_B} \tag{28-3}$$

The change in p over one generation (Δp) is equal to the new gene frequency p' minus the initial gene frequency p, so

$$\Delta p = p' - p \tag{28-4}$$

and substituting (28-3) for p',

$$\Delta p = \frac{pR_A}{pR_A + qR_B} - p \tag{28-5}$$

which may be simplified to

$$\Delta p = \frac{pq(R_A - R_B)}{pR_A + qR_B} \tag{28-6}$$

(see Note 1 at the end of this chapter). Because $pR_A + qR_B = \bar{R}$, equation 28-6 may be re-expressed as

$$\Delta p = \frac{pq}{\bar{R}}(R_A - R_B) \tag{28-7}$$

In words, the rate of change in allele frequency (p) is equal to the product of the allele frequencies (pq) times the difference in fitness between the alleles ($R_A - R_B$), divided by the average fitness of the population ($\bar{R}$).

When p changes, q must change by the same amount in the opposite direction because $p + q = 1$. Hence $\Delta q = -\Delta p$. If the fitness of the A allele were greater than that of the B allele, $R_A - R_B$ would be positive, and the frequency of the A allele would increase (that is, Δp is positive). If R_A were less than R_B, Δp would be negative, and the frequency of the B allele would increase at the expense of A. In the numerical example presented on page 447, the fitnesses of A and B were 1.2 and 0.5, respectively, and their initial frequencies were $p = 0.5$ and $q = 0.5$. Thus, the initial average fitness of the population was 0.85. Substituting these values in equation 28-7, we have

$$\Delta p = \frac{0.5 \times 0.5}{0.85}(1.2 - 0.5)$$

$$= \frac{0.25}{0.85}(0.7)$$

$$= 0.21$$

which is exactly how much the frequency of A increased in the next generation (0.50 to 0.71).

The equation for change in allele frequency can be applied if one knows the fitnesses of two alleles and their proportions in the populations. As we shall see below, equation 28-7 can be extended over many generations to predict long-term changes in allele frequency that could not be directly observed. Equation 28-7 also tells us the conditions under which allele frequency does not change—the equilibrium conditions giving $\Delta p = 0$. For the haploid model with nonoverlapping generations, allele frequency remains constant under selection only when $p = 0$ or $q = 0$ (there is no genetic variation in the population) or when $R_A - R_B = 0$ (the alleles are equally fit). Because it is unlikely that two alleles have exactly the same expression in the phenotype, selection tends, in the haploid case, to eliminate one allele or the other.

In the numerical example on page 447, we saw that selection increased the average fitness of the population. We can prove the point mathematically and, at the same time, provide a quantitative expression for the rate at which fitness increases $\Delta \bar{R}$ owing to natural selection. An indirect, but simple derivation follows from the relationship

$$\Delta \bar{R} = \frac{\Delta \bar{R}}{\Delta p} \cdot \Delta p \tag{28-8}$$

We have already derived an expression for Δp (equation 28-7) and we can show that

$$\frac{\Delta \bar{R}}{\Delta p} = R_A - R_B \tag{28-9}$$

(see Note 2 at the end of this chapter). Now, substituting equations 28-7 and 28-9 in equation 28-8, we obtain

$$\Delta \overline{R} = \frac{pq}{R}(R_A - R_B)(R_A - R_B)$$

$$= \frac{pq}{R}(R_A - R_B)^2 \qquad (28\text{-}10)$$

$\Delta \overline{R}/\Delta t$ is always positive (p, q, and $\overline{R}$ are positive, and the square of any rational number is positive), so natural selection always increases the average fitness of the population.

Because R is a valid estimate of fitness only for populations with nonoverlapping generations and discrete breeding seasons, it is desirable to reformulate the equation for natural selection in terms of the exponential growth rate r (Haldane 1927, Charlesworth 1974). For alleles A and B, with initial frequencies p and q and fitnesses r_A and r_B, we have:

	Individuals with		All
	allele A	allele B	individuals
Initial proportion of each allele	p	q	1
Initial number of each allele	Np	Nq	N
Fitness (exponential growth rate)	r_A	r_B	$\bar{r}$

The number of A alleles (N_A) in the population after a period (t) is

$$N_A(t) = Npe^{r_A t} \qquad (28\text{-}11)$$

according to the equation for exponential growth $N(t) = N(0)e^{rt}$ (see page 426). Similarly, the number of B alleles may be expressed as

$$N_B(t) = Nqe^{r_B t} \qquad (28\text{-}12)$$

The new frequency of allele A at time t ($p(t)$) is equal to the number of A alleles at time t divided by the total number of alleles in the population at time t,

$$p(t) = \frac{N_A(t)}{N_A(t) + N_B(t)} \qquad (28\text{-}13)$$

Substituting equations 28-11 and 28-12 for N_A and N_B, respectively, in 28-13, we obtain

$$p(t) = \frac{Npe^{r_A t}}{Npe^{r_A t} + Nqe^{r_B t}} \qquad (28\text{-}14)$$

Dividing numerator and denominator by N *gives*

$$p(t) = \frac{pe^{r_A t}}{pe^{r_A t} + qe^{r_B t}} \tag{28-15}$$

which may be arranged in the form

$$p(t) = \frac{1}{1 + \left(\dfrac{q}{p}\right)e^{(r_A - r_B)t}} \tag{28-16}$$

Equation 28-16 may be solved for the rate of change in allele frequency with respect to time as a function of p to give

$$\frac{dp}{dt} = pq(r_A - r_B) \tag{28-17}$$

(The solution of the equation requires calculus and will not be given here. The notation for the rate of change in gene frequency in a continuously growing population becomes dp/dt, rather than Δp.)

The expression for dp/dt is similar to that for Δp, except that the term $1/\bar{R}$, is absent. The mathematical properties of the two equations are also similar, particularly when the selection differential $(r_A - r_B)$ is small. We can also show that selection always increases fitness because $d\bar{r}/dt = pq(r_A - r_B)^2$.

In diploid populations, the fitness of each allele is affected by its expression in the heterozygote phenotype.

The mathematical description of natural selection derived for the haploid case in the last section is more complicated for diploid populations because the fitness of each allele is determined by the fitness of both the homozygous genotype (AA) and the heterozygous genotype (AB). To calculate the fitness of an allele, one must first know what proportion of the alleles occurs in homozygous and heterozygous genotypes. These proportions change with allele frequency. If we assume that gametes containing A and B alleles mate with each other at random, and furthermore that neither chance occurrences nor selection alters allele frequencies, the frequency of each genotype, AA, AB, and BB, may be calculated from the laws of probability. The chance that two independent events or objects occur together is the product of the probabilities of each occurring independently. In a diploid population with two alleles, the male and female gametes, which are haploid, each have probability p of containing an A allele. Therefore, the proportion of AA genotypes among zygotes will be $p \times p = p^2$. Similarly, the proportion of BB genotypes is q^2. Heterozygotes (AB) occur with frequencies $2pq$ because the heterozygote can be formed two ways: the A allele may come from either the male or the female parent, as shown in Figure 28-1. The frequencies of the

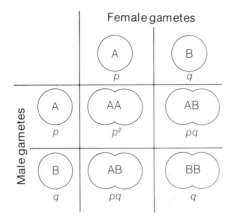

Figure 28-1   Possible combinations and their frequencies of two alleles in zygotes formed by the random union of male and female gametes.

three genotypes formed by two alleles in a diploid population, assuming no selection, mutation, or chance influence, are:

Genotype	AA	AB	BB	Total
Frequency	p^2	$2pq$	q^2	1

The frequencies of the genotypes add to 1, as they must: $p^2 + 2pq + q^2 = (p + q)^2 = (1)^2 = 1$. The relationship between allele frequency and genotype frequency is shown graphically in Figure 28-2.

Knowing that genotype frequencies vary with allele frequency, we may readily calculate the proportions of each allele in the homozygous and heterozygous state. The total number of A alleles in the homozygous genotype is the proportion of that genotype in the population, p^2, times the

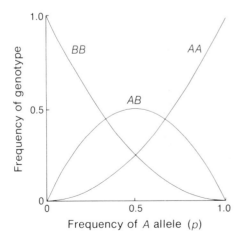

Figure 28-2   The relationship between genotype frequency and allele frequency in a large, randomly mating diploid population. The proportion of an allele in the homozygous state is equal to the frequency of that allele in the population.

number of individuals in the population, N, times the number of A alleles per individual, 2. Because the total number of alleles in the population is $2N$, the total number of A alleles in the population is $2Np$. The proportion of the A alleles that are in the homozygous phenotype is simply

$$\frac{A \text{ alleles in homozygotes}}{\text{total } A \text{ alleles in population}} = \frac{2Np^2}{2Np} = p \qquad (28\text{-}19)$$

Similarly, the number of A alleles in the heterozygous state is equal to the proportion of that genotype, $2pq$, times the number of individuals in the population, N, times the number of A alleles in each heterozygous individual, 1. Hence the proportion that is in the heterozygous genotype is $2Npq/2Np = q$. The proportion of A alleles that occur in the homozygous state is identical to the proportion of all genes that are the A allele. By analogy, proportion q of the B alleles occurs in BB homozygotes and proportion p is the heterozygous state. In practice, when selection acts on a population, the gametes produced are no longer a random sample of the gametes produced by the previous generation because some genotypes are selectively removed from the population. But so long as selection is weak, the proportions of A and B alleles in homozygous and heterozygous genotypes do not deviate enough from the proportions derived above to affect our calculations substantially.

We may now redefine the fitnesses of the alleles, r_A and r_B, in terms of the fitnesses of the three genotypes, r_{AA}, r_{AB}, and r_{BB}:

$$r_A = pr_{AA} + qr_{AB} \qquad (28\text{-}20)$$

because proportion p of the A alleles occurs in the AA genotype, and proportion q occurs in the AB genotype. Similarly,

$$r_B = pr_{AB} + qr_{BB}. \qquad (28\text{-}21)$$

Substituting these expressions for r_A and r_B in the equation for selection, $dp/dt = pq(r_A - r_B)$, gives

$$\frac{dp}{dt} = pq(pr_{AA} + qr_{AB} - pr_{AB} - qr_{BB})$$

which, rearranged in a more convenient form, is

$$\frac{dp}{dt} = pq[p(r_{AA} - r_{AB}) + q(r_{AB} - r_{BB})] \qquad (28\text{-}22)$$

The rate of change in allele frequency can be simplified for the special cases of complete dominance, intermediate expression, and complete recessiveness of one allele. When A is dominant, $r_{AA} = r_{AB}$ because the dominant homozygote and heterozygote are indistinguishable. It follows that $r_{AA} - r_{AB} = 0$, and equation 28-22 simplifies to

$$\frac{dp}{dt} = pq^2(r_{AA} - r_{BB}).$$

When A is recessive, $r_{AB} = r_{BB}$, $r_{AB} - r_{BB} = 0$, and equation 28-22 becomes

$$\frac{dp}{dt} = p^2q(r_{AA} - r_{BB}).$$

When the fitness of genotype AB is intermediate between the fitnesses of genotypes AA and BB,

$$r_{AA} - r_{AB} = r_{AB} - r_{BB} = \tfrac{1}{2}(r_{AA} - r_{BB}),$$

and so

$$\frac{dp}{dt} = \frac{pq}{2}(r_{AA} - r_{BB}),$$

which is one-half the rate of change in gene frequency in the haploid case (equation 28-17).

Heterozygote superiority results in the stable coexistence of alleles (genetic polymorphism).

Selection in diploid and haploid populations differs greatly because alleles can influence each other's expression in the heterozygote genotype. All possible relationships between genotype fitnesses and their influence on selection are compared in Table 28-1. When the fitness of the heterozygote is equal to that of one of the homozygotes (complete dominance), or when it is intermediate, the allele whose homozygous genotype is the more fit will continually increase until it replaces the other allele. When the fitness of the heterozygote is greater than that of both homozygote genotypes (variously referred to as *overdominance, heterosis,* or *heterozygote superiority*), either allele will increase when it is rare (when it occurs more commonly in heterozygotes than in homozygotes), and it will decrease when abundant (when it occurs more commonly in homozygotes than in heterozygotes). Heterozygote superiority therefore leads to a stable (balanced) polymorphism, in which both alleles coexist in a population.

When the fitness of the heterozygote is inferior to that of both homozygotes, the rarer allele decreases because it occurs primarily in the heterozygous state. Depending on the initial allele frequency, one allele or the other will be eliminated from the population. One important consequence of heterozygote inferiority is that a mutation that produces a homozygote more fit than that of the normal allele but an inferior heterozygote, cannot be selected in a population of any appreciable size. Relatively few alleles of this type have been detected in natural populations, but heterozygote inferiority may occur commonly in crosses between genetically distant populations. Such inferiority could be responsible for postmating reproductive isolation of parental populations and reduce or prevent introgression between them.

The outcome of selection can be determined graphically from the relationship between the average fitness of the population ($\bar{r}$) and allele frequency (p) because natural selection always acts to increase $\bar{r}$ (Figure 28-3). When the fitness of the heterozygote is equal to that of either homozygote

TABLE 28-1

Outcome of natural selection for various allele interactions in diploid populations. The outcome is determined by the sign (+ or −) of the equation $dp/dt = pq[p(r_{AA} - r_{AB}) + q(r_{AB} - r_{BB})]$.

Situation	Allele interaction	Fitness	$p(r_{AA} - r_{AB})$	$q(r_{AB} - r_{BB})$	Outcome (dp/dt)
AA homozygote superior	A dominant to B	$r_{AA} = r_{AB} > r_{BB}$	0	+	p increases
	heterozygote intermediate	$r_{AA} > r_{AB} > r_{BB}$	+	+	p increases
	A recessive to B	$r_{AA} > r_{AB} = r_{BB}$	+	0	p increases
AA homozygote inferior	A dominant to B	$r_{AA} = r_{AB} < r_{BB}$	0	−	p decreases
	heterozygote intermediate	$r_{AA} < r_{AB} < r_{BB}$	−	−	p decreases
	A recessive to B	$r_{AA} < r_{AB} = r_{BB}$	−	0	p decreases
Heterozygote superior to both homozygotes		$r_{AA} < r_{AB} > r_{BB}$	−	+	p increases when low, decreases when high; rare gene increases: stable polymorphism
Heterozygote inferior to both homozygotes		$r_{AA} > r_{AB} < r_{BB}$	+	−	p increases when high, decreases when low; common gene increases: unstable equilibrium point

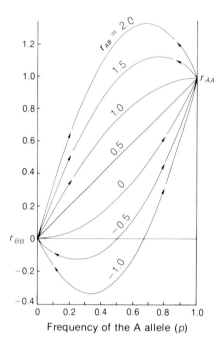

Figure 28-3   Average fitness of a population ($\bar{r}$) as a function of allele frequency and the fitness of the heterozygote. In all cases, $r_{AA} = 1$ and $r_{BB} = 0$. Curves are presented for heterozygote fitnesses (r_{AB}) of 2.0, 1.5 (heterozygote superiority), 1.0 (A dominant of B), 0.5 (heterozygote intermediate), 0.0 (B dominant to A), -0.5, and -1.0 (heterozygote inferiority). Arrows indicate the direction of changes in gene frequency. Maximum points on the curves are stable equilibria; minimum points are unstable equilibria.

(complete dominance), or is intermediate, $\bar{r}$ is greatest at either $p = 1$ or $q = 1$, depending on which homozygote is more fit, and selection will proceed to eliminate one allele. When the fitness of the heterozygote is greater than either homozygote, the maximum average fitness of the population occurs at some intermediate allele frequency. Because natural selection always increases the average fitness of the population, allele frequency will converge toward the point ($\hat{p}$) at which r is greatest. When p is greater than the equilibrium allele frequency ($\hat{p}$), $\bar{r}$ will increase only when p decreases; when p is less than $\hat{p}$, $\bar{r}$ will increase only when p increases. The equilibrium allele frequencies ($\hat{p}$) in Figure 28-3, where $r_{AA} = 1$ and $r_{BB} = 0$, are approximately 0.75 for $r_{AB} = 1.5$, and 0.66 for $r_{AB} = 2.0$. If $r_{AA} = r_{BB}$ and the heterozygote were superior, equilibrium would occur at $p = 0.5$.

When the heterozygote is inferior to both homozygotes, a minimum in the fitness curve occurs at some intermediate allele frequency. You can see from Figure 28-3 that when the A allele is rare, selection will tend to remove it from the population whereas when it is common (above the unstable equilibrium point), it will be selected.

The equilibrium value of p can be calculated because we know that at $\hat{p}$ there is no net selection on allele frequency and $dp/dt = 0$. Therefore, we may evaluate $\hat{p}$ by solving the equation

$$\frac{dp}{dt} = pq[p(r_{AA} - r_{BB}) + q(r_{AB} - r_{BB})]$$

for p when $dp/dt = 0$. Aside from the trivial cases in which either p or q is equal to zero, the solution requires that

$$\hat{p}(r_{AA} - r_{AB}) = -\hat{q}(r_{AB} - r_{BB}) \tag{28-23}$$

which may be rearranged to give

$$\frac{\hat{p}}{\hat{q}} = \frac{r_{BB} - r_{AB}}{r_{AA} - r_{AB}} \tag{28-24}$$

Employing the equation $q = 1 - p$ and a lot of algebra, we get

$$\hat{p} = \frac{r_{BB} - r_{AB}}{(r_{AA} - r_{AB}) + (r_{BB} - r_{AB})} \tag{28-25}$$

For one of the cases of heterosis shown in Figure 28-3, where $r_{AA} = 1$, $r_{AB} = 2$, and $r_{BB} = 0$, $\hat{p}/\hat{q} = (0 - 2)/(1 - 2) = -2/-1 = 2/1$ and so $\hat{p} = 0.67$ (the estimate from the graph was 0.66). We can also calculate p from equation 28-26 as $(0 - 2)/((0 - 2) + (1 - 2)) = -2/-3 = 0.67$.

One of the best-known, indeed, one of the few cases of heterozygote superiority is that of the sickle-cell trait in man (see page 333). The sickle-cell allele (S) is widespread throughout many tropical and subtropical regions of the Mediterranean area and Asia. When the sickle-cell allele is homozygous, it produces a highly deleterious form of anemia in the phenotype, whose fitness is about two-tenths of the normal homozygote phenotype (Allison 1956, Cavalli-Sforza and Bodmer 1971). But, in spite of the extreme deleteriousness of the homozygous sickle-cell genotype (SS), the allele is maintained in many populations at frequencies of 20 per cent or more, because in the heterozygous state (AS) the sickle-cell allele confers protection against malaria. Allison estimated that where malaria is prevalent and virulent, the fitness of heterozygote individuals may be 25 per cent greater than the fitness of individuals who are homozygous for the normal allele. If the fitnesses of AA, AS, and SS genotypes are 1.0, 1.25, and 0.2, respectively, the equilibrium frequency of the A allele is

$$\hat{p} = \frac{0.2 - 1.25}{(1 - 1.25) + (0.2 - 1.25)} = \frac{-1.05}{-1.30} = 0.81$$

(The fitnesses are actually estimates of the influence of the sickle-cell allele on net reproductive rate (R_0), but the equation for equilibrium gene frequency is the same whether relative or absolute measures of fitness are used.)

If malaria were completely eradicated, the heterozygote genotype would confer no greater advantage than the homozygous normal genotype, and the maximum fitness of the population would occur when the frequency of A is 1. Hence, selection would eliminate the sickle-cell allele from the population. Apparently, this is happening in the black population of the United States, which originally carried a fairly high proportion of the sickle-cell allele, perhaps 20 per cent. The allele frequency (q) has been reduced to its present level of 4.5 per cent in about twelve generations (250 to 300 years), partly through interbreeding with the white population of the United States, and partly through selective elimination of the sickle-cell allele.

Genetic polymorphisms can be maintained by mechanisms other than heterosis.

Up to this point, we have considered only the simplest possible models of population genetics. We have used these to make the point that selection inevitably follows from the influence of genetic variation on the demography of a population. In the haploid case, one allele is always favored over another and selection continually erodes the genetic variation in a population. Add a little complexity, such as the interaction between alleles in a diploid organism and it becomes possible for selection to maintain genetic polymorphism. But here the complexities in the genetic structure of populations only begin. We have not even touched upon the population genetic consequences of inbreeding, sex-linked inheritance, linkage groups, mutation, population structure, and, probably most important, variation in the fitness of an allele with environment, frequency in the population, and genetic background. These phenomena are difficult to attack theoretically and experimentally—often the case when complexities are added to a system—and much remains to be learned. We shall examine the effects of frequency-dependent fitness and mutation in more detail below, partly because they are more easily treated than the problems incurred by higher order genome interactions like linkage and environmental control over fitness. It is worth mentioning in passing that selection of different alleles in males and females, or in gametes and zygotes, can maintain a balanced polymorphism in the population, although genes of this type probably are not common.

Mutation pressure can maintain a deleterious allele in equilibrium with selection.

Alleles can be added to or removed from populations by mutation as well as by selection. When allele A mutates to B, one B allele is added to the population and one A allele is removed. As far as anyone knows, mutation rate is independent of the frequency of an allele, so the number of alleles that mutate from A to B is directly proportional to the frequency of A in the population. It follows that if A were common and B were rare, more A alleles would mutate to B than B to A (provided their mutation rates were similar), and the frequency of B would increase in the absence of selection. This can be demonstrated mathematically as follows: We will define the mutation rate of A to B as μ_A (the Greek letter mu), and from B to A as μ_B. Now, because A alleles are removed from the population by mutation to B at rate $p\mu_A$, and A alleles are added to the population by mutation from B at rate $q\mu_B$, the rate of change in A (dp/dt) due to mutation alone is

$$\frac{dp}{dt} = -p\mu_A + q\mu_B$$

If mutation were the only force acting to change allele frequency, an equilibrium would be reached when $\mu_A p = \mu_B q$, which can be solved to give

$$\hat{p} = \frac{\mu_B}{\mu_A + \mu_B} \qquad (28\text{-}26)$$

Hence, in the absence of selection, allele frequency is maintained in a stable polymorphism by the balance between forward and back mutation. If the two rates were identical, the equilibrium value $\hat{p}$ would be 0.5.

Mutation pressure is usually weak compared to selection. While mutation rates are measured in thousandths or millionths, selection coefficients are often expressed in hundredths or even tenths. But the strength of selection diminishes toward zero as allele frequency approaches zero. We can see this in the equation for selection $dp/dt = pq(r_A - r_B)$; regardless of the fitness differential between A and B, as either p or q approaches zero, the product pq, hence dp/dt, also approaches zero. The strength of selection is greatest at intermediate allele frequencies where the product pq is largest. In contrast to selection, mutation pressure to increase the frequency of an allele increases as the allele becomes rarer. In the region of allele frequencies close to zero, a deleterious allele is maintained in a population by mutation.

By itself, a rare deleterious allele would seemingly have little effect on a population, but when considered over the genome as a whole, mutation pressure maintains a reservoir of genetic variation which, on one hand, provides material for evolution but, on the other, may reduce the fitness of the individual.

We may derive an equation for an equilibrium frequency of a deleterious allele (B) in a haploid population in the following way. Two factors tend to remove allele B: selection, at rate $pq(r_A - r_B)$, and mutation from B to A, at rate $q\mu_B$. Only mutation from A to B, at rate $p\mu_A$, tends to increase q. An equilibrium is reached when

$$pq(r_A - r_B) + q\mu_B = p\mu_A \qquad (28\text{-}27)$$

Because this equation involves quadratic algebra, it will be easier for our purposes to demonstrate its solution graphically. In Figure 28-4, the rate at

Figure 28-4   Graphical representation of the balance between selection and mutation for a haploid case in which the difference in allele fitness ($r_A - r_B$) is 1, and the mutation rates (μ_A and μ_B) are both 0.1. The equilibrium point is about $\hat{p} = 0.91$ ($\hat{q} = 0.09$). In the absence of selection, the equilibrium produced by mutation is $\hat{p} = 0.5$. The example is not realistic numerically because the rates of mutation are much too high compared to selection. Note, however, that the effect of mutation from B to A on $\hat{q}$ is negligible, even for such a high value of μ_B.

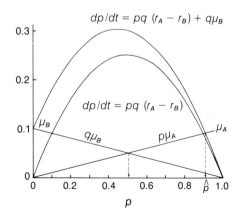

which the frequency of A increases is shown as the sum of the rates of selection and of mutation from B and A. An equilibrium point, $\hat{p}$, occurs where this curve intersects the curve for the rate of mutation from A to B.

Because selection is less efficient in diploid populations than in haploid populations, equilibrium frequencies of deleterious alleles are greater in diploid populations. Also, because selection against rare recessives is weak (page 454) mutation pressure maintains recessive deleterious alleles at higher frequencies than dominant deleterious alleles.

We can derive an approximate solution for equilibrium frequencies of deleterious alleles when the frequencies are very close to zero, if we assume that selection is much stronger than mutation, which is probably reasonable for most genes. Under these conditions, $\hat{q}$ will be very small. As we see in Figure 28-4, $q\mu_B$ becomes negligible when q is small and, therefore, the term has little quantitative effect on $\hat{q}$. When $q\mu_B$ is omitted from equation 28-27, equilibrium conditions are achieved:

$$pq(r_A - r_B) = p\mu_A$$

which may be solved to give, in the haploid case,

$$\hat{q} = \frac{\mu_A}{r_A - r_B} \tag{28-28}$$

In words, when selection is great relative to mutation rate, the equilibrium value $\hat{q}$ is approximately equal to the mutation rate divided by the difference between the fitness of the two alleles. As the mutation rate increases, so does $\hat{q}$. As the selective difference between the alleles increases, $\hat{q}$ will decrease, as intuition leads us to expect.

By applying the same simplifying approximation in the diploid case (see Note 3 at the end of this chapter), one may solve for the equilibrium frequency ($\hat{q}$) when the deleterious allele is recessive,

$$\hat{q}^2 = \frac{\mu_A}{r_{AA} - r_{BB}} \tag{28-29}$$

when the deleterious allele has intermediate fitness $r_{AA} - r_{AB} = r_{AB} - r_{BB}$,

$$\hat{q} = \frac{2\mu_A}{r_{AA} - r_{BB}} \tag{28-30}$$

and when the deleterious allele is dominant,

$$\hat{q} = \frac{\mu_A}{r_{AA} - r_{BB}} \tag{28-31}$$

At equilibrium, the proportion of deleterious genotypes in the haploid case ($\hat{q} = \mu_A/(r_A - r_B)$) is identical to the frequency of recessive homozygote genotypes in the diploid case ($\hat{q}^2 = \mu_A/(r_{AA} - r_{BB})$). Hence, the same number of deleterious alleles is exposed to selection in both cases. Similarly, when B is dominant, most of the deleterious alleles will be expressed in the heterozygote form (AB) whose frequency is $2\hat{p}\hat{q}$. Since $\hat{p}$ is close to 1, and only half the alleles in the heterozygote are B, the number of deleterious alleles exposed to selection is also $q = \mu_A/(r_{AA} - r_{BB})$. When the heterozygote is intermediate in fitness, $\hat{q}$ is twice as great as when B is dominant, but although twice as many alleles are exposed to selection, selection is only about half as strong (we again assume that relatively few homozygote recessives appear). The greater strength of selection in removing deleterious alleles in haploid than in diploid populations is demonstrated in *Drosophila* by comparing the sex chromosome (X) with the autosomes. The genotype of the male is XY (the Y is genetically relatively inert), so alleles that are recessive in the XX females are expressed in the males. Dobzhansky (1951) showed that the X chromosome harbors fewer deleterious mutations than the autosomes (see page 354).

Regardless of the specific properties of the allele, the number of deleterious alleles exposed to selection times the rate of selection per allele is a constant equal to the mutation rate,

$$\hat{q}(r_A - r_B) = \mu_A \qquad (28\text{-}32)$$

The fact that the rate of selective removal of an allele is equal to the mutation rate is frequently used to estimate mutation rates of genes whose frequency and fitness are known. Suppose, for example, that in a diploid population a recessive allele occurred in the homozygous form in one out of 10,000 individuals $(q^2 = 0.0001)$ and that its fitness was one-tenth less than the fitness of the normal gene $(r_{AA} - r_{BB} = 0.1)$. Its mutation rate thus could be calculated from the formula (rearranged from equation 28-29)

$$\mu_A = q^2(r_{AA} - r_{BB})$$
$$= 0.0001 \times 0.1$$
$$= 0.00001$$

(see Note 4 at the end of this chapter). In this example, one A allele in 100,000 mutates to B each generation, and the same number are removed by selection.

Frequency-dependent selection can maintain genetic polymorphism.

In the case of heterosis, the fitness of each allele varied with its frequency in the population. Rare alleles occur more frequently in heterozygotes than in homozygotes and thus have a higher fitness than common alleles, provided the heterozygote genotype is more fit than either homozygote. The principle of frequency-dependence has also been found to apply to the fitness of genotypes themselves in some cases (Ayala and Campbell 1974). Frequency-dependent selection and, as a result, balanced polymorphism occur whenever rare genotypes are favored because of their rareness. We first encountered this phenomenon in apostatic selection (page 363), which is the result of predators keying in on the most abundant genotype of the prey. When these genotypes affect the way the prey appears to the predator, apostatic selection can result in a population polymorphic for color and pattern.

One feature of frequency-dependent selection is that fitnesses of all genotypes are approximately equal at equilibrium allele frequency. This can be shown by the simple numerical example from Lewontin (1974: 257). Suppose the fitnesses of three genotypes are

$$
\begin{array}{ccc}
AA & AB & BB \\
2(1-p) & 1 & 2(1-q)
\end{array}
$$

The fitness of the heterozygote is constant at 1; the fitnesses of the homozygotes vary with allele frequency p and q. When A is common (p near 1), the fitnesses approach 0, 1, and 2; when B is common (q near 1) the fitnesses are reversed: 2, 1, 0. At the equilibrium allele frequency ($\hat{p} = 0.5$ in this case), the fitnesses of the genotypes are 1, 1, and 1. Hence, at equilibrium there is no selective differential between genotypes, but if the allele frequency is displaced from the equilibrium point, selection differentials quickly appear to restore the equilibrium.

There is abundant evidence for frequency-dependent effects in laboratory populations, the most striking of which is the mating advantage possessed by the rarer male genotype in populations of *Drosophila* (Ehrman 1967, Spiess 1968). The effect depends on olfactory cues that can be picked up from a food site by the larvae as well as being determined genetically by a wide variety of genes and chromosome rearrangements. Lewontin suggests that such a general behavioral effect may serve to enhance outcrossing within a population.

Kojima and his co-workers at the University of Texas identified frequency-dependent effects in genotypes of enzymes with electrophoretically distinguishable alleles, *esterase-6* and *alcohol dehydrogenase* (Kojima and Yarbrough 1967, Kojima and Tobari 1969). The interaction between genotypes appears to operate through substances exuded into the growth medium by larvae, perhaps a mechanism of direct competition by chemical interference. But despite Kojima's (1971) claim, probably reasonable, that constant fitness is the exception in natural populations (even though it is the rule in population genetics theory), frequency-dependent effects have received little attention so far.

Are neutral mutations an important factor in evolution?

Until recently, genetic polymorphisms in natural populations, other than those for obvious traits like eye color and banding patterns on shells, have been difficult to identify. Recently developed electrophoretic techniques for identifying protein polymorphism have, however, shown that 30 to 60 per cent of all loci are polymorphic, and individuals are heterozygous for perhaps 5 to 15 per cent of their genes. One's first reaction to such a revelation is to think of the sickle-cell gene and explain most polymorphic loci as the result of heterosis. The obvious difficulty with this theory is that although heterosis is fine if one happens to carry the heterozygote genotype, sexual reproduction continually produces a high frequency of homozygotes with reduced fitness. If, for example, the frequencies of two alleles of a polymorphic gene were 0.5 and 0.5, half the genotypes would be heterozygous and half, the less fit homozygotes genotype. Multiply this situation over a thousand genes and one begins to appreciate the potential magnitude of fitness depression in a multiple homozygote.

For a population in which mating occurs at random and loci are not closely linked, the laws of probability dictate that few individuals would have more or fewer heterozygous gene loci than the average and so the fitness differences between individuals would not be great. A simple calculation will make this point. Suppose a population has 1,000 polymorphic loci at each of which the frequency of heterozygous genotypes is 0.5 and the difference between the fitness of the heterozygous and homozygous genotypes is 0.01. That is, the homozygote has 99 per cent of the fitness of the heterozygote. Each individual will have 500 heterozygous loci on average with a population variance of 250 (see page 324). According to this statistical distribution, only one individual in ten will have fewer than 470 or more than 530 heterozygous loci, and only one in a hundred will have fewer than 460 or more than 540 heterozygous loci. If the fitnesses at each gene locus multiply to determine the fitness of the individual, then one individual out of a hundred will have a fitness less than 68 per cent (0.99^{40}) or greater than 149 per cent (1.01^{40}) of the population average. So far, so good. These calculations do not seem in the least bit unreasonable. Now suppose, however, that we inbreed the population in the laboratory until we obtain a strain that is entirely homozygous. The fitness of such individuals should be reduced by inbreeding depression to about 0.7 per cent (0.99^{500}) of the mean value in the wild population. It is this value, which seems unreasonably low, that has made the presence of large numbers of enzyme polymorphisms such a controversial topic (Lewontin 1974). The data on inbreeding depression do not seem compatible with predicted loss of fitness due to homozygosity.

One way out of this dilemma is to postulate that electrophoretically detectable variations in proteins do not differ in function and so have no differential effect on fitness (Kimura 1968a, King and Jukes 1969). The substitution of one amino acid for another at one position in a large molecule may affect its mobility in a starch gel under electric current, but it might have no detectable effect on the function of the molecule. When two alleles have identical fitness, they are said to be *neutral alleles*. Such genes represent the special case of the equation for natural selection under which $r_A = r_B$ and selection has no effect on allele frequency ($dp/dt = 0$ at all allele frequencies). In the absence of selection, allele frequency comes under the influence of sampling properties of finite populations. Allele frequencies drift due to random events, such as accidental death, that have nothing to do with the expression of the gene itself. Proponents of the neutral allele hypothesis envision neutral alleles appearing as mutations, sometimes becoming common and eventually replacing the predominant allele. Proteins are seen to change in this manner, resulting in a kind of non-Darwinian evolution and accounting for differences between species in protein structure (see page 342). Moreover, the neutral allele hypothesis would explain the widespread genetic polymorphism in populations as transient states in the replacement of alleles by random drift, rather than as stable states maintained by balancing selection (heterosis or frequency-dependent fitness). The resolution of this controversy will have important consequences for evolutionary ecology. If most of the alleles turn out to be selectively neutral, they are of little consequence to ecology and evolution other than as indices to phylogenetic relationship. If most of the alleles have selective value and are maintained in balanced polymorphisms, they would constitute an immense reservoir of genetic variation for response to long-term environmental change. Levels of polymorphism may then serve as important indices to environmental heterogeneity and population structure and assume an important role in the future development of ecological genetics. The controversy is by no means resolved, and in the end the answer may be that some alleles are selectively neutral and others are not. Let us review some of the evidence for each side of the argument.

Kimura (1968a) showed theoretically that the rate at which selectively neutral alleles are fixed in a population through random drift is equal to the average mutation rate. In a diploid population whose size is N there are $2N$ alleles, and the number of new mutants of a particular gene each year will be equal to $2N\mu$, if the mutation rate of a particular locus is μ. Kimura argued that at some point in time, which may be very far in the future, all the alleles at a particular locus in a population just by chance will have descended from a single allele. Of course, the chances of this occurring are extremely small, especially in large populations, but the probability is finite. The probability that the descendants of an individual new mutant will make up the entire population at some time in the future is its proportion in the total population ($1/2N$). Because new mutations are produced in the population at the rate $2N\mu$ (the rate of fixation of new mutants in the population) the rate of change in the gene pool is $(2N\mu) \times (1/2N) = \mu$ (the mutation rate).

It should be possible to determine whether the rate of allele replacement is consistent with the mutation rate from biochemical studies on the rate of change in sequences of amino acid sequences in proteins. Such rates can be determined by comparing proteins in species that have evolved separately for known periods. King and Jukes (1969) summarized data on rates of amino acid substitutions in proteins during mammalian evolution. For example, the alpha chain of the hemoglobin molecule has been compared among man, the horse, and the mouse. The three possible pairs of species (human-horse, human-mouse, and horse-mouse) have 18, 17, and 23 differences, respectively, in amino acids at homologous sites along the protein chain. If one assumes that the primate, equine, and rodent lines diverged about 75 million years ago, the average number of substitutions per amino acid codon per year would be on the order of 10^{-9}. For seven proteins surveyed in a similar manner by King and Jukes, rates of substitution varied between 3.3 and 49.2×10^{-10}, with an average of 16×10^{-10} substitutions per codon per year. This is much less than the observed average—for a class of genes with different phenotypic effects—of about 10^{-5} mutations per year (Table 22-1), which suggests that evolution is more conservative than the neutral mutation theory predicts and, hence, that new mutations are at a disadvantage. But other biochemical evidence presented by King and Jukes favors a neutral mutation interpretation.

The theory of neutral mutation suggests that the distribution of polymorphisms should not show any consistent pattern with respect to the environment. Therefore, additional evidence for the existence or absence of neutral alleles might be obtained from geographical patterns of allele frequencies of various genes. Persistent ecological trends or temporal stability of allele frequencies would suggest that the alleles were maintained in a polymorphic state by heterosis or frequency-dependent fitness.

Patterns of allele distributions can be found both to support and to reject the neutral allele hypothesis and most are open to alternative explanation (for example, see Berger 1971, Koehn *et al.* 1973, Schopf and Gooch 1971, Ayala *et al.* 1972). Further evidence has been sought by physiological and biochemical studies of the proteins themselves (Yamazaki 1971, Merritt 1972). Harris (1971) reported that of 23 polymorphic enzyme loci in man, 15 show differences in physiological function.

Despite numerous attempts to reconcile the large number of polymorphic loci in the genome with conventional models of selection (*e.g.*, King 1967, Sved *et al.* 1967, Tracey and Ayala 1974), geneticists still do not understand the general relationship between allele interactions and fitness. We are also ignorant of the extent of epistatic interactions between gene loci, particularly for the many genes with small additive effects on the phenotype (*e.g.*, Mukai 1969). Perhaps most important are the poorly understood effects of environmental variation on the interaction between alleles and the fitness of genotypes (Gillespie and Langley 1974).

Selection for quantitative traits.

Fitness does not reside in the gene; it is a property of the genotype. Whole animals live, reproduce, and die; the fate of an individual gene depends on the rest of the genotype. Animal and plant breeders appreciated this fact long before the nature of genes was understood. They were concerned with the phenotype itself, it being the sum of all genetic factors plus environmental influences.

In the chapter on inheritance (Chapter 22), we defined heritability (h^2) as the ratio of additive genetic variance to total phenotypic variance in a metrical trait. The change in a quantitative trait under selection (R) depends on the deviation of selected individuals from the mean value of the population (S) and the heritability of the trait, according to the relationship

$$R = h^2S \tag{28-33}$$

If, for example, h^2 is 0.5 and males and females ten size units larger than the population on average are mated, their progeny will be an average of five size units above the population average. The greater the heritability of the trait, the more rapidly it will respond to selection. Large values of h^2 indicate a close correspondence of phenotype and genotype, and between parent and offspring. Selection cannot produce any result when the heritability is zero because the phenotype bears no relationship to genotype. As we saw earlier, most traits have heritabilities in the range of 0.1 to 0.6 and, thus, are amenable to change by selection. As we shall see in the next chapter, however, the results of selection for quantitative traits reveal many complex interactions within the genotype.

In summary, mathematical analysis of natural selection makes several important points. Some of these are intuitive and the equations only serve to quantify relationships; in some cases, however, predictions based on the mathematical treatment of selection are not apparent before writing the equations; in still others, the theoreticians must await the tests of their models by experimentalists and further intuitions. We can say five things: First, selection generally increases the average fitness of a population. Second, selection on allele frequency in diploid populations has only half the strength found in haploid populations because the removal of a heterozygote has a smaller influence on gene frequency than the removal of a homozygote. Third, when the fitness of the heterozygote is superior to that of either homozygote, the average fitness of the population is greatest at an intermediate allele frequency, and selection preserves a balanced polymorphism. Fourth, mutation balances selection against deleterious alleles and creates an equilibrium allele frequency at which the rate of selective removal of alleles is exactly equal to the mutation rate. Fifth, and most important, we have a poor understanding of the interactions between alleles and gene loci that together make up the unit of natural selection, the genotype of the organism. In Chapter 29, we shall examine fitness values and evolution in natural—and some not-so-natural—populations.

Notes

1. By the following algebraic steps:
We start with

$$\Delta p = \frac{pR_A}{pR_A + qR_B} - p$$

then put both terms over a common denominator

$$= \frac{pR_A - p(pR_A + qR_B)}{pR_A + qR_B}$$

and multiply through

$$= \frac{pR_A - p^2R_A - pqR_B}{pR_A + qR_B}$$

Next we collect terms in R_A

$$= \frac{(p - p^2)R_A - pqR_B}{pR_A + qR_B}$$

and factor $p - p^2$

$$= \frac{p(1 - p)R_A - pqR_B}{pR_A + qR_B}$$

Since $1 - p = q$

$$= \frac{pqR_A - pqR_B}{pR_A + qR_B}$$

and factoring again we obtain

$$= \frac{pq(R_A - R_B)}{pR_A + qR_B}$$

2. In the following way: we ask how a change in $p(\Delta p)$ will affect $\overline{R}$. Adding or removing individuals with allele A will change $\overline{R}$ by the frequency change in $A(\Delta p)$ times the fitness of A individuals (R_A). Because $\Delta q = -\Delta p$, the effect of a change in p will also change $\overline{R}$ by $-\Delta pR_B$. Thus

$$\Delta \overline{R} = \Delta pR_A - \Delta pR_B$$

$$\Delta \overline{R} = \Delta p(R_A - R_B)$$

and

$$\frac{\Delta \overline{R}}{\Delta p} = R_A - R_B$$

3. If we ignore the term $q\mu_B$, the equilibrium equation is

$$pq[p(r_{AA} - r_{AB}) + q(r_{AB} - r_{BB})] = p\mu_A$$

The ps cancel to give

$$q[p(r_{AA} - r_{AB}) + q(r_{AB} - r_{BB})] = \mu_A \qquad (28\text{-}34)$$

When B is recessive, $r_{AB} = r_{AA}$ and equation 28-34 becomes

$$q^2(r_{AB} - r_{BB}) = \mu_A$$

When B is intermediate in fitness $r_{AA} - r_{AB} = r_{AB} - r_{BB} = \frac{1}{2}(r_{AA} - r_{BB})$, so equation 28-34 becomes

$$\frac{q}{2}(p + q)(r_{AA} - r_{BB}) = \mu_A$$

The term $p + q = 1$, so $q(r_{AA} - r_{BB}) = 2\mu_A$. When B is dominant, $r_{AB} = r_{BB}$, and equation 28-34 becomes

$$qp(r_{AA} - r_{BB}) = \mu_A$$

and we again assume that $p = 1$.

4. To solve this equation when one allele is either lethal or renders the organism sterile, r must be replaced by λ or R_0 as a measure of fitness. An instantaneous rate of change cannot be defined for an allele that prevents reproduction, because the life span of such organisms, from an evolutionary standpoint, is zero.

29

Fitness and Evolution in
Natural Populations

Genetic variation and natural selection are intimately tied to the ecology of populations in two ways. First, the genetic characteristics of a population, and the changes in these characteristics, reflect the genotype-environment interaction, the outcome of which is the fitness of the individual. Second, the genetic mechanism and the stored genetic variation of the population determine the ability of a population to respond to change in the environment. We experience the impact of evolutionary response within our own lifetimes in the adaptation of insects and microorganisms to pesticides and antibiotics, and in the adaptation of plants to the heavy-metal contaminated soils of mine dumps. In such cases, the favored trait often has a selective advantage of 0.5 or more (Antonovics 1971) and has a simple genetic basis of inheritance.

Unless the environment changes very rapidly, as it often does in the wake of human meddling, genes with extremely deleterious effects on fitness normally would be rare, having long since been weeded out of the population. The environment is thought to change slowly relative to generation time and the evolutionary potential of populations. Yet, most phylogenetic lines that have ever begun to wend their way up the evolutionary tree of life are now extinct. Somewhere along the line their evolutionary potential failed them, and they were replaced by more successful forms.

In this chapter we shall examine what little we know about rates of evolutionary change in natural populations in an effort to understand constraints on evolution. We may look to insufficient genetic variation and correlated adaptations of other traits for major constraints on rate of evolution. But I must emphasize that the gap between our understanding of short-term genetic change at individual gene loci and the long-term evolution of the phenotype is formidable and beyond crossing at present.

Rate of gene replacement depends on selective value and initial allele frequency.

One of the earliest attempts to relate population genetics to the more general course of evolutionary change was made by Haldane (1957), who, beginning with the equations for the rate of allele substitution (see Chapter 28), derived expressions for the length of time (t) required for allele replacement (see Note 1 at the end of this chapter). In the haploid case

$$t = \frac{1}{r_A - r_B} \left[\log_e \left(\frac{q(0)}{p(0)} \cdot \frac{p(t)}{q(t)} \right) \right]$$ (29-1)

where $p(0)$ is the initial frequency and $p(t)$ the final frequency of allele A, and $q(0)$ and $q(t)$ are the initial and final frequencies of allele B. The time required for the evolutionary substitution of an allele is inversely related to the difference in fitness between the two alleles; with greater fitness, less time is required for substitution. In a diploid population, the time required for the incorporation of an advantageous, but recessive, allele is

$$t = \frac{1}{r_{AA} - r_{BB}} \left[\log_e \left(\frac{q(0)}{p(0)} \cdot \frac{p(t)}{q(t)} \right) + \frac{1}{p(0)} - \frac{1}{p(t)} \right]$$ (29-2)

and for an advantageous, dominant allele, it is

$$t = \frac{1}{r_{AA} - r_{BB}} \left[\log_e \left(\frac{q(0)}{p(0)} \cdot \frac{p(t)}{q(t)} \right) + \frac{1}{q(t)} - \frac{1}{q(0)} \right]$$ (29-3)

Allele substitution in the diploid case takes longer than it does in the haploid case, by the term $[1/p(0) - 1/p(t)]$ for a recessive allele and by $[1/q(t) - 1/q(0)]$ for a dominant allele.

For a sample calculation, consider a mutant recessive allele (A) whose fitness (r) when homozygous is 0.1 per year in excess of the wild type allele (B) in the population (thus $r_{AA} - r_{BB} = 0.1$). If A initially has a frequency $(p(0))$ of 0.01, how long will it take to increase to a frequency of 99 per cent of the gene pool? Substituting 0.1 for $r_{AA} - r_{BB}$, 0.01 for $p(0)$, and 0.99 for $p(t)$ in equation 29-2, we find that the gene replacement will take 1,082 years.

Drawing upon a real example—the replacement of the typical form of the peppered moth by the melanistic form *carbonaria* in polluted English woods—we can compare the calculated number of years with the observed replacement period, about a century in most areas. We must use a relative measure of fitness (W) rather than r because Kettlewell recorded differential survival only for a brief part of the adult portion of the life cycle (page 93). In the polluted woods, typical and melanistic moths were marked and released: 34.1 per cent of the melanistic individuals, but only 16.0 per cent of the light, or typical, individuals were recaptured. Thus the fitness (W) of the typical form was 0.160/0.341, or 0.47 times that of the melanic form (see Note 2 at the end of this chapter), and the difference in fitness between the two alleles would be $1 - 0.47$, or 0.53, if we arbitrarily assigned the melanistic allele a fitness of 1. Because *carbonaria* is dominant, we must use equation 29-3: the calculated time required for *carbonaria* to increase from $p = 0.05$ to 0.95 is approximately 47 generations, and from 0.01 to 0.99, about 204 generations. These estimates seem to correspond closely to the observed time for the substitution of the allele for light color by *carbonaria*.

Records of trapped red and silver foxes kept by the Moravian mission posts in Labrador for a hundred years, 1834-1933, show that the percentage of silver foxes in the catch declined from about 0.15 to 0.05 (Elton 1942). What would be the selective force required to produce such a change? If we assumed that silver coat is a homozygous recessive phenotype with genotype frequency p^2, then p would have decreased from 0.39 to 0.22. Substituting these values into equation 29-2, we obtain the expression 100 years = $3.5/(r_{AA} - r_{BB})$. The difference between the fitnesses of red and silver foxes apparently must have been about $3\frac{1}{2}$ per cent per year ($r_{AA} - r_{BB} = 3.5/100$). It does not seem unreasonable that human hunting practices, combined with the greater demand for the silver fox, could cause an annual mortality for silver foxes 3 per cent in excess of the mortality of red foxes. The actual fitnesses in this example are difficult to determine because they depend on the slyness of the foxes, the patience of the hunters, and fluctuations in the fur market.

How do interactions among alleles influence the rate of allele substitution? When rare, the frequency of a dominant allele changes much faster than that of a recessive allele because the dominant is exposed to selection as a heterozygote. The rate of change is faster for a recessive allele when the allele frequency is above 0.5 because only homozygotes, bearing two of the alleles, are selected. An allele substitution involving a symmetrical change in frequency from $p(0)$ to $1 - p(0)$ requires the same amount of time, whether the gene is dominant or recessive. Suppose a deleterious gene were maintained at some frequency $p(0)$ by the balance between mutation and selection. If the coefficient of selection were suddenly reversed, that is, if the environment changed so an allele became advantageous to the same degree that it was previously deleterious, would the rate of substitution be more rapid for a dominant than for a recessive allele? At equilibrium between mutation and selection (page 460), recessive alleles are maintained at a higher frequency in the population ($\sqrt{(\mu_A/(r_{AA} - r_{BB}))}$) than dominant alleles ($\mu_A/(r_{AA} - r_{BB})$). Therefore, although the rate of increase of a rare recessive is slower than that of a rare dominant, the recessive allele has an initial head start. Because the initial and final equilibrium frequencies of alleles maintained by mutation pressure depend on the degree of dominance of the allele, the substitution of dominant and recessive alleles takes exactly the same time.

Do selective deaths place an upper limit on the rate of evolution?

For a given environment, only one genotype is most fit. Genetic variation in a population results in genotypes with reduced fitness that are continually being weeded out of the population. Thus, one effect of genetic variation is to produce a *genetic load* on the population, which can be measured in terms of the selective deaths sustained by the population (Crow 1970). A selective death is one that is wholly attributed to the deleterious effects of a genotype, or as Maynard-Smith (1968b) put it, "... deaths of individuals who would have survived had they had the optimum genotype...." In this sense, "death" also includes reduced fecundity—individuals that "die" before they are potentially born.

The rate of selective death in a population is the proportion of the population with a deleterious genotype multiplied by the reduction in fitness caused by that genotype. Thus, in a haploid population, the rate of selective death (D) is

$$D = q(r_A - r_B) \qquad (29\text{-}4)$$

when the A allele is the more fit. For the diploid case

$$D = 2pq(r_{AA} - r_{AB}) + q^2(r_{AA} - r_{BB}) \qquad (29\text{-}5)$$

when AA is the most fit genotype. If A is dominant to B, $r_{AA} - r_{AB} = 0$ and

$$D = q^2(r_{AA} - r_{BB}) \qquad (29\text{-}6)$$

In 1957, J. B. S. Haldane suggested that the total number of selective deaths sustained during an allele replacement could be used as a measure of a total cost of evolution. Haldane found that this cost is approximately

$$C = -\log_e p(0) \qquad (29\text{-}7)$$

(see Note 3 at the end of this chapter) where $p(0)$ is the initial frequency of the allele at the beginning of gene substitution; $p(t)$ is assumed to be close to 1 and thus drops out of the equation ($\log_e 1 = 0$). The cost of evolution is independent of the strength of selection, and it is identical for haploids and fully dominant alleles in diploid populations. For an allele that produces a heterozygote with intermediate fitness, the cost of evolution is twice as high as for a dominant allele. This result is not immediately obvious, but it may be understood intuitively if one remembers that selection on heterozygotes removes only one allele for each selective death. So the rate of allele substitution is only half as fast as it is for a dominant allele, and twice the number of genetic deaths are sustained. Most of the selective deaths in a population occur when a beneficial allele is rare and most of the population has the less fit genotype. Because the time required for the incorporation of a recessive allele can be very long, the number of genetic deaths accumulated during the course of allele replacement is correspondingly high, approximately $1/p(0)$ rather than $-\log_e p(0)$.

There must be a limit to the number of selective deaths that a population can withstand per unit of time without becoming extinct, which, in turn, must set an upper limit to the rate of evolution of a population. For example, the incorporation of an allele that has an initial frequency of 0.01 requires 4.5, 9.2, or more than 100 selective deaths per individual in a diploid population, depending on whether the allele is dominant, intermediate, or recessive. Haldane estimated that most gene substitutions require about thirty selective deaths per individual, although he had to make many assumptions about the expression of alleles and their initial frequencies to arrive at this figure. Haldane reasoned that if a population could sustain a tenth of a selective death per individual each year, it could evolve at the rate of one allele substitution every 300 years.

Haldane's calculations should be regarded only as rough estimates, and his figure for the upper limit to the rate of allele substitution may be entirely too low (Moran 1970, O'Donald 1969, Maynard-Smith 1968b, Sved 1968). The opposing view to Haldane's is that allele replacement imposes no genetic load because selection does not cause deaths over and above other nonselective deaths in a population. All populations produce more offspring than are required to maintain the population at a constant level. The surplus is pared down through inclement weather, predation, and starvation. If population size were limited by competition for resources, the death of an individual, whether as a result of accident or genetic deficiency, would only reduce the number of individuals that must be eliminated by other causes; it would not add to the burden of death sustained by the population. In populations that produce many eggs or newborn, perhaps thousands or even millions per individual, there is room for many genetic deaths. Some populations could conceivably sustain hundreds or even thousands of selective deaths per individual each generation, if they were caused by alleles expressed at relatively early stages of development.

A population's ability to respond by evolutionary change would be greater than Haldane's estimate, if the effects of several deleterious alleles multiplied to determine the fitness of the individual. Haldane assumed that the replacement of alleles at two gene loci would require twice the number of selective deaths as a single substitution. But if the fitness of an individual with both deleterious alleles were less than that of an individual with either, the deleterious alleles would be preferentially removed in pairs, thereby reducing the total selective deaths required for both substitutions (see Maynard-Smith 1968b). Crow (1970) presented a new set of calculations, based on the variance in fitness within a population rather than the number of selective deaths, which suggests a lower cost for gene substitution that is less sensitive to initial allele frequency, $p(0)$.

We are led to wonder whether the "cost of natural selection" has any real bearing on the rate of evolution. Regardless of the absolute number of selective deaths required per allele substitution, animals such as elephants, which have low reproductive rates, seemingly could sustain fewer genetic deaths than most fish, insects, or bacteria. Yet, paleontological evidence shows that large mammals have evolved every bit as rapidly as small invertebrates (see Chapter 26). One might argue, therefore, that the rate of evolution is ruled by some master other than the price of selective death.

The cost of natural selection actually is a misnomer, for natural selection continually acts to increase the fitness of a population. As Brues (1964) pointed out, the cost to a population would be much greater if it did not evolve. A better term for the selective deaths that occur during allele replacement might be the "cost of environmental change." Allele substitutions are, after all, a direct consequence of environmental change, and the level of selective deaths sustained by a population depends upon the rate of environmental change (Maynard-Smith 1976). The more drastic the change, the greater the difference in fitness between an allele and the one that will eventually replace it. The number of selective deaths that a population can sustain may provide a measure of its ability to respond to environmental change. Productive populations presumably can withstand rapid environmental change. We must not forget, however, that the greater regulatory capacities of large organisms tend to reduce the influence of environmental change on fitness. The cost of environmental change can also be spread over a long period through phenotypic flexibility and subsequent genetic assimilation.

Artificial selection of quantitative traits illustrates some properties of evolution in natural populations.

Artificial selection usually is accomplished with large fitness differences applied over short periods, and in this way, it is probably unlike the less intense selection that occurs in natural populations. The results of artificial selection nonetheless suggest some of the kinds of evolutionary response that one might observe under natural conditions. It should be pointed out at first that responses to artificial selection are generally repeatable, even with different genetic lines (Figure 29-1), although one occasionally runs across a particularly stubborn or genetically impressionable population.

As we saw in the last chapter, evolutionary response (R) is equal to the phenotypic selection differential (S) times the heritability. The stronger the selection, the stronger the response. R and S are measured in terms of distance from the population mean. If the measurements of individuals in the population are normally distributed, R and S can be expressed in terms of the standard deviation of measurements from the mean (see Figure 29-2). Each value expressed in standard deviation units corresponds to a particular percentage of the population above and below that value. Zero standard deviation units (SD) corresponds to the population mean; half the individuals lie above the mean and half lie below; 31 per cent exceed 0.5 SD; 16 per cent exceed 1.0 SD; 7 per cent exceed 1.5 SD; only 2.3 per cent exceed 2.0 SD. Hence, the more intense the selection and the larger the value of S, the smaller the number of individuals selected and the smaller the number of

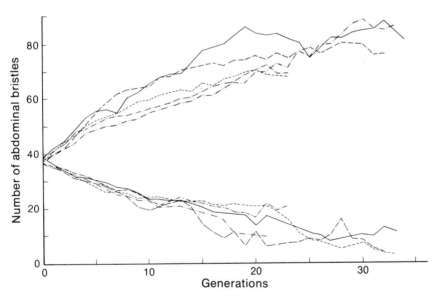

Figure 29-1   Selection for abdominal bristle number in *Drosophila melanogaster* with five replicates each for high and low selected lines (after Clayton *et al.* 1957).

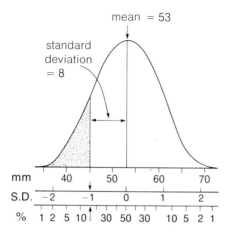

mean = 53

standard deviation = 8

mm   40   50   60   70
S.D. −2   −1   0   1   2
%   1 2 5 10   30 50 30   10 5 2 1

Figure 29-2   A schematic dia-gram of variation in a metric trait showing the relationship between the measurement scale, the standard deviation, and the propor-tion of the population with phenotypic values more extreme than a particular value. For exam-ple, 16 per cent of the population have phenotypic values greater than one standard deviation above or below the mean (shaded por-tion).

resulting progeny for the next generation of selection. If selection were too strong, the population would dwindle, eventually to extinction. Even in arti-ficial selection, the strength of selection that can be applied is limited by the size of the stock population and the reproductive rate of the selected indi-viduals.

The relationship between selection intensity (per cent of individuals selected), phenotypic selection differential (S), and response (R) can be illustrated by a program of selection for rate of egg laying in the flour beetle *Tribolium castaneum* (Ruano *et al.* 1975). In the stock population, the mean number of eggs laid from day seven to eleven after adult emergence (the phenotypic trait investigated) had a mean of 19.03, a standard deviation of 11.8, and a heritability (h^2) of 0.30. One control line and five lines with different levels of selection (ranging from 50 to 5 per cent) were established (Table 29-1). Knowing the variability, heritability, and selection differential,

TABLE 29-1

Selection procedure in six lines of *Tribolium castaneum* under selection for fecundity (after Ruano *et al.* 1975).

Line	Number of families scored per generation	Number of females scored per family	Total scored	Total selected	Selection intensity (per cent removed)	Selection differ-ential (S)*
A	10	20	200	10	95	2.0
B	20	10	200	20	90	1.8
C	40	5	200	40	80	1.4
D	66	3	198	66	67	1.1
E	100	2	200	100	50	0.8
F	200	1	200	200	0	0.0

* Standard deviation units.

one can estimate the initial response of the population to selection. For example, in the C line a selection intensity of 20 per cent corresponds to a selection differential of 1.4 SD (Falconer 1960: 193), or 16.5 eggs (1.4 × 11.8). With a heritability of 0.30, the response to selection should be about 5.0 eggs per generation ($R = h^2S = 0.30 \times 16.5$). The observed response fell somewhat short of this prediction (about 3.0 eggs per generation), probably because the estimate of heritability included components like maternal and dominance effects as well as additive genetic variation. But the beetle population did behave as predicted in that the rate of response varied in direct proportion to the intensity of selection (Figure 29-3).

A common result of selection is asymmetry of response. That is, selection works faster in one direction than the other. This type of result is illustrated by Mather and Harrison's (1949) experiments, discussed at the beginning of this book, in which large and small numbers of abdominal bristles were selected in *Drosophila melanogaster* (Figure 8-4). The high-selected line increased much more rapidly than the low-selected line decreased. Falconer (1960: 212) suggested many causes for asymmetry of response, including influence of correlated traits, genetic dominance of alleles, and artifacts caused by the scale of measurement used (*e.g.,* whether arithmetic or logarithmic).

With continued selection pressure on experimental populations, the response to selection eventually stops, as seen in Figures 29-1 and 29-3. The slowed response is the result of two factors: the erosion of genetic variation

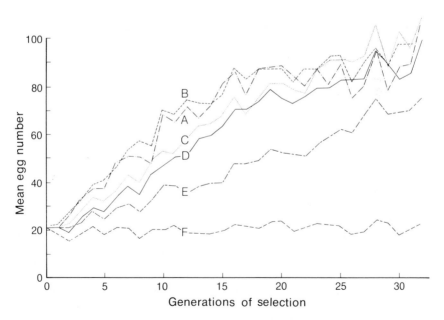

Figure 29-3   Change in rate of egg laying in *Tribolium castaneum* lines exposed to different levels of selection (from Ruano *et al.* 1975).

by selection and the application of opposing selection by correlated changes in other traits. Selection works only when the population has genetic variability. When all unfit alleles are removed by selection, evolution ceases until new mutations or gene combinations appear. Under intense selection, populations also lose variability through inbreeding because population size is reduced and all progeny are produced by relatively few matings. In the Ruano et al. (1975) experiment on Tribolium, the level of inbreeding in the A line (5 per cent of the females selected each generation) reached 0.57 (out of a possible 1.0). In lines with less intense selection, inbreeding coefficients were correspondingly smaller.

Loss of genetic variability under continued selection is reflected in reduced heritability because the additive genetic component of variation is reduced, while environmental components are generally unchanged. This cannot, however, explain the leveling off of the response in all cases, for if all genetic variation were removed, further selection in either forward or reverse direction would not be successful. Yet reverse selection, applied long after the selection response has reached a plateau, often will produce an immediate response. Furthermore, when selection is merely relaxed, so that selection coefficients are zero, a selected trait will sometimes return toward the preselection measurement apparently by itself. The most reasonable explanation for these results is that selection applied to one trait causes changes in other traits that affect the fitness of the organism. Increase in rate of egg laying, for example, may cause physiological or morphological changes that reduce viability and thus oppose the artificial selection regime.

Correlated responses to selection constrain evolutionary response.

The effects of selection are not limited to the character selected. The parts of organisms are highly integrated through both development and function. Few traits, particularly among those involving size or rate of growth and production, are independent of all others. Animal and plant breeders have estimated genetic correlations between traits in many domestic species (Falconer 1960: 316). For example, in poultry the genetic correlation between body weight and egg weight is 0.50 and between body weight and egg production it is −0.16. In other words, one cannot apply selection to body weight without also obtaining a relatively rapid increase in egg size and a slower, but steady, decrease in rate of laying. In mice, selection for body weight will also result in mice with longer tails, among other things; in pigs, selection for a longer body tends to reduce backfat thickness as well.

When Dawson (1966) selected for fast and slow larval development in Tribolium castaneum and T. confusum, he found a number of correlated responses that decreased fitness regardless of the direction of selection. Selection for rapid development resulted in decreased size and, in T. castaneum, decreased larval survival and an increase in the incidence in adult abnormalities. Selection for slow development resulted in increased adult weight, increased frequency of adult abnormalities, and decreased fertility of females.

Dawson assessed the fitness of each of the selected strains of flour beetles by placing them in competition with unselected stocks of the other species (Dawson 1967). Both the fast and slow selected lines suffered decreased fitness during early stages of selection, which suggests that development rate is optimized with respect to its effect on fitness. As Dawson continued his selection experiments, the fitness of one of the selected lines began to improve, apparently because of increased cannibalism by the selected larvae. Evolution always has surprises for us.

Results similar to Dawson's were obtained by Verghese and Nordskog (1968), who examined correlated responses in reproductive fitness in selected lines of chickens. Selection for both increase and decrease in body weight and egg weight caused a decline in reproductive fitness, measured as the product of rate of egg production, hatch rate, and survival of chicks to nine months. Regardless of the character or direction of selection, fitness in the selected lines varied from 54 to 85 per cent of the nonselected line. These and other experiments re-emphasize the fact that natural populations are well balanced genetically, and their adaptations are both well tuned to the environment and finely adjusted to each other.

Rate of evolution is slow in natural populations.

The failure of artificial selection to produce general improvement in fitness, other than in the narrowly prescribed conditions of the experimental environment, illustrates the stability of the gene pool and the general conservatism of evolution. Yet, the evidence of the fossil record is for change, both fast and slow by geologic time standards, perhaps not so fast measured against a human life span. In the natural world, all characters, not just the one or the few traits selected by the plant or animal breeder, contribute to fitness. To avoid the correlated responses that usually reduce fitness, organisms may require structural or genetic reorganization. Sometimes two genes must be separated by chromosome inversion or translocation before one can undergo allele substitution.

How fast do organisms evolve? Unfortunately, we are left with a scanty record of evolutionary change. One line of evidence comes from amino acid sequences of homologous proteins in different species. When the age of divergent evolutionary lines is known, the average rate of allele substitution can be calculated for the summed periods of independent evolution of the two lines. Uzzell and Corbin (1972) used this principle to calculate evolutionary rates for the alpha and beta chains of the hemoglobin molecule in mammals. This rate was found to vary between 1.0 and 1.7 allele substitutions per amino-acid position per billion years for the alpha chain, and 0.5 and 2.2 substitutions per position per billion years in the beta chain. Multiplied by 141 and 146 amino acids in the alpha and beta chains, these rates fall in the range 0.07 to 0.22 changes per million years in each chain, and perhaps twice that rate for the hemoglobin molecule as a whole. But hardly fast enough to turn one's head. One could argue that the hemoglobin molecule had long since been perfected and that the homeostatic capacities of mammals have regulated the blood environment closely for the last 75 million years.

Kurtén (1959) got closer to the heart of evolution when he calculated rates of change in metric characters of mammals through fossil lineages. Kurtén used the *darwin* as his unit of change, as suggested by Haldane (1949). One darwin is equal to a change of 0.1 per cent (1/1,000) per 1,000 years. When the darwin is treated as an exponential rate, it is equivalent to increase by a factor of e (2.72) per million years. The distribution of observed rates of change in many lineages is shown in Figure 29-4. Most of the data appear to fall within three ranges of evolutionary rates: approximately 0.01 to 0.1, 0.2 to 2, and 5 to 20 darwins, although additional data may fall in between these peaks. The slower rates of evolution, averaging about 0.03 darwins, characterize long-term sequences from the Tertiary Period and probably represent the normally sustainable rates of evolutionary change. Faster rates (0.5 darwins) characterized the bear and human evolutionary lines during the ice ages, but these rates were and still are being greatly exceeded (10–20 darwins) by mammal lines during the 10,000 years since the last glaciation. Evolutionary rates obtained in short-term artificial selection experiments may run into the millions of darwins.

Clearly, the rate of evolution sustained depends on the duration of the selection regime. Rapid evolutionary change apparently is not maintained for long periods, either because of inherent constraints on the rate of evolution (Maynard-Smith 1976) or because rates of change in the environment average out to lower values over longer periods. Kurtén was unable to find evidence that rapid evolution eroded genetic variability; phenotypic variation at the end of phases of rapid evolution was no less than at the beginning. He suggested instead that very rapid evolution resulted from phases of rapid environmental change, such as occurred after the retreat of glaciers, when a

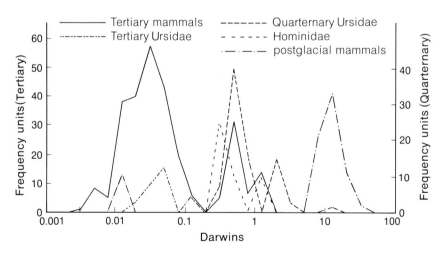

Figure 29-4   Frequency of rates of phenotypic change, measured in darwins, for a variety of phylogenetic lines of mammals (from Kurtén 1959).

population colonizes a new region, or when an area is invaded by new competitors or predators. At present, however, we must leave the problem of evolutionary rate largely unresolved.

This chapter has provided a glimpse at evolution in natural populations and in some domesticated and laboratory stocks. We left the world of the theoretician behind, carrying little more than a sharpened intuition into the more empirical world of phenotypic change. Such is the gap between population genetics theory and evolutionary change in nature. In the contrived environment of the laboratory or farm, artificial selection can accomplish rapid change in most phenotypic traits, but correlated responses in other characters and inherent constraints of the genetic architecture ultimately set limits to the results of selection. In fact, most selected lines initially lose fitness in the environment of the unselected stock. Evolution in the laboratory can be made to move thousands of times faster than the most rapid change observed in nature. The slower pace of evolution in nature apparently allows ample time for replenishment of genetic variability and for reorganization of the genome, and thereby permits continual progress. Perhaps the ultimate determinant of evolutionary change in nature is the rate of change in the environment itself.

Notes

1. The derivation of these expressions requires calculus; only a brief outline will be given here. To find the time (t) required for a gene substitution from allele frequency $p(0)$ to $p(t)$, we first rearrange the basic equation for the rate of increase of an allele

$$\frac{dp}{dt} = pq(r_A - r_B) \tag{29-8}$$

in the form

$$dt = \frac{dp}{p(1 - p)(r_A - r_B)} \tag{29-9}$$

and integrate both sides of the equation. The result for the haploid case is

$$t = \frac{1}{r_A - r_B} \int_{p=p(0)}^{p(t)} \frac{dp}{p(1 - p)}$$

$$= -\frac{1}{r_A - r_B} \left[\log_e \left(\frac{1 - p}{p} \right) \right]_{p=p(0)}^{p(t)}$$

$$= \frac{1}{r_A - r_B} \left[\log_e \left(\frac{1 - p(0)}{p(0)} \right) - \log_e \left(\frac{1 - p(t)}{p(t)} \right) \right] \tag{29-10}$$

which may be rearranged to give equation 29-1. Derivation of t for the diploid case is similar.

2. There are several problems with calculating fitness in this way. First, the alleles for wing color in the adults may also affect fitness during other stages of the life cycle. Second, measurement of the relative fitnesses of the two genotypes depends on the length of the mark-release-recapture experiment. Suppose, for example, that in a polluted woods, the mortality rate of the melanistic form is 20 per cent per day and the mortality rate of the typical form is 50 per cent per day. If one were to measure the difference in fitness between the two genotypes by releasing equal numbers into the woods and recapturing them one day later, one would find 80 per cent of the melanistic moths and 50 per cent of the typical moths still alive. Hence, the calculated fitness of typicals would be 0.625 (0.5/0.8) times that of melanistic moths. After two days, 64 per cent of the melanistic moths would be recaptured but only 25 per cent of the typicals, indicating a fitness for typicals of only 0.39 that of the *carbonaria* allele. After five days, the relative fitness of the typical phenotype, judged on this basis, would be only 10 per cent of *carbonaria*. As the length of the experiment increases, the estimate of fitness becomes more realistic.

3. Haldane derived the cost of evolution (C) as follows: The increase in the total cost of evolution during an interval of time is equal to the selective deaths sustained during that interval so, for the haploid case,

$$\frac{dC}{dt} = q(r_A - r_B)$$

or

$$dC = q(r_A - r_B) \, dt \tag{29-11}$$

We know that by rearranging the equation for selection, $dp/dt = pq(r_A - r_B)$, we can obtain the expression

$$dt = \frac{dp}{pq(r_A - r_B)} \tag{29-12}$$

By substituting dt in equation (29-11), we obtain

$$dC = \frac{1}{p} \, dp \tag{29-13}$$

which may be solved for C by integrating both sides

$$C = \int_{p(0)}^{p(t)} \frac{1}{p} \, dp$$

$$= \log_e p(t) - \log_e p(0) \tag{29-14}$$

Because we are interested in complete gene substitution, $p(t)$ will be near 1, and thus $\log_e p(t)$ is near zero and can be omitted. Hence, the total number of selective deaths per individual for the substitution of a gene in a haploid population is

$$C = -\log_e p(0) \tag{29-15}$$

In the diploid case, C is derived in a similar manner from the appropriate equations for D (29-5) and dp/dt (28-22)

$$C = \int_{p(0)}^{p(t)} \left[\frac{2pq(r_{AA} - r_{AB}) + q^2(r_{AA} - r_{BB})}{pq[p(r_{AA} - r_{AB}) + q(r_{AB} - r_{BB})]} \right] \qquad (29\text{-}16)$$

If allele A is fully dominant, $r_{AA} - r_{AB} = 0$ and equation 29-16 simplifies to

$$C = \int_{p(0)}^{p(t)} \frac{1}{p} \, dp \qquad (29\text{-}17)$$

which is identical to the haploid case (equation 29-14). If allele A is exactly intermediate in fitness, then all the fitness differences involving heterozygotes become $(r_{AA} - r_{BB})/2$ and equation (29-16) simplifies to

$$C = \int_{p(0)}^{p(t)} \frac{2}{p} \, dp$$

which, assuming $p(t)$ is close to 1,

$$= -2 \log_e p(0) \qquad (29\text{-}18)$$

If allele A is recessive, $r_{AA} - r_{BB} = 0(r_{AB} = r_{BB})$ and equation 29-16 simplifies to

$$C = \int_{p(0)}^{p(t)} \frac{(p + 1)}{p^2} \, dp$$

$$= \left[\frac{1}{p} + \log_e p \right]_{p(0)}^{p(t)} \qquad (29\text{-}19)$$

which, assuming $p(t)$ is close to 1,

$$= -\log_e p(0) + \frac{1}{p(0)} \qquad (29\text{-}20)$$

30

The Evolution of
Population Characteristics

Most adaptations are apparent in the individual. Form, physiology, and behavior are expressed in the phenotype, and the purpose of such adaptations is clearly to promote the survival and productivity of the organism. Yet the selective value of some adaptations is not always obvious in the individual. We have seen such a case in genetic heterosis where two alleles are maintained in a balanced polymorphism owing to the superior fitness of the heterozygote, but the selective advantage of this arrangement is certainly not obvious in either homozygote genotype and would not be guessed easily from the heterozygote alone. The selective basis for polymorphism in the sickle-cell trait (page 457) or in the mimicry complex of the African swallow-tailed butterfly *Papilio dardanus* (page 362) can be understood only in terms of the variety of phenotypes in the population as a whole. I do not want to imply that these are population characteristics, only that the selective value of an allele or a genotype depends upon its interaction with others in the population, hence on the frequency of alternative traits.

Many adaptations of the life history of an organism can be understood only by considering many individuals together. The sex ratio of a population can be determined only by sampling many individuals, yet we shall see below that sex ratio is a trait whose selective basis is the fitness of the individual. Sex itself is an adaptation that obviously requires several phenotypes (except in plants) in the population, but its adaptive basis is not understood at all. Probability of survival is a statistical measure of a population; each individual only dies once, at least according to conventional thinking. Still, we shall see that the survival curve of a population is an expression of the adaptations of individuals.

This chapter is devoted to a discussion of several topics in evolutionary biology—sex, the sex ratio, reproductive effort, and age-dependent effects of selection—that not only are major problems of adaptation, but also will serve to sharpen our understanding of population and genetic processes. In our discussion of population genetics, we derived mathematical models to predict the rate of change in allele frequency and conditions for genetic equilibrium. Studies of population growth and demography are concerned with

482

predicting the age structure of a population and the trajectory of its growth. Recently, population genetics and demography have been melded into a new endeavor concerned with the evolution of life-history adaptations, many of which are evident only on the population level (Cole 1954). In most of these studies, the genetic basis for evolutionary change is assumed to be sufficient to produce any adaptation that would increase fitness. The object of such studies is to express adaptations in demographic terms and to formulate models that optimize the relationship between adaptations by conferring on the individual the highest fitness possible. Hence the study of life-history patterns and population phenomena is a study of adaptation in which the effects of correlated traits on fitness are of central importance.

What use is sex?

That this most fundamental attribute of almost all living systems has defied compelling explanation shows how poorly we understand natural systems (Maynard-Smith 1970, Baker and Parker 1973, Cavalier-Smith 1975). The adaptive basis for sex has been discussed widely in recent years; the theories that have been advanced fall into three categories: (a) sex generates genetic variability, which enables populations to respond to environmental change (Crow and Kimura 1965, 1969, Maynard-Smith 1968, 1971, Stanley 1975); (b) sex creates variability within a brood of progeny and thereby increases the probability of offspring with superior fitness in unpredictable environments (Williams and Mitton 1973, Williams 1975); (c) sex makes possible independent assortment of chromosomes and recombination, both of which increase the evolutionary independence of gene loci (Crow and Kimura 1965).

Although sex may confer genetic advantage at the individual or population level, it also carries a cost. In a population in which the number of offspring is limited by the fecundity of females, males are a liability to individual fitness. Because the male parent contributes one-half of each individual's genes in a sexual population, the female's genetic contribution to her own progeny is only one-half what it would be if the young were produced asexually. This is often referred to as the cost of meiosis, and it must be balanced by whatever advantage accrues from sexual reproduction. When males help to rear offspring, as they do in many birds and mammals, they can increase the fecundity of the female and thereby reduce this cost of sexual reproduction. Reproduction in birds and mammals is unvaryingly sexual, although parthenogenetic strains have been isolated in poultry (Sarvella 1970). For most organisms, however, males play little or no role in rearing the young. Incidents of asexual reproduction are more frequent in lower vertebrates (see page 359), but are sporadically distributed among genera. Obligatory asexual reproduction is most widespread among simple organisms, including the lower plants, fungi, and unicellular animals and microorganisms (Ainsworth and Sussman 1969, Grell 1967, Hayes 1964).

None of the postulated advantages to sex is truly convincing, because the models on which advantage is based are either unrealistic or of limited application to nature. Sex is so widespread that its evolutionary basis must be very broad and pervasive, yet the absence of sex from many groups shows that it is not inherently necessary to life processes. If the primary function of sex is to generate variability or increase the independence of genetic loci, why do unvarying phylogenetic lines in seemingly constant environments retain sexual reproduction? Why is sexual reproduction least frequent in simple organisms, which apparently have the lowest rates of mutation per generation (see Table 22-1)? The phenomenon of sex will undoubtedly test our knowledge of evolution, ecology, and our imagination for years to come.

The 1:1 sex ratio maximizes the fitness of the individual in outcrossing populations.

Setting aside the problem of sex itself, let us turn our attention to the ratio of the sexes in natural populations. Various theories have been suggested to explain the evolutionary advantage of the 1:1 sex ratio, characteristic of most populations. Kalmus and Smith (1960) argued that an equal sex ratio maximizes an individual's chances of finding a mate. True, the probability that a *random* encounter between two individuals will involve a male and a female is greatest when the sex ratio is 1:1, but this could be important only if there were a limited number of chances of meeting another individual. In most populations individuals meet frequently, and, therefore, mating and reproduction can occur efficiently regardless of sex ratio.

Edwards (1960) and Kolman (1960) suggested that the most valid approach to the sex ratio problem is that of Fisher (1930), who argued that an equal probability of producing male or female offspring maximized the fitness of the individual, not the population. Imagine a population of 25 males and 25 females, in which each female gives birth to two offspring, a population total of 50. Each male contributes his genes to an average of two progeny, as does each female, hence the fitnesses of the two sexes are equal. Now suppose that in another population, males are less common, say 10 males to 25 females. While each of the females still contributes her genes to an average of two progeny, each of the males contributes his genes to five (50 progeny/10 males). In this situation, an individual's genes are transmitted to future generations at a higher rate if his or her offspring are mostly males. Therefore, any genetic change that alters the sex ratio toward a predominance of males will be favored. If males became more common than females, the female sex would be the more efficient avenue of gene transmission, and the proportion of females would be increased by natural selection until the 1:1 sex ratio was restored. The rare sex is favored by evolution because it competes less with individuals of the same sex during reproduction. This frequency-dependence makes the even sex ratio a stable genetic equilibrium.

Fisher's model predicts that the sex ratio should be 1:1 at the age at which parental care ceases, but not necessarily during other stages of the life cycle. In fact, if males and females had different probabilities of survival, the initial sex ratio at conception, or birth, would have to be sufficiently biased to ensure that the numbers of males and females in the population were roughly equal by the age at which the young became independent. In man, the sex ratio is about 114 males to 100 females at conception. Differential mortality of embryos and children shifts the sex ratio to 106 males to 100 females at birth and to 100 males to 100 females by ten years of age. Because males suffer higher mortality than females throughout most of their lives, females predominate in older age groups.

Although an equal sex ratio maximizes individual fitness, it may do so at the expense of the overall fitness of a population that is growing rapidly. When the only function of the male is to contribute gametes to the zygote, males are a costly ecological and demographic luxury for a population, to the extent that they utilize resources that would otherwise be used more productively by females. The negative impact of males on population growth is sometimes lessened by skewed sex ratios favoring females, as we shall see below. In addition, where the male remains with his mate but does not participate in reproduction beyond fertilizing the eggs, as in most spiders, the size of the male is sometimes greatly reduced, perhaps to minimize competition for resources with his mate and progeny.

Meiotic drive can unbalance the sex ratio.

In higher organisms, sexual dimorphism is usually based upon differences within a single pair of morphologically and genetically distinguishable chromosomes, the X and Y chromosomes. In most species, an individual that has paired X chromosomes is a female, and the heterogametic sex (XY) is the male, but there are many exceptions to this rule, notably birds and butterflies, in which the female is the heterogametic sex. Because gametes are haploid, only one sex chromosome can be included in each gamete. If either the X or Y chromosome has genes that favor the production of X- or Y-containing gametes during meiosis (*meiotic drive,* Sandler and Novitski 1957), the sex ratio will be biased. Only one of the four haploid chromosome complements produced by meiosis in the female becomes an egg. Meiotic drive mechanisms control which chromosome is included in the viable product of meiosis. In the male, all four meiotic products normally become functional sperm; meiotic drive can be caused by genetically controlled chromosome damage. Either the X or the Y chromosome can be the driving chromosome and gain the advantage over the other, but consequences for the population are different in each case (Hamilton 1967). When the X chromosome gains an advantage, the number of females in the population increases because female gametes (all X) are more likely to be fertilized by an X-containing sperm (producing a female zygote) than by a Y-containing sperm (producing a male zygote). As the number of females in the population increases, the reproductive rate of the population also increases, at least until so few males remain that mating efficiency is impaired. The Y-driving mechanism produces males superabundantly, and as the number of females declines, the overall fecundity of the population declines, and the population decreases rapidly toward extinction. For this reason, the Y-driving situation is much more dangerous to a population. In addition, superior Y chromosomes are incorporated into a population more rapidly than superior X chromosomes: a Y-driving chromosome competes with an X chromosome in every meiosis because the Y chromosome always occurs in a heterogametic pair (XY), normally in the male sex. The two X chromosomes in the female do not compete with Y chromosomes during meiosis, therefore the chance that an X-driving chromosome will compete with a Y chromosome is only one in three.

The fact that Y-driving genetic mechanisms can endanger populations prompted Hamilton (1967) to suggest that the relative genetic inertness of the Y chromosome (more generally, the heterogametic chromosome) in most species is an adaptation to suppress Y-driving meiotic mechanisms. Under normal conditions, both the X and Y chromosomes presumably have modifier genes to suppress the driving effects of genes on the other chromosome, thereby keeping the driving tendencies of both chromosomes in balance and preserving the 1:1 sex ratio.

The population effects of Y-driving genes have been exploited for population control programs (Hickey and Craig 1966). For example, in the mosquito *Aedes aegypti*, which is a malaria-carrying species, some populations have Y-driving mutants, which are kept under control by modifier genes. When the Y-driving alleles are introduced into other strains of the mosquito lacking the modifiers, a strong meiotic drive results, leading to the rapid decline of the population. One would not expect this mechanism to produce long-lasting effects because modifiers to suppress the introduced alleles would soon be incorporated in the population. But continuous backcrossing of different Y-driving genes into a population could keep its size at a considerably reduced level.

Superior X chromosomes are initially beneficial to a population because they increase the number of reproductive females. X-driving chromosomes are known to occur in stable equilibria in wild populations of *Drosophila* (Novitski 1947). To be maintained in the population, the advantages of the X-driving genes must be balanced by some disadvantage for the individuals that have them.

The phenomenon of meiotic drive—a disgenic force that occurs within the individual and which eventually reduces its fitness by biasing the sex ratio of its progeny—demonstrates that the principles of competition and coexistence of genetically unique units apply even on the level of the chromosome within the cell.

Biased sex ratios are favored where outcrossing is difficult or impossible.

When outcrossing is made difficult owing to wide dispersal of individuals in a population, mating (if it takes place at all) occurs primarily among siblings because they are born and raised in the same area. The maintenance of a balanced sex ratio is contingent upon random mating, and inbreeding decreases the evolutionary advantage of producing male offspring. When a species is restricted to a single, sparsely-occurring host or food plant, it is more efficient for a parent to produce female progeny that are inseminated near their hatching place by their brothers before they disperse to find hosts, than it is to produce a lot of male progeny that disperse to find mates. One would expect the sex ratios in such cases to be strongly biased in favor of females, and indeed, this is frequently so in species that parasitize organisms that are rare and widely dispersed. Biased sex ratios reinforce the idea that sex ratio is an adaptation of the parent to maximize the number of its grandchildren.

Parasitic wasps have a wide variety of adaptations to solve the basic problems created by parasitizing widely dispersed hosts (Hamilton 1967). In most of these species, females greatly predominate in broods. Furthermore, sex determination is *arrhenotokous*: that is, males develop only from unfertilized eggs and are haploid. All fertilized eggs become females. Arrhenotoky has the general property that, when males are abundant, most eggs are fertilized and the progeny are females. If males are rare, however, many eggs are not fertilized, and these produce males, thereby tending to restore the optimum sex ratio in the population.

Species of wasps that reproduce solely by sib-mating often produce one, and only one, male in each brood; the male fertilizes all his sisters, so that additional males are an encumbrance. In most of these species, the brood develops gregariously, remaining together on the host until they mate. The males of some species have no wings, thereby ensuring that they will not leave the area before mating with their sisters—only inseminated females disperse. The tendency toward reduced dispersibility in the male has progressed so far in several pyemotid wasps that males are retained within their mother's body where they fertilize the developing females before the brood is born.

Extraordinary sex ratios do, in fact, prove the rule that an even sex ratio evolves only when there is free outcrossing within the population. Though extraordinary sex ratios are not common in nature, they do provide clues to the evolutionary meaning of the more common balanced sex ratio.

An abnormal sex ratio is one manifestation of another unusual reproductive system in animals that deserves mention here: the parthenogenetic reproduction of waterfleas (*Cladocera*) and aphids. Males usually are produced only at the end of the growing season; during most of the reproductive season, females form diploid eggs that develop without fertilization into more females (see page 359). The selective advantage of biased sex ratios to cladocera and aphids cannot be based on the difficulty of locating males because their populations rarely are sparse. As with most small organisms, cladocera and aphid populations are reduced considerably during the winter, so they undergo a phase of rapid growth throughout most of the summer. Individuals in an increasing population may enhance their fitness by temporarily abandoning sexual reproduction for efficient parthenogenetic reproduction, in which the only offspring produced are females that contribute directly to population growth by laying more eggs.

Under what conditions can a temporarily parthenogenetic female have an evolutionary advantage in a sexual, outcrossing population? When can parthenogenesis become established? Assuming that males play no role in reproduction other than to fertilize the eggs and, further, that males are as expensive to produce as females, the slightest level of inbreeding will reduce the effective genetic contribution to future populations through male off-spring and will favor the production of female progeny, eventually leading to parthenogenesis. In populations that do not disperse quickly, pockets of closely related individuals could form when population growth is rapid, caus-ing inbreeding, and thus promoting the evolution of parthenogenesis. It is relevant to this argument that in many planktonic species of cladocera, in which individuals mix freely, males are produced throughout the growing season. Parthenogenesis is more frequent in sedentary benthic species.

Current thinking about sex ratio has progressed beyond the simple ideas outlined above. Fisher (1930) and Leigh (1970) point out that sex ratio is optimized in an outcrossing population when parental investment in each sex is equal on average, not when equal numbers of males and females are produced by the population. Thus, parental investment in the young by each sex and sexual selection become important aspects of the evolution of sex ratio (Trivers 1972). Trivers and Willard (1973) have suggested that under some conditions, females would be expected to vary the sex ratio of their offspring in accordance with their own condition and hence their ability to make a parental investment. For example, if the reproductive capacity of males varied greatly, as in harem-forming species like deer, and if males required greater than average parental investment to be successful, females in good physical condition should produce more male progeny on average, but females in poor condition would be better off to invest their production in female offspring.

Perennial life histories are favored by high and relatively constant adult survival.

An important controversy in the evolution of life-history adaptations has centered on the number of breeding seasons that the individual survives. Most plants and animals may be divided neatly into those that survive to reproduce during a single season and die (*annuals*) and those that have the potential to reproduce over a span of many seasons (*perennials*). Population biologists have pondered the relative advantages of each habit since Cole's (1954) pioneering mathematical treatment of the problem. The solution to the problem rests in the trade-off between survival probability and fecundity. To survive its nonreproductive period, a perennial plant must allocate consider-able resources to storage of materials in roots and formation of freeze-resistant buds, all at the expense of production. But when do the advantages of the perennial habit outweigh the cost in fecundity? The problem was formulated as a simple algebraic model by Charnov and Schaffer (1973), which we shall cast in terms of a comparison between annual and perennial strategies in plants. Suppose a population of plants is faced with the evolu-tionary decision between producing a large number of seeds at the end of the first growing season and then dying (annual), and producing fewer seeds but surviving through the winter to reproduce in subsequent growing seasons (perennial). We shall simplify the model by assuming that annual and perennial plants would have the same probability of survival during their first (in the annual's case, only) growing season. We shall use the following symbols:

P = the number of progeny (including the parent in the perennial case) that survive to reproduce

B = the number of seeds produced

S_1 = survival during the first year of growth

S_2 = survival of perennials during subsequent years

The subscript a shall refer to annuals, p to perennials.

The number of progeny of an annual plant is equal to the number of seeds produced (B_a) times their survival to reproduction (S_1), or

$$P_a = B_a S_1$$

The number of progeny of a perennial plant also is equal to the number of seeds B_p times their survival (S_1), but we must add to this number the parent, if it survives, to obtain the total productivity. Because the probability of survival of the parent is S_2,

$$P_p = B_p S_1 + S_2$$

The annual strategy is favored when the productivity of the annual (P_a) exceeds that of the perennial (P_p),

$$P_a > P_p$$

or

$$B_a S_1 > B_p S_1 + S_2$$

Dividing both sides of the equation by S_1, we obtain

$$B_a > B_p + S_2/S_1$$

or

$$B_a - B_p > S_2/S_1$$

This model suggests that the annual life history would be favored if the number of seeds produced by the annual exceeded the fecundity of the perennial by the ratio S_2/S_1. If survival from one breeding season to the next (S_2) were very low or required a large decrease in fecundity (B_p), the annual habit would be favored. Where year-to-year survival is great once an individual is established, but seedling survival is low (high S_2/S_1), an annual must be extremely fecund to be favored. It is no wonder that annuals predominate among the floras of deserts, where survival through drought periods is low, and perennials predominate among the floras of the tropics, where competition and predator pressure reduce the survival of seedlings to near zero.

Adding complexity to the model, by incorporating growth from year-to-year and annual variation in survival probabilities, would not destroy the basic qualitative conclusion that the life-history strategy is determined primarily by the ratio of adult survival to juvenile survival (Stearns 1976).

Optimum reproductive effort must weigh the trade-off between fecundity and adult survival.

Any superficial survey of life-history patterns reveals a tremendous diversity of adaptations in brood size, size of young, adult survival, and age distribution in reproductive effort (Stearns 1976). Even within such a uniform group of animals as birds, we find extremes of life history ranging from the wandering albatross, which lays one egg every other year and may live fifty years or more, to many small sparrows that lay several clutches of four eggs each year, but live only a couple of years on average (see page 280). Williams (1966) attempted to explain life-history patterns in terms of the trade-off between fecundity and adult survival. He argued that if mortality outside the breeding season were high, the probability of surviving to reproduce in the future would be low regardless of whatever additional risks might attend reproduction. Under these conditions, the optimum strategy would be to make a small sacrifice in future expectation of reproduction by attempting to raise a large number of offspring in each breeding season. In contrast, if survival probability outside the breeding season were great, as it is in albatrosses, which have no natural predators and can search widely for food, risk of mortality incurred by reproduction could greatly reduce expectation of future reproduction.

Elaborations on Williams's model have been the subject of a variety of theoretical and empirical papers published since 1966, including those by Gadgil and Bossert (1970), Cody (1971a), Murphy (1968), Goodman (1974), and Schaffer (1974). All reinforce Williams's original conclusion that fecundity, reproductive effort, and precocity of sexual maturation should vary in direct relation to the mortality of adults. Donald Tinkle and his co-workers have explored the optimization of the trade-off between growth and reproduction in lizards (Tinkle 1969, Hirschfield and Tinkle 1975, Tinkle and Hadley 1975, Tinkle et al. 1970). Because fecundity is directly related to body size in lizards, as it is in fish (page 288), to allocate energy and resources to eggs rather than to growth is to reduce next year's fecundity. If the probability of surviving from one breeding season to the next were small, little would be lost by sacrificing growth to increase fecundity.

Tinkle and Ballinger (1972) undertook a systematic study of the eastern fence lizard Sceloporus undulatus at three localities in search for evidence that the trade-off between growth and reproduction varied according to adult mortality. The patterns in life history were consistent with theory (Table 30-1). Texas populations suffered high adult mortality (at least judged by the proportion of individuals with damaged tails: predator escapees), they matured early, and they produced the largest number of eggs per season. In Ohio, adult mortality was low, reproduction was delayed to the end of the second year, and relatively few eggs were produced. South Carolina populations were intermediate in all respects.

However consistent these patterns are with the theory of demographic optimization, other explanations are equally plausible. Instead of being optimized with respect to other life-history characteristics, each aspect of the life history may be adapted independently to different factors in the environ-

TABLE 30-1

Life-history characteristics of populations of the eastern fence lizard *Sceloporus undulatus* from three localities (data from Tinkle and Ballinger 1972).

	Texas	South Carolina	Ohio
Growth	rapid	intermediate	slow
Adult mortality	high	intermediate	low
Age at maturity (months)	9	9–10	20
Average size of breeding females (mm)*	57	63	75
Average clutch size	9.5	7.4	11.8
Broods per year	3–4	3	2
Annual production (eggs)	33	22	24
Egg weight (g)	0.22	0.33	0.35
Ratio of clutch weight to body weight	0.27	0.23	0.25
Proportion of births by age of female (per cent)			
1 year	82	48	0
2 years	18	30	57
3 years	0	15	22
4 years	0	7	21

* The conventional size measurement is the snout-vent length, the distance between the tip of the snout and the cloaca.

ment. For example, in Tinkle and Ballinger's study of *Sceloporus*, reproductive rate was viewed as an adaptation to adult mortality, not to factors affecting growth and fecundity directly. But, in the Texas population, high mortality could be caused by a relative abundance of predators and a long activity season during which the lizards are exposed to predation; multiple clutches could be related to the long growing season; large clutch size and rapid prereproductive growth could be related to an abundant food supply and warm temperatures; sexual maturity could come early because the rapidly growing young were able to reach some threshold size for reproduction before the end of their first year. All these ideas are amenable to experimental work, and differences in opinion will undoubtedly be resolved in the future. For now, alternate hypotheses should remind us not to jump to conclusions too rapidly. In birds, for example, the direct relationship between reproductive rate and adult mortality rate has been accepted as evidence of Williams's general hypothesis. Yet a moment's reflection will convince us, that, in a population with constant size, any change in mortality must be balanced by a change in recruitment, a component of which is fecundity. Therefore, adult mortality and fecundity may be linked through the effect of the population on its food supply, rather than through demographic optimization. Reduced mortality can lead to increased population size and fewer resources per individual for reproduction. The life histories of birds are well enough known that I have been able to estimate that between 10 and 25 per cent of the relationship between fecundity and adult mortality is attributable to demographic optimization; the rest, between 75 and 90 per cent, is due to ecological feedback of adult population size on resources available for reproduction (Ricklefs 1977).

Life-history patterns vary according to the growth rate of the population.

Reproductive rate has been linked to the growth rate of populations to explain latitudinal variation in fecundity (Cody 1966, Skutch 1949, MacArthur and Wilson 1967). The argument runs as follows. In temperate and arctic regions, populations are periodically reduced by catastrophic weather and individuals die with little regard to their genotype. Population crashes are followed by longer periods of population increase during which adaptations that increase intrinsic population growth rate (r)—including increased fecundity and early maturity—are selected. In stable tropical environments, where populations fluctuate little, populations remain near the limit imposed by resources, and adaptations that improve competitive ability and efficiency of resource utilization are selected.

The distinction between temperate and tropical patterns has been heralded as the r- and K-selection alternative, with all intermediates possible (Pianka 1970). The r refers to the growth capacity (exponential growth rate) of the population, and K denotes the carrying capacity of the environment for the population—the upper resource limit to population size. Although the naming of the concept has set off a minor semantic battle among population ecologists (Hairston et al. 1970, Pianka 1972, Wilbur et al. 1974), r- and K-selection occupies an important place in current thinking about life-history patterns (Stearns 1976).

Over and above basic assumptions about environmental variability and fluctuations in population size, r- and K-selection theory depends on a trade-off between adaptations favored under conditions of high population growth rate and those favored under conditions of crowding and low resources. The trade-off can be visualized on a graph portraying the fitness of traits as a function of population density (Figure 30-1). The exponential rate of increase of virtually all genes decreases as population density approaches its resource limitation and the growth rate of the population slows. One of the traits illustrated in Figure 30-1 has superior fitness at low density and is said to be r-selected; the other is more fit at high density and is said to be K-selected. If the population were to remain near its carrying capacity for a long period, r-selected traits would be selectively removed from the population; under conditions of fluctuating population size, r-selected traits persist.

The proponents of r- and K-selection theory have yet to demonstrate correlations between population fluctuation and adaptation or to show

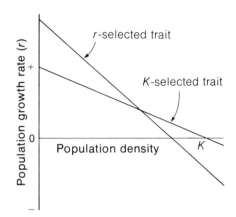

Figure 30-1   The relationship of the fitness of a trait (*r*) to population density, illustrating the difference between *r*-selected (low density) and *K*-selected (high density) traits.

trade-offs between *r*-selected and *K*-selected traits. At present, the *r-K* continuum is mostly conjecture and species are placed on the continuum according to their adaptations rather than measured fluctuations in their population size. In other words, there has been no test of the theory. Pianka (1970) lists a variety of traits that could be considered as either *r*-selected or *K*-selected (Table 30-2). He places insects at the *r*-selected end of the continuum, and mammals and birds at the *K*-selected end. Although it is true that insect populations fluctuate more and insects exhibit traits listed in the *r*-selected column of Table 30-2, this does not demonstrate causal relationship. The body sizes of mammals and insects differ by several orders of magnitude, and accompanying such difference in body size we find differences in the time and power scale of all physiological processes (Schmidt-Nielsen 1975). Small organisms move more rapidly relative to body length, use more energy relative to body weight, and have more rapid development and shorter generations than large animals. These traits may be inherently correlated to size, just as the swinging rate of a pendulum is inversely related to its length, and thus largely insensitive to environmental influence. I would be more convinced of the importance of *r*- and *K*-selection theory if one could demonstrate expected differences in adaptations between two related species with measured differences in population fluctuation. A further

TABLE 30-2
Traits that have been associated with the *r*- and *K*-selection continuum (after Pianka 1970).

r-selected traits	*K*-selected traits
rapid development	slow development
early reproduction	reduced resource requirement
small body size	delayed reproduction
semelparity (single reproduction)	large body size
	iteroparity (repeated reproduction)

difficulty with the interpretation of adaptations according to r- and K-selection theory is that the traits attributed to different levels of population fluctuation are identical to those predicted for different levels of adult mortality and population turnover, even among populations with constant size. These difficulties only emphasize our ignorance of population processes under natural conditions.

The concept of reproductive value enables us to examine age-specific patterns of life-history traits.

Up to this point, we have treated individuals either as adult or immature, depending on their capacity for reproducing. But life-history values of survival and reproduction do not remain constant after the onset of reproduction. Fecundity tends to reach a maximum and mortality a minimum during the middle of the adult life span in most species. In those that grow continuously throughout their lives, reproduction and survival rate tend to increase steadily with age. Changes in life-history pattern with age suggest that selection acts differently with age, shifting the optimum compromise between life-history functions as the individual matures and grows old.

Fisher (1930) derived a useful expression for the evolutionary value of an individual as a function of age. This function, referred to as the *reproductive value*, is calculated for an individual at age a by the equation

$$V_a = \frac{\lambda^a}{l_a} \sum_{x=a}^{\infty} \lambda^x \, l_x b_x \qquad (30\text{-}1)$$

for populations with discrete reproductive seasons, and

$$V_a = \frac{e^{ra}}{l_a} \int_{x=a}^{\infty} e^{-rx} \, l_x b_x \qquad (30\text{-}2)$$

for continuously growing populations, where l_x is survival and b_x is fecundity at age x, r is the exponential growth rate and λ the geometric growth rate. Reproductive value is the relative contribution of an individual of age a to future generations, or conversely, the diminution of future population increase by removing an individual of age a.

The reproductive value of an individual is derived from its expectation of future offspring. The contribution of a birth to the reproductive value of its parent decreases with time (age of the parent) in growing populations and increases with time in declining populations. An individual born in an expanding population at time 0 (the present) will contribute proportionately more offspring to the population in the future than an individual born at time 1, because the second is a smaller proportion of the population into which it is born. In a population whose growth rate is λ,

a birth at time (t)	0	1	2	3	. . .
will represent proportion	1	λ^{-1}	λ^{-2}	λ^{-3}	. . .

in the population relative to a birth at time 0. A newborn at time 0 is proportion $1/N(0)$ of the entire population. After one unit of time, the population will have grown to $N(0)\lambda$ individuals of which a newborn at time 1 will be as the proportion $1/N(0)\lambda$. Its proportion in the population relative to a newborn at time 0 is

$$\frac{1/N(0)\lambda}{1/N(0)} = \lambda^{-1}$$

Similarly, a newborn at time t is $1/N(0)\lambda^t$ proportion of the population, or λ^{-t} relative to a newborn at time 0.

The probability that an individual alive at age a will live to age $a + 1$ is l_{a+1}/l_a and that it will live to age $a + t$ is l_{a+t}/l_a. Thus, the total expected value of offspring born during the remaining lifetime of an individual of age a is the sum over each age of the expectation of offspring times their relative value,

$$V_a = \left(\frac{l_a}{l_a} b_a\right) + \left(\lambda^{-1} \frac{l_{a+1}}{l_a} b_{a+1}\right) + \left(\lambda^{-2} \frac{l_{a+2}}{l_a} b_{a+2}\right) + \cdots + \left(\lambda^{-t} \frac{l_{a+t}}{l_a} b_{a+t}\right)$$

which, written in summation notation, is

$$V_a = \frac{1}{l_a} \sum_{t=0}^{\infty} \lambda^{-t} l_{a+t} b_{a+t} \tag{30-3}$$

The age of the organism (x) is equal to $a + t$; therefore, $t = x - a$ ($t = 0$ when $x = a$). Substituting $x - a$ for t in equation 30-2, we get

$$V_a = \frac{1}{l_a} \sum_{x=a}^{\infty} \lambda^{-(x-a)} l_x b_x$$

$$= \frac{\lambda^a}{l_a} \sum_{x=a}^{\infty} \lambda^{-x} l_x b_x \tag{30-4}$$

(Note that $a + t$ becomes $a + x - a$, which is x.)

In a population with constant size ($\lambda = 1$), the reproductive value becomes

$$V_a = \frac{1}{l_a} \sum_{x=a}^{\infty} l_x b_x$$

which is the expected remaining reproduction at age a.

The reproductive value of a newborn individual (age 0) is

$$V_0 = \sum_{x=0}^{\infty} \lambda^{-x} l_x b_x$$

which is the equation for the exact calculation of λ or r, derived on page 439. Hence $V_0 = 1$ and we see that reproductive value is a measure of the contribution an individual alive at age a can make to the growth rate of the population relative to the contribution of a newborn individual. The reproductive value of an individual increases until the age of first reproduction. All the terms in equation 30-4 that include b_x are 0 until the beginning of reproduction, so the sum $\Sigma \lambda^{-x} l_x b_x$ does not change until the onset of reproduction. Because survival to age a (l_a) decreases with age, the reproductive value of each individual increases prior to the onset of reproduction, as shown for a human population in Figure 30-2. After the onset of reproduction, reproductive value may increase or decrease depending on whether fecundity increases faster than the expectation of further life decreases. Eventually, as the individual approaches the maximum life expectancy of the species, its reproductive value declines toward zero (see Figure 30-2).

A useful property of reproductive value is that it may be partitioned into a sum of present fecundity and expected future reproduction

$$V_a = b_a + \frac{l_{a+1}}{l_a \lambda} V_{a+1} \tag{30-5}$$

(Taylor *et al.* 1974). The second term of the expression is the *residual reproductive value* (*RRV*); it includes the survival from one year to the next (l_{a+1}/l_a) and the inverse of population growth rate ($1/\lambda$). As you can see, any trade-off between fecundity (b_a) and adult survival (l_{a+1}/l_a) is influenced by future reproductive value and by population growth rate. In a rapidly growing population, residual reproductive value is discounted by factor $1/\lambda$ and selection to maximize V_a will tend to increase b_a at the expense of survival, all other things being equal. This, of course, is one of the predictions for the *r*-selected (high population growth rate) case.

The trade-off between fecundity and future reproduction may be visualized on a graph relating *RRV* to b_a (Figure 30-3). The shape of the curve relating fecundity to residual reproductive value depends upon the type of trade-off between the two. Each point on the curve represents some possible combination of values of b_a and *RRV*. If the curve is convex (curves a and b) the effect of reproduction on survival is small for low fecundity, but increases rapidly as fecundity increases toward some upper limit determined by resources and physiological constraints. Where the curve meets the horizontal (b_a) axis, reproductive effort is all-consuming and the organism dies as a result (*RRV* = 0). When the curve is concave (c), the act of reproduction itself is costly, owing perhaps to preparatory activities like migration and territory defense, but to increase brood size further is relatively inexpensive in terms of residual reproductive value. How do we determine the optimum reproduc-

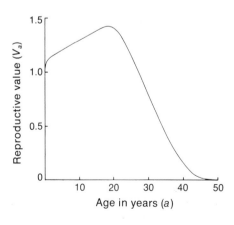

Figure 30-2   Reproductive value as a function of age for Australian women in 1911. Reproductive value increases until the onset of reproduction, after which it steadily declines (after Fisher 1930).

Figure 30-3   Hypothetical relationships between fecundity and residual reproductive value at age *a*.
The diagonal dashed lines each
represent a different value of reproductive value V_a, which increases
with a distance from the origin of
the graph. The optimum trade-off
between present fecundity and expectation of future reproduction is
indicated for each curve by the dot,
which is the maximum value of V_a
on the curve (after Pianka and
Parker 1975).

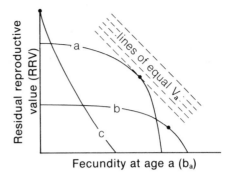

tive pattern along the continuum represented by the b_a-*RRV* curve? Rearranging equation 30-5 yields the expression $RRV = V_a - b_a$, which describes a
family of straight lines with slope -1 whose position is determined by V_a
(Figure 30-3). The optimum reproductive pattern is that point on the b_a-*RRV*
curve that is just touched by the line $RRV = V_a - b_a$; this point is the combination of fecundity and residual reproductive value that results in the highest
reproductive value at age *a*. For a convex curve relating *RRV* to b_a, an
intermediate level of fecundity will always be selected, and that level will vary
depending on how the curve changes with age (*e.g.*, from *a* to *b*). For a
concave curve, selection will favor either delayed reproduction ($b_a = 0$) or an
all-out reproductive effort ($RRV = 0$), the so-called big bang strategy of salmon, agave, and bamboo (see page 291).

Survival rate is an evolved population characteristic.

The trade-off between survival and reproduction is only one factor that
influences the survival curve of a population. The probability of survival at a
given age also depends upon adaptations for avoiding predators, homeostatic responses, stage of development, and the degree of senescence or
physiological deterioration with age. Death from any cause is, however, a
unique event for the individual; the probability of death at a given age is a
statistical property of the population as a whole that can be measured only by
observation of many individuals.

Demographers characterize probabilities of living and dying by survival
and mortality curves for a population. Survival to age *x* (l_x) is a relatively
straightforward concept described in Chapter 27. It is simply the probability
that a newborn individual will be alive at age *x*. Mortality (m_x) is the probability
that an individual will die during the age interval *x* to *x* + 1, expressed as a
proportion of those alive at age *x*. Mortality is properly calculated from
survival data by the expression

$$m_x = \frac{l_x - l_{x+1}}{l_x} \qquad (30\text{-}6)$$

To illustrate the relationship between survival and mortality, we shall turn to data on survival of Dall mountain sheep in Table 27-3. We find, for example, that survival to eight years of age was 0.571 and to nine, 0.439. Substituting these values into equation 30-6, we obtain

$$m_8 = \frac{0.571 - 0.439}{0.571}$$

$$= \frac{0.132}{0.571}$$

$$= 0.231$$

That is, on the basis of Murie's sample one would expect about 23 per cent of eight-year-old mountain sheep to die before their next birthday. The relationship between survival and mortality in the Dall mountain sheep is shown in Figure 30-4.

When graphed, the survival curve is usually plotted as the logarithm of the number or decimal fraction of survivors as a function of age, so the slope of the survival curve at a particular age is equal to the mortality rate at that age. If a population sustained a constant rate of mortality, regardless of age, its survival curve would be a straight line when plotted on a logarithmic scale. For example, if the annual mortality rate were 0.50, one-half of the population would remain after one year, one-half of those alive at one year of age (one-quarter of all newborn) would reach the end of their second year, one-half of those (one-eighth of all newborn) would reach the end of their third year, and so on (Figure 30-5).

Most survival curves have three phases: a juvenile period, during which the growing organism is particularly vulnerable to predators and inclement weather; an adult period, during which the individual is fully developed

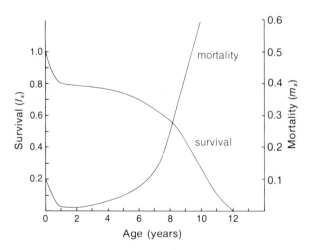

Figure 30-4   Survival (l_x) and mortality (m_x) curves for the Dall mountain sheep (*Ovis dalli*), based on data in Table 27-3.

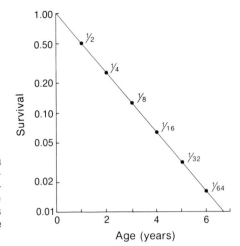

Figure 30-5   Survival curve for a hypothetical population with an annual mortality of 50 per cent. Survival (S) to age x is expressed by the formula $S_x = (1 - m)^x$, where m is the annual mortality and therefore $1 - m$ is the annual survival rate.

physically and in its prime reproductive years (generally a period of low mortality); and a senescent period of declining physiological function and increasing mortality (Figure 30-6). The differences between survival curves of various species reflect the relative emphasis of each of these three periods in the life history of the organism. In organisms with intense parental care, like birds and mammals, the juvenile phase usually has reduced impact on the survival curve as a whole (see Figure 30-7 and Deevey 1947, Caughly 1966, Spinage 1972, Ricklefs 1973). In fish and invertebrates, most of which lay vast numbers of eggs, the young suffer great mortality from predation and physical factors (Sette 1943, Beverton and Holt 1957). Plant survival curves are also characterized by high juvenile mortality, both before and after seedling germination (e.g., Hett and Loucks 1971).

Mortality rate during the adult stage of the life cycle, generally its lowest point, is determined by the adaptations of the organism to the physical and

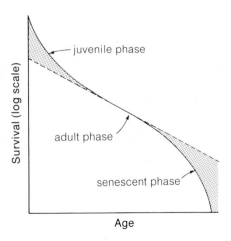

Figure 30-6   General form of the survival curve for all animal populations. Variations among species generally reflect quantitative, rather than qualitative, differences. The hatched areas represent costs of development and senescence.

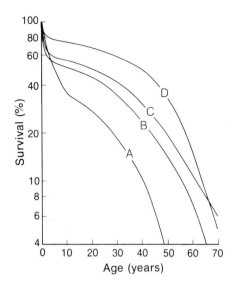

Figure 30-7   Survival curves for four human or subhuman populations calculated· from skeletal remains. A, *Sinanthropus* (Peking man); B, Maghreb culture (late Paleolithic); C, first to fourth century Roman agriculturalists; D, agriculturalists of the late Roman era (after Dumond 1975).

biological environment. Small organisms tend to suffer higher mortality rates—and hence have shorter life spans—than large organisms because their high surface-to-volume ratio reduces homeostatic capability and because they are smaller than a greater number of predatory species than are large organisms. And so the difference between large and small organisms that results in differential mortality is not unlike the growth of an individual organism that leads to a decrease in mortality.

Once the organism matures sexually, its adaptations have approached or reached their final expression and, ignoring seasonal variation, we would expect mortality rate to assume a constant level. In some species, most birds for example, this is the case (*e.g.*, Figure 30-8, Deevey 1947, Hickey 1951, Lack 1954). In species that continue to grow after becoming sexually mature, mortality rate may continue to drop during adult life as the individual outgrows its predators. In vertebrates, experience may also play an important role. But in most animals, mortality reaches its lowest point close to the time of sexual maturity and then increases, sometimes rapidly, throughout the adult portion of the life span. This type of mortality curve is shared by man (Figure 30-7 and 30-8), the Dall mountain sheep (Figure 30-4), adults of many species of moths (Figure 19-10), and laboratory populations of rotifers (Figure 30-8), rats and voles (Caughley 1966), fruit flies (Bodenheimer 1958), especially starved fruit flies (Deevey 1950), and others.

Senescence is a gradual increase in mortality resulting from deterioration of physiological function.

Old age brings about deterioration of physiological function in most organisms, even when living under the best conditions (Kohn 1971, Rockstein 1974). For example, the rates of most physiological functions in man decrease in a roughly linear fashion between the ages of 30 and 85 years, to 80 to 85 per cent of the value of 30-year-olds in nerve conduction and basal metabolism, 40 to 45 per cent in the volume of blood circulated through the kidneys, and 37 per cent for maximum breathing capacity (Mildvan and Strehler 1960). Senescence and death in old age do not result from drastic physiological change. Rather, the demographic consequences of senescence result from a gradual decrease in physiological function with age. Such changes are found throughout the animal kingdom (Comfort 1956, Strehler 1960).

The manifestations of senescence are not limited to physiological parameters. Man's ability to learn rises until about age eleven, remains fairly high until about thirty, and then begins to decline (Inglis, Ankus, and Sikes 1968). Effects of senescence may also appear as malformations in the progeny of aging individuals. Birth defects generally occur with increasing prevalence in mothers progressively older than thirty years (Table 30-3). Senescence may also cause fertility to decline or even cause reproduction to cease entirely in old age.

How can senescence evolve? Why is not senescence eliminated by selection when survival presumably is advantageous to an individual at any age? Imagine a population with no senescence, in which mortality rate is constant and individuals are *potentially* immortal. If the annual survival rate of the population were 10 per cent, the proportion of the population alive would be one-tenth after ten years, one-hundredth after twenty years, one-thousandth after thirty years, and so on. If genes were expressed at different ages, relatively fewer deleterious mutants expressed at old age could be removed by selection before they were passed on to an individual's progeny. For example, a thousandth of the deleterious mutations expressed at age thirty would be exposed to selection in each generation. Any progeny born to an

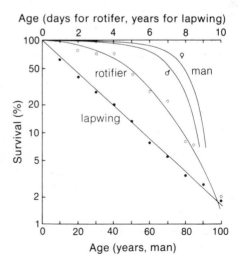

Figure 30-8   Survival curves for the white human population of the United States (from Keyfitz and Flieger 1968); for the sessile rotifer, *Floscularia conifera* (shown in inset, based on data in Edmondson 1945, quoted by Deevey 1947); and for a shorebird, the lapwing, *Vanellus vanellus* (after Lack 1943b, quoted by Deevey 1947).

TABLE 30-3
Number of defects per 10,000 human births (from Milham and Gittelsohn 1965, quoted by Emlen 1970).

	Age of Mother (years)						
	15–19	20–24	25–29	30–34	35–39	40–44	45+
Anencephalus	10	9	7	8	9	10	6
Spina bifida	12	11	10	10	13	13	16
Hydrocephalus	18	7	7	8	10	15	25
Microcephalus	1	1	1	1	1	13	—
Heart defects	17	18	17	17	24	28	50
Cleft lip, palate	10	11	11	10	12	16	—
Mongolism	2	2	2	3	11	33	89
Club foot	15	14	11	11	12	24	18
Total	85	73	66	68	92	142	202

individual before age thirty would be as likely to carry the mutant as not. If a gene were not expressed until age sixty, only one phenotype in a million would live long enough to express the trait. Therefore, the strength by which natural selection removes deleterious alleles from a population declines as age of expression increases, allowing mutation pressure and drift to increase the frequency of an allele to a high level. If this process occurred at many gene loci, it could easily burden the gene pool of the population with a heavy load of deleterious alleles expressed in old age and could account for the progressive increase in mortality rate with age that is characteristic of senescence (Medawar 1957, Williams 1957, Hamilton 1966).

The age at which senescence is first expressed and its increase with further age depend partly upon how much mortality would occur regardless of senescence. The minimum age-specific mortality rate of the population probably estimates nonsenescent mortality reasonably well. Where the minimum mortality rate is high, few individuals are long-lived, even in the absence of senescence, and the physiological and demographic effects of senescence appear early. Conversely, in a species with a low minimum mortality rate, more individuals survive to each age, and selection delays the encroachment of senescence on survival (Figure 30-9).

The evolution of senescence is often phrased in terms of the reproductive value of individuals at a given age. Because the reproductive value of an individual measures its contribution to future populations (see page 494), the sum of the reproductive values of individuals of age x is the contribution of that age group to future generations and, therefore, measures the proportion of the gene pool expressed at age x that is available to selection.

Reproduction may be considered as the transferral of reproductive value from one generation to the next. The total reproductive value of a cohort of individuals remains constant up to the age of first reproduction, and then steadily decreases as young are born. The strength of selection on prereproductive individuals, therefore, is high and constant, and thus mortality

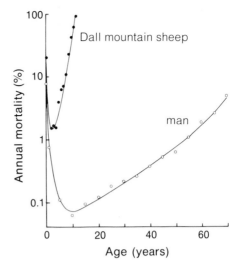

Figure 30-9   Relationship between annual mortality rate and age
in the Dall mountain sheep (from
Deevey 1947) and in a human population (Costa Rican females, 1963;
from Keyfitz and Flieger 1968),
showing that senescence encroaches more rapidly in the population whose minimum adult
mortality is higher.

related to senescence does not increase until after the age of first reproduction. With increasing age, the reproductive value of a cohort is transferred to
their progeny. After their last offspring have been born, the total reproductive
value of a cohort is zero, and selection can no longer remove mutants
expressed after that age. The mortality of postreproductive individuals
should therefore increase through the accumulation of deleterious mutations
in the gene pool. Of course, the decrease and cessation of reproduction is
itself a manifestation of senescence.

One might argue that the human female contradicts the notion that
postreproductive mortality should increase rapidly. Women rarely have children beyond the age of 45 years, yet their average life span is more than
seventy years in most industrialized countries and exceeds male life span by
seven years in the United States (Figure 30-10). But the transferral of reproductive value to offspring includes a period of parental care that corresponds
approximately to the lag between the cessation of reproduction in the human
female and rapid decrease in female survival. Closer correspondence
between the decrease in survival and the reproductive capacity of males
(Figure 30-10) suggests that, at least for the conditions under which fecundity
and mortality patterns evolved, the male was not primarily responsible for
the well-being of his children. Such family organization is reminiscent of
some species of baboons and other primates in which the female cares
primarily for her own young, but males defend all the young in the troop
(page 86).

In this chapter, we have attempted to relate adaptations in life-history
patterns to the environment and to the nature of the genetic mechanism. In
the case of the sex ratio, selection on outcrossing populations favors individuals that produce progeny of the rarer sex. Such frequency-dependent
selection leads to an equilibrium in the ratio of the sexes in a population, but

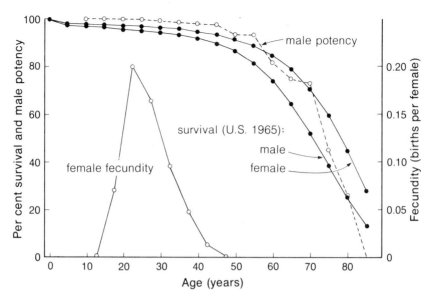

Figure 30-10   Survival and reproductive capacity for males and females in the population of the United States. (Survival and female fecundity are from Keyfitz and Flieger 1968; male potency from Kinsey *et al.* 1948).

the position of this equilibrium is influenced by the mating structure of the population, particularly by the degree of outcrossing.

Senescence emphasizes the inability of selection to perfect the gene pool. Because selection cannot efficiently remove deleterious alleles that are expressed at old age, mutation pressure can cause the frequency of such alleles in the population to increase to a level at which they adversely affect survival. The waning effect of selection at old age demonstrates the importance of age structure in population dynamics.

This chapter has been the last of a series on the evolutionary responses of populations to their environment. In the next part of the book we shall investigate demographic responses of populations to their own density and the impact of competitors, predators, and prey on the growth of populations and the regulation of population size. These interactions are analogous to the homeostatic responses of organisms and determine the steady-state properties of density and age structure of the population. Remember that the demographic properties of populations are adaptations subject to evolutionary change as environmental pressures shift.

Part 8

Ecology of Populations

31

Population Growth and Regulation

A knowledge of the growth and regulation of populations is a prerequisite to understanding the structure and dynamics of the biological community and the ecosystem. All species have a high potential for population growth under optimum conditions. In *On the Origin of Species,* Darwin wrote: "There is no exception to the rule that every organic being naturally increases at so high a rate that, if not destroyed, the earth would soon be covered by the progeny of a single pair." This potential is never realized fully by natural populations as their numbers are regulated, often within narrow limits. The factors that influence population growth rate and tend to stabilize population size are examined in this chapter.

What is a population?

Ecologists apply the term population size very loosely to pragmatically defined assemblages of individuals of one species. The walls of cages and aquaria delimit laboratory populations. Edges of study areas conveniently define populations in nature, for which *density* (individuals per unit area) is used as an index to population size. For species whose numbers and geographical ranges are restricted, and for species that are easily counted, it is possible to measure the size of the entire population. In practice, this is accomplished only for populations of endangered species (whooping cranes, grizzly bears) and commercially important species (ducks, deer, and so on). Usually, the ecologist must restrict his studies to the local population within a well-defined study area or to the artificial population in the laboratory. So long as the movement of individuals into and out of the study area is small compared with the replacement of individuals by reproduction within the area, local studies may afford reasonable insight into the processes that influence population size. In this chapter, we shall draw upon evidence from a variety of laboratory and field studies to evaluate the many factors that have been suggested as regulating population size. One point will become clear above all others: Although ecologists have sought fundamental principles applicable to all populations, their diverse views of population regulation have been determined by the systems they have studied. Students of bird, mammal, and insect populations have argued bitterly over which *single* factor is most important to all populations when, in fact, all populations are influenced by the same set of factors, each expressed to a different degree in different populations.

The growth potential of populations is great.

We can best appreciate the capacity of a population for growth by following its rapid increase when introduced to a new region with a suitable environment. The number of colonists is at first so low that crowding and depletion of resources do not hinder population growth. In 1859, twelve pairs of the European rabbit (*Oryctolagus cuniculus*) were released on a ranch in Victoria, Australia. Within six years, the population of rabbits increased so rapidly that 20,000 were killed in a single hunting drive. Even by conservative estimate, the population must have increased by a factor of at least 10,000 in six years, or a geometric rate of increase (λ) of 4.7 per year. Two male and six female ring-necked pheasants introduced to Protection Island, Washington, in 1937 increased to 1,325 adults within five years (Einarsen 1942, 1945). The 166-fold increase represents a 180 per cent annual rate of increase ($\lambda = 2.80$). When domestic sheep were introduced to Tasmania, a large island off the coast of Australia, the population increased from less than 200,000 in 1820 to more than 2 million in 1850 (Davidson 1938). The tenfold increase in thirty years is equivalent to an annual rate of increase of 8 per cent ($\lambda = 1.08$).

Even such an unlikely creature as the elephant seal, whose population had been all but obliterated by hunting during the nineteenth century, increased from twenty individuals in 1890 to 30,000 in 1970 ($\lambda = 1.096$; Bonnell and Selander 1974). If you are unimpressed, consider that another century of unrestrained growth would find 27 million elephant seals crowding surfers and sunbathers off Southern California beaches. Before the end of the next century, the shore lines of the Western Hemisphere would give lodging to a trillion elephant seals.

Elephant seal populations do not hold any growth records. Life tables of populations maintained under optimum conditions in the laboratory have shown that potential annual growth rates (λ) may be as great as 24 for the field vole (Leslie and Ranson 1940), 10 billion (10^{10}) for flour beetles (Leslie and Park 1949), and 10^{30} for the water flea *Daphnia* (Marshall 1962). Rapid growth rates are more conveniently expressed in terms of the time required for the population to double in number. The relationship between geometric growth rate (λ) and doubling time (t_2) can be derived from the equation for geometric growth

$$N(t) = N(0)\lambda^t$$

Because the ratio of $N(t_2)$ to $N(0)$ is 2, this equation may be restated as $\lambda^{t_2} = 2$. We can rearrange this expression to obtain

$$t_2 = \frac{\log_e 2}{\log_e \lambda}$$

$$= \frac{0.69}{\log_e \lambda}$$

Hence for the field vole ($\lambda = 24$),

$$t_2 = 0.69/\log_e (24)$$

$$= 0.69/3.18$$

$$= 0.22 \text{ years}$$

$$= 79 \text{ days}$$

Corresponding doubling times are 7.6 years for the elephant seal, 8 months for the pheasant, 80 days for the vole, 10 days for the flour beetle, and less than 3 days for the water flea. Populations of microorganisms (bacteria and viruses) and many unicellular plants and animals can double in a day or a few hours.

Even though natural populations occasionally attain their maximum growth potential, such rates of increase never prevail for long, and population growth is, fortunately, brought under control.

Fluctuation in size, not constancy, is the rule for natural populations.

Because birth and death are sensitive to changes in climate and food supply, population fluctuations follow directly upon variation in the environment. The degree of variation in the size of a population depends both on the magnitude of fluctuation in the environment and on the inherent stability of the population. After sheep became established on Tasmania, their population varied irregularly between 1,230,000 and 2,250,000 individuals—less than a twofold range—over nearly a century (Figure 31-1). Short-lived organisms

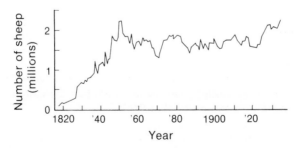

Figure 31-1   Number of sheep on the island of Tasmania since their introduction in the early 1800s (after Davidson 1938).

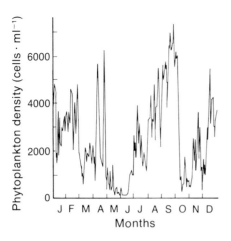

Figure 31-2   Variation in the number of phytoplankton in samples of water from Lake Erie during 1962 (after Davis 1964).

are more sensitive to short-term variation in the environment, and their populations often fluctuate dramatically (Figure 31-2). Because sheep live for several years, the population at any given time includes individuals born over a long period, and the effects of short-term fluctuations in birth rate on population size are thereby evened out. But because the life span of the single-celled algae that constitute the phytoplankton of a lake is measured in days, the turnover rate of algal populations is rapid and populations are thus vulnerable to the capriciousness of the environment. The cold snap that causes sheep a little discomfort may kill most of a population of insects.

Not all species' populations respond to the same environmental factors, even though they may be otherwise similar ecologically. For example, fluctuations in the densities of four species of moths, whose larvae feed upon pine needles, were followed for sixty years in a managed pine forest in Germany (Figure 31-3). The populations were sampled by counting the number of pupae (or hibernating larvae in *Dendrolimus*) per square meter of forest floor. The populations fluctuated by factors of between one hundred and ten thousand in a few years. Furthermore, the peaks and troughs of the four species' populations did not coincide closely, suggesting that even though the species fed on the same resource in the same forest, their populations were controlled independently by different factors. The lack of correlation among the population trends eliminates weather and food supply as major factors in population control, raising the possibility that specialized predators or parasites may have been involved.

In seasonal environments, reproduction occurs only when climate and resources combine to produce favorable conditions. Seasonal changes in temperature, moisture, and nutrients additionally affect mortality, either directly or indirectly. When life span is so short that many generations occur within a year, seasonality can greatly influence population size. Near Adelaide, South Australia, populations of the tiny insect pest *Thrips imaginis,* which infests roses and other cultivated plants, undergo regular cycles of increase during seasonally favorable conditions, followed by rapid decline

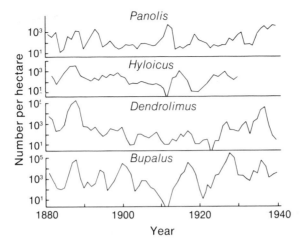

Figure 31-3   Fluctuations in the number of pupae of moth species (hibernating larvae in *Dendrolimus*) in a managed pine forest in Germany for sixty consecutive midwinter counts (after Varley 1949).

when the environment becomes either too dry or too cold to support population growth (Figure 31-4). Adelaide has a Mediterranean climate: winters are cool and rainy; summers, hot and dry. The winter months are generally too cold to sustain growth of thrips populations; summers are too dry. The spring (October through December in the Southern Hemisphere) brings an ideal combination of moisture, warmth, and plant flowering for population growth.

Thrips subsist mainly on plant pollen, whose abundance varies seasonally with the production of flowers. The climate of Adelaide is mild enough that some flowers are always available and thrips are active all year. During the winter, however, the depressing effect of cool temperature on develop-

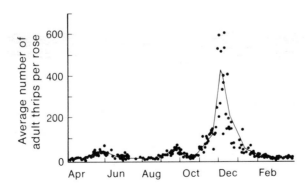

Figure 31-4   Number of *Thrips imaginis* per rose from April 1932 through March 1933 near Adelaide, Australia. Dots indicate daily records; the curve is a moving average for fifteen days (after Davidson and Andrewartha 1948).

ment rate and fecundity (Table 31-1) brings a marked decline in the number of thrips. Population growth is further checked by high mortality during the immature stages; the period from egg to adult is so long during the winter that most flowers wither and fall off before the thrips reach maturity. The warm weather of spring increases the net reproductive rate of the thrips, while shortening the mean generation time. Under these conditions, populations rapidly increase to infestation levels. Increasing mortality caused by ensuing summer drought halts population growth, and results in rapid population decline in late November or early December.

Some populations of long-lived organisms, particularly mammals and birds, fluctuate with great regularity; peaks and troughs in their numbers occur at intervals of anywhere from three to ten years. The most remarkable cycles are exhibited by some mammal populations of the New World arctic, where the regularity of population fluctuations is precise enough to predict population size several years in advance. For example, the lynx populations of Canada have a cycle of approximately ten years, closely following cyclic changes in the population of snowshoe hares (Figure 31-5).

Ecologists are not completely agreed on the cause of population cycles. The populations vary much too regularly for the cycles to be caused solely by variation in the environment. The most reasonable explanation ties the cycles to inherent population properties of predator-prey interactions. In one version of this hypothesis, an increase in the hare population is followed by an increase in its principal predator the lynx, which eventually becomes so numerous that the hare population can no longer sustain the predation rate and declines. Lynx follow suit as their food supply diminishes. When the lynx become scarce, the hare population can increase and the cycle begins again. This hypothesis does not adequately explain the lynx and hare population cycles because there are too many bits of contrary evidence. First, the reproductive potential of the hare is so much greater than that of the lynx that the lynx population could not increase fast enough to exterminate the hares unless some other factor, perhaps insufficiency of food, slowed the growth

TABLE 31-1

Influence of temperature on development rate, life span, and fecundity of *Thrips imaginis* (from Andrewartha 1935, summarized in Andrewartha and Birch 1954: 569).

Parameter	Temperature (C)	
	8 to 12	23 to 25
Length of adult life (days)	250	46
Total eggs laid per female*	192	252
Daily egg production	1.4	5.6
Development period, egg to adult (days)	44	9

* With pollen (a protein source) in their diet. If the adults are raised without pollen, their lifetime egg production falls to 20 at 24 C and adult life span increases to 77 days.

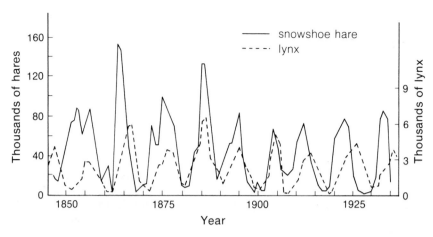

Figure 31-5   Population cycles of the lynx (*Lynx canadensis*) and the snow-shoe hare (*Lepus americanus*) in the Hudson's Bay region of Canada, as indicated by fur returns to the Hudson's Bay Company (after MacLulich 1937).

rate of the hare population. Second, lynx population peaks occasionally coincide with or precede hare population peaks rather than following them by a year or so (Leigh 1968). Third, on some islands from which lynx are absent, snowshoe hare populations fluctuate just as much as they do on the mainland (Elton and Nicholson 1942b, Keith 1963). Perhaps population crashes are caused by periodic decline in quality or quantity of plants on which the hares feed, which in turn causes a decline in the hare population (and the lynx population), and subsequently allows the plants a chance to recover from overeating by hares. After all, snowshoe hares are predators, too. The fact that their food is plant rather than animal does not mean that we cannot treat it as prey.

After analyzing hare and lynx population data, Gilpin (1973) suggested that hares were predators on lynx, not in the conventional sense of eating them but perhaps by serving as vectors and reservoirs of their diseases. Of course, the greatest pox on the lynx in recent times has been the fur trapper. If trappers waited for hare populations to increase before going out to trap the lynx, they possibly could have driven the entire cycle.

Population growth rate is influenced by both stabilizing and nonstabilizing factors.

Any general ecological theory of population regulation must be able to explain three basic patterns of population growth. First, many populations tend to remain, within narrow limits, near an equilibrium level, because *stabilizing* factors restrict the tendencies of populations to vary from this level. Second, *nonstabilizing* factors change population size without regard to its equilibrium size. In some populations, the force of nonstabilizing influences completely obliterates any trace of a stable population level. Third, some populations exhibit regular *oscillations*. To begin with, we shall focus our attention on stabilizing and nonstabilizing factors.

Stabilizing factors tend to reduce population density when it exceeds its equilibrium level, and to increase population density when it is below the equilibrium level. Hence, the influence of stabilizing factors varies with the density of the population. For this reason, stabilizing factors are usually called *density-dependent,* a distinction that was first made by Howard and Fiske (1911). Nonstabilizing factors act upon individuals without regard to population density, and thus have been called *density-independent.* Such influences do not tend to restore a population to any particular level. Whether a heavy frost will kill an individual insect is often independent of the density of the population of which it is a member. Hot, dry air robs an individual of its life-sustaining moisture without regard to the number of individuals in the population.

The distinction between density-dependent and density-independent factors, and their relative roles in regulating density, has been vigorously argued by many authors (Nicholson 1933, Smith 1935, Varley 1947, Andrewartha and Birch 1954, Lack 1954). Confusion over the terms has been heightened by their frequent association with biotic and physical factors, or with competitive and noncompetitive influences, respectively (Lack 1966). Andrewartha and Birch argued, quite rightly, that the killing effects of a frost, or other density-independent influences, depend on how sheltered the place the individual inhabits is. If the number of well-protected living places is limited, the influence of climatic factors may well be density-dependent. Clearly, the individuals in a population are not equally affected by a frost: some live, some die.

Density-dependent influences on individuals are an expression of competition.

Stabilizing influences vary with the density of the population. Death from starvation or predation is often density-dependent. The larger the number of individuals in the population, the less food for each individual. Competition for limited resources, perhaps food or water, in turn affects the survival and reproduction of the individual and the growth rate of the population. When suitable habitats or hiding places are limited to prey populations, predation may increase greatly at high population density. In this case, individuals either compete successfully for a suitable place to live or risk being eaten.

Space may limit populations of plants and sessile aquatic animals. Individual barnacles clearly compete for space, as none is left uncovered (page 239), but Andrewartha and Birch (1954) would argue that the density of the barnacle population is limited by the space itself, not by the influence of competition for space on population parameters. The population is limited to a certain number of individuals per square meter regardless of the intensity of competition.

The most straightforward approach to understanding population regulation is to start with the following axiomatic argument. The size of a stable population is maintained in equilibrium by the balance between two opposing forces: the inherent growth potential of the population and limits to population growth imposed by the environment, whose action varies according to population size. In a closely regulated population, the birth rate must, on average, exceed the death rate at low population density, and the death rate must exceed the birth rate at high population density (Figure 31-6). At some intermediate population density, the birth and death rates exactly balance and the population will remain at that level until thrown out of equilibrium by some nonstabilizing influence. Because the birth-rate and

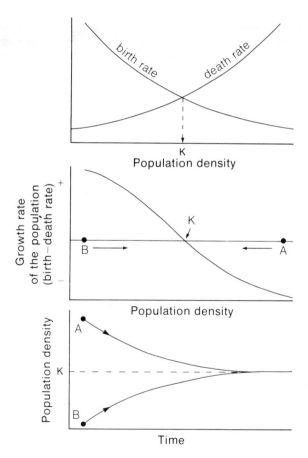

Figure 31-6   Diagramatic representation of birth and death rates in a regulated population (top), the resulting growth rate of the population with respect to density (middle), and the approach of population size to the equilibrium level (*K*) with time (bottom). A and B represent the densities of two populations whose birth and death rates are not in equilibrium.

death-rate curves vary with changing environmental conditions, equilibrium population size will also vary. For example, cold weather can decrease the reproductive rate of insects by extending their development period and reducing their egg production (Table 31-1). If the death-rate curve remained unchanged, this would shift the equilibrium population size to the left. Seasonal changes in the structure of the habitat, the quality of food resources, the activity of predators, and so on, shift the position of the equilibrium, as do such inherent characteristics of the population as age structure, acclimation to local conditions, and genetic makeup.

It is important to recognize that the factors that influence population growth can interact, regardless of whether their effect is stabilizing or not. Predators are more likely to capture individuals that are weak from starvation. Weak or diseased individuals are usually more susceptible than healthy individuals to conditions of drought or cold. Factors that are supposedly density-independent can thus accentuate the expression of stabilizing influences.

How do stabilizing factors affect population processes?

If populations were regulated by density-dependent influences, we would expect growth rates to decrease with increasing population density. Above the equilibrium population size, the growth rate of a population should be negative; below the equilibrium, it should be positive. Does the behavior of natural populations conform to this pattern? We should only have to measure the growth rate of a population at different densities to verify or reject the hypothesis. In practice, however, such data are difficult to obtain. A valid test of the hypothesis can include data gathered only under identical conditions so as to eliminate the effect of variation in density-independent factors on population size (Maelzer 1970). Density is meaningful only in the context of resource availability. To be certain resources remain unchanged, we must restrict our observations on density-dependence either to laboratory populations or to manipulated populations in natural habitats (Amant 1970).

Introductions of populations can provide such manipulated situations. The rapid increase of the pheasant population on Protection Island (mentioned on page 508) is such a case (Figure 31-7). For six years after the introduction, the pheasants were counted in the spring, prior to reproduction, and again in the fall, after the reproductive season had ended and the young were fully grown. When population size is plotted on an arithmetic scale, the absolute rate of population growth ($\Delta N/\Delta t$) appears to increase steadily in geometric fashion. But when plotted on a logarithmic scale, the data reveal that the relative growth rate ($1/N \cdot \Delta N/\Delta t$) steadily decreased with increasing population size.

The slowing of population growth rate was related to a steady decline in the percentage rate of increase of the population from spring to fall (Figure 31-8). As the number of eggs laid by females did not change, most of the decline in production was caused by reduced nesting success and increased mortality of the young. The number of pheasants always decreased between

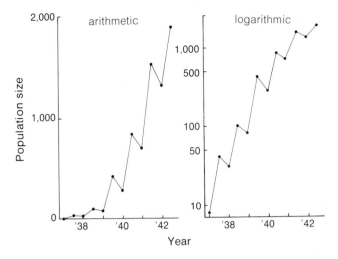

Figure 31-7   Arithmetic and logarithmic growth of the ring-necked pheasant population on Protection Island following its introduction in 1937 (data from Einarsen 1942, 1945; quoted by Lack 1954).

the fall and spring censuses because there were no births during the winter period, only deaths. The percentage drop in population size from fall to spring remained fairly constant, at about 20 per cent. By subtracting the fall-to-spring decrease in population size from the spring-to-fall increase, one can estimate the annual rate of increase of the population and relate it to population size (Figure 31-8). This line intersects the horizontal axis (zero population growth rate) at about 1,600 individuals, suggesting that the population would be regulated at about that size.

Because the environment constantly changes, it is difficult to describe precisely the relationship of population processes to density in natural populations, particularly if population size varies within narrow limits. Laboratory populations maintained at different densities, but under identical conditions, demonstrate the response of birth rate and death rate to density. In a series of experiments on the water flea *Daphnia pulex* (Frank, Boll, and Kelly 1957), initial population densities varied between 1 and 32 individuals per milliliter of water. The populations were grown in small beakers with cultures of green algae provided for food. Survival and fecundity of females were noted for two months after the beginning of the experiment, and the data were used to construct life tables for populations at each density, shown graphically in Figure 31-9. Fecundity decreased markedly with increasing population density. Survival actually increased at densities up to eight individuals per milliliter; high fecundity apparently reduces survival, indicating a trade-off between reproductive effort and adult survival. At densities of eight or more individuals per milliliter the body growth of individual water fleas was stunted, suggesting that depletion of food resources ultimately limited birth rates and survival in the dense cultures.

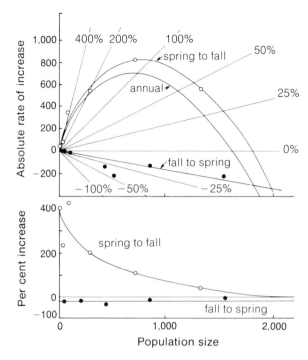

Figure 31-8   Absolute rate of increase in the ring-necked pheasant population of Protection Island (above). The approximate annual increment in population size is obtained by subtracting the curve for the fall-spring decrease from that for the spring-fall increase. Percentage increase values (below) are calculated by the expression $(N_t - N_0)/N_0$, where N_0 and N_t are initial and final population sizes, respectively, for the interval 0 to t. The percentage increase is zero (hence population size is in equilibrium) at about 1,600 individuals.

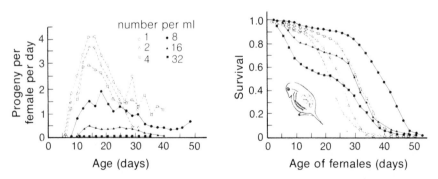

Figure 31-9   Fecundity and survival in laboratory populations of *Daphnia pulex* at different densities (from Frank, Boll, and Kelly 1957).

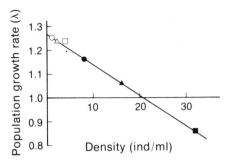

Figure 31-10   Values of λ calculated from the life table data for
Daphnia pulex portrayed in Figure
31-9. Population growth rate decreases as a function of density. A
density of 20 individuals per milliliter represents the apparent carrying capacity of the environment, at
which point λ is equal to 1.0 and
the population just maintains itself
(after Laughlin 1965).

Population growth rate (λ), calculated from life table data for the water
fleas, decreases with increasing density and falls below 1.0 (equilibrium
population size) at a density of about twenty individuals per milliliter (Figure
31-10). Under the conditions of temperature, light, water quality, and food
availability provided in the laboratory, water flea populations would attain a
stable equilibrium of twenty individuals per milliliter, representing the ability
of the environment to support the population, regardless of the initial density
of the culture.

Although nonstabilizing factors influence population growth rate, they cannot regulate population size.

Strictly speaking, factors whose influence on the survival and fecundity
of individuals are independent of the density of the population cannot regulate population size. But this amounts to saying that populations are regulated by factors that regulate populations, and critics quickly point out the
circularity of the argument. Andrewartha and Birch (1954: 649) wrote that
"the usual generalizations about 'density-dependent factors,' when they refer
to natural populations, have a peculiar logical status. They are not a general
theory, because ... they do not describe any substantial body of empirical
facts. Nor are they usually put forward as a hypothesis to be tested by
experiment and discarded if they prove inconsistent with empirical fact. On
the contrary, they are usually asserted as if their truth were axiomatic."

Andrewartha and Birch argue that most populations, particularly those of
insects and other small invertebrates, are influenced primarily by "density-
independent" factors, and that the period in which environmental conditions
are favorable for population growth ultimately controls the size of the population:

> The numbers of animals in a natural population may be limited in three ways: (a)
> by shortage of material resources, such as food, places in which to make nests,
> etc.; (b) by inaccessibility of these material resources relative to the animals'
> capacities for dispersal and searching; and (c) by shortage of time when the rate
> of increase *r* is positive. Of these three ways, the first is probably the least, and
> the last is probably the most, important in nature. Concerning *c*, the fluctuations

in the value of r may be caused by weather, predators, or any other component of environment which influences the rate of increase (Andrewartha and Birch 1954: 660).

Thrips imaginis is cited by Andrewartha and Birch as an example of a population whose size is determined by nonstabilizing factors (see page 513). Variation in the maximum size of the thrips populations in the spring was analyzed by Davidson and Andrewartha (1948) with respect to variation in climate. By far the most important of these was the temperature from the beginning of the growing season (date of the first winter rains) until August 31. Rainfall during September and October, just prior to the build-up of the thrips population, also partly determined maximum population size, but temperature during this period apparently was not critical—it was always sufficiently high for maximum reproduction and growth. Davidson and Andrewartha suggested that year-to-year variation in the growth of the thrips population, and its decline at the beginning of the dry season, is determined largely by nonstabilizing factors. The most important factor, temperature several months *prior* to the seasonal increase in the thrips population, probably influences population growth through its effects on plant growth and the production of pollen; the pollen food resource could, however, exert a stabilizing, rather than a nonstabilizing, influence on thrips population growth.

Even if the rapid population growth of the thrips in the spring were related to nonstabilizing influences, the peak populations attained would depend in large part on the population size at the beginning of the growth phase. In exponential growth, described by the equation $N(t) = N(0)e^{rt}$, the initial population size $N(0)$, is an important component of population size in the future. For a population to be maintained within narrow limits, stabilizing forces need operate only during a portion of the year that is long enough to readjust the population to its equilibrium level each year.

The thrips population data may be interpreted in the context of density-dependent factors by supposing that the equilibrium population size (K) varies seasonally, becoming so high during the spring that the thrips population cannot grow to K before it is forced back to its low dry season and winter level (Figure 31-11). Under these conditions, spring population size is determined by the regulated population size in winter ($N(0)$) and the rate of population growth in spring (r). Because population size in spring remains far below the ultimate equilibrium level, r is influenced primarily by density-independent factors like temperature and rainfall, at this time. As a result, the peak size of the population in spring appears to be determined by density-independent factors. But the fact that spring populations remain within certain limits over many years is due primarily to density-dependent regulation in winter. Davidson and Andrewartha's thrips data do not support any one theory of population regulation better than any other because they ignored population processes during the critical winter months (Nicholson 1958, Smith 1961, 1963, Andrewartha 1963). The data do illustrate the influence of density-independent factors on population processes, but they are not sufficient to reject the hypothesis of density-dependent regulation overall (Varley 1975).

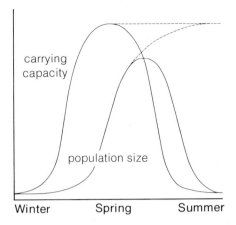

Figure 31-11   Graphical interpretation of the seasonal growth and decline of thrips populations; population changes are related to, but lag behind, seasonal trends in *K*. The behavior of the population if *K* were maintained at a new high value for a long period is indicated by the dashed line.

Some populations exhibit mosaic patterns of local extinction and recolonization.

One attempt to reconcile density-independent influence and population regulation has its roots in the patchy distribution of individuals into local subpopulations. If a population were influenced overwhelmingly by nonstabilizing factors, local subpopulations would dwindle to extinction rather frequently. Although extinction would not normally be considered as a positive force in population regulation, Ehrlich and Birch (1967) have attempted to recast the density-independent arguments for population regulation in exactly these terms. They suggested that populations are mosaics of numerous subpopulations that receive occasional immigrants from other areas but are otherwise independent. If a local population became extinct by chance, it could be re-established by colonization from surrounding subpopulations. Thus a population could be regulated far below any resource limitation through the balance of local extinction and recolonization. This hypothesis echoes Cockerell's (1934) observations on scale insects (*Coccidae*) in New Mexico:

> Certain species occur on the mesquite and other shrubs which exist in great abundance over many thousands of square miles of country. Yet the coccids are only found in isolated patches here and there. They are destroyed by their natural enemies, but the young larvae can be blown by the wind or carried on the feet of birds, and so start new colonies which flourish until discovered by predators and parasites. This game of hide-and-seek doubtless results in frequent local extermination, but the species are sufficiently widespread to survive in parts of their range, and so continue indefinitely.

Ehrlich and Birch's mosaic model of population regulation is not strictly density-independent because, while extinction may occur by chance, the colonizing ability of a subpopulation depends on how many areas of habitat are unpopulated. When the population is sparse, many local areas are available for recolonization; when it is dense, few areas are unoccupied. Therefore, the tendency of the population to increase through colonization of new areas is density-dependent. The mosaic model differs from more conventional density-dependent models only by incorporating a spatial component that separates death processes (extinction) from birth processes (recolonization). The Ehrlich-Birch model does, however, suggest that it may be impossible to identify density-dependent factors acting locally in a population that is well regulated over a large area.

Can density-dependence be demonstrated in nature?

As we have seen, density-dependent influences on populations can be demonstrated in introduced populations that expand quickly up to an equilibrium point and in laboratory populations for which all environmental factors, other than population density, are rigidly controlled. Natural populations cannot be manipulated so easily as laboratory populations, and ecologists have had to resort to elaborate statistical analysis of population trends to identify the existence of density-dependent regulation and to identify the factors influencing population growth.

A simple way to look for density-dependent regulation would be to compare population size for successive time intervals. In Figure 31-8, we examined the relationship of percentage increase $[N(t + 1) - N(t)]/N(t)$ to population size $N(t)$ for pheasants on Protection Island. In this case, reproductive rate appeared to decrease with increasing population size. As Watt (1964a) pointed out, however, this technique is not strictly valid as a test for density-dependence because if ΔN were independent of population size, the percentage increase in population size would be inversely related to $N(t)$. For example, if a population increased by ten individuals per year, λ would be 2 if $N(t)$ were 10, but only 1.1 if $N(t)$ were 100.

In order to circumvent this problem, Morris (1959, 1963a) suggested that $\log_e N(t + 1)$ should be compared with $\log_e N(t)$. According to the logarithmic form of the population growth equation,

$$\log_e N(t + 1) = \log_e N(t) + r.$$

If the growth rate of the population were independent of density (r constant), $\log_e N(t + 1)$ and $\log_e N(t)$ should be linearly related with a slope of 1. If r decreased with increasing population size, the slope of the relationship would be less than 1.

Tanner (1966) used Morris's procedure to analyze the trends of seventy populations of a wide variety of species of animals. In 63 of them the slope of $\log_e N(t)$ was significantly less than 1. Tanner then checked the slope of r upon $N(t)$ in these populations to verify the density-dependence of the population growth rate r. Only sixteen of these populations failed to show a significant negative slope for r upon $N(t)$. When density-dependence cannot be demonstrated, however, it is impossible to distinguish whether it is absent or merely undetectable at significant levels. But Maelzer (1970) and Amant (1970) additionally have pointed out that under certain conditions, even random density-independent population trends can give spurious statistical indications of density-dependence. Additional problems with the method are discussed by Benson (1973) and Ito (1972).

Regardless of whether regression analysis is valid for a particular population, it fails to provide insight about the mechanisms of population regulation. Density-dependence is more than a statistical property of populations, it is the expression of biological processes. Several studies of population regulation in economically important insect pests have attempted more detailed investigation of population regulation. Two of these studies (involving the European spruce sawfly and the diamondback moth) are of particular interest from the standpoint of the sampling procedures and statistical methods employed, and provide models for other studies.

The spruce sawfly (*Diprion hyrcyniae*) was first introduced to Quebec from Europe in about 1930. By 1938, it had spread throughout most of Quebec, the maritime provinces of Canada, and New England, causing widespread damage, including complete defoliation of spruce forests. Most sawflies are female—fewer than one in a thousand are male—and reproduction is predominantly asexual. Females lay eggs singly in slits cut in spruce needles; the larvae feed on the foliage after they hatch (Figure 31-12). After the larvae are fully grown, they drop to the ground and burrow a few inches into moss, where they spin a tough cocoon around themselves and metamorphose into adults. The sawfly population goes through several generations each summer, but toward fall, pupae (the developmental stage between larval and adult forms) enter into a diapause state in their cocoons to spend the winter. The Canadian government introduced several parasites in an attempt to control the sawfly, but eventually a virus, accidentally introduced with one of the parasites, proved to be the most effective control agent.

Canadian foresters and entomologists initiated a census program during the height of the sawfly outbreak in the late 1930s. The program was con-

Figure 31-12   The larch sawfly, a close relative of the European spruce sawfly, feeding on needles of western larch (courtesy U.S. Forest Service).

tinued for more than twenty years. Larvae were sampled by spreading large canvas sheets on the ground under a tree and vigorously shaking the limbs directly above it. The dislodged larvae were collected and reared individually in vials to determine rates of disease and parasitism in the population. A census of cocoons was taken each year. Wooden trays were filled with moss gathered from an area free of sawflies. The trays were then set out in an infested forest during July, before the first generation larvae began to drop to the forest floor, to serve the larvae in place of naturally occurring moss. The following May, after most sawflies had emerged from their cocoons, the trays were brought back to the laboratory and the cocoons classified as either (a) sound but unemerged, (b) emerged normally, (c) parasitized, (d) preyed upon by wireworms, or (e) preyed upon by small mammals, mostly rodents. Each type of predator and parasite leaves a distinguishing mark on the tough and leathery cocoons.

The size of the sawfly population varied by between tenfold and a hundredfold during a period of a few years (Figure 31-13). Rates of parasitism and disease also varied greatly. Statistical analysis of population trends in relation to biotic and climatic factors identified factors responsible for fluctuations in population size; the analysis also distinguished density-dependent factors from density-independent factors (Neilson and Morris 1964).

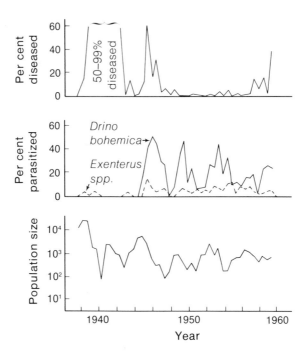

Figure 31-13   Population density, percentage of diseased individuals, and percentage of parasitized individuals of fifth-instar larvae of the European spruce sawfly (*Diprion hyrcyniae*) on a study area of black spruce forest. *N* is the mean number of larvae per tree sample times 1,000 (after Neilson and Morris 1964).

When the logarithm of the population in each year was plotted as a function of the logarithm of the population during the previous year, the slope of the relationship (0.82) was less than 1.0, suggesting that stabilizing influences were acting upon the population. To determine the possible role of parasites and disease organisms in the regulation of population size, Neilson and Morris plotted each year's population size, not as a function of the entire population of the previous year, but as a function of the number of larvae surviving parasitism or disease, or both. If either, or both, of these factors together were entirely responsible for the stabilizing influences on the population, the slope of the relationship between successive population sizes after these factors had been taken into account should have been 1. Accounting for disease did not significantly alter the relationship between the size of successive populations (Table 31-2), thus disease had little stabilizing influence on population growth. (Note that disease was, however, a major mortality factor for the sawfly and was largely responsible for bringing serious outbreaks under control.) But taking parasitism into account increased the slope of the relationship to 0.89 and also improved the significance of the correlation (72 per cent of the variance is accounted for). Therefore, parasites exerted a stabilizing influence on the population.

Neilson and Morris found that part of the deviation from the regression line relating successive population sizes was correlated with rainfall during May and June. Abnormally high precipitation tended to increase the sawfly population, whereas low precipitation reduced numbers below expected values. By eliminating the deviations owing to variation in rainfall, the correlation between population size in successive years, based on individuals surviving parasitism and disease, was improved considerably (to 85 per cent). The slope of the regression did not increase above 0.89, however, which suggests that rainfall did not act in a density-dependent manner. Some other

TABLE 31-2

Relationship of population size in the European spruce sawfly (*Diprion hyrcyniae*) to population size during the previous year: uncorrected, and corrected for disease, parasitism, and rainfall (after Neilson and Morris 1964).

Previous year's population base	Slope of relationship	Percentage of variation accounted for
Number of individuals	0.82	59
Number surviving parasitism	0.89	72
Number surviving disease	0.83	61
Number surviving both parasitism and disease	0.89	74
Number surviving both parasitism and disease with a correction for May-June rainfall	0.89	85

factor, not discovered by Neilson and Morris, must also have exerted a stabilizing influence on the sawfly populations.

The sawfly study provided several insights into the regulation of insect populations. First, although disease and weather were important components of mortality, neither exerted a stabilizing effect on population size. (Disease was, however, largely responsible for bringing serious outbreaks under control.) Second, parasites that attacked larval stages were found to be the only factor that acted in a density-dependent manner. Third, additional unidentified density-dependent factors must have acted at other stages of the life history because larval mortality could not account for all the regulatory influences.

Harcourt (1963) used a somewhat different approach from that of Neilson and Morris to analyze changes in population size in the diamondback moth, a worldwide pest of cabbage and related crops. The aim of his *key factor* method was to determine which causes of mortality were most important in determining *changes* in population size, regardless of whether their action was stabilizing or nonstabilizing (Varley and Gradwell 1960, Kuno 1971, Luck 1971).

In southern Ontario, the diamondback moth usually has four, and in exceptional years six, generations during a single season. During the course of each growing season, Harcourt sampled populations at each stage of the moth's life cycle (egg, several larval instars, pupa, and adult). Rate of parasitism was determined from extensive rearings of larvae and pupae. Fecundity was estimated by raising adult moths in the laboratory. Mortality of the adults was measured indirectly by comparing the number of eggs present on the cabbage plants at the beginning of a generation with the number expected if all the adults from the previous generation had survived to lay eggs. Although some of his data were obtained indirectly, Harcourt was able to construct a reasonable life table for the moth during each generation (for example, the life table for the second generation of 1961 is shown in Table 31-3). Mortality during each stage varied from more than 70 per cent of small larvae and adults to less than 2 per cent during the egg stage. The major causes of mortality also varied from one stage of the life history to the next. Small larvae were particularly vulnerable to heavy rainfall, whereas large larvae and pupae were heavily parasitized by wasps. In the generation described in Table 31-3, most of the adult mortality was caused by inclement weather during the oviposition period. Fecundity was only one-quarter of the normal level because of the short photoperiod during this early generation. Harcourt equated the depression in fecundity to adult death.

The relative importance of each mortality factor in determining the size of the next generation of the diamondback moth was calculated by key factor analysis. The key factor method is based on the premise that the number of eggs laid in one generation is equal to the number of eggs laid in the previous generation, times the product of the survival rates during each stage of the life history, times the fecundity of the adults raised in the previous generation. Harcourt calculated a trend index, which is the number of eggs in a generation divided by the number of eggs in the preceding generation, for eighteen successive generations within four years. The trend index for the

TABLE 31-3

Life table for the second generation of the diamondback moth (*Plutella maculipennis*) on early cabbage, in southern Ontario 1961 (modified after Harcourt and Leroux 1967).

Stage	Number per 100 plants	Survival (per cent)	Cause	Number per 100 plants	Percentage of stage
				MORTALITY	
Egg	1,580	100	Infertility	25	1.6
Larva					
Period 1	1,555	98.4	Rainfall	1,199	77.1
Period 2	356	22.5	Rainfall	36 ⎤	24.7
			Parasitism by M. plutellae	52 ⎦	
Period 3	268	18.1	Parasitism by H. insularis	69	25.7
Pupa	199	12.6	Parasitism by D. plutellae	92	46.2
Adult	107		Inclement weather	20	18.7
Reduction in fecundity			Photoperiod		73.6

overwintering generation was not calculated and so, like Davidson and Andrewartha's study on thrips, only the summer season of population increase was considered. Harcourt then compared variation in the trend index with variation in the mortality caused by each factor during the life cycle. If variations in a particular factor were relatively unimportant in determining the overall trend of the population, the correlation between variation in that factor and variation in the trend index over many generations would be negligible.

For the diamondback moth, adult mortality, responsible for only 16 per cent of the total mortality of individuals, was nonetheless responsible for more than 70 per cent of the variation in the trend index (Table 31-4). Rainfall during the first larval period caused the death of 55 per cent of the individuals in the population, but accounted for less than 10 per cent of the total variation in the trend index: The most important causes of death are not necessarily those with the greatest influence on variation in population size.

Harcourt and Leroux (1967) pointed out that the key factor is usually one that causes high, but variable, mortality from generation to generation. The key factor for the diamondback moth, adult mortality, was related to weather and thus was probably nonstabilizing. Stabilizing influences may have operated on the population as a result of parasitism during the late larval and

TABLE 31-4
Combined life-table and key-factor analysis for eighteen generations of the diamondback moth (*Plutella maculipennis*) in southern Ontario (from Harcourt 1963).

Stage of life history	Survival to beginning of stage	Death during stage	Mortality rate during stage (per cent)	Major factor	Percentage of total regulation*
Egg	100	1	1	Infertility	3
Larva 1	99	55	55	Rainfall	8
2	44	9	20	Parasitism	1
3	35	12	34	Parasitism	17
Pupa	23	5	22	Parasitism	1
Adult	18	16	89	Inclement weather	73

* The total is slightly greater than 100 per cent because of interactions among the factors.

pupal stages, or during the overwintering stage of the moth. Since Harcourt's study did not include population trends between years, it is not surprising that the key factor was density-independent. Trend indices within growing seasons ought to reflect seasonal changes in climatic parameters more strongly than they would exhibit stabilizing influences.

Harcourt and Leroux (1967) summarized key-factor analyses for twelve Canadian agricultural and forest insect pests. In these diverse insects, the critical stage at which key factors acted, as well as the nature of the key factor itself, varied from species to species. In only one of the studies was food supply found to cause substantial *variation* in population size. Weather was the most important factor in three of the species, and disease, parasitism, or emigration from local populations ranked highest in the rest.

How are oscillations maintained in populations?

The cyclic fluctuation of many arctic mammal and bird populations, first described in detail by Charles Elton (1924), has stimulated considerable interest in the processes that regulate the size of such populations (Elton 1942, Krebs *et al.* 1973). Analysis of fur returns from various regions has revealed certain generalizations concerning these cycles. Many of the cycles seem to be closely aligned with one another. We have already discussed the correlation between the cycles of the Canadian lynx and its principal prey, the snowshoe hare (see page 512). The fluctuations of animal populations that eat the same cyclic food supplies also show similar cycles. For example, the approximately nine-year cycles in the fur returns of the colored fox and the lynx from Manitoba are closely related, and both species prey heavily on snowshoe hares in that area. Several kinds of game birds, particularly the prairie chicken and ruffed grouse, have well-marked nine- or ten-year population cycles which coincide with each other, and also with the lynx and the coyote in the same region (Keith 1963). Populations of the introduced Hungarian partridge follow the same cycle.

The period of the cycle varies from species to species, and even within a species. In Canada, most cycles have periods either of nine to ten years or of four years. Although the colored fox exhibits a nine- to ten-year cycle over most of its range, it has a pronounced four-year cycle in Labrador and on the Ungava Peninsula (Elton 1942). This short cycle may be related to the fact that the principal prey of the fox in the eastern provinces are lemmings, which also have a short population cycle.

Small herbivores, such as voles and lemmings, commonly exhibit four-year cycles; large herbivores, such as the snowshoe hare, muskrat, ruffed grouse, and ptarmigan have nine- to ten-year cycles. Predators that feed on short-period herbivores (arctic fox, rough-legged hawk, snowy owl, and great northern shrike) themselves have short population cycles. Predators that feed on the larger herbivores (red fox, lynx, marten, fisher, mink, goshawk, and horned owl) have longer cycles (Dymond 1947). The length of the cycle also seems to be related to habitat: long cycles occur predominantly in forest species; short cycles in tundra species.

Population fluctuations are neither purely oscillatory nor purely periodic, in the strict mathematical sense, because the amplitude of cycles varies considerably and peaks are not spaced at equal intervals. For example, the colored fox returns of the Moravian missions between 1834 and 1932 revealed two peaks after intervals of two years, five peaks after five years, and one after six years, the remainder following four-year intervals—an average interval of almost four years. A string of seven four-year periods between 1847 and 1880 led Elton (1942) to remark that an Eskimo hunter ". . . might have reflected that his good luck and his bad luck chased each other with sufficient regularity to amount to a natural law." But variations in the period between peaks at one time led observers to believe that population fluctuations were merely the expression of random variation and not cycles at all (Moran 1952, 1953, Cole 1951, Finerty 1971). For such species as the lynx and snowshoe hare, whose populations exhibit evenly spaced peaks and troughs (Figure 31-5), such a belief seems unwarranted. In other species, cycles are less pronounced.

Cole (1951) derived an expected frequency distribution for the intervals between peaks based on a series of random numbers and compared this distribution with that obtained from population data (Table 31-5). To derive the random distribution, Cole reasoned that turning points in the trend of a random number series occur with a probability of one-half for each time interval. That is, the probability that a random number reverses the trend of the previous two numbers in a series is one-half. Furthermore, one-half of the turning points are maxima (peaks). Using probability theory, Cole calculated the theoretical proportions of intervals that will be two, three, four, and so on, years. The expected mean interval of the random series is precisely three

TABLE 31-5
Frequency distribution of intervals between maxima for a random series, for the width of growth rings in Douglas fir, and for populations of the colored fox (from Cole 1951, and data of Elton 1942).

Interval between maxima (years)	Frequency of intervals (per cent)		
	Random series	Douglas fir tree rings	Colored fox population
2	39	37	8
3	34	33	20
4	17	19	56
5	7	8	12
6	2	1	4
7	1	1	0
8 or more	0	1	0
Mean interval (years)	3.0	3.1	3.8

years. Mean intervals between peak years in many of the fluctuating animal populations analyzed by Cole were too long to be attributed to random fluctuations in population size. The distribution of intervals between peak widths of growth rings in the Douglas fir, however, suggests that the annual growth increment may vary randomly (Table 31-5).

The discovery of nonrandom population fluctuations has been limited primarily to economically important species in arctic and boreal habitats and, in a few cases, in arid regions. Populations of the desert locust, for example, undergo great fluctuations, the peaks of which occasionally are climaxed by mass migrations over broad areas (Dempster 1963). The general correlation between the occurrence of population cycles and the simplified ecological communities of arctic and desert environments suggests a basic relationship between the simplicity of a community and its stability. Although there may be theoretical grounds for this hypothesis, as we shall see in a later chapter, too few empirical data are available to make a valid generalization. Arctic mammal populations have been sampled intensively, largely because of the economic value of their thick fur. Few observations on changes in the populations of tropical species exist, but as such data are beginning to accumulate, they reveal increasingly that, at least among insects, populations may undergo marked variation in size from year to year (Smith 1971, Galindo et al. 1956). Whether these variations take the form of regular cycles remains to be seen. Fleming (1975) has suggested that populations of tropical rodents are relatively stable; rodent plagues have occurred only in areas of intense cultivation.

In the long run, observations on population size will probably tell us relatively little about the biological factors that cause regular fluctuations in populations. As Kendall (1948) has pointed out, "Experience seems to indicate that few things are more likely to mislead in the theory of oscillatory series than attempts to determine the nature of the oscillatory movement by mere contemplation of the series itself."

Oscillations in laboratory populations of the sheep blowfly are caused by time lags in density-dependent responses.

We expect density-dependent factors to restore population size to an equilibrium level, and yet cyclic populations never reach a single equilibrium point, rather they fluctuate around that point. If we could determine why density-dependent responses fail to damp population cycles, we might come closer to understanding predator-prey cycles. Experiments by Australian ecologist A. J. Nicholson (1958) on population regulation in the sheep blowfly demonstrate that when the action of density-dependent factors is delayed, population cycles will occur.

Nicholson kept one group of blowflies under conditions that forced the larvae to compete strongly for food, which caused marked periodic fluctuations in population size (Figure 31-14). Food provided for the larvae (liver) was limited to fifty grams each day, and food for the adults was unlimited. The number of adults varied from a maximum of about 4,000 per population to a minimum of zero (all the individuals were either eggs or larvae). The period of oscillation varied between thirty and forty days, approximately the maximum life span of an individual.

In this experiment, regular fluctuations of the blowfly populations were clearly caused by a time lag in the response of fecundity and mortality to the density of adults in the cages. When adults were numerous, many eggs were laid, resulting in strong larval competition for the limited food supply. Virtually none of the larvae hatched from eggs laid during adult population peaks survived, primarily because they did not grow large enough to pupate. Therefore, large adult populations gave rise to few adult progeny, and because adults live a maximum of four weeks the population soon began to decline. Eventually so few eggs were laid on any particular day that most of the larvae survived, and the size of the adult population began to increase again. Because of a time lag in the response of the population to density, fluctuations in population size did not damp out with time and come into equilibrium. Density-dependent mortality did not act immediately on the population of adults, but was felt a week or so later when the progeny were larvae. This mortality was not expressed in the adult population until those larvae emerged as adults about two weeks after eggs were laid. The adult flies did not compete directly with one another; competition was restricted to their progeny.

A direct test of the hypothesis that population fluctuations are caused by time lags would be to eliminate the time lag in the density-dependent response, that is, to make the deleterious effects of competition at high density felt immediately. Nicholson was able to do this by adjusting the amount of food provided so that adults were as severely limited by food as the larvae. Because adults must have protein to produce eggs, restricting the liver for

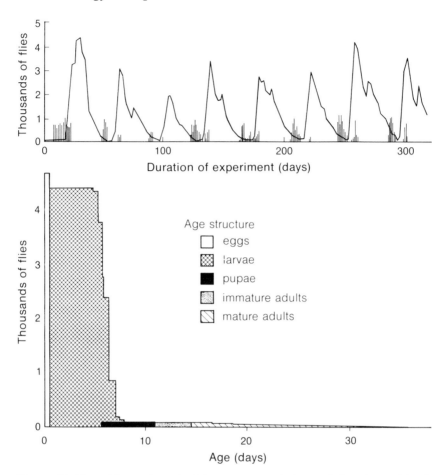

Figure 31-14   Fluctuations in laboratory populations of the sheep blowfly (*Lucilia cuprina*). Larvae were provided with fifty grams of liver per day; adults were given unlimited supplies of liver and water. Continuous line represents the number of adult blowflies in the population cage. The vertical lines are the number of adults that eventually emerged from eggs laid on the days indicated by the lines. The average age structure of the population is shown below (after Nicholson 1958).

adults to one gram per day curtailed egg production to a level determined by the availability of liver, rather than by the number of adults in the population. Under these conditions, the recruitment of new individuals into the population was determined at the egg-laying stage by the influence of food supply on fecundity (nearly all the larvae survived); as a result, fluctuations in the population all but disappeared (Figure 31-15).

Fluctuations in populations of the western tent caterpillar are governed by the same principles as in the sheep blowfly. A four-year survey on Van-

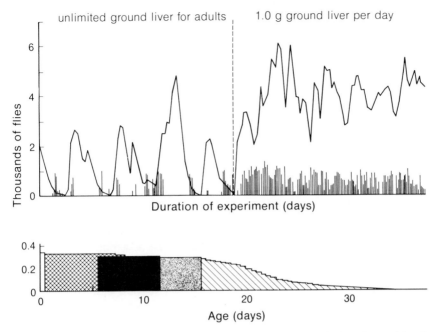

Figure 31-15   Effect on fluctuations in a population of the sheep blowfly (*Lucilia cuprina*) of limiting the amount of liver available to adults. The experiment was similar to that depicted in Figure 31-14 in all other respects. Average age structure is shown for the latter half of the experiment (after Nicholson 1958).

couver Island by Wellington (1960) included a peak year, 1956, followed by a rapid decline through 1959. The larvae of the tent caterpillar could be classified either as "active" or "sluggish" (see page 296). Broods of tent caterpillars composed primarily of active larvae differed from those with a higher proportion of sluggish larvae in that they constructed more tents with a more elongate structure, foraged over greater distances along the limbs of the tree, ate more leaves, and, as a result of their greater food intake, developed more rapidly. Active caterpillars survived better than sluggish caterpillars.

The proportion of active larvae in a particular brood was influenced by the past history of the infestation. Broods in areas in which the infestation had been low or absent in previous years were likely to contain more active caterpillars than broods from areas of recent high levels of infestation. The activity level of the larvae depended on the nutrition provided by the eggs from which they developed. The number of eggs laid by a female moth was relatively constant, regardless of her ability to provision each individual egg, and the amount of food that a moth could use to produce eggs depended upon how much she consumed as a caterpillar. When food was relatively scarce, few eggs were properly provisioned with nutrients and broods contained a relatively low proportion of active caterpillars. Since only the active forms were likely to pupate successfully, the recruitment of adult moths into the population was considerably reduced.

The quality of the caterpillars in a brood in a particular year did not depend on the food availability during that year, rather it was related to the availability of food to the female parent when she was a caterpillar during the previous year. Because the activity level of the caterpillar also partly determined how much food it consumed, the food supply of its parent during the previous year affected the success of its progeny during the following year. As a result, tent caterpillar populations were shown to have a time lag of a year or more in their response to food availability.

The causes of fluctuation in the insect populations outlined above suggest that time lags may be a general feature of regular biological fluctuations (May 1973, Auslander et al. 1974). Time lags might be expected to occur primarily in stabilizing processes involving reproduction rather than death, because death is an immediate response to environmental change. The effects of environmental conditions on reproduction may be transmitted to, and expressed only by, the offspring. Furthermore, a population can be reduced much more rapidly by death than it can be built up by reproduction. As we continue to consider how density-dependent factors stabilize populations, several possible sources for time lags will become apparent.

Stabilizing factors may act directly through physiological impairment of survival and reproduction.

Direct physiological effects of food availability upon reproduction were shown in a study of white-tailed deer by Cheatum and Severinghaus (1950). Proportion of females pregnant and average number of embryos per pregnant female were directly related to range conditions (Table 31-6). The number of corpora lutea in the ovary indicates the number of eggs ovulated, hence the reproductive potential of the female. The difference between the number of corpora lutea and number of embryos showed that reduced reproduction in females in poor range areas was due partly to secondary resorption of embryos, probably caused by poor nutrition of the pregnant female. In the central Adirondack area, where the condition of the habitat for deer was poor, ovulation was greatly reduced. Taber (1956) and Taber and Dasmann (1957) found similar patterns for mule deer in various successional chaparral communities in California, where variation in fecundity closely paralleled variation in ovulation rate. Clark (1972) related the proportion of females pregnant and litter size of coyotes directly to the abundance of their principal food, the jackrabbit. In most studies of this kind, reproductive performance and population trends directly parallel the condition of the habitat. Many studies on deer have additionally shown that high populations can lead to deterioration of the habitat through overeating. In one protected deer population in Ohio, a deteriorating range caused by overgrazing led to decrease in growth rate, adult body weight, ovulation rate, and production of fawns (Nixon 1965). The influence of variation in range condition on repro-

TABLE 31-6
Reproductive parameters of adult white-tailed deer (*Odocoileus virginianus*) in five regions of New York State, 1939–1949, arranged by decreasing suitability of range (from Cheatum and Severinghaus 1950).

Region	Per cent of females pregnant	Embryos per female	*Corpora lutea* per ovary
Western (best range)	94	1.71	1.97
Catskill periphery	92	1.48	1.72
Catskill central	87	1.37	1.72
Adirondack periphery	86	1.29	1.71
Adirondack center (worst range)	79	1.06	1.11
DeBar Mountain (1939–1943)	57	0.71	0.60
(1947)	100	1.78	1.86
Moose River (1939–1943)	91	1.00	0.98
(1947)	69	1.00	1.13

duction is usually most pronounced in young females. For example, in white-tailed deer in New England states, regional differences in range quality caused variation between 0 to 60 per cent in the pregnancy rate of one-year-old does, but only between 61 and 92 per cent in two-year-old does, and between 88 and 100 per cent in older females (Day 1964).

Range deterioration caused by overgrazing is often reversed by selective hunting in dense populations. Such programs frequently result in reduced populations, but improved range and reproductive performance. During the course of Cheatum and Severinghaus's (1950) study of white-tailed deer in New York, one of the worst ranges in the state, located in the Adirondack Mountains, was opened to hunting. Two study areas, DeBar Mountain and Moose River, were investigated between 1939 and 1943, and again in 1947; the areas were opened to deer hunting in 1943 and 1945. Because Moose River is inaccessible, few deer were shot by hunters; but in the more readily accessible DeBar Mountain area, 250 deer were removed from the 8,000-acre range in 1954 alone. Reproduction by the deer population in the DeBar Mountain area improved greatly after the hunting seasons, presumably following the reduced population of deer in the area, whereas reproductive performance in the Moose River area changed little (Table 31-6). Such studies emphasize the fact that reproduction and population growth depend on the relationship between the population and its resources. Different areas have different capacities to support populations. Within an area, the reproductive performance of a population depends on its density relative to the supportive capacity of the habitat.

Territoriality stabilizes population size.

The regulation of population size in deer occurs through adjustment of reproduction and mortality according to range conditions. The response time of the population is slow, and deer frequently increase in number so as to exceed the capacity of the environment to support them. In many areas, deer populations go through cycles of expansion followed by rapid crashes, as the habitat follows a complementary cycle of deterioration and improvement.

The failure of deer and habitat to achieve a stable equilibrium can be traced to a time lag in the response of each to the other. Other species, however, have behavioral mechanisms that adjust population size to resources at any given moment, thereby eliminating lags between the population and its environment and damping fluctuations in population size. The defense of territories is the most important of these behavioral mechanisms. To the extent that the area each individual defends is adjusted to resource availability at that time, the population as a whole will be adjusted to the ability of the habitat to support it.

One consequence of territorial behavior is that resources are distributed unevenly among individuals in a population. Under strong territorial or hierarchical systems, individuals either do or do not have sufficient resources for reproduction (Watson and Moss 1970). Environmental fluctuations are reflected more by the proportion of individuals that reproduce at all than by the rate of reproduction or survival of individuals. For example, the attainment of sexual maturity in the red vole largely depends upon the ability of individuals to procure territories (Sadleir 1969). The proportion of individuals that breeds during the year of their birth depends upon the size of the population at the beginning of the reproductive season. At the beginning of the season when population density is low, most of the yearling voles are able to reproduce. But toward the end of the season, after populations have increased to fairly large numbers, most young of the current year, particularly males, are inhibited from reproduction by the aggressive behavior of older individuals (Table 31-7).

By limiting the occupation of suitable habitats, territoriality exerts a strong stabilizing influence on population size. Among species that hold territories during a limited season, density-dependence can be demonstrated only when the species is territorial (Tanner 1966). Exclusion of individuals from reproduction by territoriality is common in higher organisms, particularly vertebrates (Brown 1969, Krebs 1971). Stewart and Aldridge (1951) and Hensley and Cope (1951) accidentally discovered evidence that populations of breeding birds in northern coniferous forests were limited by territorial behavior. In the course of a study designed to investigate the effects of predation by birds on forest insect pests, they shot all the males of several species in a small experimental area. As soon as territory-holding males were removed, they were readily replaced by others, which apparently had been prevented from establishing territories previously (see page 251). Other removal experiments have had similar results, indicating that territorial limitation of breeding populations is quite general (red-backed sandpiper, Holmes 1966; blackbirds of the genus *Agelaius*, Orians 1961; red grouse,

TABLE 31-7

Variation in attainment of puberty in the year of birth by voles in various habitats, early and late in the season (from Koshkina, quoted by Sadleir 1969).

Type of habitat	Early season (June 25–July 13)			Late season (August 2–19)		
	Relative density*	Percentage reproducing in year of birth		Relative density*	Percentage reproducing in year of birth	
		Male	Female		Male	Female
Mature coniferous forest	14.4	68	72	35.3	13	25
Second-growth coniferous forest	14.5	66	74	29.5	22	39
Aspen forest	9.2	68	73	16.0	32	59

* Number of voles per 100 trap nights.

Watson 1967, Watson and Jenkins 1968; ptarmigan, Watson 1965; Cassin's auklet, Manuwal 1974; bank voles, Smyth 1968; dragonflies, Jacobs 1955, Moore 1964; and pomacentrid fish, Clarke 1970).

Territoriality not only limits the proportion of individuals in a population that reproduce, but it may expose individuals that do not hold territories to greater risk of death (Errington 1946, Jenkins, Watson, and Miller 1963). Territorial individuals are often forced into suboptimal habitats where their chances of succumbing to starvation, disease, inclement weather, or predators are increased. Errington (1946) has suggested, furthermore, that the burden of stabilizing factors falls upon nonterritorial individuals.

Territoriality stabilizes populations by limiting the recruitment of new individuals into the reproducing population. Because the response of territorial behavior to the density of the population and to the resources of the habitat is immediate, its effects have no time lag. We might, therefore, expect to find periodic fluctuations in population size in species that are nonterritorial during part of the year. Indeed, most of the herbivorous mammals in arctic tundra and boreal forest habitats whose populations fluctuate regularly are nonterritorial during the winter. Furthermore, only species that remain active throughout the winter have marked population cycles. Species that hibernate (for example, chipmunks, squirrels, bears, and raccoons) do not exhibit regular population fluctuations (Keith 1963, Elton and Nicholson 1942a). The persistence of territoriality throughout the year in tropical regions could be responsible for the apparent lack of periodic fluctuations in populations of large animals there.

Territoriality and behavioral dominance force individuals, particularly the young, to disperse from areas of high population. In a sense, dispersal is a safety valve for the population. The pressure of recruitment into the population is sufficient to maintain the population close to the level that the habitat can support; excess individuals are shunted off into a nonbreeding reservoir. Because these individuals retain reproductive potential, they may obtain territories and mates in the event of a population decline, and thereby help to stabilize the population (Krebs *et al.* 1973, Lidicker 1975, Manuwal 1974).

Can populations be self-regulating?

Territorial behavior and dispersal undoubtedly stabilize population size and, in doing so, would seem to benefit some individuals in the population. But we wonder about those excluded from reproduction. Is their behavior the expression of adaptations to enhance the fitness of others, and if so, how could such behavior evolve by natural selection?

We normally view patterns of reproduction as adaptations to increase evolutionary fitness, even if an adaptation might sometimes result in a short-term reduction of fecundity. For example, the drain on energy and nutrient resources caused by reproduction under poor environmental conditions could increase the mortality rate of the parents enough to offset any immediate gain in fitness from the additional offspring, if they should survive. Delayed maturity in response to poor conditions could reduce an individual's drive to acquire territory, thereby foregoing the risks of territorial conflict with older, more experienced individuals. Territoriality itself assuredly evolved to preserve the fitness of dominant individuals. Ecologists usually interpret such adaptations as evolutionary responses that maximize the fitness of the individual over a wide range of environmental conditions.

This view was challenged in a fascinating book by V. C. Wynne-Edwards (1962), *Animal Dispersion in Relation to Social Behavior.* Wynne-Edwards agreed with most ecologists that resources, particularly food, ultimately set the level of population density, but he denied that stabilizing factors act directly, through the physiology of the organism, on population processes. Instead, he suggested that such manifestations of poor environmental conditions as reduced fecundity and delayed maturity are adaptations to improve the fitness, not of the individual, but of the population. We have maintained in this book that evolution proceeds to maximize the fitness of the individual, even though this might result in the production of more progeny than are needed to balance adult mortality. Population excesses are removed by behavioral exclusion of some individuals from breeding and by external stabilizing factors. The strategy of improving individual fitness can be wasteful; it often leads to high mortality of juvenile animals and plants. Wynne-Edwards rejected this view of evolution and suggested instead that evolution has proceeded to minimize wastefulness in populations by limiting the reproductive recruitment of young individuals to a level that just balances adult mortality. Wynne-Edwards argued that the regulation of population size is internally mediated through behavioral controls rather than directed through external factors that control population parameters. This view has also been espoused by Kalela (1954, 1957) and Skutch (1949, 1967), among others, who assume that efficient regulation increases the competitive ability of a population with respect to populations that utilize resources less efficiently.

Wynne-Edwards suggested that animals assess the density of their populations through various group-oriented displays, such as flocking and territorial singing. He presumed that such behavior has, in fact, evolved to facilitate the estimation of population density, whereby organisms adjust their fecundity. Wynne-Edwards's hypothesis need not imply that organisms are consciously aware of population densities and of controlling their own reproductive effort, merely that individuals respond directly to population density or to intensity of social interaction.

Wynne-Edwards's theory, reasonable as it may seem, became the most controversial topic in ecology during the 1960s and stimulated a barrage of criticism aimed at demolishing the idea utterly. The difficulties with Wynne-Edwards's idea may be summed up as follows: First, the predictions of his hypothesis are virtually identical to predictions made on the basis of selection for maximum reproduction, hence patterns in life history cannot determine the validity of either hypothesis. Second, the mechanism required to select traits that benefit the group at the expense of the individual must be based on competition between groups or populations rather than between individuals.

Wynne-Edwards (1962, 1964, 1965, 1970) suggested that populations with low adult mortality should have evolved mechanisms for reducing reproduction and recruitment of new individuals into the population. He pointed to delayed reproduction and small broods or litters in animals with long life spans as examples of such adaptations. He envisioned organisms relinquishing a portion of their reproductive potential so the population would not exceed the ability of the habitat to support it. Wynne-Edwards further suggested that through the evolution of territoriality and dominance, some individuals in the population abstain altogether from reproducing to reduce the reproductive performance of the population. But all these adaptations are to be expected on the basis of adaptation to increase individual fitness, and so their existence does not allow us to choose between theories based alternatively on internal and external regulation of populations.

The crux of the controversy centered over the mechanism of evolution. Wynne-Edwards maintained that organisms give up potential individual fitness for the sake of other members of the population. He contended that such behavior can evolve as a result of *group selection,* which is an extension of natural selection in which small groups, rather than individuals, are the units of adaptation (Wynne-Edwards 1963). The hypothesis of group selection has two major difficulties: First, there is little evidence for the existence of populations with the requisite group structure (Maynard-Smith 1964). Second, groups are more persistent in time than individuals, thus the rate of evolution by group selection must be much slower than by individual selection (Wiens 1966). Where group fitness and individual fitness conflict, therefore, the latter normally will prevail (Williams 1966b, Lack 1966).

Although few ecologists take seriously Wynne-Edwards's view of population regulation through altruistic self-regulation of reproduction, the controversy it generated greatly advanced the study of ecology. Wynne-

Edwards's ideas stimulated others to sharpen their thinking about the evolution of adaptations involved in the regulation of population size. His ideas also led directly to a complete re-evaluation of life-history patterns as adaptations, beginning with Williams (1966a, 1966b) and Cody (1966). We should also not forget that Wynne-Edwards raised many questions about the function of such behaviors as group displays and flocking that remain unanswered. Finally, and perhaps somewhat ironically, theoreticians are now beginning to explore the possibilities of group-selection models for natural populations (Wilson 1973). Whatever else, Wynne-Edwards has provided a rich legacy of behavioral and ecological phenomena that will stimulate and challenge ecologists and evolutionary biologists for years (Wilson 1975).

Is "social pathology" a major factor in the regulation of natural animal populations?

Most theories of population regulation consider population density as important only with respect to the ability of an environment to support the population. Social arrangements such as territoriality and dominance replace resources with space or rank as the object of competition, but the expression of such behavior remains closely related to environmental resources. In dense laboratory populations, social stress can cause abnormal physiological responses (*social pathology*), often lumped under the term *general adaptive syndrome,* that result in reduced fertility and survival even though food, water, and nesting opportunities are provided in excess of requirements. Some physiologists have attempted to expand observations on abnormally dense laboratory populations into a general theory to explain the regulation of population density in natural situations (Christian 1950, 1963, Christian and Davis 1964). Before we consider the implications of this theory for natural populations, we should look at some of the laboratory studies on which the theory is based.

Laboratory populations provided with ample resources typically increase to a high density, then physiological functions begin to break down. Population growth slows to a halt or may even begin to decline to extinction. A typical example, from the work of Christian, Lloyd, and Davis (1965) on house mice, is presented in Figure 31-16. The population initially grew rapidly and the reproductive rate of mature females (more than two months old) increased during the early phase of population growth. After the population had exceeded fifty individuals, however, birth rate declined rapidly, and when the population exceeded one hundred individuals infant survival was severely curtailed, in spite of more than adequate provision of food and nest boxes (Lloyd and Christian 1969, Lloyd 1975). As population density increased, the proportion of females increased, and because young did not survive well, the age structure of the population became progressively older.

A series of studies on populations of house mice was conducted during the late 1940s and early 1950s by investigators at the University of Wisconsin. Strecker and Emlen (1953) described demographic patterns in populations established in large escape-proof rooms. Populations were started by intro-

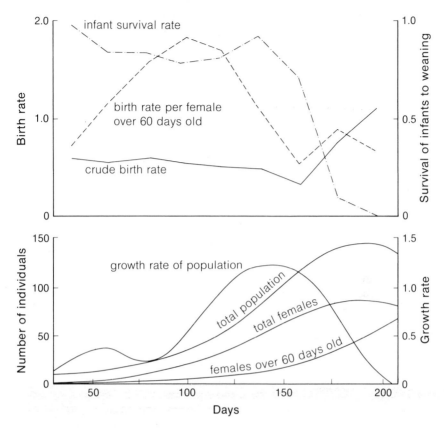

Figure 31-16   Birth, death, and population growth in a confined population of house mice (*Mus musculus*) supplied with abundant resources and nesting places (after Christian *et al.* 1965).

ducing equal numbers of adult male and female mice, trapped from wild populations in the town of Madison. Water and nesting sites were provided in abundance, but food was limited to a constant ration each day. In this experiment, the populations grew rapidly until the mice consumed all the food provided each day, at which time reproduction ceased abruptly. In another experiment, Strecker (1954) established a population of house mice in a large room in the subbasement of a University of Wisconsin building. The room had several natural escape routes, evidently by way of ventilating shafts that ran throughout the building, so the mice could, and did, emigrate from the population. This population was also provided with limited food and its demographic progress was followed in detail. Within eight months, the population had become limited by food supply, and the emigration of mice from the population, determined by trapping in various offices and laboratories throughout the building, rose to a high level. Both subadult and adult animals emigrated from the population, and the sexes were about equally represented in the exodus. As a result of emigration from the population, reproductive activity remained high even after food had become a limiting factor. When the whole colony was trapped at the end of the study, 20 per cent of the individuals were immature, and 29 per cent of the females were pregnant. Social pressure from within the population maintained the population at a sufficiently low level with respect to the food supply that reproduction could continue even when food was limited.

Southwick (1955a) worked with confined populations of house mice provided with unlimited food. Under these conditions mice continued to reproduce regardless of the density of the population, but the survival of embryos and young was so reduced that the population leveled off in spite of unlimited food resources (Figure 31-17). The populations were regulated primarily by increased mortality of preweaned young (and also of some adults) rather than by cessation of reproduction, as Strecker and Emlen had observed in the limited food situation. The mortality of preweaned mice was directly related to the presence of mice other than the mother, especially males, in the nest box (Brown 1953, Southwick 1955b). With increasing density of house mouse populations, Southwick (1955b) found that aggression, measured by the number of fights observed per hour, rose sharply (Figure 31-17), and that this was highly correlated with the decreased survival of young to weaning. The number of adult mice, mostly males, that were wounded or diseased also increased with density (Figure 31-17).

Behaviorally mediated population regulation, in the absence of resource limitation, has been observed frequently in crowded laboratory populations. Social stress can lead to a variety of abnormal physiological symptoms, collectively referred to as the general adaptive syndrome: (a) the adrenal glands enlarge considerably in size, (b) growth and reproduction are curtailed in both sexes, (c) sexual maturation is delayed or inhibited, (d) spermatogenesis in the male is delayed, (e) the estrus cycle of the female is prolonged and rates of ovulation and implantation are diminished, (f) mortality of embryos within the uterus is greatly increased (see Christian 1961), (g) lactation is often inadequate, leading to stunting of the young at weaning, and (h) susceptibility to disease may increase (Christian and Davis 1964).

Christian and Davis suggested, along with Calhoun (1949) and others, that the general adaptive syndrome may be a factor in the regulation of natural populations. They further suggested that this behavioral and physiological mechanism restricts populations considerably below the ultimate carrying capacity of their environment determined by limits of food availability, and that in this way ". . . mammals avoid the hazard of destroying their environment, and thus the hazard of their own extinction."

The theory of population regulation proposed by Christian and Davis raises the same evolutionary problems that proved to be so damaging to Wynne-Edwards's hypothesis. The curtailment of reproduction under conditions of extreme social contact can hardly be explained on the basis of individual selection, for such behavior could not benefit the individual.

Because natural populations rarely achieve the densities that are attained in the laboratory, the general adaptive syndrome would seem to be a

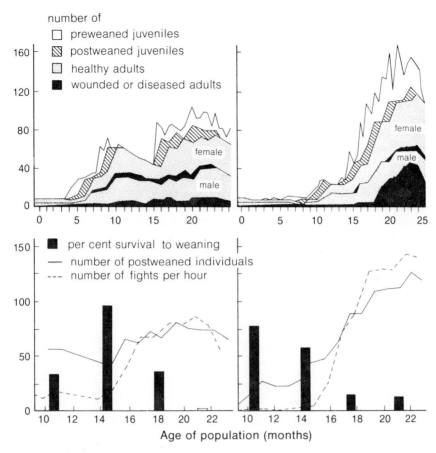

Figure 31-17   Growth of two confined house mouse (*Mus musculus*) populations supplied with unlimited food, showing the age structure, level of aggression, survival of young to weaning, and the incidence of diseased or wounded
adults (after Southwick 1955a, 1955b).

physiological response that occurs only under abnormally dense conditions,
to which selection has not had the opportunity to adjust the population. If the
environment changed so as to allow an extremely dense population to build
up, would not natural selection favor those individuals whose behavioral and
physiological response to high population density excluded the curtailment
of reproduction? In some natural populations of colonial organisms, reproductive activities proceed normally at densities far beyond levels that would
be tolerated by other species. Zebra finches nest in small colonies in the wild;
several pairs will reproduce perfectly well in a small room. Two pairs of
robins, normally a territorial species, would not, however, peaceably tolerate
confinement in an area a hundred times larger. That reproductive behavior

can evolve to function normally at high population densities can be best appreciated by a visit to a colony of prairie dogs or seabirds, where young are successfully reared in conditions that seem to be utter bedlam (Figure 31-18).

Some observations on natural populations suggest that during periods of high population density reproduction may be curtailed in a manner reminiscent of the general adaptation syndrome, or other physiological impairment may appear (for example, see Green et al. 1939, for the snowshoe hare). Adverse physiological responses to high density can be accounted for in several ways, none of which has actually been demonstrated. For example, by delaying attainment of sexual maturity, individuals could possibly extend their life spans, waiting for population density to decrease to a level at which reproduction is more likely to be successful. Newsome (1969) suggested that in a natural population of house mice in Australia, nonreproducing individuals may have a higher probability of survival through the following winter than sexually active house mice, although his evidence was rather weak. Abstinence from reproduction or delayed maturity when reproduction has a minimal chance of success may, in the long run, be advantageous to the individual. Such conditions could occur when food is extremely limited or

Figure 31-18   A dense breeding colony of royal terns on the coast of North Carolina. The social behavior of the terns has evolved to allow tolerance of high population density (courtesy U.S. Bureau of Sport Fisheries and Wildlife).

when the individual occupies a low position in the social hierarchy. In that context, curtailment of reproductive activity at high density would make good evolutionary sense.

Behavior that seemingly reduces the individual's fitness could also arise at high densities in fluctuating populations. If reproductive physiology were adapted to low population densities, the adaptations might also happen to restrict reproduction at high population density (Williams 1975). In other words, behavioral and physiological mechanisms may respond appropriately to most environmental situations, but for the rare condition of an extremely dense population, they may be entirely inappropriate. One must be careful to assess the value of an adaptive response for the whole range of conditions encountered, not for only one set of circumstances.

One additional hypothesis, concerned primarily with regulation in fluctuating populations, has been outlined by Dennis Chitty (1960, 1967). Finding no evidence of starvation, disease, increased predation, or stress at high densities in fluctuating populations of voles, Chitty suggested that population declines are brought about through a deterioration in the "quality" of individuals in the population. Chitty observed that populations of field voles declined even in the absence of adverse conditions. He also demonstrated that diseases, the most important of which was tuberculosis, played no role (Chitty 1954). Similarly, Krebs (1963) showed that food shortage could not have been responsible for the decline of lemmings near Baker Lake, Canada, from 1960 to 1962. By fencing off parts of the tundra to exclude lemmings (Table 31-8), Krebs found that the lemmings reduced the forage crop outside the enclosures by only about 25 per cent during peak and decline years. Disease did not seem to be important and there was no change in the occurrence of scars (a measure of aggression) on animals through the cycle. Krebs (1966) also found evidence that traditional factors (predation, starvation, disease, and stress) were not responsible for fluctuations in populations of voles in California, although Pearson (1966) claimed that predators had a major effect on fluctuations of the same vole population, and Batzli and Pitelka (1971) found evidence for change in diet quality associated with the population cycle. While some field evidence points to changes in the quality

TABLE 31-8

Abundance of lemmings and the standing forage at the end of the summer during a peak and subsequent decline in lemming population (from Krebs 1963).

| Year | Relative abundance | | Dry weight of forage ($g \cdot m^{-2}$) | |
	Lemmus	Dicrostonyx	Enclosed areas	Open areas
1959	1	2	92	78
1960	43	29	127	97
1961	3	6	154	117
1962	1	7	160	149

of the population, rather than of the environment, as the reason for fluctuations in mammal and bird populations, the underlying causes for the change in quality, and even the basic meaning of quality, are not at all well understood (Krebs 1970, Krebs *et al.* 1973, Krebs and Myers 1974, Batzli 1975).

Attempts to dissect the population cycles of small mammals into their working parts by field observation and experimentation have met with little success. Lemmings have been studied intensively at Barrow, Alaska, where population density varies in a three- to five-year cycle by factors of 100 or more between trough and peak years (Pitelka 1957a, 1957b, 1973). Predators undoubtedly exert a strong depressing influence on population growth because most of the increase in number of lemmings occurs during the winter beneath a protective mantle of snow. Nevertheless, the reduction of lemming population size in summer by predators is small compared with longer-term changes in population size. The only habitat characteristic that appears to parallel the lemming cycle at Barrow is the quality (not quantity) of the vegetation (Table 31-9). During one peak year, vegetation contained 22 per cent protein, but during the next population trough it contained only 14 per cent protein. If lemming populations and plant quality have parallel cycles, the two may be interrelated: the total available nutrients in the tundra ecosystem may be of such small quantity that they are mostly transferred to the bodies of lemmings during peak years, thereby depressing plant growth and nutrient quality. Nutrients would then be restored to the cycle only after large numbers of lemmings die and their remains decompose (Schultz 1964).

The debate over the cause or causes of cycles in populations of small mammals emphasizes the fact that although factors having influence on population size are understood in general terms, we frequently have difficulty applying these principles to particular populations. You may have noticed that no mention has been made of populations of plants in this chapter. The reason is simple. Although plant ecologists have recently begun to consider population processes (*e.g.,* Palmblad 1968, Putwain *et al.* 1968, Harper and

TABLE 31-9

Features of a lemming population cycle, including vegetation characteristics, at Barrow, Alaska (after Mullen 1968).

	Year			
	1960	1961	1962	1963
Relative peak density	125	0.5	1 to 10	50
Male body weight (g)	92	47	69	59
Litter size	7.6	7.0	7.3	6.7
Breeding season (days)	58	80	73	83
Green vegetation (lbs·A^{-1})	111	278	115	149
Per cent protein	22	14	17	19
Protein in plants (lbs·A^{-1})	24	40	20	28

White 1970), their studies have not yet yielded significant insights into population processes beyond what is known of animal populations. Some aspects of plant population ecology, particularly the interaction of plants with herbivores, will, however, be discussed later.

We have examined the balance of forces that regulate population size. Some forces are density-dependent. The level of resources limits the inherent capacity of a population to grow and keeps the population at a level for which the resources are adequate. When the availability of resources changes, the size of the population changes in an appropriate manner. Other forces—cold, drought, and the like—may be density-independent. That is, they may limit the growth of a population without regard to its density or to its resources. Changes in population size are compensated by homeostatic population responses, including reproduction and dispersal. But the response of a population to its resources can lead to cyclic fluctuations in population size, if the influence of the response is delayed.

Inseparable from the notion of density-dependent forces and responses of populations to them is the fact that individuals compete with one another for resources. In the next chapter, we shall examine the influence of competition, whether between individuals of the same or of different species, on population processes and on the coexistence of populations of different species within the community.

32

Competition

The tendency of populations to increase exponentially expresses the unrestricted biotic potential of the individual. The realization of this potential diminishes as population density increases; this is the essence of population regulation. The growth rate of a population is dependent upon the fecundity and longevity of individuals. The depressing influence of density on growth rate reflects the detrimental effect of competition on the survival and reproductive performance of individuals (see Chapter 31).

How does competition operate? If the availability of a particular food resource were reduced by the feeding activities of each individual that used the resource, such activities would reduce the ability of other individuals in the population to obtain the resource. Organisms whose resource requirements are most similar will compete most intensely, and so competition between individuals of the same species is usually more intense than competition between individuals of different species.

Laboratory experiments demonstrate that two species cannot coexist if they require similar resources.

Field observations and laboratory experiments convey opposite impressions of nature. We frequently observe many ecologically similar species coexisting in nature, clearly using many of the same resources. Closely related species of trees grow in the same habitat, all needing sunlight, water, and soil nutrients. Coastal estuaries and inland marshes harbor a variety of fish-eating birds, including egrets, herons, terns, kingfishers, and grebes (Figure 32-1). As many as six kinds of warblers (small insectivorous birds), having similar morphology and feeding habits, may be found in one locality in forest habitats of the northeastern United States. Eight species of large predatory snails frequent the shallow waters of Florida's Gulf Coast.

By contrast, closely related species rarely coexist in the laboratory. If two species are forced to live off the same resource, inevitably one persists and the other dies out. Reconciling these observations has been a major task for

Figure 32-1   Three closely related species of egrets (common, reddish, and snowy) feeding in the Aransas National Wildlife Refuge, Texas (courtesy U.S. Bureau of Sport Fisheries and Wildlife).

ecologists over the past half century, since the Russian biologist G. F. Gause tried to make two similar species coexist in the laboratory.

Gause (1934) established laboratory cultures of closely related species of protozoa, of the genus *Paramecium,* in the same nutritive medium. The species flourished when separate, but in mixed cultures only one species survived (Figure 32-2). Similar experiments with fruit flies, mice, flour beetles, and annual plants have always produced the same result: One species persists and the other dies out, usually after thirty to seventy generations (Miller 1967).

The results of laboratory competition experiments led to the formulation of the *competitive exclusion principle* (Hardin 1960), also called Gause's principle (Lack 1944), which states that two species cannot coexist on the same limiting resource (Gause 1934, Harper 1961). This principle was first appreciated by the American naturalist Joseph Grinnell (1904, 1943), and its existence was implicit in mathematical formulations of competition by Lotka (1925) and Volterra (1926). The earliest experimental demonstration of the principle was made by the English plant ecologist A. G. Tansley (1917). But it remained for Gause's extensive laboratory experiments to raise the principle to the level of general recognition by ecologists.

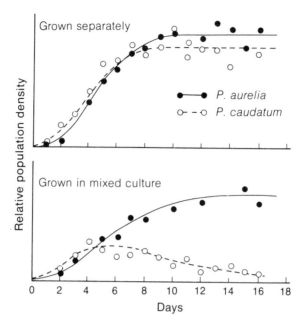

Figure 32-2   Increase in populations of two species of *Paramecium* when grown in separate cultures (above) and when grown together (below). Although both species thrive when grown separately, *P. caudatum* cannot survive when grown with *P. aurelia* (after Gause 1934).

The word "limiting" is included in the statement of the competitive exclusion principle because only resources that limit population growth can provide the basis for competition. Nonlimiting resources, like atmospheric oxygen, are superabundant compared to the needs of organisms, and their use by one organism does not make them less available to others.

How can the principle of competitive exclusion, verified by laboratory experiments, be reconciled with observations of similar species coexisting in nature? The coexistence of many similar species in nature (see page 233) suggests that competition between species generally is much weaker in natural communities than in laboratory populations. Species avoid competition by partitioning resources and habitats among themselves (Schoener 1974), a response that species in simplified laboratory environments cannot make. But the existence of resource partitioning in nature does not mean that competition does not occur. The inhibiting influence of populations upon each other's growth has been demonstrated in many instances by the response of a population to the removal of ecologically similar species from the habitat—for example, in competition for space among barnacles and competition for light and moisture among tropical forest trees (Chapter 16). In general, however, when species can avoid extensive overlap with other species in the use of resources, competition can be reduced sufficiently to permit indefinite coexistence. It seems that only in laboratory situations, where species are forced to exploit a single resource, is competition so intense that only one population can persist.

Does competitive exclusion ever occur in nature? Does competition influence the number of species that can coexist in a community? Do the numerous ways in which species exploit the environment, and thus avoid competition, allow any conceivable number of species to be crammed into a community? Or is there some degree of ecological similarity between species that cannot be exceeded without leading to extinction, and are natural communities saturated with species whose similarities just approach this level? Ecologists have not resolved these problems; the factors that determine the number of species in a community constitute an important area of ecological research. As we shall see, the analysis of competition has provided a useful approach.

Elimination of competitors following introduction of species demonstrates the population effects of competition.

The species we find coexisting in nature are the successful ones. What of the species that failed? Fossils of extinct species prove that populations have died out. Was their demise caused by superior competitors?

Competitive exclusion is a transient phenomenon. The evidence of exclusion having taken place is lost when the poorer competitor is eliminated. We can observe competitive exclusion in the laboratory because we can mix populations according to our whims and follow the course of their interaction. The closest natural analogy to the laboratory experiment is the accidental or intended introduction of species by man. When new immigrants are superior competitors to resident species, the immigrants can reduce or eliminate local populations. The explosion of rabbit populations following their introduction to Australia (see Chapter 31) worsened range conditions and thereby reduced populations of many native marsupial herbivores, such as kangaroos and wallabies. (Rabbit-control programs were aimed more at preserving the range for another introduced competitor—the sheep—than preserving the native fauna.)

Competition between introduced species has produced some vivid demonstrations of the competitive exclusion principle. When many species of parasites are introduced simultaneously to control a weed or insect pest, the control species are brought together in the same locality to prey on, or parasitize, the same resource. We should not be surprised that competitive exclusion has occurred frequently under these conditions. Between 1947 and 1952, the Hawaii agriculture department released 32 potential parasites to combat several species of fruit pests, including the Mediterranean fruit fly (Bess et al. 1961). Thirteen of the species became established, but only three kinds of braconid wasps proved to be important parasites of fruit flies. Populations of these species, all closely related members of the genus *Opius,* successively replaced each other from early 1949 to 1951, after which only *Opius oophilus* was commonly found to parasitize fruit flies (Figure 32-3). As each parasite population was replaced by a more successful species, the

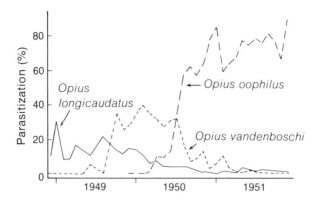

Figure 32-3   Successive change in predominance of three species of wasps of the genus *Opius,* parasitic on the Oriental fruit fly (after Bess *et al.* 1961).

level of parasitism of fruit flies by wasps also increased, suggesting superior competitive ability on the part of the most recently established wasp population.

A similar pattern of competitive replacement involving wasps that are parasites of scale insects, which are pests of citrus groves, has been thoroughly documented in southern California (DeBach and Sundby 1963, DeBach 1966). With the failure, owing to the evolution of resistance by pests, of chemical pesticides to provide adequate long-lasting control, agricultural biologists turned to the importation of insect parasites and predators (De-Bach 1974). Yellow scales have infested California citrus groves since oranges and lemons were first planted there. In the late 1800s, the red scale was accidentally introduced and has replaced yellow scale almost completely, perhaps itself a case of competitive exclusion. Of the many species introduced in an effort to control citrus scale, tiny parasitic wasps of the genus *Aphytis* (from the Greek *aphyo,* to suck) have been most successful. One species, *A. chrysomphali,* was accidentally introduced from the Mediterranean region and became established by 1900.

The life cycle of *Aphytis* begins when adults lay their eggs under the scaly covering of hosts. The newly hatched wasp larva uses its mandibles to pierce the body wall of the scale and proceeds to consume nearly all the body contents. After the wasp pupates and emerges as an adult, it continues to feed on scales while producing eggs. Each female can raise 25 to 30 progeny under laboratory conditions, and the development period is so short (egg to adult in fourteen to eighteen days at 27 C or 80 F) that populations may produce eight to nine generations per year in the long growing season of southern California.

In spite of its tremendous population growth potential, *A. chrysomphali* did not effectively control scale insects, particularly not in the dry interior valleys. In 1948, a close relative from southern China, *A. lingnanensis,* was introduced as a control agent. This species increased rapidly and almost

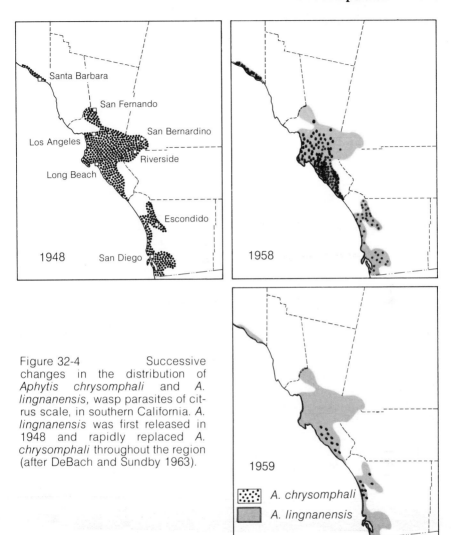

Figure 32-4   Successive changes in the distribution of *Aphytis chrysomphali* and *A. lingnanensis,* wasp parasites of citrus scale, in southern California. *A. lingnanensis* was first released in 1948 and rapidly replaced *A. chrysomphali* throughout the region (after DeBach and Sundby 1963).

completely replaced *A. chrysomphali* within a decade (Figure 32-4). When both species were grown in the laboratory, *A. lingnanensis* was found to have the higher net reproductive rate, whether the two species were placed separately or together in population cages.

Although *A. lingnanensis* had replaced *A. chrysomphali* throughout most of southern California, it still did not provide effective biological control of scale insects in the interior valleys because cold winter temperatures periodically reduced parasite populations. Wasp development slows to a standstill at temperatures below 16 C (60 F), and adults cannot tolerate temperatures below 10 C (50 F). Pupae are most resistant to cold, but average winter pupal

Figure 32-5   Distribution of three species of *Aphytis* in southern California in 1961. *A. melinus* predominates in the interior valley, while *A. lingnanensis* is more abundant near the coast (after DeBach and Sundby 1963).

mortality was 42 per cent, with extremes of about 80 per cent in the cold interior valleys.

In 1957, a third species of wasp, *A. melinus,* was introduced from areas in India and Pakistan where temperatures range from below freezing in winter to above 40 C in summer. As expected, *A. melinus* spread rapidly throughout the interior valleys of southern California, where temperatures resemble the wasp's native habitat, but did not become established in coastal areas (Figure 32-5).

Competition between *A. lingnanensis* and *A. melinus* was studied under controlled laboratory conditions—27 C (80 F) and 50 per cent relative humidity—that more closely resembled the climate of coastal areas than that of the inland valleys. The wasps were provided with oleander scale grown on lemons as hosts, and new scale-infested lemons were added every seventeen days. Populations of wasps were counted just before new food was added. When cultured separately, both species parasitized about 40 per cent of the scales, and wasp populations averaged 6,400 for *A. lingnanensis* and 8,300 for *A. melinus.* When grown together, *A. melinus* was reduced in four months from 50 per cent to less than 2 per cent of the total wasp population. During the process of competitive exclusion, the combined population of the two species (average 7,400) remained at approximately the same level as that of either species grown separately. We may infer that population size was limited by host availability and that under the conditions of the experiment, *A. lingnanensis* was the superior competitor.

Competition in plant populations.

Plants differ from animals in two ways that bear on the sudy of competition (Harper 1961, 1967, Harper and White 1970). First, few terrestrial species have generation times of less than a year. Plant ecologists, therefore, often cannot continue experiments for a long enough period to demonstrate competitive exclusion. Second, plant growth, as well as survival, is greatly affected by the variety of conditions under which the plant may live. In particular, plants grow slowly when crowded and do not attain their full stature, even though they may produce seed. In contrast, animal populations usually respond to crowding with increased mortality and reduced fecundity, not stunted growth.

When horseweed (*Erigeron*) seed is sown at a density of 100,000 seeds per square meter (equivalent to about ten seeds in the area of your thumb nail), the young plants compete vigorously. As the seedlings grow, many die and the density of surviving seedlings decreases (Figure 32-6). At the same time, the growth of surviving plants exceeds the decline of the population, and the total weight of the planting increases. Over the entire growing season, the hundredfold decrease in population density is more than balanced by a thousandfold increase in the average weight of each plant.

The relationship between plant weight and density, shown in Figure 32-6, is often called a *self-thinning curve* (Harper 1967). When obtained from very dense plantings, self-thinning curves reflect the maximum capacity of the soil to support plant growth. Combinations of density and size lying outside the self-thinning line never occur because they are beyond the carrying capacity of the environment. If horseweed were sown at a density of 10,000 seeds per square meter, rapid growth with little early mortality would carry the population to the self-thinning line as shown in Figure 32-6. Sown at densities of

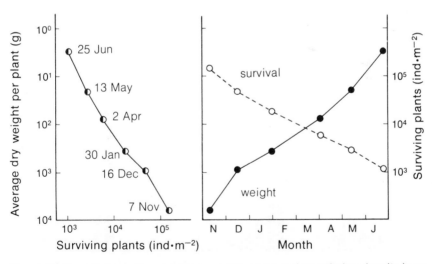

Figure 32-6   Progressive change in plant weight and population density in an experimental planting of horseweed (*Erigeron canadensis*) sown at a density of 100,000 seeds per square meter ($10 \cdot cm^{-2}$). The relationship between plant density and plant weight as the season progressed is shown at left, where the self-thinning curve of a planting with a density of 10,000 seeds per square meter is shown by the dashed line (after Harper 1967).

1,000 per square meter, few of the horseweed plants would die before reaching maturity.

Each species has a characteristic thinning curve under particular conditions of soil, light, temperature, and moisture. Regardless of the density of the initial planting, the eventual harvest of mature plant biomass is surprisingly uniform (Palmblad 1968). The self-thinning curve does not, however, tell the full story of within-species competition. When sown at low initial density, most individuals grow vigorously. At high density, only a few individuals reach large size. Other survivors do poorly because they are crowded by individuals that obtained an initial growth advantage owing to a favorable site for germination (Figure 32-7).

Because of the flexible growth response of plants and their long generation times, botanists usually assess competition between plant populations by comparing total plant weight or number of seeds produced in an experimental plot. These indexes were used to measure competition between two closely related species of oats, *Avena barbata* and *A. fatua* (Marshall and Jain 1969). Plants were grown in watered pots at six densities: 8, 16, 32, 64, 128, and 256 individuals per pot. When the species were grown separately, density had relatively little effect on germination, establishment of seedlings, or survival to maturity. Growth responses were, however, markedly different: At high densities, individual plants attained smaller average weight and height and produced fewer seeds.

To measure the influence of interspecific competition on plant growth, pots were sown at each of several densities with different proportions of the two species. In experiments with seed densities of 128 per pot, for example, a group of pots would be planted with 128 *barbata* and 0 *fatua*, 112 *barbata* and 16 *fatua*, 64 *barbata* and 64 *fatua*, and so on, each totaling 128 seeds. For each species, the pots created conditions ranging from pure intraspecific competition (128 *vs.* 0), to strong interspecific competition (16 *vs.* 112).

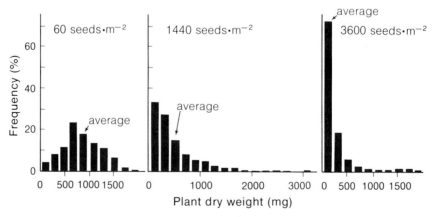

Figure 32-7   The distribution of dry weights of individuals in populations of flax plants sown at different densities (after Harper 1967).

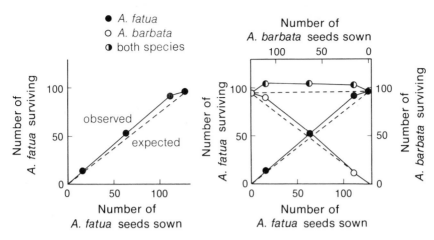

Figure 32-8   Replacement series diagrams representing the outcome of competition between *Avena fatua* and *A. barbata* grown at a density of 128 seeds per pot. The left-hand diagram shows the expected response of *fatua* survival if the effects of intra- and interspecific competition did not differ. Actual results closely match this expectation. At right, the results of *A. fatua* and *A. barbata* are plotted on the same diagram. The expected total survivorship of both species is added to the diagram (after Marshall and Jain 1969).

When grown separately at densities of 128 seeds per pot, both species showed survival rates of about 96 individuals per pot (75 per cent). If the survival of a *fatua* individual were affected by competition with *barbata* individuals no differently than from competition with other *fatua* individuals, we would expect 75 per cent of *fatua* seeds to survive regardless of the ratio of the two species in the pot. Thus 12, 48, and 84 individuals would survive from initial plantings of 16, 64, and 112 seeds. The relationship between the initial planting and the final outcome is often depicted on a graph called a *replacement series diagram* (Figure 32-8). The left-hand figure represents the expected number of *fatua* surviving to maturity if interspecific competition and intraspecific competition had equal influence on survival. A straight line drawn between 96 individuals on the right-hand axis (128 *fatua* seeds per pot) and 0 on the left-hand axis (0 *fatua* seeds per pot) represents 75 per cent survivorship of *fatua* seeds. If, on one hand, interspecific competition depressed the survival of *fatua* seedlings, relative to the effect of intraspecific competition, the outcomes of the experiment in the mixed species pots would fall below the line of equal competitive effect. If, on the other hand, *fatua* grew better under conditions of strong interspecific competition than it did under strong self-inhibition, the points would fall above the line of equal competitive effect.

Survival of both *fatua* and *barbata* exhibited no differential response to intraspecific and interspecific competition. Weight and seed production did, however, display differences (Figure 32-9). Compared with the outcomes of single-species cultures, *A. fatua* produced more seeds per plant when grown

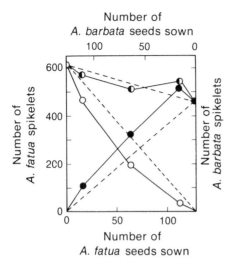

Figure 32-9   Replacement series diagram for seed production (measured as number of flower spikelets per pot) in *Avena fatua* and *A. barbata* (after Marshall and Jain 1969).

in competition with *barbata*; *A. barbata* produced fewer seeds per plant when grown with *fatua.* Thus *fatua* competed well against *barbata* under the conditions of the experiment, and we would expect *fatua* eventually to exclude *barbata.*

We may visualize the outcome of competition on a ratio diagram (Figure 32-10), in which the ratio of seeds produced by one species to seeds produced by the other is compared with the initial ratio of seeds planted. If the ratio of the species of seed produced from a pot were identical to the ratio of seeds sown, neither species would have a competitive advantage. This situation is represented by the diagonal line in Figure 32-10. If *fatua* were relatively more productive than *barbata*, the outcome of competition experiments would lie in the upper left-hand part of the diagram, representing an increase

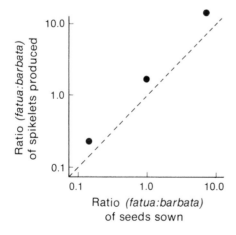

Figure 32-10   Ratio diagram of competition between *Avena fatua* and *A. barbata*, representing the outcome of the experiment depicted in Figure 32–9 (after Marshall and Jain 1969).

in the ratio of *fatua* to *barbata*. Conversely, if *barbata* were favored, the outcome would lie in the lower right-hand portion of the diagram. At high initial seed density, all experimental results favored *fatua* over *barbata*. Inasmuch as the proportion of *fatua* in mixed populations would continually increase, *fatua* eventually would exclude *barbata*.

Plant ecologists have begun to study competition between species by transplanting individuals to natural sites, with or without close competitors, and following their subsequent growth and reproduction (Cavers and Harper 1967). Smith (1975) applied this technique to the study of competition between two species of *Desmodium*, which are small herbaceous legumes common in oak woodlands in the midwestern United States. Small individuals of each species, *D. glutinosum* and *D. nudiflorum*, were planted either 10 cm from a large individual of the same species, 10 cm from a large individual of the other species, or at least 3 m from any *Desmodium* plant. To provide an index of subsequent growth, Smith measured total increase in length of all leaves, both old and new, added together. The results of the experiment (Figure 32-11) showed that both species grew most rapidly in the absence of individuals of either species. Note that the transplanted individuals of *Desmodium* were placed among numerous unrelated plant species, regardless of their proximity to other *Desmodium* individuals. It was also clear, however, that *D. glutinosum* exerted a far stronger depressing effect on the growth of *D. nudiflorum* than the reverse. All other things being equal, *D. glutinosum* should be able to replace *D. nudiflorum* in the habitats in which the transplant experiments were performed, but leaf growth does not provide a full measure of plant productivity, and the disadvantage of *D. nudiflorum* could be balanced during some other stage of the life cycle.

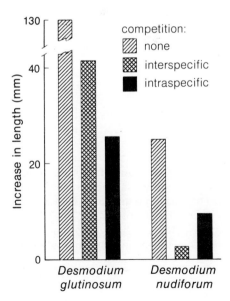

Figure 32-11   The growth responses of two species of *Desmodium* when planted near individuals of the same species (black bars), near individuals of the other species (crosshatched bars), and at a distance from individuals of either species (hatched bars) (after Smith 1975).

Experiments with laboratory and artificially manipulated natural populations show that competition is a powerful force in the ecology of populations. But the persistence of closely related species in the same habitats also tells us that competition need not lead to competitive exclusion. To determine the conditions under which competing species can coexist, population biologists have resorted to the analysis of mathematical models of population growth, including the mutual effects of competitors on the growth rates of their populations. Most of these models are built upon a simple mathematical expression for the growth of a single population limited by the availability of its resources.

The logistic equation incorporates both the intrinsic capacity of a population to grow and limitations imposed by the environment.

Several quantitative expressions of population growth have been proposed, but the most persistent has been the *logistic equation* formulated by Verhulst (1838) and independently by Pearl and Reed (1920). The logistic growth model was derived from simple biological considerations, and it provides an adequate description of the growth of many populations. Because of its simplicity and general application, it has been a springboard for many theoretical investigations of population interactions. The logistic equation characterizes the growth rate of a population in terms of two opposing biological forces: the biotic potential of the individual, and the detrimental effects of other individuals in the population on that potential, the "environmental resistance" (Chapman 1928).

Under ideal conditions, each population has a characteristic biotic potential, the innate capacity for increase (r_0), which is the rate at which each individual contributes descendants to future populations (see page 427). The unrestricted growth rate of a population under ideal conditions can be represented as the biotic potential (r_0) times the number of individuals in the population (N) or,

$$\frac{dN}{dt} = r_0 N \qquad (32\text{-}1)$$

As population density increases, however, the resources available to each individual are reduced, so the full biotic potential of the population cannot be realized. Suppose that resources are produced at rate P, and that each individual requires a certain level (M) of resources just to stay alive. A population of size N would, therefore, require resources at rate NM for the maintenance of individuals, leaving $P - NM$ resources available for reproduction. If we now assume that an individual's rate of reproduction increases in proportion to the level of resources available for that purpose, $(P - NM)/P$, we can express the productivity of each individual in the population as the biotic potential times the proportion of resources available for reproduction, that is,

$$r = r_0 \frac{(P - NM)}{P} \qquad (32\text{-}2)$$

When the population is small, that is, when N is close to zero, the proportion of resources available for reproduction is nearly 100 per cent, and each individual can realize its full biotic potential. As population size increases, the proportion of resources available for reproduction decreases, and with it the realization of the individual's reproductive potential.

The growth rate of the population is equal to the number of individuals in the population times the realization of each individual's potential for reproduction, $dN/dt = rN$. Substituting equation (32-2) for r, we obtain

$$\frac{dN}{dt} = r_0 N \left(\frac{P - NM}{P} \right) \tag{32-3}$$

If the population were to become so large that the term NM exceeded P, the resources available for reproduction would become negative, leading to a decrease in population size (a negative growth rate). Population size will remain constant ($dN/dt = 0$) when $P = NM$, hence the term $P - NM = 0$, and the number of individuals in the population N equals P/M. The total resources available divided by the minimum maintenance requirement of each individual (P/M) is the equilibrium population size—often referred to as the *carrying capacity* of the environment (K) for that population.

The differential form of the logistic equation is usually expressed with K substituted for P/M, hence

$$\frac{dN}{dt} = r_0 N \frac{(K - N)}{K} \tag{32-4}$$

or

$$\frac{dN}{dt} = r_0 N \left(1 - \frac{N}{K} \right) \tag{32-5}$$

The term N/K is equivalent to Chapman's environmental resistance. As the number of individuals in the population approaches the carrying capacity, N/K approaches 1, $(1 - N/K)$ approaches zero, and population growth ceases, as shown in Figure 32-12. If population size exceeded the carrying capacity (K), its growth rate would become negative and the number of individuals in the population would decrease toward the carrying capacity. Because carrying capacity is determined both by characteristics of the environment (P) and by characteristics of the population (M), it is subject to both environmental and evolutionary change.

The cumulative growth curve of the logistic equation—that is, the number of individuals in the population (N) as a function of time (t)—is expressed by

$$N(t) = \frac{K}{1 + [(K - N(0)/N(0)]e^{-rt}} \tag{32-6}$$

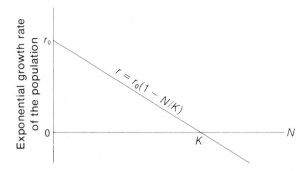

Figure 32-12   Relationship between the exponential growth rate of a population (r) and the number of individuals in the population (N), according to the logistic growth model. Compare with the growth rate of the *Daphnia pulex* population shown in Figure 31-10.

where $N(0)$ is the population size at time 0. The logistic growth curve has a sigmoid (S-shaped) form and it resembles the growth curves of many natural and laboratory populations (for example, see Figure 32-13).

That a population follows a logistic growth curve provides little information about underlying growth processes and their regulation, except that their interaction with the environment tends to maintain the population at a stable equilibrium point. We should not presume that stably regulated populations cannot have growth forms other than the logistic curve. The specific shape of the function depends on the manner in which individuals in the population interact. If individuals did not compete—that is, if the activities of

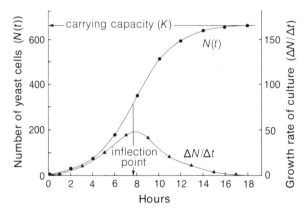

Figure 32-13   Increase in the number of cells in a laboratory culture of yeast. The number of cells, $N(t)$, was counted every two hours. The increase in the number of cells between each count is the growth rate, $\Delta N/\Delta t$. For example, between six and eight hours the population increased by 176 cells, from 175 to 351 (from data of Pearl 1925, quoted by Kormondy 1969; see Table 27-2).

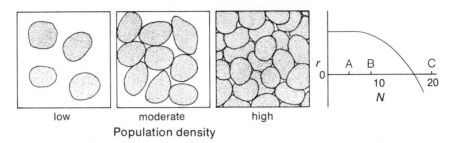

Figure 32-14   Relationship between rate of change in population size and density when populations are limited by territorial behavior. Intraspecific competition becomes more intense as density approaches the carrying capacity.

one had no influence on the reproduction and survival of others—the growth rate of the population would be unaffected by population density, and the population would continue to grow at the same rate indefinitely, all other things being equal. Although growth in all populations is eventually limited by intraspecific competition, interactions between individuals may not occur in some situations until the population has reached some threshold density, above which behavioral or other interference increases. This model could apply when a resource that can be partitioned exclusively among individuals at low population density (such as space, nesting sites, or territory) forms the basis for intraspecific competition. In the hypothetical example illustrated in Figure 32-14, the arrangement of territories defended by individuals in a population is shown for three levels of density, with the corresponding curve for r as a function of N. In this case, the carrying capacity of the area depicted is sixteen pairs of animals, which is exceeded by the dense population.

A more extreme situation could arise in a population limited by discrete, spatially isolated nesting sites. The population could grow at an undiminished exponential rate until all the nesting sites were filled, at which point the population would level off abruptly, as illustrated in Figure 32-15. If

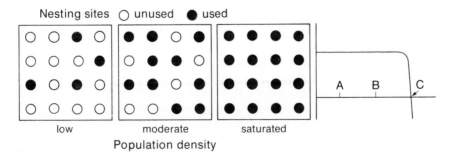

Figure 32-15   Relationship between rate of change in population size and density when populations are limited by a nonrenewable resource such as nesting sites.

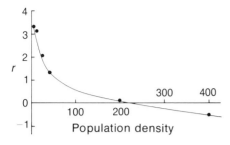

Figure 32-16   The growth rate with respect to density of a wild population of *Drosophila melanogaster* maintained in laboratory culture (after Shorrocks 1970).

competition for nest sites involved strong social interaction, other than a first-come, first-served arrangement, the population growth curve might resemble the logistic form more closely.

One can easily envision how the effects of population density upon r can increase at a disproportionately greater rate than population density, as in the preceding examples, but the opposite situation would require that interactions among individuals are most severe at low population densities, and that less severe kinds of interactions become prevalent as population size increases. This situation might occur if territoriality were maintained vigorously at low population densities, but broke down under crowded conditions, as described for several species in Chapter 31. Alternatively, limiting factors could operate during different stages of the life cycle, depending on population density. For example, intense larval competition might limit the number of adults emerging from pupae, but at high densities the population might ultimately be limited by weaker interactions among adults. The laboratory populations of *Drosophila* raised by Shorrocks (1970) may have been limited in this manner (Figure 32-16).

Whereas the logistic equation describes population growth rate as a linearly decreasing function of population size, more precise modeling of birth and death processes would almost certainly alter the predicted form of the population growth curve. Regardless of these considerations, the logistic growth function has been particularly useful to the development of population theory, as we shall see.

What is the biological meaning of carrying capacity (K)?

Although the carrying capacity of an environment for a population is partly dependent upon the resource requirements of individuals, most variation in the population density of a species in different habitats, or of similar species in a community, can be related to the availability of resources.

If individuals required a relatively fixed level of resources irrespective of their availability, population processes would balance the requirements of the population against the availability or rate of production of resources through change in population density. The concept of carrying capacity differs somewhat for renewable and nonrenewable resources. A population limited by nonrenewable resources (for example, space) will reach the carrying capacity of its environment when all of the available resource is utilized.

Populations of barnacles are limited in this way (see page 239). So are densities of forest trees. The availability of nesting sites can impose a similar limit on the number of breeding individuals, although a population could increase above this limit by accumulating nonreproducing individuals.

In contrast to nonrenewable resources, renewable resources are never completely used up. Instead, populations maintain an equilibrium resource level by the balance between production and exploitation. A population can exceed the carrying capacity of renewable resources, but because the population overexploits its resources in this situation, it will eventually decline to an equilibrium level.

A population will achieve equilibrium with resources that are continually renewed (for example, by the productivity of prey), when its resource requirements just match the renewable rate of its resources. Variation in the density of populations thus should correspond to variation in the productivity of their resources among different areas. For example, seabirds distribute themselves over the oceans so that their local density corresponds to the availability of their prey. In Figure 32-17, the number of seabirds seen from ships during a standard observation period is compared to the total volume of plankton in the same area captured in nets towed through the water, which probably is a reasonable index of resource production. The density of seabirds bore a linear relationship to the availability of food resources over a hundredfold range of plankton abundance.

We have seen that the population densities of small landbirds vary with the productivity of the habitat in which they live (page 254), and McNab (1963) has further suggested that home range size in mammals is related to body size, hence to resource requirements (page 253). McNab also pointed out that mammals that hunt for prey have much larger home ranges than species that graze on vegetation (Figure 17-2, page 253). Because hunters seek rarer food resources than those used by croppers, hunters require larger home ranges to include an amount of food production equivalent to that of grazers and browsers.

Taber (1956) directly demonstrated the influence of resource production on carrying capacity by comparing populations of mule deer in natural

Figure 32-17   Relationship between the density of seabirds in the Atlantic Ocean and the availability of their food resource (from data in a third species can invade the area between them. If the invader is efficient (large K) and overlaps its competitors greatly (small D_iH), it can outcompete both of them and reduce the number of species in the system (modified after MacArthur and Levins 1967 and May 1973).

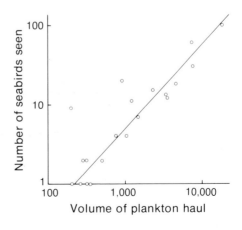

TABLE 32-1

Resource availability and population density of mule deer (*Odocoileus hemionus*) in chaparral habitat and in artificially maintained shrub habitat in northern California (from Taber 1956).

	Chaparral	Shrub
Available forage (pounds/acre)		
herbaceous species	5 (<10%)*	86 (45%)
woody species	181 (>90%)	506 (55%)
Average protein content (per cent)	9	14
Density of deer populations (ind/mi² in December)	25	60
Live weight of bucks (pounds)		
two years old	86	98
three years old	102	112
four or more years old	130	140

*Percentage of diet in parentheses.

chaparral habitat and in a nearby shrubland maintained by mechanical thinning, controlled burning, and seeding with herbaceous plants. Chaparral plants tend to be woody and much of the edible leafy vegetation is above the reach of deer. Suitable browse consisted of the young leaves of chamise and various species of oak. In the shrub habitat, the woody vegetation was thinned and herbaceous vegetation, which is highly desirable as forage for deer, was maintained artificially. As a result, the shrub area supported a population of deer more than twice that of the chaparral habitat, and the condition of the individuals, measured by their live weight, was superior in the shrub habitat (Table 32-1). The higher carrying capacity for deer in the shrub area could be attributed both to the greater quantity of available forage and to its higher quality (greater protein content).

Resource availability determines the intensity of intraspecific competition at different population densities. The relationship between competition and density, in turn, dictates the specific form of the population growth curve and its equilibrium level. Individuals of different species also compete for the same resources, and thereby exert a mutually limiting influence on population growth. Because interspecific competition plays important ecological and evolutionary roles in the organization of biological communities, population biologists have intensely investigated, both empirically and theoretically, the influence of interspecific competition on the growth of populations.

The logistic equation can be modified to incorporate interspecific competition.

Simple mathematical models based on the logistic growth equation have provided important insights into competition and the conditions that permit coexistence of competing species. Because diverse systems are cumbersome to handle mathematically (all species are independent variables in the growth equations of all other species), we shall start with the simplest case: competition between two species.

The logistic equation for population growth has the form $dN/dt = rN(1 - N/K)$, as we have seen. In the ensuing discussion, we shall make a simplifying change in notation, using $\dot{N}$ in the place of dN/dt to indicate the rate of population growth. Thus the differential form of the logistic equation becomes $\dot{N} = rN(1 \times N/K)$. The logistic equation has two component terms: rN is the unrestricted growth potential of the population in the absence of intraspecific competition; $(1 - N/K)$ describes the decrease in the realization of the growth potential caused by competition between individuals in the population. As N approaches the carrying capacity of the environment (K), N/K approaches 1 and the realized growth potential $(1 - N/K)$ decreases to zero. The term $-N/K$ expresses the degree to which the resources in the environment (K) are utilized by the population (N). To the extent that other species use the same resources, interspecific competition can also depress the growth potential of a population and influence its equilibrium level.

We may incorporate interspecific competition into the logistic by adding a second term, analogous to the term for intraspecific competition $(-N/K)$, to reduce the growth potential of the population. First, as we shall be referring to two different species in the same equation, we must attach subscripts to each variable to keep straight the species to which it refers. We shall arbitrarily call our species i and j. The logistic equation for population growth of species i in the absence of interspecific competition now becomes

$$\dot{N}_i = r_iN_i(1 - N_i/K_i) \tag{32-7}$$

The effects of interspecific competition can now be added by the term $-a_{ij}N_j/K_i$ where N_j is the number of individuals of species j and a_{ij} is the effect of an individual of species j on the population growth rate of species i. Although it need not be, the value of a_{ij} is frequently less than 1; competition is usually stronger between individuals of the same species than between individuals of different species. Strictly speaking, the term $-N_i/K_i$ expressing intraspecific competition should be multiplied by a similar coefficient a_{ii}, but a_{ii} is usually thought of as being equal to 1 and is deleted from the logistic equation.

Adding the term for interspecific competition, the equation for the population growth rate of species i becomes

$$\dot{N}_i = r_iN_i\left(1 - \frac{N_i}{K_i} - \frac{a_{ij}N_j}{K_i}\right) \tag{32-8}$$

and that for species j is

$$\dot{N}_j = r_jN_j\left(1 - \frac{N_j}{K_j} - \frac{a_{ji}N_i}{K_j}\right) \tag{32-9}$$

Thus, in a two-species system, the growth rate of each population is dependent upon the number of individuals in the populations of both species. This is shown graphically in Figure 32-18, where $\dot{N}_i$ is shown as a function of N_i and

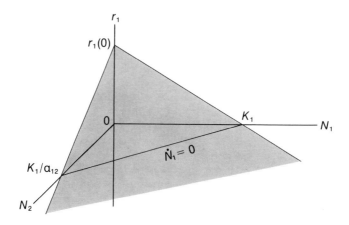

Figure 32-18   The growth rate of species 1 ($\dot{N}_1$) as a function of its population size (N_1) and the size of a competing population (N_2).

N_j. Values of $\dot{N}_i$ lie on a plane and there is a single value of $\dot{N}_i$ for every combination of N_i and N_j. If there were no individuals of species j ($N_j = 0$), the equation for the growth rate of species i (32-8) would reduce to the logistic equation 32-6; the surface of $\dot{N}_i$ values in Figure 32-18 would accordingly collapse onto a line on the N_i axis that entends from r_i at $N_i = 0$ through 0 at $N_i = K_i$. This is identical to the relationship between $\dot{N}$ and N under conditions of pure intraspecfic competition (see Figure 32-12). When N_i is small, the surface for $\dot{N}_i$ collapses onto the N_j axis, representing conditions dominated by interspecific competition.

A graphical model for coexistence of competitors is based on the conditions for equilibrium of both species.

The conditions that must be met for two competing species to coexist can be found by determining the equilibrium conditions in a two-species model—that is, the conditions under which the growth rate of both populations is zero. Equilibrium for species i occurs along a line that represents combinations of populations of species i and species j for which $\dot{N}_i$ equals zero. Since we are interested only in equilibrium values of $\dot{N}$, we can simplify the graph in Figure 32-18 by considering only the plane of zero population growth. In doing so, we revert to a two-dimensional graph with axes N_i and N_j, as shown in Figure 32-19. the line $\dot{N}_j = 0$ defines those combinations of populations of species i and j for which the growth rate is zero and the population is in equilibrium. The line also divides the graph into two areas: one toward the origin of the graph in which species i increases ($\dot{N}_i > 0$; shaded), and a second area, more distant from the origin, in which the population decreases ($\dot{N}_i < 0$; not shaded). The direction of change in population size of species i in each area is indicated by arrows. The population of species j is assumed to be constant in this case.

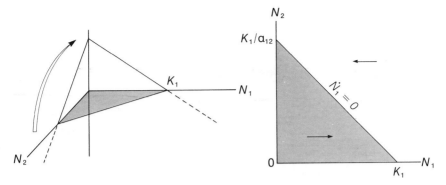

Figure 32-19   Simplification of the competition graph by considering only the plane $\dot{N}_i = 0$ bound by the N_1 and N_2 axes.

The equation for the line $\dot{N}_i = 0$ can be derived by solving equation 32-8 for $\dot{N}_i = 0$. Aside from the trivial solution $N_i = 0$ (species i does not exist), population equilibrium occurs when

$$1 - \frac{N_i}{K_i} - \frac{a_{ij}N_j}{K_i} = 0 \tag{32-10}$$

or

$$\hat{N}_i = K_i - a_{ij}N_j \tag{32-11}$$

When species i is subject to only intraspecific competition (either N_j or a_{ij} is zero), the term $a_{ij}N_j$ drops out of equation 32-11, and the equilibrium population size of species i, $\hat{N}_i$, equals K_i which is the carrying capacity of the environment for that species. We can determine the intercept of the line $\dot{N}_i = 0$ on the N_j axis by evaluating equation 32-11 when N_i is close to zero, that is, when species i is exposed to nearly pure interspecific competition. In this case, $a_{ij}N_j$ approaches K_i when species i is extremely rare, therefore, it neither increases nor decreases when $N_j = K_i/a_{ij}$.

The intercepts, K_i on the N_i axis and K_i/a_{ij} on the N_j axis, have transparent biological meaning. K_i is the number of individuals of species i that can be supported by its resources (the carrying capacity of the environment for population i). If the population were larger than K_i, its resource demands would exceed resource availability and the population would decline. Below K_i species i can increase. The intercept K_i/a_{ij} represents the number of individuals of species j that would exploit the resources of species i at a rate equivalent to that of K_i individuals of species i. When N_j is greater than K_i/a_{ij} the demands placed on the resources of species i by species j exceed the ability of the resource to support any additional individuals of species i, and the population of i declines. As the ecological overlap between species j and species i decreases, the competition coefficient a_{ij} becomes smaller and the intercept of the line $\dot{N}_i = 0$ on the N_j axis becomes higher, as shown in Figure 32-20. That is, as interspecific competition decreases, the ability of species i

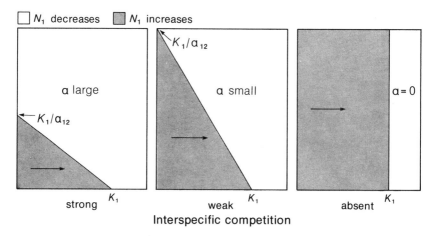

Figure 32-20   Influence of the intensity of competition (α_{ij}) on the position of the $N_i = 0$ isocline showing that when interspecific competition is weak, species j has little effect on the population growth of species i.

to increase is influenced less by the population size of species j. When there is no interspecific competition, α_{ij} is zero and the carrying capacity of the environment for species i is unaffected by species j.

Because coexistence requires the stable equilibrium of two species, we must consider both equilibrium lines, $\dot{N}_i = 0$ and $\dot{N}_j = 0$, together to determine graphically the conditions required for coexistence (Figure 32-21). The line $\dot{N}_j = 0$ is determined in the same manner as the equilibrium line for species i, and its intercepts are K_j on the N_j axis and K_j/α_{ji} on the N_i axis. Because the equilibrium lines define regions within the graph in which each species increases or decreases in number, we can determine the fate of mixed populations of species i and j graphically. If the numbers of species i and species j were represented by point A in Figure 32-21, both populations would be below their respective equilibrium lines, and so both would increase, as indicated. The point that represents the two populations would move up and to the right on the graph with time. If the populations were initially at point B, beyond the equilibrium lines of both species, both would decrease and point B would move down and to the left, toward the origin of the graph. If the populations were located between the equilibrium lines of the two species (for example, at point C), the population of species i would increase while that of species j would decrease, as indicated, and the resulting trajectory of the populations on the graph would point toward the lower right. The system portrayed in Figure 32-21 can attain equilibrium only after N_i reaches K_i and N_j has been reduced to zero; hence when species i completely eliminates species j. When the region of increase of one species completely includes that of a second, the first species is the better competitor, and it will always exclude the second. If the position of the two equilibrium lines were reversed, so $\dot{N}_j = 0$ lay outside of $\dot{N}_i = 0$, species j would outcompete species i.

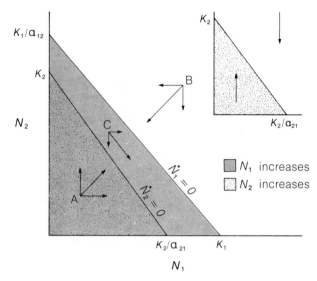

Figure 32-21   Graphical representation of the equilibrium conditions for two species where species 1 is the better competitor. The component of the population trajectory attributed to species 2 is shown in the small graph. Areas in which populations can increase are indicated by shading for species 1 and stippling for species 2.

The trajectory of two competing populations, both of which are small initially, is followed through successive time intervals in Figure 32-22. At first, both populations increase until they enter the region on the population graph between the two equilibrium lines. There, the trajectory moves down and to the right, ending at point K_i on the N_i axis. During the course of competitive elimination, the population of species i increased continuously during each interval, whereas species j initially increased and then began to decrease, until eventually it was eliminated altogether.

Two species can coexist only when the equilibrium lines for the two species cross. At their intersection, the growth rate of both populations is zero, and thus both are in equilibrium. Let us return for a moment to Figure 32-21, in which the equilibrium curve for species i lies outside of the equilibrium curve for species j. If a_{ji} decreased, that is, if the competitive influence of species i on species j were reduced, K_j/a_{ji} would increase and the equilibrium line for species j would move out along the N_i axis, as shown in Figure 32-23. As a_{ji} decreased, the exploitation of the resources of species j by species i would be reduced, permitting N_j to increase at progressively greater levels of N_i.

The population trajectories in Figure 32-23 demonstrate that the intersection of the equilibrium lines represents stable coexistence; all trajectories converge upon that point. When species j is rare, it can increase in number even when species i has reached its maximum population size. Similarly,

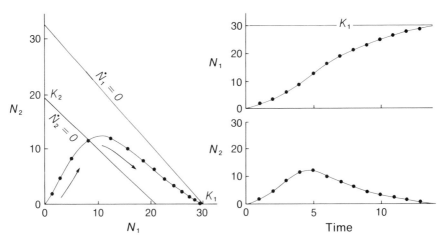

Figure 32-22   The course of competition between two populations, portrayed on a competition graph (left) and as the change in population size with respect to time (right). The time intervals are indicated on the competition graph by numerals next to the sample points.

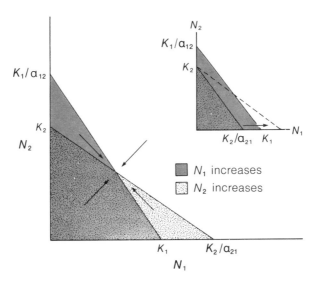

Figure 32-23   Conditions for the stable coexistence of two competing species. The graph can be obtained from Figure 32-21 by reducing the competition coefficient of species 1 (the better competitor) against species 2 as shown in the upper right. If α_{12} is decreased, the intercept K_2/α_{12} increases.

when species i is rare, it can always increase in the presence of species j. We can conclude that two species will coexist when competitive interactions between them are weak—that is, when each species is more strongly limited by intraspecific competition than by interspecific competition. Therefore, partitioning of resources differentially among species reduces interspecific competition and permits stable coexistence in a two-species system.

The conditions for coexistence between species i and j can be formalized in terms of the relationships between values of K and α. From the graph of coexistence (Figure 32-23), we see that the following inequalities must hold: $K_i < K_j/\alpha_{ji}$ and $K_j < K_i/\alpha_{ij}$. By rearranging these equalities, we obtain the conditions $\alpha_{ji} < K_j/K_i$ and $\alpha_{ij} < K_i/K_j$, and in general

$$\alpha_{ij}\alpha_{ji} < 1 \tag{32-12}$$

What happens when both populations are more strongly limited by interspecific competition than by intraspecific competition? This situation, shown in Figure 32-24, has an unstable equilibrium point, away from which populations tend to move. One species or the other will prevail, depending on which is initially more numerous. The biological conditions that would result in this situation are difficult to envision: each species must be superior to the other in a predominately intraspecific environment and inferior in a predominantly interspecific environment. This could occur if each species inhibited the population growth of the other by the release of toxic chemicals. Whichever species occurred initially in greater numbers would be likely to have a competitive edge.

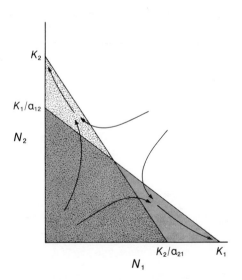

Figure 32-24   Outcome of competition between two species that are both more strongly limited by interspecific competition than by intraspecific competition (α_{12} and α_{21} are both large).

The measurement of competition coefficients in natural systems.

The general result that two species can coexist if $a_{ij}a_{ji} < 1$ and similar results from more complicated models with more than two species have stimulated attempts to measure a-values in natural communities and experimental systems. These studies have taken two approaches, one being the measurement of mutual use of resources by competing species (ecological overlap), the other being the effect of manipulating the size of one competitor population on the equilibrium size or growth rate of another.

MacArthur (1968) and Levins (1968) suggested that overlap in resource utilization by two species measures the coefficient of competition between them. Attempts to measure these overlaps (for example, Cody 1968, 1974, Hespenheide 1971, Pulliam and Enders 1971, Zaret and Rand 1971, Pianka 1973, 1974, Pulliam 1975) have produced results, to be sure, but the a-values are difficult to interpret (Schroder and Rosensweig 1975). The technical difficulties involved in measuring what animals eat have forced ecologists to rely on indirect measures of overlap: where individuals feed, their foraging techniques, the relative sizes of their mouths or bills, and so on. Furthermore, by measuring overlaps during a particular season, one implicitly assumes that the conditions of competition permitting coexistence pertain at that time. In fact, the conditions necessary for coexistence need not apply continuously, and for many species, food supplies may be so abundant during certain seasons or years that competition is greatly reduced. Finally, even if overlap in resource use were measured precisely, competition coefficients cannot be directly inferred because the a-values are population parameters that incorporate such additional factors as efficiency of resource use, predation, dispersal, and so on. Keeping these difficulties in mind, we may still use overlap values as rough estimates of competition coefficients and as measures of community organization. Overlap values suggest the manner in which species partition resources and how this partitioning may vary between taxonomic groups, geographical localities, habitats, and seasons (Schoener 1974).

Overlap indexes are calculated in a variety of ways (Horn 1966, Orloci 1975, Pielou 1977, Poole 1974), perhaps the most straightforward of which is the a index* of MacArthur and Levins (1967). Because a is a ratio between the mutual co-occurrence of individuals of different species and the mutual co-occurrence of individuals of the same species, values less than 1 suggest

---

* The strength of competition is assumed to vary with the probability that two species (i and j) will seek the same resource (h). This probability is proportional to the product of the utilization (U) of each resource by each species ($U_{ih}U_{jh}$). The overall competitive interaction between two species on all resources is $\Sigma\, U_{ih}U_{jh}$. To provide a meaningful estimate of a, however, we must compare interspecific competition to competition between individuals of the same species seeking the same resource ($U_{ih}U_{ih}$, or U_{ih}^2). Therefore, the competition coefficient of species j with species i is

$$a_{ij} = \frac{\Sigma_h\, U_{ih}U_{jh}}{\Sigma_h\, (U_{ih})^2}.$$

(32-13)

(continued on following page)

that interspecific competition is weaker than intraspecific competition for the particular resource considered.

The calculation of overlap values has revealed the basis for resource partitioning in several communities. For example, Levins (1968) studied the occurrence of five species of *Drosophila* on different kinds of fruit throughout the year. The α values calculated on the basis of food preferences (food αs) varied considerably (0.16 to 0.92), as did season αs (0.26 to 0.95), and the two measures of overlap bore no relationship to each other. One might have expected species of flies with high food αs to avoid each other seasonally, but Levins (1968) showed in another study that season is relatively unimportant compared to habitat specialization in partitioning the environment among fly species.

Zaret and Rand (1971) calculated food overlap values from stomach contents of eleven species of fish in a small stream in central Panama. For a community with $n = 11$ species there are $n(n - 1)/2 = 55$ pairwise combinations of species, hence α-values. As in Levins' study, overlap values varied widely, from 0.0 to more than 0.9, though most of the values were low. During the dry season, when stream flow was reduced and food was less abundant, food overlap values were lower than during the wet season. Whereas 56 per cent of the wet-season overlap values were greater than 0.2, and 29 per cent exceeded 0.5, comparable percentages during the dry season were only 31 and 16 per cent. These findings suggest that when food resources diminished seasonally, the fish adopted more specialized feeding behavior, and thereby avoided intense competition.

Further discussion of other studies of ecological overlap and their significance will be deferred until a later chapter on the organization of biological communities. Here we shall continue to discuss methods of estimating competition coefficients by investigating inferences that can be made from experimental perturbation of populations.

When all species in a community of competitors have achieved equilibrium population densities, the resources available to species i (K_i) are fully

---

Consider the following numerical example:

	Resource			
	A	B	C	Total
Utilization by				
Species 1	5	3	2	10
Species 2	3	7	4	14

Applying equation (32-13), we see that

$$\alpha_{12} = \frac{(5 \times 3) + (3 \times 7) + (2 \times 4)}{(5)^2 + (3)^2 + (2)^2} = \frac{15 + 21 + 8}{25 + 9 + 4} = \frac{44}{38} = 1.16.$$

Similarly, we find that

$$\alpha_{21} = \frac{(3 \times 5) + (7 \times 3) + (4 \times 2)}{(3)^2 + (7)^2 + (4)^2} = \frac{15 + 21 + 8}{9 + 49 + 16} = \frac{44}{74} = 0.59.$$

utilized by species i and by each of its competitors (j) in proportion to the coefficient of competition (a_{ij}) of that species. This relationship can be expressed by

$$K_i = \sum_{j \neq 1}^{m} a_{ij}\hat{N}_j \qquad (32\text{-}13)$$

for a community of m species. In the simplest case, the two-species community, equation 32-14 becomes

$$K_i = \hat{N}_i + a_{ij}\hat{N}_j \qquad (32\text{-}14)$$

We can determine K_i by raising species i in the absence of j or by eliminating j from the habitat and measuring the new equilibrium level of i. Knowing K_i we can arrange (32-14) to give

$$a_{ij} = \frac{K_i - \hat{N}_i}{\hat{N}_j} \qquad (32\text{-}15)$$

This procedure can easily be extended to calculate a-values in communities with more than two species (Vandermeer 1969, 1972, Wilbur 1972). A revealing set of experiments on competition within laboratory communities of four species of aquatic crustacea was performed by Neill (1974). The crustacea were maintained in 1,500 ml laboratory microcosms in which they fed on a variety of algae species. Neill determined equilibrium densities of each species after two to four months in the presence of each one of the other three species, combinations of two of the competitors, and all three competitors. Representative results are shown in Table 32-2. None of the species was grown alone, so K_i-values have to be calculated indirectly. A quick glance at the data indicates that coefficients of competition were extremely asymmetrical in many cases. On one hand, *Hyalella* was not influenced by competition from any species; its numbers did not vary outside of the range 312–330, re-

TABLE 32-2

Equilibrium population sizes in laboratory microcosms composed of different combinations of competing aquatic crustaceans (after Neill 1974).*

Experiment	Alonella	Equilibrium populations of Ceriodaphnia	Simocephalus	Hyalella
Hyalella present	417	294	48	312
Remove C	614	—	83	327
Remove S	431	331	—	325
Remove C and S	739	—	—	330
Hyalella absent	2,287	911	1,407	—
Remove C	3,624	—	1,512	—
Remove S	2,186	1,411	—	—

*Results are modified somewhat to simplify table. As a result, calculated values of a_{ij} do not correspond exactly to Neill's values.

gardless of which other species it was raised with. Hence the competition coefficients α_{Hj} were all close to zero. On the other hand, competition from *Hyalella* greatly reduced equilibrium populations of the other three species, hence values of α_{iH} were quite large. More important, the outcome of competition between two species is frequently influenced by a third. When *Ceriodaphnia* ($N = 294$) was removed from the four-species system, the population of *Alonella* increased by 197 individuals ($614 - 417 = 197$). From equation 32-16 we can calculate α_{AC} as $197/294 = 0.67$ and, similarly, α_{AS} as $(431 - 417)/48 = 0.29$. If the competitive influences of *Ceriodaphnia* and *Simocephalus* on *Alonella* population density were independent, we would expect the influences to add when both competitors were removed from the system. That is, the *Alonella* population should increase by $(614 - 417) + (431 - 417) = 211$ individuals. What Neill found, however, was that the *Alonella* population increased by $739 - 417 = 322$ individuals. In other words, coefficients of competition between any two species depend in part on other species present. To further demonstrate this point, let us recalculate the values of α_{AC} and α_{AS} in the absence of *Hyalella*. For *Ceriodaphnia* we obtain $\alpha_{AC} = (3624 - 2287)/911 = 1.47$ and for *Simocephalus,* $\alpha_{AS} = (2186 - 2287)/1407 = -0.072$, which probably does not differ significantly from zero. Neill's experiments demonstrate the complexity of competitive relationships between species and caution us from extrapolating results from two-species systems to more complex communities.

Even for two-species systems the description of competition based on the logistic equation for population growth (32-8) is often insufficient. Ayala (1970) obtained evidence that contradicted predictions of the logistic equations for competing laboratory populations of the fruit flies *Drosophila pseudoobscura,* from the western United States, and *D. serrata,* from New Guinea. At 23 C the two species were able to coexist indefinitely. *Serrata* replaced *pseudoobscura* at higher temperatures, and the reverse occurred at lower temperatures. Populations of each species maintained separately and in competition with each other are presented in Table 32-3. If we were to calculate coefficients of competition from Ayala's data, using the logistic expression for competition (equation 32-11), we would obtain $\alpha_{ps} = (K_p - \hat{N}_p)/\hat{N}_s = (664 - 252)/278 = 1.49$ and $\alpha_{sp} = (K_s - \hat{N}_s)/\hat{N}_p = (1251 - 278)/252 = 3.86$. The product $\alpha_{ps} \alpha_{sp} = 1.49 \times 3.86 = 5.75$ violates the general condition for coexistence for competitors $\alpha_{ij} \alpha_{ji} < 1$ (32-11).

Under conditions of logistic competition, the combined populations of coexisting competitors normally will be greater than that of either species grown separately. This is shown in Figure 32-25. In Ayala's experiment, however, the combined number of individuals of both species (530) was less than either species when grown alone (664 and 1,251).

The competition graph can easily be altered to accommodate Ayala's results by deforming the equilibrium lines for $\dot{N} = 0$ and $\dot{N}_s = 0$ into curved lines that are concave outward (Gilpin and Justice 1972, 1973; Figure 32-35). Note that the intersections of the equilibrium lines with the N_i axes (K_i and K_j/α_{ji}) remain unchanged. Furthermore, the basic conclusion that species can coexist only when intraspecific competition is stronger than interspecific competition remains valid. Biologically speaking, the curvature of the lines

TABLE 32-3

Outcome of competition experiments in which *Drosophila pseudoobscura* coexists with *D. serrata* (from Ayala 1970).

	Mathematical expression according to logistic equation	Adult population size
Species raised separately		
D. pseudoobscura*	$N_p = K_p$	664
D. serrata	$N_s = K_s$	1,251
Species raised together		
D. pseudoobscura	$\hat{N}_p = K_p - a_{ps}\hat{N}_s$	252
D. serrata	$\hat{N}_s = K_s - a_{sp}\hat{N}_p$	278
		Total = 530

*Arrowhead chromosome arrangement. Similar experiments with the Chiricahua chromosome arrangement had nearly identical outcomes.

detected by Ayala's experiment suggests that the coefficient of competition a_{ij} is greatest when species j is uncommon and decreases as j approaches its own carrying capacity (K_j). This might occur if the two species were specialized to feed on slightly different resources. Suppose, for example, that the habitat contained two resources, 1 and 2. Species i is more efficient than j at exploiting resource 1, and vice versa. Hence a_{ij} is greater on resource 1 than on resource 2, and a_{ji} is greater on resource 2 than on resource 1. If species i were rare, most individuals would exploit their preferred resource (1) and would, therefore, exert a strong competitive influence on species j, if it were so abundant that individuals were forced by intraspecific competition to feed on resource 1. This type of competitive interaction is a type of frequency-dependent competition (see, for example, Ayala 1970, 1971). Gilpin and Ayala (1973) suggested, alternatively, that the curvature of the isoclines on the competition graph was caused by an increase in intraspecific competition with density. They incorporated this effect into the logistic competition equation attaching an exponent to the term for intraspecific competition, hence

$$\dot{N}_i = r_i N_i \left(1 - \left(\frac{N_i}{K_i}\right)^a - \frac{a_{ij}N_j}{K_i}\right) \tag{32-16}$$

When $a = 1$, the equation reduces to the linear form of the logistic competition equation. When $a > 1$, intraspecific competition increases with density, perhaps through increasing social interference or territorial defense. Gilpin and Ayala suggest that a is more likely to exceed 1 for vertebrates, which have well-developed social interaction, than for invertebrates. Conversely, when $a < 1$, intraspecific competition decreases with density (see Figure 32-16). Although this pattern is observed in other *Drosophila*, it is difficult to account for.

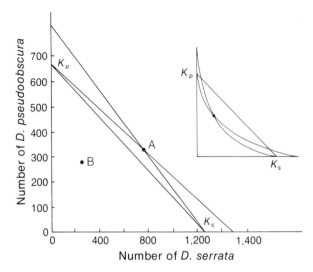

Figure 32-25   Comparison between a prediction of the logistic competition model for coexisting species (A) and the outcome of an experiment with coexisting species of *Drosophila* (B). When coexistence occurs, the model predicts that the equilibrium population (point A) must lie outside a line connecting the two carrying capacities (K_s and K_p). Ayala's result is indicated by point B. A modification of the graphical model to accommodate point B is shown to the right.

Park (1954) distinguished two types of competition—*exploitation* and *interference*—to help resolve the problem of variable competition coefficients. Park equated exploitation interactions to the linear component of the competition equations, and he suggested that interference has occurred when coexisting populations of competitors are lower than would be expected on the basis of the carrying capacities of the individual species (see Case and Gilpin 1974). Exploitation interactions occur when individuals seek the same resource but do not come into direct contact with one another. Their influence is exerted by reducing the resources available to others. Unless an individual's resource requirements change with population density, one would expect exploitation αs to remain constant.

Interference implies such direct interaction between individuals as fighting or the release of toxic chemicals (Miller 1967, 1969), and the intensity of such interaction probably increases with density. Additionally, interference usually occurs within, rather than between species (see page 243). An increase in the coefficient of intraspecific competition with increasing density of species i would be equivalent to a decrease in α_{ji}, and would tend to favor the coexistence of competing species. On the competition graph, such interference would deform the equilibrium lines in a manner consistent with Ayala's experiment (Figure 32-25).

If interspecific competition coefficients increased as a function of density, the equilibrium lines would become concave outward, rather than in-

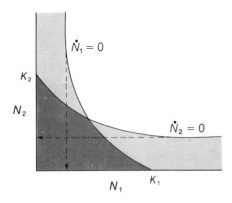

Figure 32-26   Graphical representation of competition between two species, each of which has exclusive resources. Both K_j/a_{ji} and K_i/a_{ij} are infinite. The dashed lines with arrows point to the numbers of each species that can be supported by its exclusive resources alone.

ward. This type of density dependence in competition coefficients could be brought about if competing populations poisoned each other, but not themselves, by producing metabolic waste products or toxins. Presumably, the impact of a toxin on each individual is greatest when the toxin-producing population is most dense. Substances produced by the larvae of *Drosophila melanogaster* have been observed to reduce the survival of *D. pseudoobscura* in competitive situations (Weisbrot 1966); similar results have been obtained for intraspecific competition between different genotypes of *Drosophila melanogaster* (Dawood and Strickburger 1969).

The last modification of the logistic competition equations we shall discuss is the influence of exclusive resources for one or both species (Schoener 1974). An exclusive resource is one that is not exploited by competitors. It seems reasonable that any species with exclusive resources cannot be eliminated by competition from other species. We can see this graphically in Figure 32-26, a situation in which both competitors have exclusive resources. Because K_i is always less than K_j/a_{ji}, and vice versa, the interaction between species i and j will always result in a stable equilibrium.

Although the logistic competition equations clearly do not adequately represent the nonlinear interactions that characterize some forms of competition, the basic premises of the model for competitive coexistence are undoubtedly valid. With a somewhat better perspective on our model of competitive interactions, we can now proceed to explore some of the consequences of the model for more complex situations.

The outcome of competition varies according to the environment.

Coefficients of competition depend both on the species involved and on the environment of the interaction. The competitive ability of a species, like any other demographic trait—such as birth rate and survival rate—depends on the particular environment of the individual. We have already seen this in the interaction between species of wasps parasitic on citrus scale (page 552). One species, *Aphytis melinus,* proved to be a superior competitor in the extreme environments of the interior valleys of southern California, but it was replaced by *A. lingnanensis* along the coast (Figure 32-5).

When closely related species are thrown into direct competition in the laboratory, which species turns out to be the superior competitor often depends upon the conditions of the experiment. When two species of flour beetle, *Tribolium castaneum* and *T. confusum,* were grown together in wheat flour, *castaneum* excluded *confusum* under cool, dry conditions, but itself was excluded when colonies were kept warm and moist (Figure 32-27). When grown separately, both species reached their highest densities in moist conditions. Thus a species may be a superior competitor under suboptimum conditions; the outcome of competition is not determined by the performance of a species in the absence of competition, rather it is determined by the relative productivity of species grown together (Park 1954, 1962).

We often observe the replacement of a species by a close competitor along a gradient of ecological conditions. Where one species leaves off and the other takes up, the environment evidently must switch from favoring one competitor to favoring the other. Sometimes two species overlap in a broad ecological zone, and occasionally one finds wide gaps between the ranges of two ecologically similar species (Diamond 1973, Terborgh 1971). These phenomena can be understood readily through a graphical device developed

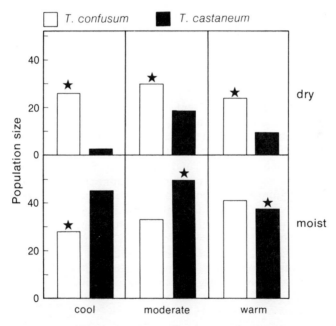

Figure 32-27   Competition between two species of flour beetles at different temperatures and relative humidity. Each section of the diagram shows the population density of the two species when grown separately (vertical bars); the superior competitor under each set of conditions is indicated by the star (Park 1954, 1962).

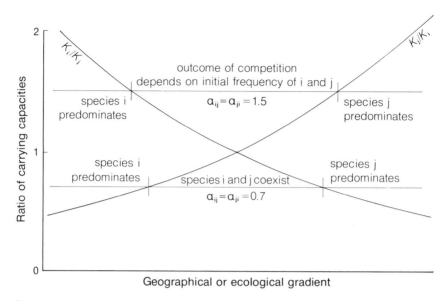

Figure 32-28   Graphical representations of conditions for coexistence of two species along a geographical or ecological gradient. When α's exceeds 1, the species cannot coexist. In the central part of the gradient, where $\alpha_{ij} > K_i/K_j$ and $\alpha_{ji} > K_j/K_i$, the species with the higher initial frequency will tend to predominate, favoring i toward the left and j toward the right. When α's are less than 1, species i and j coexist, hence their distributions overlap, in the central part of the gradient.

by Jonathan Roughgarden at Stanford University (Figure 32-28). First, we suppose that the carrying capacity (K_i) of each species varies along the environmental gradient and that the ratio K_i/K_j either increases or decreases; the ratio K_j/K_i behaves in an inverse fashion. We have seen earlier (page 573) that for coexistence $\alpha_{ij} < K_i/K_j$ and $\alpha_{ji} < K_j/K_i$. Suppose for the sake of simplicity that $\alpha_{ij} = \alpha_{ji}$ and the α-values did not change over the ecological gradient. From Figure 32-28, it becomes clear that if α_{ij} and α_{ji} are less than 1, the two species will overlap along the gradient over a narrow range of ecological conditions. That is, they will coexist locally where their carrying capacities are similar. Conversely, if α_{ij} and α_{ji} are greater than 1, the two species cannot coexist, and their ranges will abut, but not overlap. To explain wide gaps between the species, we must postulate the existence of a third species that competes effectively with the first two in the region in which they meet.

How many species can coexist?

Although mathematical and experimental studies of two-species systems have helped to increase our understanding of population interactions, they hardly comprehend natural communities in which many thousands of species can coexist. Can we extend the mathematical analysis of competition to the case of many species? Would it be profitable to do so? A few simple attempts to extend two-species models will show that when one considers coexistence of many species at the community level, methods of analysis necessarily must be changed and different questions must be asked.

We can extend the graphical model of competition (Figure 32-23) to the three-species case by adding another dimension for the third species. The equilibrium isoclines no longer remain straight lines on a two-dimensional graph; rather they become planes in a three-dimensional graph (Figure 32-29). Any three nonparallel planes will intersect at one, and only one point. If that point occurs within the boundaries of the graph, that is, if N_1, N_2, and N_3 are all positive, the three species will coexist. If the point of intersection is negative for one or more of the species, such species will be extinct at equilibrium, and coexistence will not occur. In Figure 32-23, the intersection of each pair of planes is indicated by a dashed line, all of which cross at a single point within the boundaries of the graph. Thus, Figure 32-29 represents a stable equilibrium. Notice that all Ks are less than K/αs. For example, K_1 is less than both K_2/α_{21} and K_3/α_{31}. These are the conditions for stable equilibrium in the three-species case, as they are in the two-species case, and these conditions can be extended validly to any number of species.

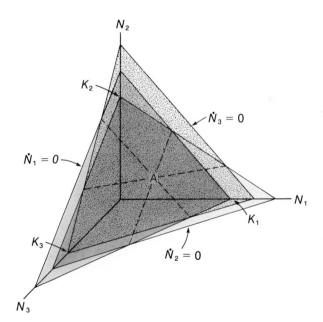

Figure 32-29   Competition graph for the three-species case. Equilibrium conditions for each species $(r = 0)$ are represented by planes whose intersection (A) designates a stable equilibrium for the community. The dashed lines represent the intersection of each pair of planes.

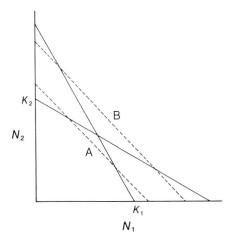

Figure 32-30   The conditions for invasion of an equilibrium two-species system by a third species. The dashed line represents the equilibrium line for the third species ($N_3 = 0$) when it is rare (N_3 close to 0). If the equilibrium line lies below the two-species equilibrium (case A), the third species cannot increase; if it lies above the equilibrium (case B), the third species can increase and may either coexist with or outcompete the other two.

Graphical analysis of the three-species system can be approached more simply by rephrasing the problem in terms of a two-dimensional model. Suppose that two species coexist in a stable equilibrium. Under what conditions can a third species increase and, therefore, not be excluded by the other two? A third species can successfully invade a two-species system when its equilibrium isocline lies outside the equilibrium point of the other two (Figure 32-30). But this condition could also result in the third species outcompeting the other two. Stable coexistence occurs only when all three species can increase when they are rare, that is, when interspecific competition is relatively weak compared to intraspecific competition. Therefore, any number of species could coexist by progressively subdividing the available resources, thereby reducing a values. The solution to the problem of competitive coexistence thus lies not so much in determining conditions for coexistence as in understanding how and to what extent species can subdivide resources.

MacArthur and Levins (1967) investigated the specific case of competition between three species lined up along a single resource dimension. They assumed that the utilization of resources by a species resembled a normal, or

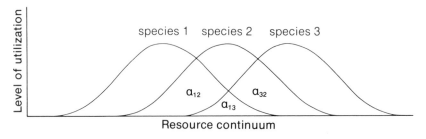

Figure 32-31   Utilization of a resource continuum by three species. The regions of overlap represent the coefficients of competition.

bell-shaped, distribution (Figure 32-31), and then asked what are the conditions that allow a third species to fit in between two others along a resource gradient? According to the logistic model of competition, the middle species (2) can increase if its carrying capacity is greater than the combined effects of competition by both species upon its population, that is, if

$$K_2 > a_{21}\hat{N}_1 + a_{23}\hat{N}_3 \tag{32-17}$$

where $\hat{N}$ is the equilibrium size of a population. MacArthur and Levins demonstrated that a third species could invade a resource continuum between two other species if the coefficients of competition of the invader with the residents were less than about 0.5 (assuming roughly equal Ks). This result can be visualized by considering the following rectangular utilization curves:

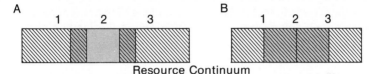

In A, each of the resident species overlaps with the invader by less than one-half, and the invading species can maintain itself on the resources that it does not share with the other two. In B, the resident species are similar enough to each other that they completely overlap the invader (a_{21} and $a_{23} = 0.5$), which cannot successfully coexist because it must share all its resources with other species.

MacArthur and Levins' analysis of the more complicated situation in Figure 32-30 showed that, as in the simpler case discussed above, the maximum values for a that permit coexistence of the third species are on the order of 0.5.* The addition of more species beyond species 1 and 3 on the

---

* Assuming that a species can invade a resource continuum between two others only when $K_2 > a_{21}\hat{N}_1 + a_{23}\hat{N}_3$ (equation 32-17), MacArthur and Levins (1967) determined maximum values of a that permit invasion. Values of $\hat{N}$ can be calculated from the equilibrium conditions of the logistic competition equations (see equation 32-12): $\hat{N}_1 = K_1 - a_{13}\hat{N}_3$, and similarly $\hat{N}_3 = K_3 - a_{31}\hat{N}_1$. Substituting $\hat{N}_3$ for N_3 in the equation for $\hat{N}_1$, we obtain

$$\hat{N}_1 = K_1 - a_{13}(K_3 - a_{31}\hat{N}_1)$$

$$= K_1 - a_{13}K_3 + a_{13}a_{31}\hat{N}_1 \tag{32-18}$$

If we now make the simplifying assumption that the system is symmetrical, that is $K_1 = K_3$ and $a_{31} = a_{13}$, equation (32-18) becomes

$$\hat{N} = K - aK + a^2\hat{N}$$

$$\hat{N}(1 - a^2) = K(1 - a)$$

$$\hat{N}(1 - a)(1 + a) = K(1 - a)$$

$$\hat{N} = \frac{K}{1 + a} \tag{32-19}$$

(continued on following page)

resource continuum does not appreciably change the equilibrium condition for species 2. From equation (32-17) we can see that the ease with which a new species can invade a system increases when the new species is very efficient (high K), or when the established species are highly specialized (low α). This is shown graphically in Figure 32-32.

MacArthur and Levins' analysis is a major theoretical step because it defines a limit to the ecological similarity of coexisting species occupying a single resource continuum. May and MacArthur (1972) have further evaluated the MacArthur-Levins model for systems in randomly varying environments. Variability makes it more difficult for a third species to invade a two-species system. In general, under varying conditions, a third species cannot invade a community if the distance between the centers of each of the two resident species' ecological distributions and the center of the ecological distribution of the invader is less than the standard deviations of their resource utilization curves ($D/H < 1$, see Figure 32-32). This value is the limiting similarity of species in the community.

There is little point in pushing the MacArthur-Levins model much beyond its present state because comparable natural situations, which could be used to test the predictions of the model, are difficult to identify and may not even exist. In natural communities, competition occurs over many resource dimensions and, as a result, large coefficients of competition can be tolerated along any one dimension. As we have seen above, α's in natural communities can be as great as 0.9 or more for a given resource dimension. We shall return to the problem of coexistence in complex communities, using more empirical approaches, in a later chapter.

The mathematical considerations of competition presented in this chapter suggest several tentative conclusions about coexistence in natural communities: First, no more than one species can coexist on a single resource; second, a limited number of species can coexist on a single resource continuum; and third, any number of species can coexist provided that sufficient numbers of discrete resources or resource dimensions are available, or that species evolve a sufficient number of ways of partitioning resources to avoid competition. These conclusions suggest that we ask other questions about the role of competition in determining the structure of natural communities. Knowing the conditions for competitive coexistence does not explain varia-

---

If we now let $\alpha_{31} = \alpha_{13} = \theta$, the conditions for invasion (equation 32-17) become

$$K_2 > \frac{2\alpha K}{1 + \theta} \tag{32-20}$$

and if $K_2 = K$, coexistence can occur when $\theta > 2\alpha - 1$. From the equation for the normal curve, MacArthur and Levins determined that $\theta = \alpha^4$ and, consequently, that α must be less than 0.544 for invasion to occur. The influence of carrying capacity and ecological overlap (interspecies distance) on the ability of a species to invade a system is shown in Figure 32-32. As one would expect, increased carrying capacity and decreased ecological overlap facilitate invasion.

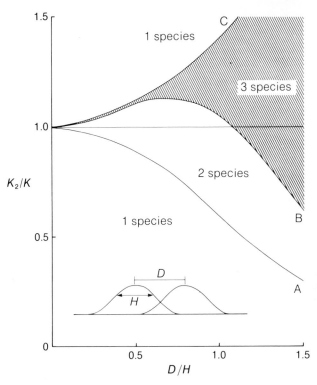

Figure 32-32   Influence of the relative carrying capacity of an invading species (K_2/K) and its ecological overlap with established species (D/H) on the coexistence of species on a resource continuum. D is the distance between modes of the utilization curves, and H is their standard deviation (breadth). Above line A, two species can coexist in competition; above line B, a third species can invade the area between them. If the invader is efficient (large K) and overlaps its competitors greatly (small D/H), it can outcompete both of them and reduce the number of species in the system (modified after MacArthur and Levins 1967 and May 1973).

tion in the number of species in natural communities. Instead, we must determine the heterogeneity of resources and the extent to which species can partition them. The problem of the number of species in communities will be considered in more detail later, after we have a better feeling for the structure and functioning of the community as a whole. In the meantime, we shall extend the analysis of species interactions to the predator-prey relationship in the next chapter.

33

Predation

Predation has received no less attention from ecologists than has competition (Huffaker 1970). In its broadest sense—food consumption—predation is the prime mover of energy through the community and defines the links in the food chain. Predation also is a basic factor in population ecology and evolution. Prey populations largely determine the growth rate of predator populations because they provide the food necessary for growth and reproduction. Conversely, predators tend to reduce the growth rate of populations of their prey. Whether predators can substantially reduce the equilibrium size of a prey population is, however, a controversial question. To the extent that predators select prey with respect to genotypic variation, they cause evolutionary changes in the prey population. These changes are manifested in part by the extensive and elaborate antipredator adaptations that occur throughout nature (Chapter 15). When predators selectively prey on individuals of different ages, they alter the age structure of the prey population and change its dynamics.

Early interest in predator-prey interactions was stimulated by discovery of regular cycles in the abundance of some arctic mammals and birds (see page 512). These cycles, whose relevance to ecological theory was first recognized by Elton (1924, 1942), have long been thought to reflect basic properties of predator-prey interactions. The early mathematical models of Lotka (1925) and Volterra (1926) predicted the presence of oscillations in predator-prey interactions. In this respect, the relationship between predator and prey populations provides a useful system for the study of population dynamics and regulation, but the relationship also is important to community and ecosystem dynamics. The influence of predators on the size of prey populations affects competition between prey species, thus making predators an integral force in shaping the structure of the community.

We may ask many questions about predator-prey relationships. To what extent do predators stabilize, or cause cyclic fluctuations in, prey populations? Do predators limit prey populations below the carrying capacity of the environment for the prey? If predators are so efficient that they can substantially reduce prey populations, how do they keep from overeating their prey? Do predators act to maximize their returns, that is, do they "manage" prey populations? Surprisingly little is known about predator and prey relations in nature, and much of what we know is a rather coarse mixture of field observations and anecdotes, experiments with laboratory systems, and mathematical models of population processes.

The impact of predation on prey population varies.

It is almost impossible to fully characterize the dynamic interactions between a single population of predators and a single population of prey in a natural habitat. The problems involved in such a task are varied and overwhelming. Measuring predation in natural populations, or even determining that the disappearance of an individual from the population is, in fact, due to predation, poses difficult technical problems. There are many causes of death, including starvation, disease, and exposure to the elements—not to mention emigration, a form of "local death." Predators are rarely caught in the act. Few predators restrict their feeding to one kind of prey, and few prey species are eaten by only a single species of predator; so additional difficulties are encountered in sorting out the interaction between two populations from the complex relationships that make up the community. .

Predators do not remove individuals from prey populations at random. In prey species that display strong territorial behavior, predators tend to capture surplus individuals that are forced into marginal habitats by intraspecific competition (Errington 1946). Jenkins, Watson, and Miller (1964) showed that the mortality of red grouse that do not hold territories and, therefore, must reside in marginal habitats is higher than the mortality of territorial individuals that occupy optimum habitats. But few of the individuals that fall prey would ever have bred, and so predators do not substantially reduce the reproductive potential of grouse populations. One would expect extreme selectivity by predators that spend much time searching for and pursuing each prey; the predators should choose the easiest catch, the individual with a relatively small chance of escape (Errington 1946, 1963). For this reason, the inexperienced young and the decrepit old are most frequently captured (*e.g.,* Mitchell *et al.* 1965). The effect of predation on healthy, reproductive-age individuals is much less. In contrast to grouse and most other vertebrates, many prey populations inhabit relatively homogeneous environments and do not establish social order. Their predators cannot be as selective about whom they eat, and they undoubtedly consume many reproductively active individuals. Individuals of such prey species usually have adopted a strategy of maximizing their production of offspring at the risk of increasing their vulnerability to predators. After all, what choice does an aphid have? If it is to suck the juices from the veins of a sycamore leaf, it must sit on a flat surface exposed to every passerby (Figure 33-1). The tiny algae of the phytoplankton have nowhere to hide. Their survival depends purely on chance. Animals that have low reproductive rates invest much more heavily in avoiding predators to shift the balance between predator and prey in their favor. Prey species are aided toward this goal when their habitat contains hiding places to which they can escape.

Figure 33-1   Adult and larvae lady beetles (family *Coccinellidae*) feeding on aphids in a laboratory culture. The flightless aphids are easy prey for the predatory beetles. Note the abundant hairs on the veins of the leaf. These help to deter the aphids from penetrating the plant and sucking its juices (courtesy U.S. Dept. Agriculture).

Do predators limit the size of prey populations below their carrying capacity?

This question is one of basic importance in community ecology. If predators reduced the size of prey populations so much that the prey did not use all the resources potentially available to them, additional species could coexist on those resources. While it can be shown that predators do greatly reduce prey populations in some cases, we cannot necessarily conclude that the rate at which the prey consume resources is similarly reduced. By limiting the number of individuals in a prey population, predators could stimulate the prey's production of young, most of which would be preyed upon before reaching maturity; if this were so, the prey might consume resources at a rate similar to that in a predator-free population, even though there were fewer adults in the population that was exposed to predators. Predator and prey should be considered together as a unit that utilizes the prey's food. Predators also consume this resource indirectly—the prey act as a go-between—

and the energy requirements of the predators must be accounted for in any attempt to determine the resources needed to support a prey population.

Hairston, Smith, and Slobodkin (1960) argued that because herbivores do not appear to consume all the food available to them (witness the abundance of green plants), and because populations of herbivores frequently increase greatly when predators are removed (for example, the Kaibab deer population, page 14), herbivore populations must be limited by predators. This logic has several flaws, which were first pointed out several years later by Murdoch (1966b). The mere presence of food in what appears to be unlimited supply does not necessarily lead to the conclusion that food does not limit herbivore populations. Leafy vegetation is relatively unsuitable as food for animals because it contains too little protein, too much indigestible cellulose, and, frequently, toxic doses of exotic organic compounds (Whittaker and Feeny 1971). Furthermore, the occurrence of large populations of herbivores could, by reasoning parallel to that of Hairston, Smith, and Slobodkin, suggest that predators are themselves limited by other predators rather than by their food resource. The weakness of the argument becomes apparent at the top of a food chain, where predators have no predators and, thus, are not also prey.

Ehrlich and Birch (1967) also argued that just because food is not completely eaten by consumer populations, one cannot infer that the consumers are not food limited. They cited the population behavior of grain weevils in food vials (Birch 1953b) to make their point. When grain is added to the vials at a specific rate, the weevil population builds up to a certain level but no further, and the food resource is never exhausted between periods of replenishment. If food is added more rapidly, the weevil population increases to a higher level, but again the food supply is not entirely used. Ehrlich and Birch argued that even though there is more food than the weevils can eat, their population size is demonstrably influenced by food. This reasoning itself may not be strictly valid. Even in a predator-limited species, a change in the prey's resources might alter its population parameters and shift the balance of the predator-prey equilibrium. In the particular case of the weevils, populations are probably limited by some sort of self-regulation, perhaps cannibalism of eggs by the larvae. As the rate of food replenishment increases, the eggs are "diluted" in the medium: the larvae run across eggs less often and more survive to augment the population. While this mechanism may be unusual, it emphasizes that the cause of population regulation probably cannot be deduced readily from observations at a particular time of the size of a population and its resources. For the weevils, space would seem to be as limiting as food, and the rate at which grain is provided influences both resources.

A line of evidence used to support the argument that populations are regulated by predators is that populations often increase dramatically when their predators are removed. Such evidence must, however, be viewed cautiously (Caughley and Burk 1970, Lowrie and Miller 1973). Reproduction by prey populations has evolved in the presence of predator pressure. The removal of predators not only reduces mortality, but it also places the prey in a new environmental context to which they are not adapted. This could upset natural population control mechanisms, temporarily resulting in abnormal population growth. The enlarged population would eventually overexploit its resources, and its size would then be forced to return to a lower level in line with its resources.

Mite-mite and other interactions demonstrate that predators may effectively limit prey populations.

Instances in which predators have been shown to depress prey populations below the carrying capacity of the environment are widely scattered. Several such studies are recounted here in detail because of the fundamental consequence of predation for community ecology. Most of these examples pertain to disturbed or artifically maintained situations in which prey populations are unusually high in the absence of predators.

The cyclamen mite is a pest of strawberry crops in California. Populations of the mites are usually kept under control by a species of predatory mite of the genus *Typhlodromus*. Cyclamen mites typically invade a strawberry crop shortly after it is planted, but their populations do not reach damaging levels until the second year. Predatory mites usually invade fields during the second year and rapidly subdue the cyclamen mite populations, which rarely reach damaging levels a second time.

Greenhouse experiments have demonstrated the role of predation in keeping the cyclamen mites in check (Huffaker and Kennett 1956). One group of strawberry plants was stocked with both predator and prey mites; a second group was kept predator-free by regular applications of parathion, an insecticide that kills the predatory species but does not affect the cyclamen mite. Throughout the study, populations of cyclamen mites remained low in plots shared with *Typhlodromus*, but their infestation attained damaging proportions on predator-free plants (Figure 33-2). In field plantings of strawberries, the cyclamen mites also reached damaging levels where predators were eliminated by parathion, but they were effectively controlled in untreated plots (a good example of an insecticide having the wrong effect). When cyclamen mite populations began to increase in an untreated planting, the predator populations quickly responded to reduce the outbreak. On the average, cyclamen mites were about twenty-five times more abundant in the absence of predators than in their presence.

The effectiveness of *Typhlodromus* as a predator owes to several factors in addition to its voracious appetite (Huffaker and Kennett 1969). Its capacity for population increase is of the same order as that of its prey. Both species reproduce parthenogenetically; female cyclamen mites lay three eggs per day over the four or five days of their reproductive life span; female *Typhlodromus* lay two or three eggs per day for eight to ten days. But even its high reproductive rate does not tell the whole success story of *Typhlodromus*. Seasonal synchrony of reproductive activities with the growth of prey populations, ability to survive at low prey densities, and strong dispersal powers all contribute to its efficiency. During the winter, when cyclamen mite popula-

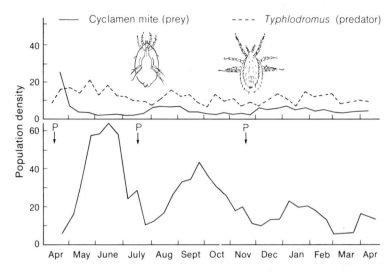

Figure 33-2   Infestation of strawberry plots by cyclamen mites (*Tarsonemus pallidus*) in the presence of the predatory mite *Typhlodromus* (above), and in its absence (below). Prey populations are expressed as numbers of mites per leaf; predator levels are the number of leaflets in 36 which had cne or more *Typhlodromus*. Parathion treatments are indicated by *p*s (after Huffaker and Kennett 1956).

tions are reduced to a few individuals hidden in the crevices and folds of leaves in the crown of the strawberry plants, the predatory mites subsist on the honeydew produced by aphids and white flies, and they do not reproduce except when they are feeding on other mites. Whenever predators are suspected of controlling prey populations, one usually finds a high reproductive capacity compared with that of the prey, combined with strong dispersal powers and the ability to switch to alternate food resources when primary prey are unavailable.

Inadequate dispersal is perhaps the only factor that keeps the cactus moth from completely exterminating its principal food source, the prickly pear cactus (Dodd 1959). When prickly pear (*Opuntia*) was introduced to Australia, it spread rapidly through the island continent, covering thousands of acres of valuable pasture and range land. After several unsuccessful attempts to eradicate the prickly pear, the cactus moth (*Cactoblastic cactorum*) was introduced from South America. The caterpillar of the cactus moth feeds on the growing shoots of the prickly pear and quickly destroys the plant—literally by nipping it in the bud. The cactus moth exerted such effective control that, within a few years, the prickly pear became a pest of the past (Figure 33-3). The cactus moth has not, however, eradicated the prickly pear because the cactus manages to disperse to predator-free areas, thereby keeping one jump ahead of the moth and maintaining a low-level equilibrium in a continually shifting mosaic of isolated patches. Indeed, one would probably not guess that the cactus moth keeps the prickly pear at its present low

Figure 33-3   Photographs of a pasture in Queensland, Australia, two months before and three years after introduction of the cactus moth to control prickly pear cactus (from Dodd 1959, courtesy of W. H. Haseler, Dept. of Lands, Queensland, Australia).

population levels; the moths are actually scarce in the remaining stands of cactus in Australia today. The same moth is probably responsible for controlling prickly pear in some areas of Central and South America, but its decisive role might have gone unnoticed if the appropriate experiment had not been performed in Australia.

The mite-mite and moth-cactus interactions prove that a predator can keep prey populations far below the capacity of the environment to support them. Yet in both situations, the prey originally occurred at unnaturally high levels in managed environments and the predators were introduced by man. These studies do not, therefore, help us to determine whether natural populations are ever controlled by predators.

Experiments on the effect of sea urchins on populations of algae have demonstrated predator control in some natural marine ecosystems. The simplest experiments consist of removing sea urchins, which feed on attached algae, and following the subsequent growth of their algae prey (Paine and Vadas 1969). When urchins are kept out of tidepools and subtidal rock surfaces, the biomass of algae quickly increases, indicating that predation reduces algal populations below the level that the environment can support. Different kinds of algae also appear after predator removal. Large brown algae flourish and begin to replace both coralline algae (whose hard shell-like coverings deter grazers) and small green algae (whose short life cycles and high reproductive rates enable algal population growth to keep ahead of grazing pressure by sea urchins). In subtidal plots kept free of predators, brown kelps formed thick forests below the ocean's surface and shaded out most small species.

Changes in kelp beds off the coast of southern California provide a striking demonstration of the role of sea urchin predation on algal populations (North 1970). The most important brown alga in the kelp beds is *Macrocystis*. Its life cycle begins when an embryo, having developed from a single cell floating in the plankton, settles to the bottom at a depth of 8 to 25 meters. As the young plant grows, it is secured to the bottom by a holdfast; its fronds extend toward the surface, buoyed up by gas-filled floats. Mature kelps form dense stands and greatly influence the overall economy of the coastal marine ecosystem. Kelps not only are major primary producers, but also provide refuge for fish populations. (Kelp beds figure in the human economy of southern California as well, because they are commercially valuable as a source of iodine and fertilizer. The 200 or so square kilometers of kelp beds between Santa Barbara and San Diego yield a harvest of 100,000 to 150,000 tons annually at a price of $20 per ton.)

Kelps are long-lived compared with other marine plants. Once a bed has become established it will last between one and ten years, depending on the locality and exposure to waves. Death occurs mostly from storms, high water temperature, and grazing by fish and sea urchins. Once a bed has been devastated, by whatever cause, it normally becomes re-established within a few years. The coastline thus witnesses the regular disappearance and persistent reappearance of kelp beds as a normal course of events in the marine community.

Beginning in 1940, kelp beds all but disappeared from areas near Los Angeles and San Diego. Disposal of sewage was suggested as a factor because deterioration of the beds started near sewage outlets and spread outward. Areas in which kelp beds had disappeared were practically devoid of all types of algae, and were instead swarming with dense populations of immature sea urchins. Normally, when a kelp bed disappears, the urchins disappear with it, giving newly settled plants a chance to regenerate. Pollutants did not affect mature kelp plants: A few beds persisted within polluted areas in sheltered spots free of wave damage. The kelp beds disappeared because no young kelp grew where normal causes of mortality had removed mature plants. The young kelp plants were eaten by the urchins.

Sea urchins apparently can obtain nourishment from suspended and dissolved organic matter in sea water. Sewage maintains an urchin population in the absence of adequate algal food in the same way that honeydew from aphids maintains predatory mites when their normal prey, the cyclamen mite, is unavailable. When young *Macrocystis* plants reinvade a devastated kelp bed, urchins are there to meet them and quickly devour newly settled plants.

When urchin populations were controlled by dumping quicklime (calcium oxide) into devastated areas, and when these areas were reseeded with young *Macrocystis* plants, the kelp beds quickly returned to their former state. The proof of the pudding came when, in 1963, San Diego stopped dumping sewage directly into the ocean. Reseeding then quickly brought kelp beds back without the need to destroy the urchins. In fact, these beds now contain fewer, but larger urchins whose grazing does not seriously depress algal growth.

Nature would not, however, leave us content with so simple a system. We must add yet another component to the dynamics of the kelp community— the sea otter. Once abundant along the west coast of the United States and Canada, sea otters were hunted to the verge of extinction during the nineteenth century by Russian and American fur traders. Under close protection and surveillance by the California Fish and Game Commission, the otters have staged a successful comeback. Wherever otter populations have reached their former densities, kelp beds also flourish. Needless to say, otters eat urchins. And a final note: Otters also eat abalone, a valuable commercial shellfish in California, and so while they are encouraged by the kelp industry, they are illegally persecuted by some abalone fishermen.

Predator and prey cycles suggest that predators exert a strong influence on prey populations.

Populations of predators and prey often vary in what appear to be closely linked cycles (see Chapter 31, page 528). The periodic fluctuations of the snowshoe hare, followed closely by fluctuations of lynx, one of the hare's major predators, is a classic example (see Figure 31-5, page 513). Because the cycles persist for long periods, they represent a stable interaction between predator and prey. This and similar relationships focused the attention of population biologists on predator-prey interactions early in the development of population ecology as a branch of science. As appears to be characteristic of most natural phenomena, however, the details of the predator-prey cycle have been difficult to pin down.

The relationship between predator and prey populations is perhaps most readily seen in laboratory experiments in which prey are provided a constant, abundant source of food. When azuki bean weevils (*Callosobruchus chinensis*) were maintained in cultures with predatory braconid wasps, the populations of predator and prey fluctuated out of phase with each other in regular cycles of population change (Figure 33-4). The weevil-braconid wasp interaction is often called a host-parasite system because the prey are consumed alive. Actually, braconid and other parasitic wasps are properly called *parasitoids* (Price 1975, Force 1974). In the life cycle of the wasp, eggs are laid on beetle larvae (Figure 33-5), which the wasp larvae proceed to consume after hatching. Abundance of prey, therefore, influences the number of adult wasps in the *following* generation, after the wasp larvae have metamorphosed into adults. This built-in time lag enhances the population fluctuation (see page 531).

Utida (1957) maintained populations of braconids and weevils together in laboratory cultures for thirty generations (about one and a half years), although some of the prey or predator populations became extinct before the end of the experiment. The course of the weevil and wasp populations can be examined with respect to each other by graphing the numbers of prey against the number of predators, as in the competition graphs (Figure 33-4). The trajectory of the single point representing both predator and prey on the population graph moves somewhat irregularly in a counterclockwise direction. When the predator and prey populations are both low (generation six), the prey increase rapidly (the trajectory moves to the right). As the prey

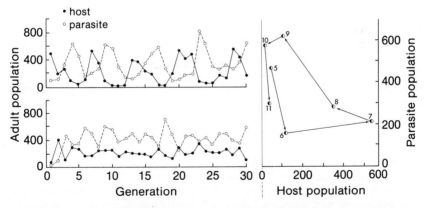

Figure 33-4.   Population fluctuations in predator–prey systems involving the Azuki Bean Weevil (*Callosobruchus chinensis*) and a slowly reproducing parasite (*Heterospilus prosopidus,* top) or a rapidly reproducing parasite (*Neocatolaccus mamezophagus,* bottom). Time course of populations is portrayed at left; the predator–prey graph at right includes six generations of the *Heterospilus–Callosobruchus* population (after Utida 1957).

Figure 33-5   A braconid wasp laying an egg in a cotton boll worm. When the egg hatches, the wasp larva will consume the boll worm. Various species of these tiny wasps attack many kinds of insect larvae and are frequently major factors in pest population control (courtesy U.S. Dept. Agriculture).

become abundant (seven), the population of wasps begins to increase and, because *Heterospilus* is a very efficient parasite, the weevil population rapidly declines as the predators increase and begin to overeat their prey (trajectory moves up and to the left to eight and nine). Eventually, the prey are nearly exterminated (trajectory approaches the predator axis, ten), and then the predator population, lacking adequate food, decreases rapidly (trajectory moves straight down, eleven) to return the cycle close to its starting point. The wasp is never so efficient that all weevil larvae are attacked, hence a small but persistent reserve of weevils always remains to initiate a new cycle of prey population growth when the predators become scarce.

The magnitude of the fluctuation in the predator-prey system is largely determined by the time lag. Although *Heterospilus* readily finds weevil larvae, its reproductive rate is low and initially it cannot keep up with the population growth of the host. The result is a considerable lag between the two. In contrast, the reproductive potential of *Neocatolaccus*, another braconid wasp used by Utida, is about six times that of *Heterospilus*. With the higher reproductive potential of *Neocatolaccus*, the time lag in the predator-prey interaction is reduced, and the system is much more stable (Figure 33-4).

With an extremely efficient predator, prey populations are often eaten to extinction, and their predators soon follow. This type of predator-prey interaction can become stable only if some of the prey can find refuges in which they can escape predators. G. F. Gause (1934) demonstrated this principle with his early studies on protozoa. He employed *Paramecium* as prey and another ciliated protozoan, *Didinium*, as predator. In one experiment, predator and prey individuals were introduced to a nutritive medium in a plain test tube. In that simple environment, the predators readily found all the prey; when they had consumed the prey population, the predators died from starvation. In a second experiment, Gause added some structure to the environment by placing glass wool at the bottom of the test tube, within which the *Paramecium* could escape predation. In this case, the *Didinium* population starved after consuming all readily available prey, but the *Paramecium* population was restored by individuals concealed from predators in the glass wool.

Gause finally achieved recurring oscillations in the predator and prey populations by periodically adding small numbers of predators—restocking the pond, so to speak. The repeated addition of individuals to the culture corresponds, in natural predator-prey interactions, to repopulation by colonists from other areas of a locality in which extinction of either predator or prey has occurred. This is reminiscent of the interaction between the cactus moth and prickly pear cactus (page 593), in which the cactus escapes complete annihilation by dispersing to predator-free areas.

C. B. Huffaker, a University of California biologist who pioneered the biological control of crop pests, attempted to produce just such a mosaic environment in the laboratory (Huffaker 1958). The six-spotted mite (*Eotetranychus sexmaculatus*) was prey; another mite, *Typhlodromus occidentalis*, was predator; oranges provided the prey's food. The experimental populations were set up on trays in which the number, exposed surface area, and dispersion of the oranges could be varied (Figure 33-6). Each tray had forty possible positions for oranges, arranged in four rows of ten each; where oranges were not placed, rubber balls of about the same size were substituted. The exposed surface area of the oranges was varied by covering the oranges with different amounts of paper; the edges of the paper were sealed with wax to keep the mites from crawling underneath. In most experiments, Huffaker first established the prey population with twenty females per tray, then introduced two female predators eleven days later. Both species reproduce parthenogenetically.

When six-spotted mites were introduced to the trays alone, their populations leveled off at between 5,500 and 8,000 mites per orange area (Figure 33-7, left-hand diagram). The predators introduced to the system increased rapidly and soon wiped out the prey population. Their own extinction followed shortly (Figure 33-7, right-hand diagram). Although predators always eliminated the six-spotted mites, the position of the exposed areas of oranges influenced the course of extinction. When the orange areas were in adjacent positions, minimizing dispersal distance between food sources, the prey reached maximum populations of only 113 to 650 individuals and were driven to extinction within 23 to 32 days after the prey were introduced to the trays. When the same amount of exposed orange area was randomly dispersed throughout the forty-position tray, the prey reached maximum populations of

Figure 33-6   (Above) One of Huffaker's experimental trays with four oranges, half exposed, distributed at random among the forty positions in the tray. Other positions are occupied by rubber balls. (Right) An orange wrapped with paper and edges sealed with wax. The exposed area is divided into numbered sections to facilitate counting the mites (from Huffaker 1958, courtesy of C. B. Huffaker).

2,000 to 4,000 individuals and persisted for 36 days. These experiments demonstrated that the survival of the prey population can be prolonged by providing remote areas of suitable habitat. The slow dispersal of predators to these areas delays the extinction of their prey.

Huffaker reasoned that if predator dispersal could be slowed further, the two species might coexist. To accomplish this, Huffaker increased the complexity of the environment and introduced barriers to dispersal. The number of possible food positions was increased to 120 and the equivalent area of six oranges was dispersed over all 120 positions. A mazelike pattern of Vaseline barriers was placed among the food positions to slow the dispersal of the predators. *Typhlodromus* must walk to get where it is going, but the six-spotted mite spins a parachutelike silk line that it can use to float on wind currents. To take advantage of this behavior, Huffaker placed vertical wooden pegs throughout the trays. The mites used the top of the pegs as jumping-off points in their wanderings. This arrangement finally produced a series of three predator-prey cycles over eight months (Figure 33-8). The distribution of the predators and prey throughout the trays continually shifted as the prey, exterminated in a feeding area, recolonized the next, one jump ahead of the predators.

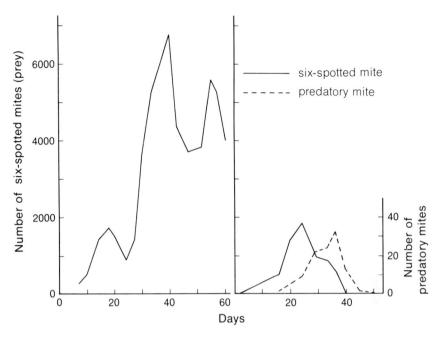

Figure 33-7   Number of six-spotted mites per orange area when raised alone (left) and in the presence of the predatory mite, *Typhlodromus* (right). The food was arranged in twenty small orange areas alternating with twenty foodless positions (after Huffaker 1958).

In another laboratory microcosm, Huffaker *et al.* (1963) placed 252 oranges in a three-tiered system and exposed one-twentieth of the orange area to the mites. In this more complex system, the predator and prey populations oscillated through longer cycles with lower amplitude. But when the orange area was increased, the productivity of the predator population at peak prey levels increased out of proportion, and the prey were eaten to extinction as they were in the simpler experimental systems.

In spite of the tenuousness of the predator-prey cycle achieved, we see that a spatial mosaic of suitable habitats allows predator-prey interactions to achieve stability. But, as we saw in Gause's experiment with protozoa, predator and prey also can coexist locally if some prey can take refuge in hiding places. And if the environment is so complex that predators cannot easily find scarce prey, stability will again be achieved.

Simple predator and prey models predict oscillations in population size.

The first mathematical formulation of the predator-prey interaction was provided by Lotka (1925) and independently by Volterra (1926), both of whom

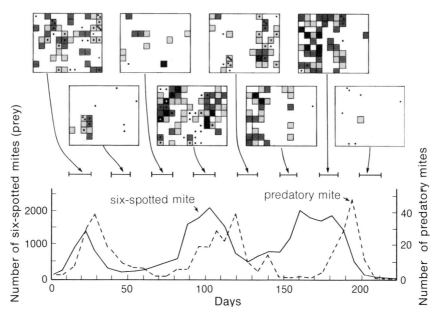

Figure 33-8   Population cycles of the six-spotted mite and the predatory mite *Typhlodromus* in a laboratory situation. The boxes above show the relative density and positions of the mites in the trays. Shading indicates the relative density of six-spotted mites; dots indicate presence of predatory mites (after Huffaker 1958).

expressed the rate of growth of predator and prey populations by simple differential equations. In this discussion, we shall designate the number of predator individuals by *P* and the number of prey by *H*. Think of predators (*P*) and herbivores (*H*) to keep these straight. Lotka and Volterra suggested that the rate of growth of the prey population, *dH/dt*, has two components. One represents the unrestricted reproductive rate of the prey population in the absence of predators, (*rH*), where *r* is the birth rate of a prey parent. The other represents the removal of prey from the population by predators, which Lotka and Volterra assumed to vary in direct proportion to the product of the prey and the predator populations (*HP*, which is proportional to the probability of a chance encounter between predator and prey). Thus the rate of increase of the prey population is

$$\frac{dH}{dt} = rH - pHP$$

or

$$\dot{H} = rH - pHP \qquad (33\text{-}1)$$

where *p* is a coefficient representing the efficiency of predation.

For the predator population, birth rate is the number of prey captured times a coefficient (a) expressing the efficiency with which food is converted to population growth. Death rate is a constant (d) times the number of predator individuals. And the growth rate of the predator populations is

$$\dot{P} = apHP - dP \qquad (33\text{-}2)$$

The model includes several simplifying assumptions: The prey do not interact with their own food supply; the relationship between predator and prey is linear; the death rate of individual predators is independent of population density.

Despite its limitations, the Lotka-Volterra model proved successful in one respect: It predicted that predator and prey populations would oscillate, as we shall see below. When both predator and prey populations are in equilibrium ($\dot{H} = 0$ and $\dot{P} = 0$), $rH = pHP$ and $apHP = dP$. These equations can be solved for

$$\hat{P} = \frac{r}{p} \quad \text{and} \quad \hat{H} = \frac{d}{ap} \qquad (33\text{-}3)$$

where $\hat{P}$ and $\hat{H}$ are the equilibrium population sizes of the predator and prey. Notice that $\hat{P}$ and $\hat{H}$ are both constant values.

When plotted on a graph of predator population size versus prey population size, the equilibrium population values $\hat{P}$ and $\hat{H}$ (the predator and prey isoclines) partition the graph into four regions (Figure 33-9). In the region below the prey isocline $\dot{H} = 0$), prey populations increase because there are few predators to eat them. In the region above the prey isocline, prey populations decrease because of overwhelming predator pressure. For the pred-

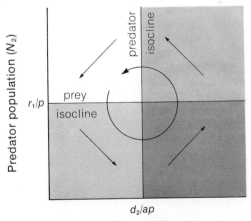

Figure 33-9   Representation of the Lotka-Volterra predator-prey model on a population graph. The trajectories of the populations show that the predator and prey will continually oscillate out of phase with each other (full circle).

ators, their population increases in the region to the right of the predator isocline $\dot{P} = 0$) where prey are abundant. To the left, predator populations decrease owing to lack of food. The change in both populations simultaneously is shown by an arrow in each of the four sections of the graph. In the lower right, for example, both predator and prey increase and the population trajectory moves up and to the right. The vectors in the four regions taken together define a counterclockwise cycling of the predator and prey population, that is, the populations cycle out of phase, with the prey population increasing and decreasing just ahead of the predator population.

The equilibrium isocline for the predator $\dot{P} = 0$) defines the minimum level of prey density ($H = d/ap$) that can sustain the growth of the predator population. That of the prey $\dot{H} = 0$) defines the greatest number of predators ($P = r/p$) that the prey population can sustain. If the reproductive rate of the prey (r) increased, the hunting efficiency of the predators (p) decreased, or both, the prey isocline (r/p) would increase—that is to say, the prey population would be able to bear the burden of a larger predator population. Similarly, if the death rate of the predators (d) increased and either the predation efficiency (p) or reproductive efficiency of the predators (a) decreased, the predator isocline (d/ap) would move to the right, indicating that more prey would be required to support the predator population. Increased predator hunting efficiency (p) would simultaneously reduce both isoclines; fewer prey would be needed to provide a given capture rate, and the prey population would be less able to support the more efficient predators.

The trajectories of predator and prey populations in Figure 33-9 show that if the populations are displaced from the equilibrium point they will continue to fluctuate, following a circular path counterclockwise. Any perturbation of the system will give the population fluctuations a new amplitude and duration until some other outside influence acts. This equilibrium is said to be neutral because no internal forces act to restore the populations to the intersection of the predator and prey isoclines. Therefore, random perturbations will eventually increase the fluctuations to the point that the trajectory strikes one of the axes of the graph, and one or both populations die out. This property in itself shows that the Lotka-Volterra equations are an oversimplified representation of nature.

As we have seen in Chapter 31 (page 531), time lags in the response of populations to their environments or to other populations can promote fluctuations in natural populations. We might ask how the introduction of a time lag into the Lotka-Volterra predator-prey model would affect the trajectory of the populations on the predator-prey graph. In the oscillating system shown in Figure 33-10, the trajectory of the populations at point one moves directly toward the right. The number of prey is increasing, and the growth rate of the predator population is momentarily zero as the latter population reaches the low point of its cycle and is about to initiate a growth phase. If the response of the predator population to the abundance of its prey has a time lag, however, perhaps owing to a long gestation period before the birth of its young, the predator population will respond at point one as if it were at some earlier point (two), and so will continue to decrease. The new trajectory produced by the time lag will spiral outward, eventually leading to the extinction of one or both populations.

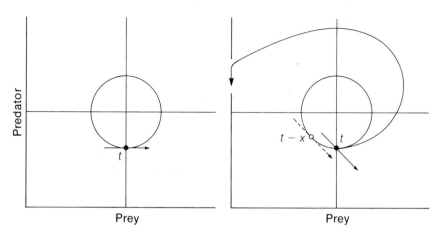

Figure 33-10   The effect of a time lag in the response of the predator population to prey density. At left, the response at time *t* is immediate and continual
oscillations are maintained. At right, the response has a time lag (*x*), which
causes the amplitude of the oscillations to increase, eventually leading to the
extinction of the predator, prey, or both.

Simple predator-prey models clearly do not adequately represent
predator-prey relationships in natural communities. Time lags are almost
certainly a feature of most interactions between predators and their prey, yet
predator-prey oscillations appear to be highly stable. It is not yet fully understood how this stability is produced.

Graphical analyses demonstrate the conditions
for stability in predator-prey systems.

We may use the population graph of the Lotka-Volterra equations (Figure
33-9) to ask how the stability of the system changes when the predator and
prey isoclines are altered. By rotating the intersection of the equilibrium
isoclines in a clockwise or counterclockwise direction, we can examine the
behavior of the system near its equilibrium point, which is the region of
primary interest. When the predator isocline is vertical and the prey isocline is
horizontal, the system is in a neutral equilibrium; in the absence of perturbation, predator and prey populations continue to cycle indefinitely (Figure
33-9). If the isoclines are rotated counterclockwise, the trajectory of the
populations will spiral outward in cycles of increasing amplitude (Figure
33-11). Hence, the system becomes unstable. When the isoclines are rotated
clockwise, however, the equilibrium becomes stable and the population
trajectory spirals in toward the intersection of the isoclines. The farther the
isoclines are rotated in a clockwise direction, the more rapidly populations
move toward the equilibrium point.

How do these graphical conditions pertain to the stability of natural
systems? The prey isocline, which, according to the Lotka-Volterra

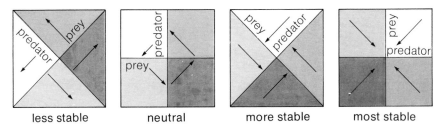

| less stable | neutral | more stable | most stable |

Figure 33-11   The effect of rotating the predator and prey isoclines on the stability of the interaction. Clockwise rotations are stabilizing, counterclockwise rotations are destabilizing.

equations, is a horizontal line, can be altered to incorporate biological properties of prey populations in several ways (Rosenzweig and MacArthur 1963). First, in the absence of predators, the prey population is limited by the carrying capacity of the environment, determined by the availability of food or other resources. As a result, the prey isocline bends down toward the prey axis as the number of prey increase, intersecting the axis at K (Figure 33-12). Two opposing factors influence the shape of the prey isocline at low densities. On one hand, small populations support fewer predators than large populations because the recruitment rate of smaller populations is lower (Rosenzweig 1969). Consequently, small populations can support fewer predators than large populations. Graphically, this is equivalent to the prey isocline bending down toward the origin of the graph at the left. On the other

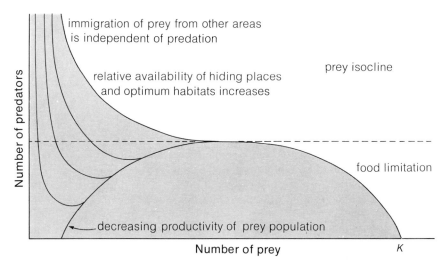

Figure 33-12   Graphical representation of the prey equilibrium isocline, incorporating several biological properties of natural systems. The isocline for the Lotka-Volterra equation is indicated by a dashed line.

hand, if prey were more difficult to locate when they were scarce (fewer prey lack good hiding places, and all live in the most suitable habitats), predation would be reduced and the prey population could persist in the presence of denser predator populations. Graphically, this corresponds to an upward shift in the prey isocline at the left. Whether the isocline swings down or up at low population densities depends on the heterogeneity of the environment of the prey. If predators easily capture only those individuals that are excluded by intraspecific competition at high prey densities from optimum habitats and the best hiding places (see page 537), the isocline will swing upward. If predators can capture prey efficiently at all densities, the isocline will swing downward toward the origin of the graph.

The predator isocline also bends under the weight of biological reality (Figure 33-13). As a population of predators increases, its food requirement increases and predators must capture more prey to maintain their population at a constant level. The predator isocline, therefore, should be rotated slightly to the right. Furthermore, social interference among predators (for example, territoriality) would reduce the efficiency with which predators utilize prey resources and would bend the predator isocline further to the right. The availability of suitable breeding sites or of alternative food resources also could limit the size of predator populations independently of the abundance of prey, causing the predator isocline to become horizontal at high prey population densities.

The relationships between a variety of predator isoclines and an idealized prey isocline are shown in Figure 33-14. In every case, the intersections of the isoclines are rotated clockwise with respect to the Lotka-Volterra model, tending to stabilize the relationship.

For cases A and B, the stability of the predator-prey interaction depends on the presence of hiding places for the prey. In the absence of such refuges,

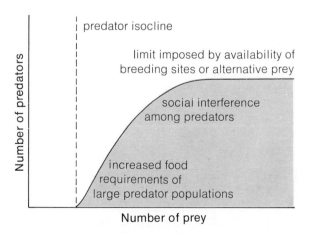

Figure 33-13   Graphical representation of the predator equilibrium isocline, incorporating several biological properties of natural systems. The isocline for the Lotka-Volterra equation is indicated by a dashed line.

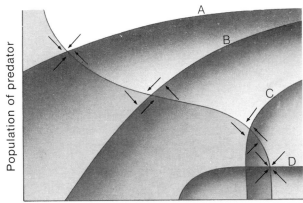

Figure 33-14   Variation in the outcome of predator-prey relationships, compared to a "standard" interaction (B), when prey-capture efficiency is reduced (C), when the predator population is limited at a low level by factors other than its food supply (D), and when the predator uses alternative food resources (A).

the prey line would bend down toward the origin at left; case B would possibly become unstable, and case A would lead to the rapid elimination of the prey population. The population trajectory is least rapidly drawn into the equilibrium point in case B (the intersection of the isoclines most closely resembles the Lotka-Volterra model). If the predators' prey-capture efficiency were reduced, the predator isocline would shift to the right (C), where the prey isocline bends down more steeply, and the stability of the interaction would thus be increased. If the predator population were limited at some low level by factors other than the availability of prey (case D), the stability of the relationship would be enhanced greatly. Suppose now that the predator could switch to alternative food resources when the prey depicted in Figure 33-14 became scarce. This strategy would move the predator isocline up along the prey axis (case A) because the predator population could increase by feeding on other species of prey. The intersection of the isoclines under these conditions indicates a highly stable equilibrium. To summarize, stability in predator and prey interactions is increased by (a) external factors that limit either population, (b) reduced prey-capture efficiency, and (c) the use of alternative food resources by the predator.

Predator-prey cycles reflect a balance between stabilizing and unstabilizing factors.

The predator-prey models suggest that predator and prey populations should be relatively constant, yet clearly they are not. Apparently, stable oscillations are the rule in many predator-prey systems (see Figure 31-5). One way in which such cycles may be obtained is through an equilibrium between stabilizing and unstabilizing forces. Rosenzweig and MacArthur's (1963) graphical model could account for oscillations in this manner (Figure 33-15), but it required that the prey curve have a hump (Rosenzweig 1969). If the predator isocline were to the left of the hump, the intersection would be unstable and the population trajectory would spiral outward. If the prey isocline swung up at very low prey densities (on account of safe hiding places, where prey could escape predators, or immigration of prey from other populations), the population trajectory, upon hitting this line, would descend along it as if it were a "safe" corridor until predators became scarce enough that the prey could increase again. This would result in a stable oscillation, because no matter where the population trajectory first strikes the safe zone, it would always continue to the same point (A) before embarking on a new cycle.

Humps have been identified in prey isoclines under experimental conditions, such as in Maly's (1969) study on the relationship between *Paramecium* and the predatory rotifer *Asplanchna*. Interactions were observed in small depression slides containing one milliliter of culture medium. By placing different numbers of predator and prey in the depressions and following population changes over short periods, Maly was able to identify the direction of population trajectories over a large area of the population graph. The resulting *Paramecium* isocline (Figure 33-16) falls off rapidly at low prey density. Furthermore, *Asplanchna* is an extremely efficient predator of *Paramecium*. Its isocline lies well to the left of the hump in the prey isocline and, as one would expect, the system is unstable.

It is difficult to determine whether, in cycling predator-prey systems, the unstabilizing component of the interaction results from the presence of a hump in the prey curve. It seems more likely that unstabilizing tendencies are caused by time lags in the response of predators to changes in the size of the population of their prey. These tendencies would be balanced by a rapid upswing in the prey isocline at low prey densities resulting from the availability of hiding places, as shown in Figure 33-17.

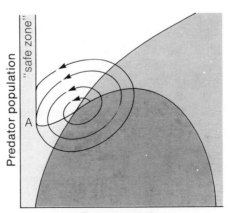

Figure 33-15   Possible stability conditions for an oscillating predator-prey system. The cycle is maintained by the balance between nonstabilizing influences at the intersection of the isoclines and the stabilizing influence of a "safe" zone.

Predator population

"safe zone"

A

Prey population

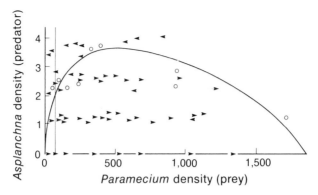

Figure 33-16   Prey (*Paramecium*) and predator (*Asplanchna*) isoclines determined experimentally in a laboratory environment. The arrows show population trajectories for *Paramecium* with varying numbers of predators. Open circles indicate populations of *Paramecium* that changed less than 5 per cent in twenty-four hours (after Maly 1969).

Recent theoretical models have shown that under some special conditions related to the geometric arrangement of the predator and prey isoclines, a hump-shaped prey isocline combined with a vertical predator isocline intersecting the prey isocline near its peak, can have internal stabilizing properties that lead to what is known as a *limit cycle* (May 1973, Gilpin 1974). The conditions that will maintain such stable oscillations in a predator-prey system are narrowly restricted, however, and therefore limit the potential application of the limit cycle concept to predator-prey interactions. It is fair to say, I think, that the most robust conditions for stability occur when two powerful forces oppose each other and bring a system into an equilibrium reflecting the balance between the two. In predator and prey systems, these forces are undoubtedly the stabilizing geometry of biologically realistic pred-

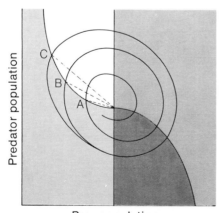

Figure 33-17   A predator-prey model demonstrating how a rise in the prey isocline at low densities can stabilize an outward spiraling population trajectory caused, for example, by a time lag in the predator response. The steepness, hence the stability of the apparent prey isocline (dashed lines) increases as the oscillations increase from A to B to C.

ator and prey isoclines and the unstabilizing influence of time lags in population response.

In any of their many forms, predator-prey models are extremely simple views of nature. An entirely different approach to predator-prey interactions has been taken by Canadian ecologist C. S. Holling and others, who have been interested primarily in practical application of theory to problems of pest control and resource management. They have attempted to dissect the predator-prey relationship into its components and then to build a precise and realistic model from these components. Much of their work has been more descriptive than theoretical but their approach has contributed greatly to our understanding of population interactions, as we shall see in the following sections.

The response of predators to prey density is not linear.

In the Lotka-Volterra models, the rate at which individuals are removed from the prey population is described by the term, pHP; for a given density of predators (P), the absolute rate of exploitation increases in direct proportion to the density of prey (H). Doubling the density of prey, for example, would double the predation rate. The assumption of linearity in the predator response is a major biological weakness of the Lotka-Volterra equations. In natural populations, the response of predation rate to prey density tends to level off with increasing prey density.

The relationship of an individual predator's rate of food consumption to prey density has been labeled the *functional response* by Holling (1959). When tested in the presence of increasing levels of prey density in the laboratory, praying mantises at first increase their rate of food consumption in proportion to prey density, but at high prey density, their feeding rate eventually levels off (Figure 33-18, curve B). Two factors dictate that the functional response of the individual should reach a plateau. First, as the predator captures more prey, the time spent handling and eating the prey cuts into hunting time. Eventually, the two reach a balance and prey-capture rate levels off. Second, predators become satiated—continually stuffed—and cannot feed any faster than they can digest and assimilate their food.

Holling (1965) identified four primary components in the functional response: rate of successful search, time available for hunting, time spent handling prey, and hunger level of the predator. The rate of successful search depends, in turn, on the speed of movement of the predator relative to its prey, the size of the field of reaction of the predator (the area within which the predator perceives the prey), and the capture success. The time required for prey handling includes the pursuit and subduction of the prey, as well as the time required to eat them. Some organisms—snakes are a good example—have a digestive pause after a meal, during which time they often will not hunt or attack.

Hunger clearly influences a predator's motivation to hunt (Holling 1966). The distance over which a praying mantis will strike at a fly depends on the time since its last feeding, which, inasmuch as one cannot ask a mantis how

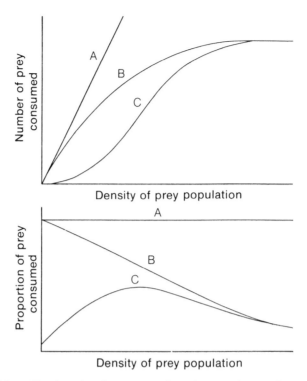

Figure 33-18   The functional response of predators to increasing prey density: (A) predator consumes a constant proportion of the prey population regardless of its density; (B) predation rate decreases as predator satiation sets an upper limit to food consumption; (C) predator response lags at low prey density owing to low hunting efficiency or absence of search image. The upper diagram portrays the functional response in terms of the *number* of prey consumed; the lower diagram, in terms of the *proportion* of the prey population consumed.

hungry it is, serves as a reasonable index of hunger. Holling (1959) also experimented with deer-mice to determine how motivation affects the functional response of a vertebrate predator. Cocoons of the pine sawfly were buried at several densities in sand on the floor of large cages; dog biscuits were provided as an alternative food. Palatability of the cocoons influenced the functional response: Deer-mice dug up many more fresh cocoons than cocoons collected during the previous year and stored prior to the experiment. Deer-mice apparently are not fond of stale cocoons. When fresh cocoons were buried deeper in the sand, the mice spent less time digging for them. The type of alternative food also influenced the functional response. With cocoons at a density of fifteen per square foot, mice consumed 200 cocoons per day when dog biscuits were provided. When sunflower seeds were added to the menu, consumption of cocoons dropped to a little more than 100 per day. Deer-mice evidently are not fond of dog biscuits either.

Holling (1959) determined the functional responses of three small mammals to the density of pine sawfly cocoons in the relatively natural habitat of pine plantations in Ontario, Canada. The eggs of the sawfly, which are laid in live pine needles in the fall, hatch early in the spring. The larvae feed on the needles of the pine. In early June, full-grown larvae fall to the ground and crawl into the litter, where they spin cocoons. The adults do not emerge to lay eggs until September. For three months during the summer, the forest floor is scattered with varying numbers of cocoons from a few thousand to more than a million per acre, depending on the level of sawfly infestation. Small mammalian predators, particularly the common shrew, the deer-mouse, and the short-tailed shrew, open the cocoons in characteristic ways enabling investigators to tally instances of predation separately for each species. Densities of the predators and prey can be assessed by trapping the rodents and sampling the forest litter for cocoons.

The functional response curve of each predator was unique. The short-tailed shrew, the least common species in the study area, increased its consumption of sawfly pupae in response to their density much more rapidly than either the common shrew or deer-mouse (Figure 33-19). Because the short-tailed shrew increased its consumption of cocoons in response to small increases in their availability, it appears to take full advantage of a highly variable food resource.

The functional response of many predators increases more slowly at low prey densities than at higher prey densities (see Figure 33-18, curve C). Two factors can cause this lag. First, hunting efficiency is decreased at low density because the few prey have the best hiding places. Second, vertebrate predators are thought to adopt their hunting behavior and prey recognition to the most worthwhile prey—usually an abundant, oft-encountered species. A preconception of what a given prey looks like and where it is found is called a *search image* (Tinbergen 1960). We use search images all the time. A lost object is easier found if we know its shape, size, and color. Predators presumably could base search images on prior experience. The more abundant

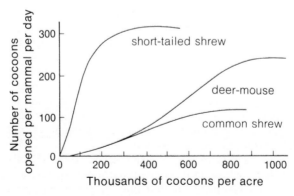

Figure 33-19   Functional response of three mammalian predators to the density of pine sawfly cocoons in the litter of pine plantations (after Holling 1959).

the prey, the more often it would be found and the better prepared the predator to find it. Conversely, search-image formation works against a predator's being able to locate uncommon prey, hence the lag in the predator's functional response at low prey density. Search images have been the subject of a number of studies (for example, Gibb 1958, 1962, Mook, Mook, and Heikens 1960, Tinbergen *et al.* 1962, Royama 1970, Murton 1971), but the phenomenon is still poorly known, and its significance to predator-prey interactions is not well understood.

The most revealing observations on the factors that influence predator choice are the experiments of the Russian biologist Ivlev (1961) on the feeding behavior of fish on invertebrate prey. All the experiments were carried out in the laboratory. Ivlev used four species of predatory fish (carp, bream, roach, and tench) and four types of prey (chironomid fly larvae, amphipods, isopods, and molluscs), which could be provided at any selected density. He determined the selectivity of a predator for each prey type by calculating an *electivity index*

$$E = (r - p)/(r + p) \qquad (33\text{-}4)$$

where p is the proportion of a prey species in the ration provided (all four prey species together sum to 100 per cent) and r is the proportion of that prey species in the diet of the predator. If, for example, 20 per cent of the prey items provided were amphipods, but 80 per cent of the roach's diet consisted of amphipods, the electivity of the roach for amphipods would be $E = (0.8 - 0.2)/(0.8 + 0.2) = 0.6/1.0 = 0.60$. Values of E can vary between -1.0, indicating complete avoidance of a particular prey, to $+1.0$, indicating exclusive preference. A value of 0.0 indicates no preference one way or the other.

Ivlev found that amphipods were generally preferred by the predators over other prey ($E = 0.10$ for carp, 0.21 bream, 0.24 roach, 0.02 tench). Comparable values for chironomid larvae, which burrowed in the silt on the bottom of the fish tanks and were less available to some of the predators, were 0.24, 0.05, -0.30, 0.27. The high electivities for chironomids of carp and tench reflect the benthic feeding habits of these predators.

Prey selection is influenced by a variety of factors. When Ivlev removed chironomid larvae as an alternative food source, electivities for amphipods increased to 0.24, 0.31, 0.31, and 0.18. Degree of satiation is also important. For example, hungry carp feed quite generally on all four prey types, but as the carp become satiated with food, molluscs, isopods, then amphipods are successively dropped from the diet; chironomid larvae are the only prey that a stuffed carp will show any interest in. Similarly, when prey are generally abundant, carp ignore molluscs and isopods to go after their preferred chironomid larvae. As the mobility of, and availability of hiding places to a certain prey type increase, electivity for that prey usually decreases. As one would expect, predators seek the easy catch.

In another series of experiments, using several different sets of predators and prey, Ivlev demonstrated that each predator has a preferred range of prey size over which its electivity for that prey is greatest (Figure 33-20). The electivity of predatory fish is usually greatest over some intermediate size range of prey; small prey are not worth eating and large prey are often too

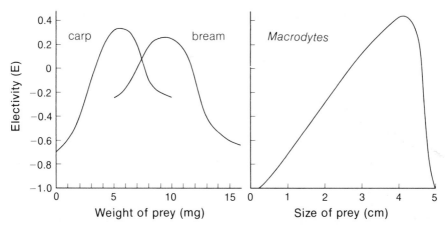

Figure 33-20   Electivity of predators as a function of prey size. Above, carp fed on chironomid larvae and bream fed on amphipods; below, larvae of *Macrodytes circumflexus* fed on young roach (after Ivlev 1961).

difficult to capture. Invertebrate predators, such as *Macrodytes*, often show increasing preference for prey as their size increases to some upper limit determined by the size of the predator's mouth.

Finally, Ivlev showed that if prey were not replenished and thus if their availability decreased as the predators ate them, each predator would consume its preferred prey first and then turn its attention to less desirable prey species. This phenomenon, called *switching* by Murdoch (1969), has also been identified in predatory snails (Murdoch 1969) and several aquatic insects (Lawton *et al.* 1974), but not in ladybird beetles, *Coccinella* (Murdoch and Marks 1973). Where it occurs, switching acts to stabilize a predator's population and, perhaps, to reduce fluctuations in the community as a whole.

Predator populations respond by growth and immigration to an increase in prey.

Individual predators can increase prey consumption only to the point of satiation. Predator response to increasing prey density above that point can be achieved only through an increase in the number of predators, either by immigration or population growth, which together constitute the *numerical response*. In Holling's (1959) study of predation on sawflies, populations of the common shrew increased from about three individuals per acre at low prey density to twenty-four per acre at medium and high prey densities. The deer mouse population exhibited a less marked numerical response, and the short-tailed shrew, none. It may be only coincidental that the short-tailed shrew had the most rapid functional response to increasing prey density, but we would nonetheless expect the functional and numerical responses to be inversely related. The increased intraspecific competition resulting from the numerical response of a population reduces the availability of prey, upon which the functional response is based.

Numerical response by growth of a local population is relatively slow, particularly when the reproductive potential of the predator is much less than that of the prey, a condition that could lengthen the time lag in a predator-prey system. Immigration from surrounding areas is a major component of the numerical response of mobile predators, many of which opportunistically congregate where resources become abundant. The bay-breasted warbler specializes on periodic outbreaks of the spruce budworm, during which its population density may reach 120 pairs per 100 acres, compared with about ten pairs per 100 acres during nonoutbreak years (Morris *et al.* 1958). The bay-breasted warbler also responds to increased prey density by laying more eggs (MacArthur 1958).

Buckner and Turnock (1965) studied the numerical responses of birds to populations of the larch sawfly in tamarack swamps in Manitoba. One of the swamps consisted of a nearly pure stand of tamarack with a well-developed shrub layer of alder, and the other area supported a mixed stand of tamarack and black spruce, with a shrub layer of dwarf birch. More than forty-three species of birds were known to eat sawfly larvae or adults, but the importance of these species as predators varied considerably. Bird censuses were taken over a period of nine years, during which the abundance of sawfly larvae varied from as few as 3,200 to more than 2 million larvae per acre. Although populations of all the species of birds varied directly with prey abundance, the numerical response was not nearly great enough to control the sawfly population. The pure tamarack stand had the greater infestation of larch sawflies, but although the density of breeding birds there was nearly twice as great as in the mixed tamarack-spruce habitat, the impact of avian predation on the sawfly population was only one-tenth as great (Table 33-1).

Three predatory birds, the pomarine jaeger, the snowy owl, and the short-eared owl, each respond in a different manner to varying densities of lemmings on the arctic tundra (Table 33-2). Lemming populations exhibit great fluctuations; high and low points in a population cycle may differ by a

TABLE 33-1

Predation by birds on larch sawfly populations in two tamarack swamp forests (from Buckner and Turnock 1965).

	Pure tamarack	Mixed tamarack-spruce
Prey population per acre		
larvae	2,138,700	40,000
adults	205,400	4,700
Predator population per acre	23.5	12.6
Potential annual predation rate		
larvae (per cent)	10,665 (0.5)	2,344 (5.9)
adults (per cent)	10,601 (5.6)	3,049 (64.9)

TABLE 33-2

Response of predatory birds to different densities of the brown lemming near Barrow, Alaska (from Pitelka et al. 1955).

	1951	1952	1953
Brown lemming (ind per acre)	1 to 5	15 to 20	70 to 80
Pomarine jaeger	Uncommon, no breeding	Breeding pairs $4 \cdot mi^{-2}$	Breeding pairs $18 \cdot mi^{-2}$
Snowy owl	Scarce, no breeding	Breeding pairs 0.2 to $0.5 \cdot mi^{-2}$ many nonbreeders	Breeding pairs 0.2 to $0.5 \cdot mi^{-2}$ few nonbreeders
Short-eared owl	Absent	One record	Breeding pairs 3 to $4 \cdot mi^{-2}$

factor of one hundred. At Barrow, Alaska, during the summer of 1951, when lemmings were scarce, none of the predatory birds bred; short-eared owls did not even appear in the area. During the following summer, one of moderate lemming density, both the jaeger and snowy owl bred, but short-eared owls again were absent. In 1953, a peak year for lemmings, all three species of bird predators bred. Jaegers were four times more abundant in 1953 than in 1952. In contrast to the jaegers, the density of snowy owls did not increase. Instead, each pair of birds raised more young. Most snowy owls laid two to four eggs during the year of moderate lemming abundance, and up to a dozen during the peak year.

A model of predator-prey equilibrium.

The diverse relations between predator and prey populations can be summarized in a diagram that compares the productivity of the prey population to the proportion of prey that are removed by predators, as both vary in relation to prey density (Figure 33-21). The two curves in the diagram represent the net addition of new prey to the population (in excess of deaths due to causes other than predation), either by reproduction or immigration (collectively called *recruitment*), and the removal of prey by predators. Both are expressed as a proportion of the prey population. The predation rate is a product of the functional and numerical responses taken together.

Recruitment and predation are analogous to birth and death. When recruitment exceeds predation, the prey population grows; when predation exceeds recruitment, prey numbers decline. Points at which the recruitment and predation curves cross are population equilibria for the prey.

The recruitment curve of the prey population declines with increasing prey density, owing to intraspecific competition for resources, and falls to zero when the prey are at the carrying capacity of the environment. In the absence of predators, prey populations are regulated at this point by the

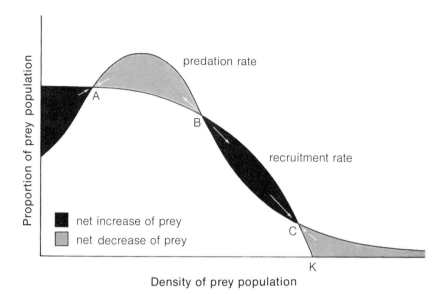

Figure 33-21   Predation and recruitment rates in a hypothetical predator-prey system. When predation exceeds recruitment, prey populations decrease, and vice versa (as shown by arrows). Points A and C are stable equilibria for the prey population; the lower point (A) represents population control by predators; the upper point (C) represents population control by food and other resources.

balance between their biotic potential and resource limitations of the environment, as we have seen in Chapter 31.

The shape of the predation curve is determined by the functional and numerical responses of the predators. Escape ability of the prey and the failure of predators to form search images limit the proportion of prey captured at low prey density. At moderate prey density, functional and numerical responses increase the effectiveness of the predators as a whole, and the proportion of prey removed increases. These responses are eventually satiated at high prey density, and although the number of prey consumed may continue to rise slowly, the proportion of individuals removed from the prey population decreases.

The recruitment and predation curves in Figure 33-21 were drawn to produce three equilibrium points for the prey population. The highest and lowest points represent stable equilibria around which populations are regulated, the middle equilibrium is unstable. The lower equilibrium point (A) corresponds to the situation in which predators regulate a prey population substantially below the carrying capacity of the environment. The upper equilibrium point (C) corresponds to the situation in which a prey population is regulated primarily by availability of food and other resources; predation exerts a minor depressing influence on population size.

Predators maintain a shaky hold on prey populations at point A. If a heavy frost or an introduced disease reduced the predator population long enough to allow the prey population to slip above point B, the prey would continue to increase to the higher stable equilibrium point (C), regardless of whether the predator population recovered. To the farmer, this means a crop pest, normally controlled at harmless levels by predators and parasites, suddenly becomes a menacing epidemic. After such an outbreak, predators could exert little control over the pest population until some quirk of the environment brought its numbers below point B, back within the realm of predator control. Outbreaks of tent caterpillars in the prairie provinces of Canada are generally preceded two to four years earlier by a year in which the winter is abnormally cold and the spring unusually warm (Ives 1973). These conditions presumably upset the normal balance between tent caterpillars and their predators and parasites. Infestations are subsequently brought under control by several cold winters that kill most of the tent caterpillar eggs (Witter *et al.* 1975).

The effectiveness of predators in maintaining prey populations at low densities depends on the relationship between predation and recruitment curves. The higher and broader the predation curve, the more effective predators are as control agents. Functional and numerical responses enhance predator effectiveness. Several species of predators attacking the same prey can control the prey population better than any one of them alone (Figure 33-22). Well-planned biological control programs take advantage of this principle in attempting to establish several species of predator and parasite, each with different predator tactics, to control pest populations.

Using the predation-recruitment rate diagram in Figure 33-21, we can examine the consequences of different levels of predation for prey population control (Figure 33-23). Inefficient predators cannot regulate prey populations at low density; they depress prey numbers slightly, but the prey population remains near the equilibrium level set by resources (upper-left diagram, point C). Increased predation efficiency at low prey density can result in predator control at point A (upper-right diagram). If functional and numerical responses are sufficient to maintain high densities of predators, or if prey are limited relative to predation by a low carrying capacity, pred-

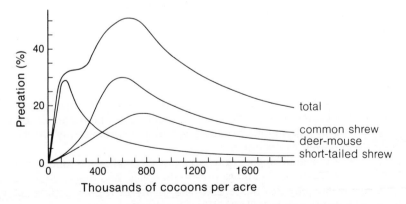

Figure 33-22   The combined functional and numerical responses of small mammals to density of sawfly cocoons showing the relative influence of several species of predator on prey populations (after Holling 1959).

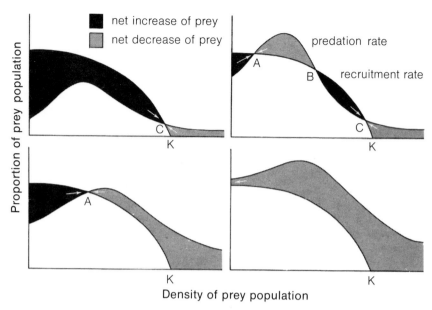

Figure 33-23   Predation and recruitment curves at different intensities of predation.

ators may effectively control prey under all circumstances, and equilibrium point C disappears (lower-left diagram). We could also envision predation being so intense at all levels of prey density that the prey are eaten to extinction (lower-right diagram, no equilibrium point). We would expect this situation only in simple laboratory systems, or when predators maintained themselves at high population levels by feeding off some alternative prey. Indeed, many ecologists have suggested that biological control of pests would be enhanced by providing parasites and predators of the pest with innocuous alternative prey.

Are predators prudent?

Predators can exert a major regulating influence on the dynamics of prey populations. There must also be some optimum level of prey population density to support the greatest number of predators. Do predators prudently manage their prey populations to maximize their own productivity? And if so, how is this achieved by evolution?

A predator that eats its prey down to a low level takes the food out of its own mouth. Predators can capture prey more easily, and therefore be more productive, when prey are numerous. The ability of a prey population to support predators varies with its density. A small prey population can support correspondingly few predators because, while each prey individual's reproductive potential may be high, the total recruitment rate of a small population is low. Prey populations near their carrying capacity also are unproductive because although numerous, each individual's reproductive potential is severely limited as a result of intraspecific competition for resources. The total recruitment rate of every prey population reaches a maximum at some density below the carrying capacity (Ricker 1954, Beverton and Holt 1957, Watt 1968). Because predators can remove a number of individual prey equivalent to the annual recruitment rate without reducing the size of the prey population, the prey population that yields the maximum recruitment also will support the greatest number of predators. Ranchers and game management biologists are clearly concerned with maintaining populations of beef cattle, deer, and geese at their most productive levels to maximize man's ability to harvest these species without reducing their populations.

The achievement of optimum yield can be illustrated with populations of guppies maintained at different densities in aquaria (Gulland 1958). Recruitment (the number of immature fish produced in three weeks) reached a peak of 33 when there were 30 adult guppies per tank, and dropped to about 7 when adult populations exceeded 100 individuals per tank (Figure 33-24). In

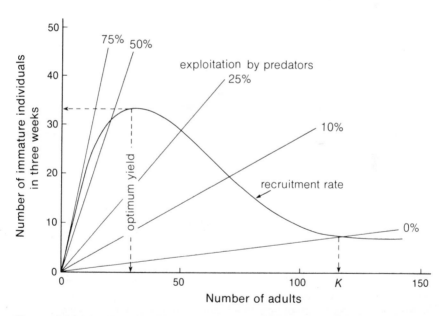

Figure 33-24   Recruitment curve and hypothetical exploitation rates for aquarium populations of guppies. In the absence of predation (0 per cent exploitation curve) the natural mortality of adult guppies would stabilize the population at about 120 individuals. The maximum exploitation rate possible is about 50 per cent per three weeks, at which point an adult population of about 30 and a yield of 33 would be maintained. A 75 per cent exploitation rate is more than the population can bear (from Gulland 1962, after data in Silliman and Gutsell 1958).

the absence of predators, the natural mortality of guppies would stabilize the population at about 120 adults, with the recruitment rate of 7 every three weeks just balancing mortality. The maximum sustainable yield would be achieved when predators removed about 40 per cent of the adult population every three weeks (about 2 per cent of the population per day).

Maximum potential daily yields in laboratory populations have been calculated as 3 per cent for flour beetles, 13 per cent for unicellular algae, 23 per cent for water fleas (*Daphnia*), and 99 per cent for the sheep blowfly (Watt 1962). Few estimates of maximum potential yield are available for natural populations of large animals. The population of ring-necked pheasants on Protection Island, Washington, could withstand removal at a rate of 1 to 3 per cent per day (Ricklefs 1973). Wolves kill about 25 per cent of the moose population of Isle Royale each year, 0.07 per cent per day (Mech 1966), and perhaps 37 per cent of the white-tailed deer population of Algonquin Park in Ontario, 0.1 per cent per day (Pimlott *et al.* 1967). The population of ring seals on Baffin Island can withstand exploitation by Eskimos at a rate of 7 per cent per year, 0.02 per cent per day (McLaren 1962).

Annual recruitment rates of adults into natural populations may be used as a crude index to the rate at which predators remove individuals from populations. These rates vary widely, from 83 to 120 per cent for plaice and haddock (Beverton and Holt 1957), 60 to 80 per cent for several species of salamanders in Virginia (Organ 1961), 40 to 60 per cent for many of our back-yard birds (Lack 1954, Farner 1955), to less than 5 per cent for some large mammals and seabirds (Ashmole 1971). Predators are not responsible for the loss of all the adults that the recruits replace, but they also capture many young before they reach reproductive age.

Could levels of exploitation observed in nature actually be maximum yields? Or would competition between individual predators cause overexploitation of their prey? Territorial animals, which exclude competitors from their feeding areas, could indeed space themselves with respect to their prey to achieve maximum yields. When the feeding areas of predators overlap, however, intraspecific competition dictates that each predator maximizes its immediate harvest at the expense of long-term yields. Man behaves no differently. Intelligently managed ranches, with fences to exclude competing livestock, can achieve maximum yields. Alas, in highly competitive situations—fishing in international waters, to name one—man has proved to be pathetically shortsighted and imprudent. Fishing and hunting practices that would attain long-term yields give way to practices that maximize today's harvest. For example, after World War II, the North Sea, between England and Norway, was fished so intensively that reducing fishing effort by 15 per cent and increasing the mesh size of nets to let more fish through actually would have *increased* the total catch by 10 to 20 per cent (Beverton and Holt 1957). Overexploitation of whale populations has similarly led to the near extinction of some species, and has virtually doomed the whaling industry (Laws 1962, McVay 1966).

We know very little about predation on natural populations, but even to show that observed rates of exploitation yield the highest possible returns to predator populations would not necessarily prove that predators were prudent by their own design. Exploitation rates depend upon the adaptations of both predator and prey. The evolution of prudence by predator populations would directly oppose the pre-eminent goal of evolution, which is to maximize the fitness of the individual predator, and would require that populations altruistically regulate their densities at an optimum level (see discussion on page 538). Selection should favor maximum exploitation of prey by each predator at the expense of optimum management. Individual prudence and population prudence have separate meanings that coincide only when the food supply of each individual is entirely under its own control. If a predator did not compete with other members of its population for the particular food items within its territory or home range, its activities alone would determine the local dynamics of the prey population and prudence would lead to the protection of long-term yields. More often than not, however, prey do not recognize the territorial boundaries of their predators, and thus they are potentially available as food to many individuals. In this case, a prudent predator should attempt to capture prey before its neighbor does, thereby maintaining its competitive status in the population.

Predators with complex behavior conceivably could exercise sufficient social feedback on each other to prevent overexploitation of food resources (Gilpin 1975). Although it seems doubtful that such behavior does occur in natural populations, the game management practices exercised by man represent social feedback mechanisms. Limits are set on the numbers of prey individuals that may be captured in a given season, based on a knowledge of the population dynamics of the prey, and fines are levied against those who exceed these limits. These practices frequently work well, but individual prudence occasionally gets the upper hand.

The level of exploitation of a prey population is not determined by prudent management practices on the part of the predator; it is determined by the ability of the predator to capture prey compared to the ability of the prey to avoid being captured. Both skills are evolved characteristics of the population. Regardless of whether predators act prudently, they tend to achieve a characteristic equilibrium with prey populations. The relationship between wolves and various prey populations in several years demonstrates this equilibrium particularly well (Table 33-3). Population ratios and, particularly, biomass ratios (one pound of wolf for each 150–300 pounds of prey) are amazingly constant despite the fact that the species and density of the principal prey of the wolf vary considerably with locality. Exploitation rates also appear to be relatively constant—18 per cent of the moose on Isle Royale (Hordan et al. 1970), and 37 per cent of the deer in Algonquin Park. Still, not all predator-prey systems achieve the same equilibrium. The population ratio of mountain lions to deer in California is 1:500–600, which is equivalent to a biomass ratio of about 1:900, and the exploitation rate is only 6 per cent (calculated from data in Longhurst et al. 1952). A mountain lion–elk–mule deer system in Idaho had a biomass ratio of 1:524, and an exploitation rate of 5 per cent for the elk population and 2.7 per cent for the mule deer population (Hornocker 1970). Evidently, wolves are more efficient predators than mountain lions, perhaps because of their social hunting habits. Where predators feed on more abundant populations of prey, as in savanna, grassland,

TABLE 33-3

The relationship between populations of predators and their prey in several localities.

Locality	Predator	Principal prey	Density of predators (ind · 100 mi⁻²)	Ratio between predator and prey populations	
				Numbers	Biomass
Jasper Nat'l Park[1]	Wolf	Elk, mule deer	1	1:100	1:250
Wisconsin[2]	Wolf	White-tailed deer	3	1:300	1:300
Isle Royale[3]	Wolf	Moose	10	1:30	1:175
Algonquin Park[4]	Wolf	White-tailed deer	10	1:150	1:150
Canadian arctic[5]	Wolf	Caribou	1.7	1:84	1:186
Utah[6]	Coyote	Jackrabbits	28	1:1000	1:100
Idaho primitive area[7]	Mt. lion	Elk, mule deer	7.5	1:116	1:524
Ngorongoro Crater, Tanzania[8]	Hyena	Ungulates	440	1:135	1:46
Nairobi Park, Tanzania[9]	Felids	Ungulates	96	1:97	1:140
Alaska[10]	Pomarine jaeger	Lemmings		1:1263	1:90

References: (1) Cowan (1947), (2) Thompson (1952), (3) Mech (1966), (4) Pimlott et al. (1967), (5) Kelsall (1968), (6) Clark (1972), Wagner and Stoddart (1972), (7) Hornocker (1970), (8) Kruuk (1969), (9) Foster and Coe (1968), (10) Maher (1970).

and tundra habitats, predators not only are more numerous, but they achieve higher biomass ratios (1:50 to 1:150, Table 33-3). Such conditions seem to enhance both prey productivity and predator efficiency. Large cats—lions and cheetahs—remove 16 per cent of prey biomass in Nairobi Park, Tanzania, where their biomass ratio is 1:140 (Foster and Coe 1968).

How do parasites affect populations of their host?

Parasites and predators use different strategies to exploit their prey or host populations. The death of a prey organism is the objective of predators, but the death of a host often causes the death of its parasite. Although predators evolve adaptations to maximize their ability to capture prey, selection acts strongly on parasites to adjust their food consumption to a level that the host can withstand. Parasite and host together evolve a balance: Adaptations of the parasite reduce its virulence, and adaptations of the host (immunity and resistance) reduce the parasite's hazard to health. The balance achieved between parasite and host is often upset when a parasite is accidentally transferred to a new host, frequently man, his livestock, or his crops, and sweeps through the population in a devastating epidemic.

Little is known about the incidence of infection or the effects of parasites on natural populations. Factors that limit populations of parasites extend far beyond the skin of their hosts (Mattingly 1969, Burnet and White 1972). Most parasites undergo complicated life cycles involving changes in form and stages of dispersal that take the parasite into the hostile exterior environment (page 221). For an organism adapted to living within the tissues of a host, the environment through which its progeny travel to reach another host, often by way of intermediate species, must be forbidding.

Infections by many parasites vary seasonally. The occurrence of the harvest mite *Thrombicula,* an external parasite of mice and voles in England, reaches a peak in the early fall and declines through the winter and spring, finally disappearing completely until the seasonal cycle is begun once more in late summer. But the pattern of abundance also varies from host to host. The harvest mite is far more persistent on the bank vole (*Clethrionomys*), than on either field mice (*Apodemus*) or meadow mice (*Microtus*), both of which are completely free of the mites by late fall (Elton 1942).

Parasite populations also fluctuate from year to year. For example, incidence of trypanosome parasites in populations of red squirrels and eastern chipmunks at Trout Lake, Manitoba, varied considerably during a three-year study (Dorney 1969). (Trypanosomes are parasitic protozoa that infect the bloodstream. In man, trypanosomes cause several fatal illnesses, of which sleeping sickness has received the most attention. The parasites are usually carried from host to host by such insects as tsetse flies.) Squirrels and chipmunks apparently are infected by different, host-specific strains of the trypanosome because the occurrence of trypanosomiasis in the two hosts fluctuated independently (4, 37, and 15 per cent in the squirrel population, and 42, 26, and 12 per cent in the chipmunk population, in 1961, 1962, and 1963). The incidence of the disease varied because adults that had previously been exposed to the disease became immune, and the number of juveniles in the population, all of which are susceptible, varied from year to year. In the eastern chipmunk, the incidence of the trypanosome in juveniles was more than four times that in adults (48 versus 11 per cent over three years), and the per cent of juveniles in the population during the summer months decreased from 68 per cent in 1961 to 29 per cent in 1963. This drop, combined with the reduction of disease incidence in adults from 19 to 4 per cent over the three years, accounted for the decrease of trypanosomiasis in the chipmunk population.

Infection by trypanosomiasis did not reduce the survival of squirrels and chipmunks. Endemic occurrences of disease organisms usually cause few detrimental effects in healthy organisms. Parasites can, however, intensify the harm caused by such stressful conditions as cold or lack of food, and thus can increase the incidence of death under these conditions. Infection by *Trypanosoma duttoni* was found to reduce the tolerance of laboratory mice to stressful conditions (Sheppe and Adams 1957). Groups of mice were given either full or half rations of food in either warm (19 to 22 C) or cold (3 to 8 C) environments. With full rations, none of the mice died over a nineteen-day period, regardless of parasitism or the temperature at which they were kept. The nonparasitized mice did, however, gain twice as much weight as the parasitized mice (14 versus 7 per cent increase). On half rations, all groups of mice had shorter average survival times in cold than in warm environments, as one would expect, but parasitism by *Trypanosoma* significantly reduced survival in both environments. Even in human populations, malnutrition often exerts its greatest effect on diseased or parasitized individuals (Latham 1975). Undernourishment also tends to lead to an increase in the incidence of disease and parasitism.

Rapid development of immunity characterizes epidemic outbreaks of most diseases. Perhaps the most famous epidemics of all times have been the outbreaks of the Black Death in human populations. The bubonic plague organism, a bacterium, is endemic in many wild populations of rodents. The disease can be spread to man by rodent fleas, particularly the rat flea. Several times in history the natural balance between and among the bacteria, rodents, and fleas has been upset so badly that the plague spread to the human population, in which it caused epidemic disease. To produce a major plague epidemic, rat populations must be so great that hordes of rats, searching for food in houses, come into close contact with humans. Furthermore, rat fleas must heavily infest rats before they will abandon their preferred hosts for humans. These conditions have occurred infrequently, even in the crowded, garbage-ridden conditions prevalent in the cities and towns of medieval Europe. But once the plague takes hold in a human population, its course runs swiftly and surely.

A typical epidemic of Black Death initially spreads rapidly, infects a large part of the population, and takes a high toll in human lives. But as susceptible individuals either die or become immune to the disease organism, the number of lethal cases and the mortality rate drop almost as rapidly as the disease first strikes. This is illustrated by data for a localized outbreak of the plague in India between 1953 and 1959 (May 1961, quoted by Watt 1968).

Year	Contracted cases	Per cent lethal
1953	20,539	70.5
1954	6,670	84.5
1955	705	23.1
1956	331	20.5
1957	44	0
1958	26	0
1959	37	0

The great plague of the fourteenth century originated in 1346 during the siege of Caffa, a small military post on the Crimean Straits. From there, the epidemic spread to Italy and the south of France by 1347, and during the next year it had reached all of Europe. The plague did not disappear entirely until 1357. It reappeared in Europe three more times during the fourteenth century, in 1361, 1371, and 1382, but with lower incidence and mortality rate (Watt 1968):

Year	Approximate percentage of population afflicted	Resultant deaths
1348	67	Almost all
1361	50	Almost all
1371	10	Many survived
1382	5	Almost all survived

The plague visited London three times during the seventeenth century, but unlike the epidemic waves that struck Europe during the fourteenth century, the effects were not attenuated during successive outbreaks (Creighton 1891):

Year	Population of London	Plague deaths	Deaths as a per cent of total population
1603	250,000	33,347	13
1625	320,000	41,313	13
1665	460,000	68,596	15

These outbreaks differed from those of the fourteenth century: Successive epidemics during the fourteenth century were separated by thirteen, ten and eleven years; during the seventeenth century, intervals were twenty-two and forty years. The longer interval increased the severity of successive epidemics in two ways. First, many more individuals lost their immunity over a twenty-year to forty-year period than over a decade. Second, the proportion of the population born since a plague epidemic, and therefore not immune, was much larger after intervals of twenty-two and forty years than after the shorter intervals during the fourteenth century. Because more humans were susceptible, the plague organism spread more rapidly through the population.

For many diseases, the frequency of outbreaks depends on the number of individuals born since the last outbreak and hence susceptible to the disease. Until the proportion of susceptible individuals reaches a certain critical level, which depends on the contagiousness and virulence of the disease, the disease organism cannot spread rapidly enough to cause an epidemic. Thus, although the smallpox virus brought to Mexico by the Spanish colonists in the sixteenth century remained at endemic levels in the native population, epidemics occurred only at eleven- to nineteen-year intervals, for example in 1520, 1531, 1545, 1564, and 1576 (Cartwright and Biddiss 1972).

The ecology of a disease vector (an organism that carries and transmits parasite organisms) frequently limits the spread of parasite populations. For example, when avian malaria and bird pox were introduced to the Hawaiian Islands in the last century, where neither disease had previously occurred in

the avifauna, highly susceptible local populations of birds were quickly destroyed and several species became extinct (Warner 1968). But malaria and
bird pox organisms are carried from host to host by a species of mosquito
that does not venture above an elevation of 600 meters, and so birds that lived
at higher altitudes completely escaped the diseases. (The mosquito is also an
introduced species in Hawaii.)

Herbivores can strongly influence plant populations.

The influence of herbivores on plant populations with animals that consume whole plants (usually seeds and seedlings) differs from the influence of
those that graze on vegetation (Slobodkin *et al.* 1967). The first are predators.
Grazers and browsers are akin to parasites. We can infer that herbivores have
an important role in the lives of plants from the elaborate morphological and
physiological defenses of plants against attack (see page 223). Plant toxins
like hypericin, digitalis, curare, strychnine, and nicotine are a fair match for
most herbivores. Thick bark, spines, thorns, and stinging hairs are also
strong deterrents.

Herbivores can stabilize plant succession on disturbed sites and prevent
the encroachment of new forms into an area. For example, after the decline
of rabbit populations in Australia following the introduction of myxamatosis
virus (see page 632), the native pine *Callitiris* regenerated extensively in New
South Wales (Harper 1969). Grazing by rabbits evidently had prevented the
growth of pine seedlings. Even though cactus moths consume only a small
portion of the net production of the prickly pear, it successfully controls the
introduced cactus population in Australia. In fact, herbivorous insects are
frequently used to control imported weeds (DeBach 1974). Klamath weed, a
European species toxic to livestock, and the source of the drug hypericin,
accidentally became established in northern California in the early 1900s. By
1944, the weed had spread over 2 million acres of range land in thirty
counties. Biological control specialists borrowed an herbivorous beetle
(*Chrysolina*) from an Australian control program. (Everything seems to become a pest in Australia.) Ten years after the first beetles were released, the
Klamath weed was all but obliterated as a range pest. Its abundance was
estimated to have been reduced by more than 99 per cent.

The impact of herbivores in plant communities, measured by the total net
primary production consumed, is least in forests, intermediate in grasslands,
and greatest in aquatic environments (Slobodkin *et al.* 1967). Herbivores
consume between 2 and 10 per cent of the net production of forests. The rest
enters detritus pathways in the community. Seed predators are much more
efficient, consuming anywhere between 10 and 100 per cent of their food
supply (Janzen 1971). Although they consume relatively little of the total
biomass of the plant, seed predators attack a vital stage in the life cycle and
can influence plant populations greatly. Ecologists are just beginning to
study their role in natural communities.

Grazing herbivores, particularly large mammals, consume 30 to 60 per
cent of grassland vegetation. The story of their influence on plant production

is particularly well told by the results of exclosure experiments. A study in California employed wire fences to exclude voles (mouselike rodents) from small areas of grassland (Batzli and Pitelka 1970). Seed production and composition of the standing crop of plants were followed for two years after the experiment began and were compared with unfenced control plots. The results, summarized by the bar diagram in Figure 33-25, show that grazing by voles in the unfenced plots reduced the abundance and seed production of food plants (mostly annual grasses) but did not affect perennial grasses and herbs absent from the vole's diet. Furthermore, competition from annual grasses in the fenced plots apparently depressed the growth of the nonfood species, suggesting a large role for the vole in determining the structure of the plant community.

Aquatic herbivores consume most of the net production of aquatic plants. The effectiveness of one type of herbivore, marine snails, as algal grazers is reflected in their abundance and annual energy flux. Representative figures for gross production (total energy assimilation) are 118 kcal·m^{-2}·yr^{-1} for the freshwater limpet *Ferrissia,* 290 kcal·m^{-2}·yr^{-1} for the periwinkle *Littorina,* and 750 kcal·m^{-2}·yr^{-1} for the turban shell *Tegula.* Comparable figures for terrestrial grazers range between 7 kcal·m^{-2}·yr^{-1} (field mouse) and 28 kcal·m^{-2}·yr^{-1} (grasshopper) (Ricklefs 1973a: 664).

Although herbivores rarely consume more than 10 per cent of forest vegetation, occasional outbreaks of tent caterpillars, gypsy moths, and other insects can completely defoliate or otherwise eradicate entire forests. Long-term studies of growth and survival of trees after defoliation by tent caterpil-

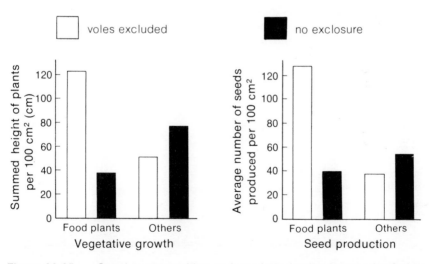

Figure 33-25   Species composition and seed production in grassland plots fenced to exclude voles and in unfenced control plots. The bar graphs present results of the experiment after two years. Food plants are mostly annual grasses; nonfood plants include perennial grasses and herbs (after Batzli and Pitelka 1970).

lars and other insects demonstrate that there may be a considerable lag between an infestation and the expression of its effects (Kulman 1971). A spruce budworm infestation on balsam fir caused varying defoliation mortality, and retardation of subsequent growth (Belyea 1952). In one area of light infestation, the defoliation exceeded 50 per cent during only three of nine years, reaching a maximum of 80 per cent in 1947 (Figure 33-26). No mortality was recorded in this area until 1951, but growth remained below one-half the normal rate for several years after the peak of the infestation. In an area of heavy infestation, defoliation exceeded 50 per cent for five years and reached 100 per cent in 1947, during the peak of the budworm outbreak. Growth was greatly suppressed and all trees in the area had died by 1951.

A Minnesota study of defoliation of quaking aspen by tent caterpillars showed that aspen usually survived defoliation and the increased intensity of insect and disease attack that inevitably followed (Duncan and Hobson 1958, Church et al. 1964). During the year of defoliation, however, growth was reduced by almost 90 per cent, and it was reduced by about 15 per cent during the following year. From studies of growth rings measured a decade later, foresters noted that among trees whose growth was normally suppressed by competition from dominant trees, mortality was independent of the history of defoliation and varied between 40 and 60 per cent following the tent caterpillar epidemic. Conversely, defoliation had a pronounced effect on survival of dominant trees. Only 2 per cent of the trees that were subjected to a single year of light defoliation died; at the other extreme, trees that had been badly defoliated three years in a row suffered almost 30 per cent mortality.

The interactions between predator and prey populations are so varied and complex as to defy summary. The examples we have looked at in this chapter lead us to the conclusion that predators, including herbivorous species, play an important role in the population processes of all species,

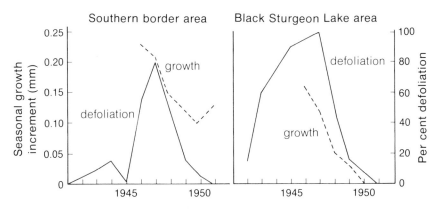

Figure 33-26   Defoliation of balsam fir trees in two areas during an infestation of the spruce budworm. In the area that was more heavily infested (right) all the trees died, whereas trees in the other area (left) had suffered no mortality by the end of the study period (after Belyea 1952).

even to the extent that efficient predators can regulate prey population density below the carrying capacity of the environment. Through their influence on prey populations, predators affect both the evolution of prey characteristics and the ability of prey to compete with other species utilizing the same resources. Evolutionary and ecological effects of predation are closely related. The efficiency of a predator as an exploiter of its prey is determined by the adaptations of both. In the next chapter, we shall examine how evolutionary responses of predator and prey populations, each to the other, influence the outcome of the predator-prey relationship.

34

Evolutionary Responses

In the preceding chapters dealing with competition and predation, we assumed that carrying capacities, competition coefficients, and predation efficiencies did not vary. But each of these population traits is the product of the adaptations of individuals in the population, and each is therefore subject to evolutionary change. Populations of predators, prey, and competitors are a part of the environment of every species. All such interacting populations select traits in each other that tend to alter their relationship.

When two populations interact, both can respond with evolutionary change. When their relationship is antagonistic, as it is between competitors and between predator and prey, the species can become locked into an evolutionary battle to increase their own fitness, each at the other's expense. Such a struggle can lead to an evolutionary equilibrium in which both antagonists continually evolve in response to each other, but the net outcome of their interaction does not change. This bubbling evolutionary pot is stirred by changes in the environment and the appearance of new species in the community, whether by natural range expansion or by human introduction.

We shall begin our discussion of evolutionary responses by describing a well-documented case of evolutionary change in a host-parasite system, that of the introduced European rabbit and myxoma virus in Australia.

The lethal myxoma virus became benign to rabbits through the rapid evolution of rabbit and virus populations.

As Western man immigrated to distant corners of Earth, he brought a host of stowaways and invited guests who exceeded their welcome. Australia was not spared. In 1859, twelve pairs of the European rabbit were released on a ranch in Victoria. Within six years, the rabbit population had increased so rapidly that 20,000 were killed in a single hunting drive. By 1900, the several hundred million rabbits distributed throughout most of the continent became a critical problem for the Australians, whose economy has always been based largely on raising sheep. Being efficient grazers, rabbits destroyed range and

632

pasture lands that otherwise would have been utilized for wool production. The Australian government tried poisons, predators, and other possible control programs—all without success. The answer to the rabbit problem seemed to be a myxoma virus, a relative of smallpox, that was discovered in populations of the related South American rabbit. Myxoma produced a small localized fibroma (a fibrous cancer of the skin). Its effect on South American rabbits was not severe, but a European rabbit infected by the virus died quickly of myxomatosis.

In 1950, the myxoma virus was introduced in one locality in Australia (Fenner and Ratcliffe 1965, Fenner 1971). A myxomatosis epidemic broke out in Victoria around Christmas time and spread rapidly. The virus was transmitted primarily by mosquitoes, which bite infected skin areas and carry the virus on their snouts. The first epidemic killed 99.8 per cent of the individuals in infected rabbit populations. During the following myxomatosis season (coinciding with the presence of mosquitoes), only 90 per cent of the remaining population was killed; and during the third outbreak, only 40 to 60 per cent of rabbits in the disease areas succumbed. At present, the myxoma virus has little effect on the rabbits, which have increased in number to the point that they are again a nuisance and an economic problem.

Several factors contributed to the decline in the virulence of the myxoma virus in the rabbits. The few rabbits that survived the disease developed immunity and were unaffected by later outbreaks of the virus. More important, immunity was conferred to the offspring of immune females through the uterus. Over and above the immune response, evolutionary changes occurred that increased the resistance of the rabbit population to the myxoma virus and reduced the virulence of the virus population. Selection was strong on both the rabbit and virus. Genetically determined immunity to a disease reduced mortality, thereby increasing the fitness of the host. At the same time, virus strains with less virulence were favored because reduced virulence lengthened the survival time of the rabbits, and thus increased the mosquito-borne dispersal of the virus. A virus organism that kills its host quickly has small chance of being carried by mosquitoes to other hosts.

Periodic samples of myxoma virus, collected in the field and tested on European rabbits with no previous exposure to the virus, demonstrated that the virulence of natural myxoma strains decreased over the years following its introduction. In 1950 and 1951, all strains of the virus collected in the field produced Grade I infections (more than 99 per cent mortality and a mean survival time of less than two weeks). By the 1958–59 season, however, no field strains of the myxoma virus had Grade I virulence; and by 1963–64, Grade II virulence (99 per cent mortality, 14-to-16-day survival time) was no longer evident. Similarly, resistance of wild rabbits to a particular grade of virus has increased steadily since 1950 (Figure 34-1). When the myxoma virus was first introduced to Australian rabbits, virtually all were mortally susceptible to myxomatosis, but by 1957 somewhat less than half the rabbits tested succumbed to virus with Grade III virulence. Studies indicated further that whether a rabbit died from myxomatosis depended largely on the inheritance of genetic immunity from its ancestors.

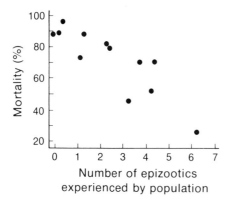

Figure 34-1 Decrease in susceptibility of wild rabbits to myxoma virus with Grade III virulence—90 per cent mortality in genetically unselected wild rabbits (after Fenner and Ratcliffe 1965).

Competition experiments show the influence of genotype on population processes.

A change in the environment, such as the introduction of a virus disease, can alter the fecundity and mortality of individuals in a population and, in doing so, change the selective pressures on the population. Evolutionary responses will follow if the appropriate genetic variation is available. We might expect many of the genetic responses to an altered environment to have such subtle effects on the phenotype, particularly on characters determined by the additive effects of many genes (see page 473), that we could not see them. But because the coexistence of two competitors and the stability of a predator-prey system can hinge on small changes in population processes, subtle genetic effects may exert a disproportionately large influence on the workings of the community. Species may drop out of, or they may be added to, a community on account of small differences in their adaptations.

The effects of subtle genetic differences can be seen in the different outcomes of competition experiments, according to the genetic strains of each species used. In competition experiments involving the flour beetles *Tribolium confusum* and *T. castaneum*, Park (1954, 1962) found that within certain ranges of temperature and humidity, the outcome was uncertain. At 29 C and 70 per cent relative humidity, for example, *castaneum* won in about five-sixths of the experiments, and *confusum* in one-sixth.

Park began all his experiments with two pairs of each species. When Lerner and Ho (1961) began experiments under identical conditions with ten pairs of each species, *T. castaneum* won twenty out of twenty times. These results suggested that in Park's experiments, genetic variation in the parent individuals might have altered the outcome of competition. Probably, the larger parental populations used by Lerner and Ho more closely resembled the average genetic makeup of the species, thereby producing more consistent results. To test this hypothesis, Lerner and Dempster (1962) developed several strains of each species that were inbred to minimize their genetic variation, and then tested the strains against each other in competition experiments. They found that if the populations were inbred for more than twelve generations, certain strains of *confusum* would consistently outcompete certain strains of *castaneum*, the complete reverse of their normal competitive relationship. These experiments indicated that changes, even minor ones, in the genetic constitution of a population can greatly influence its competitive ability (Park *et al.* 1964). Similar results have been found in competition experiments involving rice (Sakai 1961) and fruit flies, *Drosophila* (Ayala 1969, 1972a).

In Ayala's (1969) experiments on competition between *Drosophila serrata* and *D. nebulosa,* the competitive ability of *D. serrata* appeared to increase during the course of the experiment. When the two species were established in population cages at 19 C, they quickly achieved a pattern of stable coexistence with 20 to 30 per cent *D. serrata* and 70 to 80 per cent *D. nebulosa*. In one experiment, but not in a replicate, the frequency of *D. serrata* began to increase after the twentieth week and attained about 80 per cent by the thirtieth week, a complete reversal of the initial equilibrium conditions. When individuals of both species were removed from the experimental populations after the thirtieth week and tested against stocks maintained in single-species cultures, the competitive ability of each species was found to have increased after exposure to the other in the competition experiment. In one of the replicates, the competitive ability of *D. serrata* evidently had evolved much more rapidly than that of *D. nebulosa*, and their equilibrium frequencies were greatly altered. The difference between these replicates also appeared in the competitive ability of *D. serrata* tested against the unselected stocks of *D. nebulosa*.

In experiments similar to those of Ayala, Pimentel *et al.* (1965) have shown that a poor competitor can evolve a competitive advantage (judged by relative population density) over a formerly superior adversary. Haldane (1932) suggested that when populations were rare, intraspecific competition would be greatly reduced, permitting the evolution of greater efficiency in interspecific competition. To test this model, Pimentel *et al.* (1965) conducted laboratory experiments with flies to determine whether two species could coexist on one food resource by frequency-dependent evolutionary changes in their competitive ability. Moore (1952) and Lerner and Dempster (1962) had earlier shown that competitive ability could be selected for.

The housefly (*Musca domestica*) and the blowfly (*Phaenicia sericata*), which have similar ecological requirements and a comparable life cycle (about two weeks), were chosen for the experiments. Both species feed on dung and carrion in nature, and they are often found together on the same food resources. The flies were raised in small population cages at 27 C, with a mixture of agar and liver provided as food for the larvae, and sugar supplied for the adults. The outcomes of an initial series of four competition experiments between individuals from wild populations of the housefly and the blowfly were split two each. The mean extinction time for the blowfly, when the housefly won, was 92 days, and it was 86 days for the housefly when the blowfly won. The two species were certainly close competitors, but the small cages used did not allow enough time for evolutionary change before one of the populations became extinct.

To prolong the competitive interaction, Pimentel and his colleagues started a population in a sixteen-cell cage, which consisted of single cages in four rows of four cages each with connections between them (Figure 34-2).

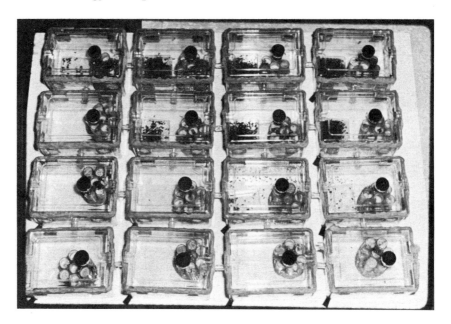

Figure 34-2   The sixteen-cell cage used by Pimentel to study competition between populations of flies. Note the vials with larval food in each cage and passageways connecting the cells. The dark objects concentrated in the upper-right-hand cells are fly pupae (courtesy of D. Pimentel, from Pimentel *et al.* 1963).

Under these conditions, populations of houseflies and blowflies coexisted for almost seventy weeks, and showed a striking reversal of number between the two species (Figure 34-3). After thirty-eight weeks, when the blowfly population was still low and just a few weeks prior to its sudden increase, individuals of both species were removed from the sixteen-cell population cage and tested in competition with each other and with wild strains of the housefly and blowfly. Captured wild blowflies turned out to be inferior competitors to wild and experimental strains of the housefly. But blowflies that had been removed from the population cage at thirty-eight weeks consistently outcompeted both wild and experimental populations of the housefly (five cases of each). Thus, the experimental blowfly population apparently had evolved superior competitive ability while it was rare and on the verge of extermination.

The tenuous competitive edge of the blowfly over the housefly was revealed in the course of competition between populations removed from the sixteen-cell experiment. In most of these populations, the blowfly eliminated the housefly in an average of 112 days, but in one of the five populations, the two species coexisted to a near draw for 519 days, during which time numerical dominance in the population cage shifted between the species several times. Apparently, the housefly had temporarily regained its competitive edge over the blowfly. Pimentel's experiment demonstrated emphatically that the

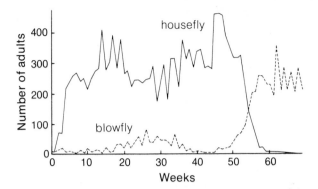

Figure 34-3   Changes in competing populations of houseflies and blowflies in a sixteen-celled cage (after Pimentel *et al.* 1965).

outcome of competition can turn upon very small genetic differences. Whether this "genetic feedback" system is important to the coexistence of species in nature is not known, however. Complex natural environments offer opportunities for species to avoid competition through divergence and ecological specialization.

Predator and prey adaptations represent a counterevolutionary equilibrium.

The outcome of competition, whether it leads to stable coexistence or to the elimination of one form, depends on a delicate balance between the genetic compositions of the competing populations. This is also true of interacting populations of predators and prey. There is abundant evidence of counterevolutionary adaptations in both (see Chapter 15). For direct evidence of evolutionary adjustments between predators and their prey, we shall turn once again to interactions between laboratory populations of insects. In Utida's (1957) host-parasite system, described on page 597, one of the populations was followed for more than eighty generations. Dramatic fluctuations that characterized the first half of the experiment were dampened considerably during the second half, and the average size of the wasp (parasite) population was lower (Figure 34-4). This change possibly could have been caused by an evolutionary response of the host.

Pimentel and Al-Hafidh (1963) explored the evolution of host-parasite relationships with the housefly and a wasp parasite, *Nasonia vitripennis* (Figure 34-5). In a control unit, *Nasonia* was allowed to parasitize a fly population that was kept at a constant level by replenishment from a stock population that had no contact with the wasp. None of the control flies that survived exposure to the wasp were returned to the population. In an experimental unit, the fly population was kept at the same constant level, but successfully emerging flies were allowed to remain in cages so the popula-

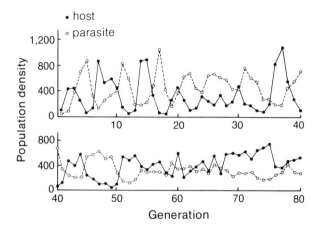

Figure 34-4   Long-term trends in a mixed population of the Azuki bean weevil (*Callosobruchus chinensis*) and its wasp parasite, *Heterospilus prosopidus* (after Utida 1957).

tion could evolve resistance to the wasp. The populations were maintained for about three years, long enough for evolutionary change to occur. In the experimental system, the reproductive rate of the wasps had dropped from 135 to 39 progeny per female, and their longevity had decreased from seven to four days. The average level of the parasite population also decreased (1,900 adult wasps, versus 3,700 in the control), and population size was more constant.

Experiments were then set up in thirty-cell cages in which the numbers of flies were allowed to vary freely. A control cage was started with flies and wasps that had no previous contact with each other, and an experimental cage was started with animals from the experimental population discussed above. In the control cage, wasps were efficient parasites, and the system underwent severe oscillations. In the experimental cage, however, the wasp population remained low and the flies attained a high and relatively constant population (Figure 34-6). This result strongly reinforced the conclusion drawn from the earlier experiments that the flies had evolved resistance to the wasp parasites.

Figure 34-5   Pupa of the housefly and the parasitic wasp *Nasonia* (courtesy of D. Pimentel, from Pimentel 1968).

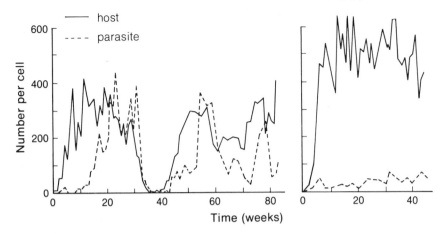

Figure 34-6   Populations of houseflies and a wasp parasite, *Nasonia vitripennis,* in thirty-celled laboratory cages. Control (left): flies had no previous experience with the wasp. Experimental (right): flies had been exposed to wasp parasitism for more than a thousand days (after Pimentel 1968).

As one might expect, parasites and diseases of economic crops have been scrutinized closely by agronomists (*e.g.,* Plank 1968, Watson 1970). Control of pathogens like wheat rust is accomplished primarily by breeding strains of wheat with genes that confer resistance to the pathogen. The defense mechanisms provided by resistance genes have proven to be easily circumvented by rapid evolutionary change in the pathogen, probably a single gene replacement. Agricultural geneticists keep track of such changes in plant disease organisms by routinely exposing different genetic strains of the crop plant to a variety of races of the pathogen and recording the virulence (Green 1971). A surprising result of such a survey in Canada of wheat rust—a fungus—was that, for a given race, virulence on strains of wheat with different resistant genes appeared and disappeared sporadically (Green 1975). Altered virulence is apparently owing to changes in single genes. For example, in 1969 race 15 B-1L of the wheat rust was virulent on strains of wheat with resistance genes (*Sr*) 8, 10, and 11, and it was avirulent on strains with genes 15 and 17. In 1970, a subrace of the rust was isolated that had become avirulent on *Sr* 11. In 1971, a new subrace appeared virulent on *Sr* 15. That same subrace lost its virulence on *Sr* 8 the following year; another lost its virulence on *Sr* 11. In 1973, a new subrace virulent on *Sr* 17 appeared. Other race groups of the wheat rust have proved to be more conservative. The significance of virulence changes for the rust is not clear. Because the rust reproduces asexually throughout most of Canada, changes in virulence must have been caused by mutation rather than recombination between previously existing genes. Furthermore, the changes in virulence were identified in samples of avirulent rusts obtained from commercial plantings of wheat varieties with complex resistance to all races of rust. The virulence on different resistance genes was determined only in experimental plantings. Hence the changes in virulence did not appear in the population of rust as a whole. What the rust story does reveal, however, is that natural populations of predators and parasites continually generate genetic variants that challenge the resistance of their prey and hosts, which presumably respond in kind. Predator and prey are clearly locked into an unending evolutionary struggle and a perpetual equilibrium.

The evolutionary responses of predator and prey populations can be depicted by a simple graphical model that relates the rates of evolution of the two to the efficiency of the predators (Rosenzweig 1973; Figure 34-7). For the prey, the rate at which new adaptations to escape or avoid predators evolve should be expected to vary in direct proportion to the predation rate. If there were no predation, there would be no selection of adaptations for predator avoidance. As predation increased, selection would increase, and so would evolutionary response, at least up to the limits set by genetic variation and balancing effects of correlated selection responses. As predation rate increased, we would expect the rate of evolution by the prey to increase up to a point and then level off. The rate of evolution by the predator should vary in opposite fashion. If a particular prey species were not heavily exploited, predators that were adapted to utilize this resource would be selected. As the rate of exploitation of the prey population increased, intraspecific competition by the predators would reduce the selective value of further increase in rate of predation on that prey. Very high rates of predation could conceivably select individuals that shifted their diets toward other prey species. Hence the rate of evolution by the predator to increase its efficiency on a particular prey could become negative.

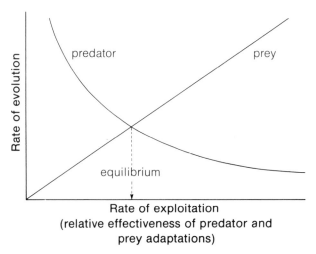

Figure 34-7   A graphical model of the evolutionary equilibrium between predator and prey adaptations which determine the level of exploitation (for explanation, see text).

In this simple model, the countervailing influences of predator and prey adaptation achieve a stable equilibrium where the two curves cross. If predator adaptations are relatively effective and the prey are exploited at a high rate, selection on the prey population will tend to improve its escape mechanisms relatively faster than selection on the predator population will improve its ability to exploit the prey. When the exploitation rate is low, the prey evolve relatively more slowly than the predators. This will lead to an equilibrium between the adaptations of predator and prey, producing a relatively constant rate of exploitation regardless of the specific predator and prey adaptations. In other words, although the predator and prey both evolve, neither gains on the other when the system is in equilibrium. The exact position of the equilibrium point will depend on potential rates of evolution of the predator and prey, and also on the availability of genetic variation and various constraints on evolutionary flexibility.

Pimentel's experiments on host-parasite interactions, described above, provide a demonstration of the predator-prey equilibrium. The housefly (host) and the parasitic wasp *Nasonia* have undoubtedly achieved an evolutionary equilibrium in their natural habitats. When brought into a simple laboratory habitat, *Nasonia* is able to exploit housefly populations at a greatly increased rate because little time is required to search out hosts. This is equivalent to shifting the exploitation rate of *Nasonia* on houseflies far above the equilibrium level in Figure 34-7. This shift increases selective pressure on the housefly to escape parasitism relative to selective pressure on the predator to further increase its exploitation rate. As a result, the ability of the housefly to escape parasitism increased and the level of exploitation by *Nasonia* decreased toward its natural equilibrium.

The equilibrium model of predator and prey counteradaptations leads one to predict that the ability of a population to exploit other populations or to avoid exploitation should be relatively similar among ecologically comparable species. Although there are few data available to test this hypothesis, it is demonstrated by the relationship between cowbirds and oropendolas in Panama (see page 79). Briefly, cowbirds, which are brood parasites, normally reduce the breeding success of oropendolas, but they may be beneficial under certain circumstances. If bot flies enter the nests to parasitize the oropendola nestlings, the cowbird young remove bot fly larvae from the young oropendolas and actually increase the nesting success of the host. If bot flies are kept away from the oropendola colony by wasps nesting in its proximity, the cowbirds perform no beneficial function but they compete for food with the oropendola young. The adaptations of both the oropendola and the cowbird are entirely different in the two settings in which they meet. But the productivity of each species is about the same in both situations. Oropendolas raise about 0.4 young per nest in both types of colonies, and cowbirds raise an average of 0.75 young per parasitized nest (Table 34-1). The values for nesting productivity in each type of colony are remarkably similar within each species, particularly when one considers that (a) in one type of colony oropendolas discriminate against cowbirds, and in the other they accept them; (b) cowbirds behave either unobtrusively or aggressively de-

TABLE 34-1

Production of oropendola (host) and cowbird (brood parasite) young in two types of oropendola colonies in which the cowbird-oropendola relationship differs strikingly (from Smith 1968).

	Colony type	
	Discriminator	Nondiscriminator
Cowbird-oropendola interaction	Parasitic	Mutualistic
Oropendolas	Reject cowbirds	Accept cowbirds
Cowbirds	Unobtrusive	Aggressive
Productivity (young/nest)		
Oropendolas	0.39	0.43
Cowbirds	0.76	0.73

pending on the colony type; (c) cowbird eggs either mimic oropendola eggs closely or have a generalized nonmimetic pattern; and (d) cowbirds parasitize only 28 per cent of nests in discriminator colonies, laying only one egg per nest, but they parasitize almost three-quarters of the nests in nondiscriminator colonies, and lay several eggs per nest. In spite of these differences in the environments of the cowbirds and the oropendolas, evolutionary responses have brought the parasitic and mutualistic systems to approximately the same equilibrium value of productivity and exploitation for both species.

The relative diversity of predators conceivably could influence the position of the exploitation equilibrium in the predator-prey counterevolutionary system (Ghiselin and Ricklefs 1970, Ricklefs 1970a). For example, suppose that a prey population is beset by two species of predators using different foraging techniques: perhaps an insect population preyed on by a mammal that feeds on the ground and runs after its prey, and by a bird that feeds by taking its prey on the wing. In the absence of birds, the insect could escape the mammalian predator by flying; when birds are present, the insect must evolve mechanisms to escape both predators. The prey, however, cannot adapt to each predator independently. Instead, it must make evolutionary compromises to adapt to both of them, and as a result, the effectiveness of the prey's counteradaptation toward the predators is reduced and the level of exploitation on the prey population increases.

Predators face similar problems when they exploit a variety of prey species, each of which can utilize different methods of avoiding predation. Ecologists generally agree that specialists can be more efficient than generalists. Applied to predators, this principle suggests that a predator with a narrow diet can attain higher levels of exploitation on each of its prey than a predator with a more catholic diet. As with many so-called ecological principles, there are few pertinent observations on natural systems to either support or refute the principle (*cf.* Zwölfer 1963).

Evolutionary responses may influence the stability of relationships between species.

In the preceding chapters on population ecology, we developed simple graphical descriptions of population regulation (Figure 31-6), interspecific competition (Figure 32-21), and predator-prey interaction (Figure 33-9). We can interpret the effect of evolutionary responses on each of these graphs. As the isoclines for each species are altered by their own evolution and by the evolution of their competitors, predator and prey equilibrium points shift, and the inherent stability of the system may change.

In the case of population regulation, we shall consider the exponential growth rate of the population as a function of population size (Figure 34-8). Competition between genotypes within a population leads to a replacement of individuals by others with higher intrinsic capacity for increase (r) at a given population size. The effect of raising the population growth rate is to shift the equilibrium population size (K) to a higher level. In a population that is limited by the availability of resources, selection will tend to increase the carrying capacity of the population by favoring adaptations that increase the efficiency with which individuals can utilize resources. This tendency is balanced by correlated results of selection having opposite effect and by the adaptations of competitors, predators, and prey. At length, selection must exhaust the genetic variation and evolutionary opportunities available for increasing K.

In the case of interspecific competition, we must remember that both species are capable of evolutionary response to each other. Selection to increase the carrying capacity (K) shifts the equilibrium isocline ($N_i = 0$) outward from the origin of the population graph and thus tends to increase competitive ability (Figure 34-9). But with selection applied equally to both species, the eventual outcome of the evolutionary relationship of two competitors would be difficult to predict. Selection may also be applied to the coefficients of competition (a_{ij}) between the two species (Gill 1974). Ecological divergence of two species tends to reduce their competition coefficients. The resulting shift in the equilibrium isoclines tends to stabilize the interaction

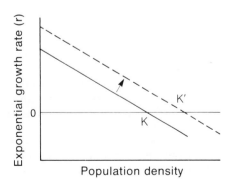

Figure 34-8   The effect of intra-specific competition on the evolution of the curve relating population growth rate to density. Selection favors individuals with higher intrinsic capacity for increase and results in an increase in the carrying capacity of the population in the absence of balancing forces.

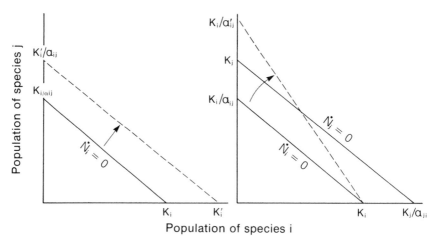

Figure 34-9   Evolution of the isoclines of competing populations. At left, selection increases the carrying capacity (K_i) and shifts the isocline away from the origin of the graph. Both intercepts $(K_i$ and $K_i/a_{ij})$ are increased by the same factor. At right, the angles of the isoclines are shifted by decreases in a_{ij} and a_{ji}, resulting from ecological divergence; this increases the probability of coexistence.

and allows the species to coexist indefinitely (Figure 34-9). In this case, evolution by both species has a similar, stabilizing influence on their interaction.

Such is not the case for the interaction between a predator and its prey (Figure 34-10). Evolution of increased predator efficiency reduces the prey isocline and shifts the predator isocline to the left. As a result, the predator isocline may fall to the left of the hump in the prey isocline and lead to increasing oscillations of predator and prey populations. If the prey were to become more efficient at exploiting its own prey, or other limiting resource, the prey population would achieve a higher (K_H) and would be able to sustain a larger predator population. The resulting shift in the prey isocline (Figure 34-10) would shift the predator isocline to the left relative to the hump in the prey isocline and would, therefore, tend to reduce the stability of their interaction. In general, increased predator efficiency tends to destabilize a predator-prey interaction. This tendency is normally balanced by the evolution of the prey to escape predation. But although an evolutionary equilibrium is inevitable, population stability is not guaranteed. Hence some evolutionary equilibria, those in which the predators are greatly favored, might never be expressed by natural predator-prey systems.

In this chapter, we have examined the role of evolutionary responses in determining the nature of interactions between populations of competitors, predators, and prey. There is ample experimental evidence that predator efficiency and competitive ability are the product of the adaptations of a

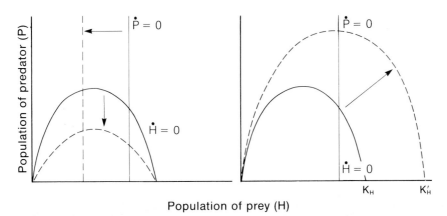

Figure 34-10   Evolution of the predator and prey isoclines. At left, an increase in predator efficiency shifts the predator isocline to the left and reduces the prey isocline, tending to reduce the stability of the system. At right, increased efficiency of the prey at exploiting its own food resource greatly enlarges the region of prey population increase within its isocline, but can also reduce the stability of the predator-prey interaction.

population and that they can be altered by applying suitable selection. Evolution in interacting populations generally works toward opposite ends and may often lead to an evolutionary equilibrium in which adaptive changes in one population are balanced by opposing changes in the other population. Evolutionary equilibrium does not necessarily imply population stability, and populations frequently fall victim to overexploitation and competitive exclusion. In spite of our understanding of population phenomena, the extinction of entire species and larger taxonomic groups remains one of the most poorly explained of all ecological and evolutionary phenomena. In the next chapter, we shall examine major patterns of extinction and some of the explanations that have been proposed for their causes.

35

Extinction

In 1810, American ornithologist Alexander Wilson observed an immense flock of passenger pigeons in the Ohio River Valley. For days, the column of birds, perhaps a mile wide, passed overhead in numbers to darken the sky. Wilson estimated that there were more than 2 billion birds. The last passenger pigeon died in the Cincinnati Zoological Garden just more than a century later, on September 1, 1914.

The demise of the passenger pigeon was caused by two unfortunate circumstances. First, the pigeons roosted and nested in huge assemblies, sometimes numbering several hundred million birds within a few hundred square miles. Second, roasted or stewed, the pigeons tasted very good. Pigeoners, as professional hunters were called, gathered in large groups to trap and slaughter them. Nesting trees were felled to collect the squabs. As farmers cleared the forests and railroads made vast areas within the pigeons' range accessible to Eastern big-city markets, the persecution increased. By the mid-1800s, increased killing disrupted breeding, and the pigeon population began to decline. By 1870, large breeding congregations were found only in the Great Lake States, at the northern edge of the pigeon's former range. The last nest was found in 1894, and the bird was last seen in the wild in 1899.

With its extinction in 1914, the passenger pigeon joined a lengthening list of species that have vanished from the Earth. Since naturalists began describing systematically the forms of plant and animal life two hundred years ago, 53 birds, 77 mammals, and a host of other animals and plants have disappeared, most of them at the hand of man. By exploiting species for food or hides or destroying them as pests, by subjecting them to depredation by domesticated animals and hangers-on like rats, and by destroying their habitats, we have managed to reduce the numbers or productivity of many populations below their self-sustaining level.

Perhaps many of these species, including the passenger pigeon, would have persisted had it not been for man's activities. But the fossil record reveals that virtually all lineages have become extinct without leaving descendants. The million or so living species of plants and animals are derived from a small fraction of those alive at any time in the distant past. It has been estimated that less than 1 per cent of all the species that have existed since the beginning of the Paleozoic Era (about 600 million years) are extant. If this were a reasonable estimate, it would mean that species have become extinct at a rate of one every six or so years on average. No doubt, therefore, that extinction is a commonplace event—in fact, that extinction is the inevitable fate of virtually all lineages.

Extinction is important in the study of ecology for two reasons. First, the extinction of local populations of a species can play a role in the regulation of population size. As we have seen, many populations exist in mosaics of habitat patches. Population size is related to the balance between local extinction and immigration of colonists to patches left empty by extinct populations (Levins 1970, Levins and Culver 1971, Levins *et al*. 1973). Second, the local extinction of one species broadens the ecological opportunity open to similar species and may enable a species previously excluded by competition to establish itself in the community.

To the extent that extinction reflects the relative inability of a population to adapt to a changing environment, it also belies the relative evolutionary success of populations better able to adapt. Hence the replacement of species by others that are more successful may be a form of selection in which the entire population rather than the individual bears the favored trait.

The study of extinction ought to elucidate many aspects of ecology and evolution, but it has largely failed to do so. This failure can be traced directly to the facts that extinction is so infrequent and difficult to predict that direct observation is not feasible, that extinction is merely the final event in a long sequence of subtle evolutionary and ecological processes leading to the demise of a population, and that the fossil record is too incomplete to record the details of a dwindling lineage. And, of course, fossils open only a tiny obscured window on biological communities of the past.

Most of what we know about extinction comes from observations of the phenomenon on three different levels. First, we can observe the extinction of small local populations when local climatic conditions become severe or when an efficient predator is introduced to the community. Such extinctions are ecological rather than evolutionary events, and would not normally lead to extermination of an entire lineage. Second, we can observe or infer the local, and sometimes global extinction of a species resulting from competition from ecologically similar species. Such replacements usually would cause a change in the structure of the community and lead to ecological and evolutionary readjustments among the remaining species. Third, fossils record the extinctions of entire groups of organisms. Dinosaurs, ammonites, and trilobites are among numerous once-successful taxonomic groups that have disappeared, sometimes abruptly, from the face of the Earth. Although the extinction of a major phylogeny must necessarily involve the extinction of all the numerous closely related species that comprise the taxon, the fossil record does not preserve so much detail and we are left guessing the ecological and evolutionary significance of the successive replacement of major phylogenetic groups.

In this chapter, we shall investigate extinction, beginning with observable, contemporaneous events that occur on a small geographic and taxonomic scale and working toward an understanding of events on a larger scale that are entwined in the evolutionary history of life on Earth.

Probability of extinction increases as population size decreases.

Birth and death rates of populations are influenced by a wide variety of ecological factors, but whether an individual dies or successfully rears one or more progeny during a particular period is largely a matter of chance. If the annual probability of death were one-half, some individuals would live and, on average, an equal number would die. There exists a finite probability, however, that all the individuals would die, just as there is a finite but small probability that ten coin tosses all would come up tails (we expect this to happen once in 1,024 trials). Changes in populations owing to chance events are called *stochastic* fluctuations, and their force is more strongly felt in small populations. This becomes clear when we consider that the probability of obtaining five tails in a row with successive tosses of a coin is one in 32, compared with the smaller chance of one in more than a thousand for twice as many flips. If we visualized each individual in the population as a coin, and if it came up tails the individual died, we could see that a population of five individuals more likely will become extinct than a population of ten.

The probability of extinction of populations has received considerable attention from probability theorists, who have derived a mathematical expression rating probability of extinction at time t ($p_0(t)$) to birth rate (b), death rate (d), and population size at present (i),

$$p_0(t) = \left[\frac{e^{(b-d)t} - 1}{(b/d)e^{(b-d)t} - 1} \right]^i \tag{35-1}$$

(Pielou 1969: 17ff). Frequently, ecologists would not be so interested in whether a population will become extinct in five years or a hundred years, as they would be in the probability of eventual extinction given essentially unlimited time. We can obtain a reasonable estimate of the probability of eventual extinction in evolutionary, or even ecological time by calculating the probability of extinction given infinite time ($p_0(\infty)$). If death rate exceeded birth rate, ($b - d$) would be negative and the exponential terms with ($b - d$) would approach 0 as t approached infinity. Hence equation 35-1 would become $p_0(\infty) = (-1/-1)^i = 1$, that is, certain extinction, regardless of initial population size. Of course, we would expect this because the population is tending to decrease on average.

If birth rate exceeded death rate, ($b - d$) would be positive and the exponential terms in equation 35-1 would become so large we could ignore the -1 terms. The equation thus would be reduced to

$$p_0(\infty) = \left(\frac{d}{b} \right)^i$$

Hence the probability of extinction decreases as the ratio of the death rate to the birth rate decreases and as population size increases. When the birth rate exceeds the death rate, the population tends to increase, and its probability of extinction decreases with time. Under such conditions, most extinctions would occur during the initial phase of population growth. This model may provide a reasonable description of events following the colonization of a habitat or island by a population. Clearly, the chances of successful establishment by a population increase in direct relation to initial population (propagule) size and the ratio of birth to death rate (MacArthur and Wilson 1967, Crowell 1973).

A more general application of equation 35-1 may be had when b and d are equal, that is, when births balance deaths and population size remains constant. We have seen in the chapter on the regulation of population size that birth and death rates are brought into balance when a population reaches its carrying capacity (K). Under these circumstances, population size usually would be large and any stochastic decrease in population size usually would be balanced by density-dependent decrease in death rate, increase in birth rate, or both. Hence it would seem unlikely that stochastic fluctuations could play a large role in extinction of most populations. This role becomes considerably strengthened, however, when one considers that efficient competing populations may reduce the realized carrying capacity of a species to a very low level. In an environment that is dominated by interspecific competition, stochastic fluctuations in population size produce little density-dependent response.

Equation 35-1 may be solved when $b = d$ to give

$$p_0(t) = \left(\frac{bt}{1 + bt}\right)^i \tag{35-2}$$

(the solution can be found by expanding the exponential terms in 35-1 into a power series). Accordingly, the probability of extinction decreases with increasing population size and it increases with larger b or d, indicating more rapid population turnover. The relationship of extinction probability to population size (i) within time period (t) is shown in Table 35-1 for a population in whch $b = d = 0.5$. These are reasonable values for adult death and recruitment in a population of terrestrial vertebrates. We see, for example, that for a population with ten individuals the probability of extinction is 0.16 within a

TABLE 35-1
Probability of extinction when birth rate = death rate = 0.5 per year, for populations of initial size i within period t. Probabilities are calculated from equation 35-2.

Population size (i)	Time (t)			
	1	10	100	1000
1	0.33	0.83	0.98	0.998
10	$<10^{-4}$	0.16	0.82	0.980
100	$<10^{-48}$	$<10^{-7}$	0.14	0.819
1000	$<10^{-99}$	$<10^{-79}$	$<10^{-8}$	0.135

ten-year period, 0.82 within a hundred-year period, and virtually certain (0.98) within 1,000 years. Even for an initial population size of 1,000, the probability of extinction is more than 10 per cent within a millennium and becomes virtually certain (0.999) within a million years.

Extinctions on islands demonstrate stochastic effects in small populations.

When populations dwindle to small size, particularly when density-dependent mechanisms no longer operate owing to intense interspecific competition, they become extremely susceptible to extinction following random fluctuations in size. By virtue of the fact that populations on small islands are restricted geographically and are not frequently augmented by immigration, they are particularly susceptible to extinction.

Extinction occurs frequently enough on small islands that probability of extinction can be determined from historical records. For example, Diamond (1969) compared species lists compiled in 1917 and 1968 for birds on the Channel Islands off the coast of Southern California. In the 51-year interval between censuses, there were numerous cases of species disappearances, varying from 70 per cent of the avifauna (seven out of ten species) on Santa Barbara Island (one square mile in area) to 17 per cent (six out of 36 species) on Santa Cruz (96 square miles). On an annual basis, these figures can be reduced to between 0.10 and 1.7 per cent of the avifauna per year, with extinction rate and island size inversely related. Comparable rates have been determined for two tropical islands: 0.20 per cent per year on Karkar, an island of 142 square miles located ten miles off the coast of New Guinea (Diamond 1971); 0.23 per cent per year on Mona, ten square miles, located between Puerto Rico and Hispaniola in the Greater Antilles (Terborg and Faaborg 1973). Lynch and Johnson (1974) presented evidence to demonstrate that reported extinctions may largely be owing to inadequate census procedures, extirpations, and habitat change, and suggested that extinction rates may be overestimated by an order of magnitude. Certainly this would be more consistent with much lower extinction rates estimated from stochastic theory (equation 35-2 and Table 35-1). It is apparent nonetheless that populations on small islands, more than populations on large islands, are susceptible to extinction, regardless of the cause.

Recent extinctions of species of birds and mammals demonstrate the susceptibility of island populations.

Such studies as those of Diamond, Terborg, and Faaborg deal with the disappearance from small islands of species with wide distributions on the mainland and, thus, do not record the extinction of entire species. As Lynch and Johnson (1974) pointed out, it is difficult to distinguish firm records of established populations on islands from sight records of vagrants or mi-

TABLE 35-2

Recorded extinctions of mammals on islands and continents (from Harper 1945).

	Number of		
	large predators	large herbivores	small species
Continents	10	17	0
Islands	1	12	42

grants as reported in the literature. After all, naturalists of fifty years ago were not interested in the problem of faunal turnover on islands and cannot be expected to have tailored their observations to suit our present needs.

When an entire species disappears, however, one can be sure that extinction has occurred. Of 53 species of birds that have become extinct in the past 300 years, only three (the crested sheldrake of eastern Asia, the passenger pigeon, and Carolina parakeet, both of eastern North America) were found on major continents. The remaining fifty species disappeared from islands the size of New Zealand and smaller. As we have seen, island populations are particularly vulnerable because of their small size and isolation. Many island species have adapted to habitats without predators and with few competitors and are unable to cope with newcomers introduced from comparable mainland habitats, or with man. Other island species that are restricted to forest habitats suffer when land is cleared for sugar cane and coconut palm plantations. Small isolated populations also are vulnerable to such perturbations in their environments as prolonged drought, hurricanes, and disease epidemics. And the local extinction of these endemic island populations means the extinction of an entire species. Prior to their extinction, many species of birds were regularly eaten by man, whose depredations led directly to the demise of ducks, pigeons, and rails (small henlike marsh birds). But small inconspicuous, forest-dwelling birds have also disappeared, indicating the general vulnerability of island populations.

In contrast to birds, mammals have become extinct almost as frequently on the major continents (27 species) as on outlying islands, including Australia (55 species). Most extinct continental species were large herbivores killed for food (ground sloths, sheep, deer, zebras) and large carnivores, killed because they competed with man for food or were thought to threaten man directly—bears, wolves, cats (Table 35-2). Losses of mammals less involved with man's well-being (small rodents, bats, shrews, small marsupials) are confined to islands (Harper 1945).

Extinctions are not random.

In tossing a coin, some trials come up heads, others come up tails, but the probability of each outcome is the same for each trial. We also observe that some species persist and others disappear, but evidence is beginning to accumulate that suggests the probability of extinction differs greatly among species. Some are close to their inevitable fate and will falter with the least ecological setback, others are resilient and productive, able to withstand the perturbations of their environments. Such differences in probability of extinction can be inferred from patterns of geographical distribution and taxonomic differentiation of populations inhabiting groups of islands, like the West Indies and Polynesian Islands.

Immigrants to islands appear to be excellent competitors initially. Colonizing species are usually abundant and widespread on the mainland; these qualities make good immigrants. Most invaders of an island exhibit ecological release: Their populations increase greatly and spread into habitats not occupied by the parent population on the mainland (Crowell 1962, Grant 1966, MacArthur et al. 1972, Cox and Ricklefs 1977, see page 692). After immigrants become established, however, their competitive ability appears to wane; their distribution among habitats becomes restricted, and local population density decreases (Ricklefs 1970, Cox and Ricklefs 1977). These trends eventually lead to extinction.

We can judge the relative ages of populations on islands by their patterns of geographical distribution and by differences in their appearance from appearance of mainland forms from which they derived. Range maps of representative species of birds in the Lesser Antilles (Figure 35-1) demonstrate the progressive changes in distribution and differentiation of species with time, called the *taxon cycle* by Wilson (1961). On the basis of such distribution patterns, we can assign populations to one of four stages (Table 35-3): expanding (I), differentiating (II), fragmenting (III), and endemic (IV). Similar patterns have been described for ants on islands in the southwestern part of the Pacific Ocean (Wilson 1961), for birds in the Solomon Islands (Greenslade 1968), and for some insects in the Solomon Islands (Greenslade 1969). Among birds of the West Indies, species in late stages of the taxon cycle exhibit a loss of interisland movement and seasonal migration, reduced

TABLE 35-3

Characteristics of distribution of species in the stages of the taxon cycle (from Ricklefs and Cox 1972).

Stage of cycle	Distribution among islands	Differentiation between island populations
I	Expanding or widespread	Island populations similar to each other
II	Widespread over many neighboring islands	Widespread differentiation of populations on different islands
III	Range fragmented due to extinction	Widespread differentiation
IV	Endemic to one island	—

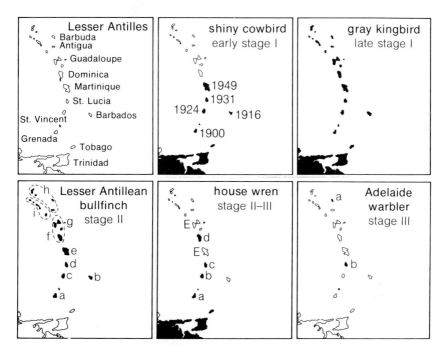

Figure 35-1   Distribution patterns and taxonomic differentiation of several birds in the Lesser Antilles, illustrating progressive stages of the taxon cycle. The shiny cowbird has expanded its range in the islands (dates of arrival are indicated), and the house wren has become extinct (E) on several islands during this century. Lower case letters designate subspecies (after Ricklefs and Cox 1972).

flocking behavior, and an increased tendency to occur in, or be restricted to, deep forest habitats, often montane forests. Similar habitat changes have been found in ants and beetles on islands in the Pacific.

Populations become more vulnerable to extinction as the taxon cycle progresses (Table 35-4). More endemic species (stage IV) of West Indian birds have become extinct since 1850, or are currently in grave danger of extinction, than widespread species (stages I to III). Almost a quarter of the endemic species either are extinct or are sufficiently reduced that they are at present in danger of extinction.

The decrease in competitive ability of island populations with age may be caused by evolutionary responses of an island's biota to new species (Ricklefs and Cox 1972). Immigrants are thought to be relatively free of parasites, predators, and efficiently specialized competitors when they colonize an island, so that their populations increase rapidly and become widespread. Having reached this stage, new immigrants constitute a larger part of the environment of many other species, which then evolve to exploit, avoid exploitation by, or outcompete the newcomer (*counteradaptation*). Apparently, a large number of species, when adapting to a single abundant new

TABLE 35-4

Rate of extinction of island populations of birds in the West Indies as a function of stage of the taxon cycle (from Ricklefs and Cox 1972).

	Taxon-cycle stage			
	I	II	III	IV
Number of recently extinct or endangered populations	0	8	12	13
Total number of island populations*	428	289	229	57
Per cent extinct or endangered	0	2.8	5.2	22.8

* A widespread species has many island populations, an endemic species only one.

population, can evolve faster than the new species can adapt to meet their evolutionary challenge. Competitive ability of the immigrants is progressively reduced by counteradaptations of island residents until the once-abundant new species becomes rare. Species are eventually forced to extinction by subsequent arrivals from the mainland that are more efficient competitors. When a species becomes rare, other species no longer gain evolutionary advantage by adapting to it, and the evolutionary pressure upon the rare species is released, similar to the effect that Pimentel observed in his experiments on competition between houseflies and blowflies (page 635). If this occurred before a species' decline had proceeded too far, the species might again increase and begin a new cycle of expansion throughout the island. This apparently has occurred many times; species distributions provide ample evidence of secondary expansions within the West Indies (for example, Figure 35-1, lower left).

Fossils record the extinction of taxonomic categories above the species level.

Observations of contemporaneous extinctions suggest that species disappear when better competitors or, less commonly, more efficient predators reduce their numbers to the point that random fluctuations quickly lead to extinction. Biological factors are, without doubt, more important than stochastic factors. Extinction by competitive exclusion results in the replacement of a species by one or more ecologically similar species that are often closely related taxonomically. Hence, although species turnover occurs, the character of the community does not change substantially. The passage of evolutionary time—measured in millions of years—reveals, however, the appearance and proliferation of new groups of organisms, often with novel structures, and the demise of once dominant groups. If we could step back 20 million years in time to the middle of the Miocene Epoch, we would see few familiar species in the fields and woods. To be sure, we could recognize many rodents and carnivores and some birds, and we could identify several of the trees, but others would be completely alien to our experience. Transported back 100 million more years into the Cretaceous Period, we would find ourselves in a world of giant reptiles and probably would hardly notice the small primitive mammals and birds that would one day inherit Earth.

The fact that extinctions of major taxonomic groups have occurred has been known for more than a century. Charles Darwin summarized what was known of extinction in the mid-1800s in *On the Origin of Species*:

> On the theory of natural selection, the extinction of old forms and the production of new and improved forms are intimately connected together. The old notion of all the inhabitants of the earth having been swept away by catastrophes at successive periods is very generally given up, even by those geologists . . . whose general views would naturally lead them to this conclusion. On the contrary, we have every reason to believe, from the study of the tertiary formations, that species and groups of species gradually disappear one after another, first from one spot, then from another, and finally from the world. In some few cases, however, as by the breaking of an isthmus and the consequent irruption of a multitude of new inhabitants into an adjoining sea, or by the final subsidence of an island, the process of extinction may have been rapid. Both single species and whole groups of species last for very unequal periods; some groups, as we have seen, have endured from the earliest known dawn of life to the present day; some have disappeared before the close of the palaeozoic period. No fixed law seems to determine the length of time during which any single species or any single genus endures. There is reason to believe that the extinction of a whole group of species is generally a slower process than their production. If their appearance and disappearance be represented, as before, by a vertical line of varying thickness the line is found to taper more gradually at its upper end, which marks the progress of extermination, than at its lower end, which marks the first appearance and the early increase in number of the species. In some cases, however, the extermination of whole groups, as of ammonites, towards the close of the secondary period, has been wonderfully sudden.

The historical record of extinction has been more recently reviewed by the late paleontologist George Gaylord Simpson (1949), who elaborated the basic points made by Darwin and others a century before. Reiterating the point that phylogenetic lines vary widely in their persistence and degree of evolutionary diversification, Simpson rejected the notion that higher taxa pass through inherently determined life stages similar to the life history of an organism. Extinction cannot in any way be construed from the fossil record as "racial senescence." He then developed the following typical profile of the history of a phylogenetic group.

(1) Periods of evolutionary activity within a line are usually accompanied by rapid diversification, geographical expansion, and adaptive radiation. As a group dies out, many lines become extinct and the remainder usually represent highly specialized forms, not a uniformly thinned-out spectrum of the group at the point of its greatest diversification.

(2) Periods of diversification reflect enlarged opportunities for the group. The reptiles diversified after they moved onto dry land; marsupials,

after they crossed the water gap to Australia; mammals, in general, after the disappearance of large reptiles; birds, after they attained the ability to fly; and so on.

(3) Major adaptive radiations were usually followed by long periods of little evolutionary change, during which extinct species were replaced by close relatives, much as we can observe or infer from present faunas and floras.

(4) The initial diversification of a group produced many false starts, innovative lines that were quickly replaced by even more successful offshoots. Thus, while diversification intensifies, so does extinction within the line.

(5) Each large taxonomic group has a characteristic average duration of species, genera, and families within it. Thus the average time span of extinct genera (that is, those for which the time span is known) is 80 million years for pelecypods (clams and their relatives) but only 8 million years for carnivore mammals.

The decline of one phylogenetic line is often accompanied by increase in another, sometimes derived from the first. In Chapter 26, we discussed the replacement of agnathe fish by placoderms, which in turn were replaced by bony fish and the sharks and rays. Surely these groups represented different approaches to exploiting the same kinds of environments. The adaptive commitments made by the agnathes and placoderms clearly did not serve them well in competition with more modern groups.

The succession of major reptilian groups by mammals provides a more familiar example. The reptiles themselves arose during the Pennsylvanian Period, perhaps 300 million years ago, from a progressive line of amphibians. The most important advance in the new reptile line was the development of an egg that could withstand desiccation—the *amniote* egg of present-day reptiles and birds—which enabled reptiles to escape water completely, and finally to launch a full-scale invasion of the terrestrial environment. From primitive beginnings in the late Paleozoic, reptiles underwent a major diversification during the Triassic Period (Figure 35-2) and produced all the major groups, including modern lizards, snakes, and turtles, by the middle of the Jurassic, 150 million years ago. The Jurassic and Cretaceous were the Age of Reptiles.

One of the ironies of the fossil record of reptiles concerns the synapsid line, which underwent a major radiation and dominated all other groups during the Permian and Triassic periods. Of the synapsids, the therapsids were the most important, but with the rise of the dinosaurs during the late Triassic, the therapsids and, with them, the synapsid line, all but dissapeared by the beginning of the Jurassic. Their only descendants eventually evolved to become what we recognize as mammals.

While the dinosaurs were having their heyday, an aberrant line of the dwindling synapsid group crossed the reptile-mammal boundary sometime during the late Triassic Period. This line then continued its precarious existence—occasionally producing abortive radiations—for nearly 100 million years, until the reptiles had finally waned, before embarking on a major period of diversification during the Paleocene Period. A major controversy

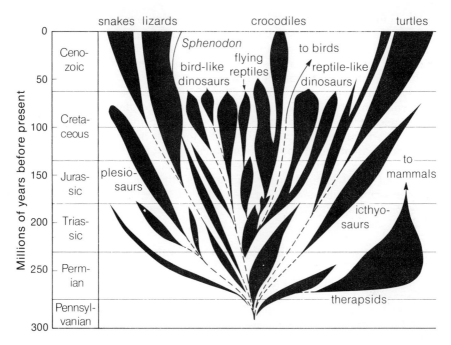

Figure 35-2   A schematic diagram of the phylogenetic tree of reptiles. The width of each line is roughly proportional to its diversity. The lines that led to birds and mammals were well-distinguished by the middle of the Jurassic Period (after Romer 1966).

exists concerning the replacement of most reptiles by mammals. Whether mammals caused the extinction of reptiles by competition and predation, or were suppressed until the reptiles succumbed to some unrelated factor.

Regardless of the sequence of events, there can be no doubt that the mammal body plan required considerable time to develop. Mammals differ from reptiles in being warmblooded, nursing their young, and abandoning the egg in favor of internal development of the embryo. Adopting a high body temperature and an active mode of life was accompanied by the development of insulating fur, a more efficient heart design, rearrangement of some parts of the skeleton, particularly the position of the appendages to elevate the body off the ground. Certain aspects of the skull and jaw became modified, and the teeth became fewer in number, more strongly rooted, and more diversified in form and function within the mouth. The manner in which the long bones of the appendages grow was also modified to allow the development of well-articulated joints in young, growing mammals. In reptiles, the bones grow at their ends by laying down cartilage that is subsequently ossified. The soft cartilage cannot be used to form an active, secure joint. In mammals, cartilage formation takes place in a region, called the *epiphysis*, under a bony cap at the end of the long bone which can assume the final shape of the joint at an early age. Mammals also developed a secondary palate at the roof of the mouth (the hard palate) that separates breathing and eating passages and allows mammals to continue breathing while they eat. This was not an important consideration for reptiles, owing to their low metabolic rates, but was a must for the more active mammals, which could suffocate in short time if breathing were cut off.

These attributes were perfected over a 100 million-year period during the Mesozoic Era, when reptiles reached their pinnacle of diversification. Then, at the end of the Cretaceous Period, most of the major lines of reptiles disappeared abruptly, followed shortly by the first major diversification of mammals. Earlier radiations of the mammal line during the Jurassic Period had been unsuccessful. None of the lines had attained any numerical importance, and only two survived the Cretaceous—the multituberculates, which became extinct in the early Eocene, and the line that gave rise to present-day marsupials and placentals.

Mass extinctions have occurred at several points in geological history.

There can be little doubt that the dinosaurs disappeared within a few million years at most. It is also reasonably certain that their demise was not caused by mammals, which did not become abundant until well after the extinction of the dinosaurs. The end of the Mesozoic also brought the simultaneous extinction of other groups, the most notable of which was the ammonites, relatives of the present-day chambered nautilus, a type of predatory mollusc. Actually, the fossil history of the ammonites shows several phases of large-scale extinction followed by rapid diversification (Figure 35-3). These phases occurred at the end of the Permian Period (230 million years ago) and at the end of the Triassic, 50 million years later. In both cases, new phylogenetic lines do not appear in the fossil record until well after most of the old lines have died out, precluding competitive exclusion as a cause of mass extinction.

The timing of major phylogenetic revolutions may be seen in Figure 35-4, where first and last appearances of families of animals are plotted as a percentage of total families existing at the time. Extinctions reached highs at the ends of the Cambrian, Devonian, Permian, Triassic, and Cretaceous periods (Newell 1967). It is no coincidence that periods of mass extinction correspond to the dividing lines of geologic periods, because geologists classified the ages of rock layers according to the fossils they contained. Any abrupt change in the animal families represented would form a natural division. It is also apparent that mass extinctions are well correlated with widespread unconformities in the geological record—lapses in the deposition of sediments. Newell (1967) has suggested that mass extinctions accompanied a general rising of the continents, resulting in the draining of vast inland seas and destruction of immense areas of habitat. If we consider that most fossils were deposited in shallow seas and that small changes in sea level would cause large changes in the area of such habitats, mass extinctions become less mysterious. We may still, however, be hard put to explain the disappearance of a primarily terrestrial group like the dinosaurs. In the final analysis,

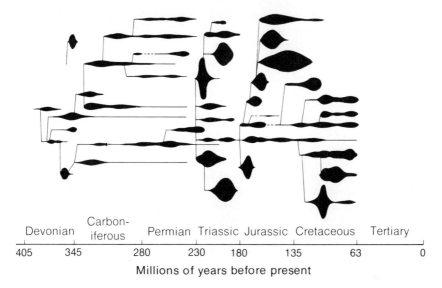

Figure 35-3   Diversification and phylogenetic relationships among superfamilies of ammonites. The group as a whole nearly went extinct at the end of the Permian and the end of the Triassic (from Newell 1967).

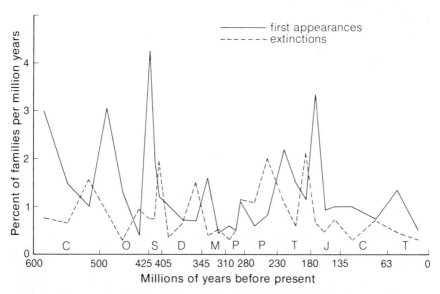

Figure 35-4   Rate of appearance and extinction of families in major groups of animals from the Cambrian (C) to the Tertiary (T). Values are expressed as a percentage of families extant during each time period (after Newell 1967).

we may never understand in detail the extinction of major taxonomic groups. Why were not some lines able to evolve rapidly enough to adapt to changing conditions? Was their extinction in any way hastened through competition with other groups or by the appearance of new, more efficient predators?

Rates of origination and extinction of taxa are approximately balanced.

Ignoring periods of mass extinction, all phylogenetic lines undergo continual replacement of taxa at all levels. A particular lineage has a characteristic rate of extinction and origination of genera, families, and so on. Although this rate may fluctuate, originations and extinctions generally keep pace so that, relative to the turnover of taxa, the net change in the number of taxa is relatively small (see Figure 26-8). Regardless of what factors fuel the evolutionary fires of a phylogenetic line, the parallel course of extinction and origination rates suggests competitive exclusion as a major cause of extinction. It is not difficult to imagine periods of rapid proliferation and diversification within a phylogenetic line producing branches that were competitively superior to older, unchanged types. But evidence of competitive exclusion is difficult to squeeze out of the fossil record.

Perhaps one of the best documented cases of extinction of a major taxonomic group is the gradual decline and final disappearance, during a 10 million-year period ending in the early Eocene, of the multituberculates, a subclass of mammals quite distinct from the marsupials and placentals. The multituberculates, primarily small herbivores, reached the peak of their diversity and numerical abundance in the early Paleocene, when they were the most abundant of mammal groups in North America and Europe. By the early Eocene they had completely disappeared, having been replaced by ecologically similar condylarths (primitive ungulates), rodents, and primates (Van Valen and Sloan 1966). It is almost universally accepted that the slow decline of the multituberculates can be attributed to competition from placental mammals (cf. Landry 1965). But if this were so, it is not clear why the multituberculates were inferior competitors. Their dentition was similar to contemporary condylarths and rodents. Of course, we know nothing about their digestive systems or reproductive behavior.

How did the multituberculates differ from the placental mammals that replaced them? Van Valen and Sloan (1966) infer several "inferior" traits from the skeletal morphology of the multituberculates. First, their hearing, particularly their pitch discrimination, may have been poor. Second, the articulation of the hip suggests that the multituberculates were capable of only limited arboreal activity. Third, the bones of the forearm did not permit sufficient rotation for effective molelike burrowing. If these characters add up to anything, it is a problem in escaping predators, not in utilizing food sources efficiently. Above all, this example shows the difficulty of attempting to understand the ecology of long gone organisms from a study of their bones.

The great American interchange illustrates the role of competition in extinction.

The history of the mammal fauna of South America reveals a series of replacements by major groups brought about at first by the evolution and diversification of new forms within South America, and later by the immigration of North American species by way of Central America (Simpson 1950). During much of the last 60 million years, South America has been isolated from other continents; Central America was underwater or appeared as a string of islands until about 5 million years ago, when a solid land bridge finally joined the two continents and began a great interchange of the faunas of North and South America. The early mammalian history of South America tells us of the diversification of notoungulates and litopterns (primitive herbivores), marsupials, and many edentates (armadillos, sloths, and anteaters). Primates, primitive rodents, and raccoonlike forms apparently crossed the water barrier from North America and diversified well before the modern land bridge formed.

Many changes occurred in the mammal fauna of South America during its extended isolation (Figure 35-5). Primitive ungulate herbivores, mostly notoungulates and litopterns, underwent a rapid diversification and reached their peak during the Eocene, 40 million to 50 million years ago. Some primitive ungulates and native marsupials evolved to fill the ecological roles of present-day rodents. Although this branch of the marsupial line did not persist, marsupials comprised most of the South American carnivorous species until the immigration of North American dogs, bears, cats, and weasels. Note that on Australia, the mammal fauna is dominated by marsupials. Few other types of mammals have ever existed there (other than bats), and Australia is still sufficiently isolated to prevent the immigration of potentially superior placental mammals.

The rodentlike ungulates of South America were gradually pushed aside by the diversification of native South American rodents, represented today by capybaras, agoutis, chinchillas, and their relatives. The success of these rodents was paralleled by a decline in the larger primitive ungulates, perhaps through competitive exclusion.

The Central American land bridge brought hordes of new immigrants from the north and caused the extinction of many South American ecological counterparts. The primitive ungulates were pushed aside by invading deer, tapirs, and camels. (Many of these are now extinct in North America. The camel family is represented in South America by llamas, vicuña, and alpaca.) Many of the early native rodents of South America vanished with the appearnace of squirrels, mice, and rabbits, but several unique forms (guinea pigs, capybaras) survived the onslaught. The marsupial carnivores were replaced by placental counterparts (dogs, cats, bears).

The modern mammal fauna of South America is a nearly equal mixture of distinctly South American and North American forms, yet few southern mammals became established in North America. In fact, porcupines, the now extinct ground sloths, and the opossum are the only distinctly South Ameri-

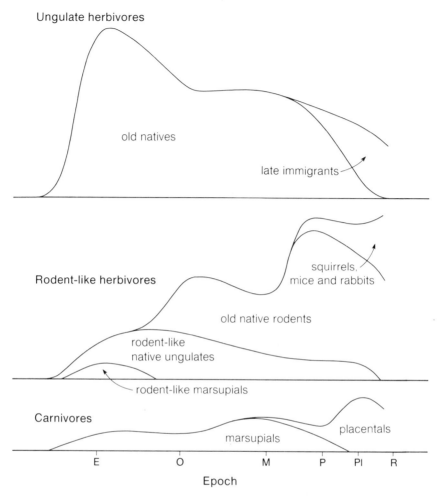

Figure 35-5   Diversification and reduction of several groups of mammals on the South American continent. Height of the graph in each case represents the relative number of genera. Time scale is only approximate: R = Recent, PI = Pleistocene (0 − 3 × 10⁶ years before present), P = Pliocene (3 − 13 × 10⁶ years), M = Miocene (13 − 25 × 10⁶ years), O = Oligocene (25 − 36 × 10⁶ years), E = Eocene (36 − 58 × 10⁶ years). Recent exchange of species between North and South America began in the Pliocene (after Simpson 1950).

can groups that have invaded the north temperate zone. Other southern forms, like armadillos, sloths, and some of the native rodents, have extended their ranges into Central America, but have not yet moved beyond the tropical zone. The failure of South American forms to colonize North America is all the more amazing in that South American mammals began to appear in North American fossil assemblages (mostly from Florida, Texas, and California) before North American forms appeared in fossil deposits in Patagonia (Webb 1976). Yet the fact remains that most of the immigrants from South America either died out or their evolution was bogged down in North America, whereas many immigrants from North America underwent extensive adaptive radiation in South America during the last 2 million years.

It seems odd that the native fauna of South America, adapted to the particular environmental conditions on that continent, should have been overwhelmed by invaders from another land. It is true that South American forms had not evolved in competition with northern forms. By the end of the Miocene period, 13 million years ago, the native fauna had partitioned most of the possible ecological roles in the community, and the major radiations of new forms were largely over. In contrast, dispersal between Asia and North America occurred regularly by way of Alaska and the Bering land bridge. The mammalian fauna of North America has continually been exposed to immigrants from other areas, challenging its competitive ability. We may be tempted to apply terms like "evolutionary vigor" to the mammals of North America and "evolutionary stagnation" to the mammals of South America, but evolutionary biologists do not yet understand how the inherent characteristics of whole groups determine competitive ability. Why, for example, did *all* the litopterns die out? What common bonds of structure and function united these species in a fatal evolutionary pact?

Evolution is basically conservative.

Once the major radiation of a group has taken place, further change in body plan or way of life evidently becomes difficult. As we have seen, most of the characteristics of present-day lungfish appeared during the first 50 million years of lungfish evolution. The line has changed relatively little in the last 250 million years. Genera of marine invertebrates and plants are recognized unchanged through hundreds of millions of years of the fossil record. These evolutionary lines have undoubtedly been buffeted by changing environments—shifts in temperature and moisture, the coming and going of predators and disease organisms—but they have retained their distinctive characteristics with a minimum of modification.

Evolutionary conservatism is well demonstrated by the floras of the southeastern regions of Asia and North America. During the late Tertiary Period, 70 million years ago, vast areas of the Northern Hemisphere were covered with broad-leaved forests. The remnants of these great forests—including primitive species like the rhododendron, tulip tree, magnolia, sweetgum, and hickory—are now localized in the southeastern United States and eastern Asia, where they have been isolated for more than 50 million years (Li 1952). Yet these plants still betray their common origin in the Tertiary flora, and although many have been given different species names, most of the genera remain unchanged. Moreover, genera with southern distributions in Asia tend to have southern distributions in North America; those with northern distributions in Asia tend to have northern distributions in North America; genera that are widespread in Asia also tend to be widespread in North America. Thus not only has form remained unchanged, but each genus has also retained a discernible portion of those physiological tolerances that determine geographical range—for more than 50 million years.

Extinction is caused by a failure to evolve in response to changing conditions.

Evolutionary conservatism in the face of environmental change opens the door to extinction. Changes in climate, competitors, predators, and disease organisms all pose serious threats. Many changes come too quickly and too abruptly for evolution to respond. Certainly many of the results of human activities fall into this category. Natural changes in the environment take a more deliberate toll.

Climatic change by itself probably never drove a species to extinction. The physical environment sets the stage for competitive struggle between populations and higher taxonomic groups. The rate of extinction of species within a particular phylogenetic line, which is related to the probability of extinction of a particular population, may depend upon the rate at which the environment changes, on one hand, and the potential rate of evolutionary change relative to that of competitors and predators, on the other. Hypotheses concerning the causes of extinction usually spring from premises of environmental change or evolutionary stagnation. And hypotheses usually distinguish mass extinctions from the more usual replacement within a phylogenetic group, and rarely attempt to encompass both.

Mass extinctions remain something of an enigma, mostly because they occur in the absence of replacement by competitors and often involve several unrelated groups. Some hypotheses invoke extraterrestrial events, such as supernovae or solar flares showering killer doses of radiation on the Earth's surface. But such events should bring more general havoc than the fossil record shows. Why should some, or most groups remain untouched, while others are disappearing? In addition, the fossil record indicates that mass extinctions were not sudden events but may have been drawn out over millions of years. The correspondence between mass extinctions and major worldwide geological changes suggests that mass extinctions were brought about by widespread change in climate or configuration of the continents and oceans that virtually eliminated certain types of habitats and the ways of life associated with them. Lipps and Mitchell (1976) have related the origination and proliferation of whales in the Eocene, followed by their near extinction in the Oligocene and subsequent reradiation in the Miocene to the availability of concentrated food resources in coastal waters. Periods of adaptive radiations in marine mammals correspond to extensive deposition of rocks formed from the silicon skeletons of diatoms, a kind of alga that supports the marine food chain. These deposits suggest periods of high marine productivity, perhaps related by some unclear steps to wind patterns and upwelling currents, and increased availability of the marine crustacea and fish upon which present-day whales and porpoises feed. That dinosaurs, ammonites, or trilobites might have died out owing to the disappearance of a critical food source seems unlikely, but perhaps only because of our ignorance of the ecology of the Earth tens and hundreds of millions of years ago.

For the study of modern ecology, extinctions involved in the replacement of species by competitors are of greater interest, for these processes play a role in shaping the structure of modern ecological communities. And the forces that lead to extinction ought to be directly observable in contemporary ecological interactions. It probably would suffice to say that we do not know why one species becomes extinct and another persists. Take two individuals, one from a population that is restricted in habitat and geographical range and contains, perhaps, only a few hundred individuals, and the other from an abundant, widespread species. One of these clearly is nearer the brink of extinction than the other, yet I would defy any biologist not familiar with the particular species, to choose with reason and certainty which individual was taken from the fading species.

Hypotheses about extinction usually are difficult to test adequately, and for each hypothesis proposed there are numerous counterexamples. Many have argued that the probability of extinction is directly related to the degree of fluctuation in the environment. But if one is willing to accept the premise that the environments of molluscs that burrow in the bottom sediments are more constant than those of molluscs that live on the surfaces of sediments, then the evidence is contrary (Thayer 1974). Rates of extinction are about the same in both groups. Bretsky and Lorenz (1970) proposed that under constant environmental conditions, selection would tend to make most genes homozygous and reduce the population's evolutionary potential. Such populations would be at greater risk of extinction should the environment suddenly change. Although there is ample evidence that species replacement rates are higher in constant environments than in fluctuating and unpredictable environments, there is no evidence that populations in constant environments are genetically impoverished (Schopf and Gooch 1972, Valentine and Ayala 1974). We should not forget, however, that ecologists cannot agree on adequate criteria for environmental constancy and genetic variability, so it is certain that these hypotheses are not close to being resolved.

Another hypothesis suggests that specialized organisms have greater probability of extinction than generalized organisms. Here again, we become bogged down trying to define "specialization." Flessa et al. (1975) have suggested that, for aquatic, free-living arthropods (trilobites, crabs, and their relatives), morphological complexity measured by the diversity of limbs (claws, swimming appendages, walking legs, etc.) on the arthropod body is a sufficient index to specialization. Accepting this definition, it turns out that specialized genera are as long-lived as generalized taxa. If we are skeptical of this index as a measure of specialization, as I am, then we are left without a conclusion.

Van Valen (1973) has followed a more general and abstract approach to the problem of extinction. He examined the survivorship curves of genera and families in dozens of taxonomic groups and discovered what appeared to be a constant probability of extinction for each age interval, just as a non-senescing organism has a constant probability of death in each age interval.

For Van Valen, this constant probability of extinction suggested that extinction is caused mainly by environmental change, not inherent evolutionary limitations, and that such changes are distributed among the environments of different organisms at random. Van Valen's hypothesis has come under considerable criticism, both for his methods of constructing taxonomic survivorship curves and for his interpretation of their pattern (Foin *et al.* 1975, Raup 1975, Salthe 1975).

As you can see, biologists have found little agreement with respect to extinction, other than that it surely has occurred in the past and, by extension, that it must be an important on-going process in the modern world. We understand so little about extinction because its time scale is too long for our own experience and too short for detailed preservation in the fossil record. The changes in population parameters that lead to extinction undoubtedly are minute on the scale of processes that influence the short-term dynamics of populations with which we are more familiar.

At this point, we shall leave extinction behind and, with it, other topics in population biology, and turn our attention to the structure and dynamics of biological communities. But just as population characteristics express the adaptations of individuals, we should not forget that community characteristics express the interrelationships between populations.

Part 9

Ecology of Communities

36

Community Ecology

In previous chapters, we have examined population processes, including the interactions between competitors, predators, and prey. The theory of population ecology and experimental tests of that theory are based on the interactions of a few species, frequently only two species. Yet, in their natural habitats, most species interact with a wide variety of others, numbering, perhaps, in the hundreds or even thousands. Many of these relationships are casual and individually exert little influence on the population. Others, such as competition with closely related species and avoidance of efficient predators and parasites, determine the carrying capacity, stability, and evolutionary course of the population. All these interactions together either permit or preclude the existence of a population at a particular locality. Summed over all species, these interactions determine the number and interrelationships of the inhabitants of a region. In this and the following three chapters, we shall study the community as a natural unit and trace the geographical, developmental, and evolutionary bases of its organization.

The community is an association of interacting populations.

The term *community* has been given such a variety of meanings by ecologists that it borders on being meaningless. It usually is restricted to the description of a group of populations that occur together, but there ends any similarity among definitions. Throughout the development of ecology as a science, the term has often been tacked on to associations of plants and animals that are spatially delimited and that are dominated by one or more prominent species or by a physical characteristic. One speaks of an oak community, a sagebrush community, and a pond community, meaning all the plants and animals found in the particular place dominated by its namesake. Used in this way, "community" is unambiguous: it is spatially defined and includes all the populations within its boundaries.

Ecologists also define communities on the basis of interactions among associated populations. This is a functional rather than descriptive use of the term. Here, the term *association* will be reserved for groups of populations that occur in the same area without regard for their interactions; community will be used to denote an association of interacting populations. We may find communities difficult to delimit because interactions among populations extend beyond arbitrary spatial boundaries. Some of the oxygen molecules that a squirrel inhales in New York might have been expired last month by a tree in the Amazon basin. Clearly, the Brazilian tree exerts a negligible influence on the squirrel, and we ought not to extend the boundaries of the community to which the squirrel belongs to the tropical rain forest. But biological interactions reach out in a complicated fashion, making it difficult to set limits—other than completely arbitrary ones—to a community.

Organisms and materials move slowly across the boundaries of most terrestrial associations of plants and animals compared to the turnover of organisms and flux of materials and energy within the association. "Community" and "association" are, therefore, nearly identical for many terrestrial habitats, so long as the boundaries of the habitats are not too narrowly set. Because most fresh-water and coastal aquatic communities receive inputs from the land, and because many organisms move freely between aquatic habitats, the boundaries of aquatic communities are more difficult to establish. Salamanders, which are aquatic as larvae and terrestrial as adults, link together the stream and forest communities; migratory birds carry a thread of interaction between many temperate and tropical communities.

Two extreme cases of associations, in which all the energy inputs come from outside, are the faunas of caves and of unlighted depths of the oceans. Strictly speaking, the inhabitants of these perpetually dark places depend upon primary producers in sunlit habitats for their source of energy, yet we may speak of competition among detritivore or carnivore members of the cave "community," if we wish. In fact, "community" may be usefully applied to any group of interacting populations—the herbivorous insect community, the oak leaf community, the kelp holdfast community—as long as community boundaries are clearly delimited.

The maintenance of community structure and functioning depends on a complex array of interactions, directly or indirectly tying all its members together in an intricate web. The influence of a population extends to ecologically distant parts of the community through its competitors, predators, and prey. Insectivorous birds do not eat trees, but they do prey on many of the insects that feed on foliage or pollinate flowers. By preying upon pollinators, birds indirectly affect the number of fruits produced, the amount of food available to animals that feed upon fruits and seedlings, the predators and parasites of those animals, and so on. The ecological and evolutionary impact of a population extends in all directions through its influence on predators, competitors, and prey, but this influence is dissipated as it passes through each successive link in the chain of interaction.

Is the local community a natural unit of organization?

We may describe the functional relationships among an association of species just as the physiologist relates the various parts of the body. The obvious analogy between community and organism led many early ecologists, notably the influential plant ecologist F. E. Clements (1916, 1936), to describe communities as discrete units with sharp boundaries. This view is reinforced by the conspicuousness of many dominant vegetation types. A forest of ponderosa pines, for example, appears distinct from the fir forests that occupy moister habitats and from the shrubby vegetation and grassland found on drier sites. The boundaries between these community types are often so sharp as to be crossed within a few yards along a gradient of climate conditions. Some community boundaries, such as that between deciduous forest and prairie in the midwestern United States are respected by most species and spanned by relatively few.

An opposite view of community organization was held by H. A. Gleason (1926, 1939), who suggested that the community, far from being a distinct unit like an organism, was merely a fortuitous association of organisms whose adaptations enabled them to live together under the particular physical and biological conditions found at a particular location.

Clements' and Gleason's concepts of community organization predict different patterns in the distribution of species over ecological and geographical gradients. On one hand, Clements believed that the species belonging to a community were closely associated with each other; the ecological limits of distribution of each species coincided with the distribution of the community as a whole. This type of community organization is commonly called a *closed community*. On the other hand, Gleason believed that each species was distributed independently of others that co-occurred in a particular association, an organization referred to as an *open community*. We would draw the boundaries of an open community arbitrarily with respect to the geographical and ecological distributions of its component species, which may extend their ranges independently into other associations of species.

The structure of closed and open communities is depicted schematically in Figure 36-1. In the upper diagram, the distributions of species in each community are closely associated along a gradient of environmental conditions—for example, from dry to moist. Closed communities represent natural ecological units with distinct boundaries. The edges of the community, called *ecotones*, are points of rapid replacement of species along the gradient. In the lower diagram, species are distributed at random with respect to each other. We may arbitrarily delimit an open community at some point, perhaps a dry forest community near the left-hand end of the moisture gradient, but some of the species included might be more characteristic of drier points along the gradient, while others reach their greatest productivity in wetter sites.

The separate concepts of open and closed communities both apply to associations of species in nature. We observe distinct ecotones between associations under two different kinds of circumstances: when there is an abrupt change in the physical environment—for example, at the transition between aquatic and terrestrial communities, between distinct soil types, or between north-facing and south-facing slopes of mountains—or when one species or life form so dominates the environment of the community that the edge of its range signals the distributional limits of many other species.

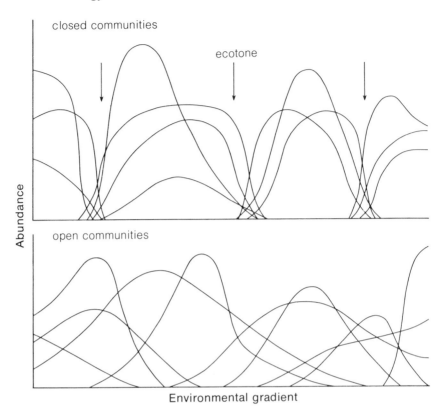

Figure 36-1   Hypothetical distributions of species organized into distinct associations (closed communities, above) or distributed at random along a gradient of environmental conditions (open communities, below). The species found at any point, or within any narrow region, along the lower gradient would be identified as an open community. Ecotones between communities in upper figure are indicated by arrows.

The transition between broad-leaved and coniferous forest is usually accompanied by an abrupt change in soil acidity. At the boundary between grassland and shrubland, or between grassland and forest, sharp changes in surface temperature, soil moisture, and light intensity result in many species replacements. Sharp grass-shrub boundaries occur because when one or the other vegetation type holds a slight competitive edge, it dominates the community (Schultz *et al.* 1955). Grasses prevent shrub seedling growth by reducing the moisture content of the surface layers of the soil; shrubs prevent the growth of grass seedlings by shading them out. Fire evidently maintained a sharp boundary between prairie and forest in the midwestern United States (Borchert 1950). Perennial grasses resist fire damage that kills tree seedlings outright, but fires do not penetrate deeply into the moister forest habitats.

Sharp physical boundaries create sharp ecotones. Such boundaries occur at the interface between most terrestrial and aquatic (especially

Figure 36-2   A sharp community boundary (ecotone) associated with an abrupt change in the physical properties of adjacent habitats. Seaweeds extend only to the high tide mark. Between the high tide mark and the spruce forest, waves wash the soil from the rocks and salt spray kills pioneering land plants, leaving the area devoid of vegetation.

marine) communities (Figure 36-2) and where underlying geological formations cause the mineral content of soil to change abruptly. The ecotone between plant associations on serpentine-derived soils and nonserpentine soils in southwestern Oregon is shown in more detail by the diagrams of soil minerals and occurrence of plant species in Figure 36-3. Levels of nickel, chromium, iron, and magnesium increase abruptly across the boundary into serpentine soils; copper and calcium contents of the soil drop off. The edge of the serpentine soil marks the boundaries of many species that are either excluded from, or restricted to, serpentine outcrops. A few species are found only within the narrow zone of transition, and others, seemingly unresponsive to variation in soil minerals, extend across the ecotone.

Plant ecologists have long recognized the influence of climate on plant associations. Shreve (1936) described the chaparral-desert transition in Baja California in relation to moisture and freezing temperatures. Desert species, particularly the cacti, do not tolerate prolonged frost and quickly drop out north of the frost line; chaparral species drop out to the south within the transition zone owing to water stress.

Turnage and Hinkley (1938) similarly related the northern edge of thorn forest communities to the occurrence of frost and the northern edge of Sonoran desert vegetation to prolonged freezes of a day or more.

Gradient analysis has revealed the open nature of plant communities.

Sharp physical boundaries often create abrupt changes in vegetation. It is difficult for species to be fence sitters in such situations; they must adapt to the conditions on one side or the other. The few species specialized to live in

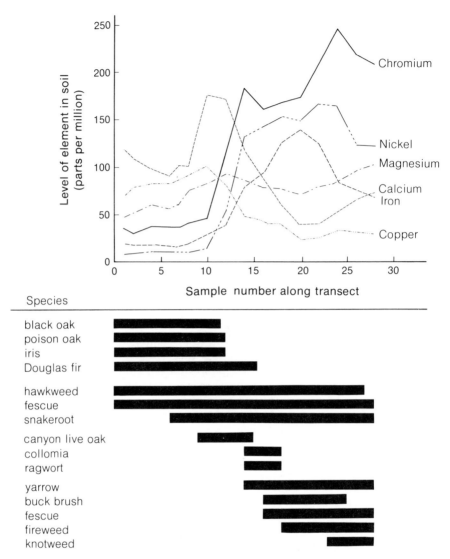

Figure 36-3   Changes in soil mineral content (above) and plant species (below) across the boundary between nonserpentine (left) and serpentine soils (right) in southwestern Oregon. The transect diagrammed here is somewhat atypical in that magnesium does not increase as abruptly as usual across the serpentine ecotone (after White 1971).

the ecotone—often referred to as *edge species*—are necessarily restricted in numbers by the scarcity of their habitat.

Most large-scale ecological changes are more gradual than the abrupt changes between land and water, forest and field, or geological formations. The major biological communities of the Earth occupy a *continuum* of gradually changing ecological conditions, varying along one dimension from cold to warm, along another from dry to moist, along a third from seasonal to moderate, and perhaps along many other minor gradients of ecological conditions. Changes along these *gradients* occur gradually over vast geographical distances; sharp physical boundaries do not intercept species ranges.

When we examine the occurrence of species along gradients of moisture or temperature, we find that the local plant community must be viewed as an open system (Whittaker 1967, Shimwell 1971). Groups of species generally are not restricted to particular associations, rather each species is distributed along environmental gradients almost without regard to the occurrence of others. The environments of the eastern United States form a continuum with a north-south temperature gradient and an east-west rainfall gradient. Species of trees found in any one region, for example those native to eastern Kentucky, have different geographical ranges, suggesting a variety of evolutionary backgrounds (Figure 36-4). Some species reach their northern limits

Figure 36-4   Geographical distribution of twelve species of trees found in plant associations in eastern Kentucky (after Fowells 1965).

in Kentucky, some their southern limits. Few species have broadly overlapping geographical ranges, hence associations of plant species found in eastern Kentucky do not represent closed communities. Each species has a unique evolutionary history, with a variable degree of association with other species in the local community.

A more detailed view of Kentucky forests would reveal that many of the tree species are segregated along local gradients of conditions. Some are found along the ridge tops, others along moist river bottoms, some on poorly developed rocky soils, others on rich organic soils. The species represented in each of these more narrowly defined associations would exhibit correspondingly closer ecological distributions, but the open community concept would still dominate our thinking about these associations.

Cornell University ecologist Robert Whittaker (1967) has examined plant distributions in several mountain ranges where moisture and temperature vary over short distances according to elevation, slope, and exposure. When Whittaker plotted the abundance of each species at sites at the same elevation distributed along a continuum of soil moisture, he found that the species occupied unique ranges with peaks of abundance scattered along the environmental gradient (Figure 36-5). The Oregon mountains have fewer species overall, but each species has a wider ecological distribution, on the average, than each Arizona species.

In the Great Smoky Mountains of Tennessee, dominant species of trees are widely distributed outside the plant associations that bear their names

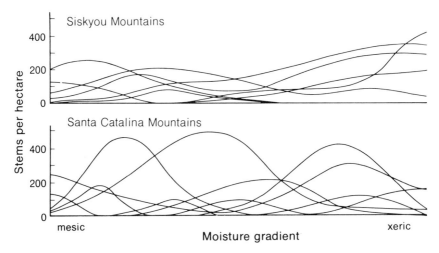

Figure 36-5   Distribution of species along moisture gradients at 460- to 470-meter elevation in the Siskyou Mountains of Oregon and at 1830- to 2140-meter elevation in the Santa Catalina Mountains of southern Arizona. Species in the more diverse Arizona flora occupy narrower ecological ranges; thus, in spite of the greater total number of species in the flora of the Santa Catalina Mountains, the Santa Catalina Mountains and Siskyou Mountains have similar number of species at each sampling locality (after Whittaker 1960, Whittaker and Niering 1965).

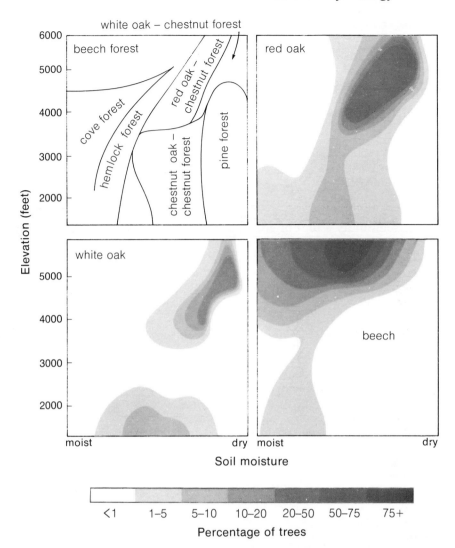

Figure 36-6   Distribution of red oak, white oak, and beech with respect to altitude and soil moisture in the Great Smoky Mountains of Tennessee. The approximate boundaries of the major forest associations are shown in the diagram at the upper left. Relative abundance, represented by degree of shading, corresponds to percentage of tree stems more than one centimeter in diameter in samples of approximately 1,000 stems (after Whittaker 1956).

(Figure 36-6). For example, red oak is most abundant in relatively dry sites at high elevation, but its distribution extends into forests dominated by beech, white oak, chestnut, and even hemlock, an evergreen coniferous species, and extends throughout the entire range of elevation in the Smoky Mountains. Beech prefers moister situations than red oak, and white oak reaches its greatest abundance in drier situations, but all three species occur together in many areas. The distributions of insect species in the same area were also independent of each other (Whittaker 1952). Similar analyses of the distribution of species of birds along an elevation gradient in Peru failed to reveal evidence of distinct ecotones between associations of species (Terborgh 1971). Even Shreve (1936), who emphasized the distinct boundary between desert and chaparral plant associations, recognized that "as is true of the meeting ground between any two great plant formations, the dominant plants of each formation are found to vary in the distance to which they extend into the other. This indicates that their habitat requirements are not so nearly identical as their close association in the midst of their respective formations would suggest."

Is the community a unit of adaptation?

The community is an association of interacting individuals. Community function is the sum of what individuals do and thus reflects the adaptations of individuals. We may ask, however, whether attributes of a community represent more than the evolved properties of individuals. Is the community itself a unit of adaptation having properties that can be interpreted only in terms of community function? Because adaptations are evolved through natural selection to improve the fitness of the individual carrying the selected trait, we would not expect to find adaptations enhancing community function that were inconsistent with this goal. What is good for the individual prevails, regardless of whether it is good for the community. We have seen, however, that evolutionary adjustments of prey to their predators, and of competitors to each other, tend to stabilize their relationships. As a result, these adaptations improve the efficiency of ecosystem function and enhance community stability.

We may now set down, at least tentatively, a basic ecological principle: Community efficiency and stability increase in direct proportion to the degree of evolutionary adjustment between associated populations. The action of this principle is shown quite clearly when foreign species are introduced to a community. In most cases they cannot successfully invade the community and so die out, but occasionally exotic species gain a foothold and rapidly come to dominate the community. Such species can upset the delicate balances achieved between members of the community and, in so doing, disrupt community function. Outbreaks of introduced pests like the European pine sawfly and gypsy moth can, in fact, almost destroy a community by defoliating its major primary producers.

The evolutionary adjustment of populations to one another depends upon their degree of association. If two species always occurred together, their interaction would exert an important influence on the evolution of each. If, however, the ecological and geographical ranges of two species were mostly nonoverlapping, each species would exert only a small portion of all the selective influences on the population of the other. A community cannot exhibit strong co-evolutionary adjustment among its members if the adaptations of its species are molded primarily by relationships in other communities.

The actual degree of association between species in a community lies somewhere between two extremes: those of obligate association, as we find among many pairs of mutually interdependent organisms (see page 226), and independence, each species being distributed randomly with respect to others. Both can be identified in natural systems, but the concensus of ecologists is that communities are open associations without clear boundaries.

The fossil record lends further support to the concept of the open community. Pollen grains deposited in lakes and bogs left by retreating glaciers in the northeastern United States record the coming and going of plant species. The composition of plant associations in the past has changed by extension and integration of new species. For example, Wright (1964) showed that the sequence of reforestation in the northeastern United States following the last major glaciation contained intermediate forest associations that are not found in the area today. The general pattern of reforestation began with spruce forest, which dominated the area until about 10,000 years ago, followed by extensive associations of pine and birch, which were later replaced by more temperate elm and oak forests.

The sequence of forest types along a transect from southern Illinois into northern Michigan differed from the normal pattern. In the southernmost area, now covered with elm and oak forest, the transition of forest associations proceeded from spruce to broad-leaved deciduous species with no intermediate stage of pine. In the northernmost area, the pollen record indicates that spruce forest yielded directly to the pine forest that covers the area at present. Wright suggested that pines migrated slowly following the recession of glaciers, and because they were restricted to refuges in the Appalachians during the height of the glaciation, they did not reach the Michigan area until after the spruce forest had deteriorated and was replaced by an elm-oak association. Farther to the west in Iowa, however, pine occurs in conjunction with spruce and fir in the earliest (more than 10,000 years old) spruce-fir forests (Durkee 1971). But there, too, pine never became a dominant component of the forest association before the conifers were replaced by oak and elm. The broad-leaved forests in the region began to disappear about 7,000 years ago with the development of prairie.

In spite of their general independence, species do adapt to the presence of other species in their environment. One would expect the degree of adaptation to be directly proportional to the degree of association between two species. Southwood (1961) provided evidence for this point by examining the association between species of insects and trees in England. He found that the more abundant a species of tree in the recent history of England's forests, measured by the number of records in fossil pollen samples, the greater the number of species of insects that can be collected from it. Species that are poorly represented in the British flora—mountain ash, hornbeam, maple—are such minor components of the environment that few insects have evolved to exploit them compared with the numbers of insects that feed on the more abundant willow, oak, and birch. Furthermore, most of the recently introduced species of trees in England have very few insect species associated with them. Similar patterns appear in the insects collected from trees in the Hawaiian Islands (Southwood 1960), leaf-mining insects collected from oaks in California (Opler 1974), and species of mites known to parasitize rodents in North America (Dritschilo *et al.* 1975). Janzen (1968, 1973) and Opler (1974) have emphasized, however, that host species may be treated as islands or patches of habitat and that the number of species associated with them is the result of a balance between colonization and extinction. Regardless of adaptations to particular host species, extinctions of herbivores and parasites should be more frequent on the least common hosts, which would therefore support the fewest species. Only further research will tell us the extent of co-evolution in these systems.

Does evolutionary history influence community structure and function?

The distribution of life forms over the surface of the Earth is by no means uniform. Some regions lack groups that are abundantly represented elsewhere. Many irregularities in distribution patterns are linked to major climatic patterns: for example, snakes and lizards cannot tolerate the cold of arctic environments. Historical accidents of distribution, caused by geographical barriers to dispersal, have also played an important role in the distribution of major groups of animals and plants. Anomalies of distribution are most obvious on islands. Australia lacks most groups of mammals except for marsupials and bats, which are highly diversified there. Few species of any kind reach small remote islands, so community structure is relatively simple compared with mainland associations.

Distributional problems are not limited to islands. The major continental land masses are sufficiently isolated to reduce the exchange of forms evolved in each area (see page 408). The terrestrial environments of the Western and Eastern Hemispheres have been joined only sporadically by a land bridge between Alaska and Siberia during the last several million years. Many groups that have evolved and radiated to prominence in one hemisphere are absent from the other. Plants, insects, and other small invertebrates disperse easily across water barriers, and few of the major groups are missing from any continent. But terrestrial vertebrates, including birds, exhibit conspicuous differences between the faunas of the continents. The Iguanidae are the most conspicuous family of lizards in the Western Hemisphere, but they are replaced by agamid lizards in most of the Old World. Many ecological types of birds are represented by unrelated families in different areas. For example, the hummingbirds (*Trochilidae*) of North and South America, which feed primarily on nectar and tiny insects, have ecological counterparts in the sunbirds (*Nectariniidae*) of Africa and Asia (Figure 36-7). Sunbirds resemble hummingbirds in their food habits, small size, bright metallic-colored plumage, strong sexual dimorphism, and mating systems. Similarly, toucans (*Ramphastidae*), which are relatives of woodpeckers and are found in Central and South America, are replaced ecologically in Africa by the hornbills

Figure 36-7   Unrelated birds from South America (left) and Africa (right), illustrating convergence of form and habits. Above, keel-billed toucan and great hornbill; below, crimson topaz hummingbird and golden-winged sunbird (after Austin 1961).

(*Bucerotidae*), which are related to the kingfishers. Hornbills not only have greatly enlarged beaks like toucans, but both are gregarious and nest in holes. The adoption of similar feeding habits by diverse families of birds causes convergence of their morphology and behavior as well.

Convergence tends to obliterate the evidence of historical accidents in the composition of a local biota. Where woodpeckers are missing from a fauna, other species may adapt to fill their role (page 411). Rain forests in Africa and South America are inhabited by plants and animals with different evolutionary origins but having remarkably similar adaptations (Keast 1972, Bourlière 1973). The rodents of North and South American deserts are more similar in morphological characteristics than one would expect from their different phylogenetic origins (Mares 1976). Many similarities have been noted in the behavior and ecology of Australian and North American lizards, in spite of the fact that they belong to different families and have been separate for, perhaps, 100 million years (Pianka 1971b). Mediterranean-climate plants in Chile and California have remarkably similar morphology and physiology (Mooney and Dunn 1970a).

Wherever one looks one finds convergence, and this reinforces our belief that community organization depends on local conditions of the environment more than it does on the evolutionary origins of the species that comprise the community. In many instances, species-for-species matchings have been made (Cody 1974, Fuentes 1976, Mares 1976), suggesting that environments may closely specify the particular characteristics of species that inhabit them and that these specifications depend only on climate and other physical factors; the plants and animals in the association have little additional modifying influence on each other by reason of their different phylogenetic backgrounds. This view is, however, much too simple. Cases of species-for-species matching do not stand up under close scrutiny (*e.g.,* Ricklefs and Travis, see p. 749). In fact, detailed studies of convergence are as likely to turn up remarkable differences between the plants and animals in superficially similar environments. Mares (1976) noted that the ancient Monte Desert of South America is the only desert region of the world lacking bipedal, seed-eating, water-independent rodents like the kangaroo rats of North America and gerbils of Asia. Among frogs and toads, however, several South American forms have carried adaptation to desert environments a step further than their North American counterparts: They construct a foam nest in which their eggs are kept from drying up (Blair 1975). Differences between the Australian agamid lizard *Amphibolurus inermis* (see page 163) and its North American iguanid analogue *Dipsosaurus dorsalis* include diet, optimum temperature for activity (see page 166), burrowing behavior, and annual cycle, even though the species appear to be dead ringers for each other at first glance (Pianka 1971a). Co-evolved relationships between species also may reveal the unique biogeographic position of each region. Unfortunately, little information has been gathered to compare such features of community organization, but one example may be drawn from the dispersal of seeds by ants. This interrelationship—a type of mutualism—is encouraged by the presence of edible appendages, called elaiosomes, on the seeds (Figure 36-8). Ants gather these, with the seeds attached, and carry them into underground

Figure 36-8   Ant-dispersed seeds of two Australian plants, *Kennedia rubicunda* (left) and *Beyeria viscosa* (right), showing the edible, light-colored appendage (elaiosome) that attracts ants (after Berg 1975).

nests, whereby they effectively plant the seeds. This seed trait is not common in most of the world and is usually restricted to trees in mesic environments. In Australia, however, the trait is exceedingly common among xerophytic shrubs, and it is associated with ecological and morphological features that are lacking in ant-dispersed plants elsewhere (Berg 1975). The cause of these particular traits in Australian plants is an unresolved subject of speculation, but we nonetheless should be cautioned about hastily drawing ecological parallels between regions. Whether the peculiarities of seed dispersal among Australian plants reflect a different environment or the particular evolutionary history of Australian plants and animals cannot be determined at this point, but the traits clearly merit a closer look.

As we examine community structure in more detail, we shall assume that community attributes are more the product of environment than the product of evolutionary history. This is conventional ecological wisdom. Still, it is better to be somewhat skeptical of the generality of community attributes described in studies of limited geographical scope.

What are the major attributes of the community?

The simplest community attributes to measure are the number and relative abundances of species. The regular arrangement of species in communities was first brought to the attention of ecologists by Raunkiaer (1934, see also Kenoyer 1927, Gleason 1929, Rommell 1930), who pointed out that some species were "dominant" and comprised a large percentage of samples, while other "subdominant" species were poorly represented. The relative abundances of species in a community often take a characteristic pattern, although the relative position of a species on a scale of abundance could vary from one community to the next. Arrhenius (1921) recognized that the number of species varied in accordance with the area of the sample.

Subsequent studies have shown that species abundance and species area curves are statistical attributes of a species association and are sensitive to sampling procedures. As such, they provide relatively little insight into the processes of community organization and regulation. Nonetheless, these patterns did stimulate ecologists to investigate the diversity and arrangement of biological associations and eventually led to studies of community dynamics.

Ecologists also recognize that the species in a community are organized in a trophic hierarchy of feeders and fed upon, and that energy moves along a one-way path up the food chain (Lindeman 1942). Plants lie at the base of all energy relationships in the community because they can assimilate the energy of sunlight. At successive trophic levels above plants are herbivores, first-level carnivores, second-level carnivores, and so on. The trophic structure of the community has many interconnections linking species into a complex food web; predators eat many different kinds of food items, and they usually compete with other predators for each of these items.

The concept of trophic structure has been taken in diverging directions by modern ecologists. Some ecologists scrutinize the structure of the web itself. What factors are responsible for the diversity of competing species on each level of the trophic structure? How is the complexity of the food web determined? What regulates the structure of the community, and what determines its stability? Other ecologists have been interested primarily in the rate and efficiency of energy and nutrient transfer through the community. How many trophic levels can be supported in a community? What is the relationship between productivity and storage of energy as biomass in the community, and how does this affect stability?

Community function also has an important space-time component. The seasonal and daily cycles of activity within the community loosely distinguish smaller groups of species that are partly independent (*e.g.,* Bider 1962). The Spanish ecologist Ramon Margalef (1968) has made a more subtle point: The spatial and temporal structure of the community is determined largely by the time and distance that an organism can move before it is captured and eaten. The community is made up of many such paths: Community structure and dynamics are the sum of the fates of the organisms in the community.

In the following chapter, we shall examine the structure of the community, concentrating on the number of species and their relative abundances as indexes to the organization of the community.

37

Community Organization

Because it is impossible to measure the interactions of each species with every other species in a community, ecologists have sought simpler indexes of community organization. The most widely studied of these is the number of species in the community. For the last two decades, ecologists have tried to understand why the number of species found in a particular habitat varies from place to place. Interest in this problem was stimulated by several essays of G. E. Hutchinson (1957b, 1959) and by Robert MacArthur's demonstration that the number of nesting birds in a habitat varied in direct proportion to the number of layers of vegetation—grass and herb, shrub, and tree (MacArthur and MacArthur 1961). Additional studies showed that more species live in a tropical habitat than in a temperate habitat with similar vegetation. These patterns were known to naturalists for more than a century, but it was not until the late 1950s that the principles of ecology and population biology were applied to understanding variation in the diversity of communities.

One may ask whether a study of the number of species can teach us anything about the organization—the inner workings—of the community. Many have argued that the number of species varies in direct proportion to the productivity of the habitat, which is influenced by temperature, sunlight, rainfall, and nutrients. If this were true, variation in number of species would seem to be a trivial problem. But this explanation, and others like it, are unsatisfactory on several counts. First, there are too many exceptions. For example, salt marshes are among the most productive habitats, yet they support relatively few species. Second, any complete theory of species number must explain how the number of species in a particular habitat is regulated; how the species are formed; where they come from; how interactions between species set an upper limit to their number. A simple correlation is not a sufficient explanation. Third, species number is too important an attribute of community structure to dismiss easily. It is possible that many species can exploit different kinds of resources more efficiently than a few species can, because each species may become specialized to fill a different ecological role. It is also likely that the environmental heterogeneity created by many species provides further opportunity for the diversification of life forms. That is, diversity tends to breed more diversity. Variation in the num-

685

ber of species brings into focus many of these related problems and has played a major role in the development of ecological thought.

In this chapter we shall examine the major patterns of variation in the number of species in a community and discuss various explanations of these patterns. Then we shall consider briefly another attribute of community organization: variation in the relative abundance of species within a community.

How do we measure the number of species?

The simplest answer to this question would seem to be, "Count them!" But because a census of species is not a complete description of a community, it may be misleading or difficult to interpret. First, species are not equally abundant. Are rare species to count the same as common species? Second, the number of species increases as one includes progressively more time, individuals, or habitats in the census. Measurements of diversity are, therefore, sensitive to the method of counting and the sampling area. Third, ecological variety does not necessarily correspond to the number of species. Many species of a similar type may have a different influence on the organization of a community from that of the same number of distinctive species.

Differences in abundance of species cause nightmarish sampling problems for the measurement of diversity. There is the purely statistical problem that rare species may be missed. If a second census is taken, some of the missing species may be recorded and others found previously will escape detection. In any event, the total number of species counted as well as the particular species included may vary from census to census. Yet, even if the populations of all species could be counted accurately, one would have to decide whether, in the functioning of the community, rare species were as important as common species.

One solution to the problem of describing diversity has been to use indexes of diversity that include relative abundance as well as the number of species. Several indexes are available (Simpson 1949, Margalef 1958, Pielou 1966a, Whittaker 1972), of which the most commonly used is one derived from information theory, referred to as the Shannon-Wiener index after its developers. The index (H) is calculated by the equation

$$H = -\sum_{i=1}^{N} p_i \log_e p_i$$

where p_i is the proportion of species i in a sample of N species. Although the Shannon-Wiener index varies directly with the number of species, rare species count relatively less than common ones. As it is formulated above, the index (H) is roughly proportional to the logarithm of the number of species. It may be more convenient, however, to rescale the diversity index so it is proportional to the number of species. We can do this by expressing H as an exponential function

$$\text{diversity} = e^H$$

The application of the diversity index may be seen in the following numerical example. Suppose a community were composed of four equally abundant species. Each would be proportion $p_i = 0.25$ of the total sample. The natural logarithm of 0.25 is -1.386, so $p_i \log_e p_i$ is $0.25 \times -1.386 = -0.347$. These terms are summed over four species to give $H = 1.386$ and $e^H = 4.00$. If another species, equally as common as the other four, were added to the community, p_i for each species would become 0.20; $\log_e p_i = -1.609$; $p_i \log_e p_i = -0.322$; $H = 1.609$; and $e^H = 5.00$. Suppose, however, that the fifth species were much less common than the other four such that p_5 were 0.04 and the other p_i's were 0.24. In this case, $H = 1.499$ and $e^H = 4.48$. It is clear that the fifth species counted less than the other four in the calculation of the diversity index. If the fifth species had amounted to only 0.1 per cent of the sample ($p_5 = 0.001$), H would have been 1.393 and e^H, 4.03. In this case, the fifth species is hardly noticed.

The relative merits of various diversity indexes and their applications have been argued back and forth (Sheldon 1969, Hurlbert 1971, Fager 1972, Peet 1975). In fact, the results of most studies are relatively insensitive to which index of diversity is applied, or to whether an index of any kind is used in place of a simple count of the species present. This indifference may be attributed to the fact that communities have characteristic patterns of relative abundance among their member species. As a result, each diversity index tends to bear a consistent relationship to all other indexes as well as to the number of species in the community. To be sure, diversity indexes, particularly when they are expressed in relation to the number of species, may provide a useful measure of the uniformity of species abundances (*equitability,* see Lloyd and Ghelardi 1964, Peet 1975). But, at present, studies of community organization are not sufficiently refined to make use of this application.

More serious problems in the measurement of diversity derive from various biological considerations, such as the size of the sampling area and duration of the sampling program, and the measurement of ecological variety. We shall touch on these problems below.

More species occur in tropical communities than in temperate and arctic communities.

In virtually all groups of organisms, the number of species increases markedly towards the equator (Fischer 1960, Stehli 1968). For example, within a small region at 60° latitude one might find as many as ten species of ants; at 40° there may be between 50 and 100 species; and within 20° of the equator, between 100 and 200 species. Greenland is inhabited by 56 species of breeding birds, New York, 105, Guatemala, 469, and Colombia, 1,395 (Dobzhansky 1950). Diversity in marine environments follows a similar trend: arctic waters harbor 100 species of tunicates, but more than 400 species are known from temperate regions, and more than 600 from tropical seas (Fischer 1960).

Some groups vary from this general pattern. Among butterflies, the milkweed butterflies (monarchs and their relatives) are more restricted to tropical latitudes than the swallow-tailed butterflies are (Figure 37-1). The number of species of benthic marine invertebrates—forms that burrow in the sand and mud on the ocean floor—does not vary markedly (Fischer 1960). The prevalent trend sometimes is reversed by small, specialized taxonomic groups. Whereas birds generally increase in variety as one approaches the equator, sandpipers and plovers are more diverse in the arctic. These shorebirds are adapted to tundra and other open habitats, which do not occur commonly in tropical areas but are extensive in the arctic.

Within a given belt of latitude around the globe, the number of species varies widely among habitats according to productivity and degree of structural heterogeneity. For example, censuses of birds in small areas (usually ten to fifty acres) of relatively uniform habitat reveal about six species of breeding birds in grasslands, fourteen in shrublands, and twenty-four in floodplain deciduous forests (Table 37-1). But because not all communities fit this pattern, other factors must influence the number of species. Marshes are productive, but they have relatively few species of birds. Deserts, by contrast, are relatively unproductive but have many more species of birds than grasslands, which are more productive but structurally much simpler. The marsh is a uniform habitat, providing little opportunity for ecological specialization; deserts often are complex, supporting a variety of plant forms. Number of species appears to increase, therefore, with the productivity and structural heterogeneity of the vegetation.

Although tropical habitats are usually more productive than temperate and, especially, arctic habitats, the factors that determine the number of species are not necessarily related to production.

In certain groups of marine crustaceans, the number of species is inversely related to the organic productivity of the marine community. For

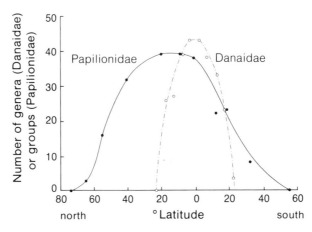

Figure 37-1   Number of genera of milkweed butterflies (Danaidae) and swallow-tail butterflies (Papilionidae) as a function of latitude (after Stehli 1964).

TABLE 37-1

Plant productivity and the number of species of birds in several representative temperate zone habitats (after Tramer 1969; productivity data from Whittaker 1970).

Habitat	Approximate productivity $(g \cdot m^{-2} \cdot yr^{-1})$	Average number of bird species
Marsh	2,000	6
Grassland	500	6
Shrubland	600	14
Desert	70	14
Coniferous forest	800	17
Upland deciduous forest	1,000	21
Floodplain deciduous forest	2,000	24

example, species of calanid crustacea in surface waters of the Pacific Ocean are more numerous in tropical regions (more than eighty species) than in temperate regions (about thirty species) and the Bering Sea (about ten species). The productivity of these habitats, indicated by the number of individuals of crustaceans per cubic meter of water, *increases* toward the north from about 100 in tropical regions to almost 3,000 in temperate and arctic regions (Fischer 1960, quoting Brodskij 1959). The number of individuals per cubic meter of water may not be a valid measure of productivity, but this example nonetheless runs counter to the general correlation between productivity and species number. It is, however, consistent with the usual arctic-tropical trend in species diversity.

Variation in the number of species may be caused by historical accidents of geography that have isolated some areas from centers of species production or, conversely, that have placed areas in ideal situations for the generation of new species or for receiving immigrants. In North America, the number of species in most groups of animals and plants increases toward tropical latitudes, but the influence on species number of geographical heterogeneity and the isolation of peninsulas is apparent. Simpson (1964) tabulated the number of species of mammals in 150-mile-square blocks distributed over the entire area of North America to the Isthmus of Panama. The number of species per block increased from 15 in northern Canada to more than 150 in Central America, following the well-known latitudinal gradient in species diversity. As one goes from west to east across the middle of the United States, one finds the greatest number of species of mammals in the western mountains, where environmental heterogeneity provides effective isolating barriers to dispersal. The ecological heterogeneity of mountains creates opportunities for allopatric speciation and allows more species to coexist in a given area. Environments in the eastern United States are more uniform and, therefore, support fewer species of mammals (50 to 75 species per block) than are found in the west (90 to 120 species per block). The number of species of breeding land birds follows a similar pattern throughout North America (MacArthur and Wilson 1967, Cook 1969), but reptile and amphibian faunas do not (Keister 1971). Reptiles are more diverse in the eastern half of the United States than in the mountainous western regions; amphibians are strikingly underrepresented in the deserts of the Southwest.

The number of species of mammals, birds, and reptiles decreases strikingly towards the end of the Florida Peninsula (Emlen 1978). Most of Florida was covered by water during the last interglacial period, so Florida may yet be in an early stage of recolonization. But it is difficult to believe that birds, at least, could not have fully repopulated Florida in a few years. Uniformity of habitats also is not a likely candidate for the cause of the so-called peninsula effect, even though the lower half of the Florida Peninsula is flat and homogeneous. The number of species in many groups of plants and animals decreases along the peninsula of Baja California, which was not submerged during the Pleistocene Epoch and which is not homogeneous. Possibly, the peninsula effect is caused by isolation from centers of speciation in western North America and the tropics.

The reduced number of species in isolated areas is exaggerated on islands. Because water gaps are more effective barriers to the dispersal of terrestrial plants and animals than are most terrestrial habitats, islands usually have many fewer species than comparable mainland areas (MacArthur and Wilson 1963, 1967). The number of species on the tops of mountains, particularly in the tropics, behaves in much the same manner as on oceanic islands (Vuilleumier 1970, Brown 1971).

Qualitative change in the composition of communities often accompanies variation in the number of species.

Variation in the number of species among areas often is the result of the selective addition or removal of particular kinds of species. We have already seen that the diversity of each taxonomic group has a characteristic geographical pattern, but what about the numbers of species that fill different ecological roles? Janzen and Schoener (1968) compared samples of insects collected during the dry season in four tropical habitats. They found that the total number of species increased greatly between dry and wet habitats, but the proportion of species on each trophic level—herbivore, predator, parasite, and scavenger—also varied. The proportion of herbivores dropped from about 65 per cent of the sample in dry (low diversity) habitats to less than 55 per cent in moist (high diversity) habitats; predators decreased from about 15 to 7 per cent, but the number of parasitic species increased from 12 to 33 per cent.

Orians (1969b) suggested that part of the increase in the number of species of birds between temperate and tropical regions is related to an increase in frugivorous and nectarivorous species, and in insectivorous species that hunt by searching for their prey while quietly sitting on perches. These types of feeding behavior are uncommon among species in temperate regions. Among mammals, the increase in number of species between temperate and tropical areas results from the addition of bats to tropical communities (Wilson 1974). Terrestrial mammals are no more diverse at the equator than they are in the United States and other temperate regions at a similar latitude, although their variety does decrease as one goes farther to the north. Not only do tropical areas have more species than temperate and arctic areas, tropical communities appear to comprise more ecological roles. This finding provides a clue to one cause of variation in species diversity, as we shall see.

Species can be added to a community by increased specialization, ecological overlap, and environmental heterogeneity.

Added species can be accommodated within a community in several ways. Consider the distribution of several species along a resource continuum—perhaps the size of prey items or the height at which feeding takes place within a forest (Figure 37-2A). If the variety of resources were increased, species could be added to the longer resource continuum without diminishing the ecological range of other species (Figure 37-2B). If the productivity of the habitat did not increase in direct proportion to the variety of resources, the added species would reduce the productivity of the others.

Without an increase in the breadth of the resource spectrum, species can be added only by increasing the degree of resource overlap between species (Figure 37-2C) or by increasing the degree of specialization (Figure 37-2D). Once more, if the productivity of the habitat did not increase in direct proportion to the number of species, each species would become less productive when another was added to the community. Whether variation in species diversity is related to resource heterogeneity, ecological overlap, degree of specialization, or a combination of these factors, is a major question in the study of community organization.

When an ecologist considers the number of species in a small patch of uniform habitat, the degree of specialization with respect to resources within

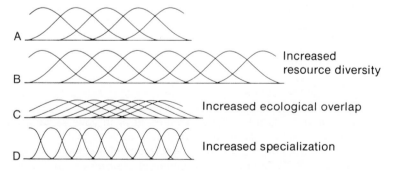

Figure 37-2   Schematic diagram showing how resource utilization along a continuum can be altered to accommodate more species.

that habitat is an important consideration. When he considers the number of species in a large region, the degree of specialization with respect to habitat is an additional consideration. Suppose a region had five distinctive habitats each of which could support no more than ten species. If all species were habitat generalists—that is, if they could live in all the available habitats—the region could support only ten species, the number that occurred in each habitat. If, however, all species were habitat specialists and could maintain populations in only one habitat, the region would support 50 species: ten different species in each of five habitats.

This simple example illustrates two kinds of diversity: local (or *alpha diversity*) and regional (or *beta diversity*). Local diversity presumably depends upon the ability of a habitat to support a variety of species and upon the degree of competition between individuals in different populations. Specialization to exploit part of a habitat is a trait of the individual—a reflection of its adaptations to the environment and to the species with which it coexists. Regional diversity depends on the replacement of populations by others in different habitats. Specialization to a particular habitat depends on the presence or absence of individuals of each species and, therefore, is a reflection of population processes as they are influenced by the physical environment, food, predators, and competitors. Because population size is more sensitive to environmental influences than is gene frequency, we might expect that where diversity increases, it does so because of habitat specialization rather than resource specialization.

Measurements of habitat specialization have shown that where many species coexist, each occurs in relatively few kinds of habitats (MacArthur, Recher, and Cody 1966). Islands usually have fewer species than comparable mainland areas, and island species often attain greater density than their mainland counterparts and expand into habitats that would normally be filled by other species on the mainland (Crowell 1962). These phenomena, called *density compensation* and *habitat expansion* (MacArthur *et al.* 1972), are collectively referred to as *ecological release.* On the island of Puerto Rico, MacArthur and his co-workers found that many species of birds occupied most of the habitats on the island. In Panama, which has a similar variety of tropical habitats, species were more narrowly restricted, often to a single type of habitat.

Surveys of bird communities in five tropical localities with different numbers of species—Panama, Trinidad, Jamaica, St. Lucia, and St. Kitts—show that where there are fewer species, each is likely to be more abundant and occur in a greater number of habitats (Table 37-2). Similar numbers of individuals of all species added together were seen in each of the five localities although the number of species differed by a factor of five between Panama and St. Kitts. In each habitat in Panama (mainland), about twice as many species were recorded and populations of each species were about half as dense as in the corresponding habitat on St. Kitts (small island).

If habitat specialization completely compensated for variation in species diversity between large regions, the number of species in small areas of uniform habitat—the local diversity—would be similar within regions of high and low diversity. The local diversity in samples of organisms obtained from tropical rivers and streams does, in fact, appear to be no greater than that

TABLE 37-2

Relative abundance and habitat distribution of birds in four tropical localities (Cox and Ricklefs 1977).*

Locality	Number of species observed (regional diversity)	Average number of species per habitat (local diversity)	Habitats per species	Relative abundance per species per habitat (density)	Relative abundance per species	Relative abundance of all species
Panama	135	30.2	2.01	2.95	5.93	800
Trinidad	108	28.2	2.35	3.31	7.78	840
Jamaica	56	21.4	3.43	4.97	17.05	955
St. Lucia	33	15.2	4.15	5.77	23.95	790
St. Kitts	20	11.9	5.35	5.88	31.45	629

* Based on 10 counting periods in each of 9 habitats in each locality. The relative abundance of each species in each habitat is the number of counting periods in which the species was seen (maximum 10); this times number of habitats gives relative abundance per species; this times number of species gives relative abundance of all species together.

in comparable temperate zone habitats, although the regional diversity of the freshwater fauna in tropical regions is greater than that of temperate regions (Patrick 1966). This example is not typical of terrestrial environments, in which local diversity usually varies in parallel to regional diversity to some degree (Table 37-2).

In streams and rivers, the number of species in most taxonomic groups increases from the headwaters to the mouth of the river. The upper region of one stream in Ontario, at a point where the water temperature averaged 9 C, was inhabited by seven species of mayflies; in lower regions, where the water temperature exceeded 20 C, between twenty and thirty species of mayflies were found. Only three species had lower distributional limits along the stream, whereas twenty-six reached upper distributional limits (Table 37-3). In other words, local diversity increased downstream by the addition

TABLE 37-3

Number of upper and lower distributional limits of species of mayflies along a stream (from Andrewartha and Birch 1954).

Collecting station	Water temperature (C)	Number of species	Number of species not found	
			Higher	Lower
1	9.0	7	0	0
2	16.3	15	8	1
3	19.5	16	2	0
4	21.5	22	8	1
5	20.5	21	2	1
6	24.0	29	6	0

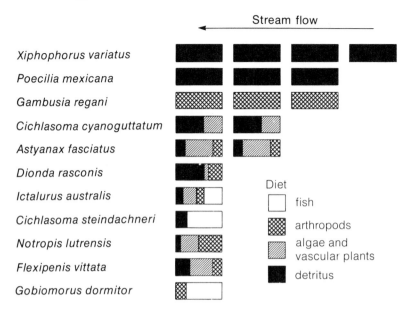

Figure 37-3   Food habits of fish species in four communities (vertical columns) from a headwater spring with one species (right) to downstream communities with up to eleven species (left). The communities sampled were in the Rio Tamesi drainage of east-central Mexico (from Darnell 1970).

of species to the set that inhabited the higher reaches of the stream. Few species were replaced by others along the length of the stream. Similar patterns appear in fish communities (Sheldon 1968, Darnell 1970). Darnell compared the food habits of fish in four communities in the Rio Tamesi drainage of east-central Mexico (Figure 37-3). A headwater spring supported only one species of platyfish, a detritus feeder. Further downstream, three species occurred: the platyfish, plus a detritus-feeding molly that preferred slightly deeper water than the platyfish, and a carnivorous mosquito fish. Species that appeared in the community farther downstream included additional carnivores and other fish that fed primarily on filamentous algae and vascular plants. None of the species dropped out of the community downstream from any of the sampling localities. Diversity increased as the stream became larger and presented more kinds of habitats and a greater variety of food items.

The examples discussed above show that variation in number of species in a region depends, in part, on the variety of habitats and the variety of resources within habitats and, in part, on the degree of specialization to particular habitats and resources. Still, we have not touched on the factors responsible for the production of species or for the regulation of their number. Regional diversity is determined by the balance between species production (speciation and immigration) and disappearance (extinction).

The number of species on an island depends on the immigration and extinction rates.

Robert MacArthur and Edward Wilson (1963, 1967) developed a simple model to explain the variation in number of species among islands. For islands that are not large enough to permit speciation through geographical isolation of populations, increase in the number of species is brought about solely by immigration from other islands or from the mainland. Whereas we know practically nothing about rates of speciation within continents, we may reasonably assume that rate of immigration to an island by new species decreases as the distance from the island to the mainland increases. Hence the advantage of the island model.

Let us consider an island located at some distance from the coast of a mainland region whose flora and fauna comprise the species pool of potential colonists for the island. The rate of immigration of new species to the island decreases as the number of species on the island increases; that is, as more and more of the potential mainland colonists are found on the island, fewer of the new arrivals constitute new species (Figure 37-4). When all the mainland species occur on the island, the immigration rate is zero. The number of extinctions per unit of time increases with the number of species present on the island. Where the immigration and extinction curves cross, the number of species on the island is at an equilibrium ($\hat{S}$).

Immigration and extinction curves probably are not strictly proportional to the number of potential colonists and the number of species established on the island. Some species are undoubtedly better colonizers than others and they reach the island first. The rate of immigration to the island initially decreases more rapidly than it would if all mainland species had equal potential for dispersal and colonization. Hence, the immigration rate follows a curved line. Competition between species on islands probably accelerates extinction, so the extinction curve rises progressively more rapidly as species diversity increases (Figure 37-4).

Figure 37-4   Equilibrium model of the number of species on islands. The equilibrium number of species ($\hat{S}$) is determined by the intersection of the immigration and extinction curves. P represents the number of potential colonists from the mainland. The immigration rate initially drops rapidly as the best colonists become established on the island. The extinction rate increases more rapidly at high species number because of increased competition between species (after MacArthur and Wilson 1963, 1967).

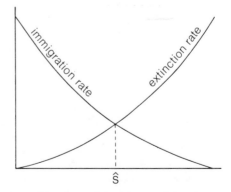

Number of species on island

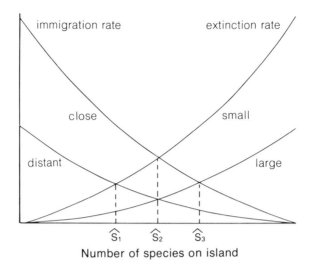

Number of species on island

Figure 37-5   Relative number of species on small, distant islands ($\hat{S}_1$) and large, close islands ($\hat{S}_3$) predicted by the MacArthur-Wilson equilibrium model. The number of species on small, close islands and large, distant islands is intermediate (after MacArthur and Wilson 1963).

If probability of extinction increased as absolute population size decreased, extinction curves for the inhabitants of small islands would be higher than for those of larger islands and small islands would have fewer species than large islands (Figure 37-5). If the rate of immigration to islands decreased with distance from mainland sources of colonists, the immigration curve would be lower for far islands than for near islands, and one would expect that the equilibrium number of species for distant islands would lie to the left of the equilibrium point for islands that are close to the mainland (Figure 37-5). These predictions have been verified for several groups of organisms on various islands throughout the world. For example, on the Sunda group of islands in the East Indies—including the Philippines, New Guinea, Borneo, and many smaller islands—the number of species of birds is closely related to the area of the island (Figure 37-6). The islands are so close to each other and to mainland sources of colonists that no distance effect is evident. On islands to the east of New Guinea, however, which are located at a greater distance in the Pacific Ocean, the number of species of birds falls below a curve representing the number of species on islands of similar size close to the mainland (Figure 37-6). In principle, the equilibrium model should apply to islands of all kinds, including patches of habitat, widely-dispersed species of plants (Janzen 1968b, Opler 1974, Strong and Levin 1975), and mountain tops (Brown 1971, Vuilleumier 1970).

If the number of species on an island behaved according to the equilibrium model, a change in the number of species would lead to a response tending to restore the equilibrium diversity. A natural test of this prediction

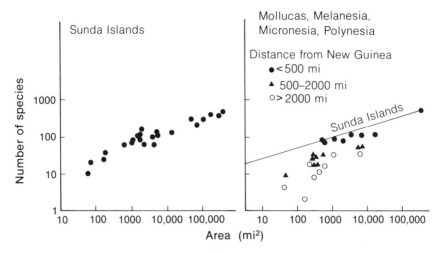

Figure 37-6   Species-area curves for land and freshwater birds on the Sunda Islands, together with the Philippines and New Guinea (left), and various islands of the Moluccas, Melanesia, Micronesia, and Polynesia (right). The latter islands show the effect of distance from the major source of colonization (New Guinea) on the size of the avifauna (after MacArthur and Wilson 1963).

was begun quite spectacularly in 1883 when the island of Krakatau, located between Sumatra and Java in the East Indies, blew up after a long period of repeated volcanic eruptions. At least half the island disappeared beneath the sea and its remaining area was covered with hot pumice and ash. The entire flora and fauna of the island was certainly obliterated. During the years that followed the explosion, plants and animals recolonized Krakatau at a surprisingly high rate: within 25 years, more than 100 species of plants and 13 species of land and freshwater birds were found there (see Table 37-4). During the ensuing 13 years, two species of birds disappeared and 16 were gained, bringing the total to 27. During the next 14-year period, the number of species on the island did not change, but five species disappeared and five new ones arrived, indicating that the number of species had reached equilibrium. The number of species of plants continued to increase. The number of species of birds on Krakatau after recolonization was about what would have been predicted from the species area curve for an island eight square miles in area in the Sunda group, where Krakatau lies (see Figure 37-6). Experimental studies, involving the colonization of glass slides by diatoms (Patrick 1967), sponges by protozoans (Cairns et al. 1967), water-filled vials by micro-organisms (Maguire 1963), and mangrove islands by arthropods (Simberloff 1969, Simberloff and Wilson 1969), reinforce the pattern of colonization seen on Krakatau.

Loss and replacement of species on islands is to be expected from MacArthur and Wilson's model; when the number of species is at equilibrium, immigration and extinction still occur and the species that inhabit an

TABLE 37-4

Recolonization of Krakatau by land and fresh water birds, and by plants, after it exploded in 1883 (from MacArthur and Wilson 1967, after Dammerman 1948, and Docters van Leeuwen 1936).

	Number of species					
	1886	1897	1908	1920	1928	1934
Plants	26	64	115	184	214	272
Birds	—	—	13	27	—	27

island are continually replaced. On Krakatau, five of the 27 species of birds present in 1920 had been replaced by other species 14 years later, indicating a turnover rate of 1.3 per cent per year [5 species / (27 species × 14 years)].

Because continual turnover, or replacement, of species is a central part of MacArthur and Wilson's equilibrium theory, ecologists have attempted to estimate turnover rates of species in a variety of island communities from historical accounts of changes in the species recorded. Most of the sources are limited to birds. Analysis of faunal changes on the Channel Islands off the coast of southern California indicated replacement rates on the order of 0.3 to 1.2 per cent per year, depending on the size of the island (Diamond 1969). Communities on small tropical islands appear to have comparable turnover rates (Diamond 1971, Terborg and Faaborg 1973). These findings have, however, been severely criticized. Lynch and Johnston (1974) thoroughly document the view that instances of extinction and colonization during historical times, usually the past century, are artifacts of inadequate census techniques and modification of the island habitats by man. Although available estimates indicating rapid turnover rates may be suspect, the equilibrium model is not less valid; it makes no prediction about the time scale of colonization and extinction. These processes are perceived, perhaps, over thousands or millions of years, not decades. Additional criticism of the equilibrium model comes from Abbott and Grant (1976), who reported increases in the number of bird species on islands near Australia, and from Lack (1976), who believed that, once established on an island, a population is entrenched; the resident has a negligible probability of extinction and it prevents the immigration of other species with similar ecological requirements. As Simberloff (1976) pointed out, critical tests of the MacArthur-Wilson model can be made only by experimenting with species that have short generation times and inhabit small islands.

The equilibrium theory can be applied to the number of species in mainland communities.

MacArthur and Wilson's approach to the number of species of organisms on islands can be applied equally well to mainland biotas by adding specia-

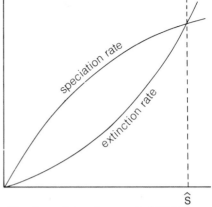

Figure 37-7   Equilibrium model of the number of species in a mainland region with a large area. When smaller areas are considered, immigration from neighboring mainland regions becomes important and the immigration-speciation curve more closely resembles that of an island with a high rate of immigration (after MacArthur 1969).

Number of species on mainland

tion within the region to the immigration rate (MacArthur 1969, Rosensweig 1975). The shape of the speciation curve for a mainland area differs from that of the immigration curve on an island. Because the island biota is derived from an established mainland source of colonists, immigration of new species is frequent when the diversity of species on the island is low. On large mainland areas, however, new species are generated from populations already present, so the rate of speciation varies in direct relation to the number of species in the area (Figure 37-7). As the number of species increases, and there are fewer opportunities for geographical expansion and subsequent isolation of subpopulations, the possibilities for speciation could decrease and result in a leveling off of the speciation rate. The shape of the extinction curve for mainland species should be similar to that for island species, but it will be lower owing to the larger area being considered.

Most of the theories that have been proposed to account for geographical variation in the number of species on mainland areas (Connell and Orias 1964, Pianka 1966, Buzas 1972) can be grouped into three categories, deriving from the equilibrium model: (1) the number of species is not in equilibrium and diversity reflects the time period in which communities have developed; (2) the combined immigration and speciation rate varies from area to area; and (3) the extinction rate varies from area to area. These hypotheses are not mutually exclusive, and all of them, or any combination of them, could describe the observed differences in number of species between areas.

Factors that could result in differences in the number of species between tropical, temperate, and arctic regions (or between any two areas within one of these regions) are outlined in Table 37-5. These hypotheses will be considered in detail below. At the outset, we must remember that any explanation of variation in diversity must ultimately rest upon variation in the physical environment—climate, edaphic factors, topographic heterogeneity, and geographical position.

TABLE 37-5
An outline of the basic hypotheses concerning species diversity, particularly the increased species diversity in the tropics compared to temperate and arctic regions.

Nonequilibrium hypothesis
    Time—the tropical habitats are older and more stable, hence tropical communities have had more time to develop.

Equilibrium hypotheses
I. Speciation rates are higher in the tropics.
    A. Tropical populations are more sedentary, facilitating geographical isolation
    B. Evolution proceeds faster due to
        1. a larger number of generations per year.
        2. greater productivity, leading to greater turnover of populations, hence increased selection.
        3. greater importance of biological factors in the tropics, thereby enhancing selection.

II. Extinction rates are lower in the tropics.
    A. Competition is less stringent in the tropics due to
        1. presence of more resources.
        2. increased spatial heterogeneity.
        3. increased control over competing populations exercised by predators.
    B. The tropics provide more stable environments, allowing smaller populations to persist, because
        1. the physical environment is more constant.
        2. biological communities are more completely integrated, thereby enhancing the stability of the ecosystem.

The time hypothesis.

Several authors (for example, Fischer 1960, and Margalef 1963, 1968) have suggested that more species live in the tropics than in temperate and arctic regions because the tropical environment is older, allowing time for the evolution of a greater variety of plants and animals. This is a nonequilibrium hypothesis: diversity is thought to increase so long as the environment remains undisturbed. The time hypothesis and the MacArthur-Wilson hypothesis are not incompatible, however. Supporters of the time hypothesis suggest that the diversity of today's biological communities is much less than the equilibrium level determined by the balance between speciation and extinction rates. Furthermore, because tropical environments are thought to be older and more stable than temperate and arctic habitats, the diversity of tropical communities is closer to the equilibrium.

The time hypothesis is not new; it was fully stated in 1878 by Alfred Russel Wallace, the co-author, with Darwin, of the theory of evolution:

> The equatorial zone, in short, exhibits to us the result of a comparatively continuous and unchecked development of organic forms; while in the temperate regions there have been a series of periodical checks and extinctions of a more

or less disastrous nature, necessitating the commencement of the work of development in certain lines over and over again. In the one, evolution has had a fair chance; in the other, it has had countless difficulties thrown in its way. The equatorial regions are then, as regards their past and present life history, a more ancient world than that represented by the temperate zones, a world in which the laws which have governed the progressive development of life have operated with comparatively little check for countless ages, and have resulted in those wonderful eccentricities of structure, of function, and of instinct—that rich variety of colour, and that nicely balanced harmony of relations which delight and astonish us in the animal productions of all tropical countries.

Evidence bearing on the age of major climate belts is sketchy and varied. Because the tropical zone girdles the Earth about its Equator—the Earth's widest point—the total area included within the tropics is much greater than that included within temperate and arctic regions. The Earth's climate has undergone several cycles of warming and cooling, which have been discovered by records, in sediments and fossils, of their influence on vegetation and ocean temperature. As the climate of the Earth warmed, as it last did during the Oligocene Epoch, the area of the tropics and subtropics expanded, reaching what is now the United States and southern Canada, and the temperate and arctic zones were squeezed into smaller areas closer to the poles. During the last 25 million years, the climate of the Earth has become cooler and drier, and the tropics have contracted. Regardless of their area, the climate of the tropics is undoubtedly influenced less by major climate trends than that of temperate and arctic regions.

Northern and tropical regions both underwent drastic fluctuations during the Ice Ages of the past 2 million years. Temperate and arctic areas witnessed the expansion and retreat of glaciers, causing major habitat zones to be displaced geographically and, possibly, disappear. Periods of glacial expansion were coupled with high rainfall in the tropics. The Amazonian rain forest, which today covers vast regions of the Amazon River's drainage basin, was repeatedly restricted to small, isolated refuges during periods of drought (Haffer 1969). Restriction and fragmentation of the rain-forest habitat could have caused the extinction of many species; conversely, the isolation of populations in patches of rain forest could have facilitated the formation of new species. Keast (1961) proposed a scheme of alternate expansion and contraction of humid habitats to account, in part, for the diversity of the present Australian avifauna. How these events have affected local and regional diversity in these areas is not known.

The only unequivocal test of the time hypothesis would be found in the number of species as it changed through time, but the fossil record is so fragmentary that this test could be applied to few taxa and would be restricted to certain types of habitats, particularly marine habitats (for example, see Stehli, Douglas, Newell 1969, and Hallam 1977). Clearly, worldwide diversity has increased greatly from the beginning of the Paleozoic Era to the present, particularly as plants and animals invaded new adaptive zones, and the end probably is not in sight (Simpson 1969). In terms of the equilibrium hypothesis, each new adaptive radiation represents a resetting of the speciation and extinction curves. To test the time hypothesis adequately on a scale that is relevant to the ecological structure of today's communities, we should measure the number of species in communities whose composition has not changed qualitatively. Such data are difficult to come by, but Buzas (1972) and Gibson and Buzas (1973) have reported that the diversity of temperate zone communities of foraminifera (marine, shelled protozoa) has not changed during the last 15 million years. We hope that more studies of this kind will be forthcoming.

Theories based on high speciation rates in the tropics.

New species would be produced at a greater rate in the tropics than in temperate and arctic regions if geographical isolation were enhanced and if species-specific characteristics evolved more rapidly in isolated popula-tions in the tropics. Tropical populations tend to be sedentary, partly because the tropical environment is relatively unseasonal and individuals, therefore, have little need to move from one locality to another to find favorable hab-itats. Temperate and arctic seasons change so drastically that many species migrate to other regions during the winter. It is widely believed that seasonal movements promote widespread gene flow and reduce local differentiation, but this notion has not been adequately tested. Whether plant populations are more sedentary in the tropics depends on how wide-ranging their dis-persal mechanisms are.

Sedentariness reduces gene flow and speeds the genetic differentiation of subpopulations. Even for such highly mobile organisms as birds, a wide river or a small mountain range can effectively block dispersal in the tropics; the ranges of many species of birds stop abruptly at major river systems in the Amazon Basin (Haffer 1969, 1974, Mayr 1969, Prance 1973); many families of deep-forest birds disperse poorly to nearby tropical islands (Ricklefs and Cox 1972). Because temperature varies so little seasonally within tropical habitats, mountains present a greater ecological barrier to dispersal in the tropics than in temperate regions (Janzen 1967b); high mountain passes differ in climate from lowland habitats more in the tropics than in temperate zones. In the tropics, seasonal changes within habitats are small compared to the differences between habitats. This allows increased habitat speciali-zation and thereby reduces the dispersal of populations across different habitats. In temperate and arctic regions, seasonal differences in climate within a habitat usually are greater than the average differences between habitats and environmental heterogeneity presents less of a barrier to dispersal.

The relationship between the biological diversity of a region and its geographical heterogeneity is obvious. Colombia has more species of birds, and probably most other types of organisms, than any other country; its geography is broken repeatedly by the Andes Mountains and by numerous valleys between the mountain ranges. Brazil and Zaire, which are topograph-ically more uniform than Colombia yet of similar size and located within the same latitudes, have fewer species. A component of Colombia's high diver-sity may be the wide range of habitats created by its mountainous topogra-phy, yet it is also possible that the many opportunities for geographical isolation have fueled the fires of species production in Colombia. These factors are, however, difficult to distinguish with the only evidence being patterns of diversity.

If evolution proceeded more rapidly in the tropics than in temperate and arctic regions, the evolutionary divergence of populations might lead more rapidly to speciation. Rapid evolution is favored by many characteristics of tropical environments. A longer growing season permits many more generations each year, which creates more opportunity for selection, particularly on adaptations of courtship and mating that are responsible for the evolution of reproductive isolation between species. Theodosius Dobzhansky (1950) suggested that, in temperate and arctic regions, physical factors in the environment select individuals for their degree of physiological adaptation to extreme conditions. In the tropics, physical factors are less important and biological agents assume a more important role in selection. Dobzhansky argued that the release of adapting systems from rigid physiological constraints allows evolution to respond more directly to the biological environment which, in turn, could enhance the evolution of divergence and reproductive isolating mechanisms through competition (see Chapter 16).

Theories based on low extinction rates in the tropics due to reduced competition.

Competition could be less among tropical species for several reasons: (1) greater resource availability, or use of fewer resources by each species; (2) greater heterogeneity of the environment, so competition can be reduced by specialization; and (3) limitation of population size by predation rather than by food resources.

If resources were more abundant in tropical regions than in temperate or arctic regions, tropical species could be more specialized without sacrificing productivity or population size; specialization would reduce interspecific competition and allow species to coexist more easily. The number of species in a community is consistently related to the organic productivity of the environment: tropical habitats are generally more productive than temperate and arctic habitats; mountains tend to be less productive and also have fewer species than lowland regions; deserts and grasslands are generally less productive, and also less diverse, than moister habitats. But if tropical and temperate habitats with similar productivity could be compared, the tropical habitat would likely have more species; differences in productivity cannot account for all variation in species number. Moreover, variation in productivity between regions is generally less than variation in organic diversity. Tropical forests are about twice as productive, on average, as temperate zone, broad-leaved forests, yet they are inhabited by perhaps five to ten times as many species of trees in small areas and have an equally greater diversity in other groups of plants and animals. Several contrary examples must also be accounted for: the productivity of abyssal oceanic habitats is low, yet the diversity of the biological communities found there tends to be high (Sanders 1969, Buzas 1972); also, as we have seen, the number of species of planktonic crustaceans is inversely related to the local productivity of the ocean (Fischer 1960, Pianka 1966; page 689).

Reduced energy or resource requirements of individuals would tend to reduce competition. In birds, fewer eggs are laid by tropical species than temperate zone species; nests are less successful; the period between broods is longer: the intensity of reproduction of tropical species, therefore, is lower than that of temperate species (MacArthur 1965, Ricklefs 1966). Connell and Orias (1964) suggested that organisms need fewer resources for homeostasis in the more constant external environment of the tropics; this could reduce the overall resource requirement of a population, but this gain also could be reduced by increased population density. Unfortunately, the productivity of habitats and the energy and resource budgets of populations in tropical and temperate regions have not been adequately measured to test Connell and Orias's hypothesis.

Heterogeneity in the environment allows specialization. No matter how productive a habitat may be, only one species could occur on each trophic level if the environment were completely uniform; competitive exclusion could not be avoided by specialization. Heterogeneity of habitats and geographical regions has not been quantified on a comparative basis in tropical and temperate regions, but tropical forests impress the casual observer with their variety of vegetational structure—including lianas, vines, and bromeliads (air plants)—each of which provides habitats for a variety of animals that could not otherwise live in the forest. For example, by trapping water at the base of their leaves, bromeliads create small aquatic environments inhabited by many types of organisms. Laessle (1961) recorded 68 species of animals and plants from water trapped in bromeliads in Jamaica; many of the species were restricted to that habitat.

The variety of flowers and fruits in tropical communities is accompanied by an equally impressive variety of frugivores, seed predators, pollinating species, and the predators and parasites that prey on them. But although much of the animal diversity in tropical habitats can be attributed to the structural heterogeneity of the vegetation, one is forced to consider the more basic problem of the diversity of plants. The number of species of trees and other plants in tropical habitats greatly exceeds that of temperate habitats, but the diversity of soils, drainage, and exposure patterns for plants probably does not differ greatly. Does environmental heterogeneity limit the number of species that can coexist in an area; or does the degree of specialization within a community reflect competition within the community that exerts an influence on resource partitioning over and above the physical structure of the habitat. The relationship between diversity and heterogeneity will be difficult to disentangle because as species are added to a community, its structural heterogeneity increases.

Theories based on reduced competition in the tropics due to predation.

Predation could reduce competition among species if predators were so efficient that they could keep prey populations below the carrying capacity of the environment, as this would permit additional species to coexist on the resources used by the prey species (Parrish and Saila 1970). We have seen that efficient predators with high reproductive rates relative to their prey are able to limit prey populations, often to a spectacular degree (Chapter 33). Many experiments have been conducted to assess the role predators play in regulating the number of prey species in a community. The most notable of these are a study on the diversity of intertidal marine faunas (Paine 1966), and experiments on the control of diversity of annual plants by rabbits (Harper 1969).

Paine worked in the intertidal region of a rocky shore habitat along the Pacific coast of Washington that was dominated by several species of barnacles, gooseneck barnacles, mussels, limpets, and chitons (a kind of grazing mollusc); these were preyed upon by the starfish *Pisaster* (Figure 37-8). One study area, eight meters in length and two meters in vertical extent, was kept free of starfish by physically removing them on a regular schedule. An adjacent control area was left undisturbed. Following the removal of the starfish, the number of prey species in the experimental plot decreased rapidly, from fifteen at the beginning of the experiment to eight at the end. Diversity declined in the experimental areas when populations of barnacles and mussels increased and crowded out many of the other species. Paine concluded that starfish were a major factor responsible for maintaining the diversity of the area. The crown-of-thorns starfish (*Acanthaster*) similarly enhances the diversity of coral reef communities near the Pacific coast of Central America by voraciously consuming one species of coral, *Pocillopora,* that otherwise crowds many species of coral out of the community (Figure 37-9; Porter 1972, 1974). On a small island off the coast of England, rabbits had grazed the vegetation heavily, creating an evenly cut turf composed of many species of plants. When the rabbits were eliminated by myxomatosis virus (page 632), the character and species composition of the island's vegetation changed dramatically, and the number of plant species decreased (Harper 1969).

In a series of experiments on communities of protozoa and other small organisms living in pitcher plants, Addicott (1974) showed that predation by mosquito larvae reduced the number of prey species, in contrast to the results of Paine's predator removal experiment. But in the simple, test tube–like environment of the pitcher plant, mosquito larvae are such efficient predators that they drastically reduce populations of all prey species, driving some to extinction. It would appear, therefore, that the outcome of experiments on the role of predation in the maintenance of diversity among prey species will vary with the complexity of the habitat and the efficiency and abundance of the predators.

Doutt (1960), Gillette (1962) and Janzen (1970) suggested that the activities of herbivores could be responsible for the large number of species of trees in tropical forests compared to temperate forests. Janzen argued that herbivores specialize on the buds, seeds, and seedlings of abundant species to such an extent that they destroy young trees and eventually reduce the density of the species. This, in turn, allows other, less common, species to grow in their place. Several lines of evidence support this hypothesis. For example, attempts to grow rubber trees in dense stands in their native

Figure 37-8   Congregation of starfish (*Pisaster*) at low tide on the coast of the Olympic Peninsula, Washington. The starfish, shown at lower left, is an important predator on mussels (lower right).

habitats in the Amazon Basin have met with singular lack of success, but rubber tree plantations are successful in Malaya. In the Amazon Basin, epidemics of plant pests rapidly destroyed the rubber trees, but in the East Indies, where the natural diseases and pests of rubber trees do not occur, the trees can be grown successfully in large stands. Attempts to grow other commercially valuable trees in single-species stands in the tropics have frequently met the same disastrous end that befell large rubber plantations in South America. In contrast, cacao, the South American plant that is the source of chocolate, is infested with no more species of insect pests where

Figure 37-9   Crown-of-thorns star-fish consuming a coral head in Panama (courtesy of J. W. Porter, 1972b).

it is planted in its native region than in plantations in Africa and Asia (Strong 1974). The difficulties of monoculture, when they occur, are not restricted to the tropics, but for unknown reasons predators and parasites appear to be much more important in tropical environments. (Different views of tropical tree species diversity were expressed by Ashton (1969), who suggested that the age and stability of tropical environments permitted extreme specialization, thereby enhancing diversity, and by Ricklefs (1977a), who suggested that conditions for seedling establishment, partly the result of disturbances to the forest, are more varied in the tropics. We shall return to this problem in the next chapter.)

Theories based on reduced extinction rates in the tropics due to greater environmental constancy.

If fluctuations in the environment increased the probability that a population would become extinct (see page 648), small populations could persist longer in environments that were relatively constant. Reduced extinction rate would allow the accumulation of rare species in a community, thereby increasing its diversity. Many ecologists believe that tropical environments are less variable than temperate and arctic environments. Temperature certainly is less variable in tropical latitudes, but rainfall can be as irregular and unpredictable as it is in temperate regions. There have been few long-term studies of tropical communities. Populations of some tropical organisms, like the moth *Urania,* undergo eruptions and cyclic changes in size reminiscent of temperate and arctic populations (for example, see Smith 1971), but populations of rodents vary little except where they are influenced by human activity (Fleming 1975).

The abyssal depths of the ocean probably come closer than any other environment to having constant conditions (George and Rowe 1973). Sanders (1968, 1969) suggested that the constancy of the deep-sea environment is responsible for the fact that more species occur there than in nearby shallow waters (Figure 37-10). According to Sanders, a rigorous physical environment requires physiological adaptation at the expense of what he calls biological accommodation within the community. This accommodation has both ecological and evolutionary components, hence Sanders' hypothesis is akin to Margalef's (1968) time hypothesis (page 700), based on the maturation of communities over long periods—evolutionary time.

Environmental stability could strengthen whatever role predators play in maintaining the species diversity of a community. Inasmuch as relatively little energy funnels up the food chain to predator trophic levels, a perturbation in the primary production of a habitat probably affects predator trophic levels more than lower trophic levels (Paine 1966). Reduced environmental variation could promote diversity in predator trophic levels and, in turn, increase the diversity of the community.

The stability of the community may be partly a function of the organization of the community itself. Ecologists generally believe that communities with complex pathways of energy flow between trophic levels are inherently more stable than simpler communities. The reasoning is simple: if a predator utilized a single prey species, its population would parallel fluctuations in the prey population; if the predator could switch to alternative prey species when the one becomes rare, it could maintain its population near a constant level by switching back and forth between prey, provided that the prey do not fluctuate in unison (see page 857). As a community develops more complex links between trophic levels, its overall stability may increase; and this would further promote greater diversity of species. Relatively little is known about the trophic organization of natural communities, and there is reason to believe, as we will see in Chapter 38, that increasing the complexity of community organization can have a destabilizing effect.

Variation in the number of species among communities has been a central issue in the study of community structure. Correlations between number of species and the physical conditions of the environment, while not directly describing the organization of the community, have suggested a number of important factors that may influence community structure. Species diversity appears to bear a direct relationship to primary production and the heterogeneity of the environment (including the activities of predators) and, perhaps, an inverse relationship to the variability of physical conditions. These relationships indicate that factors having influence on the degree of competition between species and on the ability of species to reduce competition by partitioning resources play an important role in the organization of the community.

That competition is a major organizing principle in ecology is so widely accepted (e.g., Cody and Diamond 1975) that it has achieved the status of a paradigm. The theory that diversity is regulated by competition between species is, however, difficult to prove. One can demonstrate that competition exists in natural communities by perturbation experiments. But to demon-

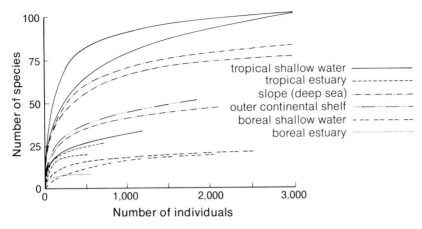

Figure 37-10   Number of species of bivalves and polychaete worms as a function of sample size for different marine environments. Tropical and deep sea faunas tend to be more diverse than faunas in less constant environments (after Sanders 1969).

strate that competition regulates diversity, one would have to show that the addition of a species to a community increased the probability of extinction of other coexisting species. Such experiments would be impossible in any reasonably complex community, even though the desired result can be obtained easily in test-tube microcosms (for example, see page 705). Ecologists are forced to study competition within communities indirectly, inferring community processes from easily measured attributes. As we have seen, the number of species in the community is one of these attributes. A second is the relative abundance of species.

Ecologists have long recognized that species are not equally abundant. Why are some species exceedingly common in samples of organisms while others are represented only by few individuals? Gleason (1929) recognized that interactions between species influence relative abundance. He suggested that the most common species of plants, which he called dominant species, were able to exclude competitors from the community. Rare species, which Gleason called non-dominants, could not compete effectively with more dominant species. From these beginnings, ecologists have since pursued the phenomenon of relative abundance from three directions: (1) statistical description of relative abundances of species in communities; (2) use of patterns of relative abundance to test hypotheses about the interactions between species; and (3) attempts to elucidate the factors that determine relative abundance of species. As we shall see below, the study of relative abundance has shed some light on the relationship between community function and community structure; the limitations of this approach also will be plainly evident.

Species abundances conform to regular patterns.

Samples of organisms from a community reveal, at the very least, two kinds of information: the number of species and the abundance of each. When the number of species with each level of abundance—one individual, two individuals, three individuals, and so on—is plotted as a histogram, the data assume characteristic patterns of distribution. An early attempt to quantify these patterns mathematically was made by Fisher, Corbet, and Williams (1943), who suggested that the pattern of relative abundance observed in a sample of English moths caught at a light trap followed a logarithmic series. The series consists of terms of the form

$$ax, \frac{ax^2}{2}, \frac{ax^3}{3}, \frac{ax^4}{4}, \cdots, \frac{ax^i}{i}$$

where x is a number between 0 and 1 and a is proportional to the number of species in the sample. Each term in the series is equal to the number of species in the sample that are represented by i individuals. The term $(ax^3/3)$, for example, is the number of species represented by three individuals. The logarithmic series provides a reasonable fit to some kinds of samples, but although Williams (1964) has used it extensively, the series has not been applied widely by others to describe relative abundance.

In 1948, Frank Preston suggested that species abundances were distributed in a regular pattern such that species with intermediate numbers of individuals were the most prevalent in communities and that very rare and extremely common species were less frequent (Figure 37-11). Preston assigned species to classes of abundance based on a logarithmic scale of numbers of individuals per species: 1–2 individuals, 2–4 individuals, 4–8

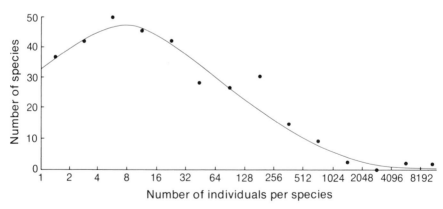

Figure 37-11   Relative abundances of species of moths attracted to light traps near Orono, Maine. Size classes (octaves), which increase by a factor of two from one class to the next, are on the horizontal axis; the number of species in each size class is plotted along the vertical axis. The distribution of abundances is hump-shaped, with a mode of 48 species in the 4–8 individuals size class (after Preston 1948).

individuals, 8–16 individuals, and so on. Preston called these classes "octaves" because each is twice as large as the preceding class. (In the musical scale, the vibration frequency of each note is twice that of the note one octave lower.) Species whose numbers in the sample fall on the class boundaries (2, 4, 8, 16, . . .) are placed half in the class above the boundary and half in the class below the boundary. In a large sample of individuals, the number of species in each abundance category tends to assume a distribution that resembles a bell-shaped, or "normal," curve. According to sampling theory, such a normal distribution would indicate that species are distributed randomly about some mean value on a logarithmic scale of abundance.

The normal curve is described by the equation

$$n_R = n_0 e^{-(1/2)(R/\sigma)^2}$$

in which n_R is the number of species whose abundance is R octaves greater or less than the modal abundance of species within the community, n_0 is the modal number of species, and σ (the standard deviation) is a measure of dispersion. These relationships can be seen in a graph of the lognormal distribution (Figure 37-12). The dispersion of the curve—whether it is narrow or broad—is proportional to the constant σ. In theory, at least, the entire lognormal distribution of species abundances in a community can never be fully sampled. There always will be some species that are too rare to be represented by one or more individuals in a sample of any size. These

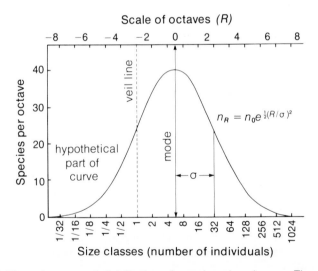

Figure 37-12   Lognormal distribution of species abundances. The part of the curve to the left of the "veil line," which corresponds to species with less than one individual in the sample, and thus not represented, is hypothetical. The scale of octaves (R) begins at the modal octave (O). One standard deviation (σ) on both sides of the mode includes about two-thirds of all the species in the sample (after Preston 1948).

species fall below the "veil line" of the distribution and their presence can be revealed only by increasing the size of the sample. If the sample is doubled, the modal abundance of species is moved one octave to the right (the abundance of all species is doubled, on the average) and additional species, each represented by one individual, appear in the distribution at the veil line. A useful feature of Preston's lognormal curve is that it takes sample sizes into account. One can predict the total number of species (N) in a community, including those not represented in the sample, knowing only the number of species in the modal abundance class (n_0) and the dispersion of the lognormal distribution (σ), by the equation

$$N = n_0\sqrt{2\pi\sigma^2} = 2.5\sigma n_0$$

For the sample of moths graphed in Figure 37-11, n_0 is 48, and σ is 3.4 octaves; therefore $N = 2.5 \times 3.4 \times 48 = 408$ species. The actual sample of over 50,000 specimens contained only 349 species, 86 per cent of the number theoretically present in the sample area. In practice, the confidence limits of N are so broad that it has dubious value.

The dispersion (σ) of lognormal curves is fairly similar for various groups of organisms. Preston (1948) obtained values of 2.3 for birds and 3.1–4.7 for moths; Patrick, Hohn, and Wallace (1954) obtained values of 2.8–4.7 for diatoms. May (1975) suggested that this range of values is to be expected from statistical sampling properties and is independent of the influence of the environment on the community. But, as MacArthur (1969) pointed out, dispersion values do vary with the environment. Working with censuses of forest birds, MacArthur obtained values of $\sigma = 0.98$ for lowland tropical localities, 1.36 for temperate localities, and 1.97 for islands. MacArthur's data indicate that there are proportionally fewer moderately common species, and more abundant species, on islands compared to temperate and, especially, to tropical mainland areas. Patrick (1963) found similar variations in the dispersion of the lognormal curve fitted to diatom samples from streams with different water conditions in the eastern United States.

Although the lognormal curve appears to describe a consistent pattern in the structure of natural communities, we must not forget that it is an empirical description, without a known biological basis. Furthermore, the curve is not applicable to all groups; one cannot assume *a priori* that a particular association of species will exhibit a lognormal distribution of species abundances. Samples of large numbers of individuals and species tend to be described well by the lognormal distribution, but for smaller samples of fewer species, the lognormal curve is less satisfactory. In addition, to obtain an accurate description of the lognormal curve, the sample must be large enough that the modal abundance class is well exposed to the right of the veil line; one cannot calculate a normal curve from one tail of its distribution.

Preston extended his analysis of species abundances to patterns in space and time in two fascinating and stimulating papers (Preston 1960, 1962), but the biological basis of the lognormal curve of abundance is so obscure and there are so many exceptions to predictions based on its properties (for example, see MacArthur and Wilson 1967) that we shall not pursue the lognormal distribution further.

MacArthur's model of relative abundance is based on competitive interactions among species.

As ecologists accumulated information about communities, they became increasingly aware that the distribution of relative abundances in samples assumed one of a few characteristic patterns, depending primarily on the kinds of organisms sampled. Inevitably, models based upon interactions among populations were developed to explain patterns of relative abundance. In effect, these models sought to determine whether the pattern of relative abundance could elucidate the organization of the community. Robert MacArthur (1957, 1960) proposed a model of relative abundance based on the biological organization of the community. He suggested that the abundance of each species was determined by the random partitioning of resources distributed along a continuum of resource types. MacArthur considered resources as if they were distributed evenly along the length of a stick. To predict the relative abundance of N species, $N - 1$ points are picked at random along the length of the stick, and the stick is broken at those points. The lengths of the resulting segments are assumed to be proportional to the relative abundances of each species. If the segments are arranged on a logarithmic scale of decreasing rank—that is, longest, second longest, third longest, and so on—the expected distribution of stick lengths assumes a nearly linear decrease (Figure 37-13). Distributions generated according to this model can be extremely variable, however, and one must examine many samples to be sure that a distribution fits the model. The model has the following biological characteristics: resources are distributed continuously along a single dimension; all resources are utilized, but those used by one species do not overlap those used by the next; the number of species is a premise of the model, not a prediction; relative abundances are determined by random apportionment of resources among species. As such, the model corresponds to a community in which species compete, each one excluding all others from the resources it exploits, and the competitive ability of each species is a random variable.

Since MacArthur published the broken-stick model, many ecologists have compared the distribution of relative abundances to samples of organisms from natural communities. MacArthur himself examined relative abundance in communities of birds, found that they were adequately described by his model, and proclaimed the correspondence as a test of the hypothesis that competition organizes natural communities. Groups of species whose relative abundances are adequately described by the broken-stick model are characterized by large body size and long life cycles—for example, birds, fish, ophiuroid worms, and predatory gastropods (King 1964). Groups that exhibit less equitably distributed abundances than the broken-stick model typically include small-bodied and relatively short-lived species—soil arthropods, nematodes, protozoa, and phytoplankton (Hairston 1959, King 1964, Batzli 1969). MacArthur (1960) had suggested that only those species whose abundances were in a fairly stable equilibrium could be expected to fit the model closely; opportunistic species, such as many insects and other small organisms, which are common during favorable periods and rare when conditions are unfavorable, could not be expected to adhere to the model.

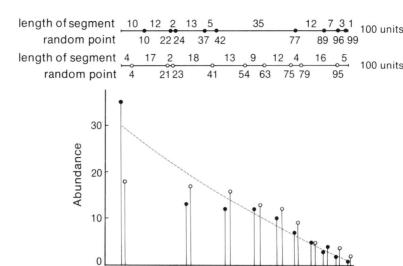

Figure 37-13   A model of the relative abundance of species in a community in which resources are distributed along a continuum and species do not overlap. The utilization of resources by n species is determined by $n - 1$ random points on the line; each segment corresponds to the abundance of a species. The species are conventionally portrayed along a logarithmic scale of ranked abundance, as shown. The expected (average) distribution of abundances according to the broken-stick model for a sample of 10 species and 100 individuals is indicated by the dashed line.

King's analysis demonstrated that species whose populations can be expected to fluctuate most, those with relatively small individuals having short life cycles, indeed are poorly fit by the broken-stick model.

The predictions of the broken-stick model can also be obtained from other models.

That many communities conform to the broken-stick model of relative abundances suggested that population size is limited by competition with species that have similar resource requirements. But the validity of MacArthur's broken-stick model as a description of community organization is greatly weakened by the fact that the same distribution of relative abundances can be predicted from models that do not include competition among their assumptions. Joel Cohen (1968) devised two alternatives: a "balls-and-boxes" model and an "exponential" model. In Cohen's balls-and-boxes model, the environment is subdivided and the abundance of each species is determined by throwing units of abundance into the subdivisions until each species has at least one subdivision to itself. In this model, the number of species cannot exceed the number of subdivisions. Competition is implicit in this model because no two species overlap completely. Resources are not distributed continuously, however, nor is there an upper limit to resource availability. The expected distribution of relative abundances is identical to that of the broken-stick model. Although the balls-and-boxes model has many unreasonable assumptions, it should caution us from drawing conclusions about underlying community processes from their superficial result.

Cohen's exponential model departs even further from the premises of the broken-stick model: species abundances fluctuate as random variables with time, independently of each other, and with a cumulative frequency distribution

$$P(x_i \geq x) = e^{-\lambda x}$$

That is, the probability that the abundance of species i (x_i) is equal to or greater than x is an exponentially decreasing function of x, where λ is constant. No dimensioning of resources, no interaction between species, no equilibrium of populations is assumed, yet Cohen's exponential model predicts the same relative abundance of species as MacArthur's broken-stick model.

Cohen considered species varying independently in time. We may also consider the abundances of species varying independently among habitats or geographical localities. The distribution of a typical species along a gradient of habitats has a peak abundance in its optimum habitat and its numbers decrease with ecological distance in either direction from the optimum. Suppose several species occur on the same environmental gradient, and their abundances are independent and randomly distributed. What is the relative abundance of each species in a given habitat along the gradient? In Figure 37-14, five hypothetical species are distributed independently along an environmental gradient and their abundances in habitats A and B are indicated by the intersection of the abundance curves and the vertical lines projected from points A and B. When one arbitrarily chooses at random the position of the species along the gradient, the height of the distribution at its maximum, or the relative dispersion of the species throughout the habitat gradient, or any combination of these variables, the relative abundances of the species in a particular habitat correspond to the broken-stick distribution, as they do in Cohen's exponential model.

Competition does not determine species abundance in the exponential model or in the open-community model illustrated in Figure 37-14. Thus the basic assumption of MacArthur's broken-stick model is not validated by observed patterns of relative abundance in communities; these patterns support models without competition equally well. Observed distributions of species abundances apparently result from the complex structure of natural communities, which causes their organization to assume a superficial element of randomness. It would seem that we have been approaching the problem of relative abundance from the wrong direction. We should not ask what patterns of relative abundance can tell us about the organization of communities; rather we must ask how the environment influences the abundance of each species and why these factors have created so much variation

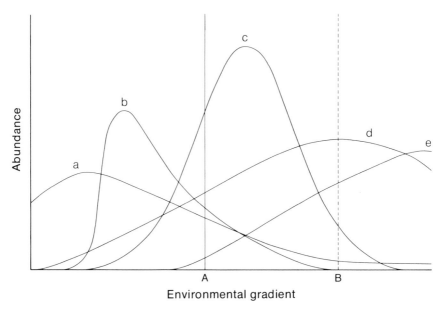

Figure 37-14   A model demonstrating how the relative abundance of species in a locality is related to the distribution curve of the species along an environmental gradient. The abundances of species a-e at localities A and B are indicated by the intersection of the species curves and lines projected vertically from the habitat gradient at points A and B.

in the relative abundance of species. Reformulated in this way, the study of community organization takes on a new dimension, namely, the integration of community patterns and population processes. If community patterns represented no more than the sum of the independent properties of many populations—the open-community concept of Gleason (page 671)—there would be no point to the study of community ecology. All the information in the community would reside at the population level. If community organization were important to the regulation of each population, population processes would take on a new dimension of interaction and communities would begin to assume some of the organizational attributes of organisms. We shall return to this problem after considering one more property of community function, the development of the community and its dynamic steady state.

38

Community Development

Communities are constantly changing. Organisms die and others are born to take their places. Energy and nutrients continually pass through the community. Yet the appearance and composition of most communities do not vary. Oaks replace oaks, robins replace robins, and so on, in continual self-perpetuation. If a community is disturbed—a forest cleared for agriculture, a prairie burned, a coral reef obliterated by a hurricane—the community is slowly rebuilt. Pioneering species adapted to the disturbed habitat are successively replaced by others until the community attains its former structure and composition. The sequence of changes on a disturbed site is called *succession* and the final association of species achieved is called a *climax.* These terms describe natural processes that caught the attention of early ecologists (Drury and Nisbet 1973, McIntosh 1974). By 1916, University of Minnesota ecologist Frederic Clements had outlined the basic features of succession, supporting his conclusions by detailed studies of change in plant communities in a wide variety of environments. In this chapter, we shall examine the course and causes of succession both from the traditional view of community development and in the light of recent studies on community succession.

Succession follows an orderly pattern of species replacements.

The creation of any new habitat—a plowed field, a sand dune at the edge of a lake, an elephant's dung, a temporary pond left by a heavy rain—invites a host of invading species to exploit the resources made available. The first colonizers are followed by others slower to take advantage of the new habitat but eventually more successful than the pioneer species. In this way, the character of the community changes with time. Successional species themselves change the environment. For example, plants shade the surface, contribute detritus to the soil, and alter soil moisture. These changes often inhibit the continued success of the species that cause them, and make the environment more suitable for other species which then exclude those responsible for the change.

Figure 38-1   Stages of secondary succession in the oak-hornbeam forest in southern Poland. From upper left to lower right, the time since clearcutting progresses from 0 to 7, 15, 30, 95, and 150 years (photos by Z. Glowacinski courtesy of O. Järvinen from Glowacinski and O. Järvinen, 1975).

The opportunity to observe succession is almost always at hand on abandoned fields of various ages (Figure 38-1). On the piedmont of North Carolina, bare fields are quickly covered by a variety of annual plants. Within a few years, most of the annuals are replaced by herbaceous perennials and shrubs. The shrubs are followed by pines, which eventually crowd out the earlier successional species, but pine forests are in turn invaded and then replaced by a variety of hardwood species that represent the end of the successional sequence (Oosting 1942). Change is rapid at first. Crabgrass quickly enters an abandoned field, hardly giving the plow furrows a chance to smooth over. Horseweed and ragweed dominate the field in the first summer after abandonment, aster in the second, and broomsedge in the third. The pace of succession falls off as slower-growing plants appear. The transition to pine forest requires 25 years. Another century must pass before the developing hardwood forest begins to resemble the natural climax vegetation of the area.

The transition from abandoned field to mature forest is only one of several successional sequences leading to the same climax. In the eastern part of the United States and Canada, forests are the end point of several different successional series, or *seres,* each having a different beginning. The sequence of species on newly formed sand dunes at the southern end of Lake Michigan differs from the sere that develops on abandoned fields a few miles away (Cowles 1899, Olson 1958). The sand dunes are first invaded by marram grass and bluestem grass. Plants of these species established in soils at the edge of a dune send out rhizomes (runners) under the surface of the sand, from which new shoots sprout (Figure 38-2; see also Figure 5-10). These grasses stabilize the dune surface and add organic detritus to the sand. Numerous annuals follow the perennial grasses onto the dunes, further enriching and stabilizing them and gradually creating conditions suitable for the establishment of shrub species. Sand cherry, dune willow, bearberry, and juniper form shrub layers before pines become established. As in the abandoned fields in North Carolina, pines persist for only one or two generations, with little reseeding after initial establishment, giving way in the end to the beech-maple-oak-hemlock forest characteristic of the region.

Succession follows a similar course on Atlantic coastal dunes, where beach grass initially stabilizes the dune surface, followed by bayberry, beach plum, and other shrubs (Oosting 1954). Shrubs act like the snow fencing frequently used to prevent the blowout of dunes; they are called dunebuilders because they intercept blowing sand and cause it to pile up around their bases (Figure 38-2). Succession in estuaries leading to the establishment of terrestrial communities begins with salt-tolerating plants and progresses as sediments and detritus build the soil surface above the water line (Redfield 1972).

Primary succession develops on newly exposed land forms.

Ecologists classify seres into two groups, according to their origin. The establishment and development of plant communities in newly formed habitats previously without plants—sand dunes, lava flows, rock bared by ero-

Figure 38-2   Initial stages of plant succession on sand dunes along the coast of Maryland. Top: beach grass on the frontal side of a dune. This grass is used widely to stabilize dune surfaces. Bottom: invasion of back dune areas by bayberry and beach plum (courtesy of the U.S. Soil Conservation Service).

Figure 38-3   A valley exposed by a receding glacier, visible at top center, in North Tongass National Forest, Alaska. The recently bared rock surfaces at the bottom of the valley just below the glacier have not yet been recolonized by shrubby thickets (courtesy of the U.S. Forest Service).

sion or exposed by a receding glacier—is called *primary succession*. The return of an area to its natural vegetation following a major disturbance is called *secondary succession*. We have already followed the course of primary succession on Lake Michigan sand dunes. The sequence of species colonizing habitats exposed by receding glaciers in the Glacier Bay region of southern Alaska is quite different (Figure 38-3). The surfaces of glacial deposits are stable but the thin clay soils are deficient of nutrients, particularly nitrogen, and pioneering plants are exposed to wind and cold stress. Here the sere involves mat-forming mosses and sedges, prostrate willows, shrubby willows, alder thicket, sitka spruce, and, finally, spruce-hemlock forest. Succession is rapid, reaching the alder thicket stage within ten to twenty years, and tall spruce forest within 100 years (Crocker and Major 1955).

The development of vegetation on bare rock, sand, or other inorganic sediments is called *xerarch succession*. The low water retention by such habitats results in *xeric* (drought) conditions during the early stages of succession. At the opposite extreme, *hydrarch* succession begins in the open water of a shallow lake, bog, or marsh. Hydrarch succession can be initiated by any factor that reduces water depth and increases soil aeration, whether natural drainage, progressive drying up, or filling in by sediments.

Changes in the vegetation of bogs illustrate hydrarch succession. Bogs form in kettleholes or dammed streams in cool temperate and subarctic regions. Bog succession begins when aquatic plants become established at the edge of the pond (Figure 38-4). Some species of sedges (rushlike plants) form mats on the water surface extending out from the shoreline. Occasionally these mats grow completely over the pond before it is filled in by

Figure 38-4   Stages of bog succession illustrated by a bog formed behind a beaver dam in Algonquin Park, Ontario. The open water in the center is stagnant, poor in minerals, and low in oxygen. Those conditions result in the accumulation of detritus from the vegetation at the edge and lead to a gradual filling-in of the bog, passing through stages dominated by shrubs and, later, black spruce.

Figure 38-5   A three-foot vertical section through a peat bed in a filled-in bog in Quebec, Canada. The layers represent the accumulation of organic detritus from plants that successively colonized the bog as it was filled in. The peat beds are probably several yards thick. Vegetation on the surface of the bog consists mostly of sphagnum, blueberry, and Labrador tea.

sediments, producing a more or less firm layer of vegetation over the water surface. The detritus produced by the sedge mat accumulates in layers of organic sediments on the pond bottom because the stagnant water of the pond contains little or no oxygen and thus does not support rapid decomposition of the organic matter by microbes. Eventually these sediments become peat, used by man as a soil conditioner and, sometimes, as a fuel for heating (Figure 38-5).

As the bog is filled in by sediments and detritus, sphagnum moss and bog shrubs, like Labrador tea and cranberry, become established along the edges, themselves adding to the development of a soil with progressively more terrestrial qualities. The shrubs are followed by a bog forest of black spruce and larch, which eventually is replaced by local climax species, including birch, maple, and fir, depending on the region.

In northern Michigan, the hydrarch succession that develops on bogs is only one of several seres of primary and secondary succession leading to a climax forest of spruce, fir, and birch (Figure 38-6). Following a fire, development of the climax follows a different course, passing through intermediate grass and aspen stages. If the soil is badly scorched and most of the humus is burned, a sere resembling that beginning on bare rock surface develops.

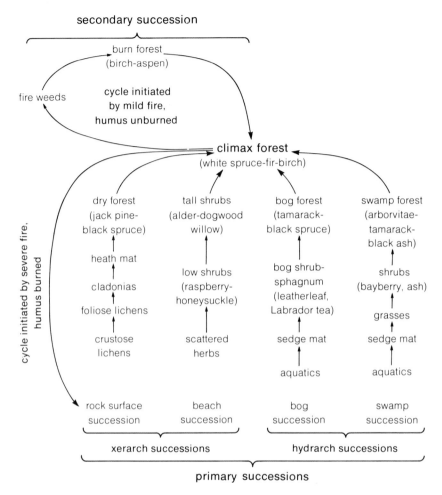

Figure 38-6   Trends of succession on Isle Royale, Lake Superior. Each habitat is characterized by a unique sere leading to the same climax. Secondary succession on burned sites follows different courses depending on the extent of the fire damage (after Oosting 1956).

Successional stages in time resemble gradients of communities in space.

The sequence of species in the sere parallels change in the physical environment as it is modified by the developing vegetation. Many of the stages in a time sequence through a sere may be found along geographical gradients in vegetation, often called *ecoclines* (Figure 38-7). For example, the xerarch succession from rock surface to forest in the eastern United States corresponds in structure, if not species composition, to the ecocline

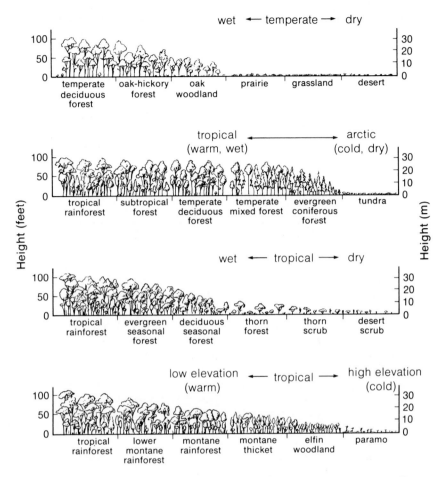

Figure 38-7   Schematic profiles of four ecoclines in vegetation type. Top, a
wet-dry gradient from the Appalachian Mountains (left) to the southwestern
United States (right); second, a warm-cold (and wet-dry) gradient from Panama
to northern Canada; third, a wet-dry gradient within the tropics; bottom, an
altitudinal gradient in the tropics from the Amazonian forest into the Andes (after
Beard 1955 and Whittaker 1970).

in vegetation from the nearly bare rock surfaces of the western deserts,
through dry grasslands, prairie, shrubby oak woodland, to tall mixed hardwood forest that occurs along an increasing moisture gradient from west to
east (Whittaker 1975). The temporal sequence of primary succession in a
particular place also follows an increasing gradient of moisture availability.
The ecocline represents a series of stages of community development
stopped at different points by lack of moisture. Each stage of the ecocline
represents the local end point of succession. The species composition of a

community along the ecocline differs from the corresponding stage of the complete sere in the eastern United States because species must be well adapted to the local conditions along the ecoclinal gradient, but, again, the general structure is the same.

Hydrarch succession on a bog is mirrored in the sequence of communities present at any one time from the open water at the center of the pond to the developing forest at its edge, where sedimentation and soil development are most advanced. Concentric bands of vegetation ring the bog, progressing from an inner circle of sedge mat outward through sphagnum and bog shrubs, larch, and finally black spruce.

The climax is the local end point of a sere.

Succession traditionally is viewed as leading inexorably toward an ultimate expression of plant development, the *climax community* (Clements 1936, Shimwell 1971). In fact, studies of succession have demonstrated that the many seres found within a region, each developing under a particular set of local environmental circumstances, progress towards the same climax (see Figure 38-6). These observations led to the concept of the mature community as a natural unit, even as a closed system. Frederic Clements stated the concept in 1916: "The developmental study of vegetation necessarily rests upon the assumption that the unit or climax formation is an organic entity. As an organism the formation arises, grows, matures, and dies. Its response to the habitat is shown in processes or functions and in structures which are the record as well as the result of these functions. Furthermore, each climax formation is able to reproduce itself, repeating with essential fidelity the stages of its development. The life history of a formation is a complex but definite process, comparable in its chief features with the life history of an individual plant."

Clements recognized 14 climaxes in North America, including two types of grassland (prairie and tundra), three types of scrub (sagebrush, desert scrub, and chaparral), and nine types of forest, ranging from pine-juniper woodland to beech-oak forest. The nature of the local climax was thought to be determined solely by climate. Aberrations in community composition caused by soils, topography, fire, or animals (especially grazing) were thought to represent interrupted stages in the transition toward the local climax—immature communities.

In recent years, the concept of the climax as an organism or unit has been greatly modified, to the point of outright rejection by many ecologists, with the recognition of communities as open systems whose composition varies continuously over environmental gradients. Whereas in 1930 plant ecologists described *the* climax vegetation of much of Wisconsin as a sugar maple-basswood forest, by 1950 ecologists placed this forest type on an open continuum of climax communities extending both over broad, climatically defined regions and over local, topographically defined areas (Whittaker 1953, McIntosh 1967, Peet and Loucks 1977). To the south, beech increased in prominence, to the north, birch, spruce, and hemlock were added to the climax community; in drier regions bordering prairies to the west, oaks became prominent. Locally, quaking aspen, black oak, and shagbark hickory, long recognized as successional species on moist, well-drained soils, came to be accepted as climax species on drier upland sites.

Mature stands of forest in Wisconsin, representing the end points of local seres, have been ordered along a *continuum index* ranging from dry sites dominated by oak and aspen to moist sites dominated by sugar maple, ironwood, and basswood (Curtis and McIntosh 1951). The continuum index for Wisconsin forests was calculated from the species composition of each forest type, and its value varied between arbitrarily set extremes of 300 for pure bur oak forest to 3,000 for a pure stand of sugar maple. Although increasing values of the continuum index correspond to seral stages leading to the sugar maple climax, they may also represent local climax communities determined by topographic or soil conditions. Thus the so-called climax vegetation of southern Wisconsin is actually a continuum of forest (and, in some areas, prairie) types (Figure 38-8). Some botanists prefer to retain the term *climatic climax* for the furthest point of vegetational succession within a region, relegating all other endpoints of seres to terminated stages of succession or *subclimaxes*. We run the risk of becoming entangled in semantics at this point without further elucidating the mechanism of plant succession. We would do well to follow one guideline in thinking about succession. If interrupted stages of succession are so prevalent and persistent in a region that species have adapted to the particular environmental conditions in these subclimax communities, these species should be recognized as climax forms even though they enter transitional seral stages elsewhere. The concept of the climax is rooted in the self-perpetuation of an association under prevalent local conditions and in the adaptations of climax species that ensure their self-perpetuation.

Succession results from variation in the ability of plants to colonize disturbed habitats and from changes in the environment following the establishment of new species.

Two factors determine the positions of species in a sere: the rate at which species invade a newly formed or disturbed habitat, and changes in the environment over the course of succession. Some species disperse slowly, or grow slowly once they have become established, and therefore become dominant late in the sequence of associations in a sere. Rapidly growing plants that produce many small seeds, carried long distances by the wind or by animals, have an initial advantage over species that are slow to disperse, and they dominate early stages of the sere. Where fire is a regular feature of a habitat, many species have fire-resistant seeds or root crowns that germinate or sprout soon after a fire and quickly reestablish their populations (Hanes 1971, Vogl and Schorr 1972, Vogl 1973).

Early colonists often change the environment in ways that favor the invasion of the community by species of superior competitors. Horseweed resists desiccation and rapidly colonizes abandoned farmland in the piedmont of North Carolina but, once established, horseweed plants modify the

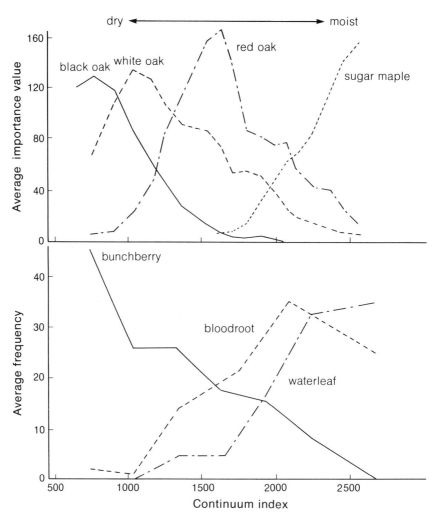

Figure 38-8   Relative importance of several species of trees (top) and herbs (bottom) in forest communities of southwestern Wisconsin arranged along a continuum index. Soil moisture, exchangeable calcium, and pH increase to the right on the continuum index (after Curtis and McIntosh 1951).

environment by shading the soil surface. Because horseweed seedlings require full sunlight, horseweed is quickly replaced by shade-tolerant species. As the community matures, the developing vegetation protects the surface layers of the soil from drying, permitting drought-intolerant species to get a foothold in the community. Progressive changes in the physical environment foster the replacement of species through seral stages.

Early stages of plant succession on old fields in the piedmont region of North Carolina demonstrate how succession is driven (Keever 1950). The first three to four years of old-field succession are dominated by a small number of species that replace each other in rapid sequence: crabgrass, horseweed, ragweed, aster, and broomsedge. The life-history cycle of each species partly determines its place in succession (Figure 38-9). Crabgrass, a rapidly growing annual, is usually the most conspicuous plant in a cleared field during the year in which the field is abandoned. Horseweed is a winter annual, whose seeds germinate in the fall. Through the winter, the plant exists as a small rosette of leaves, and it blooms by the following midsummer. Because horseweed has strong dispersal powers and develops rapidly, it usually dominates one-year-old fields. Ragweed is a summer annual; seeds germinate early in the spring and the plants flower by late summer. Ragweed dominates the first summer of succession in fields that are plowed under in the late fall, after horseweed normally germinates. Aster and broom-sedge are biennials that germinate in the spring and early summer, exist through the winter as small plants, and bloom for the first time in their second autumn. Broomsedge persists and flowers during the following autumn as well.

Horseweed and ragweed both disperse their seeds efficiently and, as young plants, tolerate desiccation. These characteristics allow them to in-vade cleared fields rapidly and produce seed before populations of compet-itors become established. Decaying horseweed roots stunt the growth of horseweed seedlings; this self-inhibiting effect, whose function and origin is not understood, cuts short the life of horseweed in the sere. Such growth inhibitors presumably are the byproduct of other adaptations that increase the fitness of horseweed during the first year of succession. If horseweed plants had little chance of persisting during the second year, owing to invasion of the sere by superior competitors, self-inhibition would have little negative selective value. At any rate, the phenomenon is fairly common in early stages of succession (Rice 1975).

Aster is a relatively successful colonist of recently cleared fields but, being a slow grower, it does not become dominant until the second year.

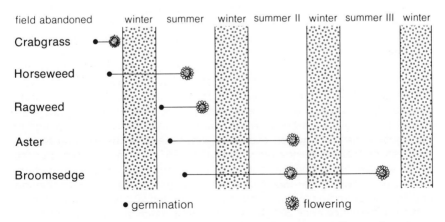

Figure 38-9   Schematic summary of the life histories of five early succes-sional species of plants that colonize abandoned fields in North Carolina.

The first aster plants to colonize a field thrive in the full sunlight, but, because aster seedlings are not shade-tolerant, they shade their progeny out of existence. Furthermore, asters do not compete effectively with broomsedge for soil moisture. To determine this, Keever (1950) cleared a circular area, one meter in radius, around several broomsedge plants and planted aster seedlings at various distances. After two months, the dry weight of asters planted 13, 38, and 63 cm from the bases of the broomsedge plants averaged 0.06, 0.20, and 0.46 gram; available soil water at these distances was 1.7, 3.5, and 6.4 grams per 100 grams of soil. Broomsedge does not, however, dominate early successional communities until the third or fourth year, in spite of its competitive edge over aster. Because their seeds do not disperse well, broomsedge plants do not increase rapidly in number until the first colonists of the field have themselves produced seeds.

When does succession stop?

Succession continues until the addition of new species to the sere and the exclusion of established species no longer change the environment of the developing community. Conditions of light, temperature, moisture, and, for primary seres, soil nutrients, change quickly with the progression of different growth forms. The change from grasses to shrubs and trees on abandoned fields bring a corresponding modification of the physical environment. Conditions change more slowly, however, when the vegetation reaches the tallest growth form that the environment can support. The final biomass dimensions of the climax community are limited by climate independently of events during succession. Once forest vegetation is established, patterns of light intensity and soil moisture are not changed by the introduction of new species of trees, except in the smallest details. For example, beech and maple replace oak and hickory in northern hardwood forests because their seedlings are better competitors in the shade of the forest-floor environment, but beech and maple seedlings probably develop as well under their own parents as they do under the oak and hickory trees they replace. At this point, succession reaches a climax; the community has come into equilibrium with its physical environment (Leak 1970, Waggoner and Stephens 1970, Horn 1975).

To be sure, subtle changes in species composition usually follow the attainment of the climax growth form of a sere. For example, a site near Washington, D.C., left undisturbed for nearly 70 years developed a tall forest community dominated by oak and beech. The community had not then reached an equilibrium because the youngest individuals—the saplings in the forest understory, which eventually would replace the existing trees—included neither white nor black oak (Dix 1957). In another century, the forest will be dominated by species with the most vigorous reproduction, namely red maple, sugar maple, and beech (Figure 38-10). The composition and age structure of a forest in northwestern Wisconsin having had minimal human disturbance over 200 years indicated a transitory state, perhaps towards the end of a sere, between oak dominance and a basswood-maple climax (Eggler 1938). At the time of that study, red oak was the commonest

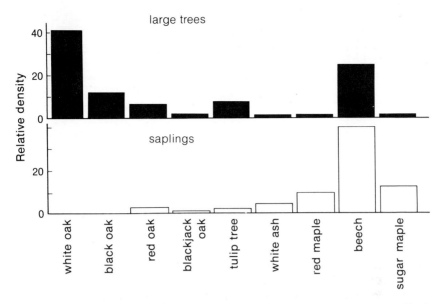

Figure 38-10   Composition of a forest undisturbed for 67 years near Washington, D.C. The relative predominance of beech and maple saplings in the forest understory foretells a gradual successional change in the community beyond the present oak-beech stage (after Dix 1957).

large tree in the forest, but basswood and, especially, maple were reproducing much more vigorously (Table 38-1). The ratios of seedlings and saplings (less than one inch diameter) to large trees (greater than ten inches diameter) were maple 186, basswood 155, red oak 18, and white oak 37. (White oak and bitternut are close to the northern edge of their ranges in northern Wisconsin and did not form a major component of the forest.)

The preponderance of red oak in the canopy of the Wisconsin forest and the evidence in the understory of successional changes yet to come indicate that the forest had been disturbed in some way, allowing seral species to enter and setting into motion the machinery of succession. The age structure of the tree populations, determined by increment borings (see page 236), suggested that fire destroyed much of the forest sometime between 1840 and 1850 (Figure 38-11). Most of the sugar maples were over 150 years old, indicating that they withstood fire damage. In fact, two-thirds of the sugar maple cores were so badly scarred by fire that their growth rings could not be counted accurately. Red oak and basswood both exhibited a period of rapid proliferation starting about 1850. Red oak gained its predominant position in the forest at that time and will not be excluded until present trees die and are replaced by basswood or maple seedlings.

The time required for succession to proceed from a cleared habitat to a climax community varies with the nature of the climax and the initial quality of the soil. Clearly, succession is slower to gain momentum when starting on bare rock than on a recently cleared field. A mature oak-hickory forest

TABLE 38-1

Number of trees of different species in a 2,500-square-meter forest area in northwestern Wisconsin. Individuals are separated into size classes (from Eggler 1938).

Species	Diameter of trunk (inches)			
	less than 1	1 to 3	4 to 9	greater than 10
Sugar maple	3,913	2	16	21
Basswood	931	22	21	6
Red oak	781	1	34	44
White oak	75	3	9	2
Bitternut	88	4	0	0
White pine	0	0	1	2
Ironwood	1,606	40	3*	0

*Maximum size class of ironwood, an understory species.

climax will develop within 150 years on cleared fields in North Carolina (Oosting 1942). Climax stages of western grasslands are reached in 20 to 40 years of secondary succession (Shantz 1917). On the basis of radiocarbon dating methods, Olson (1958) suggested that complete primary succession to a beech-maple climax forest on Michigan sand dunes requires up to 1,000 years. In the humid tropics, forest communities regain most of their climax elements within 100 years after clearcutting, provided that the soil is not abused by farming or prolonged exposure to sun and rain (Budowski 1965). But the development of a truly mature tropical forest devoid of any remnants of successional species requires many centuries.

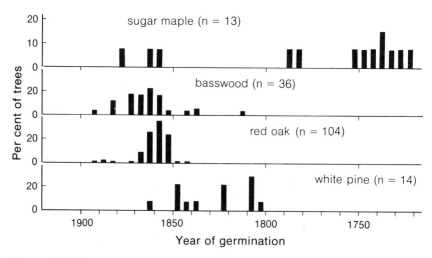

Figure 38-11   Age groups of sugar maple, basswood, and oak in a northern Wisconsin forest (after Eggler 1938).

The character of the climax is determined by local conditions.

Clement's idea that a region had only one true climax (the *monoclimax* theory) forced botanists to recognize a hierarchy of interrupted or modified seres by attaching names like subclimax, preclimax, and postclimax. This terminology naturally gave way before the *polyclimax* viewpoint, which recognized the validity of many different vegetation types as climaxes, depending on the habitat. More recently, the development of the continuum index and gradient analysis has fostered the broader *pattern-climax* theory of Robert Whittaker (1953), which recognizes a regional pattern of open climax communities whose composition at any one locality depends on the particular environmental conditions at that point.

Many factors determine the climax community, among them soil nutrients, moisture, slope, and exposure. Fire is an important feature of many climax communities, favoring fire-resistant species and excluding others that otherwise would dominate (Cooper 1961, Kozlowski and Ahlgren 1974). The vast southern pine forests in the gulf coast and southern Atlantic coast states are maintained by periodic fires. The pines are adapted to withstand scorching under conditions that destroy oaks and other broad-leaved species (Figure 38-12). Fire is even necessary to the life history of some species of pines that do not shed their seeds unless triggered by the heat of a fire passing through the understory below. After a fire, pine seedlings grow rapidly in the absence of competition from other understory species.

Any habitat that is occasionally dry enough to create a fire hazard but normally wet enough to produce and accumulate a thick layer of plant detritus is likely to be influenced by fire. Chaparral vegetation in seasonally dry habitats in California is a fire-maintained climax (see Figure 13-13) that is replaced by oak woodland in many areas when fire is prevented. The forest-prairie edge in the midwestern United States separates climatic climax and fire climax communities (Borchert 1950). Frequent burning eliminates seedlings of hardwood trees but the perennial grasses sprout from their roots after a fire. The forest-prairie edge occasionally shifts back and forth across the countryside, depending on the intensity of recent drought and the extent of recent fires. After prolonged wet periods the forest edge advances out onto the prairie as tree seedlings grow up and begin to shade out the grasses. Prolonged drought followed by intense fire can destroy tall forest and allow rapidly spreading prairie grasses to gain a foothold. Once prairie vegetation is established, fires become more frequent owing to the rapid buildup of flammable litter. Reinvasion by forest species then becomes more difficult. By the same token, mature forests resist fire and rarely become damaged enough to allow the encroachment of prairie grasses. Hence the stability of the forest-prairie boundary.

Alternating floods and drought maintain wet prairies and meadows in a quasi-equilibrium in southeastern Wisconsin. Rapid hydrarch succession on peat marshes produces wet prairie and meadow vegetation. Further encroachment of species from the local climax forest of the area is prevented by occasional flooding, alternating with fire during periods of drought. Al-

Figure 38-12   A stand of longleaf pine in North Carolina shortly after a fire. Although the seedlings are badly burned (lower left), the growing shoot is protected by the dense, long needles (shown on an unburned individual, lower right) and often survives. In addition, the slow-growing seedlings have extensive roots, which store nutrients to support the plant following fire damage.

though fire normally checks the advancement of terrestrial vegetation, its influence varies with the content of organic matter and water in the soil at the time of its occurrence. Between the turn of the century and 1930, farming, swamp drainage, and forest-clearing steadily lowered the water table in southeastern Wisconsin and caused the peat underlying the wet prairie to dry out. By 1920, a marsh in one area was dry enough that farmers could cut hay there. Following a period of below normal rainfall, a discarded cigarette ignited the marsh in August 1930. The fire persisted throughout the winter, burning the peat to an average depth of one foot and thus destroying the wet prairie vegetation, roots and all. Instead of returning to prairie and marsh vegetation, burned areas came up in aspen, evidently seeded in from a great distance inasmuch as aspen trees were not known to occur in the immediate area. By 1966, the aspen groves still remained, but their vigor was declining rapidly and many of the trees were diseased or dead. Left undisturbed, the aspen groves will eventually be invaded by hardwoods—in fact, elm had already appeared by 1940. Periodic fire, however, would preserve the aspen in a new climax because aspen resprouts from its roots following fires intense enough to kill all other hardwoods. Fire, then, plays various roles in plant succession, depending on its intensity and frequency. Although fire is usually thought of as retrogressive, returning a community to earlier seral stages, the complete destruction of prairie vegetation by an especially intense fire allowed the invasion of an area by aspen trees and moved the habitat closer to a hardwood forest climax.

Grazing pressure also can modify the climax (Harper 1969). Grassland can be turned into shrubland by intense grazing. Herbivores kill or severely damage perennial grasses and allow shrubs and cacti unsuitable for forage to establish themselves. Most herbivores graze selectively, suppressing favored species of plants and bolstering competitors that are less desirable as food. On the African plains, grazing ungulates follow a regular succession of species through an area, each using different types of forage (Vesey-Fitzgerald 1960, Gwynne and Bell 1968). By excluding wildebeest, the first of the successional species, from large fenced-off areas, McNaughton (1976) was able to show that the subsequent wave of Thompson's gazelles preferred to feed in areas previously used by wildebeest. Apparently, heavy grazing by wildebeest stimulates growth of the preferred food plants of gazelles.

Transient and cyclic climaxes develop where climax conditions are unstable.

We view succession as a series of changes leading to a climax, determined by, and in equilibrium with, the local environment. Once established, the beech-maple forest is self-perpetuating and its general appearance does not change in spite of the constant replacement of individuals within the community. Yet not all climaxes are persistent. A simple case of a *transient climax* would be the development of animal and plant communities in seasonal ponds—small bodies of water that either dry up in the summer or freeze solid in the winter and thereby regularly destroy the communities that become established each year during the growing season. Each spring the ponds are restocked either from larger, permanent bodies of water, or from spores and resting stages left by plants and animals before the habitat disappeared the previous year.

Succession recurs whenever a new environmental opportunity appears. For example, dead organisms are a resource for a wide variety of scavengers and detritus feeders. On African savannas, carcasses of large mammals are fed upon by a succession of vultures, beginning with large, aggressive species that devour the largest masses of flesh, followed by smaller species that glean smaller bits of meat from the bones, and finally by a kind of vulture that cracks open bones to feed on the marrow (Kruuk 1967). Scavenging mammals, maggots, and microorganisms enter the sere at different points and assure that nothing edible remains. This succession has no climax, however, because all the scavengers disperse when the feast is concluded. Nonetheless, we may consider all the scavengers a part of a climax, which is the entire savanna community.

In simple communities, particular life-history characteristics in a few dominant species can create a *cyclic climax.* Suppose, for example, that species A can germinate only under species B, B can germinate only under C, and C only under A. This situation would create a regular cycle of species dominance in the order A, C, B, A, C, B, A . . . with the length of each stage determined by the life span of the dominant species. Stable cyclic climaxes, which are known from a variety of localities, usually follow the scheme presented above, often with one of the stages being bare earth (Watt 1947, Forcier 1975, Sprugel 1976). Wind or frost heaving sometimes drives the cycle. When heaths suffer extreme wind damage, shredded foliage and broken twigs create an opening for further damage and the process becomes self-accelerating. Soon a wide swath is opened in the vegetation; regeneration occurs only on the protected side of the damaged area while wind damage further encroaches upon the exposed vegetation. As a result, waves of damage and regeneration move through the community in the direction of the wind. If we watched the sequence of events at any one point, we would witness a healthy heath being reduced to bare earth by wind damage and then regenerating in repeated cycles (Figure 38-13). Similar cycles occur where hummocks of earth form in windy regions around the bases of clumps of grasses. As the hummocks grow, the soil becomes more exposed and better drained. With these changes in soil quality, shrubby lichens take over the hummock and exclude the grasses around which the hummock formed. Shrubby lichens are worn down by wind erosion and eventually are replaced by prostrate lichens, which resist wind erosion but, lacking roots, cannot hold the soil. Eventually the hummocks are completely worn down and grasses once more become established and renew the cycle.

Frost action and wind erosion work together in alpine meadows of the Rocky Mountains to produce a mosaic of vegetation patches. Sedges become established in wet hollows and start the process of hummock building. As peat accumulates in the developing hummock, water is absorbed by the organic detritus. Repeated cycles of freezing and thawing begin to thrust the hummock upward. This process continues until the top of the hummock protrudes above the snow surface in winter, at which point the grasses

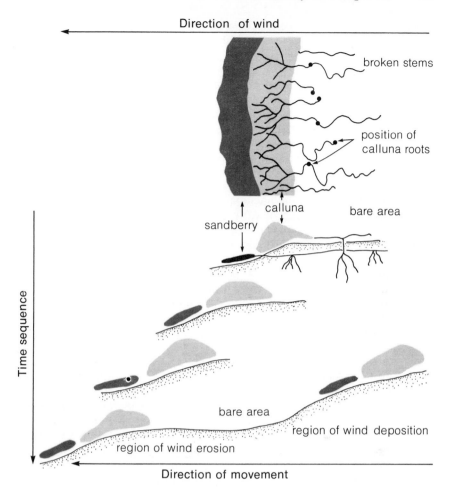

Figure 38-13   Sequence of wind damage and regeneration in the dwarf heaths of northern Scotland (after Watt 1947).

and herbaceous perennials at the crest are killed by exposure, and wind erosion begins to wear away the hummock, forming a hollow and completing the cycle.

   Mosaic patterns of vegetation types are common to any climax community where the death of individuals alters the environment. Treefalls open the forest canopy and create patches of habitat that are dry, hot, and sunlit compared to the forest floor under unbroken canopy. These openings are often invaded by early seral forms, which persist until the canopy closes (Aubreville 1938, Forcier 1975, Williamson 1975). Treefalls thus create a mosaic of successional stages within an otherwise uniform community. Indeed, adaptation by some species to grow in the particular conditions created by different-sized openings in the canopy could enhance the overall diversity of the climax community. Similar models have been developed for intertidal regions of rocky coasts, where wave damage and intense predation continually open new patches of habitat (Dayton 1971, Levin and Paine 1974, Connell 1978).

Cyclic patterns of changes and mosaic patterns of distribution must be incorporated into the concept of the community climax. The climax is a dynamic state, self-perpetuating in composition, even if by regular cycles of change. Persistence is the key to the climax. If a cycle persists, it is inherently as much a climax as an unchanging steady state.

The characteristics of dominant species change during succession.

Succession in terrestrial habitats entails a regular progression of plant forms. Plants characteristic of early stages and late stages of succession employ different strategies of growth and reproduction. Early-stage species are opportunistic and capitalize on high dispersal ability to colonize newly created or disturbed habitats rapidly. Climax species disperse and grow more slowly, but shade tolerance as seedlings and large size as mature plants gives them a competitive edge over early successional species. Plants of climax communities are adapted to grow and prosper in the environment they create, whereas early successional species are adapted to colonize unexploited environments.

Some characteristics of early and late successional stage plants are compared in Table 38-2. To enhance their colonizing ability, early seral species produce many small seeds that usually are wind dispersed (dandelion and milkweed, for example). Their seeds are long-lived, and they can remain dormant in soils of forests and shrub habitats for years until fires or treefalls create the bare-soil conditions required for germination and growth. The seeds of most climax species, being relatively large, provide their seedlings with ample nutrients to get started in the highly competitive environment of the forest floor (Salisbury 1942).

The survival of seedlings in the shaded environment of the forest floor is directly related to seed weight (Figure 38-14). The ability of seedlings to

TABLE 38-2

General characteristics of plants during early and late stages of succession.

Character	Early stage	Late stage
Seeds	Many	Few
Seed size	Small	Large
Dispersal	Wind, stuck to animals	Gravity, eaten by animals
Seed viability	Long, latent in soil	Short
Root/shoot ratio	Low	High
Growth rate	Rapid	Slow
Mature size	Small	Large
Shade tolerance	Low	High

survive the shade conditions of climax habitats is inversely related to their growth rate in the direct sunlight of early successional habitats (Grime and Jeffrey 1965). When placed in full sunlight, early successional herbaceous species grew ten times more rapidly than shade-tolerant trees. Shade-intolerant trees, like birch and red maple, had intermediate growth rates. Shade tolerance and growth rate represent a trade-off; each species reaches a compromise between those adaptations best suited for its place in the sere.

The rapid growth of early successional species is due partly to the relatively large proportion of seedling biomass allocated to leaves (Abrahamson and Gadgil 1973). Leaves carry on photosynthesis, and their productivity determines the net accumulation of plant tissue during growth. Hence the growth rate of a plant is influenced by the allocation of tissue to the root and the aboveground parts (shoot). In the seedlings of annual herbaceous plants, the shoot typically comprises 80 to 90 per cent of the entire plant; in biennials, 70 to 80 per cent; in herbaceous perennials, 60 to 70 per cent; and in woody perennials, 20 to 60 per cent (Monk 1966).

The allocation of a large proportion of production to shoot biomass in early successional plants leads to rapid growth and production of large crops of seeds. Because annual plants must produce seeds quickly and copiously they never attain large size. Climax species allocate a larger proportion of their production to root and stem tissue to increase their competitive ability; hence they grow more slowly. The progression of successional species is therefore accompanied by a shift in the compromise between adaptation for great dispersal power and adaptation for great competitive ability.

The biological properties of a developing community change as species enter and leave the sere. As a community matures, the ratio of biomass to productivity increases; the maintenance requirements of the community also increase until production no longer can meet the demand, at which point the net accumulation of biomass in the community stops (Odum 1969). The end of biomass accumulation does not necessarily signal the attainment of

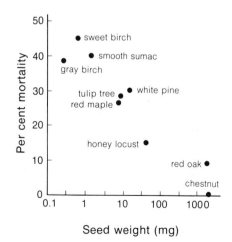

Figure 38-14   Relationship between seed weight and the survival of seedlings after three months under shaded conditions (after Grime and Jeffrey 1965).

climax; species may continue to invade the community and replace others whether the biomass of the community increases or not. The attainment of a steady-state biomass does mark the end of major structural change in the community; further changes are limited to the adjustment of details.

As plant size increases with succession, a greater proportion of the nutrients available to the community are tied up in organic materials. Furthermore, because the vegetation of mature communities has more supportive tissue, which is less digestible than photosynthetic tissue, a larger proportion of their productivity enters the detritus food chain rather than the consumer food chain. Other aspects of the community change as well. Soil nutrients are held more tightly in the ecosystem because they are not exposed to erosion; minerals are taken up more rapidly and stored to a greater degree by the well-developed root systems of forests; the environment near the ground is protected by the canopy of the forest; conditions in the litter are more favorable to detritus-feeding organisms.

Ecologists generally agree that communities become more diverse and complex as succession progresses (Margalef 1968, Odum 1969), although Whittaker (1975) suggested that, in some seres, intermediate stages of succession may be more diverse because they contain elements of early seral stages as well as elements of the climax community. It is not known whether the increase in the diversity of a community during its early stages of succession is related to increased production, greater constancy of physical characteristics of the environment, or greater structural heterogeneity of the habitat. Furthermore, there is no reason to suspect that gradients of diversity along a successional continuum are related to the same factors that determine diversity along a structurally analogous gradient of mature communities.

The study of succession leads us to formulate four basic ecological principles. First, succession is one-directional; good colonizers with rapid growth and high tolerance of conditions on disturbed or newly exposed habitats are replaced by slowly growing species with great competitive ability. Second, successional species alter the environment by their structure and activities, often to their own detriment and to the benefit of other species. Third, the climax community is not a unit, rather it represents, at any given place, a point on a continuum of possible climax formations. The nature of the climax is influenced by climate, soil, topography, fire, and the activities of animals. Fourth, the climax may be a changing mosaic of successional stages maintained by wind, frost, or other sources of mortality acting locally within the community.

Successional changes are an inherent response of vegetation to disturbance, tending to reinstate the particular plant formation characteristic of the habitat. Succession occurs by the total replacement of populations of some species by populations of others, and thus cannot be compared to the homeostatic responses of an organism. Climax communities, nonetheless, are endowed with an inherent stability of structure and function, in part the sum of organism homeostasis and population responses, but in part the unique property of community organization. We shall examine community stability further in Chapter 43.

39

The Niche Concept in
Community Ecology

Community attributes such as number of species and their relative abundances are superficial measures that reflect characteristics of a habitat and the interactions among species that live there. By relating such attributes to variation in the environment, ecologists have attempted to sort out the factors that regulate community structure and function. We have, however, become increasingly aware that simple correlations cannot elucidate community organization. Furthermore, the study of organization at the community level is made difficult by the nature of communities themselves.

Comparisons between communities in different environments rarely constitute "natural" experiments, in which all variables but one are controlled. Environments usually differ by several characteristics, making it difficult to distinguish between hypotheses based on the influences of different variables. For example, tropical and temperate regions differ in a variety of properties including seasonality and constancy of climate, average temperature, average precipitation, and organic production, each of which has been argued to influence the diversity of species in a community. As so-called natural experiments—comparisons between communities—are difficult to interpret, we might consider manipulating the environment of a single community and observing the response in its structure and function. But, in addition to being impractical, such experiments often are invalid because evolutionary adjustments of populations to the new conditions cannot be accounted for. Whereas comparisons between communities usually are not experiments, experimental manipulations of communities usually are not natural. For example, when Paine (1966) removed starfish from an experimental area of rocky coast in Washington, he observed a decrease in the number of species of their prey (see page 705). But a similar result could also have been brought about by other types of disturbance: raising the area to a higher level in the intertidal zone, artificially altering the tidal cycle, changing the average temperature of the water, or applying a poison. The response of the community to any environmental disturbance is a short-term adjustment; over a longer period, species could evolve new adaptations in response to the changed conditions, or other species could invade the community from elsewhere, and the number of species could regain its original level or, perhaps, reach a higher level. Of course, ecologists cannot wait for communities to return to evolutionary equilibrium following manipulation. Results of experiments on natural communities must be interpreted cautiously, particularly when evolved adjustments among species are tampered with.

The most fundamental difficulty for the study of community ecology arises because observed patterns of community structure are the outcome of ecological and evolutionary interactions between populations; but they in no way reveal the mechanisms of community function. Population processes are translated into community patterns by a complex array of interactions, through which most of the "information" about those processes is lost. To understand the mechanisms that underlie community patterns, questions asked about community structure must be rephrased in terms of such population processes as predation, competition, and evolution. We must begin to look to the adaptations of individuals and their expression in population dynamics for explanations of community patterns. The number of species that coexist in a community can be understood in terms of the outcome of competition between them and in terms of the factors that promote species production. The complexity of community organization—the number of links in the web of interaction between species—can be understood in terms of the variety of prey eaten by each predator and the effectiveness of the adaptations of prey to escape predators. In other words, community structure reflects the adaptations of each species in the community—adaptations that are selected in part by the activities of coexisting species.

The niche concept expresses the relationship of the individual to all aspects of its environment.

The reorientation of community ecology towards a population view of community organization has taken place around the concept of the *niche,* particularly as it has been envisioned by G. E. Hutchinson (1958) and others following in his path. Before ecology and population biology had become thoroughly fused, the niche had been given a variety of meanings including Joseph Grinnell's (1917) use of niche to describe the habitats and habits of birds and Charles Elton's (1927) use of niche to describe the species' place in the biological environment—its relationship to food and enemies. Gause (1934) made the connection between the degree to which the niches of two species overlapped and the intensity of competition between them, but the idea was not followed up until Hutchinson defined the niche concept formally. Ideally, he said, one could describe the activity range of each species along every dimension of the environment, including such physical and chemical factors as temperature, humidity, salinity and oxygen concentration, and such biological factors as prey species and resting backgrounds against which an individual may escape detection by predators. Each of these dimensions can be thought of as a dimension in space. If there are *n* dimensions, the niche is described in an *n*-dimensional space, of which it occupies a certain volume. Of course, we cannot visualize a space with more

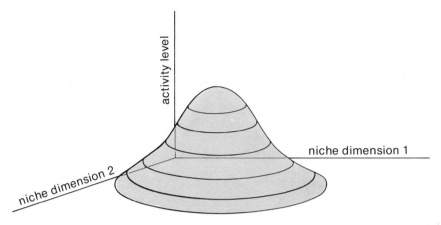

Figure 39-1   A diagram of the activity space of a species plotted on two dimensions. Contour intervals correspond to level of activity.

than three dimensions; the concept of the n-dimensional niche is therefore an abstraction. But it is possible to handle multidimensional concepts mathematically and statistically, and we can understand them with the aid of physical or graphical models in three or fewer dimensions. For example, the graphs relating biological activity to environmental gradients (see Figure 14-6, 15-1, 16-1) may be thought of as representing the distribution of a species' activity along one niche dimension. In two dimensions, a similar diagram would resemble a hill, with contours representing the various levels of biological activity (Figure 39-1).

Each species occupies a part of the n-dimensional volume—its niche—that represents the total resource space of the community. We may think of the total niche space of a community as a volume into which are fit the niches of all the species of a community, as one would pack balls of various sorts into a box. The number of species in the community depends upon the total amount of niche space and the average size of each species' niche. Community ecologists are interested in the factors that determine both these quantities. But one must first find ways to measure the niche.

Think of the niche once more as a bell-shaped curve of biological activity distributed along a single resource dimension (Figure 39-2). Viewed in this way, the most important characteristics of each species' niche are its height (the maximum level of activity or rate of resource procurement) and its breadth (variety of resources utilized). The overlap between the niches of any two species is determined by their breadths and the distances between the optima of each species. Because niches are not likely to be symmetrical, it may be more fruitful to measure niche breadth and overlap directly, and to ignore the distance between optima. This is the practice normally followed by community ecologists (Pianka 1973, Cody 1974). Before describing the results of studies of niche organization, some additional properties of resource spaces and niches should be mentioned. These properties remind us

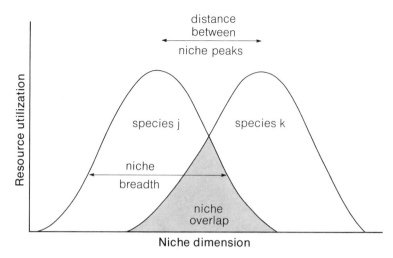

Figure 39-2   The activity curves of two species along a single resource dimension showing niche breadth, the distance between the centers of two species' niches, and niche overlap.

that niches are not smooth curves along a few simple dimensions of the environment, and that our ability to measure the niche is likely to fall as short of reality as our ability to conceptualize the niche relationships in a community.

Some resource dimensions may be continuous variables, as temperature and humidity are, but others are discrete and discontinuous. How does one place, for example, the many species of prey organisms in a community along one or more dimensions in space? Continuous dimensions do not provide equal ecological opportunities along their entire length (Schoener 1965). For example, small prey often are more abundant than large prey; the abundance of earthworms and other soil organisms varies with the acidity and moisture content of the soil; the suitability of vegetation to herbivores tends to decrease with increasing levels of cellulose, tannins, and toxic materials. Measurements of niche breadth and overlap are diminished by our ignorance of the resource dimension itself. Another difficulty in measuring the niche is that the combined resource utilization of a population is greater than that of any one individual (Roughgarden 1974). All individuals are specialized to some degree with respect to the population as a whole. Each individual spends its life within a limited geographic area and restricts its activities to a portion of the total habitat space occupied by the population. This specialization arises in part because the individual has limited time and dispersal range, but it may be based on learned or inherited behavioral variation, morphological variation based on genetic polymorphism, sexual dimorphism, or age. In addition to this between-individual component to the niche of a population, niche space shifts in a daily and seasonal rhythm, following the ever-changing resource space of the habitat. The relationship of each species in the community to every other species must also change continually. Niches are, therefore, so difficult to measure that field observations on the ecological relationships of the species in a community probably could not be relied on as adequate descriptions. At best, we can hope to obtain only fragmentary measures that indicate relative degrees of niche breadth and overlap within communities. Such glimpses are only as valuable as their ability to distinguish hypotheses about community organization based on niche theory.

Niche relationships can be investigated using ecological overlaps of species in nature.

The key to understanding the interactions between species is not in attempting to describe completely the niches of all species, which clearly is impossible, but to measure the degree to which the niches of any two species overlap. This sharing of resources or environmental niche space is often thought of as representing the degree of competition between two species (MacArthur 1968). But although competition must be based on the sharing of resources, competitive ability measured in terms of the α-values of the Lotka-Volterra equations (see page 567) depends as well on the efficiency of exploitation and conversion of resources into population biomass. For this reason, we must interpret ecological overlaps between species with caution. As we have seen in Chapter 32, the degree of competition between two species can be measured with confidence only in perturbation experiments. The relationship between the results of such manipulations and ecological measures of overlap has never been established.

Ecological overlaps usually are measured in terms of items in the diet, period of activity, microhabitat utilized for foraging, feeding method, or any combination of these. The particular dimensions of the niche included in any study depend partly on what is easy to measure and partly on inituitive notions about important niche dimensions. In most studies, these dimensions usually number no more than three or four (Schoener 1974).

In his extensive studies on lizard communities in desert habitats, Eric Pianka (1973) used time of activity (see, for example, page 166), microhabitat preference (open sun, shade; ground, grass, bush, tree), and type of prey eaten (termites, beetles, locusts, ants, etc.) to characterize the niche dimensions of each species. Niche breadths were calculated by the equation

$$B = \frac{1}{\sum\limits_{i=1}^{n} p_i^2}$$

where p_i is the proportion of records for a species in each category (i) of a particular niche dimension. B takes on values between 1 (low niche breadth) and the number of categories (n) recognized for a given niche dimension. For example, if a species of lizard ate 10 per cent beetles, 20 per cent termites, and 70 per cent grasshoppers, its niche breadth on the food dimension would be $1/(0.10^2 + 0.20^2 + 0.70^2) = 1/(0.01 + 0.04 + 0.49) = 1/0.54 = 1.85$.

Overlaps between two species (j and k, for example) were calculated by the equation

$$O_{jk} = \frac{\sum\limits_{i=1}^{n} p_{ij}p_{ik}}{\sqrt{\sum\limits_{i=1}^{n} p_{ij}^2 \sum\limits_{i=1}^{n} p_{ik}^2}}$$

where p_{ij} is the proportion of species j's activity recorded from category i of a particular resource dimension. The simplifications and potential measurement biases in a study of this kind are obvious but, recognizing the amount of effort required to do even this much analysis of community organization and the lack of more sophisticated analyses, let us see the results of Pianka's study.

Lizard communities in three continents were studied: North America, with four to eleven species per locality; the Kalahari Desert of southern Africa, with twelve to eighteen species per locality; and the Western Australian Desert, with eighteen to forty species. Average niche breadths of species were similar on the three continents (time: 5.3, 5.6, 5.1; microhabitat: 2.2, 3.4, 2.9; food: 4.4, 3.8, 3.9) with the exception that more burrowing species occurred in Australia and Africa than in North America. Because niche breadths do not appear to vary with the number of lizard species, variation in diversity would have to be accompanied either by variation in the total niche volume occupied by the community, by the degree of ecological overlap between species, or both. In fact, Pianka discovered that in the more diverse Australian communities, total niche volume was greater and ecological overlap was less than that in North American and African communities. This result suggests that the larger niche space of the Australian communities is less fully utilized by species of lizards than it is in the less diverse communities. In addition, Pianka's finding implicitly suggests that Australian lizards are limited to a lesser degree by the resource dimensions he measured. Perhaps predation is a more important component of community organization in Australia, but at this point such notions are pure speculation.

Cody (1974) analyzed niche relationships among bird communities in scrub habitats using methods similar to those of Pianka. Cody measured niche breadths and overlaps along three dimensions: microhabitat, height of feeding zone above ground, and food type as it is reflected in beak size and foraging behavior. Indices of niche overlap were calculated with formulas similar to those used by Pianka. The number of species in Cody's communities varied from five to twenty. Over this range, average niche overlaps within a community did not bear any systematic relationship to species diversity. In fact, average niche overlaps were similar in all communities, ranging between 0.43 and 0.52. A more revealing index to community organization is the overlap of each species with its nearest neighbor—that is, with the ecologically most similar species having the highest overlap value. The average overlap among nearest neighbors in a community tends to increase in direct relation to the number of species in a community, suggesting that the more diverse communities are more tightly packed with

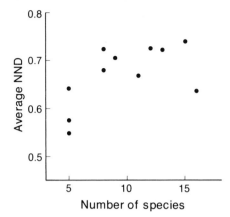

Figure 39-3   Relationship between nearest neighbor distance (NND), based on ecological overlap of species, to number of species in the community (from data in Cody 1974).

species (Figure 39-3). This finding is exactly opposite to that of Pianka, who found less overlap with increasing diversity in lizard communities.

Approaches similar to that of Pianka and Cody have been employed in a number of studies of niche organization among other kinds of organisms, including ants (Culver 1974), stem boring insects (Rathke 1976), lizards (Schoener 1968), fish (Werner 1977), and rodents (Brown 1975). As yet, these studies have failed to reveal any general patterns of community organization. Part of the problem may be that each kind of species or each type of community obeys different laws. At this point, however, a more crucial part of the problem may be one of field techniques and analysis. As they are measured by the ecology of species in their environments, niche breadth and overlap depend upon the structure of the habitat and the particular behavior of the species being considered. As a result, values of niche breadth and overlap obtained in different habitats cannot be compared properly because they are measured against different backdrops. Similarly, values for different kinds of species—like lizards and birds—cannot be compared because categories of feeding behavior, microhabitat utilization, and food type cannot be compared.

Morphological similarity provides a habitat-free measure of ecological similarity.

One way to avoid the habitat biases of ecological overlap measures is to treat an organism's structure as a mirror of its niche. The general principle that an organism's structure is adapted to the particular environment it inhabits is well established. Working from this premise, we may reasonably assume that the niche relationships of species in a community are reflected in the relationships among their adaptations. Structural measurements are independent of the habitat, so the organization of a bird or lizard community in rain forest can be compared to that in desert scrub habitats. Other advantages of morphological analysis include the lack of behavioral flexibility and its attendant sampling problems, and relative ease and consistency of measurement. Of course, there are pitfalls. One must choose morphological attributes that are important to the ecology of the species; one must assume that morphological distance bears a consistent relationship to ecological similarity; and one must assume that behavioral flexibility is not a major organizing principle in biological communities. Another drawback of morphological studies is that one cannot measure overlap or degree of ecological specialization, only ecological distance. In most cases, morphological variation within a species is much less than morphological differences between species. In some cases, however, sexual dimorphism and size variation related to age may contribute substantially to the niche space occupied by a population.

The results of morphological studies on community organization are more consistent than those of ecological studies, although both techniques are new and conclusions are still tentative. In a morphological analysis, the degree of packing of species into niche space can be measured by the average distance between nearest neighbors in morphological space. Total morphological niche volume occupied by the entire community could be estimated by the average distance between farthest neighbors or by using complicated statistical methods to describe the morphological volume of a community.

Morphological analyses of communities of birds and bats indicate that nearest neighbor distance is relatively fixed while total morphological volume increases in direct relation to number of species. If morphological volume corresponded to ecological variety, this result would suggest that there is a limiting degree of similarity beyond which competition precludes coexistence (page 586) and that species are added to communities only when ecological opportunities are increased.

In a study of bat communities in temperate and tropical localities, Fenton (1972) used only two dimensions to define morphological space: the ratio of ear length to forearm length and the ratio of the lengths of the third and fifth digits of the hand bones in the wing. The first is a measure of ear length relative to body size and is related to the bat's sonar system and hence to the type of prey it is able to echolocate. The second dimension is related to the shape of the wing—whether it is long and thin or short and broad— which is related to flight characteristics of the bat and, in turn, to the type of prey it can pursue and capture efficiently. When each species of bat in a community is plotted on a graph whose axes are the two morphological dimensions, one can visualize the niche relationships among species (Figure 39-4). In the less diverse community in Ontario, the species have similar morphology reflecting the fact that they are all small insectivorous species. In the more diverse community in Cameroun, the total morphological space occupied is greatly enlarged, corresponding to the greater variety of ecological roles played by bats there. To small insectivorous species are added fruit-eaters, nectar-eaters, fish-eaters, and large, predatory bat-eaters. More detailed analyses of bat communities by Findley (1973, 1976) have verified Fenton's finding that morphological space increases in direct proportion to number of species.

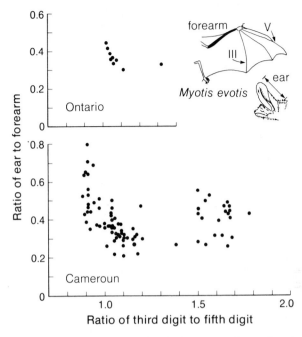

Figure 39-4   Distribution in morphological space of species in the aerial-feeding bat faunas of southeastern Ontario (above) and Cameroun (below). The horizontal axis is the ratio of the lengths of the third and fifth digits of the hand and the vertical axis the ratio of ear length to forearm length (after Fenton 1972).

Karr and James (1975) have shown the close correspondence between morphology and ecology in birds. In addition, they demonstrated that species of birds in tropical communities occupied greater morphological space than species in a less diverse temperate zone sample. Their result agrees fully with the findings of Findley and Fenton on bat communities.

With Joseph Travis, I have undertaken a study of species packing in morphological space in a variety of bird communities. One series of communities that has been analyzed is the group of scrub habitats studied by Cody (1974), for which there is a corresponding ecological analysis of community organization. In our analysis, morphological space is defined by eight dimensions, which are the logarithms of eight measurements that can be obtained from museum skins. Figure 39-5 shows how the morphological space looks when portrayed in two dimensions, in this case tarsus length and wing length. Logarithms of measurements are used for two reasons: nearest neighbor distances tend to be more uniform throughout the morphological space than they are when linear dimensions are used, and straight lines in morphological space represent ratios and products of measurements and thus correspond to size and shape. For example, a line on Figure 39-5 that runs diagonally from lower left to upper right would have the equation *log tarsus = c + log wing* which is equivalent to *c = log (tarsus/wing)*. Be-

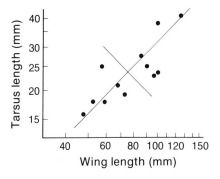

Figure 39-5   Relationship between the logarithm of wing length and the logarithm of tarsus length among passerine birds in a chaparral habitat in California.

cause the ratio of tarsus to wing is constant, the line is a size variable incorporating both wing and tarsal length. A diagonal line perpendicular to this one would have the equation *log tarsus = c − log wing* or *c = log (tarsus × wing)*. Because the product of tarsus and wing is constant, the line is a shape variable representing the ratio of tarsus to wing. On one end of the line, one would find such ground-feeding birds as thrushes with relatively long legs and short wings; on the other end, one would find such aerial species as swallows with relatively long wings and short legs.

Average nearest neighbor distances in morphological space for the eleven scrub communities studied by Cody are portrayed in Figure 39-6. The morphological analysis shows that the density of species packing does not vary with respect to number of species in the community. Hence species apparently are added to communities by expanding the total space and, presumably, variety of ecological roles. This result is consistent with other morphological analyses but differs from that of Cody's ecological analysis in which species packing increased in direct relation to number of species (Figure 39-3). Clearly the two approaches require some independent test of their validity before their differences can be resolved.

In searching for patterns of community structure, we seek to test hypotheses concerning community organization. Uniformity of species packing

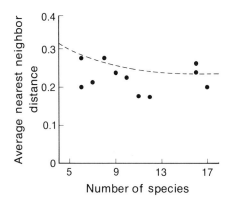

Figure 39-6   Average nearest neighbor distance in morphological space among passerine birds inhabiting eleven scrub habitats in the United States and Chile. The dashed line represents nearest neighbor distances in randomly generated communities having no interaction among species (after Ricklefs and Travis, unpubl.).

would suggest that there is some level of similarity set by competition that limits the ability of species to coexist. But what would an alternative hypothesis predict? Suppose species were assembled into communities at random, without regard to other species present and therefore without any sort of interaction between them. What would nearest neighbor patterns look like? Ricklefs and Travis constructed such communities artificially by randomly drawing species from the entire pool of species found in all eleven of the habitats in Cody's study. In these random communities, nearest neighbor distance decreases, hence species packing becomes tighter, as the number of species increases. The pattern of species packing exhibited by the natural communities differs significantly from that of the random communities and thus the existence of organization is demonstrated for natural communities.

Hypotheses based on species interaction would also predict that the spacing of individuals within morphological space would be more regular in natural communities than in randomly generated assemblages. Ricklefs and Travis measured regularity of spacing by the standard deviation of the nearest neighbor distances. As we can see in Figure 39-7, natural and random communities do not differ in this respect. The apparent lack of structure may, however, be explained by considering that although the general level of packing is determined by competition, the morphology of the species in a particular habitat is determined by their interactions both there and in numerous other habitats. The local community is a point on a continuum of open associations, as we have seen in Chapter 36. The morphology of each species has evolved independently of that of every other species in the community. We might not expect, therefore, that spacing would necessarily be regular at any particular point on the continuum.

If competition set a limiting degree of morphological similarity that permitted coexistence, nearest neighbor distances in most habitats and geographical regions would be similar. For the communities of passerine birds (songbirds) analyzed to date, this appears to be the case (Table 39-1). The habitats surveyed include temperate and tropical, island and continental, and forest and scrub, suggesting the generality of the result. It is interesting to note that on the Galapagos Islands, where the avifauna—particularly the

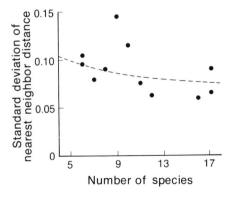

Figure 39-7   Standard deviation of average nearest neighbor distance in morphological space among passerine birds inhabiting scrub communities. The dashed line is the predicted value for randomly generated communities (after Ricklefs and Travis, unpubl.).

TABLE 39-1

Average nearest neighbor distances in morphological space for a variety of communities of birds (unpublished).

Community	Habitat	Location	Number of communities	Number of species	Average nearest neighbor distance
Passerine birds	Scrub	New World Temperate Zones	11	6–17	0.212
Passerine birds	Deciduous forest	Eastern United States	15	11–23	0.194
Passerine birds antbird flock	Rain forest	Panama	1	8	0.201
Passerine birds	Humid scrub and forest	Galapagos	13	10–13	0.200
Passerine birds	Secondary forest	St. Kitts	1	11	0.245
Hummingbird	Humid forest	Trinidad	1	9	0.167

Darwin finches—has had a long history of coexistence, nearest neighbor distances are similar to those in continental communities, about 0.20. On St. Kitts, a small island in the Lesser Antilles where all the species are recent colonists, primarily from South America and other Antillean islands, the average nearest neighbor distance is greater, about 0.25. On St. Kitts, the composition of the community may be determined more by dispersal and colonizing ability than by interactions of the species on the island. It is also worth noting that in communities of hummingbirds, which are generally thought to be food specialists, species are packed more closely in morphological space.

Escape space is a part of the niche.

The concept of the niche need not be limited to dimensions that are related to feeding. Avoidance of predation is equally important to the well-being of the population and avenues of predator escape constitute dimensions of the niche along which competition may occur. Predators should tend to focus their attention where prey resources are congregated in niche space. Where many populations use the same predator avoidance technique, predators having adaptations that enable them to exploit these prey will be strongly selected. Conversely, prey with unusual adaptations for predator escape will be strongly selected, and as a result species will tend to become uniformly distributed within the available escape space (Blest 1963, Rand 1967).

Ricklefs and O'Rourke (1975) used a morphological approach to study the packing of species of moths in escape space, in this case the variety of backgrounds against which moths conceal themselves to avoid detection by diurnal visual predators. Among cryptic species, appearance has evolved to match the background against which the species rests. Hence the morphology, or appearance, mirrors the resting place (Figure 39-8). Appearance was described by twelve characters, including morphology and position of the legs, coloration and reaction to disturbance (Table 39-2). Samples of moths were analyzed in three areas: Colorado spruce-aspen forest (43 species); Arizona Sonoran desert (51 species); and Panama lowland rain forest (203 species). The total size of the escape space was larger in Panama than in both Colorado and Arizona; that is, a greater variety of morphological characters was represented in the Panama sample (Table 39-2).

Figure 39-8   Representative species of moths from Panama, photographed against the window screens to which they were attracted by ultraviolet lights. The moths show the variety of appearances in the community (from Ricklefs and O'Rourke 1975).

TABLE 39-2

Diversity in the appearance and behavior of moths in three localities. Only three of twelve characters studied are included (after Ricklefs and O'Rourke 1975).

	Panama	Colorado	Arizona
Number of species in sample	203	43	51
Appearance of legs (per cent of species)			
hidden	29	14	8
exposed, not modified	57	81	92
exposed, modified	15	5	0
Coloration (per cent of species)			
white	10	14	4
gray	39	58	67
brown	29	12	18
green	9	0	6
yellow	7	14	4
black	5	2	2
Behavior when disturbed (per cent of species)			
stay	18	9	0
walk	13	2	14
fly	59	58	86
drop	8	30	0
Average distance to nearest neighbor	0.144	0.134	0.140

Average nearest neighbor distances based on appearance characters were nearly identical in the three samples of moths, varying between 0.134 and 0.144. These distances were calculated differently than the morphological distances in the bird communities described above and are not comparable. But the result is the same: species are added to a community by expansion of the niche space rather than by denser packing of the same space. Whether this pattern will prove to be a general one when other groups are analyzed remains to be seen. In addition, ecologists have yet to explain why some habitats present a broader range of resources than do others.

In some groups of organisms, species in more diverse communities are more specialized.

The morphological analyses of bird and bat communities discussed above indicated that, as the number of species in a community varies, niche size remains constant. To a first approximation, this means that the degree of ecological specialization is independent of species diversity. Specialization itself is difficult to define operationally and even more difficult to measure. It corresponds in some way to the range of utilization along a continuum of resources or to the variety of discrete resources exploited. Pianka's (1973) measure of niche breadth in lizard communities provides an estimate of resource specialization. In Pianka's study niche breadths did not vary noticeably in relation to diversity of species or total niche space, with the possible exception that lizards in the more diverse Australian and African communities utilized a *greater* range of microhabitats than their North American counterparts.

Scriber (1973) analyzed food-plant specialization among swallow-tail butterflies (Papilionidae) of the world. He called species generalists if the larvae had been reported feeding on plants belonging to more than one taxonomic family, and specialists if they had been recorded from only one family of food plant. Swallow-tail butterflies are most diverse in the tropics, as are most groups of organisms, and the faunas there have a low proportion

TABLE 39-3
Latitudinal gradients in specialization of larval swallow-tail butterflies on host plant families (from Scriber 1973).

Latitude belt	Number of species			Per cent generalist
	Specialists	Generalists	Total	
0–10	434	46	480	9.6
10–20	271	44	315	14.0
20–30	172	42	214	19.6
30–40	105	36	141	25.5
40–50	40	15	55	27.3
50–60	10	9	19	47.4
60–70	4	3	7	42.9

of generalists (Table 39-3). Even though the total number of generalists tends to decrease as one goes towards the poles, specialists decrease more rapidly and the proportion of generalists increases as a result. Scriber's criterion for specialization is, however, open to a number of potential biases. First, only family of food plant, not species, was recorded. Second, variation within families and differences between families of food plants may be greater in tropical areas than in temperate areas.

Degree of specialization reflects optimal breadth of prey choice.

Degree of specialization serves as an index of niche breadth in studies of community ecology like that of Scriber (1973) on swallow-tail butterflies. Specialization also derives from the feeding adaptations of predators and thus may be treated as an evolved phenomenon that influences fitness. Optimization of predator choice has been explored theoretically by a number of authors (Emlen 1966, MacArthur and Pianka 1966, Schoener 1969a, 1969b,

1971, Rapport 1971, Pulliam 1973, Cody 1974a, Pyke *et al.* 1977). All these models have one prediction in common, namely that as production of resources increases, increased specialization is favored. MacArthur and Pianka's treatment is the simplest and most general, and one aspect of their model will be considered here.

MacArthur and Pianka sought to determine by a cost-benefit analysis the number of kinds of prey that a predator should include in its diet. The optimum diet breadth was assumed to be that which minimized the time required to find and consume an individual prey. To simplify the model, we shall consider only the case in which prey species have identical nutritive value but vary in abundance and predator escape tactics. The addition of a new kind of prey to the diet has two opposing effects on the rate of consumption of prey: first, a broader diet makes more individual prey potentially available to the predator; second, the average ease of pursuit, capture, and consumption decreases provided, of course, that the predator always chooses the prey species in descending order of suitability. A predator should broaden its diet until the decrease in the average quality of its prey more than offsets any decrease in searching time due to increased abundance of prey.

When prey are ranked according to their suitability, one can graph these cost and benefit functions. In Figure 39-9, the average search time (T_s) and pursuit and handling time (T_p) per prey are graphed as a function of diet breadth. The optimum diet breadth is that which results in the lowest sum of search and pursuit time ($T_s + T_p$). The slope of the search-time curve becomes less steep with increasing diet breadth because each new prey species adds proportionately less to the total diet. The slope of the pursuit time curve increases because as the suitability of the prey decreases the capture and handling time increases more rapidly. Because of the shapes of the T_s and T_p curves, the sum of the two is always U-shaped and hence there is an optimum diet breadth.

A change in the overall abundance of prey, resulting from a change in the production of the habitat or in the number of competing species, will change the search time, but not the pursuit and handling time. Only the quantity of prey vary, not their quality. If the production of the habitat were

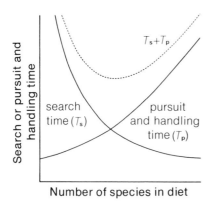

Figure 39-9   A graphical model of the optimization of diet breadth by a predator. Searching time (T_s) and pursuit and handling time (T_p) are drawn as a function of diet breadth. The sum of the two ($T_s + T_p$) is inversely related to rate of prey consumption. Hence optimum diet breadth corresponds to the minimum of the ($T_s + T_p$) curve.

to increase, the T_s curve would decrease and the optimum diet breadth would shift to the left. That is, specialization would be favored. If the level of competition increased, prey would become more scarce and the T_s curve would increase, shifting the optimum to the right and favoring increased generalization (Figure 39-10). Other effects may be considered. If the number of potential prey species were increased without increasing their variety or production, the T_s curve would be increased, the T_p curve would be reduced, and the optimum diet breadth would shift to the right. Although more species might be included in the diet, the proportion of potential prey included in the diet and the variety of prey, assessed by their predator escape characteristics, would not be altered. If the number of species were increased by increasing the variety of species, only the search time curve would increase (each prey would be less abundant) and although diet breadth would increase slightly, proportion of total prey eaten would decrease as well as rate of prey capture. In general, then, increased production leads to specialization

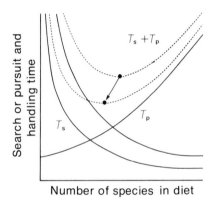

Figure 39-10   Influence   of   a change in the production of prey on optimal diet breadth. Variation in production could be related to habitat production or level of competition from other predators. As production increases, the T_s line is lowered. Rate of consumption of prey can increase and optimum diet breadth decreases.

and increased rate of consumption of prey. Increased competition is equivalent to reduced production and has an opposite effect. This presumably could lead to greater niche overlap among predators. Increased variety of prey species may lead to a somewhat broader food niche but reduces rate of prey consumption. Can we predict from these theoretical results whether tropical species should be more specialized than temperate species? Unfortunately not, because the increased production of habitats in the tropics has an effect on diet breadth opposite that of competition and increased variety of prey.

How are rate of prey consumption and diet specialization related?

In addition to predictions about diet breadth, the MacArthur-Pianka model also may be used to explore factors that influence rate of prey consumption. When diet breadth is restricted owing to increased production of prey, rate of prey consumption should increase. When diet breadth is restricted owing to more effective prey escape strategies, rate of prey consumption should decrease. There are, unfortunately, few data available to test these predictions. Among brood parasites of birds, specialized species, such as the giant cowbird (see page 80), are generally more successful, at greater expense to the host, than parasites with broad host preferences, such as the brown-headed cowbird (Nice 1937, Norris 1947, Friedmann 1963, Smith 1968). By contrast, Watt (1965) found that the number of species of trees fed upon by species of Canadian moths and butterflies has little effect on the efficiency (measured by predator population size) with which any one of the host trees is exploited. The mean population size of species of lepidoptera increased in direct proportion to the total number of tree species eaten, hence generalization does not affect exploitation. Dean and Ricklefs (1978) reached the same conclusion about the relationship between exploitation efficiency and diet breadth among species of flies and wasps parasitic on lepidoptera larvae in Ontario. In fact, the parasitism rate of wasps actually increased with greater diet breadth. Of course, none of these analyses take into account the abundances of prey species or their antipredator adaptations. Clearly the factors that influence degree of specialization merit much greater attention from ecologists than they have received.

In this chapter, we have begun to examine community properties, such as the number of coexisting species and the complexity of community organization in terms of the adaptations and ecological relationships of the species that comprise the community. Niche relationships among species have evolved against the backdrop of resources and environmental conditions from which each community ultimately derives its particular pattern of structure and function. The challenge of community ecology is to understand the evolution of ecological relationships in terms of a species' simultaneous interaction with an unresponsive physical environment and a responsive, counter-adapting and co-adapting biological environment. Clearly we are a long way from this goal but the study of niche relationships has pointed out some worthwhile directions.

Part 10

Ecology of Ecosystems

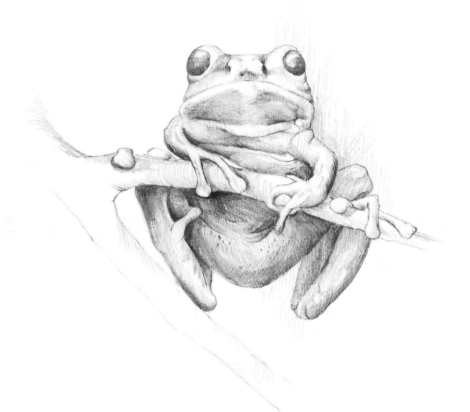

40

Primary Production

Organisms need energy to move, grow, and maintain the functions of their bodies. Energy to support these activities enters the ecosystem as light, which plants convert to chemical energy during photosynthesis. The rate at which plants assimilate the energy of sunlight is called *primary productivity.* It is important to realize that primary production underlies the entire trophic structure of the community. The energy made available by photosynthesis drives the machinery of the ecosystem. The flux of energy through populations of herbivores, carnivores, and detritus feeders, and the biological cycling of nutrients through the ecosystem are ultimately tied to the primary productivity of plants. In this chapter, we shall consider how light, temperature, water, and nutrients influence the rate of photosynthesis, and thereby determine the productivity of natural communities.

Plants assimilate energy by photosynthesis.

During photosynthesis, plants capture light energy and transform it into chemical energy in the form of organic compounds (Rabinowitch and Govindjee 1969). These compounds may be stored conveniently, and their energy later released to meet the demands of biological processes. Photosynthesis chemically unites two common inorganic compounds, carbon dioxide and water, to form glucose, a simple sugar. The overall chemical equation for photosynthesis is

$$6CO_2 \quad + \quad 6H_2O \quad \longrightarrow \quad C_6H_{12}O_6 \quad + \quad 6O_2$$

| six molecules of carbon dioxide | + | six molecules of water | yield, with an input of light energy | one molecule of glucose sugar | + | six molecules of oxygen |

Photosynthesis requires a net energy input in the form of light equivalent to 9.3 kilocalories* per gram of carbon assimilated. Because the light energy is converted to chemical energy, the metabolic breakdown of sugar into its inorganic components makes available 9.3 kilocalories per gram of carbon converted to carbon dioxide; this energy can then be utilized to perform such functions as muscle contraction and biosynthesis.

Photosynthesis is an uphill process; it requires an input of energy to drive the chemical reaction. The mere presence of the ingredients for photosynthesis is not, however, sufficient to make the reaction occur. Carbon dioxide, water vapor, and sunlight occur together in the earth's atmosphere, yet we are not deluged by a constant rain of glucose. The probability of the appropriate chemical reaction occurring there is very low. In plant cells, pigments and enzymes bring molecules and energy together so as to make the chemical reactions of photosynthesis highly probable.

All green plants have identical photosynthetic reactions. Hydrogen atoms, usually obtained from water, combine with carbon and oxygen to form a sugar molecule. Some bacteria have evolved alternative biochemical pathways for producing glucose. For example, photosynthetic sulfur bacteria obtain the hydrogen needed to form sugars from hydrogen sulfide (H_2S) rather than from water. The resulting chemical equation is

$$12H_2S + 6CO_2 \rightarrow C_6H_{12}O_6 + 6H_2O + 12S$$

Production of sugars by sulfur bacteria is a true photosynthetic process, however, because the source of energy is light.

The assimilation of energy is called production.

Plants are made up of more than just glucose. Various biochemical processes use glucose both as a source of energy and as a building block to construct other, more complex organic compounds. Rearranged and joined together, sugars become fats, oils, and cellulose, the basic structural material of the plant cell wall. Combined with nitrogen, phosphorus, sulfur, and magnesium, sugars are used to produce an array of proteins, nucleic acids, and pigments. Plants cannot grow unless they have all these basic building materials. Chlorophyll contains an atom of magnesium in addition to nitrogen and carbon atoms, just as hemoglobin contains an atom of iron. Thus, even though all other necessary materials might be present in abundance, a plant lacking magnesium cannot produce chlorophyll, and thus cannot grow. Plants clearly cannot function in the absence of water, either.

---

* The kilocalorie (kcal) is a measure of heat energy. It is defined as the energy required to raise the temperature of one kilogram (2.2 pounds or 1.06 quarts) of water one degree Celsius. The kilocalorie, sometimes referred to as the large calorie (Cal), well-known to diet watchers, equals 1,000 small calories (cal). Some familiar benchmark figures may prove useful. Average people utilize 2,000 to 3,000 kcal of energy daily. One kilowatt of power is equivalent to the expenditure of 860 kcal per hour; one horsepower is equivalent to 642 kcal per hour.

The amount of water required by photosynthesis is a drop compared to the bucket needed to balance transpiration. Water makes up the bulk of plant tissues and is the essential medium that makes nutrients available.

Thus the basic equation for production must include mineral inputs as well as the raw materials for photosynthesis. Many biological reactions are involved in production, but we can summarize production by a word equation:

carbon dioxide + water + minerals
(in the presence of light and within the proper temperature range) →
plant production + oxygen + transpired water

The sugars produced by photosynthesis are not wholly incorporated into plant biomass; that is, all do not serve to increase the size and number of plants. Some must be oxidized to release energy for biosynthesis and maintenance. Plants are no different from animals in this respect. They need energy to keep going.

We must distinguish two measures of production. *Gross production* refers to the total energy assimilated by photosynthesis. *Net production* refers to the accumulation of energy in plant biomass, that is to say, plant growth and reproduction. The difference between the two measures is accounted for by *respiration* by the plant. Gross production (photosynthesis) is allocated between respiration and net production. In most studies of plant productivity, particularly in terrestrial habitats, ecologists measure net rather than gross production because the techniques are less difficult and because net production indicates the quantity of resources available to heterotrophic consumers in the ecosystem.

Methods of measuring primary production vary with habitat and species.

Net production can be expressed conveniently as grams of carbon assimilated, dry weight of plant tissues, or the energy equivalent of dry weight. Ecologists use these indexes interchangeably. The energy content of an organic compound depends primarily on its carbon and nitrogen content. The proportion of carbon by weight in most plant tissues is close to the proportion of carbon in glucose, 40 per cent.* When plants convert sugars to fats and oils, oxygen is biochemically stripped from the molecules, thereby increasing the proportion of carbon. The fat tripalmitin ($C_{51}H_{98}O_6$), for example, contains 76 per cent carbon by weight. Fats and oils contain more than twice as much energy per gram as sugars and are, therefore, widely used by plants and animals for energy storage.

The energy content of a substance is estimated by burning a sample in a device called a bomb calorimeter (Paine 1971a). The guts of a calorimeter

---

* The relative weights of atoms of hydrogen, carbon, and oxygen are 1, 12, and 16, respectively. The proportion by weight of carbon in glucose ($C_6H_{12}O_6$) is therefore $72/180 = 0.40$.

are a small chamber where the sample is burned. Oxygen is forced under high pressure into the chamber to ensure complete combustion. The chamber is surrounded by a water jacket that absorbs the heat produced. The increase in temperature of a known amount of water in the jacket provides a direct estimate of the heat energy released by combustion.

The photosynthetic combination of carbon dioxide and water requires an energy input of 9.3 kcal for each gram of carbon assimilated (Rabinowitch and Govindjee 1969). The complete oxidation of a carbon compound to carbon dioxide and water should, therefore, release exactly 9.3 kcal per gram of carbon oxidized. In practice, the biochemical rearrangements involved in making most complex organic compounds alter energy values slightly. As a result, ecologists rely on established values for energy content obtained directly from calorimeters. Generally accepted amounts of energy released in oxidation are 4.2 kcal per gram of carbohydrate (sugars, starch, cellulose), 5.7 kcal per gram of protein, and 9.5 kcal per gram of fat metabolized (King and Farner 1961, Kleiber 1961).

The equation for production suggests several possible methods for measuring the primary productivity of natural habitats (Newbould 1967, Milner and Hughes 1968). Uptake of carbon dioxide and mineral nutrients, production of plant biomass, and release of oxygen are all proportional to production. Water flux would not provide a useful measure of photosynthesis because water is too abundant in the plant and the environment, and, depending upon soil moisture, temperature, and humidity, its uptake and transpiration vary independently of the rate of photosynthesis. Uptake of carbon dioxide and production of oxygen and organic matter can be measured more reliably.

Primary production in terrestrial ecosystems is usually estimated by the annual increase in plant biomass (net production). Yearly growth of annual plants is measured by cutting, drying, and weighing the plants at the end of the growing season. The harvest method is commonly used for crop and field plants in temperate regions, where most plants die back to the ground each year. Because root growth is usually ignored—roots are difficult to remove from most soils—harvesting measures the *net annual aboveground productivity* (NAAP), which is perhaps the most commonly used basis for comparing the producitivity of terrestrial communities.

The harvest method has several inherent problems. Herbivores harvest some of the net production. Root growth, as we have just noted, is difficult to measure, though the root systems of annual plants can sometimes be separated from the soil by painstaking washing. But the roots of perennials continue to grow each year, so that their biomass represents the accumulation of many years' growth. The difficulty in measurement created by root production in field habitats is compounded by branch and trunk growth in forests. Harvesting leaf fall and clippings of new twigs allows only a partial estimate of production. The annual growth of woody parts is often calculated by relating tree girth to total biomass, and then measuring annual increments in the girth of living trees (Ovington 1957, Ovington and Madgwick 1959, Whittaker and Woodwell 1968). To arrive at a measurement of total biomass, a series of trees of increasing size is cut down and divided into trunk, branch, and, sometimes, root components, which are then dried in large ovens and weighed. The annual increase in girth of living trees can then be converted to an increase in total weight. Leaves, flowers, and fruits, which are renewed each year, are collected at the end of the growing season and dried. Their weight is then added to the growth of woody parts to complete the estimate of production.

In aquatic habitats, plant production can be measured by gas exchange. The concentration of dissolved oxygen in water is so low that the input of oxygen by photosynthesis adds substantially to the oxygen already present. Under natural conditions, most of the oxygen produced by photosynthesis is either consumed by animals or bacteria, or escapes into the atmosphere. Ecologists evade problems by measuring production within sealed bottles. At desired depths beneath the surface of a natural body of water, samples of water containing phytoplankton are suspended in *light bottles,* which are clear and allow sunlight to enter, and *dark bottles,* which are opaque and exclude light. In the light bottles, photosynthesis and respiration occur together, and part of the oxygen released into the water in the bottles is consumed. Photosynthesis does not occur in the dark bottles, but respiration does consume oxygen. The change in oxygen concentration in the light bottle provides a measure of net production; by adding to that measure the oxygen removed from the dark bottle, we obtain a measure for gross production. The calculations are summarized: photosynthesis *minus* respiration (light bottle) *plus* respiration (dark bottle) *equals* gross production. The estimate for net production obtained from the light bottle includes the respiration of plants, animals, and bacteria. Only the estimate for gross production is really valid as a measure of plant productivity.

The light-and-dark-bottle technique is restricted to short-term measurements in small parts of an aquatic ecosystem. The technique cannot be applied easily to benthic algae or to whole systems. Ecologist Howard T. Odum (1956) partly solved this problem of measuring the production of entire stream communities. Rather than employ light and dark bottles, he compared the change in the oxygen content of the stream water during the day and the night, correcting for the exchange of oxygen between the stream and the atmosphere. By combining Odum's method with both light-and-dark-bottle techniques and conventional harvest methods (for large seaweeds; Mann 1973), ecologists have obtained reasonably accurate measurements of aquatic production (Bunt 1973).

For measuring photosynthesis in terrestrial ecosystems, carbon dioxide exchange is more useful than oxygen exchange because carbon dioxide is the rarer gas in the atmosphere. Small changes in carbon dioxide concentration are relatively easy to measure (the atmosphere contains only 0.03 per cent CO_2), and leaks in sampling chambers do not produce large errors. Measurement of production by carbon dioxide exchange resembles the light-and-dark-bottle technique. A portion of a habitat, or even of an individual plant, is enclosed in an air-tight chamber, and the decrease in carbon dioxide during the day is compared to the increase in carbon dioxide, owing to respiration alone, during the night. Gross production can be measured accurately in this way, although attempts to measure production in whole forest canopies (Odum and Jordan 1970, Woodwell 1970) have been fraught with technical difficulties, including leakage and problems of air conditioning large plastic enclosures, which have forced ecologists to fall back on more conventional harvest techniques.

The use of radioactive atoms of carbon, particularly the isotope ^{14}C, provides a useful variation on the gas exchange method of measuring productivity (E. P. Odum 1971). When a known amount of radioactive carbon is added, in the form of carbon dioxide, to an air-tight enclosure, plants assimilate the radioactive carbon atoms in the same proportion in which they occur in the air in the chamber. The rate of carbon fixation is calculated by dividing the amount of radioactive carbon in the plant by the proportion of radioactive carbon dioxide in the chamber at the beginning of the experiment. Thus, if a plant assimilates 10 milligrams of ^{14}C in an hour, and the proportion of radioactive carbon dioxide in the plant chamber is 0.05 (5 per cent), we calculate that the plant has assimilated carbon at the rate of 200 milligrams per hour (10 ÷ 0.05). Plant respiration eventually releases some of the assimilated carbon as carbon dioxide, which the plant can reassimilate. Measured over a one- to three-hour period, uptake of radioactive carbon allows a reliable estimate of gross productivity. After one to two days, uptake and release of radioactive carbon approach a steady state, and estimates represent net production more nearly than gross production (Whittaker 1975).

Plants use nutrients other than carbon dioxide and water to synthesize organic compounds. The disappearance of dissolved nitrates and phosphates from aquatic environments can sometimes be used as a relative measure of net production, but only under restricted conditions: Growth must occur rapidly, and plants must convert inorganic nutrients into biomass much more quickly than they are made available by decomposition of dead plants or by mixing with deep water. When production and decomposition balance each other in a steady state, decomposition releases inorganic nutrients at the same rate that they are assimilated by photosynthesis, and the concentration of dissolved nutrients does not change. Nor are nutrients necessarily accumulated by plants in fixed proportion relative to rates of production. Algae are known to take up more phosphorus when dissolved phosphates are plentiful than when they are scarce. Conversely, plants sometimes leak dissolved minerals into the environment. Many physical and chemical processes, particularly erosion, upwelling, and sedimentation, also influence nutrient concentrations in aquatic systems. Conditions permitting reliable estimation of productivity from the disappearance of inorganic nutrients usually occur only during algal blooms, which follow the quiescent winter period in temperate and arctic lakes and oceans.

A final method for estimating plant production is based on the idea that chlorophyll determines the rate of photosynthesis. Marine algae assimilate a maximum of 3.7 grams of carbon per gram of chlorophyll per hour (Ryther and Yentsch 1957). The productivity of a marine habitat may be estimated if the concentration of chlorophyll at different depths and the decrease in light intensity with depth are known. Although the chlorophyll method lacks the

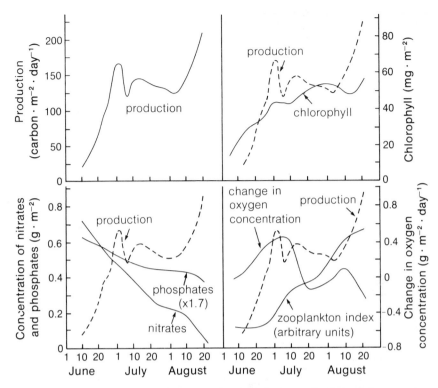

Figure 40-1   Relationship between phytoplankton production and concentrations of chlorophyll, nitrates, phosphates, oxygen, and zooplankton in Ogac Lake, Baffin Island (after McLaren 1969).

precision of gas-exchange methods, it does provide a simple and rapid index to the productivity of oceans and lakes.

Several methods for measuring the productivity of aquatic ecosystems were compared by McLaren (1969) at Ogac Lake, a landlocked fiord on Baffin Island, Canada. Primary production was measured throughout the growing season by the uptake of radioactive carbon in light bottles suspended beneath the fiord surface, but concentrations of chlorophyll, nitrates, phosphates, and dissolved oxygen were also monitored (Figure 40-1). The daily productivity of the fiord increased rapidly in early summer as ice disappeared from the surface and light began to penetrate the fiord's depths. Chlorophyll concentration paralleled the increase in productivity. Nitrate and phosphate concentrations declined throughout the summer. Dissolved oxygen increased in spring with the burst of plant production, but as the season progressed, increased zooplankton respiration obscured any direct relationship between oxygen and production. A surge in production in late summer was, curiously, unrelated to any of the factors that McLaren monitored.

Rate of photosynthesis varies in relation to light, temperature, and the availability of water and nutrients.

Light and photosynthesis.

By subjecting the photosynthetic machinery of plants to varied levels of light intensity, plant physiologists have determined the influence of light on productivity. At relatively low intensities, usually less than one-fourth the intensity of bright sunlight, the rate of photosynthesis is directly proportional to light intensity (Ryther 1956, Giese 1968). Brighter light saturates the photosynthetic pigments, however, and the rate of photosynthesis increases more slowly or levels off (Berry 1975). In many algae, very bright light reduces photosynthesis because it deactivates the photosynthetic apparatus.

The response of photosynthesis to light intensity can be characterized by two reference points. The first, the *compensation point,* is the level of light intensity at which photosynthetic assimilation of energy just balances respiration. Above the compensation point, the plant has a positive energy balance; below it, the plant suffers a net energy loss. The second reference point is the *saturation point,* above which rate of photosynthesis no longer responds to increasing light intensity.

Species differ in their response to light intensity according to their nature and the habitat in which they live. Among terrestrial plants, the compensation points of species that normally grow in full sunlight occur between 0.02 and 0.03 $kcal \cdot m^{-2} \cdot min^{-1}$; the compensation points of shade species are usually below 0.01 $kcal \cdot m^{-2} \cdot min^{-1}$. The saturation point, at which photosynthesis reaches a maximum, occurs in several groups of marine phytoplankton when light intensity falls between 0.5 and 2 $kcal \cdot m^{-2} \cdot min^{-1}$ (Figure 40-2). Above this point, photosynthesis declines rapidly. Although oak and dogwood leaves become light saturated at intensities similar to those that saturate algae, supersaturation does not depress photosynthetic activity in these species, and loblolly pine is fully light saturated only on the brightest days. In general, the saturation points of sun species occur between 0.4 and 0.6 $kcal \cdot ^{-2} \cdot min^{-1}$, and those of shade species occur between 0.09 and 0.11 $kcal \cdot m^{-2} \cdot min^{-1}$.

The sunlight that strikes the surface of a leaf is made up of a spectrum of light of different wavelengths, which we perceive as different colors (page 124). Not all colors of light are utilized in photosynthesis. Green leaves contain several pigments, particularly *chlorophylls* (green) and *carotenoids* (yellow), that absorb light and harness its energy. Carotenoids, which give carrots their orange color, absorb light primarily in the blue and green regions of the spectrum (see Figure 11-3) and reflect light with yellow and orange wavelengths. Chlorophyll absorbs light in the red and violet portions of the spectrum and reflects green, the color we perceive in leaves. The absorption spectra of whole leaves resemble the combined absorption spectra of photosynthetic pigments, but organic compounds not involved in photosynthesis evidently absorb considerable orange light (Federer and Tanner 1966). As one might expect, light under the canopy of a forest is relatively rich in the green and infrared, but poor in the red-orange and blue portions of the spectrum that are most effective in photosynthesis. Leaves of different

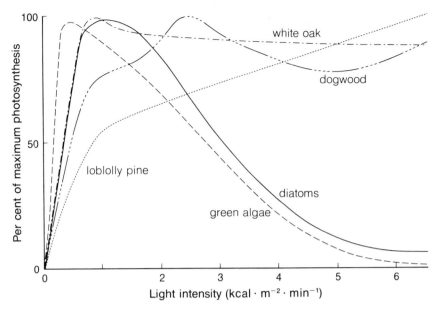

Figure 40-2   Relationship between light intensity and photosynthesis in green algae, diatoms, oak, dogwood, and loblolly pine. Photosynthetic rate is expressed as a per cent of the maximum. Light intensities above 7 kcal·m^{-2}·min^{-1} are achieved on bright summer days (after Ryther 1956, Kramer 1957).

species have different absorption spectra. Fig leaves, being thick and heavily pigmented, absorb 85 per cent of green light (550 mμ), the wavelength absorbed *least* efficiently. Tobacco leaves absorb only 50 per cent of green light.

Temperature and photosynthesis.

Temperature and light intensity normally have a close relationship in natural systems, and so their effects on photosynthesis are difficult to separate. By controlling these factors in the laboratory, one can, however, assess their separate influences. Photosynthesis is relatively insensitive to temperature at low light intensities, where light constitutes a limiting factor, but at moderate light intensity photosynthetic rate increases two to five times for each 10 C rise in temperature (Giese 1968).

Like most other physiological functions, photosynthesis is greatest within a narrow range of temperature, above which its rate declines rapidly. Because leaves absorb light, their temperatures can become great enough during the middle of the day that photosynthesis is effectively prohibited; rate of photosynthesis then reaches a peak in mid-morning and a second peak in mid-afternoon (Figure 40-3). As one would expect, the optimum temperature for photosynthesis varies with the environment, from about 16 C in many temperate species to as high as 38 C in tropical species. The opti-

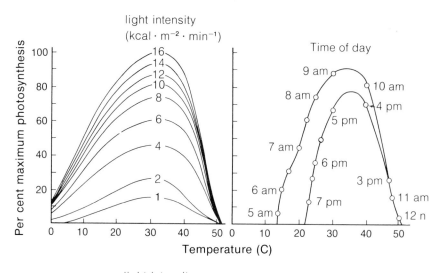

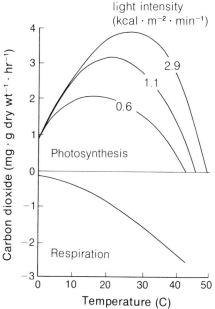

Figure 40-3   Net photosynthetic rate as a function of leaf temperature and light intensity in a typical desert bush (upper left) and the heath *Loiseleuria* (lower right). The daily course of photosynthetic rate in the desert bush (right) shows a dip in midday because the leaves become excessively hot (after Gates 1969 and Larcher *et al.* 1975).

mum temperature also varies with light intensity in some species, such as the alpine heath *Loiseleuria* in Austria. Net production depends on rate of respiration as well as rate of photosynthesis. In general, respiration increases steadily with increasing leaf temperature (Kramer 1958; Figure 40-3).

Photosynthetic efficiency is a useful index of rates of primary production of plant formations under natural conditions. The photosynthetic efficiency

TABLE 40-1

Photosynthetic efficiency (ratio of net primary production to visible light energy received) in several plant communities.

Community	Locality	Days in leaf*	Photosynthetic efficiency (per cent)	Reference
Beech	Denmark	150	1.5	Moller *et al.* 1954
Oak and pine	Long Island	365	0.9	Whittaker and Woodwell 1969
Scots pine	Britain	360	2.4	Ovington 1962
Sugar cane	Java	360	1.9	Hellmers 1964
Rice	Japan	150	2.2	Hellmers 1964
Corn	U.S.	100	1.3	Transeau 1926
Perennial grass and herbs	Michigan		1.0	Golley 1960

* Measurement of photosynthetic efficiency based on period in leaf.

is the per cent of incident visible radiation that is converted to net primary production during seasons of active photosynthesis. Where water and nutrients do not limit plant production, photosynthetic efficiency varies between 1 and 2 per cent of available light energy (Table 40-1).

Water and transpiration efficiency.

Because photosynthesis requires gas exchange across the surface of the leaf, productivity also parallels the rate of transpiration of water from the leaf surface. As the moisture content of soil decreases, plants have greater difficulty removing water from the soil and leaves must close their stomata to reduce water loss (see page 140). When soil moisture is reduced to the wilting point, leaves are effectively shut off from the surrounding air and photosynthesis slows to a standstill. Rate of photosynthesis is, therefore, closely tied to the plant's ability to tolerate water loss, to the availability of moisture in the soil, and to the influence of air temperature and solar radiation on rate of evaporation (Kramer 1969). Humid environments favor high rates of photosynthesis by reducing transpiration from leaves.

Agronomists have devised *transpiration efficiency* as an index of the drought resistance of plants. Transpiration efficiency is the ratio between net production and transpiration, expressed as grams of production per 1,000 grams of water transpired. Most plants have transpiration efficiencies of less than 2 grams of production per 1,000 grams of water transpired; some drought-resistant crops have transpiration efficiencies of 4 (Odum 1971). High efficiencies result from morphological and physiological adaptations to reduce evaporation and leaf temperature (see page 196). Regardless of their adaptations to conserve water, plants cannot escape the physical reality that if they must reduce transpiration during drought, they must also reduce the gas exchange necessary for photosynthesis, and consequently reduce their productivity.

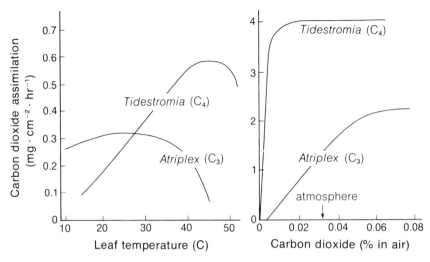

Figure 40-4   Rate of photosynthesis as a function of leaf temperature (left) and concentration of carbon dioxide (right) in a C_3 plant, *Atriplex glabriuscula,* and a C_4 plant, *Tidestromia oblongifolia* (after Berry 1975).

Solutions to the opposing problems of water retention and gas exchange have led to a variety of biochemical modifications of the assimilation of carbon dioxide by plants (Black 1973, Hochachka and Somero 1973, Laetsch 1974). In most temperate zone plants, carbon dioxide enters photosynthetic cells by diffusion. But because the level of carbon dioxide in the atmosphere is low (0.03 per cent) and the affinity of carbon dioxide for the enzymes that assimilate it in photosynthesis is weak, carbon dioxide uptake is extremely inefficient. Many plants in hot, dry climates have greatly increased affinity for carbon dioxide owing to a highly efficient carbon dioxide uptake step. As a result, these plants—called C_4 plants because carbon dioxide is first assimilated into a four-carbon carbohydrate—take up carbon dioxide from the atmosphere more rapidly and, therefore, with relatively less water loss. This is illustrated in Figure 40-4, in which rate of carbon dioxide assimilation is plotted as a function of carbon dioxide concentration in the atmosphere surrounding the plant. The honey-sweet plant *Tridestromia oblongifolia* (Amaranthus family), with C_4 metabolism, reaches maximum assimilation at less than 0.01 per cent carbon dioxide whereas the C_3 saltbush plant *Atriplex glabriuscula* does not attain its potential rate of carbon dioxide assimilation until the concentration of carbon dioxide is greater than its normal concentration in the atmosphere (0.03 per cent).

A disadvantage of C_4 metabolism is that the assimilation of carbon dioxide is low at low temperatures (Figure 40-4). The temperature optimum of C_4 plants is usually close to the maximum tolerated—about 45 C. That of C_3 plants usually lies between 20 C and 30 C. As a result, the proportions of C_4 and C_3 plants in a community vary inversely in relation to the average temperature during the growing season. C_4 plants predominate in hot climates, and C_3 species in cool climates (Teeri and Stowe 1976).

Nutrients and production.

Most habitats respond to artificial applications of fertilizers by increased primary production (Thomas 1955, Treshow 1970). No matter what the natural fertility of the habitat, nutrient availability interacts with water, temperature, and light to determine levels of production. Nutrient limitation is probably most strongly felt in aquatic habitats, particularly the open ocean, where the scarcity of dissolved minerals reduces production far below terrestrial levels. The abundance of nutrients in lakes and oceans depends on upwelling currents, water depth, proximity to coastlines and rivers, and drainage patterns of nearby land masses. Recent intense fertilization of inland and coastal water by sewage and runoff from fertilized agricultural land has greatly increased aquatic production in some inland and coastal waters. Such relationships will be discussed in greater detail in Chapter 42.

Even oxygen can limit terrestrial plant production under some circumstances. Roots require oxygen in the soil for metabolism and growth (Norman 1957, Leyton and Rousseau 1958). If the soil is nonporous, or if it is completely saturated with stagnant water, available oxygen can be reduced below the point required to support plant growth. In the waterlogged soils of swamps, many plants have structures for obtaining oxygen for their roots directly from the atmosphere. For example, the knees of cypress trees are projections of the roots above the surface of the water (Figure 40-5). Cypress knees allow free exchange of gases between the root system and the atmosphere (Kramer *et al.* 1952). The air roots, or pneumatophores, of the white mangrove *Avicennia* whose root system grows in anaerobic mud, also provide direct gas exchange between the roots and the atmosphere (Scholander *et al.* 1955).

Production in terrestrial ecosystems is determined primarily by light, temperature, and rainfall.

The favorable combination of intense sunlight, warm temperature, and abundant rainfall makes the humid tropics the most productive terrestrial ecosystem on Earth, square mile for square mile. Low winter temperatures and long nights curtail production in temperate and arctic ecosystems. Lack of water limits plant production in arid regions where light and temperature are otherwise favorable for plant growth.

The production of litter (falling leaves, fruits, and branches) provides an index to plant production in forests at various latitudes (Bray and Gorham 1964). Annual litter production varies between 900 and 1,500 grams per square meter per year ($g \cdot m^{-2} \cdot yr^{-1}$) in equatorial forests (0° to 10° latitude), 400 to 800 $g \cdot m^{-2} \cdot yr^{-1}$ in warm temperate forests (30° to 40°), 200 to 600 $g \cdot m^{-2} \cdot yr^{-1}$ in cool temperate localities (40° to 60°) and 0 to 200 $g \cdot m^{-2} \cdot yr^{-1}$ in arctic or alpine communities (60° to 70°). These data demonstrate clearly that the production of litter decreases with distance from the Equator. Litter

Figure 40-5   Cypress knees in a drained swamp in South Carolina. Under normal water levels in the swamp the knees would project above the surface, providing an avenue of gas exchange between the roots and the air (courtesy of the U.S. Forest Service).

production is known to represent about 40 per cent of total net annual above ground productivity (NAAP) in temperate forests, so aboveground primary production is about 2.5 times the litter fall.

Within a given latitude belt, where light and temperature do not vary appreciably from one locality to the next, net production is directly related to annual precipitation. Studies of net aboveground productivity in grasslands around the world show that in dry areas (Idaho, South Dakota, southwestern Africa) production varies between 90 and 200 $g \cdot m^{-2} \cdot yr^{-1}$, in moister areas (including South Carolina, Louisiana, Japan, and parts of India), where grassland is maintained by agricultural practices and is not the predominant plant association, primary production varies between 250 and 500 $g \cdot m^{-2} \cdot yr^{-1}$ (Singh 1968). Although plant production initially increases rapidly as annual precipitation increases, productivity tends to level off in temperate habitats with more than 100 cm (39 inches) of rain annually, presumably because the availability of light and nutrients becomes limiting (Figure 40-6). The production of dry habitats (less than 70 cm, or 20 inches annual rainfall) falls below the trend for moister areas; this may be the result of the more seasonal distribution of precipitation in areas of sparse rainfall.

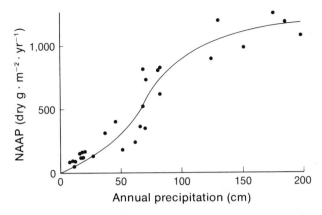

Figure 40-6   Relationship between net annual aboveground production (NAAP) and rainfall; data are primarily from temperate areas (after Whittaker 1970).

Ecologists Robert Whittaker and Gene Likens have recently estimated net primary production for representative terrestrial and aquatic ecosystems (Table 40-2). Their estimates are based on many studies using a wide variety of techniques and are probably reasonably close to true values. Production in terrestrial habitats decreases dramatically from the wet tropics to temperate regions, even more so where the climate is too dry or too cold to support forests. Swamp and marsh ecosystems occupy the interface between terrestrial and aquatic habitats, where plants are as productive as in tropical forests. Maximum rates of production in marshes have been reported to be as high as 4,000 $g \cdot m^{-2} \cdot yr^{-1}$ in temperate regions and 7,000 $g \cdot m^{-2} \cdot yr^{-1}$ in the tropics. Marsh plants are highly productive because their roots are frequently under water and their leaves extend into the sunlight and air, obtaining benefits of both aquatic and terrestrial living. In addition, rapid decomposition by bacteria of detritus that washes into marshes releases abundant nutrients.

The productivity of agricultural land usually falls somewhat below the productivity of natural vegetation in the same area because croplands are plowed each year and laid bare early and late in the growing season, when undisturbed habitats continue to be productive. Furthermore, most agricultural plantings consist of a single species, which cannot exploit the resources of the land as efficiently as a mixture of species with differing ecological requirements.

Irrigation and the application of fertilizers can increase agricultural yields two- to threefold over world averages (Wittwer 1974, 1975). Sugar cane, a common tropical crop, has a world average production of about 1,700 $g \cdot m^{-2} \cdot yr^{-1}$. Intensively cultivated sugar cane in the Hawaiian Islands has double the average world yield and a maximum productivity of about 7,000 $g \cdot m^{-2} \cdot yr^{-1}$. Poor crop management, conversely, can lead to soil deterioration and reduced production.

TABLE 40-2

Average net primary production and related dimensions of the Earth's major habitats (from Whittaker and Likens 1973).

Habitat	Net primary production ($g \cdot m^{-2} \cdot yr^{-1}$)	Biomass ($kg \cdot m^{-2}$)	Chlorophyll ($g \cdot m^{-2}$)	Leaf surface area ($m^2 \cdot m^{-2}$)
Terrestrial				
Tropical forest	1,800	42	2.8	7
Temperate forest	1,250	32	2.6	8
Boreal forest	800	20	3.0	12
Shrubland	600	6	1.6	4
Savanna	700	4	1.5	4
Temperate grassland	500	1.5	1.3	4
Tundra and alpine	140	0.6	0.5	2
Desert	70	0.7	0.5	1
Cultivated land	650	1	1.5	4
Swamp and marsh	2,500	15	3.0	7
Aquatic				
Open ocean	125	0.003	0.03	—
Continental shelf	360	0.01	0.2	—
Algal beds and reefs	2,000	2	2.0	—
Estuaries	1,800	1	1.0	—
Lakes and streams	500	0.02	0.2	—

Net primary production of temperate zone cereal crops (wheat, corn, oats, and rice), hay, and potatoes varies between 250 and 500 $g \cdot m^{-2} \cdot yr^{-1}$; sugar beets commonly attain twice that productivity (Odum 1959). These values are compared to the estimates of Whittaker and Likens for temperate forests (600 to 2,500 $g \cdot m^{-2} \cdot yr^{-1}$) and temperate grasslands (150 to 1,500 $g \cdot m^{-2} \cdot yr^{-1}$). The productivity of all agricultural land varies between 100 and 4,000 $g \cdot m^{-2} \cdot yr^{-1}$, depending on the crop, with an average of 650 $g \cdot m^{-2} \cdot yr^{-1}$.

Production in aquatic ecosystems is limited primarily by nutrients.

The open ocean is a virtual desert, where scarcity of mineral nutrients—not water—limits productivity to one-tenth or less that of temperate forests (Table 40-2). Upwelling zones (where nutrients are brought up from the depths by vertical currents) and continental-shelf areas (where exchange between shallow bottom sediments and surface waters is well developed) support greater production, averaging 500 and 360 $g \cdot m^{-2} \cdot yr^{-1}$, respectively. In shallow estuaries, coral reefs, and coastal algae beds, production approaches that of adjacent terrestrial habitats, with averages approaching

2,000 $g \cdot m^{-2} \cdot yr^{-1}$. Primary production in fresh-water habitats is similar to that of comparable marine habitats.

Availability of nutrients largely determines variation in the production of aquatic ecosystems. Light apparently does not limit production within the euphotic zone. As much as 95 per cent of the incident radiation penetrates the surface of the water and is available to plants. Variation in the depth to which light penetrates, a function of the clarity of the water, influences the depth to which photosynthesis occurs but does not affect annual production per square meter of ocean surface. In clear water, algae are spread thinly throughout the deep euphotic zone; in turbid water, they are concentrated closer to the surface.

Temperature evidently does not influence the production of marine habitats. Although the photosynthetic rates of individual plants may be depressed by cold temperatures, marine algae attain great density in cold water, enough so that arctic oceans are as productive as warm tropical seas. In cold temperate waters, large seaweeds produce as much biomass per square meter of marine habitat as in the Indian Ocean and Caribbean Sea (Mann 1973). In Nova Scotia, the alga *Laminaria* alone attains a productivity of about 1,500 $g \cdot m^{-2} \cdot yr^{-1}$, a respectable value for a temperate forest!

The annual net primary production of the Earth is about 10^{18} (one billion billion) kilocalories.

The most productive habitats attain photosynthetic efficiencies of 1 to 2 per cent, but so much of the Earth's surface lacks optimum conditions for plant growth that only one-tenth of 1 per cent of the light energy striking the Earth's surface is assimilated by plants. The energy value of sunlight reaching the outer atmosphere of the Earth directly under the Sun (the *solar constant*) is about 10^7 (10 million) $kcal \cdot m^{-2} \cdot yr^{-1}$. The angle of incident radiation varies, however, with time of day and season, and over a year's time, all points on the Earth are shrouded by night for an equivalent of six months. If the total energy reaching the outer atmosphere were spread evenly over the surface of the Earth, each square meter would receive one-quarter of the solar constant, or 2.5×10^6 $kcal \cdot m^{-2} \cdot yr^{-1}$. In fact, the Earth's surface does not actually receive that much solar energy during the year, and its distribution is not uniform. Perhaps 40 per cent of the total light income of the Earth is absorbed by the atmosphere and re-radiated back into space as heat. Part is also reflected and scattered by particles of dust in the atmosphere and reflected by surfaces of water, rock, and vegetation (Geiger 1957, Barry and Chorley 1970).

The annual energy income of a particular locality varies with latitude and cloud cover. Temperate localities usually receive between 2×10^5 and 2×10^6 $kcal \cdot m^{-2} \cdot yr^{-1}$, which represents 2 to 20 per cent of the solar constant (Phillipson 1966). Only half the light energy can be assimilated by plants; the other half lies outside the absorption spectrum of plant pigments.

Whittaker and Likens (1973) estimated the total annual primary productivity of the Earth as 162×10^{15} (million billion) grams (about 730×10^{15} kcal),

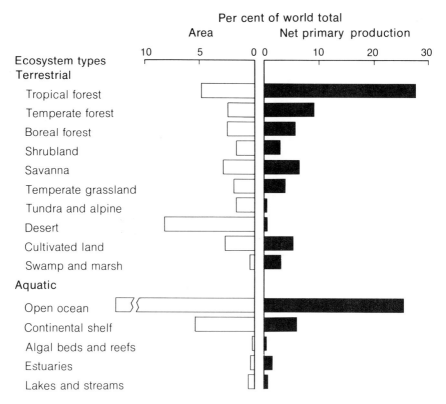

Figure 40-7   Surface area and total annual productivity of the major ecosystems. Values are expressed as a percentage of the total for the Earth (after data in Whittaker and Likens 1973).

of which terrestrial habitats are responsible for two-thirds.* The average productivity of terrestrial areas (720 g·m^{-2}·yr^{-1} or 3,200 kcal·m^{-2}·yr^{-1} represents assimilation of about 0.3 per cent of the light reaching the surface. The overall photosynthetic efficiency of aquatic habitats is less than one-quarter that of terrestrial communities.

The distribution of production among the major vegetation zones of the Earth is a function of the local productivity of these zones, determined by

* Estimates of the Earth's productivity vary considerably, but Whittaker and Likens' is probably the most soundly based. An early estimate for terrestrial productivity (Schroeder 1919) was 147 × 10^{15} kcals·yr^{-1}, and J. R. Vallentyne (quoted by Kormondy 1969:21) suggested a range of 200 to 290 × 10^{15}, compared to Whittaker's estimate of 489 × 10^{15}. Whittaker's proposed value for the oceans (248 × 10^{15} kcals·yr^{-1}) is closely matched by other estimates of 180 × 10^{15} (Ryther 1969) and 198 × 10^{15} (Steeman-Neilsen 1960), and falls within Vallentyne's range of 200 to 540 × 10^{15}. An earlier guess of 1,200 × 10^{15}, made by Riley (1938–39), reflects the fact that precise and extensive measurements of primary production have been made only within the last two or three decades (for an historical survey see Lieth 1973).

light, temperature, rainfall, and nutrients, and their total surface area (Figure 40-7). Tropical forests cover only 5 per cent of the Earth's surface, but they account for almost 28 per cent of the total production. Temperate forests and the open ocean, representing 2.4 and 63 per cent of the surface area, are responsible for 9.2 and 25 per cent of the Earth's productivity. Although inshore waters (estuaries, algal beds, and reefs) occupy only 0.4 per cent of the Earth's surface they account for 2.3 per cent of the Earth's productivity, and 6.7 per cent of all aquatic production. Swamps and marshes have similarly high productivity for their small area.

In this chapter we have examined basic patterns of primary production over the surface of the Earth. Productivity is greatest where light, warmth, water, and mineral nutrients are all abundant. Decreasing light and temperature reduce production of habitats distant from the tropical zones. Lack of moisture further restricts production in arid regions. In some aquatic habitats, scarcity of nutrients imposes the greatest limitation to production, particularly in the open ocean, but inshore areas with abundant nutrients can match or exceed nearby terrestrial communities for organic production.

Net productivity measures the overall energy input to the ecosystem. This production is eventually consumed and dissipated by herbivores, by the carnivores that eat them, and by the myriad detritus feeders that scavenge dead debris. The precise pathways that energy follows on its route through the ecosystem and the time required for its journey are the subject of the next chapter.

41

Energy Flow in the Ecosystem

Plants manufacture their own "food" from raw materials. Hence they are referred to as *autotrophs,* literally self-nourishers. Animals obtain their energy in ready-made food by eating plants or other animals. Animals are, therefore, referred to as *heterotrophs,* meaning nourished from others. The specialization of living forms as food producers and food consumers creates an energetic structure—called the *trophic structure*—in biological communities, through which energy flows and nutrients cycle. The food chain from grass to caterpillar to sparrow to snake to hawk delineates the path of organic materials and the energy and nutrient minerals they contain (Figure 41-1). Each link in the food chain, each trophic level in the community, dissipates most of the food energy it consumes as heat, motion, and, in the case of luminescent organisms, light. None of these energy forms is useful to other organisms. Hence, with each step in the food chain, the total amount of usable energy passed to the next higher trophic level becomes smaller. It is no wonder that all the grass in Africa heaped into a big pile would dwarf a like assemblage of grasshoppers, gazelles, zebras, wildebeests, rhinoceroses, and all other animals that eat grass. So much would the piles of plants and herbivores overwhelm us that we would probably not even notice the pitiful heap of lions, cheetahs, and hyenas nearby.

Energy flux is the only sound currency in the economics of ecosystem function; biomass and numbers are static descriptions of the community frozen in an instant of time. The dynamics of the community are measured in terms of change—rates of energy and nutrient transferral from organism to organism through the structure of the food web. The unique status of energy as an ecological currency has greatly stimulated the study of community energetics.

The food chain illustrated in Figure 41-1 oversimplifies nature in several ways. First, few carnivores feed on a single trophic level; second, many organisms (or their parts in the case of plants) die of causes other than predation, and their remains are consumed by detritus-feeding organisms; third, most energy assimilated by a trophic level is dissipated as heat because biological processes are energetically inefficient and organisms utilize energy to maintain themselves as well as to grow and reproduce. Energy

780

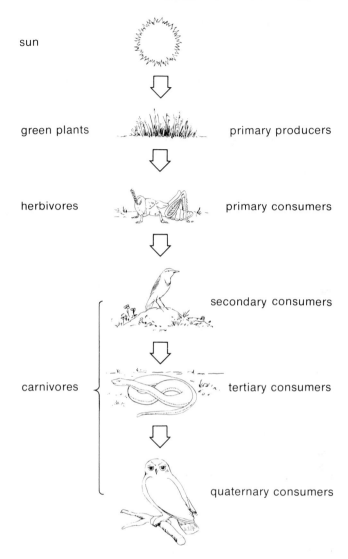

Figure 41-1   A simplified terrestrial food chain showing the sequence of trophic levels.

incorporated into growth and reproduction, which becomes the potential food of the next higher trophic level, is a small percentage of the total energy consumed and assimilated by organisms on any one trophic level.

In this chapter, we shall study the pathways of energy through the community and how the physical environment of the community influences energy flow.

A word on definitions is needed here.

Ecologists have applied many different terms to the feeding relationships of organisms. Such terms are usually introduced to clear up confusion and give precise definitions to the activity of organisms. But they often have the opposite effect. Plants and animals have been functioning perfectly well without having names applied to their activities. We should take this cue and concentrate on understanding the feeding relationships within a community, rather than try to categorize them. Many of the terms that have been used to describe trophic relationships are in the Glossary, but brief distinctions are included here as well. Lindeman (1942), Janzen (1970), and Wiegert and Owen (1971) also discuss several of these terms.

We can arrange most of the troublesome terms in pairs with opposite meanings. *Producer* and *consumer* refer to different activities performed by the same organisms, although *primary producer* is usually reserved for green plants and *primary consumer* for herbivores. Production refers to the assimilation of materials into the body of the organism, that is, growth and reproduction. The source of energy for production, whether sunlight or chemical energy in food, distinguishes primary producers from secondary, tertiary, quaternary, etc., producers. Just as all organisms produce, they also consume. Plants and animals both must metabolize assimilated substances to keep themselves going. Ecologists apparently will not consider the assimilation of sunlight by plants as an act of consumption. Herbivores, therefore, are designated primary consumers by default.

Autotroph and *heterotroph* distinguish organisms that convert inorganic forms of energy to organic forms from those whose sole source of energy is organic matter. Thus, autotrophs are primary producers. Green plants are *photo*autotrophs because they assimilate the energy of sunlight. The bacterium *Thiobacillus* is a *chemo*autotroph because it can derive energy from the oxidation of hydrogen sulfide to sulfur.

Herbivore and *carnivore* refer to eaters of plants and animals, respectively, although carnivore is sometimes reserved for meat eaters. Other terms such as insectivore and piscivore (fish eater) are used. Herbivores also come in a variety of forms—nectivores, frugivores, grazers, browsers, and so on—depending on what they eat and how they eat it. *Predator* distinguishes organisms that consume prey from *parasites,* organisms that feed on living prey, or hosts. Hence seed eaters are properly called predators because they destroy the tiny plant embryo in the seed, and many grazers and browsers are properly called parasites, because they consume leaves or buds without killing the tree. Mosquitoes and vampire bats would fit equally well in the category of browsers.

Biophage and *saprophage* distinguish organisms that eat living prey from those that go the easy route of eating dead prey. Saprophages also may be called *detritivores,* as they are in this book. Detritivores, or saprophages, are often referred to as decomposers. While detritivores *do* decompose organic compounds into simpler inorganic compounds, this function is not unique to detritivores. All organisms metabolize organic foods to obtain usable energy and release carbon dioxide and water as end products. All organisms therefore are decomposers.

Feeding relationships define food chains and food webs in the community.

Trophic levels provide a simple framework for understanding energy flow through the ecosystem. We should be able to determine the trophic level of the energy in a particular chemical bond by the number of times it has changed hands since it was first put together in a green plant. But energy takes complicated paths through the trophic structure of the ecosystem, and we cannot easily follow a particular chemical bond.

Many carnivores eat fruit and other plant materials when their normal prey is not readily available. Some carnivores eat other carnivores as well as herbivores. Whether an owl eats herbivorous field mice or insectivorous shrews is of little consequence nutritionally to the owl; both prey have about the same food value and require similar effort to capture. Large owls sometimes eat snakes or small weasels, which themselves feed on mice and shrews. Owls are not known to feed on plant material, but foxes, which normally prey on small mammals, often eat fruit. It is, therefore, not unusual for predators to feed on three or four trophic levels. And plants are not always the *victims* of animals; some partly carnivorous species are known— Venus's-flytraps, pitcher plants, and sundews. Most herbivores, however, particularly those that eat leafy vegetation, are so specialized in their food habits that they are incapable of carnivory. A cow's teeth can grind tough plant material to shreds but could not get a grip on a sheep's flesh, much less tear it apart.

The great variety of feeding relationships within the community links species into a complex *food web*. Each species feeds on many kinds of prey. Within the context of this complexity, trophic levels become abstract concepts, useful for understanding general patterns of structure and energy flow through the ecosystem, but usually quite useless as categories for assignment of individual species.

The trophic structure of a community can be described in terms of pyramids of productivity, biomass, and numbers.

The amount of energy metabolized by each trophic level decreases as energy is transferred from level to level along the food chain. Green plants, the primary producers, constitute the most productive trophic level. Herbivores are less productive, and carnivores still less. The productivity of each trophic level is limited by the productivity of the trophic level immediately below it and by the efficiency with which energy is transformed into biomass. Because plants and animals expend energy for maintenance, less and less energy is made available, through growth and reproduction, to each higher trophic level. Of the light energy assimilated by a plant, 30 to 70 per cent is used by the plant itself for maintenance functions and the energetic costs of biosynthesis. Herbivores and carnivores are more active than plants and expend even more of their assimilated energy on maintenance. As a result, the productivity of each trophic level is usually no more than 5 to 20 per

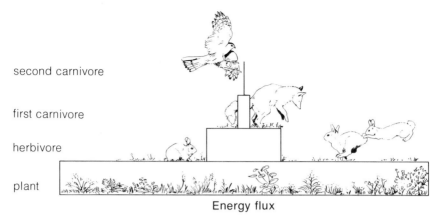

Energy flux

Figure 41-2   An ecological pyramid representing the net productivity of each trophic level in the ecosystem. This particular structure represents ecological efficiencies of 20, 15, and 10 per cent between trophic levels, but these values vary widely between communities.

cent of that of the level below it. The percentage transfer of energy from one trophic level to the next is called the *ecological efficiency* or *food chain efficiency* of the community.

The pioneer ecologist Charles Elton (1927) suggested that if each trophic level in the community were represented by a block whose size corresponded to the productivity of the trophic level, and if the blocks were then stacked on top of each other with the primary producers at the bottom, one would obtain a characteristic pyramid-shaped structure (Figure 41-2). The structure of the pyramid varies from community to community, depending on the ecological efficiencies of the trophic levels. In the particular case pictured in Figure 41-2, these efficiencies are 20, 15, and 10 per cent. Herbivore production is, therefore, 20 per cent of plant production; first-level carnivore production is 15 per cent of herbivore production and only 3 per cent of plant production (15 per cent of 20 per cent equals 3 per cent); second-level carnivore production is only 0.3 per cent of plant production. These values are probably unrealistically high compared to most natural communities, but they illustrate the universal decrease in the availability of energy at progressively higher trophic levels.

As energy availability decreases, the biomass and numbers of individuals on each trophic level usually decrease, although no law of energetics prevents a reverse trend. The biomass structure of the community resembles the pyramid of productivity in most terrestrial communities. If one were to collect all the organisms in a grassland, the plants would far outweigh the grasshoppers and ungulates that eat the plants. The herbivores in turn would outweigh the birds and large cats at the first carnivore level, and these too would outweigh their predators, if there were such. An individual lion may be heavy, but lions are spread out so thinly that they do not count for much on a gram-per-square-meter basis.

The pyramid of biomass is sometimes turned upside down in aquatic plankton communities. Algae must be more productive than the tiny animals that eat them; the laws of energetics cannot be violated. But the phytoplankton are sometimes consumed so rapidly that their numbers are kept small by herbivorous zooplankton. Intensive grazing reduces phytoplankton biomass, but the algae are so productive that they can often support a larger biomass of herbivores under optimum conditions for growth.

The pyramid of numbers is even more shaky than the pyramid of biomass. Disease organisms (parasitic bacteria and protozoa, for example), mosquitoes, ants, and others are certainly more numerous than the organisms they feed on, even though all of them together do not weigh as much as their prey or hosts. A single tree may be host to thousands of aphids, caterpillars, and other herbivorous insects.

Detritivores can be a major part of the food web.

Many ecologists attribute a distinctive role to organisms that consume dead plant and animal matter. It is frequently said that these detritus eaters are responsible for breaking down dead organic remains, which would otherwise accumulate, and for releasing their nutrients so they can be used again by plants. Detritus-consuming organisms *do* perform this function in the ecosystem, but this view of a special role in the community is misleading for two reasons. First, as we have seen, all organisms "decompose" organic matter. Terrestrial mammals consume 20 to 200 grams of food for every gram of body weight produced. The remainder is metabolized and used as a source of energy or egested. Required minerals and other nutrients are retained and incorporated into body tissue, but most ingested food is returned to the environment in an inorganic form: carbon dioxide and water are exhaled; water and various mineral salts are excreted in sweat and urine.

Second, detritivores have no special purpose different from any other organism in the overall function of the ecosystem. The individual detritivore obtains energy and nutrients in its food just like herbivores and carnivores. And just like all other organisms, detritivores leave behind undigestible remains, breakdown products of metabolism, and excess minerals that they cannot use. Detritivores eat the garbage of the ecosystem for the same reason herbivores and carnivores eat fresh food: to make a living. It is all a matter of taste.

Detritivores include such diverse species as carrion eaters—crabs, vultures, and the like, whose freshly dead food differs little from the live prey eaten by carnivores—and bacteria and fungi, which are biochemically specialized to consume certain organic materials and waste products that are particularly difficult for most organisms to digest.

From the standpoint of energy use, detritus feeders are not readily distinguishable from other kinds of consumers. Detritivores have better pickings in some communities than others. As we shall see below, terrestrial plants produce large quantities of indigestible supportive tissue, most of which is consumed, after the plant's death, by organisms of decay in the soil. More than 90 per cent of the net primary production of a forest is consumed by detritivores, and less than 10 per cent by herbivores. Aquatic plants are more thoroughly digestible by herbivores, and the detritus pathways are correspondingly less prominent.

The individual link in the food chain is the basic unit of trophic structure.

Once food is eaten, its energy follows a variety of paths through the organism (Figure 41-3). Not all food can be fully digested and assimilated. Hair, feathers, insect exoskeletons, cartilage and bone in animal foods, and cellulose and lignin in plant foods cannot be digested by most animals. These materials are either egested by defecation or regurgitated in pellets of undigested remains. Some egested wastes are substances that have been relatively unaltered chemically during their passage through an organism, but nearly all have been mechanically broken up into fragments by chewing and by contractions of the stomach and intestines and are thereby made more readily usable by detritus feeders.

Organisms use most of the food energy that they assimilate into their bodies to fulfill their metabolic requirements: performance of work, growth, and reproduction. Because biological energy transformations are inefficient,

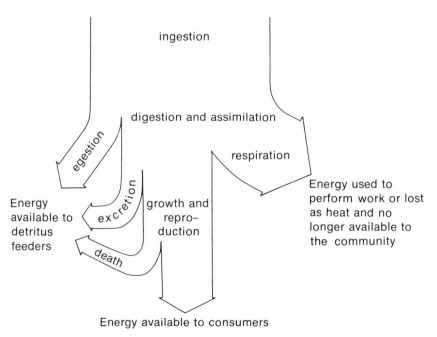

Figure 41-3   Partitioning of energy within a link of a food chain.

a substantial proportion of metabolized food energy is lost, unused, as heat. Organisms are no different from man-made machines in this respect. Most of the energy in gasoline is lost as heat in a car's engine rather than being transformed into motion. In natural communities, energy used to perform work or dissipated as heat cannot be consumed by other organisms and is forever lost to the ecosystem.

Proteins create special metabolic problems because they contain nitrogen. Nitrogen in excess of requirements for growth and body maintenance is usually excreted in an organic form—ammonia in most aquatic organisms, urea and uric acid in most terrestrial organisms. Excreted nitrogenous waste products, therefore, represent a loss of potential chemical energy. But because these waste products can be metabolized by specialized microorganisms, they enter the detritus pathways of the community. We shall examine the path of nitrogen compounds through the ecosystem in greater detail in the next chapter.

Assimilated energy that is not lost through respiration or excretion is available for the synthesis of new biomass through growth and reproduction. Populations lose some biomass by death, disease, or annual leaf drop, which then enters the detritus pathways of the food chain. The remaining biomass is eventually consumed by herbivores or predators, and its energy thereby enters the next higher trophic level in the community.

Energetic efficiencies characterize the flow of energy through the food chain.

The movement of energy through the community depends on the efficiency with which organisms exploit food resources and convert them into biomass. This efficiency is referred to as the food chain, or *ecological efficiency*. Ecological efficiencies are determined by both internal, physiological characteristics of organisms and their external, ecological relationships to the environment. To understand the biological basis of ecological efficiency, one must dissect the individual link of the food chain into its components (Figure 41-4). Ecological efficiency depends on the efficiencies of three major steps in energy flow: exploitation, assimilation, and net production (Table 41-1). The product of the *assimilation* and *net production efficiencies* is the *gross production efficiency*—the percentage of ingested food converted into consumer biomass. The product of the exploitation and gross production efficiencies is the food chain, or ecological efficiency—the percentage of available food energy in prey converted into consumer biomass.

Pigeonholing nature into a stilted and often unnatural framework by rigid categorization often produces more problems than solutions. Ecological efficiencies are no exception. Egested and excreted energy are the square pegs one tries to fit into round holes in this case. On what trophic level do we place egested and excreted detritus? If put on the prey trophic level, it must be classed as unexploited energy and figured into the exploitation efficiency. If put on the consumer trophic level, egested and excreted matter properly should be figured into gross production efficiency, thereby

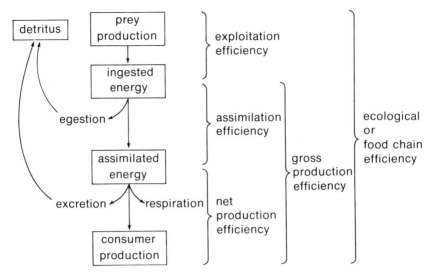

Figure 41-4   Diagram of the partitioning of energy in a link of the food chain and the energetic efficiencies associated with each metabolic step through the organism. Detritus produced by egestion and excretion of unusable food remains on the trophic level of the prey and is available for consumption by other organisms.

increasing overall ecological efficiency. Most ecologists either place detritus in a special food category all its own, belonging to no trophic level, or ignore the problem altogether. Here we shall leave detritus at the prey trophic level.

As ecologists define them, energetic efficiencies correspond to quantities that are readily measured. Practicality is often the architect of concept. The study of energy transformation by plants and animals does, however, provide a useful insight into the basis of ecological efficiency and the trophic structure of the community.

TABLE 41-1
Definitions of energetic efficiencies.

---

(1) Exploitation efficiency $= \dfrac{\text{Ingestion of food}}{\text{Prey production}}$

(2) Assimilation efficiency $= \dfrac{\text{Assimilation}}{\text{Ingestion}}$

(3) Net production efficiency $= \dfrac{\text{Production (growth and reproduction)}}{\text{Assimilation}}$

(4) Gross production efficiency $= (2) \times (3) = \dfrac{\text{Production}}{\text{Ingestion}}$

(5) Ecological efficiency $= (1) \times (2) \times (3) = \dfrac{\text{Consumer production}}{\text{Prey production}}$

Plants are terrible predators. Or perhaps we should say that light is an elusive prey. Light can pass right through a leaf, it can be absorbed by the wrong molecules, or it can be quickly converted to heat and slip away before the plant can harness its energy. As a form of energy, light differs so much from organic molecules that the conversion of energy from one form to the other is inefficient. Once light energy *has* been harnessed, plants can utilize it efficiently for production.

Animals have different problems. With their prey in hand—or in mouth— its energy can be assimilated efficiently by rearranging chemical bonds rather than by converting one form of energy to another. Yet animals expend so much energy in maintaining their delicate bodies—compared to plants— and in pursuing prey that relatively little assimilated energy ends up as production. The fate of energy in plant and animal links of the food chain will be considered in more detail below.

Most energy absorbed by plants is lost as heat or through the evaporation of water.

The absorption of light by green plants is a form of ingestion and may be used as a basis for calculating the energetic efficiency of the plant trophic level. Probably 20 to 30 per cent of all light striking the Earth's surface in productive habitats is absorbed by green plants (Rabinowich and Govindjee 1969). Most of the remainder is either reflected or absorbed by bare earth and water. The vegetation in a dense forest may absorb more than 90 per cent of the light striking leaf surfaces (Federer and Tanner 1966). About 24 per cent of the light striking Silver Springs River in Florida is absorbed by aquatic plants (Odum 1957a). The remainder is partly reflected and partly absorbed by water and the stream bottom. Because plants incorporate little absorbed energy into biomass, gross production efficiency is usually less than 2 per cent.

Most of the light energy absorbed by plants is converted directly to heat and lost by re-radiation, convection, or transpiration. In a classic study of the energy relations of a corn field, E. N. Transeau (1926) determined that only 1.6 per cent of the light energy incident on the field during one growing season (100 days) was assimilated by photosynthesis. The remainder was either converted to heat (54 per cent) or absorbed by water, either in the soil or in leaves, in evaporating and transpiring much of the 15 inches of rainfall received during the growing season (44.4 per cent). Although plants convert 1 to 2 per cent of absorbed light into chemical energy under suitable natural conditions, photosynthesis can achieve a maximum efficiency of 34 per cent under the most favorable laboratory conditions (Rabinowich and Govindjee 1969).

Plants use relatively less assimilated energy for maintenance than animals because they do not move or maintain body temperatures, and they "feed" continuously during daylight periods. Estimates of net production efficiency vary between 30 and 85 per cent depending on the habitat (Table 41-2). Rapidly growing vegetation in temperate zones, whether natural fields, crops, or aquatic plants, exhibits uniformly high net production efficiency

TABLE 41-2

Net production efficiency (net primary production/gross primary production) of several plants and plant communities.

Plant community	Locality*		Net production efficiency (per cent)
Terrestrial			
Perennial grass and herb	Michigan	(1)	85
Corn	Ohio	(6)	77
Alfalfa		(5)	62
Oak and pine forest	New York	(7)	45
Tropical grasslands		(2)	55
Humid tropical forest		(2)	30
Aquatic			
Duckweed	Minnesota	(4)	85
Algae	Minnesota	(4)	79
Bottom plants	Wisconsin	(3)	76
Phytoplankton	Wisconsin	(3)	75
Sargasso Sea	Tropical Atlantic Ocean	(5)	47
Silver Springs	Florida	(5)	42

* References numbered in parentheses following localities are (1) Golley 1960, (2) Golley and Misra 1972, (3) Juday 1940, (4) Lindeman 1942, (5) Odum 1959, (6) Transeau 1926, and (7) Whittaker and Woodwell 1969.

(75 to 85 per cent). Lower values characterize tropical vegetation. The cause of low net production efficiencies in tropical communities (30 to 55 per cent) is unclear. Bright light can deactivate the photosynthetic mechanism of aquatic plants and reduce the rate of photosynthesis without reducing respiration (see Figure 40-2). In addition, high temperature accelerates plant respiration even above the temperature optimum for photosynthesis (see Figure 40-3, or Mooney et al. 1964). The combination of these factors in tropical waters would increase the rate of respiration relative to photosynthesis and thereby lower net production efficiency.

Wet tropical forests exhibit low net production efficiencies, probably around 30 per cent (Golley and Misra 1972). (Measurements of forest productivity is exceedingly difficult and most published figures should be viewed as tentative.) Variation in energetic efficiency of terrestrial habitats roughly parallels the ratio of photosynthetic to supportive tissue in different plants. As we shall see below, leaves comprise 1 to 10 per cent of the aboveground biomass of forest trees compared to 20 to 60 per cent for shrubs and over 80 per cent for most herbs. Nonphotosynthetic living parts of plants respire, though the outer bark and dead cores of trunks and branches do not. Roots certainly contribute substantially to plant respiration.

If the ratio between photosynthetic and supportive tissue determined net production efficiency, young plants that have not yet developed extensive supportive or root tissue should exhibit higher efficiencies than large mature plants. Comparative data for different vegetation types support this hypothesis. Further evidence comes from studies of seasonal change in the net production efficiency of a field of lespedeza (a member of the pea family). The energetic efficiency fell from 75 per cent in April to 15 per cent in August, paralleling a similar decline in growth rate and increase in the proportion of root and stem biomass (Menhinick 1967).

The allocation of net production in plants determines the availability of energy to higher trophic levels.

Plant parts differ greatly as food for herbivores. Supportive tissues (trunks and branches) are mostly cellulose and lignin, which are extremely indigestible, and lack nitrogen and other necessary minerals. They are, therefore, relatively unsuitable foods for herbivores. Although roots contain more living tissue, their composition is similar to that of other supportive tissues. Leaves contain a higher proportion of nitrogen, generally between 2 and 4 per cent (Daubenmire and Prusso 1963), and are more suitable foods for animals. Leaves nonetheless contain relatively high proportions of cellulose, and in some species, accumulate tannins to make leaf proteins unavailable to herbivores (see page 224). Seeds, however, are provisioned with protein and lipids to nourish the young plant once it has germinated, and thus are highly desirable foods (Janzen 1971). The seeds of sugar pines and Jeffrey pines contain, for example, about 50 per cent oil and 30 per cent protein, with the remainder including 5 per cent sugars, 2 per cent water, and 11 per cent mineral ash. The fleshy fruit that surrounds the seeds of many species of plants often has a high food value to attract animals, which then act unwittingly as seed dispersers (Snow 1971).

Because the food value of plant parts varies, the amount of plant biomass consumed by herbivores depends on the proportions of photosynthetic (leafy) and reproductive (seed) tissues. Most of the net production of forests is allocated to roots and branches; relatively little is directly available to herbivores that eat leaves and fruits. The rest accumulates as stem and root growth eventually to be consumed by detritus feeders after death of the plant. By contrast, most of the net production of aquatic plants, which do not have supporting tissue or root systems, is available to herbivorous consumers.

The ratio of net productivity to biomass (P/B) of plants in a community provides a general index of the availability of energy to herbivores. High P/B ratios indicate that relatively little energy is accumulated in supporting tissues, whereas low ratios indicate extensive development of supporting and root tissues. As one would expect, P/B ratios are high in aquatic communities (25 to 42, based on data in Table 40-2) and low in terrestrial communities (0.4 to 0.65). Among terrestrial communities P/B ratios are lowest in mature forests (0.040 to 0.044) highest in grasslands (0.33); young forests that have accumulated relatively little biomass but are growing rapidly having relatively high P/B ratios. In a young oak and pine forest on Long Island, New York, P/B ratios were 0.12 for trees and 0.38 for shrubs (Whittaker and Woodwell

1969). At the other end of the spectrum, a mature forest with high biomass in the Great Smoky Mountains of Tennessee had a P/B ratio of 0.028 (Whittaker 1966).

Because the production of roots and stems is difficult to measure, the complete allocation of net production by plants is poorly known. Golley and Gentry (1966) determined the distribution of harvested biomass among plant parts for corn grown in heavily fertilized fields in South Carolina. Corn is an annual crop, and the biomass at the end of the growing season roughly corresponds to the allocation of net production: leaves 15 per cent, stems 30 per cent, roots 11 per cent, and seeds 44 per cent. Of course, the allocation of production to seeds is much greater in corn than it is in species in natural communities because crop strains are selected for high seed production. In the young oak-pine forest on Long Island, about one-quarter of the net production of trees was allocated to root growth, 40 per cent to stem growth, 33 per cent to the production of leaves, and only 2 per cent to the production of fruits and flowers (Figure 41-5). In the same forest, shrubs allocated more production to roots (54 per cent), and relatively less to stems (21 per cent) and leaves (23 per cent). In the mature cove forest in Tennessee, which had a lower P/B ratio owing to its greater biomass, more production was allocated to the growth of stems (50 per cent) and proportionally less to roots (19 per cent) and leaves (29 per cent). Fruits and flowers represented about 2 per cent of the total. In the forests studied by Whittaker and Woodwell, aboveground production was approximately evenly distributed among

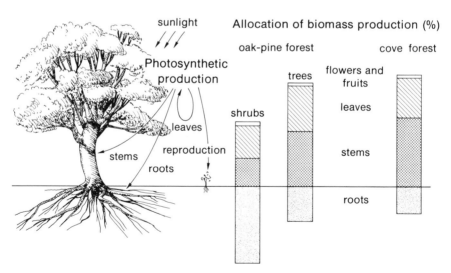

Figure 41-5   Allocation of net production in an oak and pine forest on Long Island, New York (Whittaker and Woodwell 1969), and a cove forest in the Great Smoky Mountains of Tennessee (Whittaker 1966).

stem and leaf components. In grasslands, most of the aboveground production ends up in leafy vegetation because there is relatively little supporting structure other than that in the leaves themselves.

Flower and fruit production varies greatly with habitat. Twelve per cent of the net aboveground primary production of the perennial legume lespedeza is allocated to seeds (Menhinick 1967). In a grassy savanna in Minnesota, aboveground production was allocated as follows: 70 per cent to leaves, 8 per cent to stems, and 22 per cent to fruits and flowers (Ovington et al. 1964). These data indicate that the allocation of net production to flowers and fruits is highest in grasslands, intermediate in shrubby habitats, and lowest in forests; the allocation of production to roots and stems follows an inverse trend.

The biomass dimensions of terrestrial plant communities provide additional insight into the allocation of plant production to various components (Table 41-3). The ratio of belowground biomass to aboveground biomass (the root/shoot ratio, R/S) is largest in grasslands and early stages of forest development, and least in mature forests. Local environmental factors also can affect R/S ratios. Exposure of aboveground parts to harsh conditions and scarcity of water in the soil are often associated with high R/S ratios: tundra 4.75, desert vegetation 5.3 (Rodin and Bazilevic 1964); compared to ratios of 0.2 to 0.3 for most forests. The R/S ratio of the oak-pine forest studied by Whittaker and Woodwell (1969) was relatively high (0.52), partly because the forest is young and partly because it is frequently burned. Fires destroy aboveground biomass while leaving root tissues relatively untouched. Moreover, fire-adapted plants store biomass in roots, where it is protected from fire by the soil, so the aboveground parts can quickly regenerate after being burned (see Figure 13-13).

The proportion of aboveground biomass in leaves is high in grasslands and young forests and low in mature forests (Table 41-3). Leaves comprise between 1 and 10 per cent of the aboveground biomass of forests, the percentage being higher in evergreen forests (tropical broadleaf forests and temperate needleleaf—conifer—forests) than in temperate deciduous forests. Leaf biomass varies from about 20 to 60 per cent of aboveground biomass in shrubs, and is a larger proportion in shrubby habitats than in the shrub layers of forests.

The allocation of net production by plants shows that more plant production is suitable and available to herbivores in aquatic communities and in grasslands than in shrubby, terrestrial habitats and forests. Conversely, detritus feeders consume a larger proportion of net production in forests than in grasslands and aquatic habitats. Furthermore, the storage of plant production for long periods as woody tissue and roots, which are made available to detritus feeders at a slow rate by the death of trees or parts of trees, reduces the turnover rate of production and increases the stability of the forest as a habitat for other organisms. In grasslands and aquatic communities, most primary production is immediately available to herbivores, and energy flow in these communities more strongly reflects fluctuations in environmental conditions.

TABLE 41-3

Biomass dimensions of some representative terrestrial habitats.

Habitat	Locality*		Shoot biomass ($g \cdot m^{-2}$)	Root/shoot ratio	Leaf biomass (per cent shoot)
Grassland and shrubland					
Prairie	Minnesota	(4)		3.7	
herbs			96		100.0
shrubs			94		61.0
Old field	Thailand	(2)	1	1.4	
Grass stand	Thailand	(2)	551	0.65	
Savanna	Minnesota	(4)	2,554	0.23	
herbs			3,497		100.0
shrubs			192		41.0
trees			7		4.5
Mixed savanna	Thailand	(2)	3,299	0.31	8.3
Arundo donax			5,867		
thicket	Thailand	(2)	10,361	0.57	20.0
Temperate deciduous forest					
Oakland	Minnesota	(4)	16,488	0.08	
shrubs			61		30.0
trees			16,406		1.7
Oak and pine	New York	(6)			
shrubs			159	1.93	19.0
trees			6,403	0.52	6.4
Birch	Russia	(3)	17,080	0.25	1.6
Birch	Japan	(3)	11,700		1.9
Aspen	Japan	(3)	11,520		1.9
Aspen	Minnesota	(1)	20,675		1.8
Cove forest	Tennessee	(5)	50,600	0.16	0.6
Nothofagus	New Zealand	(3)	26,950	0.15	1.0
Temperate needleleaf forest					
Douglas fir	Washington	(3)	20,470	0.06	5.9
Spruce	Sweden	(3)	10,860		8.4
Tropical evergreen forest					
Evergreen gallery	Thailand	(2)	29,049		7.2
Dipterocarp	Thailand	(2)	5,028		11.0
Temperate evergreen	Thailand	(2)	17,803		8.9
Tropical deciduous	Ghana	(3)	26,160	0.09	
Young tropical evergreen	Congo	(3)	12,290	0.26	5.3

* References numbered in parentheses following localities are 1) Bray and Dudkiewicz 1963, 2) Ogawa *et al.* 1961, 3) Ovington 1962, 4) Ovington *et al.* 1964, 5) Whittaker 1966, and 6) Whittaker and Woodwell 1969.

Assimilation efficiency is usually greater on animal food than plant food.

Assimilation efficiency is determined largely by the quality of the diet. Animal food is more easily digested than plant food; assimilation efficiencies of predatory species vary from 60 to 90 per cent (Table 41-4). Vertebrate prey are digested more efficiently than insect prey, because most insects are covered by a tough chitinous exoskeleton that is resistant to digestion. The assimilation efficiencies of insectivores vary between 70 and 80 per cent, whereas those of most carnivores are about 90 per cent.

The nutritional value of plant foods depends upon the amount of cellulose, lignin, and other indigestible materials present (Grodzinski and Wunder 1975). Assimilation efficiencies of seed eaters (kangaroo rats, for example) can be as high as 80 per cent. When young vegetation is eaten, efficiencies of 60 to 70 per cent are typical (jackrabbits and field mice). Chew and Chew (1970) found that the assimilation efficiency of rabbits on browse was 42 per cent, compared to 62 per cent on herbaceous plant material. Most grazers and browsers (elephants, cattle, grasshoppers) assimilate 30 to 40 per cent of the food they eat. Millipedes are an extreme. They eat decaying wood, which is composed almost entirely of cellulose and the microorganisms that occur in decaying wood, and their assimilation efficiency is only about 15 per cent (O'Neill 1968).

Phytoplankton are easily digestible because they lack thick cell walls. The assimilation efficiencies of zooplankton on various types of planktonic plants range from 50 to 90 per cent. The assimilation efficiency of a planktonic crustacean, *Calanus hyperboreus,* varied between 54 per cent on diatoms, 72 per cent on dinoflagellates, and 87 per cent on green algae (Conover 1966a). These foods contained progressively less mineral ash—42.6, 10.6, and 7.6 per cent, respectively.

Slightly decomposed remains of plants and animals often are assimilated with higher efficiency than live food. For example, Conover (1966b) found that the efficiency of assimilation of prey by marine zooplankton increased with the depth at which the prey occurred—48 per cent at the surface, 77 per cent at 50 meters, and 89 per cent at 100 meters—because food materials decompose as they sink below the euphotic zone. The periwinkle (*Littorina irrorata*) feeds on partly decomposed plant detritous in salt marshes and it has an assimilation efficiency of 45 per cent (Odum and Smalley 1959); the grasshopper (*Orchelimum fidicinium*) eats the same food plants but in the form of living salt marsh vegetation, and it has an assimilation efficiency of only 27 per cent (Smalley 1960).

Production efficiencies are inversely related to activity and degree of homeothermy.

Maintenance activities and, in warm-blooded animals, heat production require energy that otherwise could be utilized for growth and reproduction (King 1974, Grodzinski and Wunder 1975). Active, terrestrial, warm-blooded

TABLE 41-4
Assimilation, net production, and gross production efficiencies in animals in relation to their diet and physiology.

Species	Habitat	Diet	Efficiencies (per cent)			Reference*
			Assimilation	Production		
				Net	Gross	
Aquatic poikilotherms						
(1) *Megalops cyprinoides* (fish)	Freshwater	Fish	92	35	32	(18)
(2) Bleak (carp family)		Invertebrates		7		(11)
(3) *Nanavax* (predatory gastropod)	Intertidal	Invertebrates	62	45	28	(15)
(4) Roach (carp family)		Plants		7		(11)
(5) *Modiolus demissus* (ribbed muscle)	Salt marsh	Phytoplankton		30		(10)
(6) *Hyallela azteca* (amphipod)	Lakes	Algae	15	15	2.3	(7)
(7) *Ferrissia rivularis* (stream limpet)	Streams	Algae	8	19	1.5	(2)
(8) *Tegula funebralis* (gastropod)	Intertidal	Algae	70	24	17	(16)
(9) Nematodes	Salt marsh	Detritus		25		(22)
(10) *Littorina irrorata* (periwinkle)	Salt marsh	Detritus	45	14	6.3	(13)
(11) *Calospecta dives* (midge)	Spring	Detritus		25		(21)
(12) *Scrobicularia plana* (bivalve)	Tidal flat	Phytoplankton	61	21	13	(8)
Terrestrial poikilotherms						
(13) Harvester ant	Old field	Seeds		0.3		(6)
(14) Saltmarsh grasshopper	Salt marsh	*Spartina*	27	37	10	(20)
(15) *Melanoplus* (grasshopper)	Old field	Lespedeza	37	37	14	(25)
(16) Grasshoppers (3 species)	Old field	Vegetation	16			(12)
(17) Spittlebug	Alfalfa	Plant juices		39		(24)
(18) Woodlouse	Forest litter	Dead leaves	33			(23)
(19) Mites (oribatei)	Old field	Detritus	25	22	5.4	(4)
(20) Millipede	Forest floor	Decaying wood	15			(14)
Terrestrial homeotherms						
(21) Least weasel	Old field	Mice	96	2.3	2.2	(5)

(22)	Marsh wren	Salt marsh	Insects	70	0.5	0.35	(9)
(23)	Grasshopper mouse	Desert shrub	Insects	78	5.7	4.4	(3)
(24)	Kangaroo rat	Desert shrub	Seeds	81	5.2	4.2	(3)
(25)	Vole	Old field	Vegetation	70	3.0	2.1	(5)
(26)	Field mouse	Old field	Grass seed		1.8		(12)
(27)	Jackrabbit (*Lepus*)	Desert shrub	Vegetation	52	5.5	2.9	(3)
(28)	Cottontail rabbit (*Sylvilagus*)	Desert shrub	Vegetation	52	6.0	3.1	(3)
(29)	Uganda kob	Savanna	Vegetation		1.3		(1)
(30)	Cattle	Pasture	Vegetation	38	11	4.2	(19)
(31)	African elephant	Savanna	Vegetation	32	1.5	0.48	(18)

* 1) Buechner and Golley 1967, 2) Burky 1971, 3) Chew and Chew 1970, 4) Englemann 1961, 5) Golley 1960, 6) Golley and Gentry 1964, 7) Hargrave 1971, 8) Hughes 1970, 9) Kale 1965, 10) Kuenzler 1961, 11) Mann 1964, 12) Odum, Connell, and Davenport 1962, 13) Odum and Smalley 1959, 14) O'Neill 1968, 15) Paine 1965, 16) Paine 1971, 17) Pandian 1967, 18) Petrides and Swank 1965, 19) Phillipson 1966, 20) Smalley 1960, 21) Teal 1957, 22) Teal 1962, 23) White 1968, 24) Wiegert 1964, 25) Wiegert 1965.

animals exhibit low net production efficiencies: birds less than 1 per cent because they maintain uniformly high activity; small mammals with high reproductive rates (rabbits and mice, for example) up to 6 per cent (Table 41-4). Man maintains the net production efficiency of beef cattle at as much as 11 per cent by slaughtering them soon after, or even before, growth is completed. More sedentary, cold-blooded animals, particularly aquatic species, channel as much as 75 per cent of their assimilated energy into growth and reproduction (Welch 1968) which approaches the biochemical efficiency of tissue growth of between 70 and 80 per cent (Kielanowski 1964, Ricklefs 1974).

The efficiency of biomass production within a trophic level (gross production efficiency) is the product of assimilation efficiency and the net growth efficiency (Figure 41-6). Gross production efficiencies of few warm-blooded, terrestrial animals exceed 5 per cent, and those of some birds and large mammals fall below 1 per cent. The gross production efficiencies of insects lie within the range of 5 to 15 per cent, and some aquatic animals exhibit efficiencies in excess of 30 per cent. Net production efficiency tends to be inversely related to assimilation efficiency (Welch 1968), especially in aquatic animals, but ecologists do not fully understand the basis for this relationship.

So many factors influence the use of energy that it is impossible to make any general statements about ecological energetics and production efficiencies. We may presume, however, that because energetic efficiencies are tied to food getting, self-maintenance, and reproduction, their values in natural

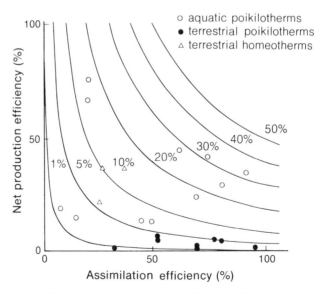

Figure 41-6   Relationships between assimilation efficiency and net production efficiency for a variety of animals. Gross production efficiencies are indicated by the curved lines on the graph (data are from Table 41-4).

ecosystems are maintained by strong selection. Some of the ways in which adaptations influence energy use are described below (also, see Golley and Beuchner 1968 and Petrusewicz and Macfadyen 1970).

Rate of metabolism per gram of body weight generally decreases with increasing body size (Kleiber 1961, Schmidt-Nielsen 1970, 1975a). If small organisms consumed food at the same rate—relative to their body weight— as large organisms, they would have less productive energy. But the rapid metabolism of small animals is compensated by their greater rate of foraging and more rapid growth and reproduction. When these factors are balanced against each other, body size exerts little influence upon the efficiency of energy transfer within a trophic level.

Maintenance metabolism reflects many functions, all of which are influenced by the environment and by adaptations of the organism to the environment. The basal (standard) metabolism of an organism represents the energy expenditure required to maintain the circulation of body fluids, muscle tone, gradients of salt concentration across cell membranes, and similar functions. Over and above this level, one must add temperature regulation, activity, and the energy required to digest and assimilate food (specific dynamic action); the latter can be 20 to 30 per cent of assimilated energy.

Temperature and activity influence the metabolic rate greatly. Warm-blooded animals increase their metabolic rate to generate heat at low temperatures (see page 148), thereby reducing net production efficiency. But adaptations of warm-blooded organisms to reduce heat loss—for example, the thickness of fur—largely compensate for differences between regions in their temperature regimes.

For most mammals, sustained peak work requires the expenditure of about eight times as much energy as resting. Over short periods, work rates may be much higher: up to twenty times the basal rate in man, and up to one hundred times the basal rate in some other mammals (Buechner and Golley 1967). Rapid flight in birds requires ten to sixteen times as much energy expenditure as standard conditions (Dolnik 1967). Most of the activity that organisms engage in during routine tasks probably requires the expenditure of energy at between two and four times basal metabolic rates (King 1974). Even social status can have a marked effect on energy expenditure. For example, the rate of energy metabolism in aggressive mice is about two and a half times that of submissive mice in laboratory populations (Catlett 1961). Activity of juvenile coho salmon similarly varies with their status in the social hierarchy. Even though dominant fish are more active and expend more energy than submissive individuals, they can also gather more food than individuals with low social status; as a result, dominant individuals grow faster (Warren 1971).

The growth of tissues requires more energy than is actually incorporated into the tissue because work is performed during biosynthesis and because energy transfers are not 100 per cent efficient. The efficiency of biosynthesis is difficult to measure directly because the energy requirements for growth cannot be isolated easily from the maintenance requirements of the organism. Growth efficiency can be estimated indirectly, however, by the increase in metabolism of organisms that are growing or depositing fat compared to individuals that are not. The biosynthesis of tissues from assimilated materials has an overall energetic efficiency of 40 to 85 per cent.

The energetic efficiency of biosynthesis sets an upper limit to the net production efficiency (growth/assimilation) of the organism. It is not surprising that few species attain net growth efficiencies exceeding 50 per cent. Even in such rapidly growing birds as nestling ring doves, production efficiency is only about 20 per cent during the first four weeks after hatching (Brisbin 1969). Production is more efficient shortly after hatching when the young grow most rapidly, but it decreases as the growth rate drops off and the young attain mature size. When the metabolism of the parents is included in the calculation of production efficiency for ring doves, the overall energetic efficiency of reproduction becomes 12 per cent when the young are one to two days old, 6 per cent at eighteen days, and 2 per cent at 36 days. If the energy expended by the parents during incubation and courtship were included, the efficiency of reproduction would be about 5 per cent. Adding the energy expenditure of the parents during the nonreproductive period would reduce the production efficiency to 1 or 2 per cent, which is consistent with wild populations of birds and mammals (Table 41-4).

The maintenance energy expenditure of most organisms does not vary greatly according to the amount of food ingested. Therefore, the surplus energy available for growth and reproduction depends partly on the rate of food consumption. For example, juvenile coho salmon must consume forty calories of food energy per kilocalorie of biomass per day just to maintain themselves without growth. When the fish eat less, they lose weight; when they eat more, their rate of growth and hence the efficiency of growth increases (Figure 41-7).

Exploitation efficiency is a population parameter.

The efficiency of energy conversion within a trophic level (gross production efficiency) does not fully describe the flow of energy from one trophic level to the next. One must also include the efficiency with which consumer populations exploit their food resources. Unless organic material is steadily accumulating in an ecosystem, exploitation efficiencies on each trophic level, including those of detritivores, account for all the net production of the next lower trophic level. Peat bogs are an exception to this rule; a large fraction of plant production sinks to the bottom of the bog where acid anaerobic conditions prevent its decay. Nonetheless, most ecosystems exhibit steady-state conditions in which all production is eventually consumed or transported out of the system by wind or water currents.

The most appropriate measure of energy flux through a population is not the number of individuals in the population or their biomass, rather it is their rate of energy consumption (Table 41-5). One of the most striking features of the data summarized in Table 41-5 is that although the density of consumer populations varies over nine orders of magnitude* and popu-

* An order of magnitude is a factor of 10. For example, 1,000 is three orders of magnitude greater than 1.

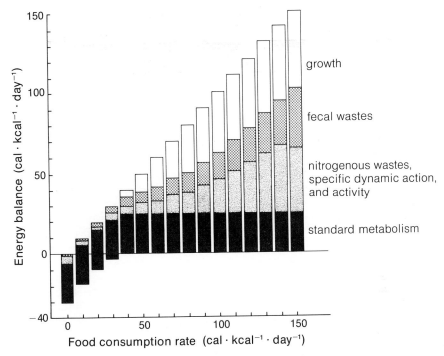

Figure 41-7   Relationship between energy consumption and expenditure of energy for growth and maintenance by juvenile coho salmon at a temperature of 17 C. Minimum metabolic requirement is 40 cal per cal of biomass per day; less food consumption results in an energy deficit (after Warren 1971).

lation biomass varies over five orders of magnitude, the rate of energy assimilation varies over only three orders of magnitude. Populations are thus more similar functionally than structurally or numerically.

Energy flow between trophic levels represents the sum of the feeding activities of many species. Individual populations usually consume only a small fraction of available food resources. Most of the data outlined in Table 41-5 were obtained from populations of abundant species; less common species contribute less to the flow of energy through the community. The most prominent grazers on the savannas of Queen Elizabeth National Park in Uganda (measured by population biomass) are the elephant (4.9 kcal·m^{-2}), hippopotamus (3.8), water buffalo (12.3), and kob (1.9). Other grazing species also occur there, such as waterbuck (0.2), reedbuck (0.1), bushbuck (0.3), and duiker (0.1), but their annual energy flow is relatively small due to their small numbers (Petrides and Swank 1966, in H. T. Odum 1971).

The exploitation efficiency varies considerably among populations. Carnivores, seed eaters, and aquatic herbivores are most efficient; the commonest species consume 10 to 100 per cent of the food available to them. Terrestrial herbivores usually consume only 1 to 10 per cent of the leafy vegetation. Consumers that eat a variety of prey usually exploit each with

TABLE 41-5A
Biomass and energy flow in consumer populations.

Species*	Body size (g)	Density (ind·m⁻²)	Biomass (kcal·m⁻²)	Annual energy flow (kcal·m⁻²·yr⁻¹)		
				Ingestion	Assimilation	Production
(5) Ribbed mussel	7.0×10^{-4}†	750	2.0	95	56	17
(6) Hyallela azteca	0.031	461	64.3	145	17	2
(7) Ferrissia rivularis	0.117†	600	487	1,071	118	11.6
(8) Tegula funebralis					750	179
(9) Nematodes				85		21
(10) Littorina irrorata	3.3×10^{-4}†		11	645	290	40.6
(11) Calospectra dives	8.3×10^{-5}		43.8		520	131
(12) Scrobicularia plana	6.6×10^{-3}	110	23.4		336	71
(13) Harvester ant		2.2	0.022		30	0.09
Leafhoppers (29)					275	
(14) Saltmarsh grasshopper			0.65	78	28	13.3
(16) Grasshoppers	1.25	0.15	0.28	71	25.6	4.0
(17) Spittlebug				0.88	0.78	0.072
(18) Woodlouse	0.021	16.1	0.50	2.8	0.93	
(19) Mites (oribatei)				2.0		0.43
(20) Millipede	2.5	0.29	1.1	13	1.91	
Mites (26)	4.0×10^{-5}	90,000	5.4		21.5	
Enchytraeid worms (28)	8.3×10^{-5}	28,600	3.4			
(21) Least weasel	60	7.4×10^{-5}	0.0067	0.58	0.56	0.013
(22) Marsh wren	10	1.1×10^{-2}	0.17	126	88	0.5
(23) Grasshopper mouse	22	1.83×10^{-4}	0.033	0.89	0.69	0.037
(24) Kangaroo rat	39.5	1.15×10^{-3}	0.23	7.1	5.8	0.3
(25) Vole	29	8.7×10^{-4}	0.038	25	17.5	0.52
(26) Field mouse	13.2	8.6×10^{-4}	0.017		6.7	0.12
(27) Jackrabbit	2,280	2.0×10^{-5}	0.23	4.6	2.3	0.12
(28) Cottontail rabbit	782	1.1×10^{-5}	0.043	1.1	0.54	0.03
Beech forest rodents			0.37		11.3	
Savanna sparrow (12)					3.6	0.04

(30) Uganda kob	77,500	1.5 × 10⁻⁵	3.1	62.4	0.81
White-tailed deer (27)	42,500	2.0 × 10⁻⁵	1.3	43.1	0.34
(31) African elephant			7.1	23.3	

* Numbers correspond to species in Table 41-4. Additional references in parentheses following species names are 26) Berthet 1963, 27) Davis and Golley 1963, 28) O'Connor 1963, 29) Smalley 1959.
† Dry weight.

TABLE 41-5B
Exploitation efficiencies of selected consumer populations.

Species*	Diet	Production of habitat ($kcal \cdot m^{-2} \cdot yr^{-1}$)	Available food ($kcal \cdot m^{-2} \cdot yr^{-1}$)	Consumption (per cent of available food)
(7) *Ferrisia rivularis* (limpet)	Algae	465–1,400	465–1,400	10.4–31.2
(8) *Tegula funebralis* (snail)	Algae	1,167	1,167	92
(11) *Calospectra dives* (midge)	Detritus	2,300		ca.50
(13) Harvester ant	Seeds	440	22	up to 100
	Leathoppers	Plant sap		4.6
(14) Saltmarsh grasshopper	Vegetation			3
(16) Grasshoppers	Vegetation	941	800	2–7
(21) Least weasel	Mice	4,950	1.87	31.0
(24) Kangaroo rat	Seeds	570	8.27	86.5
(25) Vole	Shoots	4,950	1,580	1.6
(26) Field mouse	Seeds	941–1,411	50–100	10–50
(27) Cottontail rabbit	Vegetation	570	228	2.5
Beech forest rodents	Seeds	4,400	194	5.8
Savanna sparrow	Seeds			10–50
White-tailed deer	Vegetation			4.5
(31) African elephant	Vegetation			9.6

* Numbers correspond to species in Tables 41-4. References as in Tables 41-4 and 41-5A.

different efficiency. A study in Alberta, Canada, revealed that red-tailed hawks consumed 22 to 60 per cent of the local ground squirrel population during the summer months, but removed only 0.5 to 1.6 per cent of the snowshoe hares and 1.0 to 2.8 per cent of the ruffed grouse during the same period (Luttich *et al.* 1970). Other predators—foxes, weasels, owls, snakes— also hunt the same prey and ensure the transfer of their biomass to the next trophic level. Plants and prey that escape herbivores and predators eventually die and their chemical energy enters the next trophic level by way of the detritus pathway.

Although it is nearly impossible to measure directly the energy flow through a trophic level, we may identify several general patterns concerning energy flow through the ecosystem. First, assimilation efficiency (and conversely, the proportion of food that enters detritus food chains) depends largely upon the quality of the food. Animal prey and aquatic plants, especially phytoplankton, are assimilated most efficiently. Seeds and fruits are readily digested, but the supportive tissues of terrestrial plants resist digestion and result in low assimilation efficiencies. Second, production efficiency depends largely on activity, and homeotherms are the most active animals. Reported net production efficiencies vary from 14 to 45 per cent among invertebrates and from 0.5 to 6 per cent among warm-blooded vertebrates. Third, exploitation efficiency is high (10 to 100 per cent) among predators (including seed and seedling predators) and low (1 to 10 per cent) among grazers and browsers.

Assimilation, production, and exploitation efficiencies allow us to reconstruct the flow of energy through representative populations. We expect the over-all efficiency of energy transfer from one trophic level to the next to vary from between 1 to 5 per cent for most warm-blooded species to between 5 and 30 per cent for other groups. Because the average efficiency of energy transfer is probably higher in aquatic habitats than in terrestrial habitats, aquatic communities may have longer food chains.

One predator's failure is another's success, but the relative feeding efficiency of consumers influences two important aspects of community energetics: the proportion of energy that travels through detritus pathways in the ecosystem; and the time energy remains in the system before it is dissipated as heat.

The detritus food chain is the major pathway of energy flow in most terrestrial ecosystems.

Detritus feeders consume remains of dead plants and animals, undigested or partially digested fecal matter, and excreted nitrogenous waste products of protein metabolism—any nonliving organic material that can be metabolized to provide energy. Detritus feeding is inversely related to the digestibility of fresh food materials. Detritivores are not so prominent in planktonic aquatic communities as in terrestrial communities, where they consume as much as 90 to 95 per cent of net primary production.

Large carrion eaters and scavengers—vultures and crows on land, crabs in the sea—draw one's attention to detritivores as members of natural communities, but most dead organic matter is consumed by the unnoticed worms, mites, bacteria, and fungi that teem under the litter of the forest floor and in the mucky sediments at the bottom of streams, lakes, and the sea (Birch and Clark 1953, Griffin 1972, Harley 1972).

Of all the detritus-based communities, the organisms that consume the litter of leaves and branches on the forest floor are probably best known (*e.g.,* Witkamp 1966, Minderman 1968, Gosz *et al.* 1972, 1973). Herbivores consume less than 10 per cent of the production of a broad-leaved forest— except during outbreaks of defoliating insects, like gypsy moths. The remainder of the production drops to the forest floor each year as old leaves and branches (Bray and Gorham 1964), or accumulates in roots, trunks, and branches, where it escapes consumers until the tree finally dies.

The breakdown of leaf litter occurs in three ways: 1) leaching of soluble minerals and small organic compounds from leaves by water; 2) consumption of leaf material by large detritus feeding organisms (millipedes, earthworms, woodlice, and other invertebrates); 3) eventual breakdown of organic compounds to inorganic nutrients by specialized bacteria and fungi.

Between 10 and 30 per cent of the substances in newly fallen leaves dissolve in cold water; leaching rapidly removes most of these from the litter (Daubenmire and Prusso 1963). As soil microorganisms decompose the litter further, they produce many small organic and inorganic molecules, which are also exposed to leaching if they are not first assimilated by detritivores.

The role of large organisms in the breakdown of leaf litter has been demonstrated by enclosing samples of litter in mesh bags with openings large enough to let in microorganisms and small arthropods such as mites and springtails, but small enough to keep out large arthropods and earthworms (Figure 41-8). Large detritus feeders assimilate no more than 30 to 45 per cent of the energy available in leaf litter, and even less from wood. They nonetheless speed the decay of litter by microorganisms because they macerate the leaves in their digestive tracts, and the finer particles in their egested wastes expose new surfaces for microbial feeding.

Figure 41-8   Percentage consumption of leaf area by detritus feeders. Leaves were enclosed in mesh bags with either large (7 mm) or small (0.5 mm) openings. The small openings admitted bacteria, fungi, and small arthropods, but excluded most large detritus feeders such as earthworms and millipedes (after Edwards and Heath 1963, in Phillipson 1966).

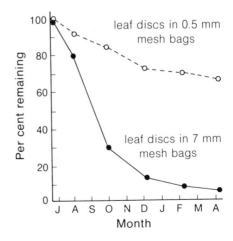

Leaves from different species of trees decompose at different rates, depending on their composition (Witkamp and van der Drift 1961, Van Cleve 1974). In eastern Tennessee, weight loss of leaves during the first year after leaf fall varies from 64 per cent for mulberry, to 39 per cent for oak, 32 per cent for sugar maple, and 21 per cent for beech (Shanks and Olson 1961). The needles of pines and other conifers also decompose slowly. Differences between species depend to a large extent upon the lignin content of the leaves. Lignin, which lends wood many of its structural qualities, is even more difficult to digest than cellulose. Conifer needles typically contain 20 to 30 per cent lignin, broad leaves 15 to 20 per cent (Table 41-6).

The toughness of some types of plant litter, particularly wood, points up the unique role of the fungi as detritivores (Harley 1972). The familiar mushrooms and shelf fungi are merely fruiting structures produced by the mass of the fungal organism deep within the litter or wood (Figure 41-9). Most fungi consist of a vast network, or mycelium, of hyphae, threadlike elements that can penetrate the woody cells of plant litter that bacteria cannot reach. Fungi are also distinguished by secreting enzymes and acids into the substrate itself, digesting organic matter even at a distance. The fungal hypha is like a biochemical blowtorch, cutting its way deep into wood and opening the way for bacteria and other microorganisms. Fungi are prominent in woody litter that is not attacked readily by larger detritus feeders. Bacteria occur more abundantly where detritus has been mechanically broken up by the earthworms and large arthropods.

Decomposition of detritus is influenced greatly by the physical conditions of the environment, particularly temperature and moisture, because these determine the buildup and activity of populations of bacteria, fungi, and other soil organisms. Decomposition rates of litter in temperate forests are related to the daily temperature cycle; microbes respire most rapidly

TABLE 41-6

Loss in weight of litter after incubation for 100 days at 25C. Chemical characteristics of fresh litter (day 0) are also given (from Daubenmire and Prusso 1963).

	Broad-leaved				Needle-leaved			
Species	Aspen		Paper birch		Western larch		Whitebark pine	
Days	0	100	0	100	0	100	0	100
Weight loss (per cent)		23		23		14		11
Composition (per cent)								
cold water soluble	29	11	11	6	16	8	17	6
cellulose	21	21	18	20	22	27	20	23
lignin	15	32	17	35	23	34	28	35
protein	3	4	4	5	2	2	3	3
ash	5	10	9	12	10	12	2	2
pH	4.8	6.4	4.8	6.3	3.7	4.7	3.7	4.2

Figure 41-9   Shelf fungi speed the decomposition of a fallen log. The brackets are fruiting structures produced by the fungal hyphae, together called the mycelium, that grow throughout the interior of the log, slowly digesting its structure (courtesy of the U.S. National Park Service).

during the middle of the day when the temperature is highest (Figure 41-10). The seasonal pattern of microbial respiration also is controlled primarily by temperature, but density of bacteria and fungi, moisture content of the litter, and number of weeks since leaf drop (hence the level of decomposition of the litter) also influence decomposition rate (Witkamp 1966). At higher altitude, litter decomposes more slowly because temperatures are lower (Shanks and Olson 1961). Loss of weight from litter owing to decomposition decreases by 1.5 to 2.5 per cent per degree centigrade decrease in the mean temperature of the environment (Mikola 1960). The chemical environment of the litter also can influence the rate of decomposition. Leaves of any species decompose more slowly in the litter of coniferous forests than in the litter of broad-leaved forests (Shanks and Olson 1961), probably because the acid conditions under conifers are unfavorable to soil organisms. Heal and French (1974) investigated factors that influenced decomposition of detritus in tundra. They found that rates of decay were greatest on sites where soil moisture is about four times the dry weight of the soil. At this level of soil moisture (80 per cent water), rate of decomposition increases about 20 per cent for each 1,000 degree-days above 0C. The influence of soil nutrients on decomposition was less clear.

Decay and microbial populations go hand in hand. In one study, the loss of weight from leaf litter samples composed of loblolly pine (35 to 45 per

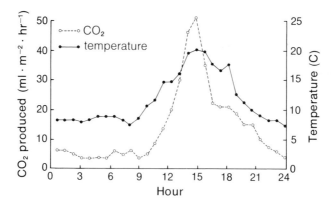

Figure 41-10   Relationship between carbon dioxide production (respiration) from the litter and soil of a pine forest and the daily cycle of temperature (after Witkamp 1969).

cent loss per year), white oak (50 to 60 per cent), redbud (70 per cent), and red mulberry (90 per cent) was highly correlated with microbial populations in the litter (Table 41-7). The relative importance of bacteria was greatest in the more easily decomposed species (redbud and mulberry), whereas fungi were more important components of the decomposer flora in oak and pine litter.

How long does energy take to flow through the community?

We have examined energy flow in terms of its utilization by each trophic level, but the dynamics of the ecosystem cannot be fully appreciated without understanding how long it takes a particular parcel of energy, and the organic compounds which contain that energy, to move through the eco-

TABLE 41-7

Average weight loss, carbon/nitrogen (C/N) ratio, and microflora in four species of decaying leaves in forest litter at Oak Ridge, Tennessee, November 1960–November 1961 (Witkamp 1966).

Species	Weight loss (per cent)	C/N ratio	Bacteria colonies (millions/g dry weight)	Fungi colonies (thousands/g dry weight)	Bacteria / Fungi
Red mulberry	90	25	698	2,650	264
Redbud	70	26	286	1,870	148
White oak	55	34	32	1,880	17
Loblolly pine	40	43	15	360	42

system. Characteristics of plants that inhibit digestion by animals slow the passage of assimilated energy through the ecosystem. For a given level of productivity, the *transit time* of energy in the community and the storage of chemical energy in living biomass and detritus are directly related: the longer the transit time, the greater the accumulation of living and dead biomass. Energy storage in the ecosystem has important implications for the stability of the biological community and its ability to withstand perturbation.

The average transit time of energy in living organisms in the community is equal to the energy stored in the system as biomass divided by the rate of energy flow through the system, or

$$\text{transit time (yrs)} = \frac{\text{biomass (g} \cdot \text{m}^{-2})}{\text{net productivity (g} \cdot \text{m}^{-2} \cdot \text{yr}^{-1})}$$

(The transit time defined by this equation is sometimes referred to as the *biomass accumulation ratio*.) According to Whittaker and Likens (1973), wet tropical forests produce an average of 2,000 grams of dry matter per square meter per year and have an average living biomass of 45,000 g·m⁻². Inserting these values into the equation for average transit time we obtain 22.5 years (45,000/2,000). Average transit times in representative ecosystems (Table 41-8) vary from more than twenty years in forested terrestrial environments to less than twenty days in aquatic plankton communities. These figures underestimate the total transit time of energy in the ecosystem, however, because they do not include the accumulation of dead organic matter in the litter.

An estimate of the transit or residence time of energy in accumulated litter can be obtained by an equation analogous to the biomass accumulation ratio

$$\text{transit time (yrs)} = \frac{\text{litter accumulation (g} \cdot \text{m}^{-2})}{\text{rate of litter fall (g} \cdot \text{m}^{-2} \cdot \text{yr}^{-1})}$$

TABLE 41-8
Average transit time of energy in living plant biomass (biomass/net primary production) for representative ecosystems (from data in Whittaker and Likens 1973).

System	Net primary production (g·m⁻²·yr⁻¹)	Biomass (g·m⁻²)	Transit time (yrs)
Tropical rain forest	2,000	45,000	22.5
Temperate deciduous forest	1,200	30,000	25.0
Boreal forest	800	20,000	25.0
Temperate grassland	500	1,500	3.0
Desert scrub	70	700	10.0
Swamp and marsh	2,500	15,000	6.0
Lake and stream	500	20	0.04 (15 days)
Algal beds and reefs	2,000	2,000	1.0
Open ocean	125	3	0.024 (9 days)

The average transit time for energy varies from a minimum of three months in wet tropical forests to one to two years in dry and montane tropical forests, four to sixteen years in pine forests of the southeastern United States, and more than a hundred years in montane coniferous forests (Olson 1963). Warm temperature and abundance of moisture in lowland tropical regions create optimum conditions for rapid decomposition. Because most of the energy assimilated by forest communities is dissipated by detritus feeders (most of a tree being unavailable to herbivores for the several reasons we have seen), the average transit time of energy in the litter must be added to the transit time in living vegetation to obtain a complete estimate of the persistence of assimilated energy in the ecosystem.

More direct estimates of the rate of energy flow can be obtained by using radioactive tracers. Energy itself cannot be followed directly, but organic compounds containing energy can be labeled with a radioactive element, and their movement followed. In radioactive tracer studies, plants (or water if an aquatic system is being studied) are labeled with a radioactive isotope, usually of phosphorus (^{32}P), applied in a phosphate solution. Consumer species are collected at intervals after the initial labeling, and they are examined with a radiation counter.

Eugene Odum and his co-workers at the University of Georgia have followed the movement of radioactive phosphorus through components of an old-field community (Odum 1962, Odum and Kuenzler 1963). They labeled the dominant plant—telegraph weed (*Heterotheca*)—with drops of radioactive phosphate solution placed directly on the leaves, and then collected insects, snails, and spiders at intervals of several days for about five weeks (Figure 41-11). Certain herbivores, notably crickets and ants, began to accumulate radioactive phosphorus within a few days of its initial application, and attained peak amounts within two weeks. Other herbivorous insects and snails accumulated peak amounts of the tracer at two to three weeks. Ground-living, detritus-feeding insects (carabid beetles, tenebrionid beetles, and gryllid crickets) and predatory spiders did not accumulate peak amounts of tracer until three weeks after the start of the experiment. Thus, the phosphorus label appeared in herbivores first and in detritivores and predators later, as the investigators undoubtedly had hoped.

The movement of radioactive phosphorus through components of the old-field community gives some indication of the time required by labeled substances to reach various trophic levels. Most of the energy assimilated into a trophic level is dissipated by respiration before it reaches the next level. Respired energy, therefore, has a shorter residence time in the ecosystem than the food energy that eventually reaches higher trophic levels. Energy appears to move between trophic levels via the herbivorous insect pathway in an average of a few weeks. Most plant production in terrestrial communities is not, however, consumed by herbivores, rather it is stored and consumed by detritus feeders over a prolonged period. Based on the biomass accumulation ratio, a minimum estimate of average transit in temperate grasslands is on the order of three years (Table 41-8). Seasonal changes in the environment can delay the decomposition of energy stored in litter. In temperate forests, most leaf fall occurs in the late autumn, just before the coldest period of the year when detritus-feeding organisms are

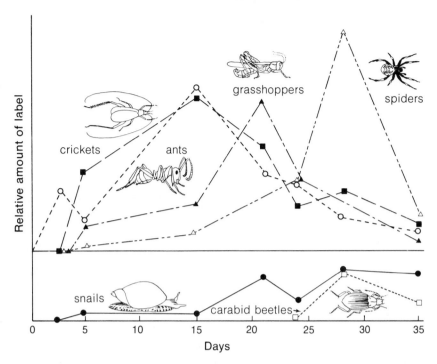

Figure 41-11   Accumulation of radioactive phosphorus by animals on different trophic levels after initial labeling of plants. The experiment was performed
in an old-field ecosystem (after Odum 1962).

less active. In moist tropical forests litter fall is more evenly distributed
throughout the year (Bray and Gorham 1964), and dead leaves are attacked
by detritus feeders immediately. In seasonally dry tropical forests, however,
leaf fall is concentrated at the beginning of the dry season, and litter accumulates on the forest floor until the return of moist conditions stimulates
the activity of bacteria and other detritus feeders.

Where primary production is seasonal, detritus feeding can be the major
energetic mainstay of a community during periods of reduced photosynthetic
activity. For example, although the production of *Spartina* grass in Georgia
salt marshes occurs primarily between March and September, remains of
dead *Spartina* are consumed at a more steady rate throughout the year. As
a result, herbivorous organisms, such as the grasshopper *Orchelimum,* are
active only during the summer months, whereas detritus feeders like the
periwinkle *Littorina* remain active throughout the year (Odum and Smalley
1959).

A radioactive tracer experiment performed on a small trout stream in
Michigan gave results comparable to those of Odum's old-field study (Ball
and Hooper 1963). Radioactive phosphate was added to the stream at one
point and its accumulation in plants and animals downstream from the
release site was monitored for two months. The median time for each pop-

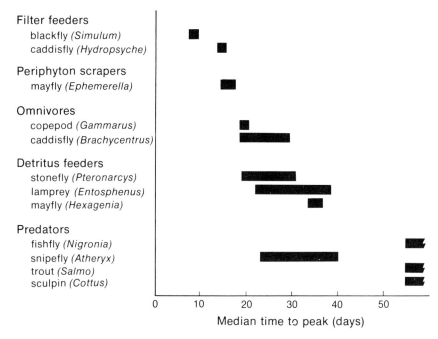

Figure 41-12   Median time required for the accumulation of peak levels of radioactive phosphorus in various components of a stream ecosystem after addition of the tracer (after Ball and Hooper 1963).

ulation to accumulate its maximum concentration of radioactive phosphorus varied from a few days for aquatic plants to one to two weeks for filter feeders and other herbivores, three to four weeks for omnivores, four to five weeks for detritus feeders, and four weeks to more than two months for most predators (Figure 41-12). The results suggest that most of the energy assimilated in aquatic ecosystems is dissipated within a few weeks, although a small portion may linger for months in the predator food chain, and perhaps for years in organic sediments on the bottoms of streams and lakes. The decomposition of organic matter in streams has received considerable attention recently owing to the effects of organic pollutants on aquatic ecosystems (King and Ball 1967, Minshall 1967, Cummins et al. 1972, Fisher and Likens 1972). We shall consider this topic in more detail in the next chapter.

Energetics can be used to summarize the trophic structure of the community.

The measurement of energy flow through an entire community is a complex and virtually impossible task. Yet the total flux of energy and the efficiency of its movement determine the basic trophic structure of the community: number of trophic levels, relative importance of detritus and predatory feeding, steady-state values for biomass and accumulated detritus, and turnover rates of organic matter in the community. Each of these properties, in turn, influences the inherent stability of the community. Detritus pathways stabilize communities because detritivores do not affect the abundance of their "prey" directly. Accumulation of biomass and detritus, coupled with low turnover rates, stabilizes communities by ironing out short-term fluctuations in the physical environment that affect energy flux. Ecologists have not yet agreed on the influence of predation and competition at higher trophic levels on community stability. Some argue that the increased complexity of a diverse and trophically stratified community adds stability to the system. Others contend that biological interactions build lag times into the response of the system to the physical environment, thereby promoting instability. Ecosystem stability will be considered more fully toward the end of this book. In any event, the trophic structure of the community constitutes an important component of the regulation of ecosystem function.

The flow of energy through a community is not unlike the flow of money through a checking account. Income must equal or exceed expenditure over a long period to keep the checking account open and the community viable. Just as the community undergoes regular daily and seasonal fluctuations in energy income and expenditure, the balance undergoes normal periodic fluctuations corresponding to pay day and bills due. Just as a hurricane or cold snap shakes the community, an unexpected disaster shakes the bank account. Under such circumstances, a large balance and hockable assets are as helpful as stored biomass and detritus in meeting disaster.

The red and black ink of community energy balance can be summarized by the following sources of income and expenditure:

Income	Expenditure
Assimilated light energy	Respiration
Transport in	Transport out

All the energy assimilated or transported into the community is either dissipated by respiration or is carried out of the system by wind, water currents, or gravity.

It is difficult to avoid drawing boundaries when we discuss community structure and function, particularly when we wish to distinguish aquatic and terrestrial communities. The lake community clearly extends to the edge of the lake, and no farther. Smaller distinctions of communities within the aquatic or terrestrial realms are drawn more arbitrarily. We may find it difficult to point to the exact place where the desert shrub community stops and the grassland community takes up. We can be sure, however, of the boundaries around a study area, and ecologists often define the community pragmatically as a place with a characteristic vegetation type.

Regardless of the kind of boundary we imagine around a community, energy frequently finds its way in and out of communities across these boundaries. Movement, or transport, of energy between communities accounts for a varying proportion of the community's income and expenditures. Terrestrial communities receive most of their energy income from light, although they lose energy when plant and animal detritus falls, blows, or washes into rivers and lakes. Transport of material between different terrestrial communities is probably negligible.

Some isolated communities rely primarily on detritus produced elsewhere and transported into the community. For example, in Root Spring, near Concord, Massachusetts, herbivores consume 2,300 kcal·m⁻²·yr⁻¹, but only 655 kcal are produced by aquatic plants; the balance is transported into the spring by leaf fall from nearby trees (Teal 1957). Life in caves and abyssal depths of the oceans, to which no light penetrates, subsists entirely on energy transported into the system by wind, water currents, and sedimentation.

The relative importance of predatory- and detritus-based food chains varies greatly among communities. Predators are most important in plankton communities; detritus feeders consume the bulk of production in terrestrial communities. As we have seen earlier, the proportion of net production that enters the herbivore-predator segment of the food web depends on the relative allocation of plant tissue between structural and supportive functions on one hand and growth and photosynthetic functions on the other. Herbivores consume 1.5 to 2.5 per cent of the net production of temperate deciduous forests, 12 per cent in old-field habitats, and 60 to 99 per cent in plankton communities (Table 41-9). The low level of herbivory in old fields may not truly represent natural habitats because most large North American herbivores became extinct during the last 15,000 years. Large grazing mammals today consume between 28 and 60 per cent of the net production of African grasslands; beef cattle consume 30 to 45 per cent of the net production of managed range lands in the United States.

In 1942, Raymond Lindeman made one of the earliest attempts to describe the energy flow in an entire community. He chose for his study Cedar Bog Lake, Minnesota, a relatively small, self-contained system. Lindeman estimated energy flow from the harvestable net production of each trophic level and from laboratory determinations of respiration and assimilation. The animals and plants collected at the end of the growing season constituted the net production of the trophic levels to which each species was assigned:

Trophic level	Harvestable production (kcal·m⁻²·yr⁻¹)
Primary producers (green plants)	704
Primary consumers (herbivores)	70
Secondary consumers (carnivores)	13

Lindeman estimated the energy dissipated by respiration from ratios of respiratory metabolism to production measured in the laboratory: 0.33 for aquatic plants, 0.63 for herbivores, and 1.4 for the more active carnivores in the lake. The gross production of carnivores was calculated as the sum of their harvestable production (13 kcal·m⁻²·yr⁻¹) and respiration (13 × 1.4 = 18 kcal·m⁻²·yr⁻¹), or 31 kcal·m⁻²·yr⁻¹ (Table 41-10). Lindeman determined that predation on secondary consumers was negligible. The gross production of

TABLE 41-9

Exploitation of net primary production as living vegetation by primary consumers (from Wiegert and Owen 1971; desert shrub from Chew and Chew 1970).

Community	Characteristics of primary producers	Exploitation by herbivores (per cent)*
Mature deciduous forest	Trees; large amount of non-photosynthetic structure; low turnover rate	1.5 to 2.5
Thirty-year-old Michigan field	Perennial forbs and grasses; medium turnover rate	1.1
Desert shrub	Annual and perennial herbs and shrubs; low turnover rate	5.5†
Georgia salt marsh	Herbaceous perennial plants; medium turnover rate	8
Seven-year-old South Carolina fields	Herbaceous annual plants; medium turnover rate	12
African grasslands	Perennial grasses; high turnover rate	28 to 60†
Managed rangeland	Perennial grasses, high turnover rate	30 to 45†
Ocean waters	Phytoplankton; very high turnover rate	60 to 99

* Aboveground production only for terrestrial systems.
† Grazing by mammals only.

TABLE 41-10

An energy flow model for Cedar Lake Bog, Minnesota (from Lindeman 1942).

Energy $(kcal \cdot m^{-2} \cdot yr^{-1})$	Trophic level		
	Primary producers	Primary consumers	Secondary consumers
Harvestable production*	704	70	13
Respiration	234	44	18
Removal by consumers			
assimilated	148	31	0
unassimilated	28	3	0
Gross production (totals)	1,114	148	31

* Does not include net production removed by consumers. Actual net production, including removal by consumers, was 879 $kcal \cdot m^{-2} \cdot yr^{-1}$ for primary producers, 104 $kcal \cdot m^{-2} \cdot yr^{-1}$ for primary consumers, and 13 $kcal \cdot m^{-2} \cdot yr^{-1}$ for secondary consumers.

primary consumers was similarly calculated as the sum of their harvestable production (70 kcal·m⁻²·yr⁻¹), respiration (70 × 0.63 = 44 kcal·m⁻²·yr⁻¹), and the consumption of primary consumers by secondary consumers (34 kcal·m⁻²·yr⁻¹). (Lindeman assumed that because secondary consumers have assimilation efficiencies of 90 per cent they must consume 3 kcal·m⁻²·yr⁻¹ over and above their gross production of 31 kcal·m⁻²·yr⁻¹.) Therefore, the gross production of primary consumers was 148 kcal·m⁻²·yr⁻¹, which corresponded in turn to the removal of net primary production by herbivores. Assuming the assimilation of herbivores feeding on plant material to be 84 per cent, Lindeman calculated that herbivores consumed, but did not assimilate (0.16/0.84) × 148 = 28 kcal·m⁻²·yr⁻¹, making the gross primary production 1,114 kcal·m⁻²·yr⁻¹.

Studies of community energetics have come a long way since Lindeman's extraordinary pioneering venture. A more recent study by Howard T. Odum (1957a) on another small aquatic ecosystem at Silver Springs, Florida, employed more refined techniques. Gross production of aquatic plants was estimated by gas exchange rather than by the harvest method. Odum also accounted for the inflow of energy in the form of detritus from tributary streams and the surrounding land. The community energetics of Cedar Bog Lake and Silver Springs are compared in Table 41-11. The more southern location and warmer temperatures of Silver Springs probably accounts for its greater primary production and the lower net production efficiencies of

TABLE 41-11

A comparison of energy flow models for Cedar Lake Bog, Minnesota (Lindeman 1942) and Silver Springs, Florida (Odum 1957a).

	Cedar Lake Bog	Silver Springs
Incoming solar radiation (kcal·m⁻²·yr)	1,188,720	1,700,000
Gross primary production (kcal·m⁻²·yr)	1,113	20,810
Photosynthetic efficiency (per cent)	0.10	1.20
Net production efficiency (per cent)		
Producers	79.0	42.4
Primary consumers	70.3	43.9
Secondary consumers	41.9	18.6
Exploitation efficiency (per cent)*		
Primary consumers	16.8	38.1
Secondary consumers	29.8	27.3
Ecological efficiency (per cent)		
Primary consumers	11.8	16.7
Secondary consumers	12.5	4.9

* Based on assimilated energy rather than ingested energy. Assimilation efficiencies were probably above 80 per cent, so values are not much below actual exploitation efficiencies.

its inhabitants. Herbivores consumed little of the net primary production of Cedar Bog Lake; most was deposited as organic detritus in lake sediments. Consequently, the exploitation efficiency of primary consumers was lower in Cedar Bog Lake than in Silver Springs. In spite of high respiratory energy losses in Silver Springs and quantities of production transported out of the system at Cedar Bog Lake, exploitation efficiencies varied in both locations between 15 and 40 per cent and the overall ecological efficiency of energy transfer between trophic levels varied between 5 and 17 per cent.

Ecological efficiencies are usually lower in terrestrial habitats, and a useful rule of thumb states that the top carnivores in terrestrial communities can feed no higher than the third trophic level on the average, whereas aquatic carnivores may feed as high as the fourth or fifth level. This is not to say that there can be no more than three links in a terrestrial food chain; some energy may travel through a dozen links before it is dissipated by respiration. These high trophic levels probably do not, however, contain enough energy to fully support a predator population.

We can crudely estimate the average length of food chains in a community from the net primary production, average ecological efficiency, and average energy flux of predator populations. Because the energy reaching a given trophic level is the product of the net primary production and the intervening ecological efficiencies, the appropriate equation for calculating the average trophic level that a community can support is

$$\text{trophic level} = 1 + \frac{\log (\text{predator ingestion} \div \text{net primary production})^*}{\log (\text{average ecological efficiency})}$$

Using this equation and some rough estimates for the values needed on the right-hand side of the equation, we can calculate average number of trophic levels as about seven for marine plankton communities, five for inshore aquatic communities, four for grasslands, and three for wet tropical forests (Table 41-12). These estimates should be taken with a grain of salt, to be sure, but they do indicate how measurements of energetics for individual

---

* We note that the energy $E(n)$ available to a predator on the nth trophic level may be calculated by the equation

$$E(n) = NPP \cdot Eff^{n-1}$$

where NPP = net primary production, Eff is the geometric mean ecological efficiency of trophic levels 1 to n, and $n - 1$ is the number of links in the food chain between trophic levels 1 and n. We now rearrange the equation above in the following form

$$Eff^{n-1} = \frac{E(n)}{NPP}$$

take the logarithm of both sides of the equation

$$(n - 1) \log (Eff) = \log (E[n]/NPP)$$

and rearrange to obtain

$$n = 1 + \frac{\log (E[n]/NPP)}{\log (Eff)}$$

TABLE 41-12

Community energetics and the average number of trophic levels in various communities. Values for production, predator energy flux, and ecological efficiencies are rough estimates based on many studies.

Community	Net primary production $(kcal \cdot m^{-2} \cdot yr^{-1})$	Predator ingestion $(kcal \cdot m^{-2} \cdot yr^{-1})$	Ecological efficiency (per cent)	Number of trophic levels
Open ocean	500	0.1	25	7.1
Coastal marine	8,000	10.0	20	5.1
Temperate grassland	2,000	1.0	10	4.3
Tropical forest	8,000	10.0	5	3.2

species and trophic levels can be used to determine the overall trophic structure of the community.

We have seen how the quality of food and allocation of energy to various functions by organisms create patterns of energy flow through communities. These patterns differ most between aquatic and terrestrial environments because of basic differences in the adaptations of organisms to each of these realms. In aquatic ecosystems, energy flows rapidly and is transferred efficiently between trophic levels, thereby permitting long food chains. In terrestrial ecosystems, some energy is dissipated rapidly, making energy transfer between trophic levels relatively inefficient, and the remainder is stored for long periods as supportive tissue in plants and as organic detritus in the soil.

Each parcel of energy assimilated by plants travels through the ecosystem only once. Energy is dissipated as heat, a form that plants cannot harness for primary production. But as we shall see in the next chapter, mineral nutrients are continually recycled through the ecosystem. The cycles differ greatly among different chemical elements, but they share two fundamental properties: first, the cycles are tied to, and are driven by, energy flux through the ecosystem; second, nutrients alternate between inorganic and organic forms through the complimentary processes of assimilation and decomposition.

42

Nutrient Cycling

Nutrients, unlike energy, are retained within the ecosystem where they are continually recycled between living organisms and the physical environment. Because plants and animals can use only those nutrients that occur at or near the surface of the Earth, persistence of life requires that the materials assimilated by organisms eventually become available to other organisms. Each chemical element follows a unique route in its cycle through the ecosystem, as we shall see below, but all cycles are driven by energy, and their elements alternate between organic and inorganic forms as they are assimilated and excreted by plants and animals.

Nutrient cycles of communities sometimes become unbalanced, and nutrients accumulate in or are removed from the system. For example, during periods of coal and peat formation, dead organic materials accumulate in the sediments of lakes, marshes, and shallow seas where anaerobic conditions prevent their decomposition by microorganisms. In other instances, under intensive cultivation or after removal of natural vegetation, erosion can wash away nutrient-laden layers of soil that took years to develop. Most ecosystems exist in a steady state in which outflow of nutrients from the system is balanced by inflow from other systems, from the atmosphere, and from the rock beneath the system. Furthermore, gains and losses are usually small compared to the rate at which nutrients are cycled within the system.

The movement of nutrients parallels energy flow through the community.

Exchanges of nutrients between living organisms and inorganic pools are about evenly balanced in most communities. Carbon and oxygen are recycled by the complementary processes of photosynthesis and respiration. Nitrogen, phosphorus, and sulfur follow more complex paths through the ecosystem, aided along the way by microorganisms with specialized metabolic capabilities.

We may describe the ecosystem as being divided into *compartments* through which material passes and within which material may remain for varying periods (Figure 42-1). Most of the mineral cycling in the ecosystem involves three active compartments: living organisms (or biomass), dead organic detritus, and available inorganic minerals. Two additional compartments—indirectly available inorganic minerals and organic sediments—are peripherally involved in nutrient cycles, but exchange between these compartments and the rest of the ecosystem occurs slowly compared to exchange among active compartments.

The processes responsible for the movement of mineral nutrients within the ecosystem are indicated in Figure 42-1. Assimilation and production cause elements to move from the inorganic to the organic compartment: Primary production by plants is the most important component of this step in the cycling of carbon, oxygen, nitrogen, phosphorus, and sulfur, but

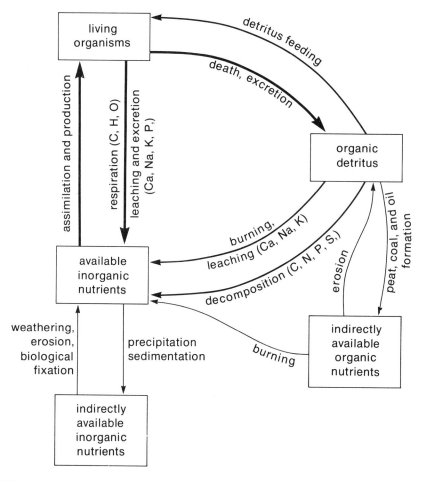

Figure 42-1   A compartment model of the ecosystem with some of the most important routes of mineral exchange indicated.

animals also assimilate many such essential minerals as sodium, potassium, and calcium directly from the water they drink.

Respiration returns some carbon and oxygen directly to the pool of available inorganic nutrients, perhaps after being cycled within the living biomass compartment several times in predator food chains. Calcium, sodium, and other mineral ions are excreted or leached out of leaves by rainfall or the water surrounding aquatic organisms and are also recycled rapidly. Most of the carbon and nitrogen assimilated into living biomass is transferred by death and excretion into the detritus compartment. Some nutrients in detritus may be returned to the biomass compartment by detritus feeders, but all eventually are returned to the pool of available inorganic minerals by leaching and decomposition. Exchange between the actively cycled pools of minerals and the vast reservoirs of indirectly available nutrients locked up in the atmosphere, limestone, coal, and the rocks forming the crust of the Earth, occurs slowly, primarily by geological processes.

Elements usually occur in different forms in the air, soil, water, and living organisms (often referred to as the atmosphere, lithosphere, hydrosphere, and biosphere). For example, oxygen occurs as oxygen molecules (O_2) and as carbon dioxide (CO_2) both in gaseous form in the atmosphere and in dissolved form in water, but it also combines with hydrogen to form water (H_2O). Oxygen appears in the form of oxides (iron oxide, Fe_2O_3) and salts (calcium carbonate, $CaCO_3$) in the lithosphere. The rate at which an element is transferred between its inorganic forms and thus its availability in inorganic forms to living organisms vary greatly. The largest pool of oxygen—more than 90 per cent of the oxygen near the surface of the Earth—occurs as calcium carbonate in sedimentary rocks, particularly limestone. Except for minute quantities released by volcanic activity, the oxygen in limestone and other sedimentary rock is virtually unavailable to the biosphere (Cloud and Givor 1970). In contrast to oxygen, nitrogen is most abundant in its gaseous form (N_2) in the atmosphere, but plants assimilate nitrogen primarily from nitrates (NO_3^-) in the soil or in water. Despite its abundance, atmospheric nitrogen plays a minor role in short-term nutrient cycling.

The assimilation and decomposition processes that cycle nutrients through the biosphere are closely linked to the acquisition and release of energy by organisms. The paths of nutrients, therefore, parallel the flow of energy through the community. (As we saw in the last chapter, radioactive elements incorporated into organic compounds can be used to follow the path of energy through the community.) The carbon cycle is most closely linked to the transformation of energy in the community because organic carbon compounds contain most of the energy assimilated by photosynthesis. Most energy-releasing processes, of which respiration is the most important, release carbon as carbon dioxide. When organisms metabolize organic compounds containing nitrogen, phosphorus, and sulfur, these elements are often retained in the body for the synthesis of structural proteins, enzymes, and other organic molecules that make up structural and functional components of living tissue. Consequently, nitrogen, phosphorus, and sulfur pass through each trophic level somewhat more slowly than the average transit time of energy.

Movement of oxygen and hydrogen in the ecosystem is overwhelmingly influenced by the water cycle. Organisms lose water rapidly by evaporation and excretion; body water may be replaced hundreds or even thousands of times during an organism's lifetime. When discussing the oxygen cycle, ecologists usually distinguish pathways involving chemical assimilation of oxygen into organic compounds and those involving movement of water.

The water, or hydrological, cycle does, however, demonstrate the basic features of all nutrient cycles—that they are approximately balanced on a global scale, and that they are driven by energy—and so we shall let it provide a model.

Precipitation balances evaporation and transpiration in the water cycle.

Although water is chemically involved in photosynthesis, most of the water flux through the ecosystem occurs through evaporation, transpiration, and precipitation (Barry and Chorley 1970). Evaporation and transpiration correspond to photosynthesis in that light energy is absorbed and utilized to perform the work of evaporating water and lifting it into the atmosphere. The condensation of water vapor in the air, which eventually causes rainfall, releases the potential energy in water vapor as heat, much as respiration by plants and animals releases energy. The water cycle is outlined in Figure 42-2.

Because more than 90 per cent of the Earth's water is locked up in rocks in the core of the Earth and in sedimentary deposits near the Earth's surface,

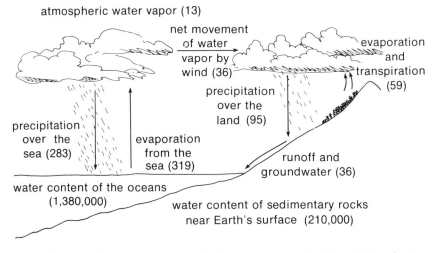

Figure 42-2   The water cycle, with its major components expressed on a global scale. All pools and transfer values (shown in parentheses) are expressed as billion billion (10^{18}) grams, and billion billion grams per year.

this water enters the hydrological cycle in the ecosystem very slowly through volcanic outpourings of steam (Hutchinson 1957a). Hence the great reservoirs of water in the Earth's interior contribute little to the movement of water near the Earth's surface.

Precipitation over the land surface of the Earth exceeds evaporation and transpiration from terrestrial habitats. Whereas 23 per cent of the Earth's precipitation occurs over land surfaces, only 16 per cent of the water vapor in the Earth's atmosphere comes from the continents. The oceans exhibit a corresponding deficit of rainfall compared to evaporation. Much of the water vapor that winds carry from the oceans to the land condenses over mountainous regions where rapid heating and cooling of the land create vertical air currents. The net flow of atmospheric water vapor from ocean to land areas is balanced by runoff from the land into ocean basins.

We can estimate the role of plant transpiration in the hydrological cycle from the organic productivity of terrestrial habitats and the transpiration efficiency of plant production (see page 771). The primary production of terrestrial habitats is about 1.1×10^{17} grams (g) of dry material per year (Whittaker and Likens 1973), and approximately 500 g of water are transpired for each gram of production. Terrestrial vegetation, therefore, transpires 55×10^{18} g of water annually, nearly the total evapotranspiration from the land. Although this figure may overestimate transpiration somewhat, plants clearly play the major role (Stanhill 1970, Swank and Douglas 1974). The influence of vegetation on water movement is best shown by removing it. At Hubbard Brook, New Hampshire, experimental cutting of all trees from small watersheds increased the flow of streams draining the clear-cut areas by more than 200 per cent (Pierce 1969). This excess would normally have been transpired, as water vapor from leaves, directly into the atmosphere.

We may calculate the amount of energy required to drive the hydrological cycle by multiplying together the energy required to evaporate 1 g of water (0.536 kcal) and the total annual evaporation of water from the Earth's surface (378×10^{18} g). The product, about 2×10^{20} kcal, represents about one-fifth of the total energy income in light striking the surface of the Earth. The remaining energy is absorbed and reradiated as heat. Of the total incident radiation in temperate forests, about 40 per cent is absorbed by plants, and two-thirds of the absorbed energy (one-fourth of the incident energy) is dissipated through evapotranspiration (Stanhill 1970).

Evaporation, not precipitation, determines the flux of water through the hydrological cycle. Ignoring the relatively minor inputs of energy to create wind currents and to heat water vapor in the atmosphere, the absorption of light energy by liquid water is the major point at which an energy source is geared to the water cycle. Furthermore, the ability of the atmosphere to hold water vapor is limited. An increase in the rate of evaporation of water into the atmosphere eventually results in an equal increase in precipitation.

The water vapor in the air at any one time corresponds to an average of 2.5 centimeters (1 inch) of water spread evenly over the surface of the Earth. An average of 65 cm (26 in) of rain falls each year, which is twenty-five times the average amount of water vapor in the atmosphere. The steady-state content of water vapor in the atmosphere, referred to as the atmospheric pool, is therefore recycled twenty-five times each year. Conversely, water has an average transit time (see page 809) of about two weeks. The water content of soils, rivers, lakes, and oceans is a hundred thousand times greater than that of the atmosphere. Rates of flux through both pools are the same, however, because evaporation equals precipitation. The average transit time of water in its liquid form at the Earth's surface (about 3,650 years) is, therefore, 100,000 times longer than in the atmosphere.

The oxygen cycle is balanced by photosynthesis and respiration.

Next to nitrogen, oxygen is the most abundant element in the atmosphere, accounting for 21 per cent of its weight. But because oxygen is abundant and ubiquitous in the terrestrial environment, ecologists do not pay as much attention to the oxygen cycle as they do to the cycles of scarcer nutrients—carbon, nitrogen, phosphorus, and so on. The oxygen cycle is relatively simple, but it exhibits the basic characteristics of nutrient cycling in the ecosystem.

The atmosphere contains about 1.1×10^{21} g of oxygen. Much more is bound up in water molecules, mineral oxides, and salts in the Earth's rocky crust, but this tremendous pool of oxygen is not directly available to the ecosystem. Terrestrial plants probably assimilate close to 10^{17} g of carbon in gross production. Photosynthesis releases two atoms of oxygen for each atom of carbon fixed. Because oxygen weighs 16/12 as much as carbon, atom for atom, green plants release about 2.7×10^{17} g of oxygen each year. This amount corresponds to about 1/2,500 of the oxygen in the atmosphere, and therefore the cycling (transit) time of oxygen in the atmosphere would be about 2,500 years, if exchange of oxygen between the atmosphere and surface water were ignored.

The oxygen cycle is more complicated than the complementary equations for photosynthesis and respiration suggest. Water enters into the complex biochemistry of photosynthesis, and although it is released in equal amounts by respiration, water molecules do not survive these processes intact. The oxygen molecule (O_2) produced by photosynthesis derives one atom from carbon dioxide and one from water; the oxygen molecule consumed in respiration supplies one atom to carbon dioxide and the other to water.

The carbon cycle is complementary to the oxygen cycle.

The biological cycling of carbon in the ecosystem is more direct than the cycling of oxygen. The carbon cycle involves only organic compounds and carbon dioxide (Figure 42-3). Photosynthesis and respiration fully complement each other. Photosynthesis assimilates carbon entirely into carbohydrate; respiration converts all the carbon in organic compounds to carbon dioxide. Large inorganic pools of carbon—atmospheric carbon dioxide, dissolved carbon dioxide, carbonic acid, and carbonate sediments—enter the

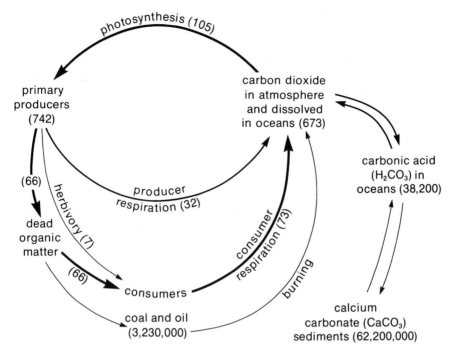

Figure 42-3   The global carbon cycle, with some estimated pools and annual transfer rates. Values are in million billion (10^{15}) grams (after Waldron and Ricklefs 1973, based on many sources).

carbon cycle to different degrees (Figure 42-3). The carbon in igneous rocks, calcium carbonate (limestone) sediments, coal, and oil is exchanged with other more active pools so slowly that these sources have little influence on the short-term functioning of the ecosystem.

Plants assimilate about 105×10^{15} g of carbon each year, of which about 32×10^{15} g are returned to the carbon dioxide pool by plant respiration. The remainder, 73×10^{15} g, supports the respiration and production of animals, bacteria, and fungi in herbivore- and detritus-based food chains. Anaerobic respiration (without oxygen) produces a small quantity of methane (CH_4) that is converted to carbon dioxide by a photochemical reaction in the atmosphere. Plants and animals annually cycle between 0.25 and 0.30 per cent of the carbon present in carbon dioxide and carbonic acid in the atmosphere and oceans, hence the total active inorganic pool is recycled every 300 to 400 years ($1 \div 0.003$ to $1 \div 0.0025$). Because the atmosphere and oceans exchange carbon dioxide slowly, they may be considered as separate pools over short periods. Terrestrial ecosystems annually cycle about 12 per cent of the carbon dioxide in the atmosphere; the transit time of atmospheric carbon is, therefore, about eight years ($1 \div 0.12$).

The combustion of coal and oil adds carbon dioxide to the atmosphere. Although man's present use of fossil fuels amounts to less than 2 per cent of the carbon cycled through the ecosystem each year, combustion of fuels adds carbon dioxide to the atmosphere over and above that produced by photosynthesis. The carbon dioxide content of the atmosphere has in fact risen during this century, and it can be expected to rise even more rapidly in the future. Scientists cannot agree on the implications of increased atmospheric carbon dioxide for air temperature or for plant production, yet man is unquestionably shifting the steady-state balance of the ecosystem.

The nitrogen cycle has many chemical steps.

The path of nitrogen through the ecosystem differs from that of carbon in several important respects. First, the immense pool of nitrogen (N_2) in the atmosphere (3.85×10^{21} g) cannot be assimilated by most organisms. Second, nitrogen is not directly involved in the release of chemical energy by respiration; its role is linked to protein molecules and nucleic acids, which provide structure and regulate biological function. Third, the biological breakdown of nitrogenous organic compounds to inorganic forms requires many steps, some of which can be performed only by specialized bacteria (Figure 42-4). Fourth, most of the biochemical transformations involved in the decomposition of nitrogeneous compounds occur in the soil, where the solubility of inorganic nitrogen compounds influences the availability of nitrogen to plants.

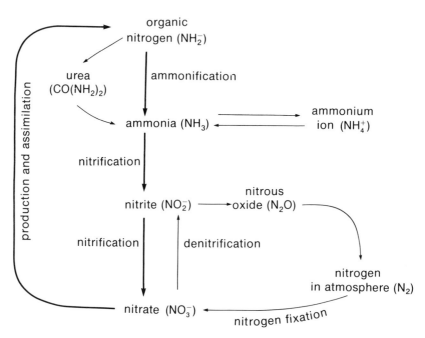

Figure 42-4   Basic biochemical steps in the nitrogen cycle.

TABLE 42-1

Percentage distribution of nitrogen among active pools, and annual transfer rates. The pools taken together contain about 10^{18} grams of nitrogen (based partly on Kormondy 1969).

Pool	Nitrogen (per cent)	Transfer rate (per cent per year)
Organic forms		
Plants	342⎤	25
Animals	11⎦	
Detritus	6,100	1.4
Inorganic forms in soils and oceans		
Ammonia (NH_3)	286	30
Nitrites (NO_2^-)	138	63
Nitrates (NO_3^-)	4,180	2.1

Living tissues contain slightly more than 3 per cent of the nitrogen in active pools in the ecosystem. The rest is distributed between detritus and nitrates (NO_3^-) in the soils and ocean, with smaller amounts in the intermediate stages of protein decomposition—ammonia, NH_3, and nitrites, NO_2^- (Table 42-1). Plants assimilate 86×10^{14} g of nitrogen annually, less than 1 per cent of the active pool; the overall cycling time of nitrogen, therefore, exceeds one hundred years.

The nitrogen cycle involves the stepwise breakdown of organic nitrogen compounds by many kinds of organisms until nitrogen is finally converted to nitrate. Of the forms of nitrogen in the soil that are available to plants, ammonia (NH_3) or the ammonium ion (NH_4^+) would seem to be the most desirable because their conversion to organic compounds requires the least chemical work. Ammonia is, however, unsuitable as a source of nitrogen in the soil because it is toxic to plant tissues in high concentrations and is not persistent in the soil (see Chapter 5). Ammonia dissolves easily in water and is quickly leached out of soil. Under acid soil conditions, ammonia is converted to the ammonium ion (Brady 1974). Although the positively charged ion can adhere to the surface of the clay-humus micelle (see page 56), it is easily displaced by hydrogen ions in acid soils, and thus it is readily leached out by water. (Some types of clay minerals actually adsorb ammonium ions into their crystal framework so tightly that they are inaccessible both to leaching and to plants.) The soil ammonia that escapes leaching is readily attacked by specialized bacteria, which obtain energy when they oxidize the nitrogen in ammonia to nitrites and nitrates. Negatively charged nitrite and nitrate ions do not bind to clay particles at all, hence they are susceptible to leaching. Once nitrates are produced in the soil they are quickly assimilated by plant roots. Storage of nitrogen in terrestrial ecosystems occurs primarily in organic detritus. Most of the nitrogen in aquatic ecosystems occurs as dissolved nitrates and in the detritus in sediments.

The biochemical reactions of nitrogen compounds are highly varied because nitrogen can combine with other elements in several different ways (Delwiche 1970). The most important processes in the nitrogen cycle are the breakdown of organic nitrogen compounds by *ammonification* and *nitrification,* the reduction of nitrate and nitrite to nitrogen (N_2) and its release into the atmosphere by *denitrification,* and the biological assimilation of atmospheric nitrogen by *nitrogen fixation* (Table 42-2).

Denitrification removes nitrogen from active pools in the soil and surface waters and releases it into the atmosphere; nitrogen fixation brings atmospheric nitrogen back into active circulation through the ecosystem. Although these processes are minor compared to the overall cycling of nitrogen in the ecosystem, nitrogen fixation can assume local importance where soils lack adequate nitrogen for normal plant growth, as we shall see below.

In its organic form, nitrogen occurs in amine groups (NH_2), or some variation, combined with other organic molecules. Animals rid their bodies of excess nitrogen by detaching the amines from organic compounds and excreting them relatively unchanged, primarily as ammonia, NH_3, or urea, $CO(NH_2)_2$. Soil microorganisms readily convert urea to ammonia by hydrolysis

$$CO(NH_2)_2 + H_2O \rightarrow 2NH_3 + CO_2$$

This reaction does not, however, release energy to perform biological work.

Some specialized but ubiquitous bacteria can release the chemical energy contained in the amine group by a series of nitrifying steps that require oxygen. *Nitrosomonas* transforms the ammonia ion to nitrite; *Nitrobacter* completes the nitrification process by oxidizing nitrite to nitrate.

Nitrification represents a critical step in the nitrogen cycle, ultimately determining the rate at which nitrates become available to green plants and thereby influencing the productivity of the habitat. Any soil condition that inhibits bacterial activity—high acidity, poor soil aeration, low temperature, and lack of moisture—also inhibits nitrification. Slow nutrient release in the soil of cold and dry regions may depress plant productivity beyond the direct effect of cold and dryness on photosynthesis. Furthermore, if the organic detritus in the soil has a low nitrogen content compared to its content of carbon, bacteria assimilate all the nitrogen into their cell structure rather than using some as a substrate for metabolism. Nitrogen thus becomes tied up in bacteria biomass rather than being made available to plants. The ratio of carbon to nitrogen in detritus influences the rate of bacterial decomposition. At one extreme, mulberry leaves (C/N ratio = 25) support an abundant microflora of bacteria and fungi and decompose rapidly; at the other extreme, loblolly pine (C/N ratio = 43) inhibits the activity of microorganisms and decomposes slowly (see Table 41-7).

Denitrification transforms nitrate to nitrogen in a series of steps

$$NO_3^- \rightarrow NO_2^- \rightarrow N_2O \rightarrow N_2$$

each of which releases oxygen. Nitrous oxide (N_2O) and nitrogen molecules (N_2) escape into the atmosphere and leave the active nitrogen pools. Denitrification also occurs purely by chemical means, independently of micro-

TABLE 42-2

The biochemical processes involved in the nitrogen cycle, their biological occurrence, and their energy yield (from Delwiche 1970).

Process*	Organism	Yield (kcals/mole)
Respiration†		
(1) $C_6H_{12}O_6 + 6O_2 \rightarrow 6CO_2 + 6H_2O$	Virtually universal	686
Denitrification		
(2) $C_6H_{12}O_6 + 6KNO_3 \rightarrow 6CO_2 + 3H_2O + 6KOH + 3N_2O$	*Pseudomonas denitrificans*	545
(3) $5C_6H_{12}O_6 + 24KNO_3 \rightarrow 30CO_2 + 18H_2O + 24KOH + 12N_2$	*Pseudomonas denitrificans*	570
(4) $5S + 6KNO_3 + 2CaCO_3 \rightarrow 3K_2SO_4 + 2CaSO_4 + 2CO_2 + 3N_2$	Anaerobic sulfur bacteria	132
Ammonification		
(5) $C_2H_5NO_2 + 1\frac{1}{2}O_2 \rightarrow 2CO_2 + H_2O + NH_3$	Many bacteria, most plants and animals	176
Nitrification		
(6) $NH_3 + 1\frac{1}{2}O_2 \rightarrow HNO_2 + H_2O$	*Nitrosomonas bacteria*	66
(7) $KNO_2 + \frac{1}{2}O_2 \rightarrow KNO_3$	*Nitrobacter*	17.5
Nitrogen fixation		
(8) $2N + 3H_2 \rightarrow 2NH_3$	Some blue-green algae, *Azotobacter*	−147.2

* $C_6H_{12}O_6$ = glucose; CO_2 = carbon dioxide; $C_2H_5NO_2$ = glycine (an amino acid); $CaSO_4$ = calcium sulfate; $CaCO_3$ = calcium carbonate; HNO_2 = nitrous acid; KNO_2 = potassium nitrite; KNO_3 = potassium nitrate; KOH = potassium hydroxide; NH_3 = ammonia; N_2O = nitrous oxide; S = sulfur.
† Included for comparison.

organisms. For example, the reaction of nitric acid (HNO_3) and urea to release nitrogen occurs in acid soils

$$2\ HNO_3 + CO(NH_2)_2 \rightarrow CO_2 + 3\ H_2O + 2\ N_2$$

Nitrogen fixation is energetically expensive and requires considerable chemical work, although it requires no more energy than does conversion of an equivalent amount of nitrate, NO_3^-, to ammonia, NH_3 (Hardy and Havelka 1975). For the price of the chemical energy in a glucose molecule ($C_6H_{12}O_6$), some blue-green algae and the bacterium *Azotobacter* can assimilate eight nitrogen atoms (ignoring the inefficiency of biochemical transformations). Nitrogen fixation requires specialized biochemical machinery that is apparently unavailable to higher plants. Nonetheless, many such legumes as alfalfa and peas, plus scattered species in other plant groups, have entered into symbiotic relationships with nitrogen-fixing bacteria (Quispel 1974). Even some marine algae (Wiebe 1975), lichens (Milbank and Kershaw 1969), and shipworms (Carpenter and Culliney 1975) have symbiotic, nitrogen-fixing bacteria or blue-green algae. Peas develop clusters of nodules throughout their root system that become infected by *Azotobacter* (Figure 42-5). The relationship benefits both parties: the plant furnishes glucose to the bacteria, and the bacteria assimilate nitrogen from the soil atmosphere for plant uptake.

In nitrogen-deficient habitats, nitrogen fixation is a critical factor in plant production. The nitrogen-fixing capabilities of certain plants have become widely exploited in agriculture to restore soil fertility after farmland has been planted with soil-depleting crops like corn. Nitrogen-accumulating plants (usually peas or alfalfa) are planted in rotation with corn, and then plowed under the soil, increasing its nitrogen and humus content and water retention. Hardy and Havalka (1975) have estimated that about 175×10^{12} g of atmospheric nitrogen are fixed annually, about half of it in agricultural land. The total amounts to about 2 per cent of the nitrogen assimilated by plants annually.

Assimilation of nitrogen from the atmosphere probably constitutes a more important agent promoting soil fertility than its annual rate indicates. Nitrogen-fixing microbes are widespread in natural habitats, even on the leaves of trees (Last and Warren 1972). If we stop to consider for a moment that parental rocks underlying most soils are completely devoid of nitrogen, we realize that most of the nitrogen in active pools in the ecosystem must have originated through nitrogen fixation.

The movement of phosphorus is closely tied to oxygen and acidity.

Oxygen, carbon, and nitrogen cycles demonstrate the basic features of mineral cycles in the ecosystem, but other elements—particularly phosphorus, potassium, calcium, sodium, sulfur, magnesium, and iron—play important roles in ecosystem function. Other elements—such as cobalt, aluminum,

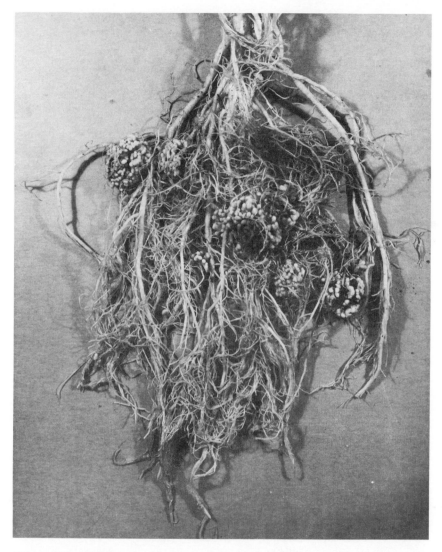

Figure 42-5   The root system of an Austrian winter pea plant, showing the clusters of nodules that harbor symbiotic nitrogen-fixing bacteria (courtesy U.S. Soil Conservation Service).

and manganese—may influence ecosystem dynamics in ways still undiscovered.

Ecologists have studied the role of phosphorus in the ecosystem most intensively because organisms require phosphorus at a high level (about one-tenth that of nitrogen) as a major constituent of nucleic acids, cell membranes, energy-transfer systems, bones, and teeth. Phosphorus is important for a number of other reasons. It is thought to limit plant productivity in many aquatic habitats, and the influx of phosphorus to rivers and lakes in the form of sewage (particularly from phosphate detergents) and runoff from fertilized agricultural lands stimulates the production of aquatic habitats to undesirably high levels. Also, ecologists can measure concentrations of phosphorus easily and use one of its isotopes as a radioactive tracer in the ecosystem.

The phosphorus cycle has fewer steps than the nitrogen cycle: plants assimilate phosphorus as phosphate ion ($PO_4^\equiv$) directly from the soil or water; animals eliminate excess organic phosphorus in their diets by excreting phosphorus salts in urine; phosphatizing bacteria convert the organic phosphorus in detritus to phosphate in the same way. Phosphorus does not enter the atmosphere in any form other than dust. The phosphorus cycle therefore involves only the soil and water of the ecosystem.

In spite of the relative simplicity of the phosphorus cycle, many environmental factors influence the availability of phosphorus to plants. In the presence of abundant dissolved oxygen, phosphorus readily forms insoluble compounds that precipitate and remove phosphorus from the pool of available nutrients. If such conditions persist, deposits of phosphate accumulate and eventually form phosphate rock, which returns to active pools in the ecosystem very slowly by erosion—or by artificial fertilization of crops and disposal of phosphate detergents in sewage (fertilizers and detergents themselves deriving their phosphorus content from phosphate rock).

Acidity also affects the availability of phosphorus to plants. Phosphate compounds of sodium and calcium are relatively insoluble in water. Under alkaline conditions phosphate ions ($PO_4^\equiv$) readily combine with sodium or calcium ions to form insoluble compounds. Under acid conditions, phosphate is converted to highly soluble phosphoric acid. At intermediate levels of acidity, phosphate ions form compounds with intermediate solubility, as shown below:

	⟶ Increasing acidity ⟶			
Ionic form	$PO_4^\equiv$ →	$HPO_4^\equiv$ →	$H_2PO_4^-$ →	H_3PO_4
	↓	↓	↓	
Salt	Na_3PO_4	$CaHPO_4$	NaH_2PO_4	None
Solubility	(slightly soluble)	(insoluble)	(soluble)	(very soluble)

Although acidity increases the solubility of phosphate in the laboratory, high acidity reduces the availability of phosphorus in the ecosystem, due to its reaction with other minerals. In acid environments, aluminum, iron, and manganese become soluble and reactive, forming chemical complexes that bind phosphorus and thereby remove it from the active pool of nutrients. The acid conditions in bogs and in soils of cold, wet regions remove phosphorus in this way, and reduce the fertility of these habitats. Phosphorus is most readily available in a narrow range of acidity, just on the acid side of neutrality.

Phosphorus is a major factor in eutrophication.

Ecologists classify natural bodies of water between two extremes on the basis of their nutrient content and organic productivity (Beeton 1965, National Academy of Science 1969). On one hand, *oligotrophic* (from the Greek, meaning little-nourished) habitats have low nutrient content and harbor relatively little plant and animal life. The water in oligotrophic lakes and rivers is clear and unproductive. On the other hand, *eutrophic* (from the Greek, meaning well-nourished) lakes and rivers are rich in nutrients and support an abundant flora and fauna. Eutrophic habitats are excellent fisheries because of their high productivity. In fact, lakes and ponds are often artificially fertilized to increase fish production: Harvests vary between one and seven pounds of fish per acre per year (0.2 to 1.6 kcal$\cdot$m$^{-2}\cdot$yr^{-1}) from the oligotrophic Great Lakes; small eutrophic lakes in the United States yield up to 160 pounds per acre per year (36 kcal$\cdot$m$^{-2}\cdot$yr^{-1}); in Germany and the Philippines, where ponds are artificially fertilized, yields of 1,000 pounds per acre per year (200 kcal$\cdot$m$^{-2}\cdot$yr^{-1}) are not uncommon. Primary production of the phytoplankton increases with eutrophy in a similar progression: 7 to 25 g carbon$\cdot$m$^{-2}\cdot$yr^{-1} in oligotrophic lakes, 75 to 250 g carbon$\cdot$m$^{-2}\cdot$yr^{-1} in naturally eutrophic lakes, and 350 to 700 g carbon$\cdot$m$^{-2}\cdot$yr^{-1} in lakes polluted by sewage and agricultural runoff.

Eutrophication does not by itself constitute a major problem for the aquatic ecosystem. High rates of production and rapid nutrient cycling are to be expected where the mineral resources for production are abundant. In naturally eutrophic lake and stream ecosystems, most of the energy and minerals come from within the system—*autochthonous* inputs—in the form of primary production. In culturally eutrophic lakes and streams, external sources of mineral nutrients and organic matter—*allochthonous* inputs—contribute to the productivity of the system. (In the course of attaching special names to almost everything, ecologists have frequently turned to Greek or Latin roots for terms. *Chthonos* is Greek for "of the earth," *auto* means "the same," and *allo,* "other" or "different," referring to internal and external inputs.)

Whereas naturally eutrophic systems are usually well balanced, the addition of artificial nutrients can upset the natural workings of the community and create devastating imbalances in the ecosystem (Likens 1972). Algal blooms are among the most noticeable of these effects. The combination of high nutrient loads and favorable conditions of light, temperature, and carbon dioxide stimulate rapid algal growth. Algal blooms are a natural response of algae to their environment. But when the environment changes and no longer can support dense algal populations, the algae that accumulate during the bloom die and begin to decay. The ensuing rapid decomposition of organic detritus by bacteria robs the water of its oxygen, sometimes so thoroughly depleting the water of oxygen that fish and other aquatic animals suffocate. The nutrient imbalance in culturally eutrophic systems stems from the addition of nutrients at seasons when nutrients are less available in naturally eutrophic waters, primarily during the summer peak of plant production. During less productive seasons, phosphorus is readily absorbed by benthic bacteria and sediments at the bottoms of lakes, and its concentration in lake water is thus quickly reduced.

Early studies of eutrophication suggested that algal blooms occurred only when the concentration of phosphorus was greater than 0.01 milligrams per liter of water (Sawyer 1946). These findings strongly implicated phosphorus as the prime factor limiting the productivity of many aquatic communities (Serruya and Berman 1975). Although ecologists have since argued this point, recent experiments on small Canadian lakes demonstrate that adding carbon (as sucrose) and nitrogen (as nitrate) does not stimulate algal blooms without simultaneously adding phosphate (Schindler 1973, 1974, Schindler and Fee 1974; Figure 42-6).

In spite of the disturbing effects of outside enrichment, culturally eutrophied lakes can recover their original condition if inputs are shut off. Apparently the sediments at the bottom of most lakes have a high affinity for phosphates and quickly remove them from active pools in the surface waters of the lake. Diversion of sewage from Lake Washington, in Seattle, quickly reversed the eutrophication process (Edmondson 1970). Similar recovery could be expected from other bodies of water if inputs from sewage and agricultural runoff were reduced.

In their normal development, lakes usually proceed from oligotrophic to eutrophic stages (Hutchinson 1969). New lake basins are deep and devoid of nutrients. As sediments wash into a lake, nutrients are added to the water. As the lake begins to fill in and become shallow, exchange between the bottom and surface water accelerates, returning the nutrients to the active pool near the surface and enriching the water. In extreme old age, lakes fill in completely and are succeeded by the local terrestrial vegetation.

Cation exchange is not directly tied to energy flux.

Calcium, potassium, sodium, and magnesium are not chemically incorporated into organic compounds, although they are abundant as dissolved *cations* in cellular and extracellular fluids. The cycles of cations in the ecosystem are only loosely associated with the assimilation and release of energy, but they are tremendously important to cell function.

Cation cycles have been studied most intensively in temperate forests, within which the ions exhibit remarkable mobility. The forest is an open system, one which freely exchanges minerals with other parts of the environment (Figure 42-7). Leached cations are washed out of the system through runoff, both in streams and groundwater. Minerals enter the system in precipitation, in wind-born dust and organic debris, and by weathering of the parent rock over which the forest lies.

Detailed cation budgets have been obtained for small watersheds by measuring the inputs in rainwater collected at various locations in the watershed area (Figure 42-8) and the outputs in water leaving the watershed by way of the stream that drains it (Figure 42-9). Care must be taken to select

Figure 42-6   Experimental lake demonstrating the crucial role of phosphorus in eutrophication. The near basin, fertilized with carbon (in sucrose) and nitrogen (in nitrates), exhibited no change in organic production. The far basin, separated from the first by a plastic curtain, received phosphate in addition to carbon and nitrogen and was covered by a heavy bloom of blue-green algae within two months (courtesy D. W. Schindler, from Schindler 1974).

a watershed with impervious bedrock to eliminate the problem of ground-water movement into and out of the watershed.

Several watersheds in the Hubbard Brook Experimental Forest, New Hampshire, have been studied intensively during the past decade to establish patterns of water and nutrient cycles and the effects of disturbances, pri-

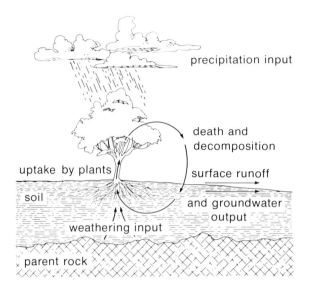

Figure 42-7   A generalized cation cycle for a terrestrial ecosystem. If the system is in a steady state, losses owing to surface flow and ground-water drainage are balanced by gains from precipitation and weathering.

Figure 42-8   Rain gauges installed in a ponderosa pine stand in California to intercept precipitation falling through the canopy of the forest and running down the trunks of trees. Analysis of the nutrient content of water collected in sampling programs like this helps determine the overall cation budget of the forest and the specific routes of mineral cycles (courtesy U.S. Forest Service).

Figure 42-9   A stream gauge at the lower end of a watershed at the Coweeta Hydrological Laboratory, North Carolina. The V-shaped notch regulates the flow of water through the weir in such a way that the flow rate is proportional to the water level in the basin (courtesy U.S. Forest Service).

marily clear-cutting, on these cycles (Likens *et al.* 1967, 1970, 1971, 1977, Bormann and Likens 1969, Eaton *et al.* 1972). The annual distribution of precipitation and runoff for the Hubbard Brook Forest (Figure 42-10) shows a fairly uniform distribution of rainfall during the year, typical of moist temperate locations. Precipitation exceeds runoff during the cold winter months because of snow accumulation. This pattern is reversed in spring as melting snow swells the streams. The difference between precipitation and runoff during the summer months, seen in Figure 42-10, is accounted for by the evaporation and transpiration of water from the watershed.

Cation budgets have been calculated for the entire watershed from the concentrations of the minerals in precipitation and stream flow (Figure 42-11). Only potassium exhibited a net gain in the watershed. Net losses of other cations would have to be balanced by weathering of the bedrock for the system to be in a steady state (Klausing 1956). The cation budgets of watersheds around the world vary tremendously with total precipitation, soil acidity, and the relative abundance of minerals in rainfall and the soil. Where sodium is abundant in the soil or in rainwater, as in many coastal areas,

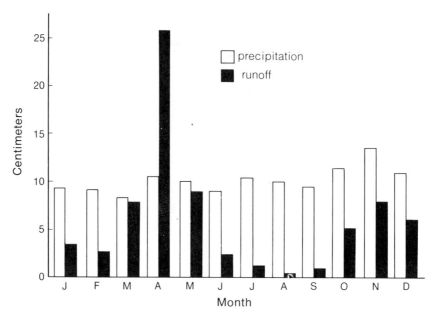

Figure 42-10   Average annual distribution of precipitation and runoff for the Hubbard Brook Experimental Forest, New Hampshire, 1955–1969 (after Likens *et al.* 1971).

sodium output in stream flow is great (Ingham 1950, Art *et al.* 1974). Similar patterns apply to other cation budgets.

The uptake of cations by plants closely parallels their availability in the soil and hence their general mobility in the ecosystem (Table 42-3). Calcium is most rapidly cycled, magnesium least. In general, uptake by vegetation is one to ten times the annual loss of cations in stream flow; therefore the average transit time (see page 809) of cations in the ecosystem is probably one to ten years (Duvigneaud and Denayer-de Smet 1970). Considering the great mobility of cations in the soil, the long transit times of cations indicate that plants rapidly assimilate free ions in the soil before they are leached out of the system by runoff and groundwater. The role of vegetation in nutrient cycling is dramatized by experiments in which entire watersheds are denuded of trees and shrubs (Figure 42-12). Clear-cutting of small watersheds as in the Hubbard Brook Forest increased stream flow several times owing to removal of transpiring leaf surface; losses of cations increased three to twenty times over comparable undisturbed systems. The nitrogen budget of the cut-over watershed sustained the most striking change. Plants assimilate available soil nitrogen so rapidly that the forest usually gains nitrogen at the rate of 1 to 3 kilograms per hectare per year ($kg \cdot ha^{-1} \cdot yr^{-1}$, a hectare = 2.47 acres). In the clear-cut watershed net loss of nitrogen as nitrate (NO_3^-) soared to 54 $kg \cdot ha^{-1} \cdot yr^{-1}$, which is comparable to the annual turnover of nitrogen by vegetation. Precipitation brought only 7 $kg \cdot ha^{-1} \cdot yr^{-1}$ of nitrogen into the

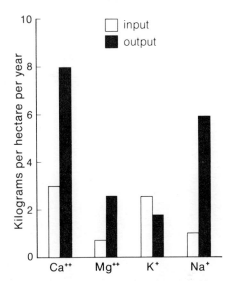

Figure 42-11   Cation budgets for the Hubbard Brook Experimental Forest, New Hampshire, during 1963–1964, expressed as kg·ha⁻¹·yr⁻¹ (after Likens *et al.* 1967).

system, and thus the loss of nitrate represented nitrification of organic nitrogen sources at the normal annual rate by soil microorganisms without simultaneous rapid uptake by plants. These experiments demonstrate the important role of vegetation in maintaining the fertility of the soil, and they further emphasize that the physical and biological components of the ecosystem cannot exist apart.

Nutrients are cycled more rapidly in tropical forests than in temperate forests.

Because of the year-round warm temperatures of tropical climates, leaf litter decomposes rapidly throughout the year and does not accumulate in

TABLE 42-3
Summary of cation budgets for representative temperate forest ecosystems (from Carlisle *et al.* 1966, Likens *et al.* 1967, Duvigneaud and Denayer-de Smet 1970).

	Range of values (kg·ha⁻¹·yr⁻¹)*			
	Precipitation input	Stream outflow	Net loss	Uptake by vegetation
Calcium	2–8	8–26	3–18	25–201
Potassium	1–8	2–13	−1–5	5–99
Magnesium	1–11	3–13	2–4	2–24
Sodium	1–58	6–62	4–21	—

* Kilograms per hectare per year. A hectare is equal to 10,000 square meters, or 2.47 acres.

Figure 42-12   Clear-cut watershed at the Coweeta Hydrological Laboratory, North Carolina, employed in studies of evapotranspiration and runoff in forest ecosystems (courtesy U.S. Forest Service).

tropical forests to the extent that it does in temperate forests (see page 810). Warm temperatures and abundant rainfall additionally result in rapid leaching of nutrients from the soil by ground water. The combination of rapid decomposition and rapid leaching is thought to cause many tropical soils to have lower nutrient contents than temperate soils. The rapid mobilization of minerals in tropical soils is balanced in part by rapid uptake by plants, whose roots actively assimilate water and nutrients throughout the year. As a result, minerals are cycled through the soil of tropical forests quickly but are retained by the vegetation.

Of the total biomass of forests, including dead branches and leaves, litter on the forest floor comprises an average of about 20 per cent in temperate coniferous forests, 5 per cent in tropical hardwood forests, and 1 to 2 per cent in tropical rain forests (Ovington 1965). Because measured litter fall in forests consists mostly of dead leaves, forests may be more meaningfully compared by considering the ratio of accumulated litter on the forest floor to the biomass of living leaves. In temperate forests this ratio is between 5:1 and 10:1, whereas in tropical forests it is less than 1:1. These data are consistent with the notion that dead organic material is decomposed rapidly in the tropics and does not form a substantial nutrient reservoir, as it does in temperate regions.

In northern coniferous forests more than 50 per cent of the organic carbon occurs in the soil and litter, whereas in tropical rain forests less than 25 per cent occurs in soil and litter (Kira and Shidei 1967). Most of the carbon in tropical forests is accumulated in the woody parts of trees. This pattern prevails only for carbon, however, and does not necessarily reflect the distribution of other elements. Nonliving woody materials in the roots, trunks, and branches of trees contain abundant carbon, but relatively little of other elements such as nitrogen and phosphorus. Furthermore, because carbon enters the terrestrial ecosystem from the air by way of photosynthesis, its distribution in the ecosystem must be considered apart from mineral nutrients that plants obtain from the soil.

It is difficult to make a general statement about the relative proportion of nutrients in the soil compared to the living vegetation in tropical forests. The distribution of potassium, phosphorus, and nitrogen in two temperate forests and one tropical forest are compared in Table 42-4. Two points seem to be clear. First, the accumulation of nutrients in vegetation, on a weight for weight basis, is slightly greater in the tropical forest. For example, the

TABLE 42-4

The distribution of mineral nutrients in the soil and vegetation components of representative temperate and tropical forest ecosystems (from Ovington 1965 and Duvigneaud and Denayer-de Smet 1970).

Forest (and locality)	Biomass (tons·ha^{-1})	Nutrients (kg·ha^{-1})		
		Potassium	Phosphorus	Nitrogen
Ash and oak (Belgium)	380			
Living		624	95	1,260
Soil		767	2,200	14,000
Soil/living		1.2	23.1	11.1
Oak and beech (Belgium)	156			
Living		342	44	533
Soil		157	900	4,500
Soil/living		0.5	20.5	8.4
Tropical deciduous (Ghana)	333			
Living		808	124	1,794
Soil (30 cm)		649	13	4,587
Soil/living		0.8	0.1	2.0

total dry weight of living vegetation in the Belgian ash-oak forest exceeds that of the tropical deciduous forest in Ghana by 14 per cent, but the accumulation of the three elements per gram of dried vegetation is 32 to 38 per cent lower in the Belgian forest than in the Ghana forest. Second, levels of some nutrient elements in tropical forest soils can be as high as, or higher than, in temperate soils (also see Sanchez and Buol 1975). We should remember, however, that among different forests, even in the same region, the availability of nutrients varies greatly. For example, in the soils of deciduous forests in Belgium, the amount of nitrogen varied between 4,480 and 13,760 kilograms per hectare (Duvigneaud and Froment 1969). Similar studies on a large series of forests in Bavaria revealed between 1,905 and 15,929 kilograms of nitrogen per hectare (Emberger 1965). We must be cautious, therefore, in drawing conclusions from a limited number of study areas.

The forest vegetation itself plays a major role in maintaining the fertility of tropical soils against the leaching effects of rainfall. In very sandy regions, where the weathering of the substrate produces poor soils with little nutrient content, the fertility of the soil is built up by nitrogen fixation and by plants trapping and retaining nutrients imported by rainfall. The importance of the forest vegetation in maintaining nutrient concentrations is strikingly demonstrated by the fact that when tropical forests are cleared and planted with annual crops, soil fertility declines rapidly unless mulch or fertilizers are applied artificially (Table 42-5).

We do not know to what extent soil nutrients limit the productivity of temperate or tropical forests, but the forest thinning experiments of Schultz (see page 237) and experiments in which fertilizers are added to forests (Ovington 1965) suggest that nutrients limit primary production to some extent. Whether tropical soils are poorer than temperate soils is also open to question. The well-known difficulties encountered in applying temperate agricultural methods to tropical regions may be caused more by the agricultural practices themselves than by the intrinsic fertility of the soil. After all, no other ecosystem can match the productivity of a mature tropical forest.

TABLE 42-5
Influence of mulching cut vegetation into the soil and of adding inorganic fertilizers on soil fertility in continuously planted cotton crops in tropical Africa. Data are kilograms of harvested cotton per acre (from Jurion and Henry 1969).

	Clean weeded		Mulched	
Year	Without fertilizer	Fertilized since 1953*	Without fertilizer	Fertilized since 1953*
1947–48	1,032	—	1,127	—
1953–54	200	440	1,117	1,434
1955–56	186	797	1,464	1,977
1956–57	124	706	986	1,344

* Fertilizer applied: 150 kilograms per hectare (kg./ha^{-1}) bicalcium phosphate, 250 kg./ha^{-1} sodium nitrate, and 50 kg./ha^{-1} potassium sulphate.

In this chapter, we have traced the cycles of several elements through the ecosystem. The pattern of movement depends partly on the chemical properties of the element and partly on how it is used by living organisms. The cycles of the various elements are similar only to the extent that the decomposition activities of organisms return them to forms that can be assimilated by plants. The rate of flow of a nutrient at each step in its cycle is determined by the ecology of the organisms involved in the step. If a substance is utilized slowly by organisms, it tends to accumulate and form a large reservoir behind the bottleneck in the cycle. Because minerals are cycled, a disruption at any one step in the cycle can be reinforced through successive cycles, leading to instability. For this reason, those who apply ecology to solve environmental problems probably should pay more attention to nutrient budgets than to energy budgets of ecosystems. The disturbance of normal ecosystem function leading to such problems as eutrophication and nutrient depletion can probably be traced to disruption of a critical step in one or more nutrient cycles.

In the preceding chapters, we have examined some overall measures of ecosystem function—primary reproduction, energy flow, and nutrient cycling—and we have seen how adaptations of organisms to their particular environments determine patterns in these community characteristics. In the next and final chapter, we shall examine the stability of structure and function in the biological community.

43

Stability of the Ecosystem

Stability is the inherent capacity of a system to tolerate or resist changes caused by outside influences. Suppose, for example, that precipitation fell 50 per cent below its long-term average, but plant production decreased by only 25 per cent and herbivore populations decreased by only 10 per cent. The dampening of the environmental fluctuation as it passed up the food chain is a measure of the internal stability of the system—its capacity to resist change. In this case, stability may be derived from storage of water in the soil, physiological responses of plants to drought, and if the drought lasts long enough, partial replacement of drought-sensitive herbs by drought-resistant species. The stability of a community depends upon the homeostatic responses of its constituent species.

We can visualize the concept of stability by considering a small ball placed in a bowl. If we nudge the ball, sending it a little way up the side of the bowl, the force of gravity quickly returns the ball to the bottom of the bowl. The steeper the sides of the bowl, the more powerful is the stabilizing force of gravity. This tendency of a system to be restored to a particular condition is referred to as a *stable equilibrium.* If the bowl itself represented the environmental factors acting on a population, the weight of the ball could correspond to the homeostatic capacities of the individuals in the population. A nudge of a given force would displace a steel ball bearing much less than it would a table tennis ball, just as a cold snap has less effect on a population of bears than on a population of flies.

A ball placed on a level table represents a system having no forces tending either to maintain the position of the ball or to change it. This condition is known as a *neutral equilibrium.* A nudge applied to the ball sends it across the table in one direction or the other, and its movement is unchanged until some outside force intercedes. Rarely is nature engineered like a level table. All systems have equilibrium points toward which they tend when displaced. Disturbances can, however, push a system beyond its capabilities of response—a ball sent careening out of the bowl; an insect population escaping predator controls and rising to outbreak levels.

When a system moves away from an equilibrium point, as when a ball rolls down the outside of an inverted bowl, the situation is referred to as an *unstable equilibrium.* The highest point on the outside surface of the inverted bowl is an equilibrium point because it is possible to balance the ball there. But the slightest disturbance sends the ball on its way.

The amount of movement a nudge of given force causes in a ball in a bowl depends then on the weight of the ball and the configuration of the bowl. Similarly, the amount of fluctuation in a population or a community caused by an external disturbance depends on the inherent stability of the system.

We may define the inherent stability of a system as the ratio between variation in the environment and variation in the system itself, but this definition is difficult to apply to populations or communities. Which aspect of environmental variation should we measure? Which component of the system gives the best indication of adequate function and continued persistence? Do we judge the stability of a community by the constancy of its function (production, ecological efficiency) or its structure (diversity, species turnover)? Moreover, change itself is often the best response to change. Hibernation and diapause represent a near total shutdown of biological activity in response to environmental change, yet dormancy allows the population to persist.

The biological significance of stability is even more elusive than its description and measurement. Constancy in the natural world is desirable to man because it enables him to predict conditions in advance and plan his activities accordingly. If weather and insect pests did not vary from year to year, farming would be simplified and a reasonably constant crop yield would be assured each year.

Virtually all human activity disturbs natural communities. Natural biological communities do not yield enough harvestable food to support dense human populations, and man selects the most "desirable" components of natural communities, alters their evolution through artificial selection to suit his purposes, and maintains these populations of crops, livestock, pulp trees, city parks, and backyards in a continually disturbed state, the populations constantly exposed to conditions they are not adapted to cope with. The price man pays for exploitation of natural resources is the price of maintaining their stability by constant management: curbing pest infestations, maintaining soil fertility, and cleaning out weeds.

Our concern with the constancy of the natural world and with the basis for stability in natural communities is understandable. This concern is shared, unconsciously, by all species. Constancy of weather, resources, predation, and competition reduces the cost of self-maintenance (homeostasis) and increases the allocation of energy and nutrients to production. Most organisms would stand to gain from a more constant world, but because their competitors and predators would also gain, we cannot predict the net benefit of constancy to an individual or to a species.

Constancy of the environment and the community both undoubtedly enhance production and ecological efficiency because few resources are used for homeostasis, materials do not accumulate behind bottlenecks caused by population fluctuations, and adaptations of organisms become more finely tuned to the environment. But we must ask whether communities possess inherent stabilizing mechanisms in addition to the homeostatic responses of organisms and the growth responses of populations. If we are to reject the notion of community adaptations, we must also reject many contemporary ideas about community stability: that it is desirable for the community to be stable because constancy increases the efficiency of energy flow and nutrient cycling; that natural selection leads to increased complexity and diversity within the community so as to enhance the inherent stability of trophic structure and improve the ability of the community to resist perturbation; that many adaptations of organisms, such as large size, long life, low reproductive rates, and low productivity-biomass ratios, are adaptations to increase the stability of the community rather than to increase the evolutionary fitness of the organism (Dunbar 1960).

Measures of community structure and function—number of species, number of trophic levels, rates of primary production, energy flow, and nutrient cycling—reflect ecological interactions among populations and between individuals and their physical environments. The ability of the community to resist ecological perturbations no doubt reflects the homeostatic mechanisms of individuals and the growth responses of populations. But beyond these separate homeostatic capacities, some properties of organization influence community stability and the efficiency of community function; that is, the homeostatic capacity of a community transcends the summed properties of its constituent populations. We might imagine, for example, that competition and ecological release help to stabilize the function of each trophic level. If the population of one species were suppressed by climate or disease, a competitor could respond by using the first population's leftover resources, and thus maintain the total production of the trophic level. We might also view predation as a destabilizing influence, magnifying population fluctuations in lower trophic levels.

In this chapter, we shall examine the inherent stability of ecological systems, and how stability of community function might be influenced by the structure of the community itself.

Some terms need defining.

Ecologists use stability and related terms in so many different ways that it will be necessary to provide explicit definitions here:

stability is the intrinsic ability of a system to withstand or to recover from externally caused change

constancy is a measure of the degree of variation of some attribute of a system

predictability is a measure of regularity in patterns of change

Seasonal fluctuation in the environment usually is predictable; day-to-day variation often is not.

The degree of fluctuation in the ecosystem is determined by three factors, each of which will be considered separately below: (a) the constancy and predictability of the physical environment; (b) the homeostatic mechanisms of organisms and growth responses of populations as subunits of community stability; (c) that component of community stability uniquely contributed by the feeding and competitive relationships of populations within the community—in other words, the trophic organization of the community.

Variability in the physical environment is inversely correlated with biological production.

Temperature and rainfall have been measured for long periods at weather monitoring stations throughout the world. Although local climatological data do not adequately measure the environment of any particular population, they do indicate the overall constancy of the environment. Three components of the environment are important to organisms: regular seasonal fluctuations, variation about seasonal norms, and the predictability of short-term variations. All habitats change daily and seasonally; coastal marine environments vary, in addition, with a lunar rhythm. Daily, lunar, and seasonal fluctuations reflect regular cycles in the physical world, as we have seen in Chapter 4. Many years of measurement would reveal the average conditions for each day of the year and each time of day. But the environment rarely exhibits average conditions. Irregularities in climate related to changing wind patterns and random meteorological events cause the environment to vary around its norm. For example, one hundred years of weather records in Philadelphia, Pennsylvania, show that while the average July rainfall was 4.2 inches, precipitation was less than half the average in nine years and more than twice the average in four years (Figure 43-1).

Diurnal and seasonal patterns are more predictable than short-term variations because they are tied to precise physical cycles, such as the daily light-dark cycle and the seasonal change in daylength during the year. But the unreliability of weather forecasts, particularly for several days ahead, attests to the lack of predictability of short-term variations in climate; the further one is removed in time from an event of brief duration like a rainstorm, the less predictable it becomes. Rain can be predicted a few minutes or a few hours before a thunderstorm by the appearance of the sky. A change in wind direction during certain seasons signals the passage of a front, often accompanied by precipitation and temperature change. But it is virtually impossible to know in January, or even in May, whether June will bring drought or deluge.

Rainfall is generally most variable from year to year where it is least abundant (Figure 43-2). At a given place, dry-season precipitation is more variable and less predictable than wet-season precipitation. Similarly, temperature variations are greatest when the average temperatures are lowest—geographically in polar regions and seasonally during winter (see Figure 4-8). These patterns of variability indicate that in the tropics, the physical

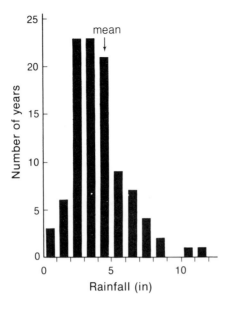

Figure 43-1   July   precipitation for one hundred years at Philadelphia, Pennsylvania.

environment is more constant than in temperate and arctic regions; the tropics are warm the year around, and the climate of most tropical areas is relatively wet. The generally observed pattern could be misleading. Dry-season rainfall in some tropical areas is less than that in the driest months in many temperate localities; rainfall during the tropical dry season can be correspondingly variable and unpredictable.

Every region, no matter how constant its environment, is subject to infrequent extremes. In a wet environment, an extreme condition encountered only once in hundreds of years may be one that differs from the normal by a factor of two. In a dry locality, an extreme condition that is encountered equally infrequently may differ from the normal by a factor of ten. The homeostatic capabilities of organisms are adapted to the range of conditions normally encountered. Regardless of the degree of fluctuation in the environment, infrequent "extreme" conditions impose a stress on organisms in any region. The homeostatic mechanisms of tropical populations might, in fact, be much more poorly developed than those of temperate and arctic populations because the environment usually varies within a narrower range.

Some types of environmental variation are so drastic that they cannot be accommodated by the homeostatic mechanisms of organisms. Such events—environmental catastrophes—include hurricanes, tornadoes, fires, and hard freezes. Although many species are adapted to prosper in the aftermath of such disasters—the weedy species that colonize disturbed habitats, and others adapted to a regular cycle of minor fires—natural catastrophes completely destroy the fabric of most communities. Their structure must be rebuilt gradually over long periods by succession. Many human disturbances create equally catastrophic conditions—beyond the limits of

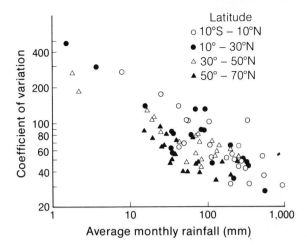

Figure 43-2   Year-to-year variability in monthly rainfall as a function of average monthly rainfall. The coefficient of variation (the standard deviation divided by the mean) is proportional to the percentage deviation from the mean. For a given amount of precipitation, tropical localities exhibit greater year-to-year variability than temperate localities. Based on twenty-five years of records for the months of February, May, August, and November at twenty localities, five in each latitude belt (data from Clayton 1944).

variation normally encountered by organisms—completely destroying communities and sometimes preventing the re-establishment of community structure.

Stability at the individual level depends on body size and homeostatic capacity.

The constancy of population size or organism activity reflects the interplay between environmental fluctuation and intrinsic stability. The outcome of this interaction can be seen by examining the growth rings of trees (Fritts 1966). A core of wood contains a record of a tree's growth rate from the sapling stage on. Because trees produce one ring each year, the rings can be dated easily, and the variation in their width compared to variation in temperature and rainfall. The sensitivity of a tree to climate fluctuation depends on where it grows. In moist habitats, water may never become sufficiently limiting to affect tree growth adversely even in the driest years. Where moisture levels are marginal for a species, drought can exert a profound effect on growth. This point is illustrated by growth ring chronologies in two populations of bristlecone pine in the White Mountains of California (Figure 43-3). Trees in a moist grove with abundant accumulation of winter snow exhibited little year-to-year variation in growth rate compared to stunted individuals growing on a dry, windswept rocky ridge. Although variation in

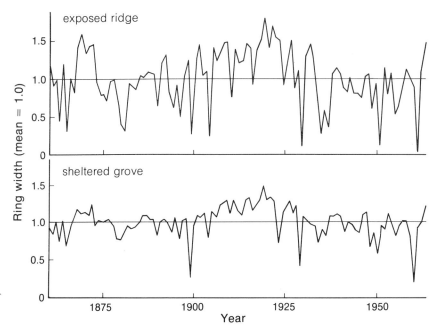

Figure 43-3   Variation in ring width of bristlecone pines growing on a rocky ridge (top) and in a protected grove with abundant moisture (below). The ring-width axis is scaled in such a way that the mean width at each locality equals 1 (after Fritts 1967).

ring width at each site paralleled variation at the other, moisture deficits severely depressed growth in the moist site only three times during the 104-year chronology—1899, 1929, and 1959.

Most of the ring-width variation in the bristlecone pine is related to the moisture level of the habitat during June, a hot month with little or no rainfall and, therefore, with a large evapotranspiration deficit. Autumn temperatures and winter moisture also influence growth during the subsequent season. The climates of different areas vary in their effect on tree growth (Figure 43-4). Moisture stress at the beginning of the growing season is more important to the bristlecone pine in California; Douglas fir responds equally to winter and spring moisture; growth of the piñon pine is determined primarily by winter moisture, indicating that it relies heavily on accumulated ground water for growth during the arid summer months. In dry habitats in Illinois, the growth of white oak responds to conditions of moisture and temperature during the growing season but, in addition, late summer drought depresses growth during the following year, perhaps by reducing storage of food in the roots or by interfering with the formation of leaf buds.

As one might expect, much of what we know about variability in communities comes from economically important species (van Emden and Williams 1974, Goodman 1975, Thompson 1975). Williams et al. (1975) related

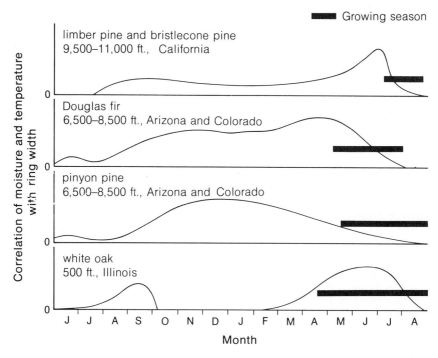

Figure 43-4   Correlation of tree-ring width with temperature and moisture during each of fifteen months prior to and including the growing season (after Fritts 1967).

variation in annual production of wheat and other crops in the Prairie Provinces of Canada to variation in weather. Variability from year to year was expressed as ·the coefficient of variation (CV), the standard deviation (SD) expressed as a percentage of the mean value. Crops were studied on three soil types: brown, dark brown, and black (Table 43-1). Variation in wheat production was greatest on brown soils (CV = 35 per cent) and least on black soils (CV = 24 per cent). Furthermore, most of the variation in crop production was related to soil moisture deficits during the growing season. These deficits were higher on brown soils (142 mm ± 71 mm SD for the season) than on black soils (43 ± 82 mm). And as we would expect, variation had more influence on crop growth where moisture was least abundant. On the brown soils, 86 per cent of the variation in crop production could be related statistically to weather and soil moisture. On black soils, only 57 per cent of the variation in crop production was so related. This example shows the potential for objectively measuring stability but, defined as such, stability includes a component of the environment itself. Clearly we are dealing with a complex phenomenon.

Tree-ring and cereal crop data show that organisms do not completely compensate for variation in the environment, but we are left guessing about

TABLE 43-1

Variation in annual wheat production and moisture properties on three soil types in the Prairie Provinces of Canada (after Williams *et al.* 1975).

	Soil type					
	Brown		Dark brown		Black	
	mean	CV	mean	CV	mean	CV
Wheat crop (kg·ha⁻¹)	1,301	35	1,587	28	1,640	24
Moisture reserves in soil May 1 (mm)	113	19	117	16	147	17
Potential evapotranspiration for June (mm)	138	11	134	13	134	10
Moisture deficit (mm)						
June	69	62	64	72	65	71
July	111	32	88	44	65	53
Season	142	50	106	70	43	190
Variation in crop explained by climate and soil moisture (per cent)	86		76		57	

the inherent stability of a plant's growth processes. Some general principles concerning stability are apparent from our earlier study of homeostasis and population growth (Chapters 12 and 31). Large organisms have small surface-to-volume ratios; their internal environments are, therefore, more independent of the external environment than are the internal environments of small organisms. Large accumulated biomass and low turnover rate of individuals in populations also increase buffering against environmental variation (page 509). Immature organisms are generally more susceptible to environmental change than adults because they are smaller, less experienced in the case of animals, and physiologically less mature. Populations with many immature individuals tend to be less stable than populations of predominantly mature individuals.

Reproduction and adult survival are sometimes interrelated so as to tend to stabilize population size. When unfavorable conditions during the reproductive period lead to reduced fecundity, increased survival can partly compensate for the loss of production (see page 517). This phenomenon is exhibited by natural populations of carabid beetles of the genus *Agonium,* which inhabit the ground litter of marshes and fens (Murdoch 1966). Adults breed during May, June, and July in the northeastern United States, laying between fifty and a hundred eggs. The larvae emerge as adults in the autumn, hibernate over the winter, and breed during the following year. Adults that survive their first summer may live to reproduce in the following year. Murdoch noticed a general inverse correlation between reproductive rate and the survival of adults between the beginning of the reproductive period and the onset of hibernation in the winter. In one population, the mean number of mature eggs in the oviducts of females collected during the summer varied between 7.0 to 9.5; the post-reproductive survival was on the order of 5 per cent, and only 4 per cent of females did not breed (no mature eggs in the oviducts). During another summer, female oviducts held an average of 4.0 eggs and 24 per cent of the population did not breed, but survival after the reproductive season was much higher, between 17 and 68 per cent.

These observations prompted Murdoch to perform some simple experiments. He put adults into wire bags in a marsh at the beginning of the reproductive period, thereby excluding competitors and predators. Some of the bags contained two beetles of each sex, so the females reproduced. Other bags contained only four females which, lacking mates, produced no mature eggs. The post-reproductive survival of the nonbreeding females was twice that of the breeding females in these experiments. Murdoch added four males to some of the bags containing four females at the end of the reproductive period to demonstrate that the presence of males had no effect on the survival of the females. In spite of competition, or any other interaction that might have occurred between females and males, the survival of the females remained high. Therefore, Murdoch concluded, the reduced survival of breeding females probably was caused by the nutritional drain of egg laying. The holdover of nonreproducing females to the following year allows the population to respond to environmental conditions during the reproductive season and at the same time to maintain a less variable population size.

Some adaptations of life history, like the inverse relation between egg laying and survival, may fortuitously enhance population stability. In other cases, adaptations may have evolved indirect response to environmental variation or unpredictability. Murphy (1968) has shown that where the environment of a particular life-cycle stage is less predictable, selection favors shortening that stage and lengthening others whenever there are trade-offs between stages. Gillespie (1974, 1977) has shown that all other factors being equal, a few offspring should be produced repeatedly rather than many produced at once. Putting eggs in many baskets reduces the variance in offspring number, which confers selective value on the parent. The importance of adaptations to reduce variance in production has not, however, been evaluated in the context of other constraints on reproduction.

Population stability reflects turnover and potential population growth rates.

The response time of the population to fluctuations in the environment constitutes an important component of stability. Small organisms reproduce more rapidly than large organisms and their populations can respond more rapidly to change in the environment. Furthermore, small organisms may choose among several avenues of response: developmental and evolutionary responses are practical ways of coping with short-term environmental change only for small organisms with short life spans (see Chapters 28 and 31). Size and biomass-to-productivity ratios have two opposing influences on stability. The number of individuals in populations of large organisms change little but respond slowly to change; the opposite is true of populations of small organisms (see Figures 31-1 and 31-2). Which is more stable? Which is better attuned to fluctuation in the physical environment? The answers are not clear.

Populations of birds have been studied in detail for long periods (Lack 1966), and it is possible to examine the degree of variation in different parts of the life cycle (Table 43-2). Coefficients of variation in local population density are mostly between 30 and 50 per cent when populations are counted once each year. The stage of the life cycle that is most sensitive to change in the environment is the survival of young from independence to maturity, although in some species annual adult survival rates may be quite variable. In studies of this kind, only local populations are dealt with. In fact, local variation in population size may be considerably reduced by the movement of individuals between localities; and we might expect to see larger fluctuations in populations that are more isolated. Variation in a single population also overestimates variation in the community where populations fluctuate independently or inversely to one another. For example, in a study of deciduous forest birds in the eastern United States, Brewer (1963) found that the average of the coefficients of variation of populations within a habitat varied between 23 and 37 per cent, whereas the coefficients of variation in the total number of birds in each habitat varied between 8 and 18 per cent.

In the long run, populations that persist are stable, regardless of their fluctuations. The ultimate measure of instability is extinction (Preston 1969). Populations with few individuals are disadvantaged compared to larger populations, and they are more likely to pass into oblivion following perturbations in their environment (see page 648). What factors determine the size of a population? The study of taxon cycles (page 652) indicates that the relative success of the adaptations of a species, compared to the adaptations of its predators, parasites, competitors, and prey, determines the degree of resource specialization and influences the size of the species' population. In an evolutionary struggle to achieve superior adaptations, species beset by few kinds of exploiters and competitors fare better than species that must confront the adaptations of many antagonists. For example, a grasshopper

TABLE 43-2

Relative degree of variation in annual measures of population size, reproduction, and survival of various species of birds (based on Ricklefs 1973).

Property	Range of coefficients of variation (per cent) in most studies
Population size	30–50
Clutch size	5–15
Nesting success per clutch	15–22
Young fledged per pair	25–55
Survival of fledged young	20–80
Survival of adults	10–50

attacked by a shrew can escape by flight. If grasshoppers are preyed upon by both shrews and sparrow hawks, each predator compromises the grass-hopper's escape from the other.

The number of species in communities is greatest in the humid tropics and decreases toward the poles. Populations are surely confronted by a greater diversity of competing populations in the tropics than elsewhere, and average population size tends to be smaller. Whether tropical species are eaten by a greater variety of predators is not known. Many ecologists have asserted that species of predators are more numerous relative to spe-cies of prey in the tropics than in temperate and arctic communities. Insect communities appear to bear this out (Table 43-3). The ratio of predators and parasites to herbivores and detritivores (the predator ratio) is greater in tropical samples than in temperate zone samples. Furthermore, within the tropics both diversity and the predator ratio increase along a moisture gra-dient from dry hillsides (ratio = 0.38) to wet lowland and river-bottom forest (ratio = 0.77). The greater predator ratio of moist tropical habitats is due primarily to an increase in the number of parasitic species, which tend to specialize their attack upon a few host species.

If high diversity and complexity of community organization reduced the average population size of tropical species and increased the probability of their extinction, the composition of tropical communities could be intrinsi-cally less stable than temperate or arctic communities. These factors could be balanced, of course, by low variability of the physical environment and a preponderance of mutualistic associations in the tropics. Direct measure-ments of the life spans of species in tropical and temperate populations would have to be based on the fossil record, which is meager at best. Some groups of marine organisms are well enough represented in the fossil record that the persistence of taxa can be compared between regions (Stehli, Doug-las, and Newell 1969). For example, genera of planktonic foraminifers (small protozoans with calcareous shells) that lived during the Cretaceous Period persisted longer in oceans north of 50° latitude than in the warm-water regions closer to the Equator (Figure 43-5). But the differences between

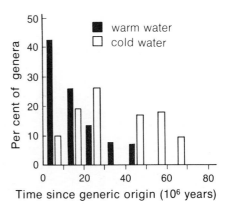

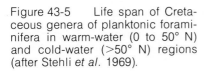

Figure 43-5   Life span of Creta-ceous genera of planktonic forami-nifera in warm-water (0 to 50° N) and cold-water (>50° N) regions (after Stehli *et al.* 1969).

TABLE 43-3

Percentages of trophic groups in samples of insects from various localities and habitats, arranged by decreasing latitude. In general, the tropical samples contain the most parasites and the fewest detritivores.

Locality*	Habitat	Herbivores	Detritivores	Predators	Parasites	Predator ratio[†]
Arctic coast (4)	Whole fauna	47	27	14	10[‡]	0.32
Connecticut (4)	Whole fauna	49	19	16	12	0.41
New Jersey (4)	Whole fauna	52	19	16	10	0.37
South Carolina (3)	Old field	68	1	19	9	0.41
Great Smoky Mountains (5)	Average for 15 habitats	41	31	12	15	0.38
Florida Keys (1)	Mangrove islands	41	28	22	7	0.42
Costa Rica (2)	Average for 4 habitats	59	4	9	26	0.56

* References in parentheses are 1) Heatwole and Levins 1972, 2) Janzen and Schoener 1968, 3) Menhinick 1963, 4) Weiss 1924, and 5) Whittaker 1952.
[†] Proportion of predators plus parasites divided by proportion of herbivores plus detritivores.
[‡] Totals do not add to 100 per cent because some species are classified as "miscellaneous."

aquatic and terrestrial environments are great enough to render this evidence slender indeed.

The relationship between diversity, complexity, and community stability is uncertain.

If high diversity speeded individual populations to extinction, thereby reducing the stability of community composition, diversity and complexity of trophic organization could enhance the stability of community function (MacArthur 1955). This principle may be stated simply: Where predators eat many kinds of prey organisms, they can specialize momentarily on whichever prey species are most abundant (Murdoch 1969). This *switching* behavior makes predators less sensitive to variation in the abundance of any one prey species. In simpler communities, predators are restricted to eating few kinds of prey—perhaps only one—so their populations follow variations in their prey populations more closely (Figure 43-6). Studies of population cycles of fur-bearing mammals in arctic North America, where communities are relatively simple, are consistent with a direct rela-

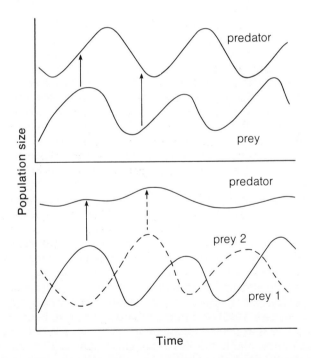

Figure 43-6   A diagram of population cycles of predator species feeding on either one (top) or two (bottom) kinds of prey. The predator in the two-prey system switches its feeding from one species to the other depending on which is more abundant.

tionship between diversity and stability (see page 707). Populations of tropical rodents fluctuate much less than populations of temperate and arctic species (Fleming 1975).

One may, however, take a different view of the relationship between diversity, complexity, and stability. In rigorous climates, either too dry or too cold to support diverse communities, the physical environment affects most species directly and at the same time. A drought or cold snap depresses biological activity throughout the community. Milder conditions return the community as a whole to its original state. With physical conditions exerting dominant control over fluctuations in the community, all species are linked directly to the dominant cause of the fluctuation. In the less harsh tropics, population trends are determined more by interactions with other populations than by the physical environment. Like a ripple moving across the surface of a pond, perturbations can be passed through many links of species interaction. A population in a diverse and complex tropical community can be removed by several links from an external source of disturbance. In such circumstances, time lags in population response and time lags between the links may increase rather than dampen the effects of perturbation, as we have seen in Chapters 31 and 33.

An example will emphasize the point. Many species of trees in tropical forests flower during the dry season when flying insect pollinators are most abundant. Occasionally, heavy rains fall after the beginning of the dry season. Trees caught with their flowers out may not be pollinated because rain delays the season of activity of the insects. When flowers are not pollinated, trees do not produce fruit during their normal fruiting period several months later; mammals and birds that rely on fruit when other food sources are less abundant starve, curtail reproduction, or turn to novel sources of food; rodents and other gnawing mammals severely damage trees and kill saplings by chewing off bark to get at the nutritious growing layers underneath; herbivorous insects that normally would have eaten the tree seedlings, and predators and parasites of these insects, are affected in turn. In this way, the initial physical perturbation is maintained in the tropical community. In simpler communities, response to physical perturbation tends to be rapid, direct, and short-lived.

Biological complexity may either dampen fluctuations (MacArthur 1955), or exaggerate perturbations to communities (May 1973, Saunders and Bazin 1975a, 1975b). Where theory is equivocal, we must turn to direct observation and experimentation. A few studies are pertinent to the diversity-stability problem. The extensive literature on crop pests has been reviewed by van Emden and Williams (1974) and found not to support the hypothesis that diversity leads to stability. In addition, some species of tropical insects are known to exhibit extensive year-to-year variation in population size (Galindo *et al.* 1956, Ricklefs 1975) or infrequent large-scale eruptions (Smith 1972). Watt (1964b, 1965) examined fluctuations in populations of Canadian forest lepidoptera (moths and butterflies), comparing species that fed on few species of trees with those that had broader diets. Watt's analysis (Table 43-4) showed that population size and the relative degree of fluctuation both increase as the number of tree species eaten increases. Hence diversity of

TABLE 43-4
Relationship of number of tree species eaten to the abundance and constancy of Canadian forest moth and butterfly populations (from Watt 1965).

Number of tree species eaten*	Number of lepidoptera species	Relative population size	Index of variation in population size†
1.5	179	5	0.23
4.2	173	47	0.30
10.9	134	52	0.34
24.7	34	442	0.37

* Means of four groups, gregarious species excluded.
† Variation relative to the size of the population: mean of the standard error of the logarithms of counts for each species.

food resources does not appear to enhance the stability of butterfly and moth populations.

Disturbance and simplification tend to reduce community stability.

If we wished to determine the ability of a community to resist disturbance, a good approach would be to disturb the community and watch its response. We would have difficulty predicting the results of our experiment beforehand, owing to the great variety of effects dependent on the nature of the disturbance and the adaptations of organisms in the community. A fire in a longleaf pine plantation in Mississippi depresses tree growth for only one year, then tree production returns to normal (Figure 43-7). When New Mexican grasslands are overgrazed, shrubs replace blue gramma and other grasses completely, and permanently alter the character of the community.

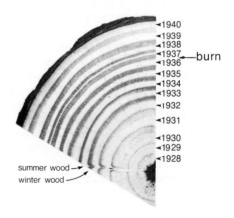

Figure 43-7   Cross section of a longleaf pine tree burned in January 1937 and cut in the fall of 1940 in Harrison County, Mississippi. Narrow growth rings were produced during the summer, 1937—light wood—and the winter, 1937–1938—dark wood (courtesy of the U.S. Forest Service).

Figure 43-8   Root systems of blue gramma grass (left) and snakeweed (*Gutierrezia,* right). Overgrazing on grasslands can kill grasses and allow shrubs with little forage value and poor soil-holding qualities to take over (courtesy of the U.S. Forest Service).

Whereas grasses have extensive shallow, fibrous root systems, which hold the topsoil firmly in place, the root systems of many shrubs do not prevent soil erosion (Figure 43-8).

One of the most consistent effects of disturbance on the structure of communities is to reduce the total number of species while allowing some of the survivors to reach abnormally high population levels (Patrick 1963, Cairns *et al.* 1971). For example, polluted streams, compared to natural streams, tend to have fewer species, but some of these are extremely abundant (Figure 43-9). Strong pollution creates conditions that are lethal to most species in the community, but which are extremely favorable for a few.

When the number of plant species in a habitat is reduced by planting crops, for example, the number of species on all trophic levels decreases. In these simplified communities, the abundance of some herbivore species increases to outbreak levels in the absence of effective control by predators (Pimentel 1961). Herbivorous insects and disease organisms parasitic on plants are usually specialized to occur on a single host plant. Some species, adapted to feed on a particular crop, do spectacularly well in agricultural habitats, much to the farmer's dismay. On the other hand, predators require

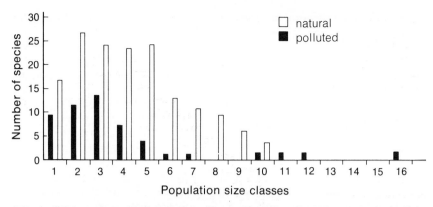

Figure 43-9   Distribution of abundances of diatoms in samples taken from a natural stream community and from a polluted river. Population-size classes increase from left to right by a factor of two, *e.g.*, 1–2, 2–4, 4–8, 8–16, and so on. The largest population in the polluted sample falls into the range 32,768–65,536 (after Patrick 1963).

a more complex habitat and do poorly in single species stands of crops. Spruce budworms cause more damage to solid stands of balsam fir than to fir trees well spaced among other species (Morris *et al.* 1956, Morris 1963b). Infestations are worse when all the trees in the forest are the same age. Susceptibility to budworm infestation increases with age; if a forest consists entirely of old trees—often the case in managed woodlands—most of the trees become infected, easing the spread of the pest and leaving fewer unaffected trees to replace those killed by the budworm.

The relationship between community simplification and population outbreaks has been explored by Pimentel (1961). Insect populations on two groups of collard plants were compared. One group of collards was planted in a field that had been uncultivated for fifteen years. In the old field, collards were planted nine feet apart among vegetation that contained about 300 species of plants, including five in the same family as collards (Cruciferae, the mustard family). A second group of collard plants was planted in a dense stand having no other species. The abundances of several species of insects—particularly aphids and flea beetles—reached outbreak proportions in the single species plantings. The same species of insects were kept under control by predators in the old field. The number of predators and parasitic insects collected from collard plants was also higher in the single species planting than in the old field, but they appeared too late in the season to exercise effective control over herbivore populations. Also, because Pimentel collected samples only from the collard plants in both plantings, he may, in the old field, have missed the many species on other types of plants. Such species could also have preyed on herbivorous insects on the collards. This often happens. Predators on the diamondback moth, a cabbage pest, survive the winter by eating other prey species that attack hawthorn. Many of the alternate hosts of gypsy moth parasites live in the forest understory, not in the canopy; the presence of forest undergrowth is, therefore, indirectly important to the growth of gypsy moth populations. Unfortunately, one could point to as many examples of weed or hedgerow plants maintaining populations of crop pests as populations of their predators and parasites. The role of diversity in maintaining stability is thus cast into doubt, perhaps because crop systems are so simple that a few populations, perhaps pests, perhaps their predators, can dominate a community just by chance.

In an experiment similar to Pimentel's, arthropod populations were followed in a field that was planted with millet one year and then left to develop a natural successional community during the second year (Odum, Barrett, and Pulliam, in Odum 1971). Populations were sampled ten times each summer at regular intervals. Each trophic group of insects and other arthropods was represented by more species in the second year than in the first. Populations of predatory and parasitic species were also greater in the second year, but numbers of herbivores were reduced, presumably by the predators and parasites (Table 43-5). An interesting result of the study was that variation in the numbers of predatory insects was greater during the second year in spite of the increased diversity of prey species (see page 642). Once more, observations do not support a direct relationship between diversity and constancy.

In another study near Syracuse, New York, fertilizer was applied to abandoned hay fields of different ages (Hurd et al. 1971). The experiment was designed to measure the effect of a perturbation—nutrient enrichment—on the composition of the community and to learn how this perturbation was transmitted through the trophic structure. The seventeen-year-old field was more diverse and had greater biomass of vegetation than the six-year-old field. Presumably the older field should have resisted perturbation to a greater degree, if increased diversity and biomass enhance stability. In fact, the older field was more sensitive to enrichment, and the effects were felt

TABLE 43-5

Diversity, density, and variation of arthropod populations in unharvested grain crop compared to the natural successional community which replaced it during the following year (from Odum 1971).

		Herbivorous insects	Predatory insects	Spiders	Parasitic insects	Total arthropods
Diversity*	1966	7.2	2.8	1.6	6.3	15.6
	1967	10.6	6.0	7.2	12.4	30.9
Density	1966	482	64	18	24	624
(ind·m⁻²)	1967	156	79	38	51	355
Variation						
(per cent of	1966	24	44	33	57	19
mean density)†	1967	28	49	67	40	32

* Calculated by the equation, diversity $= (S - 1)/\log N$, where S is the number of species in the sample and N is the number of individuals.
† Coefficient of variation (per cent) $=$ (standard deviation/mean) $\times$ 100.

TABLE 43-6

Percentage change in diversity and productivity of producer, herbivore, and carnivore trophic levels as a result of fertilization of six-year-old and seventeen-year-old abandoned hay fields (from Hurd *et al.* 1971).

Trophic level	Diversity		Productivity*	
	6 years	17 years	6 years	17 years
Producers	−7	+3	+96	+70
Herbivores†	+24	+51	+31	+201
Carnivores†	+27	+74	−6	+108

* $Mg \cdot m^{-2} \cdot day^{-1}$.
† Diversity and productivity measured during the earlier of two growing seasons.

more strongly at the herbivore and carnivore levels than among the primary producers, which were affected directly by the fertilizer (Table 43-6). It is arguable whether the increased diversity and production reflected failure of internal stabilizing mechanisms or a favorable response to enrichment, further increasing stability. The point is well made, however, that in a successionally more mature community, perturbations were transferred more efficiently throughout the trophic structure.

As we have seen, ecologists are terribly ignorant of stability in natural systems—the pertinent internal mechanisms of communities and how they work, the relative stability of different communities. Even recognizing stability where it occurs is difficult. Why has this single property of nature been so elusive? Stability is the culmination of all ecological interrelationships; it is the sum of all the components and interactions that make up the community, the union of all lower-order properties of community, population, and organism. To understand stability, we must understand ecological and evolutionary responses and interrelationships at all levels. The science of ecology is not yet mature enough to mold its diverse knowledge and concepts into a unified theory of stability. It is unlikely such a theory will emerge before another decade has passed. However provocative the challenge, not every ecologist should direct his research entirely or even partly toward understanding stability. Other challenges remain, other puzzles must be solved, other speculation will engage us. Still, the significance of new knowledge and new ideas surely will be judged by their contribution to our understanding of the ecological synthesis—the stability of natural systems.

A Survey of Biological Communities

The photographs on the pages that follow illustrate most of the important plant formations in temperate North America and tropical Central and South America. They are arranged to show the influence of temperature and moisture on vegetation structure. In addition, there are photographs of many important fresh-water and marine communities.

Environments Without Life

Life can gain a foothold in regions with almost any combination of temperature and moisture found on Earth, providing the moisture is available and other nutrients are present. But life is excluded from a few environments. The extreme cold on the slopes of Mount McKinley, Alaska (below), freezes life to a standstill. Water occurs only as ice and is therefore unavailable to

plants. Water is a problem as well on the shifting sand dunes of Death Valley, California (below). The little rain that falls either evaporates or percolates through the coarse sand. Temperatures at White Sands, New Mexico (above), are favorable for life, and the region's rainfall supports desert shrubs in the surrounding valley, but the pure gypsum sand (calcium sulfate) does not contain the nutrients needed to support life (courtesy of W. J. Smith and U.S. National Park Service).

The Humid Tropics

Year-round warm temperatures and plentiful moisture in the humid tropics create conditions for the most luxuriant and diversified communities in the world. Vegetation forms include vines that drape the trees in a lowland forest in Panama (right), and air plants that clothe trees in a mistenshrouded cloud forest in Guatemala (above). Because soils are impoverished of nutrients except near the surface, root systems of tropical trees tend to be shallow and the trunks of many trees are buttressed for support (left; courtesy W. J. Smith).

Tropical Mountains

Temperature decreases about 6 C for each 1,000-meter increase in elevation. Plant productivity parallels the lower temperatures of montane habitats, creating cold and almost barren deserts in the tropics. The mean annual temperature and rainfall would support a forest or woodland in seasonal temperate climates with warm summers, but the year-round cold of tropical mountains does not permit such luxurious growth. On the paramo of the high Andes in Colombia at about 3,700 meters (above left), the temperature hovers around 5 C throughout the year. One is struck by the paucity of life forms and by the silence, broken only by the relentless wind. At the same elevation in Costa Rica on the Cerro de la Muerte (Mountain of Death, below left), the ever-present fog slips among dwarfed plants, whose small thick leaves are clustered tightly around the plant stem for protection from the cold wind (above). The bare patch of rock shows the thinness of the soil layer in the tropical montane habitat.

Subtropical Deserts

    Belts of seasonally hot, dry climate girdle the Earth at about 30° north and south of the Equator (see Figures 4-4, 4-5). These are harsh environments where only a few drought-adapted species of plants and animals thrive. Whereas light and nutrients are critical in the humid tropics, the bare ground exposed in deserts testifies that these resources go wanting where rainfall limits plant growth. Cacti have greatly reduced leaves to decrease water loss. Their thick, succulent stems have taken over the function of photosynthesis. Numerous thorns hinder desert animals from getting at their stored water. Desert shrub habitats of the Sonoran Desert of Arizona and northern Mexico (above left) are among the most diverse vegetation types of arid regions. Giant saguaro cacti and paloverde trees dominate the landscape. The joshua tree, a treelike yucca, occurs primarily in the Mohave Desert of southern California (below left). An extremely dry habitat near the Gulf of California in northern Sonora, Mexico (below), supports only two kinds of large plants. Note the wide spacing between individuals. Desert plants do not tolerate close proximity because their extensive root systems compete for water (courtesy U.S. Fish and Wildlife Service and U.S. National Park Service).

Temperate Woodland and Shrubland

In temperate habitats with better water relations and lower summer temperatures than deserts, succulent cacti are replaced by bushes, shrubs, and small trees. The wide spacing and low growth form of plants in the Great Basin region exemplified in Zion National Park (above) indicate that water is still a critical factor. At higher elevations, in Coconino National Forest of Arizona, an open woodland dominates the landscape (above right). Juniper woodland develops at about 2,000-meters (6,000 to 7,000 feet) elevation in this area, where snow covers the ground for much of the winter and summers are cool. The milder Mediterranean climate of the southern California coast, characterized by warm, dry summers and cool, moist winters (see Figure 4-11), supports a characteristic dense shrubland called chaparral (below right). In moist canyons and valleys, oak woodland tends to replace chaparral species, but frequent fires often prevent this natural succession and maintain the fire-adapted chaparral vegetation (courtesy U.S. Forest Service and U.S. Soil Conservation Service).

Temperate Forests

Tall forests of broad-leaved, deciduous trees occur throughout the temperate zone where rainfall is plentiful and winters are cold. Oak, beech, maple, hickory, and other hardwoods dominate temperate forests. Seasonal patterns of summer activity and winter dormancy are characteristic. The stand of Indiana hardwoods dominated by white oak has a well-developed understory of sugar maple and smaller shrubs (above). In the Appalachian Mountains of West Virginia, red spruce and hemlock occur with broad-

leaved trees to form mixed forests (below left). In the southeastern United States, sandy soils are too poor for broad-leaved trees. Pines are widely distributed in vast forests that are managed and harvested for paper pulp. In Florida, the palmetto frequently forms a dense understory (above). In the northern United States and Canada, and in mountainous regions of the west, birch and aspen, frequently mixed with spruce and fir, represent the farthest incursion of broad-leaved forests into cold regions (below; courtesy U.S. Forest Service and J. Lane, Archbold Biological Station).

Temperate Grasslands

Grasslands occur under a variety of temperate climates with cold winters and summer drought. True prairie, remnants of which can be found in Kansas (above), Texas (below), and other midwestern states, is characterized by grasses with extensive root systems. Tall grass prairies grow on fertile

soil and are probably maintained by periodic fires that keep trees from becoming established. Farther to the west, lower rainfall supports sparser vegetation, the shortgrass prairies to the east of the Rocky Mountains (above) and in western interior valleys (below). These grasslands are delicate and they are sensitive to plowing and overgrazing (courtesy U.S. National Park Service, U.S. Department of Agriculture, and U.S. Soil Conservation Service).

Fresh-Water Habitats

Fresh water covers a small fraction of the Earth's surface, yet fresh-water habitats display remarkable diversity. Variation in water movement, mineral and oxygen content of the water, and size and shape of the stream or lake basin all contribute to this variety. Communities in deep lakes and fast-moving streams consist mostly of phytoplankton and thin layers of diatoms on the surfaces of rocks. Vegetation shows above the surface only where water is shallow and still, as in an artificially flooded marsh in Maine (left), or a cattail marsh in New York (above). Floating water hyacinths choke a deeper channel in Louisiana, buoyed up by gas trapped in their stems (right; courtesy U.S. Department of Agriculture).

Temperate Montane Environments

Montane habitats are much colder and are often drier than the surrounding lowlands. Trees reach their upper limit of elevation at about 3,000 meters (10,000 feet) in the Cascade Mountains of Oregon (above left). Above timberline, snow persists well into summer in habitats that can support only the low grassy vegetation characteristic of the alpine tundra, as in the Rocky Mountains of Colorado at 3,700 meters (12,000 feet) elevation (below). Lichens are the first plants to colonize bare rock surfaces in these habitats (right) and start the slow process of soil formation. Wind-driven ice strips bark and branches from trees near the timberline in Colorado (below left; courtesy U.S. Forest Service).

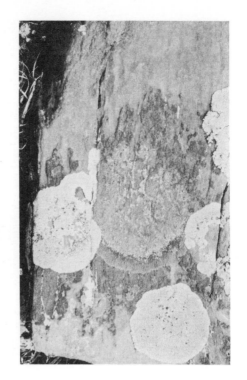

Temperate Conifer Forests

Forests of pine, spruce, fir, hemlock, redwood, and others grow under a variety of temperature, soil, moisture, and fire conditions that favor drought-resistant needle-leaf species over less tolerant broad-leaved trees. Poor soils and frequent fires favor pines throughout much of the southeastern United States. Dry summers and cold winters characterize the environments of coniferous forests at high elevations in western mountains. Pinyon pine-juniper-cedar woodland is found near Flagstaff, Arizona, where the climate is too dry to support a closed forest (below left). A moister site in Inyo National Forest, California, is dominated by tall Jeffrey pines (above left). Undergrowth is sparse in the dry, acid soil. Abundant winter rainfall and cool, foggy summers create ideal conditions for redwoods in the temperate rain forests of northern California (above). They bear little resemblance to the humid forests of the tropics because they lack diversity in species and plant forms and are relatively unproductive (courtesy U.S. Forest Service).

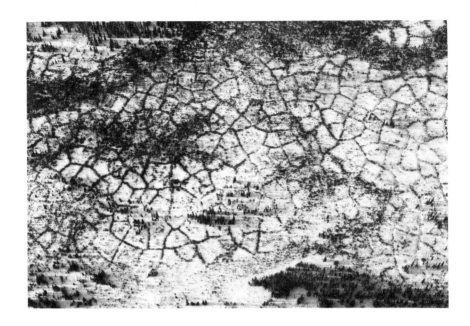

The Arctic Tundra

Permanently frozen soils underlie the arctic tundra habitat. Warm summer temperatures thaw the ground to a depth of a few inches or feet, briefly creating a shallow, often waterlogged layer of soil on which arctic vegetation develops. Repeated freezing and thawing creates characteristic polygonal patterns in the ground surface of some areas (above). At Cold Bay, Alaska,

lichens, mosses, and grasses are found on the hummocky, frost-heaved soil (below left). Kettle lakes formed by the melting of large blocks of ice, left by retreating glaciers, are a prominent feature of the Kuskokwim River Delta, Alaska (above). Montane tundra in the arctic is better drained than lowland habitats and spruce trees occasionally get a foothold in protected valleys, such as these near Mount McKinley, Alaska (below; courtesy U.S. Soil Conservation Service).

The Land Meets the Sea

The topography of coastal areas often determines the character of plant communities at the edge between the land and the sea. Shallow, sloping, sandy shores, like those of Cape Cod, Massachusetts (above), create shifting dune habitats colonized by a few species of plants that stabilize the dune and allow other species to establish a foothold. Salt marshes develop in more protected bays and in river estuaries, as at Barnstable Harbor on Cape

Cod (below left). In the tropics, such protected habitats are usually invaded by mangrove trees, creating forests in standing water at the edge of the ocean, as in Biscayne Bay, Florida (above). At the rockbound coast of Maine, an abrupt meeting of land and sea tolerates little intermingling of the two environments (below; courtesy F. B. Bowles, Marine Biological Laboratories, P. J. Tzimoulis, American Littoral Society, W. J. Smith, and U.S. National Marine Fisheries Service).

The Marine Environment

   The little explored mantle of water covering most of the surface of the
Earth contains a wide variety of habitats and life forms. Open waters create
a vast realm for the tiny phytoplankton and zooplankton, the fish that exploit
them (above left), and the sea birds that eat the fish (below). Other fish are
more like terrestrial grazers and predators, feeding on algae and small animals near the bottom and among corals (above right). Subtidal tropical
habitats on the Caribbean coast of Panama are dominated by corals, including the elk horn coral in shallow, rough water (above far right) and the
important reef-building coral at a depth of 15 meters (50 feet; below right).
Reef-building corals are restricted to sunlit depths because they rely on
symbiotic green algae in their tissues for much of their nutrition (courtesy
James Porter, from Porter 1974).

Glossary

Acclimation. A reversible change in the morphology or physiology of an organism in response to environmental change.

Adaptability. Capacity for evolutionary change. Adaptability may depend on the phenotype's tolerance of environmental change as well as on the genetic variability of the population.

Adaptation. A genetically determined characteristic that enhances the ability of an organism to cope with its environment.

Adaptive radiation. Evolutionary diversification of species derived from a common ancestor into a variety of ecological roles.

Adaptive zone. A particular type of environment requiring unique adaptations. Species in different adaptive zones usually differ by major morphological or physiological characteristics.

Aggressive mimicry. Resemblance of predators or parasites to harmless species to deceive potential prey or hosts.

Allele. One of several alternative forms of a gene.

Allelopathy. Direct inhibition of one species by another using noxious or toxic chemicals.

Allochronic speciation. Separation of a population into two or more evolutionary units as a result of reproductive isolation because two subpopulations in the same area mate at different times.

Allochthonous. Referring to materials transported into a system, particularly minerals and organic matter transported into streams and lakes.

Allopatric. Occurring in different places; usually refers to geographical separation of populations.

Allopatric speciation. Separation of a population into two or more evolutionary units as a result of reproductive isolation caused by geographical separation of two subpopulations.

Allopolyploidy. Increase in the number of chromosome sets in a fertilized egg which receives two unlike sets of chromosomes because of hybrid parentage.

Alpha diversity. The variety of organisms occurring in a particular place or habitat. Often called local diversity.

Altruism. In an evolutionary sense, enhancing the fitness of an unrelated individual by acts that reduce the evolutionary fitness of the altruistic individual.

Amino acid. One of about thirty organic acids containing the group NH_2 which are the building blocks of proteins.

Ammonification. Breakdown of proteins and amino acids with ammonia as an excretory by-product.

Anaerobic. Without oxygen.

Annual. Referring to an organism that completes its life cycle from birth or germination to death within a year.

Apomixis. Reproduction without sexual union, as in parthenogenesis.

Aposematism. Conspicuous appearance of an organism warning that it is noxious or distasteful; warning coloration.

Apostatic selection. Selective predation on the most abundant forms in a population, regardless of their appearance, leading to balanced polymorphism, the stable occurrence of more than one form in a population.

Arena. A display area used by many males in concert for courting females. Also called a lek.

Arrhenotoky. A sex determination system in which males develop only from unfertilized haploid eggs, and thus have only one chromosome complement.

Artificial selection. Intentional manipulation by man of the fitnesses of individuals in a population to produce a desired evolutionary response.

Aspect diversity. Variations in the outward appearance of species that live in the same habitat and are eaten by visually hunting predators.

Assimilation. Incorporation of any material into the tissues, cells, and fluids of an organism.

Assimilation efficiency. A percentage expressing the proportion of energy ingested that is absorbed into the bloodstream.

Association. A group of species living in the same place.

Autecology. The study of organisms in relation to their physical environment.

Autochthonous. Referring to materials produced within a system, particularly organic matter produced, and minerals cycled, within streams and lakes.

Autopolyploidy. Spontaneous increase in chromosome number often resulting from abnormal meiosis.

Autotroph. An organism that assimilates energy from either sunlight (green plants) or inorganic compounds (sulfur bacteria). *See also* Heterotroph.

Average fitness of a population. ($\bar{r}$). The average fitness of the genotypes in a population weighted by their frequencies.

Balanced polymorphism. Maintenance of more than one allele in a population by the selective superiority of the heterozygote over both types of homozygotes.

Barren. An area with sparse vegetation owing to some physical or chemical property of the soil.

Basal metabolism. The energy expenditures of an organism that is at rest, fasting, and in a thermally neutral environment.

Batesian mimicry.   Resemblance of an edible species (mimic) to an unpalatable species (model) to deceive predators.

Benthic.   Bottom dwelling in rivers, lakes, and oceans.

Beta diversity.   The variety of organisms summed over all habitats in a particular region. Often called regional diversity.

Biomass.   Weight of living material, usually expressed as a dry weight, in all or part of an organism, population, or community. Commonly expressed as weight per unit area, a biomass density.

Biomass accumulation ratio.   The ratio of weight to annual production, usually applied to vegetation.

Biota.   Fauna and flora together.

Biotic environment.   Biological components of an organism's surroundings that interact with it, including competitors, predators, parasites, and prey. Interactions within a population are subclassified as the social, sexual, and parent-offspring environment.

Boreal.   Northern. Often refers to the coniferous forest regions that stretch across Canada, northern Europe, and Asia.

Broken-stick model.   A model of relative abundance of species obtained by random division into segments of a line representing the resources of an environment.

Calcification.   Deposition of calcium and other soluble salts in soils where evaporation greatly exceeds precipitation.

Caliche.   An alkaline salt deposit on the soil surface, usually occurring in arid regions with ground water close to the surface.

Carnivore.   An organism that consumes mostly flesh.

Carrying capacity.   Number of individuals that the resources of a habitat can support.

Cation.   A part of a dissociated molecule carrying a positive electrical charge, usually in an aqueous solution (e.g. Ca^{++}, Na^+, NH_4^-, H^+).

Cation-exchange capacity.   The ability of soil particles to absorb positively charged ions, such as hydrogen (H^+) and calcium (Ca^{++}).

Character convergence.   Evolution of similar appearance or behavior in unrelated species for the purpose of facilitating direct interaction between individuals. Also called social mimicry.

Character displacement.   Divergence in the characteristics of two otherwise similar species where their ranges overlap, caused by the selective effects of competition between the species in the area of overlap.

Character divergence.   Evolution of differences between similar species occurring in the same areas, caused by the selective effects of competition.

Chemoautotroph.   An organism that oxidizes inorganic compounds (often hydrogen sulfide) to obtain energy for synthesis of organic compounds; e.g. sulfur bacteria.

Chromosome.   Any of several small bodies found in the nucleus of the cell that bear the genetic material.

Cleaning symbiosis.   A mutualistic relationship in which one species benefits from having parasites removed by a cleaning species, which gains a food resource.

Climatic climax.   The steady-state community characteristic of a particular climate.

Climax. The end point of a successional sequence; a community that has reached a steady state under a particular set of environmental conditions.

Climograph. A diagram on which localities are represented by the annual cycle of their temperature and rainfall.

Cline. Change in population characteristics over a geographic area, usually related to a corresponding environmental change.

Closed-community concept. The idea, popularized by F. C. Clements, that communities are distinctive associations of highly interdependent species.

Coadaptation. Evolution of characteristics of two or more species to their mutual advantage.

Coarse-grained. Referring to qualities of the environment that occur in large patches, with respect to the activity patterns of an organism and, therefore, among which the organism can select.

Codon. Three consecutive bases in a nucleic acid chain whose sequence is complementary to a particular amino acid in the protein encoded by the nucleic acid chain.

Coevolution. Development of genetically determined traits in two species to facilitate some interaction, usually mutually beneficial. *See* Counterevolution.

Coexistence. Occurrence of two or more species in the same habitat; usually applied to potentially competing species.

Community. An association of interacting populations, usually defined by the nature of their interaction of the place in which they live.

Compensation point. Depth of water at which respiration and photosynthesis balance each other; the lower limit of the euphotic zone.

Competition. Use or defense of a resource by one individual that reduces the availability of that resource to other individuals.

Competitive exclusion principle. The hypothesis that two or more species cannot coexist on a single resource that is scarce relative to the demand for it.

Congeneric. Belonging to the same genus.

Conspecific. Belonging to the same species.

Constancy. A measure of the variability of a system.

Continuum. A gradient of environmental characteristics or of change in the composition of communities.

Continuum index. An artificial scale of an environmental gradient based on changes in community composition.

Convergent evolution. Development of characteristics with similar functions in unrelated species that live in the same kind of environment but in different places.

Counter adaptation. Evolution of characteristics of two or more species to their mutual disadvantage.

Counterevolution. Development of traits in a population in response to exploitation, competition, or other detrimental interaction with another population.

Cross-resistance. Resistance or immunity to one disease organism resulting from infection by another, usually closely related organism.

Crypsis. An aspect of the appearance of organisms whereby they avoid detection by others; usually applied to the prey of visually hunting predators.

Cyclic climax. A steady-state, cyclic sequence of communities, none of which by itself is stable.

Death rate. The percentage of newborn dying during a specified interval. *See* Mortality.

Decomposition. Metabolic breakdown of organic materials; the by-products are released energy and simple organic and inorganic compounds. *See* Respiration.

Denitrification. The reduction by microorganisms of nitrate and nitrite to nitrogen.

Density-dependent. Having influence on individuals in a population that varies with the degree of crowding in the population.

Density-independent. Having influence on individuals in a population that does not vary with the degree of crowding in the population.

Deoxyribonucleic acid (DNA). A long, chainlike organic molecule composed of four kinds of purine and pyrimidine bases whose order determines the genetic code and which resides in the chromosome.

Detritivore. An organism that feeds on freshly dead or partially decomposed organic matter.

Detritus. Freshly dead or partially decomposed organic matter.

Developmental response. Acquisition of one of several alternative forms by an organism, depending on the environmental conditions under which it grows.

Diapause. Temporary interruption in the development of insect eggs or larvae, usually associated with a dormant period.

Diffuse competition. The sum of weak competitive interactions with species that are ecologically distantly allied.

Dimer. A molecule composed of two distinct subunits.

Dimorphism. Occurrences of two forms of individuals in a population.

Dioecy. In plants, the occurrence of reproductive organs of both sexes on different individuals. *See* Monoecy.

Diploid. Pertaining to cells or organisms having two sets of chromosomes. *See* Ploidy.

Direct competition. Exclusion of individuals from resources by aggressive behavior or use of toxins.

Directional selection. Selection favoring individuals at one extreme of the distribution of phenotypes in a population.

Dispersal. Movement of organisms away from the place of birth or from centers of population density.

Dispersion. Pattern of spacing of individuals in a population.

Disruptive selection. Evolutionary disadvantage of individuals in a population that have intermediate traits, which leads to the divergence of distinct subpopulations with different extreme traits.

Diversity. The number of species in a community or region. Alpha diversity refers to the diversity of a particular habitat, beta diversity to the species added by pooling habitats within a region. Also, a measure of the variety of species in a community that takes into account the relative abundance of each species.

Dominance (genetic). Ability of a genetic trait (allele) to mask the expression of an alternative form of the same gene when both are present in the same cell (that is, in a heterozygote).

Dominance hierarchy. Orderly ranking of individuals in a group, based on the outcome of aggressive encounters.

Ecocline. A geographical gradient of vegetation structure associated with one or more environmental variables.

Ecological efficiency. Percentage of energy in the biomass produced by one trophic level that is incorporated into biomass produced by the next highest trophic level.

Ecological isolation. Avoidance of competition between two species by differences in food, habitat, activity period, or geographical range.

Ecological release. Expansion of habitat and resource utilization by populations in regions of low species diversity, resulting from reduced interspecific competition.

Ecosystem. All the interacting parts of the physical and biological worlds.

Ecotone. A habitat created by the juxtaposition of distinctly different habitats; an edge habitat; a zone of transition between habitat types.

Ecotype. A genetically differentiated subpopulation that is restricted to a specific habitat.

Ectoparasite. A parasite, like a tick, which lives on, or attached to, the host's surface.

Edaphic. Pertaining to, or influenced by, soil conditions.

Edge species. Species preferring the habitat created by the abutment of distinctive vegetation types.

Egestion. Elimination of undigested food material.

Electivity index (E). A value varying between 1 and -1 expressing degree of preference for or avoidance of a particular item in the diet.

Eluviation. The downward movement of dissolved soil materials carried by percolating water.

Endemic. Confined to a certain region.

Endoparasite. A parasite that lives within the tissues or bloodstream of its host.

Environment. Surroundings of an organism, including the plants and animals with which it interacts.

Environmental gradient. A continuum of conditions ranging between extremes, as the gradation from hot to cold environments.

Enzymes. Organic compounds in living cells that accelerate specific biochemical transformations without themselves being affected.

Epilimnion. The warm, oxygen-rich surface layers of a lake or other body of water.

Epiphyte. A plant that lives wholly, but nonparasitically, on other plants, usually above ground.

Epistasis. Interaction of two or more genes to produce a particular phenotypic trait.

Equilibrium isocline. A line on the population graph designating combinations of competing populations, or predator and prey populations, for which the growth rate of one of the populations is zero.

Equilibrium model. Any hypothetical or actual representation of a system in which two balancing forces act to maintain stability in the system.

Equitability. Uniformity of abundance in an assemblage of species. Equitability is greatest when all species are equally numerous.

Euphotic zone. Surface layer of water to the depth of light penetration at which photosynthesis balances respiration. *See also* Compensation point.

Eusocial. Referring to the complex social organization of termites, ants, and many wasps and bees, dominated by an egg-laying queen that is tended by nonreproductive offspring.

Eutrophic. Referring to a body of water with abundant nutrients and high productivity.

Eutrophication. Enrichment of bodies of water, often caused by sewage and runoff from fertilized agricultural land.

Evapotranspiration. The sum of transpiration by plants and evaporation from the soil. Potential evapotranspiration is the amount of evapotranspiration that would occur, given the local temperature and humidity, if water were superabundant.

Evolutionary opportunism. The principle that adaptations are based on whatever genetic variation is available in a population.

Exploitation. Removal of individuals or biomass from a population by predators or parasites.

Exploitation efficiency. The percentage of potential prey or food plants that are consumed by predators and herbivores.

Exploitation equilibrium. Stability between predator and prey populations in which counterevolutionary responses of one just balance those of the other.

Exponential rate of increase (r). Rate at which a population is growing at a particular instant, expressed as a proportional increase per unit of time.

Facultative. Being able to live under different conditions or to respond to particular conditions in a variety of ways.

Fall bloom. The rapid growth of algae in temperate lakes following the autumnal breakdown of thermal stratification and mixing of water layers.

Fall overturn. Vertical mixing of water layers in temperate lakes in autumn following breakdown of thermal stratification.

Fecundity. Rate at which an individual produces offspring, usually expressed only for females.

Field capacity. The amount of water that soil can hold against the pull of gravity.

Fine-grained. Referring to qualities of the environment that occur in small patches with respect to the activity patterns of an organism, and among which the organism cannot usefully distinguish.

Fitness. Genetic contribution by an individual's descendants to future generations of a population.

Floristic. Referring to studies of the species composition of plant associations.

Food chain. An abstract representation of the passage of energy through populations in the community.

Food chain efficiency. *See* Ecological efficiency.

Food web. An abstract representation of the various paths of energy flow through populations in the community.

Founder principle. The principle that a population started by a small number of colonists contains only a small fraction of the genetic variation of the parent population.

Functional response. Change in the rate of exploitation of prey by an individual predator as a result of a change in prey density. *See also* Numerical response.

Gamete. A haploid cell that fuses with another haploid cell of opposite sex during fertilization to form the zygote. In animals the male gamete is called the sperm and the female gamete the egg, or ovum.

Gametophyte. Haploid stage in the life cycle of plants arising by mitosis from a spore. The dominant stage in many algae and mosses.

Gene. Generally, a unit of genetic inheritance. In biochemistry, gene refers to the part of the DNA molecule that encodes a single enzyme or structural protein unit.

Gene flow. Exchange of genetic traits between populations by movement of individuals, gametes, or spores.

Gene locus. Segment of a chromosome on which a gene resides.

General adaptive syndrome. The set of abnormal physiological responses to the stress of social interaction in dense populations.

Generalist. A species with broad food or habitat preferences, or both.

Generation time. Average age (T_c) at which a female gives birth to her offspring, or average time (T) for a population to increase by a factor equal to the net reproductive rate.

Genetic drift. Change in allele frequency due to random variations in fecundity and mortality in a population.

Genetic feedback. Evolutionary response of a population to the adaptations of competitors, predators, or prey.

Genetic load. Selective deaths sustained by a population due to genotypes that deviate from the genotype with the maximum fitness.

Genome. The entire genetic complement of the individual.

Genotype. All the genetic characteristics that determine the structure and functioning of an organism; often applied to a single gene locus to distinguish one allele, or combination of alleles, from another.

Geometric rate of increase (λ). Factor by which the size of a population changes over a specified period. *See* Exponential rate of increase.

Gross production. The total energy or nutrients assimilated by an organism, a population, or an entire community. *See also* Net production.

Gross production efficiency. The percentage of ingested food utilized for growth and reproduction by an organism.

Group selection. Elimination of a group of individuals with a detrimental genetic trait, caused by competition with other groups lacking the trait. Often called intergroup selection.

Habitat. Place where an animal or plant normally lives, often characterized by a dominant plant form or physical characteristic (*i.e.,* the stream habitat, the forest habitat).

Habitat expansion. Increase in average breadth of habitat distribution of species in depauperate biotas, especially on islands, compared with species in more diverse biotas.

Habitat selection. Preference for certain habitats.

Haploid. Referring to a cell or organism that contains one set of chromosomes.

Heat.   A measure of the kinetic energy of the atoms or molecules in a substance.

Heat of fusion.   Amount of heat energy removed to make a substance freeze or solidify. Conversely, the energy added to melt a substance.

Heat of vaporization.   Amount of energy added to vaporize a substance.

Herbivore.   An organism that consumes living plants or their parts.

Heredity.   Genetic transmission of traits from parents to offspring.

Heritability.   The proportion of variance in a phenotypic trait due to the effects of additive genetic factors.

Hermaphrodite.   An organism that has the reproductive organs of both sexes.

Heterogeneity.   The variety of qualities found in an environment (habitat patches) or a population (genotypic variation).

Heterosis.   Situation in which the heterozygous genotype is more fit than either homozygote. Also called overdominance.

Heterostyly.   Variation in the spatial arrangement of sexual organs in flowers that promotes outcrossing.

Heterotroph.   An organism that utilizes organic materials as a source of energy and nutrients. *See also* Autotroph.

Heterozygote superiority.   Superiority of the fitness of the heterozygote over that of either homozygote. Also called heterosis and overdominance.

Heterozygous.   Containing two forms (alleles) of a gene, one derived from each parent.

Homeostasis.   Maintenance of constant internal conditions in the face of a varying external environment.

Homeothermic.   Able to maintain constant body temperature in the face of fluctuating environmental temperature; warm-blooded.

Home range.   An area, from which intruders may or may not be excluded, to which an individual restricts most of its normal activities. *See* Territory.

Homologous chromosomes.   Corresponding chromosomes from the male and female parents which pair during meiosis.

Homology.   The condition of having similar evolutionary or developmental origin.

Homozygous.   Containing two identical alleles at a gene locus.

Hybridization.   Crossing of individuals from genetically different strains, populations, or, sometimes, species.

Hybrid zone.   Region of reproduction among individuals of different species, often occurring where the ranges of two close relatives abut.

Hydrarch succession.   Progression of terrestrial plant communities developing in an aquatic habitat such as a bog or swamp.

Hyperosmotic.   Having a salt concentration greater than that of the surrounding medium.

Hypolimnion.   The cold, oxygen-poor part of a lake or other body of water that lies below the zone of rapid change in water temperature. *See also* Epilimnion.

Hypo-osmotic.   Having a salt concentration less than that of the surrounding medium.

Illuviation.   The accumulation of dissolved substances within a soil layer.

Inbreeding.   Mating among related individuals.

Inclusive fitness.   The total fitness of an individual and the fitness of its relatives, the latter weighted according to degree of relationship.

Independent assortment.   The separate inheritances of genes occurring on different chromosomes. Whether a particular trait is inherited from the male or female parent is not influenced by the parental origin of other traits. *See* Linkage groups.

Indirect competition.   Exploitation of a resource by one individual that reduces the availability of that resource to others. *See also* Direct competition.

Inheritance.   Genetic transmission of traits from parent to offspring.

Innate capacity for increase (r_0).   The intrinsic growth rate of a population under ideal conditions without the restraining effects of competition.

Interference.   Direct antagonism between individuals whether by behavioral or chemical means.

Interspecific competition.   Competition between individuals of different species.

Intertidal zone.   The region between the high and low tide marks on the shore, which is alternately covered by water and exposed to the air with each tidal cycle.

Intraspecific competition.   Competition between individuals of the same species.

Intrinsic rate of increase (r_m).   Exponential growth rate of a population with a stable age distribution.

Introgression.   Incorporation of genes of one species into the gene pool of another species.

Inversion (chromosomal).   Reversal of the order of gene loci within a segment of a chromosome.

Ion.   The dissociated parts of a molecule, each of which carries an electrical charge.

Karyotype.   Characteristic chromosomes of a particular species.

Key factor analysis.   A statistical treatment of population data designed to identify factors most responsible for change in population size.

Kin selection.   Differential reproduction among lineages of closely related individuals. May counteract individual selection when there is genetic variation for social behavior among individuals within the lineages.

Laterite.   A hard substance rich in oxides of iron and aluminum; frequently formed when tropical soils weather under alkaline conditions.

Laterization.   Leaching of silica from soil, usually in warm, moist regions with an alkaline soil reaction.

Leaching.   Removal of soluble compounds from leaf litter or soil by water.

Lek.   A communal courtship area on which several males hold courtship territories to attract and mate with females; sometimes called an arena.

Life form.   Characteristic structure of a plant or animal.

Life table.   A summary by age of the survivorship and fecundity of female individuals in a population.

Life zone.   A more or less distinct belt of vegetation occurring within, and characteristic of, a particular range of latitude or elevation.

Limit cycle.   Stable oscillation of predator and prey populations occurring when their interaction meets certain conditions.

Limiting resource.   A resource that is scarce relative to demand for it. *See* Resource.

Linkage group.   Genes on a single chromosome.

Loam.   Soil that is a mixture of coarse sand particles, fine silt, clay particles, and organic matter.

Logistic equation.   Mathematical expression for a particular sigmoid growth curve in which the percentage rate of increase decreases in linear fashion as population size increases.

Lower critical ambient temperature.   Surrounding temperature below which warm-blooded animals must generate heat to maintain their body temperature.

Malthusian parameter (r_m).   The intrinsic exponential growth rate of a population.

Mating system.   Pattern of matings between individuals in a population, including number of simultaneous mates, permanence of pair bond, degree of inbreeding, and so on.

Meiosis.   A series of two divisions by cells destined to produce gametes, involving pairing and segregation of homologous chromosomes, and reducing chromosome number from diploid to haploid. Recombination occurs during meiosis. *See* Mitosis.

Meiotic drive.   Preferential inclusion of chromosomes with one type of allele, or of one of the sex chromosomes, in gametes.

Melanism.   Occurrence of black pigment, usually melanin.

Mesic.   Referring to habitats with plentiful rainfall and well-drained soils.

Metamorphosis.   An abrupt change in form during development that fundamentally alters the function of the organism; often called complete metamorphosis. Incomplete metamorphosis refers to more gradual change.

Micelle.   A complex soil particle resulting from the association of humus and clay particles, with negative electric charges at its surface.

Microhabitat.   The particular parts of the habitat that an individual encounters in the course of its activities.

Mimicry.   Resemblance of an organism to some other organism or object in the environment, evolved to deceive predators or prey into confusing the organism and that which it mimics. *See* Batesian mimicry, Müllerian mimicry, Aggressive mimicry.

Mitosis.   Division of somatic or germ line cells to yield two daughter cells that are genetically identical to the parent cell. *See* Meiosis.

Monoclimax theory.   The idea that all successional sequences led ultimately to one of a few distinctive community types, depending on the climate of the region.

Monoecy.   In plants, the occurrence of reproductive organs of both sexes on the same individual, either in different flowers (hermaphrodite) or in the same flowers (perfect flowers; see Dioecy).

Morph.   A specific form, shape, or structure.

Morph ratio cline.   A gradual geographical change in the frequency of morphs in a population, usually associated with a gradual change in ecological conditions.

Mortality (m_x).   Ratio of the number of deaths of individuals to the population, often described as a function of age (x). *See* Death rate.

Müllerian mimicry.   Mutual resemblance of two or more conspicuously marked, distasteful species to enhance predator avoidance.

Multiple gene inheritance. Determination of a quantitatively varying trait by the additive effects of more than one allele.

Mutation. Any change in the genotype of an organism occurring at the gene, chromosome, or genome level; usually applied to changes in genes to new allelic forms.

Mutation rate (μ). Probability that an allele will be altered to a different form in a generation or other time period.

Mutualism. Relationship between two species that benefits both parties.

Mycorrhizae. Close association of fungi and tree roots in the soil that facilitates the uptake of minerals by trees.

Myoglobin. A large protein with an iron-heme group, associated with storage of oxygen in muscle, and similar to the subunits of hemoglobin.

Natural selection. Change in the frequency of genetic traits in a population through differential survival and reproduction of individuals bearing those traits.

Negative feedback. Tendency of a system to counteract externally imposed change and return to a stable state.

Net aboveground productivity (NAP). Accumulation of biomass in aboveground parts of plants (trunks, branches, leaves, flowers, and fruits), over a specified period; usually expressed on an annual basis.

Net production. The total energy or nutrients accumulated by the organism by growth and reproduction; gross production minus respiration.

Net production efficiency. The percentage of assimilated food utilized for growth and reproduction by an organism.

Net reproductive rate. Number of offspring that females are expected to bear on average during their lifetimes.

Neutral alleles. Alleles of the same gene having identical fitness.

Neutral equilibrium. The particular state of a system that has no forces acting upon it.

Niche. All the components of the environment with which the organism or population interacts.

Nidicolous. Referring to young birds that hatch naked, with their eyes closed, and unable to move about (also called altricial).

Nidifugous. Referring to young birds that hatch covered with down, with their eyes open, and able to move about (also called precocial).

Nitrification. Breakdown of nitrogen-containing organic compounds by microorganisms, yielding nitrates and nitrites.

Nitrogen fixation. Biological assimilation of atmospheric nitrogen to form organic nitrogen-containing compounds.

Nonevolutionary responses. Adaptive changes made by the organism in response to changes in the environment. *See* Regulatory, Acclimatory, and Developmental responses.

Nonstabilizing factors. Influences on population growth that are independent of the size of the population.

Normalizing selection. Selection favoring individuals in the middle of the distribution of phenotypes in a population and disfavoring the extremes. Often called stabilizing selection.

Numerical response. Change in the population size of a predatory species as a result of a change in the density of its prey. *See also* Functional response.

Nutrient. Any substance required by organisms for normal growth and maintenance. (Mineral nutrients usually refer to inorganic substances taken from soil or water.)

Nutrient cycle. The path of an element through the ecosystem including its assimilation by organisms and its release in a reusable inorganic form.

Obligatory. Referring to a way of life or response to particular conditions without alternatives.

Oligotrophic. Referring to a body of water with low nutrient content and productivity.

Omnivore. An organism whose diet is broad, including both plant and animal foods.

Open-community concept. The idea, advocated by H. A. Gleason and R. H. Whittaker, that communities are the local expression of the independent, geographic distributions of species.

Opportunistic species. A species that takes advantage of temporary or local conditions. (Populations of opportunistic species usually fluctuate markedly.)

Oscillation. Regular fluctuation through a fixed cycle above and below some mean value.

Osmoregulation. Regulation of the salt concentration in cells and body fluids.

Osmosis. Diffusion of substances in aqueous solution across the membrane of a cell.

Overdominance. Superiority of the fitness of the heterozygote over that of either homozygote. Also called heterosis.

Overturn. Vertical mixing of layers of large bodies of water caused by seasonal changes in temperature.

Pangenesis. The now discredited belief, held by Darwin among others, that inherited traits were transmitted by a fluid derived of minute secretions from every cell in the body.

Parasite. An organism that consumes part of the blood or tissues of its host, usually without killing the host. Parasites may live entirely within the host (endoparasites) or on its surface (ectoparasites).

Parasitoid. Any of a number of so-called parasitic insects whose larvae live within and consume their host, usually another insect.

Parthenogenesis. Reproduction without fertilization by male gametes, usually involving the formation of diploid eggs whose development is initiated spontaneously.

Pattern-climax theory. The hypothesis that succession reaches a wide variety of nondiscrete climax communities depending on local climate, soil, slope, grazing pressure, and so on.

Perennial. Referring to an organism that lives for more than one year.

pH. A scale of acidity or alkalinity.

Phenotype. Physical expression in the organism of the interaction between the genotype and the environment; outward appearance of the organism.

Pheromones. Chemical substances used for communication between individuals.

Photoautotroph. An organism that utilizes sunlight as its primary energy source for the synthesis of organic compounds.

Photoperiod.   Length of the daylight period each day.

Photoperiodism.   Seasonal response by organisms to change in the length of the daylight period. (Flowering, germination of seeds, reproduction, migration, and diapause are frequently under photoperiodic control.)

Photosynthesis.   Utilization of the energy of light to combine carbon dioxide and water into simple sugars.

Photosynthetic efficiency.   Percentage of light energy assimilated by plants; based either on net production (net photosynthetic efficiency) or on gross production (gross photosynthetic efficiency).

Phyletic evolution.   Genetic changes that occur within an evolutionary line.

Phytoplankton.   Microscopic floating aquatic plants.

Plankton.   Microscopic floating aquatic plants (phytoplankton) and animals (zooplankton).

Pleiotrophy.   Influence of one gene on the expression of more than one trait in the phenotype.

Ploidy.   Number of chromosome complements, or sets, contained by a cell: one—haploid, two—diploid, three—triploid, four—tetraploid, and so on.

Podsolization.   Breakdown and removal of clay particles from the acidic soils of cold, moist regions.

Poikilothermic.   Unable to regulate body temperature; cold-blooded.

Polyclimax theory.   The hypothesis that succession leads to one of a variety of distinct climax communities, depending on local environmental conditions.

Polygamy.   A mating system in which a male pairs with more than one female at one time (polygyny) or a female pairs with more than one male (polyandry).

Polymorphism.   Occurrence of more than one distinct form of individuals in a population.

Potential evapotranspiration.   The amount of transpiration by plants and evaporation from the soil that would occur, given the local temperature and humidity, if water were not limited.

Precocial.   Referring to birds that hatch covered with down, with their eyes open, and able to move about (also called nidifugous).

Predator.   An animal (rarely a plant) that kills and eats animals.

Predictability.   A measure of the regularity of variation in a system.

Primary consumer.   An herbivore, the lowermost eater on the food chain.

Primary producer.   A green plant that assimilates the energy of light to synthesize organic compounds.

Primary productivity.   Rate of assimilation (gross primary productivity) or accumulation (net primary productivity) of energy and nutrients by green plants and other autotrophs.

Primary succession.   Sequence of communities developing in a newly exposed habitat devoid of life.

Protandry.   Course of development of an individual during which its sex changes from male to female.

Protogyny.   The sequence of sexual change by an individual from female to male.

Proximate factors.   Aspects of the environment that organisms use as cues for behavior; for example, daylength. (Proximate factors are often not directly important to the organism's well-being.)

Rain shadow. Dry area on the leeward side of a mountain range.

Recessiveness. Failure of an allele to influence the phenotype when present in heterozygous form.

Recombination. Exchange of genetic information between paired homologous chromosomes during meiosis.

Recruitment. Addition of new individuals to a population by reproduction.

Regulatory response. A rapid, reversible physiological or behavioral response by an organism to change in its environment.

Relative abundance. Proportional representation of a species in a sample or a community.

Replacement series diagram. A diagram showing the outcome of competition between two species in experiments in which the initial ratio of the two species was varied.

Reproductive isolation. Prevention of successful interbreeding between individuals of opposite sex. By definition such individuals belong to different species.

Reproductive value (v_a)**.** The expected reproductive output of an individual at a particular age (a) relative to that of a newborn individual at the same time.

Residual reproductive value (RRV). The reproductive value of an individual discounted by its expected reproductive output at present.

Resource. A substance or object required by an organism for normal maintenance, growth, and reproduction. If the resource is scarce relative to demand, it is referred to as a limiting resource. Nonrenewable resources (such as space) occur in fixed amounts and can be fully utilized; renewable resources (such as food) are produced at a rate that may be partly determined by their utilization.

Respiration. Use of oxygen to break down organic compounds metabolically for the purpose of releasing chemical energy.

Ribonucleic acid (RNA). A long chainlike organic molecule similar to DNA, primarily an intermediary in the transcription of proteins from DNA.

Riparian. Along the bank of a river or lake.

r- and K-selection. Alternative expressions of selection on traits that determine fecundity and survival to favor rapid population growth at low population density (r) or competitive ability at densities near the carrying capacity (K).

Salinization. Accumulation of salts, such as sodium chloride and calcium sulfate, in surface layers of soil, usually in arid climates.

Saprophage. An organism that consumes detritus.

Saturation point. With respect to primary production, the amount of light that causes photosynthesis to attain its maximum rate.

Search image. A behavioral selection mechanism that enables predators to increase searching efficiency for prey that are abundant and worth capturing.

Secondary substances. Organic compounds produced by plants, not directly involved in metabolic pathways, that are implicated as chemical defenses.

Secondary succession. Progression of communities in habitats where the climax community has been disturbed or removed entirely.

Segregation. Separation of alleles derived from the male and female parent into different gametes during meiosis.

Selection. *See* Artificial selection, Group selection, Natural selection, Sexual selection, Sibling selection, Social selection.

Selective death. A death attributed to the deleterious expression of a genotype; a death that would have been avoided had the individual had the optimum genotype.

Selfing. Mating with oneself, applicable, of course, only to individuals having both male and female sexual organs.

Self-thinning curve. In populations of plants limited by space or other resources, the characteristic logarithmic relationship between number and biomass.

Senescence. Gradual deterioration of function in an organism leading to an increased probability of death; aging.

Sere. A series of stages of community change in a particular area leading toward a stable state. *See also* Succession.

Sex chromosomes. A pair of nonhomologous chromosomes, called *X* and *Y* in mammals, which determine the sex of the organism (*XX* is female and *XY* is male in mammals.)

Sexual selection. Selection by one sex for specific characteristics in individuals of the opposite sex, usually exercised through courtship behavior.

Sibling selection. Selection within a brood of offspring for interaction, both competitive and cooperative, that increases their evolutionary fitness.

Sibling species. Species with similar appearance that are incapable of breeding with each other.

Social adaptation. Any adaptation that facilitates behavioral interactions among individuals of the same species; usually does not include reproductive activities.

Social behavior. Any direct interaction among distantly related individuals of the same species; usually does not include courtship, mating, parent-offspring, and sibling interactions.

Social facilitation. Enhancement of any behavior by association with other individuals engaged in similar behavior.

Social feedback. Direct interaction by which some individuals exercise control over the activities of other individuals so as to regulate population processes.

Social group. A group of individuals of the same species formed by mutual attraction of individuals to each other and within which individuals are interdependent to some degree for their well-being.

Social mimicry. Evolution of similar appearance or behavior in unrelated species to facilitate direct interaction between individuals. Also called character convergence.

Social pathology. A syndrome of physiological and behavioral disturbances, caused by crowding, that lead to reduced fecundity and increased mortality.

Society. A group of organisms of the same species characterized by specialization of individual roles, divisions of labor, and mutual dependence.

Soil. The solid substrate of terrestrial communities resulting from the interaction of weather and biological activities with the underlying geological formation.

Soil horizon. A distinctive zone of soil formed at a characteristic depth by weathering and organic contributions to the soil.

Soil skeleton. Physical structure of mineral soil, referring principally to sand grains and silt particles.

Specialization. Restriction of an organism's or a population's activities to a portion of the environment; a trait that enables an organism (or one of its organs) to modify (or differentiate) in order to adapt to a particular function or environment.

Speciation. Separation of one population into two or more reproductively isolated, independent evolutionary units.

Species. A group of actually or potentially interbreeding populations that are reproductively isolated from all other kinds of organisms.

Specific heat. Amount of energy that must be added or removed to change the temperature of a substance by a specific amount. By definition, one calorie of energy is required to raise the temperature of one gram of water by one degree Celsius.

Sporophyte. Diploid stage in the life cycle of plants arising by mitosis from the zygote. The dominant stage in ferns and seed-bearing plants.

Spring overturn. Vertical mixing of water layers in temperate lakes in spring as surface ice disappears.

Stability. Inherent capacity of any system to resist change.

Stabilizing factors. Factors that tend to restore a system to its equilibrium state; specifically, the class of density-dependent factors that act to restore populations to equilibrium size.

Stabilizing selection. Selection favoring individuals in the middle of the distribution of phenotypes in a population and disfavoring the extremes. Often called normalizing selection.

Stable age distribution. Proportion of individuals in various age classes in a population that has been growing at a constant rate.

Stable equilibrium. The particular state to which a system returns if displaced by an outside force.

Stochastic. Referring to patterns resulting from random factors.

Subclimax. A stage of succession along a sere prevented from progressing to the climatic climax by fire, soil deficiencies, grazing, and similar factors.

Subspecies. Subpopulations within a species that are distinguishable by morphological characteristics and, sometimes, by physiological or behavioral characteristics.

Succession. Replacement of populations in a habitat through a regular progression to a stable state.

Supergene. A series of genes, often with related functions, so closely placed on a chromosome that virtually no recombination occurs between them.

Survival (l_x). Proportion of newborn individuals alive at a given age.

Switching. Change in diet to favor items of increasing suitability or abundance.

Symbiosis. Intimate, and often obligatory, association of two species, usually involving coevolution. Symbiotic relationships can be parasitic or mutualistic.

Sympatric. Occurring in the same place, usually referring to areas of overlap in species distributions.

Sympatric speciation. Speciation without geographical isolation; reproductive isolation occurring between segments of a single population.

Synecology. The relationship of organisms and populations to biotic factors in the environment.

Taxon cycle. Cycle of expansion and contraction of the geographical range and population density of a species or higher taxonomic category.

Temperature profile. The relationship of temperature to depth below the surface of water or the soil, or the height above the ground.

Territory. Any area defended by one or more individuals against intrusion by others of the same or different species. *See* Home range.

Tetramer. A molecule composed of four distinct subunits. Hemoglobin is one such molecule.

Tetraploid. Referring to a cell or organism containing four sets of chromosomes. *See* Ploidy.

Thermal conductance. Rate at which heat flows through a substance.

Thermocline. The zone of water depth within which temperature changes rapidly between the upper warm water layer (epilimnion) and lower cold water layer (hypolimnion).

Time lag. Delay in response to a change.

Torpid. Having lost the power of motion and feeling, usually accompanied by greatly reduced rate of respiration.

Transient climax. The end of a successional sequence terminated because the appropriate habitat ceases to exist. Examples are the plant communities in seasonal pools and the series of scavengers that feed on carcasses.

Transit time. Average time that a substance or quantum of energy remains in the biological realm; ratio of biomass to productivity.

Translocation. Switching of a segment of a chromosome to another chromosome.

Transpiration. Evaporation of water from leaves and other plant parts.

Transpiration efficiency. The ratio of net primary production to transpiration of water by a plant, usually expressed by grams per kilogram of water.

Triploid. Referring to a cell or organism containing three sets of chromosomes. *See* Ploidy.

Trophic. Pertaining to food or nutrition.

Trophic level. Position in the food chain determined by the number of energy-transfer steps to that level.

Trophic structure. Organization of the community based on feeding relationships of populations.

Ultimate factors. Aspects of the environment that are directly important to the well-being of an organism (for example, food).

Unstable equilibrium. The particular state of a system upon which forces are precisely balanced, but away from which the system moves when displaced.

Upwelling. Vertical movement of water, usually near coasts and driven by onshore winds, that brings nutrients from the depths of the ocean to surface layers.

Variance. A statistical measure of the dispersion of a set of values about its mean.

Wilting capacity.   The minimum water content of the soil at which plants can obtain water.

Xerarch succession.   Progression of terrestrial plant communities developing in habitats with well-drained soil.

Xeric.   Referring to habitats in which plant production is limited by availability of water.

Zooplankton.   Tiny floating aquatic animals.

Zygote.   Diploid cell formed by the union of male and female gametes during fertilization.

Bibliography

Abrahamson, W. G., and Gadgil, M. 1973. "Growth form and reproductive effort in goldenrods (*Solidago,* Compositae)." *Amer. Nat.,* 107: 651–661.

Abrams, L. 1951. *Illustrated Flora of the Pacific States, Vol. III.* Stanford, California, Stanford. Univ. Press.

Addicott, J. F. 1974. "Predation and prey community structure: an experimental study of the effect of mosquito larvae on the protozoan communities of pitcher plants." *Ecology,* 55: 475–492.

Ahmadjian, V. 1967. *The Lichen Symbiosis.* Waltham, Mass., Blaisdell.

Ainsworth, G. C., and Sussman, A. S. (eds.). 1969. *The Fungi: An Advanced Treatise.* 3 vols. New York, Academic Press.

Alcock, J. 1972. "The evolution of the use of tools by feeding animals." *Evolution,* 26: 464–473.

Alexander, R. D. 1961. "Aggressiveness, territoriality, and social behavior in field crickets (Orthoptera: Gryllidae)." *Behaviour,* 17: 130–223.

———— 1974. "The evolution of social behavior." *Ann. Rev. Ecol. Syst.,* 5: 325–383.

———— and Bigelow, R. S. 1960. "Allochronic speciation in field crickets, and a new species *Acheta veletis.*" *Evolution* 14: 334–346.

Alexander, R. D., and Otte, D. 1967. "The evolution of genetalia and mating behavior in crickets (Gryllidae) and other Orthoptera." *Misc. Publ. Mus. Zool. Univ. Mich.,* 133: 1–62.

Alexander, R. M. 1968. *Animal Mechanics.* Seattle, Univ. of Washington Press.

———— 1971. *Size and Shape.* London, Edward Arnold.

Allard, R. W. 1975. "The mating system and microevolution." *Genetics,* 79: 115–126.

Allee, W. C. 1926. "Measurements of environmental factors in the tropical rainforest in Panama." *Ecology,* 7: 273–302.

———— Emerson, A. E., Park, O., Park, T., and Schmidt, K. P. 1949. *Principles of Animal Ecology.* Philadelphia, Saunders.

Allen, J. A. 1877. "The influence of physical conditions in the genesis of species." *Radical Rev.,* 1: 108–140.

Allen, J. A., and Clarke, B. 1968. "Evidence for apostatic selection by wild passerines." *Nature,* 220: 501–502.

Allison, A. C. 1956. "Sickle cells and evolution." *Sci. Amer.,* 195: 87–94.

Amadon, D. 1947. "Ecology and evolution of some Hawaiian birds." *Evolution,* 1: 63–68.

Amant, J. L. S. St. 1970. "The detection of regulation in animal populations." *Ecology,* 51: 823–828.

Andrewartha, H. G. 1935. "Thrips investigation No. 7: On the effect of temperature and food upon egg production and the length of adult life of *Thrips imaginis, Bagnall*." *J. Council Sci. Indust. Res. Australia,* 8: 281–288.

——— 1963. "Density-dependence in the Australian thrips." *Ecology,* 44: 218–220.

——— and Birch, L. C. 1954. *The Distribution and Abundance of Animals.* Chicago, Univ. of Chicago Press.

Andrewartha, H. G., and Browning, T. O. 1961. "An analysis of the idea of 'resources' in animal ecology." *J. Theoret. Biol.,* 1: 83–97.

Antonovics, J. 1968. "Evolution in closely adjacent plant populations, VI: Manifold effects of gene flow." *Heredity,* 23: 507–524.

——— 1968a. "Evolution in closely adjacent plant populations, V: Evolution of self-fertility." *Heredity,* 23: 219–238.

——— 1971. "The effects of a heterogeneous environment on the genetics of natural populations." *Amer. Sci.,* 59: 593–599.

——— and Bradshaw, A. D. 1970. "Evolution in closely adjacent plant populations, VIII: Clinal patterns at a mine boundary." *Heredity,* 25: 349–362.

Applebaum, S. W. 1964. "Physiological aspects of host specificity in the Bruchidae—I: General considerations of developmental compatibility." *J. Insect Physiol.,* 10: 783–788.

——— Gestetner, B., and Birk, Y. 1965. "Physiological aspects of host specificity in the Bruchidae—IV: Developmental incompatibility of soybeans for *Callosobruchus*." *J. Insect Physiol.,* 11: 611–616.

Arrhenius, O. 1921. "Species and area." *J. Ecol.,* 9: 95–99.

Art, H. W., Bormann, F. H., Voigt, G. K., and Woodwell, G. M. 1974. "Barrier island forest ecosystem: Role of meteorologic nutrient inputs." *Science,* 184: 60–62.

Ashmole, N. P. 1971. "Sea bird ecology and the marine environment," in D. S. Farner and J. R. King (eds.), *Avian Ecology.* New York and London, Academic Press.

——— and Tovar, S. H. 1968. "Prolonged parental care in royal terns and other birds." *Auk,* 85: 90–100.

Ashton, P. S. 1969. "Speciation among tropical forest trees: Some deductions in the light of recent evidence." *Biol. J. Linnean Soc.,* 1: 155–196.

Assem, J. van den. 1967. "Territory in the three-spined stickleback (*Gasterosteus aculeatus*)." *Behaviour,* Suppl. 16: 1–164.

Aubréville, A. 1938. "La forêt coloniale: les forêts de l'Afrique occidentale francaise." *Ann. Acad. Sci. Colon., Paris,* 9: 1–245.

Auslander, D. M., Oster, G. F., and Huffaker, C. B. 1974. "Dynamics of interacting populations." *J. Franklin Inst.,* 297: 345–376.

Austin, G. T. 1974. "Nesting success of the cactus wren in relation to nest orientation." *Condor,* 76: 216–217.

Ayala, F. J. 1965. "Evolution of fitness in experimental populations of *Drosophila serrata*." *Science,* 150: 903–905.

——— 1968. "Biology as an autonomous science." *Amer. Sci.,* 56: 207–221.

——— 1969. "Evolution of fitness. IV. Genetic evolution of interspecific competitive ability in *Drosophila*." *Genetics,* 61: 737–747.

——— 1970. "Competition, coexistence, and evolution," in M. K. Hecht and W. C. Steere (eds.), *Essays in Evolution and Genetics.* New York, Appleton-Century-Crofts.

——— 1971. "Competition between species: Frequency dependence." *Science,* 171: 820–824.

——— 1972. "Darwinian *versus* non-Darwinian evolution in natural populations of *Drosophila*." *Proc. 6th Berkeley Symposium Math. Stat. Prob.,* 5: 211–236.

——— 1972a. "Competition between species." *Amer. Sci.,* 60: 348–357.

—— 1975. "Scientific hypotheses, natural selection and the neutrality theory of protein evolution," in F. M. Salzana (ed.), *The Role of Natural Selection in Human Evolution.* Amsterdam, North-Holland Publ. Co.

—— and Anderson, W. W. 1973. "Molecular evolution—evidence for natural selection." *Nature New Biol.,* 241: 274–276.

Ayala, F. J., Powell, J. R., Tracey, M. L., Mourão, and Perez-Salas, S. 1972. "Enzyme variability in the *D. willistoni* group. IV. Genic variation in natural populations of *D. willistoni.*" *Genetics,* 71: 113–39.

Ayala, F. J., and Campbell, C. A. 1974. "Frequency-dependent selection." *Ann. Rev. Ecol. Syst.,* 5: 115–138.

Baerends, G. P., and Baerends-van Roon, J. M. 1950. "An introduction to the ethology of cichlid fishes." *Behav. Suppl.,* 1: 1–243.

Baker, H. G. 1963. "Evolutionary mechanisms in pollination biology." *Science* 139: 877–883.

—— 1965. "Characteristics and modes of origin of weeds," in H. G. Baker and G. L. Stebbins (eds.), *Genetics of Colonizing Species.* New York, Academic Press.

Baker, J. R. 1938. "The evolution of breeding seasons," in G. R. de Beer (ed.), *Evolution: Essays on Aspects of Evolutionary Biology.* London and New York, Oxford Univ. Press.

Baker, R. R., and Parker, G. A. 1973. "The origin and evolution of sexual reproduction up to the evolution of the male-female phenomenon." *Acta Biotheoretica,* 22: 1–77.

Balda, R. P., and Bateman, G. C. 1971. "Flocking and annual cycle of the piñon jay, *Gymnorhinus cyanocephalus.*" *Condor,* 73: 287–302.

—— 1973. "The breeding biology of the piñon jay." *Living Bird,* 11: 5–42.

Ball, R. C., and Hooper, F. F. 1963. "Translocation of phosphorus in a trout stream ecosystem," in V. Shultz and A. W. Klement (eds.), *Radioecology.* New York, Reinhold.

Balon, E. K. 1975. "Reproductive guilds of fishes: a proposal and definition." *J. Fish. Res. Bd. Can.,* 32: 821–864.

Banta, A. M., and Brown, L. A. 1939. "Control of male and sexual-egg production," in A. M. Banta (ed.), *Studies on the Physiology, Genetics, and Evolution of Some Cladocera.* Washington, D.C., Carnegie Inst. Wash. Publ. No. 513.

Barkham, J. P., and Norris, J. W. 1970. "Multivariate procedures in an investigation of vegetation and soil relations of two beech woodlands, Cotswold Hills, England." *Ecology,* 51: 631–639.

Barnes, R. D. 1968. *Invertebrate Zoology,* 2d ed. Philadelphia, Saunders.

Barr, T. C. 1968. "Cave ecology and the evolution of troglobytes." *Evol. Biology,* 2: 35–102.

Barron, D. H., and Meschia, G. 1954. "A comparative study of the exchange of the respiratory gases across the placenta." *Cold Spring Harbor Symp. Quant. Biol.,* 19: 93–101.

Barry, R. G., and Chorley, R. J. 1970. *Atmosphere, Weather, and Climate.* New York, Holt, Rinehart and Winston.

Bartholomew, G. A. 1942. "The fishing activities of double-crested cormorants on San Francisco Bay." *Condor,* 44: 13–21.

—— Lasiewski, R. C. and Crawford, E. C. 1968. "Patterns of panting and gular flutter in cormorants, pelicans, owls, and doves." *Condor,* 70: 31–34.

Batzli, G. O. 1969. "Distribution of biomass in rocky intertidal communities on the Pacific coast of the United States." *J. Anim. Ecol.,* 38: 531–546.

——— 1975. "The role of small mammals in arctic ecosystems," in F. B. Golley, K. Petrusewicz, and L. Ryszkowski (eds.), *Small Mammals: Their Productivity and Population Dynamics.* Cambridge, Cambridge Univ. Press.

——— and Pitelka, F. A. 1970. "Influence of meadow mouse populations on California grassland." *Ecology,* 51: 1027–1039.

——— 1971. "Condition and diet of cycling populations of the California vole, *Microtus californicus.*" *J. Mammal.,* 52: 141–163.

Bawa, K. S. 1974. "Breeding systems of tree species of a lowland tropical community." *Evolution,* 28: 85–92.

Beals, E. W., and Cope, J. B. 1964. "Vegetation and soils in eastern Indiana woods." *Ecology* 45: 777–792.

Beard, J. S. 1955. "The classification of tropical American vegetation types." *Ecology,* 36: 89–100.

Beck, S. D. 1965. "Resistance of plants to alkaloids." *Ann. Rev. Entomol.,* 10: 207–232.

Beckman, W. A., Mitchell, J. W., and Porter, W. P. 1973. "Thermal model for prediction of a desert iguana's daily and seasonal behavior." *J. Heat. Transfer,* May, 1973: 257–262.

Beeton, A. M. 1965. "Eutrophication of the St. Lawrence Great Lakes." *Limnol. Oceanogr.,* 10: 240–254.

Belyea, R. M. 1952. "Death and deterioration of balsam fir weakened by spruce budworm defoliation in Ontario." *J. Forestry,* 50: 729–738.

Benson, J. F. 1973. "Some problems of testing for density-dependence in animal populations." *Oecologia,* 13: 183–190.

Benson, W. W. 1971. "Evidence for the evolution of unpalatability through kin selection in the Heliconiinae (Lepidoptera)." *Amer. Nat.,* 105: 213–226.

Berg, R. Y. 1975. "Myrmechochorous plants in Australia and their dispersal by ants." *Aust. J. Bot.,* 23: 475–508.

Berger, E. M. 1971. "A temporal survey of allelic variation in natural and laboratory populations of *Drosophila melanogaster.*" *Genetics,* 67: 121–136.

Bergmann, C. 1847. "Ueber die Verhaltnisse der Warm eökonomie des Thiere zu ihrer Grösse." *Göttinger Studien,* 3: 595–708.

Berkner, L. V., and Marshall, L. C. 1964. in P. J. Brancazio and A. G. W. Cameron, eds., *The Origin and Evolution of Atmospheres and Oceans.* New York, Wiley.

——— 1965. "History of major atmospheric components." *Proc. Natl. Acad. Sci.,* 53: 1215–1226.

Berlin, B., Breedlove, D. E., and Raven, P. H. 1966. "Folk taxonomies and biological classification." *Science,* 154: 273–275.

Bernal, J. D. 1967. *The Origin of Life.* London, Weidenfeld and Nicolson.

Berry, J. A. 1975. "Adaptation of photosynthetic processes to stress." *Science,* 188: 644–650.

Berthet, P. 1963. "Mesure de la consommation d'oxygène des oribatides (Acariens) de la litière des forêts," in J. Doekson and J. Van Der Drift (eds.), *Soil Organisms.* Amsterdam, North-Holland Publ. Co.

Bess, H. A., van den Bosch, R. and Haramoto, F. A. 1961. Fruit fly parasites and their activities in Hawaii." *Proc. Haw. Ent. Soc.,* 17: 367–378.

Beverton, R. J. H., and Holt, S. J. 1957. "On the dynamics of exploited fish populations." *Fish. Invest.,* 19: 1–533.

Bider, J. R. 1962. "Dynamics and the temporo-spatial relations of a vertebrate community." *Ecology,* 43: 634–646.

Billings, W. D. 1964. *Plants and the Ecosystem.* Belmont, California, Wadsworth.

——— and Mooney, H. A. 1968. "The ecology of arctic and alpine plants." *Biol. Rev.,* 43: 481–529.

Birch, L. C. 1948. "The intrinsic rate of natural increase of an insect population." *J. Anim. Ecol.*, 17: 15–26.

—— 1953a. "Experimental background to the study of the distribution and abundance of insects, I: The influence of temperature, moisture and food on the innate capacity for the increase of three grain beetles." *Ecology*, 34: 698–711.

—— 1953b. "Experimental background to the study of the distribution and abundance of insects, II: The relation between innate capacity for increase in numbers and the abundance of three grain beetles in experimental populations." *Ecology*, 34: 712–726.

—— and Clark, D. P. 1953. "Forest soil as an ecological community with special reference to the fauna." *Quart. Rev. Biol.*, 28: 13–36.

Black, C. C. 1973. "Photosynthetic carbon fixation in relation to net CO_2 uptake." *Ann. Rev. Plant Physiol.*, 24: 253–286.

Blair, W. F. 1975. "Adaptation of anurans to equivalent desert scrub of North and South America," in D. W. Goodall (ed.), *Evolution of Desert Biota.* Austin, Univ. Texas Press.

Blanchard, B. 1941. "The white-crowned sparrows (*Zonotrichia leucophrys*) of the Pacific seaboard: Environment and annual cycle." *Univ. Calif. Pub. Zool.*, 46: 1–178.

Blest, A. D. 1957. "The function of eye-spot patterns in Lepidoptera." *Behaviour*, 11: 209–256.

—— 1960. "A study of the biology of saturniid moths in the Canal Zone biological area." *Smithson. Rept.*, 1959: 447–464.

—— 1963. "Longevity, palatability and natural selection in five species of New World Saturniid moth." *Nature*, 197: 1183–1186.

—— 1963. "Relations between moths and predators." *Nature*, 197: 1046–1047.

—— 1964. "Protective display and sound production in some New World arctiid and ctenuchid moths." *Zoologica*, 49: 161–181.

Bodenheimer, F. S. 1958. "Animal ecology today." *Monogr. Biol.*, 6: 1–276.

Bodmer, W. F., and Parsons, P. A. 1962. "Linkage and recombination in evolution." *Adv. Genet.*, 2: 1–100.

Bogert, C. M. 1949. "Thermoregulation in reptiles, a factor in evolution." *Evolution*, 3: 195–211.

Bolin, B. 1970. "The carbon cycle." *Sci. Amer.*, 223: 125–132.

Bonaventura, J., Bonaventura, C., and Sullivan, B. 1974. "Urea tolerance as a molecular adaptation of elasmobranch hemoglobins." *Science*, 186: 57–59.

Bonnell, M. L., and Selander, R. K. 1974. "Elephant seals: Genetic variation and near extinction." *Science*, 184: 908–909.

Borchert, J. R. 1950. "The climate of the central North American grassland." *Ann. Assoc. Amer. Geogr.*, 40: 1–39.

Borgstrom, G. 1969. *Too Many: A Study of the Earth's Biological Limitations.* New York, Macmillan.

Bormann, F. H. 1958. "The relationships of ontogenetic development and environmental modification of photosynthesis in *Pinus taeda* seedlings," in K. V. Thimann (ed.), *The Physiology of Forest Trees.* New York, Ronald Press.

Boss, W. R. 1943. "Hormone determination of adult characters and sex behavior in Herring Gulls (*Larus argentatus*)." *J. Exptl. Zool.*, 94: 181–209.

Bossert, W. H., and Wilson, E. O. 1963. "The analysis of olfactory communication among animals." *J. Theoret. Biol.*, 5: 443–469.

Bourlière, F. 1973. "The comparative ecology of rainforest mammals in Africa and tropical America: Some introductory remarks," in B. J. Meggars, E. S. Ayensu, and W. D. Duckworth (eds.), *Tropical Forest Ecosystems in Africa and South America: A Comparative Review.* Washington, D.C., Smithsonian Inst.,

Bovbjerg, R. V. 1970. "Ecological isolation and competitive exclusion in two crayfish (*Orconectes virilus* and *Orconectes immunis*)." *Ecology*, 51: 223–236.

Bowler, P. A., and Rundel, P. W. 1975. "Reproductive strategies in lichens." *Bot. J. Linn. Soc.*, 70: 325–340.

Brady, N. C. 1974. *Nature and Property of Soils*, 8th ed. New York, Macmillan.

Brawn, V. M. 1961. "Aggressive behavior in the cod (*Gadus callarias* L.)." *Behaviour*, 18: 107–147.

Bray, J. R., and Dudkiewicz, L. A. 1963. "The composition, biomass and productivity of two *Populus* forests." *Bull. Torrey Bot. Club*, 90: 298–308.

Bray, J. R., and Gorham, E. 1964. "Litter production in forests of the world." *Adv. Ecol. Res.*, 2: 101–157.

Breder, C. M., Jr. 1967. "On the survival value of fish schools." *Zoologica*, 52: 25–40.

Bretsky, P. W., and Lorenz, D. M. 1970a. "An essay on genetic-adaptive strategies and mass extinctions." *Geol. Soc. Amer. Bull.*, 81: 2449–2456.

—— 1970b. "Adaptive response to environmental stability: A unifying concept in paleoecology." *North Amer. Paleont. Conv. Proc.*, Pt. E, 522–550.

Brett, J. R. 1956. "Some principles in the thermal requirements of fishes." *Quart. Rev. Biol.*, 31: 75–87.

—— 1971. "Temperature—animals—fishes," in O. Kinne (ed.), *Marine Ecology*. Vol. I. New York, Wiley-Interscience.

Brewbaker, J. L. 1964. *Agricultural Genetics*. Englewood Cliffs, New Jersey, Prentice-Hall.

Brewer, R. 1963. "Stability in bird populations." *Occ. Pap. C. C. Adams Cent. Ecol. Stud.*, 7: 1–12.

Brian, A. D. 1957. "Differences in the flowers visited by four species of bumblebees and their causes." *J. Anim. Ecol.*, 26: 71–98.

Briggs, D., and Walters, S. M. 1969. *Plant Variation and Evolution*. New York, McGraw-Hill, World Univ. Library.

Briggs, L. J., and Schantz, H. L. 1914. "Relative water requirements of plants." *J. Agric. Res.*, 3: 1–63.

Brinkhurst, R. O. 1959. "Alary polymorphism in the Gerroidea (Hemiptera-Heteroptera)." *J. Anim. Ecol.*, 28: 211–230.

Brisbin, I. L., Jr. 1969. "Bioenergetics of the breeding cycle of the ring dove." *Auk*, 86: 54–74.

Brock, T. D. 1970. "High temperature systems." *Ann. Rev. Ecol. Syst.*, 1: 191–220.

—— 1975. "The effect of water potential on photosynthesis in whole lichens and their liberated algal components." *Planta*, 124: 13–23.

—— and Darland, G. K. 1970. "Limits of microbial existence: Temperature and pH." *Science*, 169: 1316–1318.

Brockway, B. F. 1965. "Stimulation of ovarian development and egg-laying by male courtship vocalization in budgerigars (*Melopsittacus undulatus*)." *Anim. Behav.*, 13: 575–578.

Brodskij, A. K. 1959. "Leben in der Tiefe des Polarbeckens." *Naturwiss. Rundschau*, 12: 52–56.

Brody, S. 1945. *Bioenergetics and Growth*. New York, Reinhold.

Broekhuysen, G. J. 1941. "A preliminary investigation of the importance of desiccation, temperature and salinity as factors controlling the vertical distribution of certain intertidal marine gastropods in False Bay, South Africa." *Trans. R. Soc. South Africa*, 28: 255–292.

Brower, J. V. Z. 1958. "Experimental studies of mimicry in some North American butterflies: Part I, the monarch, *Danaus plexippus* and viceroy, *Limenitis archippus;* Part II, *Battus philenor* and *Pipilio troilus*, *P. polyxenes* and *P. glaucus;* Part

III, *Danaus gilippus berenice* and *Limenitis archippus floridensis.*" *Evolution,* 12: 32–47, 123–136, 273–285.

———— 1960. "Experimental studies of mimicry: Part IV, The reactions of starlings to different proportions of models and mimics." *Amer. Nat.,* 94: 271–282.

———— and Brower, L. P. 1962. "Experimental studies of mimicry: Part 6, The reaction of toads (*Bufo terrestris*) to honeybees (*Apis mellifera*) and their dronefly mimics (*Eristalis vinetorum*)." *Amer. Nat.,* 96: 297–308.

Brower, L. P. 1969. "Ecological chemistry." *Sci. Amer.,* 220: 22–29.

———— 1970. "Plant poisons in a terrestrial food chain and implications for mimicry theory," in K. L. Chambers (ed.), *Biochemical Evolution* (*Proc. Ann. Biol. Coll.* 1968). Corvallis, Oregon, Oregon State Univ. Press.

———— Brower, J. V. Z., Stiles, F. G., Croze, H. J., and Hower, A. S. 1964. "Mimicry: Differential advantage of color patterns in the natural environment." *Science,* 144: 183–185.

Brower, L. P., Cook, L. M., and Croze, H. J. 1967. "Predator responses to artificial Batesian mimics released in a neotropical environment." *Evolution,* 21: 11–23.

Brower, L. P., Ryerson, W. N., Coppinger, L. L., and Glasier, S. C. 1968. "Ecological chemistry and the palatability spectrum." *Science,* 161: 1349–1351.

Brown, J. H. 1971. "Mammals on mountaintops: Nonequilibrium insular biogeography." *Amer. Nat.,* 105: 467–478.

———— 1975. "Geographical ecology of desert rodents," in M. L. Cody and J. M. Diamond (eds.), *Ecology and Evolution of Communities.* Cambridge, Mass., Belknap Press.

Brown, J. L. 1963. "Aggressiveness, dominance and social organization in the stellar jay." *Condor,* 65: 460–484.

———— 1969. "Territorial behavior and population regulation in birds." *Wilson Bull.,* 81: 293–329.

———— 1970. "Cooperative breeding and altruistic behavior in the Mexican jay, *Aphelocoma ultramarina.*" *Anim. Behav.,* 18: 366–378.

———— 1974. "Alternate routes to sociality in jays—with a theory for the evolution of altruism and communal breeding." *Amer. Zool.,* 14: 63–80.

Brown, R. Z. 1953. "Social behavior, reproduction, and population changes in the house mouse." *Ecol. Monogr.,* 23: 217–240.

Brown, W. L., Jr., 1957. "Centrifugal speciation." *Quart. Rev. Biol.,* 32: 247–277.

———— and Wilson, E. O. 1956. "Character displacement." *Syst. Zool.,* 5: 49–64.

Brues, A. M. 1964. "The cost of evolution *vs.* the cost of not evolving." *Evolution,* 18: 379–383.

———— 1969. "Genetic load and its varieties." *Science,* 164: 1130–1136.

Brues, C. T. 1924. "The specificity of food plants in the evolution of phytophagous insects." *Amer. Nat.,* 58: 127–144.

Brussard, P. F., Ehrlich, P. R., and Singer, M. C. 1974. "Adult movements and population structure in *Euphydryas editha.*" *Evolution,* 28: 408–415.

Buck, J. B. 1937. "Studies on the firefly, II: The signal system and color vision in *Photinus pyralis.*" *Physiol. Zool.,* 10: 412–419.

Buckner, C. H., and Turnock, W. J. 1965. "Avian predation on the larch sawfly, *Pristiphora erichsonii* (Htg.) (Hymenoptera: Tenthredinidae)." *Ecology,* 46: 223–236.

Budowski, G. 1965. "Distribution of tropical American rain forest species in the light of successional processes." *Turrialba,* 15: 40–42.

Buechner, H. K. 1961. "Territorial behavior in the Uganda kob." *Science,* 133: 698–699.

———— and Golley, F. B. 1967. "Preliminary estimation of energy flow in Uganda kob. (*Adenota kob thomasi* Neumann)." *Secondary Product. Terrest. Ecosyst.,* 243–253.

Bumpus, H. C. 1898. "The elimination of the unfit as illustrated by the introduced sparrow, *Passer domesticus.*" *Biol. Lect., Mar. Biol. Lab., Woods Hole,* 11: 209–226.

Bünning, E. 1967. *The Physiological Clock* (rev. 2d ed.). New York, Springer-Verlag, Heidelberg Science Library.

Bunt, J. S. 1973. "Primary production: Marine ecosystems." *Human Ecol.,* 1: 333–345.

Bunting, B. T. 1967. *The Geography of Soil* (rev. ed.). Chicago, Aldine.

Buol, S. W., Hole, F. D., and McCracken, R. J. 1973. *Soil Genesis and Classification.* Ames, Iowa, Iowa State Univ. Press.

Burk, C. J. 1973. "The Kaibab deer incident: A long persisting myth." *Bio. Science,* 23: 113.

Burkhardt, D., Schleidt, W., and Altner, H. 1967. *Signals in the Animal World,* trans. by K. Morgan. New York, McGraw-Hill.

Burky, A. J. 1971. "Biomass turnover, respiration, and interpopulation variation in the stream limpet, *Ferrissia rivularis* (Say)." *Ecol. Monogr.,* 41: 235–251.

Burnet, M., and White, D. O. 1974. *Natural History of Infectious Disease* (4th ed.). London, Cambridge Univ. Press.

Bush, G. L. 1969. "Sympatric host race formation and speciation in frugivorous flies of the genus *Rhagoletis* (Diptera, Tephretidae)." *Evolution,* 23: 237–251.

Buxton, P. A., and Lewis, D. J. 1934. "Climate and tsetse flies: Laboratory studies upon *Glossina submorsitans* and *G. tachinoides.*" *Phil. Trans. R. Soc. London, B,* 224: 175–240.

Buzas, M. A. 1972. "Patterns of species diversity and their explanation." *Taxon,* 21: 275–286.

Cain, A. J., and Sheppard, P. M. 1950. "Selection in the polymorphic land snail *Cepaea nemoralis.*" *Heredity,* 4: 275–294.

——— 1954. "Natural selection in *Cepaea.*" *Genetics,* 39: 89–116.

Cairns, J., Dahlberg, M. L., Dickson, K. L., Smith, N., and Waller, W. T. 1969. "The relationship of fresh-water protozoan communities to the MacArthur-Wilson equilibrium model." *Amer. Nat.,* 103: 439–454.

Cairns, J., Crossman, J. S., Dickson, K. L., and Herricks, E. E. 1971. "The recovery of damaged streams." *Assoc. Southeastern Biol. Bull.,* 18: 79–106.

Calder, W. A. 1968. "Nest sanitation: A possible factor in the water economy of the roadrunner." *Condor,* 70: 279.

Calhoun, J. B. 1949. "A method for self-control of population growth among mammals living in the wild." *Science,* 109: 333–335.

Cambell, B. G. 1966. *Human Evolution.* Chicago, Aldine.

Campbell, J. W. (ed.). 1970. *Comparative Biochemistry of Nitrogen Metabolism.* Vols. I and II. New York, Academic Press.

Cantlon, J. E. 1969. "The stability of natural populations and their sensitivity to technology." *Brookhaven Sym. Biol.,* 22: 197–205.

Carey, F. G., Teal, J. M., Kanwisher, J. W., and Lawson, K. D. 1971. "Warm-bodied fish." *Amer. Zool.,* 11: 137–145.

Carey, M., and Nolan, V. 1975. "Polygyny in indigo buntings: A hypothesis tested." *Science,* 190: 1296–1297.

Carlisle, A., Brown, A. H. F., and White, E. J. 1966. "The organic matter and nutrient elements in the precipitation beneath a sessile oak (*Quercus petraea*) canopy." *J. Ecol.,* 54: 87–98.

Carlquist, S. 1966. "The biota of long distance dispersal, IV: Genetic systems in the floras of oceanic islands." *Evolution,* 20: 433–455.

Carpenter, E. J., and Culliney, J. L. 1975. "Nitrogen fixation in marine shipworms." *Science,* 187: 551–552.

Carrick, R. 1963. "Ecological significance of territory in the Australian Magpie, *Gymnorhina tibicen.*" *Proc. XIII Inter. Ornith. Congr.,* 740–753.

Carson, H. L. 1955. "The genetic characteristics of marginal populations of *Drosophila.*" *Cold Spring Harbor Symp. Quant. Biol.,* 20: 276–287.

—— 1956. "Marginal homozygosity for gene arrangement in *Drosophila robusta.*" *Science,* 123: 630–631.

—— 1958. The population genetics of *Drosophila robusta.*" *Adv. Genet.,* 9: 1–40.

—— 1959. "Genetic conditions which promote or retard the formation of species." *Cold Spring Harbor Symp. Quant. Biol.,* 24: 87–105.

Carter, M. A. 1967. "Selection in mixed colonies of *Cepaea nemoralis* and *Cepaea hortensis.*" *Heredity,* 22: 117–139.

—— 1968. "Thrush predation of an experimental population of the snail *Cepaea nemoralis* (L.)." *Proc. Linn. Soc. Lond.,* 179: 241–249.

Cartwright, F. F. and Biddiss, M. D. 1972. *Disease and History.* New York, Crowell.

Case, T. J., and Gilpin, M. E. 1974. "Interference competition and niche theory." *Proc. Nat. Acad. Sci.,* 71: 3073–3077.

Catlett, R. H. 1961. "An evaluation of methods for measuring fighting behavior with special reference to *Mus musculus.*" *Anim. Behav.,* 9: 8–10.

Caughley, G. 1966. "Mortality patterns in mammals." *Ecology,* 47: 906–918.

—— 1970. "Eruption of ungulate populations, with emphasis on Himalyan thar in New Zealand." *Ecology,* 51: 53–72.

Cavalier-Smith, T. 1975. "The origin of nuclei and of eukaryotic cells." *Nature,* 256: 463–468.

Cavalli-Sforza, L. L., and Bodmer, W. F. 1971. *The Genetics of Human Populations.* San Francisco, W. H. Freeman.

Cavers, P. B., and Harper, J. L. 1967. "Studies in the dynamics of plant populations, I. The fate of seed and transplants introduced into various habitats." *J. Ecol.,* 55: 59–71.

Center, T. D., and Johnson, C. D. 1974. "Coevolution of some seed beetles (Coleoptera: Bruchidae) and their hosts." *Ecology,* 55: 1096–1103.

Cerami, A., and Manning, J. M. 1971. "Potassium cyanate as an inhibitor of the sickling of erythrocytes *in vitro.*" *Proc. Natl. Acad. Sci.,* 68: 1180–1183.

Chapman, R. N. 1928. "The quantitative analysis of environmental factors." *Ecology,* 9: 111–122.

Charnov, E. L., and Schaffer, W. M. 1973. "Life history consequences of natural selection: Cole's result revisited." *Amer. Nat.,* 107: 791–793.

Cheatum, E. L., and Severinghaus, C. W. 1950. "Variations in fertility of white-tailed deer related to range conditions." *Trans. North Amer. Wild. Conf.,* 15: 170–189.

Chesher, R. H. 1969. "Destruction of Pacific corals by the sea star *Acanthaster planci.*" *Science,* 165: 280–283.

Chew, R. M., and Chew, A. E. 1970. "Energy relationships of the mammals of a desert shrub." *Ecol. Monogr.,* 40: 1–21.

Chinnici, J. P. 1971. "Modification of recombination frequency in *Drosophila.* II: The polygenic control of crossing over." *Genetics,* 69: 85–96.

Chitty, D. 1954. "Tuberculosis among wild voles: With a discussion of other pathological conditions among certain mammals and birds." *Ecology,* 35: 227–237.

—— 1960. "Population processes in the vole and their reference to general theory." *Canad. J. Zool.,* 38: 99–113.

—— 1967. "The natural selection of self-regulatory behavior in animal populations." *Proc. Ecol. Soc. Australia,* 2: 51–78.

Christian, J. J. 1950. "The adreno-pituitary system and population cycles in mammals." *J. Mamm.,* 31: 247–259.

—— 1963. "Endocrine adaptive mechanisms and the physiologic regulation of population growth," in W. V. Mayer and R. G. van Gelder, eds., *Physiological Mammalogy,* Vol. I: *Mammalian Populations.* New York, Academic Press.

—— and Davis, D. E. 1964. "Endocrines, behavior and population." *Science,* 146: 1550–1560.

Christian, J. J., Lloyd, J. A., and Davis, D. E. 1965. "The role of endocrines in the self-regulation of mammalian populations." *Recent Prog. Hormone Res.,* 21: 507–578.

Christiansen, K., and Culver, D. 1968. "Geographical variation and evolution in *Pseudosinella hirsuta.*" *Evolution,* 22: 237–255.

Churchill, G. B., John, H. H., Duncan, D. P., and Hobson, A. C. 1964. "Long-term effects of defoliation of aspen by the forest tent caterpillar." *Ecology,* 45: 630–633.

Clark, F. W. 1972. "Influence of jackrabbit density on coyote population change." *J. Wildl. Mgmt.,* 36: 343–356.

Clarke, B. 1966. "The evolution of morph-ratio clines." *Amer. Nat.,* 100: 389–402.

—— 1969. "The evidence for apostatic selection." *Heredity,* 24: 347–352.

Clarke, C. A., and Sheppard, P. M. 1959. "The genetics of *Papilio dardanus,* Brown, I: Race *cenea* from South Africa." *Genetics,* 44: 1347–1358.

—— 1960a. "The genetics of *Papilio dardanus,* Brown, II: Races *dardanus, polytrophus, meseres,* and *tibullus.*" *Genetics,* 45: 439–457.

—— 1960b. "The evolution of dominance under disruptive selection." *Heredity,* 14: 73–87.

—— 1960c. "The evolution of mimicry in the butterfly *Papilio dardanus.*" *Heredity,* 14: 163–173.

—— 1960d. "Super-genes and mimicry." *Heredity,* 14: 175–185.

Clarke, T. A. 1970. "Territorial behavior and population dynamics of a pomacentrid fish, the garibaldi, *Hypsypops rubicunda.*" *Ecol. Monogr.,* 40: 189–212.

Clausen, J., Keck, D. D., and Hiesey, W. M. 1948. "Experimental studies on the nature of species, III: Environmental responses of climatic races of *Achillea.*" *Carnegie Inst. Wash. Publ.,* 581: 1–129.

Clayton, G. A., Morris, J. A., and Robertson, A. 1957. "An experimental check on quantitative genetical theory. I: Short-term responses to selection." *J. Genet.,* 55: 131–151.

Clayton, H. H. 1944. "World weather records." *Smithsonian Misc. Coll.,* 79: 1–1199.

—— and Clayton, F. L. 1947. "World weather records 1931–1940." *Smithsonian Misc. Coll.,* 105: 1–646.

Clements, F. E. 1916." Plant succession: Analysis of the development of vegetation." *Carnegie Inst. Wash. Publ.,* 242: 1–512.

—— 1936. "Nature and structure of the climax." *J. Ecol.,* 24: 252–284.

Cloud, P. 1968. "Pre-metazoan evolution and the origins of the metazoa," in E. T. Drake (ed.), *Evolution and Environment.* New Haven, Conn., Yale Univ. Press.

—— and Gibor, A. 1970. "The oxygen cycle." *Sci. Amer.,* 223: 111–123.

Cloud, P. 1974. "Evolution of ecosystems." *Amer. Sci.,* 62: 54–66.

Cockerell, T. D. A. 1934. "'Mimicry' among insects." *Nature,* 133: 329–330.

Cody, M. L. 1966. "A general theory of clutch size." *Evolution,* 20: 174–184.

—— 1968. "On the methods of resource division in grassland bird communities." *Amer. Nat.,* 102: 107–147.

—— 1969. "Convergent characteristics in sympatric species: A possible relation to interspecific competition and aggression." *Condor,* 71: 222–239.

—— 1970. "Chilean bird distribution." *Ecology,* 51: 455–464.

—— 1971. "Finch flocks in the Mohave desert." *Theoret. Pop. Biol.,* 2: 142–158.

——— 1971a. "Ecological aspects of reproduction," in D. S. Farner and J. R. King (eds.), *Avian Biology.* Vol. I. New York, Academic Press.

——— 1974. *Competition and the Structure of Bird Communities.* Princeton, N.J., Princeton Univ. Press.

——— 1974a. "Optimization in ecology." *Science,* 183: 1156–1164.

——— and Brown, J. H. 1969. "Song asynchrony in neighboring bird species." *Nature,* 222: 778–780.

Cody, M. L., and Diamond, J. M. (eds.). 1975. *Ecology and Evolution of Communities.* Cambridge, Mass., Harvard Univ. Press.

Coe, W. R. 1944. "Sexual differentiation in molluscs, II: Gastropods, amphineurans, scaphopods, and cephalopods." *Quart. Rev. Biol.,* 19: 85–97.

——— 1953. "Influence of association, isolation, and nutrition on the sexuality of snails of the genus *Crepidula.*" *J. Exptl. Zool.,* 122: 5–19.

Cogger, H. G. 1974. "Thermal relations of the mallee dragon *Amphibolurus fordi* (Lacertilia: Agamidae)." *Aust. J. Zool.,* 22: 319–339.

Cohen, J. E. 1966. *A Model of Simple Competition.* Cambridge, Mass., Harvard Univ. Press.

——— 1968. "Alternate derivations of a species-abundance relation." *Amer. Nat.,* 102: 165–172.

Cole, L. C. 1951. "Population cycles and random oscillations." *J. Wildl. Mgmt.,* 15: 233–252.

——— 1954. "The population consequences of life history phenomena." *Quart. Rev. Biol.,* 29: 103–137.

Comfort, A. 1956. *The Biology of Senescence.* New York, Rinehart.

——— 1962. "Survival curves of some birds in the London Zoo." *Ibis,* 104: 115–117.

Conder, P. J. 1949. "Individual distance." *Ibis,* 91: 649–655.

Connell, J. H. 1961. "The influence of interspecific competition and other factors on the distribution of the barnacle *Chthamalus stellatus.*" *Ecology,* 42: 710–723.

——— 1978. "Diversity in tropical rainforests and coral reefs." *Science,* 199: 1302–1310.

——— and Orias, E. 1964. "The ecological regulation of species diversity." *Amer. Nat.,* 98: 399–414.

Conover, R. J. 1966a. "Assimilation of organic matter by zooplankton." *Limnol. Oceanogr.,* 11: 338–345.

——— 1966b. "Factors affecting the assimilation of organic matter by zooplankton and the question of superfluous feeding." *Limnol. Oceanogr.,* 11: 346–354.

Cook, L. M., Askew, R. R., and Bishop, J. A. 1970. "Increasing frequency of the typical form of the peppered moth in Manchester." *Nature,* 227: 1155.

Cook, R. E. 1969. "Variation in species density in North American birds." *Syst. Zool.,* 18: 63–84.

Cooke, F. and McNally, C. M. 1975. "Mate selection and colour preferences in lesser snow geese." *Behaviour,* 53: 151–170.

Cooper, C. F. 1961. "The ecology of fire." *Sci. Amer.,* 204: 150–160.

Cooper, R. A., and Uzmann, J. R. 1971. "Migration and growth of deep-sea lobsters, *Homarus americanus.*" *Science,* 177: 288–290.

Cope, E. D. 1896. *Primary Factors of Organic Evolution.* Chicago, Open Court Publ. Co.

Corben, C. J., Ingram, G. J., and Tyler, M. J. 1974. "Gastric brooding: Unique form of parental care in an Australian frog." *Science,* 186: 946–947.

Cott, H. B. 1940. *Adaptive Coloration in Animals.* London, Methuen.

Coulson, J. C. 1968. "Differences in the quality of birds nesting in the center and on the edges of a colony." *Nature,* 217: 478–479.

Cowles, H. C. 1899. "The ecological relations of the vegetation on the sand dunes of Lake Michigan." *Bot. Gaz.,* 27: 95–117, 167–202, 281–308, 361–391.

Cowles, R. B., and Bogert, C. M. 1944. "A preliminary study of the thermal requirements of desert reptiles." *Bull. Amer. Mus. Nat. Hist.,* 83: 265–296.

Cox, G. W., and Ricklefs, R. E. 1977. "Species diversity, ecological release, and community structuring in Caribbean land bird faunas." *Oikos,* 29: 60–66.

Crane, J. 1949a. "Comparative biology of salticid spiders at Rancho Grande, Venezuela, Part III: Systematics and behavior in representative new species." *Zoologica,* 34: 31–52.

————— 1949b. "Comparative biology of salticid spiders at Rancho Grande, Venezuela, Part IV: An analysis of display." *Zoologica,* 34: 159–214.

Cranmer, M. F., and Turner, B. L. 1967. "Systematic significance of lupine alkaloids with particular reference to *Baptisia* (Leguminosae)." *Evolution,* 21: 508–517.

Creighton, C. 1891. *A History of Epidemics in Britain from A.D. 664 to the Extinction of Plague.* Cambridge, Cambridge Univ. Press.

Croat, T. B. 1969. "Seasonal flowering behavior in central Panama." *Ann. Missouri Bot. Gard.,* 56: 295–307.

Crocker, R. L. 1952. "Soil genesis and the pedogenic factors." *Quart. Rev. Biol.,* 27: 139–168.

————— and Major, J. 1955. "Soil development in relation to vegetation and surface age at Glacier Bay, Alaska." *J. Ecol.,* 43: 427–448.

Croghan, P. C. 1958. "The osmotic and ionic regulation of *Artemia salina* (L.)." *J. Exptl. Biol.,* 35: 219–233.

Crook, J. H. 1964. "The evolution of social organisation and visual communication in the weaver birds (*Ploceinae*)." *Behav. Suppl.,* 10: 1–178.

————— 1966. "Gelada baboon herd structure and movement." *Symp. Zool. Soc. Lond.,* 18: 237–258.

————— 1970. "The socio-ecology of primates," in J. H. Crook (ed.), *Social Behaviour in Birds and Mammals.* London and New York, Academic Press.

Crosby, J. L. 1949. "Selection of an unfavorable complex." *Evolution,* 3: 212–230.

————— 1959. "Outcrossing on homostyle primrose." *Heredity,* 13: 127–131.

Crow, J. F. 1957. "Genetics of insect resistance to chemicals." *Ann. Rev. Entomol.,* 2: 227–246.

————— 1970. "Genetic loads and the cost of natural selection," in K. Kojima (ed.), *Mathematical Topics in Population Genetics.* New York, Springer-Verlag.

————— and Kimura, M. 1965. "Evolution in sexual and asexual populations." *Amer. Nat.,* 99: 439–450.

————— 1969. "Evolution in sexual and asexual populations: A reply." *Amer. Nat.,* 103: 89–91.

————— 1970. *An Introduction to Population Genetics Theory.* New York, Harper and Row.

Crowell, K. L. 1962. "Reduced interspecific competition among the birds of Bermuda." *Ecology,* 43: 75–88.

————— 1973. "Experimental zoogeography: Introductions of mice to small islands." *Amer. Nat.,* 107: 535–558.

Cruz, A. A. de la, and Wiegert, R. G. 1967. "32-phosphorus tracer studies of a horse-weed–aphid–ant food chain." *Amer. Midl. Nat.,* 77: 501–509.

Cullen, J. M., Shaw, E., and Baldwin, H. A. 1965. "Methods for measuring the three-dimensional structure of fish schools." *Anim. Behav.,* 13: 534–543.

Culver, D. C. 1974. "Species packing in Caribbean and north temperate ant communities." *Ecology,* 55: 974–988.

Cummins, K. W., Klug, J. J., Metzel, R. G., Petersen, R. C., Suberkropp, K. F., Manny, B. A., Wuycheek, J. W., and Howard, F. O. 1972. "Organic enrichment with leaf leachate in experimental lotic ecosystems." *Bio. Science,* 22: 719–722.

Curtis, J. T., and McIntosh, R. P. 1951. "An upland forest continuum in the prairie-forest border region of Wisconsin." *Ecology,* 32: 476–496.

Damian, R. T. 1964. "Molecular mimicry: Antigen sharing by parasite and host and its consequences. *Amer. Nat.,* 98: 129–147.

Dammerman, K. W. 1948. "The fauna of Krakatoa. 1883–1933." *Verhandel. Koninkl. Akad. Wetenschap. Afdel. Natuurk.,* 44: 1–594.

Dansereau, P. 1957. *Biogeography—An Ecological Perspective.* New York, Ronald Press.

Darlington, P. J., Jr. 1957. *Zoogeography: The Geographical Distribution of Animals.* New York, Wiley.

Darnell, R. M. 1970. "Evolution and the ecosystem." *Amer. Zool.,* 10: 9–15.

Darwin, C. R. 1859. *On the Origin of Species.* London, Murray.

—— 1877. *The Different Forms of Flowers on Plants of the Same Species.* London, Murray.

Dasmann, R. F. 1964. *Wildlife Biology.* New York, Wiley.

Daubenmire, R. 1968. "Ecology of fire in grasslands." *Adv. Ecol. Res.,* 5: 209–266.

—— 1968. *Plant Communities: A Textbook of Plant Synecology.* New York, Harper and Row.

—— and Prusso, D. C. 1963. "Studies of the decomposition rates of tree litter." *Ecology,* 44: 589–592.

Davidson, J. 1938. "On the growth of the sheep population in Tasmania." *Trans. R. Soc. South Australia,* 62: 342–346.

—— and Andrewartha, H. G. 1948. "The influence of rainfall, evaporation, and the atmospheric temperature on fluctuations in the size of a natural population of *Thrips imaginis* (Thysanoptera)." *J. Anim. Ecol.,* 17: 193–199.

Davis, C. C. 1964. "Evidence for the eutrophication of Lake Erie from phytoplankton records." *Limnol. Oceanogr.,* 9: 275–283.

Davis, D. E. 1942. "The phylogeny of social nesting habits on the Crotophaginae." *Quart. Rev. Biol.,* 17: 115–134.

—— and Golley, F. B. 1963. *Principles in Mammology.* New York, Reinhold.

Davis, H. C. 1958. "Survival and growth of clam and oyster larvae at different salinities." *Biol. Bull.,* 114: 296–307.

Dawood, M. M., and Strickburger, M. W. 1969. "The effect of larval interaction on viability in *Drosophila melanogaster.*" *Genetics,* 63: 201–220.

Dawson, P. S. 1966. "Correlated responses to selection for developmental rate in *Tribolium Genetica.*" *Genetica,* 37: 63–77.

—— 1967. "Developmental rate and competitive ability in *Tribolium.* II: Changes in competitive ability following further selection for developmental rate." *Evolution,* 21: 292–298.

Day, B. W. 1964. "The white-tailed deer in Vermont 1964." *Vermont Fish Game Dept. Wildl. Bull.,* 641: 1–24.

Dayton, P. K. 1971. "Competition, disturbance, and community organization: The provision and subsequent utilization of space in a rocky intertidal community." *Ecol. Monogr.,* 41: 351–389.

DeBach, P. 1966. "The competitive displacement and coexistence principles." *Ann. Rev. Entomol.,* 11: 183–212.

—— 1974. *Biological Control by Natural Enemies.* London, Cambridge Univ. Press.

—— and Sundby, R. A. 1963. "Competitive displacement between ecological homologues." *Hilgardia,* 34: 105–166.

Deevey, E. S., Jr. 1942. "A re-examination of Thoreau's 'Walden.'" *Quart. Rev. Biol.,* 17: 1–11.

—— 1947. "Life tables for natural populations of animals." *Quart. Rev. Biol.,* 22: 283–314.

—— 1950. "The probability of death." *Sci. Amer.,* 182: 58–60.

Deleurance-Glaçon, S. 1963. "Recherches sur les coléoptères troglobies de la sous-famille des Bathysciinae." *Ann. Sci. Nat. (Zool.),* Sér. 12, 5: 1–172.

Delwiche, C. C. 1970. "The nitrogen cycle." *Sci. Amer.,* 223: 137–146.

Dempster, J. P. 1963. "The population dynamics of grasshoppers and locusts." *Biol. Rev.,* 38: 490–529.

Denton, E. 1960. "The buoyancy of marine animals." *Sci. Amer.,* 203: 119–128.

Dessauer, H. C., and Nevo, E. 1969. "Geographic variation of blood and liver proteins in cricket frogs." *Biochem. Genet.,* 3: 171–188.

DeVore, I., and Washburn, S. L. 1963. "Baboon ecology and human evolution," in F. C. Howell and F. Bourliere (eds.), *African Ecology and Human Evolution.* Chicago, Aldine.

DeVries, A. L. 1971. "Glycoproteins as biological antifreeze agents in Antarctic fishes." *Science,* 172: 1152–1155.

DeWitt, C. B. 1967. "Precision of thermo-regulation and its relation to environmental factors in the desert iguana, *Dipsosaurus dorsalis." Physiol. Zool.,* 40: 49–66.

Diamond, J. M. 1966. "Zoological classification system of primitive people." *Science,* 151: 1102–1104.

—— 1969. "Avifauna equilibria and species turnover rates on the Channel Islands of California." *Proc. Natl. Acad. Sci.,* 67: 1715–1721.

—— 1971. "Comparison of faunal equilibrium turnover rates on a tropical island and a temperate island." *Proc. Nat. Acad. Sci.,* 68: 2742–2745.

—— 1973. "Distributional ecology of New Guinea birds." *Science,* 179: 759–769.

Dickson, R. C. 1940. "Inheritance of resistance to hydrocyanic acid fumigation in the California red scale." *Hilgardia,* 13: 515–522.

—— 1949. "Factors governing the induction of diapause in the Oriental fruit moth." *Ann. Ent. Soc. Amer.,* 42: 511–537.

Dijkgraaf, S. 1963. "The functions and significance of the lateral line organs." *Biol. Rev.,* 38: 51–106.

Dilger, W. C. 1960. "The comparative ethology of the African parrot genus *Agapornis. Zeit. Tierpsychol.,* 17: 649–685.

—— 1962. "The behavior of lovebirds." *Sci. Amer.,* 206: 88–98.

Dillon, L. S. 1966. "The life cycle of the species: An extension of current concepts." *Syst. Zool.,* 15: 112–126.

Dingle, H. 1968. "Life history and population consequences of density, photoperiod, and temperature in a migrant insect, the milkweed bug, *Oncopeltus." Amer. Nat.,* 102: 149–163.

—— 1972. "Migration strategies of insects." *Science,* 175: 1327–1335.

Dix, R. L. 1957. "Sugar maple in forest succession at Washington, D.C." *Ecology,* 38: 663–665.

Dixon, A. F. G. 1959. "An experimental study of the searching behaviour of the predatory coccinellid beetle *Adalia decempunctata* (L.)." *J. Anim. Ecol.,* 28: 259–281.

Dobzhansky, T. 1937. *Genetics and the Origin of Species.* New York, Columbia Univ. Press.

—— 1947. "Adaptive changes induced by natural selection in wild populations of *Drosophila." Evolution,* 1: 1–16.

—— 1948. "Genetics of natural populations, XVI: Altitudinal and seasonal changes produced by natural selection in certain populations of *Drosophila persimilis." Genetics,* 33: 158–176.

—— 1950. "Evolution in the tropics." *Amer. Sci.,* 38: 209–221.

—— 1951. *Genetics and the Origin of Species* (rev. 3d ed.). New York, Columbia Univ. Press.

—— 1970. *Genetics of the Evolutionary Process.* New York, Columbia Univ. Press.

—— and Ayala, F. J. 1973. "Temporal frequency changes of enzyme and chromosomal polymorphisms in natural populations of *Drosophila." Proc. Nat. Acad. Sci.,* 70: 680–683.

Dobzhansky, T., and Epling, C. 1944. "Contributions to the genetics, taxonomy, and ecology of *Drosophila pseudoobscura* and its relatives." *Carnegie Inst. Wash. Publ.,* 554: 1–183.

Dobzhansky, T., and Pavlovsky, O. 1957. "An experimental study of interaction between genetic drift and natural selection." *Evolution,* 11: 311–319.

Dobzhansky, T., and Spaasky, B. 1963. "Genetics of natural populations, XXXIV: Adaptive norm, genetic load and genetic elite in *Drosophila pseudoobscura.*" *Genetics,* 48: 1467–1485.

Docters van Leeuwen, W. M. 1936. "Krakatau, 1883–1933." *Ann. Jard. Botan. Buitenzorg,* 56–57: 1–506.

Dodd, A. P. 1959. "The biological control of prickly pear in Australia," in A. Keast, R. L. Crocker, and C. S. Christian (eds.), *Biogeography and Ecology in Australia. Monogr. Biol.,* Vol. VIII.

Dodson, C. H., Dressler, R. L., Hills, H. G., Adams, R. M., and Williams, N. H. 1969. "Biologically active compounds in orchid fragrances." *Science,* 164: 1243–1249.

Dodson, C. H., and Frymire, G. P. 1961. "Natural pollination of orchids." *Missouri Bot. Gard. Bull.,* 49: 133–152.

Dolinger, P. M., Ehrlich, P. R., Fitch, W. L., and Breedlove, D. E. 1973. "Alkaloids and predation patterns in Colorado lupine populations." *Oecologia,* 13: 191–204.

Dolnik, V. R. 1967. "Metabolism and bird migration." *Coll. Inter. C. N. R. C. Photoreg. Reprod., Montpellier,* 16–22/7/67.

Dorf, E. 1964. "The use of fossil plants in paleoclimate interpretations," in P. S. Martin (ed.), *Problems in Paleoclimatology.* London, Interscience.

Dorney, R. S. 1969. "Epizootiology of trypanosomes in red squirrels and eastern chipmunks." *Ecology,* 50: 817–824.

Doutt, R. L. 1960. "Natural enemies and insect speciation." *Pan-Pacific Entomol.,* 36: 1–13.

Downhower, J. F., and Armitage, K. B. 1971. "The yellow-bellied marmot and the evolution of polygamy." *Amer. Nat.,* 105: 355–370.

Dressler, R. L. 1968. "Pollination by euglossine bees." *Evolution,* 22: 202–210.

Dritschilo, W., Cornell, H., Nafus, D., and O'Connor, B. 1975. "Insular biogeography: Of mice and mites." *Science,* 190: 467–469.

Drury, W. H., and Nisbet, I. C. T. 1973. "Succession." *J. Arnold Arboretum,* 54: 331–368.

Dubinin, N. P., and Tiniakov, G. G. 1946. "Structural chromosome variability in urban and rural populations of *Drosophila funebris.*" *Amer. Nat.,* 80: 393–396.

Dublin, L. I., and Lotka, A. J. 1935. *Length of Life: A Study of the Life Table.* New York, Ronald Press.

Dumond, D. E. 1975. "The limitation of human population: A natural history." *Science,* 187: 713–721.

Dunbar, C. O. 1960. *Historical Geology,* 2d ed. New York, Wiley.

Dunbar, M. J. 1960. "The evolution of stability in marine environments: Natural selection at the level of the ecosystem." *Amer. Nat.,* 94: 129–136.

Dunbar, R. I. M., and Crook, J. H. 1975. "Aggression and dominance in the weaverbird, *Quelea quelea.*" *Anim. Behav.,* 23: 450–459.

Duncan, D. P., and Hobson, A. C. 1958. "Influence of the forest tent caterpillar upon the aspen forests of Minnesota." *For. Sci.,* 4: 71–93.

Dunn, L. C., and Dobzhansky, T. 1946 (rev. 1957). *Heredity, Race and Society.* Mentor Books. New York, New American Library.

Durkee, L. H. 1971. "A pollen profile from Woden Bog in north-central Iowa." *Ecology,* 52: 837–844.

Duvigneaud, P., and Denaeyer-de Smet, S. 1970. "Biological cycling of minerals in temperate deciduous forests," in D. E. Reichle (ed.), *Analysis of Temperate Forest Ecosystems.* New York, Springer-Verlag.

Duvigneaud, P., and Froment, A. 1969. "Recherches sur l'ecosystème forêt, Serie E: Forêts de Haute-Belgique. Eléments biogènes de l'édaphotope et phytocénose forestière." *Bull. Inst. R. Sci. Nat. Belg.,* 45: 1–48.

Duxbury, A. C. 1971. *The Earth and Its Oceans.* Reading, Mass., Addison-Wesley.

Dymond, J. R. 1947. "Fluctuations in animal populations with special reference to those of Canada." *Trans. R. Soc. Canada,* 41: 1–34.

East, E. M. 1910. "A Mendelian interpretation of variation that is apparently continuous." *Amer. Nat.,* 44: 65–82.

Eaton, J. S., Likens, G. E., and Bormann, H. F. 1972. "Throughfall and stemflow chemistry in a northern hardwood forest." *J. Ecol.,* 61: 495–508.

Eberhard, M. J. W. 1975. "The evolution of social behavior by kin selection." *Quart. Rev. Biol.,* 50: 1–39.

Edmonds, V. W. 1968. "Survival of the cutthroat finch (*Amadina fasciate*) under desert conditions without water." *Auk,* 85: 326–328.

Edmondson, W. T. 1945. "Ecological studies of sessile rotatoria, Part II: Dynamics of populations and social structures." *Ecol. Monogr.,* 15: 141–172.

—— 1955. "Factors affecting productivity in fertilized salt water." *Pap. Mar. Biol. Oceanogr., Deep-Sea Res. Suppl,* 3: 451–465.

—— 1970. "Phosphorus, nitrogen, and algae in Lake Washington after diversion of sewage." *Science,* 169: 690–691.

Edwards, A. W. F. 1960. "Natural selection and the sex ratio." *Nature,* 188: 960–961.

Edwards, C. A., and Heath, G. W. 1963. "The role of soil animals in breakdown of leaf material," in J. Doeksen and J. van der Drift (eds.), *Soil Organisms.* Amsterdam, North-Holland Publ. Co.

Eggler, W. A. 1938. "The maple-basswood forest type in Washburn County, Wis." *Ecology,* 19: 243–263.

Ehrlich, P. R. 1961. "Intrinsic barriers to dispersal in checkerspot butterfly." *Science,* 134: 108–109.

—— 1965. "The population biology of the butterfly *Euphydrias editha,* II: The structure of the Jasper Ridge colony." *Evolution,* 19: 327–336.

—— and Birch, L. C. 1967. "The 'balance of nature' and 'population control.'" *Amer. Nat.,* 101: 97–107.

Ehrlich, P. R., and Holm, R. W. 1963. *The Process of Evolution.* New York, McGraw-Hill.

Ehrlich, P. R., and Raven, P. H. 1969. "Differentiation of populations." *Science,* 165: 1228–1232.

Ehrlich, P. R., White, R. R., Singer, M. C., McKechnie, S. W., and Gilbert, L. E. 1975. "Checkerspot butterflies: A historical perspective." *Science,* 188: 221–228.

Ehrman, L. 1967. "Further studies on genotype frequency and mating success in *Drosophila.*" *Amer. Nat.,* 101: 415–424.

Eickmeier, W., Adams, M., and Lester, D. 1975. "Two physiological races of *Tsuga canadensis.*" *Canad. J. Bot.,* 53: 940–951.

Einarsen, A. S. 1942. "Specific results from ring-necked pheasant studies in the Pacific northwest." *Trans. North Amer. Wildl. Conf.,* 7: 130–145.

—— 1945. "Some factors affecting ring-necked pheasant population density." *Murrelet,* 26: 39–44.

Eisner, T., and Meinwald, J. 1966. "Defensive secretions of arthropods." *Science,* 153: 1341–1350.

Eisner, T., Silberglied, R. E., Aneshansley, D., Carrel, J. E., and Howland, H. C. 1969. "Ultraviolet video-viewing: The television camera as an insect eye." *Science,* 166: 1172–1174.

Eisner, T., Johnessee, J. S., Carrel, J., Hendry, L. B., and Meinwald, J. 1974. "Defensive use by an insect of a plant resin." *Science,* 184: 996–999.

Elton, C. 1924. "Periodic fluctuations in the numbers of animals: Their causes and effects." *Brit. J. Exptl. Biol.,* 2: 119–163.

——— 1927. *Animal Ecology.* New York, Macmillan.

——— 1942. *Voles, Mice and Lemmings. Problems in Population Dynamics.* Clarendon Press. Oxford, Oxford Univ. Press.

——— and Nicholson, M. 1942a. "Fluctuations in numbers of the muskrat (*Ondatra zibethica*) in Canada." *J. Anim. Ecol.,* 11: 96–125.

——— 1942b. "The ten-year cycle in numbers of the lynx in Canada." *J. Anim. Ecol.,* 11: 215–244.

El-Wailly, A. J. 1966. "Energy requirements for egg-laying and incubation in the Zebra finch, *Taeniopygia castanotis.*" *Condor,* 68: 582–594.

Emberger, S. 1965. "Die Strickstoffvorräte bayrischer Waldböden." *Forstwiss. Cbl.,* 84: 156–193.

Emerson, R., and Lewis, C. M. 1942. "The photosynthetic efficiency of phycocyanin in *Chroococcus,* and the problem of carotenoid participation in photosynthesis." *J. Gen. Physiol.,* 25: 579–595.

Emerson, S. 1939. "A preliminary survey of the *Oenothera organensis* population." *Genetics,* 24: 524–537.

Emlen, J. M. 1966. "The role of time and energy in food preference." *Amer. Nat.,* 100: 611–617.

——— 1970. "Age specificity and ecological theory." *Ecology,* 51: 588–601.

Emlen, J. T. 1978. "Density anomalies and regulatory mechanisms in land bird populations on the Florida peninsula." *Amer. Nat.,* 112: 265–286.

Emlen, S. T. 1967a. "Migratory orientation in the indigo bunting, *Passerina cyanea,* Part I: Evidence for use of celestial cues." *Auk,* 84: 309–342.

——— 1967b. "Migratory orientation in the indigo bunting, *Passerina cyanea,* Part II: Mechanism of celestial orientation." *Auk,* 84: 463–489.

——— 1969. "Bird migration: Influence of physiological state upon celestial orientation." *Science,* 165: 716–718.

——— and Ambrose, H. W., III, 1970. "Feeding interactions of snowy egrets and red-breasted mergansers." *Auk.,* 87: 164–165.

Endler, J. A. 1973. "Gene flow and population differentiation." *Science,* 179: 243–250.

Engelmann, M. D. 1961. "The role of soil arthropods in the energetics of an old field community." *Ecol. Monogr.,* 31: 221–238.

Epling, C., and Dobzhansky, T. 1942. "Genetics of natural population, VI: Microgeographic races in *Linanthus parryae.*" *Genetics,* 27: 317–332.

Epling, C. H., Lewis, H., and Ball, F. M. 1960. "The breeding group and seed storage: A study in population dynamics." *Evolution,* 14: 238–255.

Erickson, R. O. 1945. "The *Clematis fremontii* var. *riehlii* population in the Ozarks." *Ann. Mo. Bot. Gard.,* 32: 413–460.

Erpino, M. J. 1968. "Nest-related activities of black-billed magpies." *Condor,* 70: 154–165.

Errington, P. L. 1946. "Predation and vertebrate populations." *Quart. Rev. Biol.,* 21: 144–177, 221–245.

——— 1963. "The phenomenon of predation." *Amer. Sci.,* 51: 180–192.

Espenshade, E. B., Jr., ed. 1971. *Goode's World Atlas,* 13th ed. Chicago, Rand-McNally.

Eyre, S. R. 1968. *Vegetation and Soils: A World Picture,* 2d ed. Chicago, Aldine.

Faegri, K., and van der Pijl, L. 1971. *The Principles of Pollination Ecology,* 2d ed. New York, Pergamon.

Fager, E. W. 1972. "Diversity: A sampling study." *Amer. Nat.,* 106: 293–310.

Falconer, D. S. 1960. *Introduction to Quantitative Genetics.* New York, Ronald Press.

Falls, J. B. 1969. "Functions of territorial song in the white-throated sparrow, in R. A. Hinde (ed.), *Bird Vocalisations.* London and New York, Cambridge Univ. Press.

Farner, D. S. 1955. "Bird banding in the study of population dynamics," in A. Wolfson (ed.), *Recent Studies in Avian Biology.* Urbana, Univ. of Illinois Press.

—— 1964. "The photoperiodic control of reproductive cycles in birds." *Amer. Sci.,* 52: 137–156.

Farner, D. S., and King, J. R. (eds.). 1971. *Avian Biology,* Vol. I. New York and London, Academic Press.

Feder, H. M. 1966. "Cleaning symbiosis in the marine environment," in M. S. Henry (ed.), *Symbiosis,* Vol. I. New York, Academic Press.

Fedejer, C. A., and Tanner, C. B. 1966. "Spectral distributions of light in the forest." *Ecology,* 47: 555–560.

Feeney, P. P. 1968. "Effect of oak leaf tannins on larval growth of the winter moth *Operophtera brumata.*" *J. Insect Physiol.,* 14: 805–817.

—— 1969. "Inhibitory effect of oak leaf tannins on the hydrolysis of proteins by trypsin." *Phytochem.,* 8: 2119–2126.

—— 1970. "Seasonal changes in oak leaf tannins and nutrients as a cause of spring feeding by winter moth caterpillars." *Ecology,* 51: 565–581.

—— 1975. "Biochemical evolution between plants and their insect herbivores," in L. E. Gilbert and P. H. Raven (eds.), *Coevolution of Animals and Plants.* Austin, Univ. of Texas Press.

—— and Bostock, H. 1968. "Seasonal changes in the tannin content of oak leaves." *Phytochem.,* 7: 871–880.

Fenner, F. 1971. "Evolution in action: Myxomatosis in the Australian wild rabbit," in A. Kramer, ed., *Topics in the Study of Life. The Bio Source Book.* New York, Harper and Row.

—— and Ratcliffe, F. N. 1965. *Myxomatosis.* London and New York, Cambridge Univ. Press.

Fenton, M. B. 1972. "The structure of aerial-feeding bat faunas as indicated by ears and wing elements." *Canad. J. Zool.,* 50: 287–296.

Findley, J. S. 1973. "Phenetic packing as a measure of faunal diversity." *Amer. Nat.,* 107: 580–584.

—— 1976. "The structure of bat communities." *Amer. Nat.,* 110: 129–139.

Finerty, J. P. 1971. "Cyclic fluctuations in biological systems: A reevaluation." Unpublished Ph.D. dissertation, Yale University.

Fischer, A. G. 1960. "Latitudinal variation in organic diversity." *Evolution,* 14: 64–81.

Fisher, J., and Hinde, R. A. 1949. "The opening of milk bottles by birds." *Brit. Birds,* 42: 347–357.

Fisher, R. A. 1928. "The possible modifications of the response of the wild type to recurrent mutations." *Amer. Nat.,* 62: 115–126.

—— 1930. *The Genetical Theory of Natural Selection.* London, Oxford Univ. Press.

—— 1931. "The evolution of dominance." *Biol. Rev.,* 6: 345–368.

Fisher, S. G., and Likens, G. E. 1972. "Stream ecosystem: Organic energy budget." *Bio-Science,* 22: 33–35.

Fleming, T. H. 1975. "The role of small mammals in tropical ecosystems," in F. B. Golley, K. Petrusewicz, and L. Ryszkowski (eds.), *Small Mammals: Their Productivity and Population Dynamics.* Cambridge, Cambridge Univ. Press.

Flessa, K. W., Powers, K. V., and Cisne, J. L. 1975. "Specialization and evolutionary longevity in the Arthropoda." *Paleobiol.,* 1: 71–81.

Flohn, H. 1968. *Climate and Weather.* World Univ. Press. New York, McGraw-Hill.

Foin, T. C., Valentine, J. W., and Ayala, F. J. 1975. "Extinction of taxa and Van Valen's law." *Nature,* 257: 514–515.

Force, D. C. 1974. "Ecology of insect host-parasite communities." *Science*, 184: 624–632.

Forcier, L. K. 1975. "Reproductive strategies and the co-occurrence of climax tree species." *Science*, 189: 808–809.

Ford, E. B. 1971. *Ecological Genetics*, 3d ed. London, Chapman and Hall.

Fowells, H. A. 1965. *Silvics of Forest Trees of the United States*. Agric. Handbook No. 271. Washington, D.C., U.S. Dept. Agric.

Fox, R., Lehmkuhle, S. W., and Westendorf, D. H., 1976. "Falcon visual acuity." *Science*, 192: 263–266.

Fox, W. 1951. "Relationships among the garter snakes of the *Thamnophis elegans* Rassenkreis." *Univ. Calif. Publ. Zool.*, 50: 485–530.

Fraenkel, G. S. 1959. "The *raison d'être* of secondary plant substances." *Science*, 129: 1466–1470.

——— 1969. "Evaluation of our thoughts on secondary plant substances." *Ent. Exptl. Appl.*, 12: 474–486.

Fraenkel, G., and Gunn, D. L. 1940. *The Orientation of Animals: Kineses, Taxes and Compass Reactions*. Clarendon Press. Oxford, Oxford Univ. Press.

Frank, P. W., Boll, C. D., and Kelly, R. W. 1957. "Vital statistics of laboratory cultures of *Daphnia pulex* De Geer as related to density." *Physiol. Zool.*, 30: 287–305.

Fretter, V., and Graham, A. 1962. *British Prosobranch Molluscs. Their Functional Anatomy and Ecology*. London, Ray Society.

Friedmann, H. 1963. "Host relations of the parasitic cowbirds." *Bull. U.S. Natl. Museum*, 233: 1–276.

Frisancho, A. R. 1975. "Functional adaptation to high altitude hypoxia." *Science*, 187: 313–319.

Frith, H. J. 1967. *Waterfowl in Australia*. Sydney, Australia, Angus and Robertson.

Fritts, H. C. 1966. "Growth-rings of trees: Their correlation with climate." *Science*, 154: 973–979.

——— 1967. "Growth-rings of trees: A physiological basis for their correlation with climate," in *Ground Level Climatology*. Washington, D.C., Amer. Assoc. Adv. Sci.

Fry, C. H. 1972. "The social organisation of bee-eaters (Meropidae) and co-operative breeding in hot-climate birds." *Ibis*, 114: 1–14.

Fry, F. E. J., Brett, J. R., and Clawson, G. H. 1942. "Lethal limits of temperature for young goldfish." *Rev. Can. Biol.*, 1: 50–56.

Fry, F. E. J., and Hart, J. S. 1948. "Cruising speed of goldfish in relation to water temperature." *J. Fish. Res. Bd. Can.*, 7: 169–175.

Fryer, G. 1959. "Some aspects of evolution in Lake Nyasa." *Evolution*, 13: 440–451.

Fryxell, P. A. 1957. "Mode of reproduction in higher plants." *Bot. Rev.*, 23: 135–223.

Fuentes, E. R. 1976. "Ecological convergence of lizard communities in Chile and California." *Ecology*, 57: 3–18.

Gadgil, M., and Bossert, W. H. 1970. "Life historical consequences of natural selection." *Amer. Nat.*, 104: 1–24.

Gadgil, M., and Solbrig, O. T. 1972. "The concept of *r*- and *K*-selection: Evidence from wild flowers and some theoretical considerations." *Amer. Nat.*, 106: 14–31.

Gaines, M. S., and Krebs, C. J. 1971. "Genetic changes in fluctuating vole populations." *Evolution*, 25: 702–723.

Galindo, P., Trapido, H., Carpenter, S. J., and Blanton, F. S. 1956. "The abundance cycles of arboreal mosquitos during six years at a sylvan yellow fever locality in Panama." *Ann. Entomol. Soc. Amer.*, 49: 543–547.

Gans, C. 1974. *Biomechanics: An Approach to Vertebrate Biology*. Philadelphia, Lippincott.

Gates, D. M. 1963. "The energy environment in which we live." *Amer. Sci.,* 51: 327–348.

—— 1969. "Climate and stability." *Brookhaven Symp. Biol.,* 22: 115–126.

Gause, G. J. 1934. *The Struggle for Existence.* Baltimore, Williams and Wilkins.

Geiger, R. 1957. *The Climate Near the Ground,* 2d ed., revised by M. N. Stewart. Cambridge, Mass., Harvard Univ. Press.

Geist, V. 1966. "The evolution of hornlike organs." *Behaviour,* 27: 175–214.

—— 1971. *Mountain Sheep: A Study in Behavior and Evolution.* Chicago, Univ. of Chicago Press.

George, R. Y., and Rowe, G. T. 1973. *Abyssal Environment and Ecology of the World Oceans.* New York, Wiley.

Ghiselin, J., and Ricklefs, R. E. 1970. "Prey population: A parsimonius model for evolution of response to predator species diversity." *Science,* 170: 649–651.

Ghiselin, M. T. 1975. "The rationale of pangenesis." *Genetics,* 79: 47–57.

Gibb, J. A. 1958. "Predation by tits and squirrels on the eucosmid *Ernarmonia conicolana* (Heyl.)." *J. Anim. Ecol.,* 27: 375–396.

—— 1962. "L. Tinbergen's hypothesis of the role of specific search images." *Ibis,* 104: 106–111.

Gibbs, G. W. 1967. "The comparative ecology of two closely related sympatric species of *Dacus* (Diptera) in Queensland." *Aust. J. Zool.,* 15: 1123–1139.

—— 1968. "The frequency of interbreeding between two sibling species of *Dacus* (Diptera) in wild populations." *Evolution,* 22: 667–682.

Gibson, E. S., and Fry, F. E. J. 1954. "The performance of the lake trout, *Salvelinus namaycush,* at various levels of temperature and oxygen pressure." *Canad. J. Zool.,* 32: 252–260.

Gibson, T. G., and Buzas, M. A. 1973. "Species diversity: Patterns in modern and Miocene foraminifera of the eastern margin of North America." *Geol. Soc. Amer. Bull.,* 84: 217–238.

Geise, A. C. 1968. *Cell Physiology,* 3d ed. Philadelphia, Saunders.

Gifford, C. E., and Odum, E. P. 1965. "Bioenergetics of lipid deposition in the bobolink, a transequatorial migrant." *Condor,* 67: 383–403.

Gill, D. E. 1974. "Intrinsic rate of increase, saturation density, and competitive ability, II: The evolution of competitive ability." *Amer. Nat.,* 108: 103–116.

Gillespie, J. 1974. "Polymorphism in patchy environments." *Amer. Nat.,* 108: 145–151.

—— 1974a. "Natural selection for within-generation variance in offspring number." *Genetics,* 76: 601–606.

Gillespie, J. H. 1977. "Natural selection for variance in offspring numbers: A new evolutionary principle." *Amer. Nat.,* 111: 1010–1014.

—— and Kojima, K. 1968. "The degree of polymorphism in enzymes involved in energy production compared to that in nonspecific enzymes in two *Drosophila ananassae* populations." *Proc. Natl. Acad. Sci.,* 61: 582–585.

Gillespie, J. H., and Langley, C. H. 1974. "A general model to account for enzyme variation in natural populations." *Genetics,* 76: 837–884.

Gillette, J. B. 1962. "Pest pressure, an underestimated factor in evolution." *Syst. Assoc. Publ.,* 4: 37–46.

Gilliard, E. T. 1956. "Bower ornamentation versus plumage characters in bowerbirds." *Auk,* 73: 450–451.

—— 1962. "On the breeding behavior of the cock-of-the-rock (Aves, *Rupicola rupicola*). *Bull. Amer. Mus. Nat. Hist.,* 124: 35–68.

—— 1963. "The evolution of bowerbirds." *Sci. Amer.,* 209: 38–46.

Gilpin, M. E. 1973. "Do hares eat lynx?" *Amer. Nat.,* 107: 727–730.

—— 1974. "A model of the predator-prey relationship." *Theoret. Pop. Biol.,* 5: 333–344.

—— 1975. *Group Selection in Predator-Prey Communities.* Princeton, N.J., Princeton Univ. Press.

—— and Ayala, F. J. 1973. "Global models of growth and competition." *Proc. Nat. Acad. Sci.,* 70: 3590–3593.

Gilpin, M. E., and Justice, K. E. 1972. "Reinterpretation of the invalidation of the principle of competitive exclusion." *Nature,* 236: 273–274, 299–301.

—— 1973. "A note on nonlinear competition models." *Math. Biosci.,* 17: 57–63.

Gleason, H. A. 1926. "The individualistic concept of the plant association." *Torrey Bot. Club Bull.,* 53: 7–26.

—— 1929. "The significance of Raunkiaer's law of frequency." *Ecology,* 10: 406–408.

—— 1939. "The individualistic concept of the plant association." *Amer. Midl. Nat.,* 21: 92–110.

Glowacinski, Z., and Järvinen, O. 1975. "Rate of secondary succession in forest bird communities." *Ornis Scand.,* 6: 33–40.

Glynn, P. W. 1970. "On the ecology of the Caribbean chitons *Acanthopleura granulata* Gmelin and *Chiton tuberculatus* Linné: Density, mortality, feeding, reproduction, and growth." *Smithson. Contrib. Zool.,* No. 66.

Goddard, S. V., and Board, V. V. 1967. "Reproductive success of red-winged blackbirds in north central Oklahoma." *Wilson Bull.,* 79: 283–289.

Goldschmidt, R. B. 1948. "Ecotype, ecospecies and macroevolution." *Experientia,* 4: 465–472.

—— 1952. "Evolution as viewed by one geneticist." *Amer. Sci.,* 40: 84–98.

Golley, F. B. 1960. "Energy dynamics of a food chain of an old-field community." *Ecol. Monogr.,* 30: 187–206.

—— and Buechner, H. K. 1968. *A Practical Guide to the Study of the Productivity of Large Herbivores.* IBP Handbook No. 7. Oxford, Blackwell.

Golley, F. B., and Gentry, J. B. 1964. "Bioenergetics of the southern harvester ant, *Pogonomyrmex badius.*" *Ecology,* 43: 217–225.

—— 1966. "A comparison of variety and standing crop of vegetation on a one-year and a twelve-year abandoned field." *Oikos,* 15: 185–199.

Golley, F. B., and Misra, R. 1972. "Organic production in tropical ecosystems." *Bio. Science,* 22: 735–736.

Goodman, D. 1974. "Natural selection and a cost ceiling on reproductive effort." *Amer. Nat.,* 108: 247–268.

—— 1975. "Stability and diversity." *Quart. Rev. Biol.,* 50: 237–266.

Gordon, H. and Gordon, M. 1957. "Maintenance of polymorphism by potentially injurious genes in eight natural populations of the platyfish, *Xiphophorus maculatus.*" *J. Genetics,* 55: 1–44.

Gordon, M. S. 1968. *Animal Function: Principles and Adaptations.* New York, Macmillan.

Gosz, J. R., Likens, G. E., and Bormann, F. H. 1972. "Nutrient content of litter fall on the Hubbard Brook Experimental Forest, New Hampshire." *Ecology,* 53: 769–784.

Gould, H. 1952. "Studies on sex in the hermaphrodite mollusk *Crepidula plana.* II: Internal and external factors influencing sex development." *J. Exptl. Zool.,* 119: 93–163.

Goulden, C. E. 1969. "Temporal changes in diversity." *Brookhaven Symp. Biol.,* 22: 96–102.

Grant, P. R. 1966. "The density of land birds on the Tres Marias Islands in Mexico, II: Distribution of abundance in the community." *Canad. J. Zool.,* 44: 1023–1030.

—— 1972. "Interspecific competition among rodents." *Ann. Rev. Ecol. Syst.,* 3: 79–106.

Grant, V. 1951. "The fertilization of flowers." *Sci. Amer.* 184: 52–56.

—— 1963. *The Origin of Adaptations.* New York, Columbia Univ. Press.

—— 1975. *Genetics of Flowering Plants.* New York, Columbia Univ. Press.

—— and Grant, K. A. 1965. *Flower Pollination in the Phlox Family.* New York, Columbia Univ. Press.

Grassé, P.-P. 1951. *Traité de Zoologie.* Vol. X. *Insects Supérieurs et Hemipteroïdes. Part II.* Paris, Masson.

Gray, I. E. 1954. "Comparative study of the gill area of marine fishes." *Biol. Bull.,* 107: 219–225.

Green, G. J. 1971. "Physiologic races of wheat stem rust in Canada from 1919 to 1969." *Canad. J. Bot.,* 49: 1575–1588.

—— 1975. "Virulence changes in *Puccinia graminis* f. sp. *tritici* in Canada." *Canad. J. Bot.,* 53: 1377–1386.

Green, R. G., Larson, C. L., and Bell, J. F. 1939. "Shock disease as the cause of the periodic decimation of the snowshoe hare." *Amer. J. Hyg.,* 30, Sec. B: 83–102.

Greenberg, B. 1947. "Some relations between territory, social hierarchy, and leadership in the green sunfish, (*Lepomis cyanellus*)." *Physiol. Zool.,* 20: 267–299.

—— and Noble, G. K. 1944. "Social behavior of the American chameleon (*Anolis carolinensis* Voight)." *Physiol. Zool.,* 17: 392–439.

Greenslade, P. J. M. 1968. "Island patterns in the Solomon Islands bird fauna." *Evolution,* 22: 751–761.

—— 1969. "Land fauna: Insect distribution patterns in the Solomon Islands." *Phil. Trans. R. Soc. B,* 255: 271–284.

Grell, K. G. 1967. "Sexual reproduction in the protozoa," in T. Chen (ed.), *Research in Protozoology,* Vol. II. Oxford, Pergamon.

Griffin, D. M. 1972. *Ecology of Soil Fungi.* London, Chapman and Hall.

Grime, J. P., and Jeffrey, D. W. 1965. "Seedling establishment in vertical gradients of sunlight." *J. Ecol.,* 53: 621–642.

Grinnell, A. D. 1968. "Sensory physiology," in M. S. Gordon, *Animal Function: Principles and Adaptations.* New York, Macmillan.

—— 1970. "Comparative auditory neurophysiology of neotropical bats employing different echolocation signals." *Z. vergl. Physiol.,* 68: 117–153.

Grinnell, J. 1904. "The origin and distribution of the chestnut-backed chickadee." *Auk,* 21: 364–382.

—— 1917. "The niche-relationships of the California thrasher." *Auk,* 34: 427–433.

—— 1943. *Philosophy of Nature.* Berkeley, Univ. of California Press.

Grodzinski, W., and Wunder, B. A. 1975. "Ecological energetics of small mammals," in F. B. Golley, K. Petrusewicz, and L. Ryszkowski (eds.), *Small Mammals: Their Productivity and Population Dynamics.* Cambridge, Cambridge Univ. Press.

Gross, A. D. 1947. "Cyclic invasions of the snowy owl and the migration of 1945–1946." *Auk,* 64: 584–601.

Gross, F., and Zeuthen, E. 1948. "The buoyancy of plankton diatoms: A problem of cell physiology." *Proc. Roy. Soc. London B,* 135: 382–389.

Gulland, J. A. 1962. "The application of mathematical models to fish populations," in E. D. LeCren and M. W. Holdgate (eds.), *The Exploitation of Natural Animal Populations.* New York, Wiley.

Gunn, D. L. 1960. "The biological background of locust control." *Ann. Rev. Entomol.,* 5: 279–300.

Guppy, J. C. 1967. "Insect parasites of the army worm *Pseudaletia unipuncta* (Lepidoptera: Noctuidae) with notes on species observed in Ontario." *Canad. Entomol.,* 99: 94–106.

Guthrie, R. D. 1968. "Paleoecology of a late Pleistocene small mammal community from interior Alaska." *Arctic,* 21: 223–254.

—— and Petocz, R. G. 1970. "Weapon automimicry among mammals." *Amer. Nat.,* 104: 585–588.

Gwynne, M. O., and Bell, R. H. V. 1968. "Selection of vegetation components by grazing ungulates in the Serengeti National Park." *Nature,* 220: 390–393.

Hadley, N. F. 1970. "Desert species and adaptation." *Amer. Sci.,* 60: 338–347.

—— 1972. "Micrometeorology and energy exchange in two desert arthropods." *Ecology,* 51: 434–444.

Haffer, J. 1967. "Speciation in Colombian forest birds west of the Andes." *Amer. Mus. Novit.,* No. 2294: 1–57.

—— 1969. "Speciation in Amazonian forest birds." *Science,* 165: 131–137.

—— 1974. "Avian speciation in tropical South America." *Publ. Nuttall Ornithol. Club,* No. 14.

Hahn, W. E., and Tinkle, D. W. 1965. "Fat body cycling and experimental evidence for its adaptive significance to ovarian follicle development in the lizard *Uta stansburiana.*" *J. Exptl. Zool.,* 158: 79–86.

Hainsworth, F. R., and Wolf, L. L. 1970. "Regulation of oxygen consumption and body temperature during torpor in a hummingbird, *Eulampis jugularis.*" *Science,* 168: 368–369.

Hairston, N. G. 1959. "Species abundance and community organization." *Ecology,* 40: 404–416.

—— 1965. "On the mathematical analysis of schistosome populations." *Bull. World Health Org.,* 33: 45–62.

—— Smith, F. E., and Slobodkin, L. B. 1960. "Community structure, population control, and competition." *Amer. Nat.,* 94: 421–425.

Hairston, N. G., Tinkle, D. W., and Wilbur, H. M. 1970. "Natural selection and the parameters of population growth." *J. Wildl. Mgmt.,* 34: 681–690.

Haldane, J. B. S. 1927. "A mathematical theory of natural and artificial selection. Part IV." *Proc. Camb. Phil. Soc.,* 23: 607–615.

—— 1932. *The Causes of Evolution.* New York, London, and Toronto, Longmans, Green and Co.

—— 1939. "The theory of the evolution of dominance." *J. Genet.,* 37: 365–374.

—— 1957. "The cost of natural selection." *J. Genet.,* 55: 511–524.

—— and Huxley, J. 1927. *Animal Biology.* Clarendon Press. Oxford, Oxford Univ. Press.

Hale, M. E. 1967. *The Biology of Lichens.* London, Arnold.

Halkka, O., Raatikainen, M., and Halkka, L. 1974. "The founder principle, founder selection, and evolutionary divergence and convergence in natural populations of *Philaenus.*" *Hereditas,* 78: 73–84.

Hall, K. R. L. 1965. "Behaviour and ecology of the wild patas monkey, *Erythrocebus patas,* in Uganda." *J. Zool., Lond.,* 148: 15–87.

Hallam, A. (ed.). 1977. *Patterns of Evolution as Illustrated by the Fossil Record.* Amsterdam and New York, Elsevier.

Hamilton, A. G. 1950. "Further studies on the relation of humidity and temperature to the development of two species of African locusts—*Locusta migratoria migratorioides,* R. and F., and *Schistocerca gregaria,* Forsk." *Trans. Entomol. Soc. London,* 101: 2–56.

Hamilton, T. H., and Barth, R. H., Jr. 1962. "The biological significance of seasonal change in male plumage appearance in some New World migratory bird species." *Amer. Nat.,* 96: 129–144.

Hamilton, W. D. 1964. "The genetical evolution of social behaviour." *J. Theoret. Biol.,* 7: 1–52.

—— 1966. "The moulding of senescence by natural selection." *J. Theoret. Biol.,* 12: 12–45.

—— 1967. "Extraordinary sex ratios." *Science,* 156: 477–488.

—— 1971. "Geometry for the selfish herd." *J. Theoret. Biol.,* 31: 295–311.

—— 1972. "Altruism and related phenomena, mainly in social insects." *Ann. Rev. Ecol. Syst.,* 3: 193–232.

Hammel, H. T., Caldwell, F. T., and Abrams, R. M. 1967. "Regulation of body temperature in the blue-tongued lizard." *Science,* 156: 1260–1262.

Hanes, T. L. 1971. "Succession after fire in the chaparral of southern California." *Ecol. Monogr.,* 41: 27–52.

Harcourt, D. G. 1963. "Major mortality factors in the population dynamics of the diamondback moth *Plutella maculipennis* (Curt.)." *Mem. Entomol. Soc. Can.,* 32: 55–66.

—— and Leroux, E. J. 1967." Population regulation in insects and man." *Amer. Sci.,* 55: 400–415.

Hardin, G. 1960. "The competitive exclusion principle." *Science,* 131: 1292–1297.

Hardy, R. W., and Havelka, U. D. 1975. "Nitrogen fixation research: A key to world food?" *Science,* 188: 633–642.

Hargrave, B. T. 1971. "An energy budget for a deposit feeding amphipod." *Limnol. Oceanogr.,* 16: 99–103.

Harley, J. L. 1969. *The Biology of Mycorrhyzae.* London, Hill.

—— 1972. "Fungi in ecosystems." *J. Anim. Ecol.,* 41: 1–16.

Harper, F. 1945. *Extinct and Vanishing Mammals of the Old World.* New York, Amer. Comm. Int. Wildl. Prot.

Harper, J. L. 1961. "Approaches to the study of plant competition." *Symp. Soc. Exptl. Biol.,* 15: 1–39.

—— 1967. "A Darwinian approach to plant ecology." *J. Ecol.,* 55: 247–270.

—— 1969. "The role of predation in vegetational diversity." *Brookhaven Symp. Biol.,* 22: 48–62.

—— and White, J. 1970. "The dynamics of plant populations." *Proc. Adv. Study Inst. Dynamics Numbers Popul.,* Oosterbeek: 41–63.

Harper, J. L., Williams, J. T., and Sagar, G. R. 1965." The behaviour of seeds in soil." *J. Ecol.,* 51: 273–286.

Harris, H. 1966. "Enzyme polymorphisms in man." *Proc. R. Soc. London B,* 164: 298–310.

—— 1970. *Principles of Human Biochemical Genetics.* Amsterdam, North-Holland Publ. Co.

—— 1971. "Annotation: Polymorphism and protein evolution: The neutral mutation–random drift hypothesis." *J. Med. Gen.,* 8: 444–452.

—— and Hopkinson, D. A. 1972. "Average heterozygosity in man." *J. Hum. Genet.,* 36: 9–20.

Harrison, A. T., Small, E., and Mooney, H. A. 1971. "Drought relationships and distribution of two Mediterranean-climate California plant communities." *Ecology,* 52: 869–875.

Hart, J. S. 1957. "Climatic and temperature induced changes in the energetics of homeotherms." *Rev. Canad. Biol.,* 16: 133–171.

Haxo, F. T., and Blinks, L. R. 1950. "Photosynthetic action spectra of marine algae." *J. Gen. Physiol.,* 33: 389–422.

Hayes, F. R. 1949. "The growth, general chemistry, and temperature relations of salmonid eggs." *Quart. Rev. Biol.,* 24: 281–308.

Hayes, W. 1964. *The Genetics of Bacteria and Their Viruses.* New York, Wiley.

Heal, O. W., and French, D. D. 1974. "Decomposition of organic matter in tundra," in E. A. J. Holding *et al.* (eds.), *Soil Organisms and Decomposition in Tundra.* Stockholm, Tundra Biome Steering Committee.

Heath, J. E. 1965. "Temperature regulation and diurnal activity in horned lizards." *Univ. Calif. Publ. Zool.,* 64: 97–136.

—— 1967. "Temperature responses of the periodical '17-year' cicada, *Magicicada cassini." Amer. Midl. Nat.,* 77: 64–76.

—— and Adams, P. A. 1967. "Regulation of heat production by large moths." *J. Exptl. Biol.,* 47: 21–33.

Heath, J. E., and Wilkin, P. J. 1970. "Temperature responses of the desert cicada, *Diceroprocta apache* (Homoptera, Cicadiidae)." *Physiol. Zool.,* 43: 145–154.

Heatwole, H. 1970. "Thermal ecology of the desert dragon *Amphibolurus inermis." Ecol. Monogr.,* 40: 425–457.

—— and Levine, R. 1972. "Trophic structure, stability and faunal change during recolonization." *Ecology,* 53: 531–534.

Heinrich, B., and Bartholomew, G. A. 1971. "An analysis of pre-flight warm-up in the sphinx moth, *Manduca sexta." J. Exptl. Biol.,* 55: 223–239.

Hellmers, H. 1964. "An evaluation of the photosynthetic efficiency of forests." *Quart. Rev. Biol.,* 39: 249–257.

—— Horton, J. S., Juhren, G., and O'Keefe, J. 1955. "Root systems of some chaparral plants in southern California." *Ecology,* 36: 667–678.

Henderson, L. J. 1913. *The Fitness of the Environment.* New York, Macmillan.

Henny, C. J. 1972. "An analysis of the population dynamics of selected avian species." *Bur. Sport Fish. Wildl., Wildl. Res. Rept.,* 1: 1–99.

Hensley, M. M., and Cope, J. B. 1951. "Further data on removal and repopulation of the breeding birds in a spruce-fir forest community." *Auk,* 68: 483–493.

Hespenheide, H. A. 1971. "Food preference and the extent of overlap in some insectivorous birds, with special reference to the Tyrannidae." *Ibis,* 113: 59–72.

Hess, E. H. 1959. "Imprinting." *Science,* 130: 133–141.

—— 1964. "Imprinting in birds." *Science,* 146: 1128–1139.

Hett, J. M., and Loucks, O. L. 1971. "Sugar maple (*Acer saccharum* Marsh.) seedling mortality." *J. Ecol.* 59: 507–520.

Hickey, J. J. 1951. "Survival studies of banded birds." *U.S. Dept. Interior Spec. Scientific Report: Wildlife,* 15: 1–177.

Hickey, W. A., and Craig, G. B., Jr. 1966. "Genetic distortion of sex ratio in a mosquito, *Aedes aegypti." Genetics,* 53: 1177–1196.

Hiesey, W. M., and Milner, H. W. 1965. "Physiology of ecological races and species." *Ann. Rev. Plant Physiol.,* 16: 203–216.

Hildemann, W. H. 1959. "A cichlid fish, *Symphysodon discus,* with unique nuture habits." *Amer. Nat.,* 93: 27–34.

Hilden, O. 1965. "Habitat selection in birds: A review." *Ann. Zool. Fenn.,* 2: 53–75.

Hinde, R. A. 1956. "The biological significance of the territories of birds." *Ibis,* 98: 340–369.

—— and Fisher, J. 1952. "Further observation on the opening of milk bottles by birds." *Brit. Birds,* 44: 393–396.

Hirschfield, M. F., and Tinkle, D. W. 1975. "Natural selection and the evolution of reproductive effort." *Proc. Nat. Acad. Sci.,* 72: 2227–2231.

Hochachka, P. W., and Somero, G. N. 1973. *Strategies of Biochemical Adaptation.* Philadelphia, Saunders.

Hogan-Warburg, A. J. 1966. "Social behavior of the ruff, *Philomachus pugnax* (L.)." *Ardea,* 54: 109–225.

Hohn, E. O. 1969. "The phalarope." *Sci. Amer.,* 220: 105–111.

Holdridge, L. 1967. *Life Zone Ecology.* San Jose, Costa Rica, Tropical Science Center.

Holling, C. S. 1959. "The components of predation as revealed by a study of small mammal predation of the European pine sawfly." *Canad. Entomol.,* 91: 293–320.

―――― 1965. "The functional response of predators to prey density and its role in mimicry and population regulation." *Mem. Entomol. Soc. Can.,* 45: 5–60.

―――― 1966. "The functional response of invertebrate predators to prey density." *Mem. Entomol. Soc. Can.,* 48: 1–85.

Holm, C. H. 1973. "Breeding sex ratios, territoriality, and reproductive success in the red-winged blackbird (*Agelaius phoeniceus*)." *Ecology,* 54: 356–365.

Holmes, R. T. 1966. "Breeding ecology and annual cycle adaptations of the redbacked sandpiper (*Calidris alpina*) in northern Alaska." *Condor,* 68: 3–46.

Horn, H. S. 1966. "Measurement of 'overlap' in comparative ecological studies." *Amer. Nat.,* 100: 419–424.

―――― 1968. "The adaptive significance of colonial nesting in the Brewer's blackbird (*Euphagus cyanocephalus*)." *Ecology,* 49: 682–694.

―――― 1975. "Markovian properties of forest succession," in M. L. Cody and J. M. Diamond (eds.), *Ecology and Evolution of Communities.* Cambridge, Mass., Harvard Univ. Press.

Hornocker, M. G. 1970. "An analysis of mountain lion predation upon mule deer and elk in the Idaho Primitive Area." *Wildl. Monogr.,* 21: 3–39.

Howard, B. H. 1966. "Intestinal microorganisms of ruminants and other vertebrates," in M. S. Henry (ed.), *Symbiosis,* Vol. II. New York, Academic Press.

Howard, H. E. 1920. *Territory in Bird Life.* London, Murray.

Howard, L. O., and Fiske, W. F. 1911. "The importation into the United States of the parasites of the gypsy moth and the brown-tailed moth." *U.S. Dept. Agric. Bur. Ent. Bull.,* 91: 1–312.

Hubby, J. L., Lewontin, R. C. 1966. "A molecular approach to the study of genic heterozygosity in natural populations, I: The number of alleles at different loci in *Drosophila pseudoobscura.*" *Genetics,* 54: 577–594.

Hudson, J. W. 1962. "The role of water in the biology of the antelope ground squirrel, *Citellus leucurus.*" *Univ. of Calif. Publ. Zool.,* 64: 1–56.

Huey, R. B. 1974. "Behavioral thermoregulation in lizards: Importance of associated costs." *Science,* 184: 1001–1003.

Huffaker, C. B. 1958. "Experimental studies on predation: Dispersion factors and predator-prey oscillations." *Hilgardia,* 27: 343–383.

―――― 1970. "The phenomenon of predation and its roles in nature." *Proc. Adv. Study Inst. Dynamics Numbers Pop.,* Oosterbeek: 327–343.

―――― and Kennett, C. E. 1956. "Experimental studies on predation: Predation and cyclamen-mite populations on strawberries in California." *Hilgardia* 26: 191–222.

―――― 1959. "A ten-year study of vegetational changes associated with biological control of Klamath weed." *J. Range Mgmt.,* 12: 69–82.

―――― 1969. "Some aspects of assessing efficiency of natural enemies." *Canad. Entomol.,* 101: 425–447.

Huffaker, C. B., Shea, K. P., and Herman, S. G. 1963. "Experimental studies on predation: Complex dispersion and levels of food in an Acarine predator-prey interaction." *Hilgardia,* 34: 305–330.

Hughes, R. N. 1970. "An energy budget for a tidal-flat population of the bivalve *Scrobicularia plana* (Da Costa)." *J. Anim. Ecol.,* 39: 357–381.

Hunsaker, D., II. 1962. "Ethological isolating mechanisms in the *Sceloporus torquatus* group of lizards." *Evolution,* 16: 62–74.

Hunt, E. G., and Bischoff, A. I. 1960. "Inimical effects on wildlife of periodic DDD applications to Clear Lake." *Calif. Fish and Game,* 46: 91–106.

Hurd, L. E., Mellinger, M. V., Wolf, L. L., and McNaughton, S. J. 1971. "Stability and diversity at three trophic levels in terrestrial successional ecosystems." *Science,* 173: 1134–1136.

Hurlbert, S. H. 1971. "The nonconcept of species diversity: A critique and alternative parameters." *Ecology,* 52: 577–586.

Hutchinson, G. E. 1957a. *A Treatise on Limnology,* Vol. I: *Geography, Physics, and Chemistry.* New York, Wiley.

—— 1957b. "Concluding remarks." *Cold Spring Harbor Symp. Quant. Biol.,* 22: 415–427.

—— 1959. "Homage to Santa Rosalia, or Why are there so many kinds of animals?" *Amer. Nat.,* 93: 145–159.

—— 1965. *The Ecological Theater and the Evolutionary Play.* New Haven, Conn., Yale Univ. Press.

—— 1969. *Eutrophication, past and present,* in *Eutrophication: Causes, Consequences, Correctives.* Washington, National Academy of Sciences.

Hutchison, V. H., Dowling, H. G., and Vinegar, A. 1966. "Thermoregulation in a brooding female Indian python, *Python molurus bivittatus.*" *Science,* 151: 694–696.

Hyman, L. H. 1967. *The Invertebrates,* Vol. VI: *Mollusca I.* New York, McGraw-Hill.

Ingham, G. 1950. "The mineral content of air and rain and its importance to agriculture." *J. Agric. Sci.,* 4: 55–61.

Inglis, J., Ankus, M. N., and Sykes, D. H. 1968. "Age-related differences in learning and short-term memory from childhood to the senium." *Human Devel.,* 11: 42–52.

Ingram, V. M. 1963. *The Hemoglobins in Genetics and Evolution.* New York, Columbia Univ. Press.

Irvine, W. 1955. *Apes, Angels, and Victorians.* New York, McGraw-Hill.

Irving, L. 1966. "Adaptations to cold." *Sci. Amer.,* 214: 94–101.

—— 1969. "Temperature regulation in marine mammals," in H. T. Anderson (ed.), *The Biology of Marine Mammals.* New York, Academic Press.

—— Krog, H., and Monson, M. 1955. "The metabolism of some Alaskan animals in winter and summer." *Physiol. Zool.,* 28: 173–185.

Isley, F. B. 1944. "Correlation between mandibular morphology and food specificity in grasshoppers." *Ann. Entomol. Soc. Amer.,* 37: 47–67.

Istock, C. A. 1967. "The evolution of complex life-cycle phenomena: An ecological perspective." *Evolution,* 21: 592–605.

Itô, Y. 1972. "On the methods for determining density-dependence by means of regression." *Oecologia,* 10: 347–372.

Ives, W. G. H. 1973. "Heat units and outbreaks of the forest tent caterpillar, *Malacosoma disstria* (Lepidoptera: Lasiocampidae)." *Canad. Entomol.,* 105: 529–543.

Ivlev, V. S. 1961. *Experimental Ecology of the Feeding of Fishes.* New Haven, Conn., Yale Univ. Press.

Jackson, G. J., Herman, R., and Singer, I. (eds.). 1969–1970. *Immunity to Parasitic Animals,* Vols. I and II. New York, Appleton-Century-Crofts.

Jacobs, M. E. 1955. "Studies on territorialism and sexual selection in dragonflies." *Ecology,* 36: 566–586.

Jain, S. K., and Bradshaw, A. D. 1966. "Evolutionary divergence among adjacent plant populations, 1: The evidence and its theoretical analysis." *Heredity,* 20: 407–441.

Janzen, D. H. 1966. "Coevolution of mutualism between ants and acacias in Central America." *Evolution,* 20: 249–275.

—— 1967a. "Interaction of the bull's-horn acacia (*Acacia cornigera* L.) with an ant inhabitant (*Pseudomyrmex furruginea* F. Smith) in eastern Mexico." *Univ. Kansas Sci. Bull.,* 47: 315–558.

—— 1967b. "Why mountain passes are higher in the tropics." *Amer. Nat.,* 101: 233–249.

—— 1967c. "Synchronization of sexual reproduction of trees within the dry season in Central America." *Evolution,* 21: 620–637.

—— 1968a. "Reproductive behavior in the *Passifloraceae* and some of its pollinators in Central America." *Behavior,* 32: 33–48.

—— 1968b. "Host plants as islands in evolutionary and contemporary time." *Amer. Nat.,* 102: 592–595.

—— 1969a. "Seed-eaters versus seed size, number, toxicity and dispersal." *Evolution,* 23: 1–27.

—— 1969b. "Birds and the ant × acacia interaction in Central America, with notes on birds and other myrmecophytes." *Condor,* 71: 240–256.

—— 1970. "Herbivores and the number of tree species in tropical forests." *Amer. Nat.,* 104: 501–528.

—— 1971a. "Euglossine bees as long-distance pollinators of tropical plants." *Science,* 171: 203–206.

—— 1971b. "Seed predation by animals." *Ann. Rev. Ecol. Syst.,* 2: 465–492.

—— 1973. "Host plants as islands, II: Competition in evolutionary and contemporary time." *Amer. Nat.,* 107: 786–790.

—— 1975. "*Pseudomyrmex nigropilosa:* A parasite of mutualism." *Science,* 188: 936–937.

—— 1976. "Why bamboos wait so long to flower." *Ann. Rev. Ecol. Syst.,* 7: 347–391.

—— and Schoener, T. W. 1967. "Differences in insect abundance and diversity between wetter and drier sites during a tropical dry season." *Ecology,* 49: 96–110.

Jenkins, D., Watson, A., and Miller, G. R. 1963. "Population studies on red grouse *Lagopus lagopus scoticus* (Lath) in northeast Scotland." *J. Anim. Ecol.,* 32: 317–376.

—— 1964. "Predation and red grouse populations." *J. Appl. Ecol.,* 1: 183–195.

Jenny, H. 1958. "Role of the plant factor in the pedogenic function." *Ecology,* 39: 5–16.

Jermy, T., Hanson, F. E., and Dethier, V. G. 1968. "Induction of specific food preferences in lepidopterous larvae." *Entomol. Exp. Appl.,* 11: 203–211.

Jespersen, D. 1924. "On the frequency of birds over the high Atlantic Ocean." *Nature,* 114: 281–283.

—— 1929. "On the frequency of birds over the high Atlantic Ocean." *Verh. VI Int. Orn. Kongr.,* 1926: 163–172.

Johnson, W. S., Gigon, A., Gulmon, S. L., and Mooney, H. A. 1974. "Comparative photosynthetic capacities of intertidal algae under exposed and submerged conditions." *Ecology,* 55: 450–453.

Johnston, D. W., and Odum, E. P. 1956. "Breeding bird populations in relation to plant succession on the Piedmont of Georgia." *Ecology,* 37: 50–62.

Johnston, R. F., Niles, D. M., and Rowher, S. A. 1972. "Hermon Bumpus and natural selection in the house sparrow *Passer domesticus.*" *Evolution,* 26: 20–31.

Johnston, R. F., and Selander, R. K. 1964. "House sparrows: Rapid evolution of races in North America." *Science,* 144: 548–550.

Jones, F. R. H. 1968. *Fish Migration.* London, Edward Arnold.

Jones, J. S. 1973. "Ecological genetics and natural selection in molluscs." *Science,* 182: 546–552.

Jordan, P., and Webbe, G. 1969. *Human Schistosomiasis.* London, Heinemann Medical Books.

Jordan, P. A., Botkin, D. B., and Wolfe, M. L. 1970. "Biomass dynamics in a moose population." *Ecology,* 52: 147–152.

Juday, C. 1940. "The annual energy budget of an inland lake." *Ecology,* 21: 438–450.

Jurion, F., and Henry, J. 1969. *Can Primitive Farming Be Modernized?* Congo, National Institute for Agricultural Studies.

Kale, H. W., II. 1965. "Ecology and bioenergetics of the long-billed marsh wren in Georgia salt marshes." *Publ. Nuttall Ornith. Club,* No. 5: 1–142.

Kalela, O. 1954. "Über den Revierbesitz bei Vögeln und Säugetieren als Population-sökologischer Faktor." *Ann. Zool. Soc. "Vanamo,"* 16: 1–48.

——— 1957. "Regulation of reproductive rate in subarctic populations of the vole *Clethrionomys rufocanus* (Sund.)." *Ann. Acad. Scien. Fenn.,* Ser. A, IV, No. 34.

Kalmus, H., and Smith, C. A. B. 1960. "Evolutionary origin of sexual differentiation and the sex-ratio." *Nature,* 186: 1004–1006.

Karr, J. R., and James, F. C. 1975. "Eco-morphological configurations and convergent evolution in species and communities," in M. L. Cody and J. M. Diamond (eds.), *Ecology and Evolution of Communities.* Cambridge, Mass., Belknap Press.

Kazacos, K. R., and Thorson, R. E. 1975. "Cross-resistance between *Nippostrongylus brasiliensis* and *strongyloides ratti* in rats." *J. Parasitol.,* 61: 525–529.

Keast, A. 1961. "Bird speciation on the Australian continent." *Bull. Mus. Comp. Zool.,* 123: 305–495.

——— 1965. "Interrelationships of two zebra species in an overlap zone." *J. Mammol.,* 46: 53–66.

——— 1966. *Australia and the Pacific Islands.* New York, Chanticleer Press.

——— 1972. "Ecological opportunities and dominant families, as illustrated by the Neotropical Tyrannidae (Aves)." *Evol. Biol.,* 5: 229–277.

Keen, R. E. 1967. "Laboratory population studies of two species of Chydoridae (*Cladocera, Crustacea*)." Unpublished M.S. thesis, Michigan State Univ.

Keever, C. 1950. "Causes of succession on old fields of the Piedmont, North Carolina." *Ecol. Monogr.,* 20: 230–250.

Keister, A. R. 1971. "Species density of North American amphibians and reptiles." *Syst. Zool.,* 20: 127–137.

Keith, L. B. 1963. *Wildlife's Ten-Year Cycle.* Madison, Univ. of Wisconsin Press.

Kelsall, J. P. 1968. *The Migratory Barren-ground Caribou of Canada.* Ottawa, Canadian Wildlife Service.

Kendall, M. G. 1948. *Advanced Theory of Statistics,* 4th ed. London, Charles Griffin.

Kennedy, J. S., and Stroyan, H. L. G. 1959. "Biology of aphids." *Ann. Rev. Entomol.,* 4: 139–160.

Kenoyer, L. A. 1927. "A study of Raunkiaer's law of frequency." *Ecology,* 8: 341–349.

Ketchum, B. H., Lillick, J., and Redfield, A. C. 1949. "The growth and optimum yields of unicellular algae in mass culture." *J. Cell. Comp. Physiol.,* 33: 267–279.

Kettlewell, H. B. D. 1955. "Selection experiments on industrial melanism in the lepidoptera." *Heredity,* 10: 287–301.

——— 1956. "Further selection experiments on industrial melanism in the lepidoptera." *Heredity,* 10: 287–301.

——— 1958. "A survey of the frequencies of *Biston betularia* and its melanic forms in Great Britain." *Heredity,* 12: 51–72.

——— 1959. "Darwin's missing evidence." *Sci. Amer.,* 200: 48–53.

——— 1961. "The phenomenon of industrial melanism in Lepidoptera." *Ann. Rev. Entomol.,* 6: 245–262.

——— 1965. "Insect survival and selection for pattern." *Science,* 148: 1290–1296.

Keyfitz, N., and Flieger, W. 1968. *World Population: An Analysis of Vital Data.* Chicago and London, Univ. of Chicago Press.

Kidwell, M. G. 1972. "Genetic change of recombination value in *Drosophila melanogaster,* I: Artificial selection for high and low recombination and some properties of recombination-modifying genes." *Genetics,* 70: 419–432.

Kielanowski, J. 1964. "Estimates of the energy cost of protein deposition in growing animals," in K. L. Blaxter (ed.), *Energy Metabolism.* New York, Academic Press.

Kikkawa, J. 1961. "Social behaviour of the white-eye *Zosterops lateralis* in winter flocks." *Ibis,* 103a: 428–442.

Kimura, M. 1968a. "Evolutionary rate at the molecular level." *Nature,* 217: 624–626.

—— 1968b. "Genetic variability maintained in a finite population due to mutational production of neutral and nearly neutral isoalleles." *Genet. Res. Cambr.,* 11: 247–269.

—— and Ohta, T. 1971a. *Theoretical Aspects of Population Genetics.* Princeton, N.J., Princeton Univ. Press.

—— 1971b. "Protein polymorphism as a phase of molecular evolution." *Nature,* 229: 467–469.

King, C. E. 1964. "Relative abundance of species and MacArthur's model." *Ecology,* 45: 716–727.

King, D. L., and Ball, R. C. 1967. "Comparative energetics of a polluted stream." *Limnol. Oceanogr.,* 12: 27–33.

King, J. L. 1967. "Continuously distributed factors affecting fitness." *Genetics,* 55: 403–413.

—— and Jukes, T. H. 1969. "Non-Darwinian evolution." *Science,* 164: 788–797.

King, J. R. 1974. "Seasonal allocation of time and energy resources in birds," in R. A. Paynter, Jr. (ed.), *Avian Energetics.* Cambridge, Mass., Nuttall Ornithol. Club.

—— and Farner, D. S. 1961. "Energy metabolism thermoregulation and body temperature," in A. J. Marshall (ed.), *Biology and Comparative Physiology of Birds,* Vol. II. New York, Academic Press.

King, M. C., and Wilson, A. C. 1975. "Evolution at two levels in humans and chimpanzees." *Science,* 188: 107–116.

Kinsey, A. C., Pomeroy, W. B., and Martin, C. E. 1948. *Sexual Behavior in the Human Male.* Philadelphia, Saunders.

Kira, T., and Shidei, T. 1967. "Primary production and turnover of organic matter in different forest ecosystems of the western Pacific." *Jap. J. Ecol.,* 17: 70–87.

Klausing, O. 1956. "Untersuchungen über den Mineralumsatz in Buchenwäldern auf Granit und Diorit." *Forstwiss. Cbl.,* 75: 18–32.

Kleiber, M. 1961. *The Fire of Life.* New York, Wiley.

Klomp, H. 1970. "The determination of clutch-size in birds: A review." *Ardea,* 58: 1–124.

Klopfer, P. H. 1962. *Behavioral Aspects of Ecology.* Englewood Cliffs, N.J., Prentice-Hall.

—— and MacArthur, R. H. 1961. "On the causes of tropical species diversity: Niche overlap." *Amer. Nat.,* 95: 223–226.

Knutson, R. M. 1974. "Heat production and temperature regulation in eastern skunk cabbage." *Science,* 186: 746–748.

Koch, A. 1966. "Insects and their endosymbionts," in M. S. Henry (ed.), *Symbiosis,* Vol. II. New York, Academic Press.

Koehn, R. K., Turano, F. J., and Mitton, J. B. 1973. "Population genetics of marine pelecypods, II: Genetic differences in microhabitats of *Modiolus demissus.*" *Evolution,* 27: 100–105.

Koford, C. B. 1963. "Group relations in an island colony of rhesus monkeys," in C. H. Southwick, ed., *Primate Social Behavior.* Princeton, N.J., Van Nostrand.

Kohn, A. J. 1966. "Food specialization in *Conus* in Hawaii and California." *Ecology,* 47: 1041–1043.

Kohn, R. R. 1971. *Principles of Mammalian Aging.* Englewood Cliffs, N.J., Prentice-Hall.

Kojima, K. 1971. "Is there a constant fitness value for a given genotype? No!" *Evolution,* 25: 281–285.

—— Gillespie, J., and Tobari, Y. N. 1970. "A profile of *Drosophila* species' enzymes by electrophoresis, I: Number of alleles, heterozygosities, and linkage disequilibrium in glucose-metabolizing systems and some other enzymes." *Biochem. Gen.,* 4: 627–637.

Kojima, K., and Tobari, Y. N. 1969. "The pattern of viability changes associated with genotype frequency at the alcohol dehydrogenase locus in a population of *D. melanogaster.*" *Genetics,* 61: 201–209.

Kojima, K., and Yarbrough, K. M. 1967. "Frequency dependent selection at the ester-ase-6 locus in *D. melanogaster.*" *Proc. Nat. Acad. Sci.,* 57: 645–649.

Koller, D. 1969. "The physiology of dormancy and survival of plants in desert environments." *Symp. Soc. Exptl. Biol.,* 23: 449–469.

Kolman, W. A. 1960. "The mechanism of natural selection for the sex ratio." *Amer. Nat.,* 94: 373–377.

Komai, T., Chino, M., and Hosino, Y. 1950. "Contributions to the evolutionary genetics of the lady beetle, *Harmonia,* I: Geographic and temporal variations in the relative frequencies of the elytral pattern types and in the frequency of the elytral ridge." *Genetics,* 35: 589–601.

Komai, T., and Emura, S. 1955. "A study of population genetics on the polymorphic land snail *Bradybaena similaris.*" *Evolution,* 9: 400–418.

Koplin, J. R., and Hoffmann, R. S. 1968. "Habitat overlap and competitive exclusion in voles (*Microtus*)." *Amer. Midl. Nat.,* 8;: 494–507.

Kormondy, E. J. 1969. *Concepts of Ecology.* Englewood Cliffs, N.J., Prentice-Hall.

Korringa, P. 1947. "Relation between the moon and periodicity in the breeding of marine animals." *Ecol. Monogr.,* 17: 349–381.

Kozhov, M. 1963. *Lake Baikal and Its Life.* The Hague, W. Junk.

Kozlowski, T. T., and Ahlgren, C. E., eds. 1974. *Fire and Ecosystems.* New York, Academic Press.

Kramer, G. 1961. "Long distance orientation," in A. J. Marshall, ed., *Biology and Comparative Physiology of Birds,* Vol. II. New York, Academic Press.

Kramer, P. J. 1958. "Photosynthesis of trees as affected by their environment," in K. V. Thimann, ed., *The Physiology of Forest Trees.* New York, Ronald.

—— 1969. *Plant and Water Relationships: A Modern Synthesis.* New York, McGraw-Hill.

—— Riley, W. S., and Bannister, T. T. 1952. "Gas exchange of cypress knees." *Ecology,* 33: 117–121.

Krebs, C. J. 1963. "Lemming cycle at Baker Lake, Canada, during 1959–1962." *Science,* 146: 1559–1560.

—— 1966. "Demographic changes in fluctuating populations of *Microtus californicus.*" *Ecol. Monogr.,* 36: 239–273.

—— 1970. "Genetic and behavioral studies on fluctuating vole populations." *Proc. Adv. Study Inst. Dynamics Numbers Popul.,* Oosterbeek: 243–256.

—— 1972. *Ecology. The Experimental Analysis of Distribution and Abundance.* New York, Harper and Row.

—— Gaines, M. S., Keller, B. L., Myers, J. H., and Tamarin, R. H. 1973. "Population cycles in small rodents." *Science,* 179: 35–41.

Krebs, C. J., and Myers, J. 1974. "Population cycles in small mammals." *Adv. Ecol. Res.,* 8: 267–399.

Krebs, J. R. 1971. "Territory and breeding density in the great tit, *Parus major* L." *Ecology,* 52: 2–22.

Krogh, A. 1941. *The Comparative Physiology of Respiratory Mechanisms.* Philadelphia, Univ. of Pennsylvania Press.

Kruckeberg, A. R. 1951. "Intraspecific variability in the response of certain native plants to serpentine soil." *Amer. J. Bot.,* 38: 408–419.

—— 1954. "The ecology of serpentine soils: A symposium, III: Plant species in relation to serpentine soils." *Ecology,* 35: 267–274.

Kruijt, J. P., de Vos, G. J., and Bossema, I. 1972. "The arena system of black grouse." *Proc. XV Intern. Ornithol. Congr.:* 399–423.

Kruuk, H. 1967. "Competition for food between vultures in East Africa." *Ardea,* 55: 171–193.

—— 1969. "Interactions between populations of spotted hyenas *Crocuta crocuta* (Erxleben) and their prey species," in A. Watson, ed., *Animal Populations in Relation to Their Food Resources.* Oxford, Blackwell.

Küchler, A. W. 1949. "A physiognomic classification of vegetation." *Ann. Assoc. Amer. Geogr.,* 39: 201–210.

—— 1964. "Potential natural vegetation of the conterminus United States." *Amer. Geogr. Soc., Spec. Publ. No. 36.*

—— 1967. *Vegetation Mapping.* New York, Ronald Press.

Kuentzel, L. E. 1969. "Bacteria, carbon dioxide, and algal blooms." *J. Water Poll. Control Fed.,* 41: 1737–1747.

Kuenzler, E. J. 1961. "Structure and energy flow of a mussel population." *Limnol. Oceanogr.,* 6: 191–204.

Kulman, H. M. 1971. "The effect of defoliation on tree growth." *Ann. Rev. Entomol.,* 16: 289–324.

Kummer, H. 1968. *Social Organization of Hamadryas Baboons.* Chicago, Univ. of Chicago Press.

—— 1971. *Primate Societies.* Chicago, Aldine-Atherton.

Kuno, E. 1971. "Sampling error as a misleading artifact in 'key factor analysis.'" *Res. Pop. Ecol.,* 13: 38–45.

Kurtén, B. 1959. "Rates of evolution in fossil mammals." *Cold Spring Harbor Symp. Quant. Biol.,* 24: 205–215.

—— 1965. "The carnivora of the Palestine caves." *Acta Zool. Fenn.,* 107: 1–74.

Lack, D. 1942. "Ecological features of the bird faunas of the British small islands." *J. Anim. Ecol.,* 11: 9–36.

—— 1943a. *The Life of the Robin.* London, Witherby.

—— 1943b. "The age of some more British birds." *Brit. Birds,* 36: 193–197, 214–221.

—— 1944. "Ecological aspects of species formation in passerine birds." *Ibis,* 86: 260–286.

—— 1945. "The ecology of closely related species with special reference to cormorant (*Phalacrocorax carbo*) and shag (*P. aristotelis*)." *J. Anim. Ecol.,* 14: 12–16.

—— 1947a. *Darwin's Finches.* Cambridge, Cambridge Univ. Press.

—— 1947b. "The significance of clutch-size, Parts 1 and 2." *Ibis,* 89: 302–352.

—— 1948. "Natural selection and family size in the starling." *Evolution,* 2: 95–110.

—— 1954. *The Natural Regulation of Animal Numbers.* London, Oxford Univ. Press.

—— 1966. *Population Studies of Birds.* Clarendon Press. Oxford, Oxford Univ. Press.

—— 1971. *Ecological Isolation in Birds.* Cambridge, Mass., Harvard Univ. Press.

—— 1976. *Island Biology.* Cambridge, Cambridge Univ. Press.

—— and Arn, H. 1947. "Die Bedeutung der Gelegegrösse beim Alpensegler." *Ornith. Beob.,* 44: 188–210.

Lack, D., and Lack, E. 1951. "The breeding biology of the swift *Apus apus.*" *Ibis.,* 93: 501–546.

Laessle, A. M. 1961. "A micro-limnological study of Jamaican bromeliads." *Ecology,* 42: 499–517.

Laetsch, W. M. 1974. "The C_4 syndrome: A structural analysis." *Ann. Rev. Plant Physiol.,* 25: 27–52.

Lagler, K. F., Bardach, J. E., and Miller, R. R. 1962. *Ichthyology.* New York, Wiley.

Lamb, R. J., and Pointing, P. J. 1972. "Sexual morph determination in the aphid, *Acyrthosiphon pisum.*" *J. Insect Physiol.,* 18: 2029–2042.

Landahl, J. T., and Root, R. B. 1969. "Differences in the life tables of tropical and temperate milkweed bugs, genus *Oncopeltus* (Hemiptera: Lygaeidae)." *Ecology,* 50: 734–737.

Landau, B. R., and Dawe, A. R. 1958. "Respiration in the hibernation of the 13-lined ground squirrel." *Amer. J. Physiol.,* 194: 75–82.

Landry, S. 1965. "The status of the theory of the replacement of the Multituberculata by the Rodentia." *J. Mammol.,* 46: 280–286.

Lange, R. T. 1966. "Bacterial symbiosis with plants," in M. S. Henry, ed., *Symbiosis,* Vol. I. New York, Academic Press.

Larcher, W., Cernusca, A., Schmidt, L., Grabherr, G., Nötzel, E., and Smeets, N. 1975. "Mt. Patscherkofel, Austria," in T. Rosswall and O. W. Heal, eds., *Structure and Function of Tundra Ecosystems.* Stockholm: Swedish Natural Science Research Council, Ecol. Bull. 20.

Lasiewski, R. C., and Snyder, G. K. 1969. "Responses to high temperature in nestling double-crested and pelagic cormorants." *Auk,* 86: 529–540.

Last, F. T., and Warren, R. C. 1972. "Non-parasitic microbes colonizing green leaves: Their form and function." *Endeavour,* 31: 143–150.

Latham, M. C. 1975. "Nutrition and infection in national development." *Science,* 188: 561–565.

Laughlin, R. 1965. "Capacity for increase; a useful population statistic." *J. Anim. Ecol.,* 34: 77–91.

Laws, R. M. 1962. "Some effects of whaling on the southern stocks of baleen whales," in E. D. LeCren and M. W. Holdgate, eds., *The Exploitation of Natural Animal Populations.* New York, Wiley.

—— and Clough, G. 1966. "Observations on reproduction in the hippopotamus, *Hippopotamus amphibius,* Linn." *Symp. Zool. Soc. London,* 15: 117–140.

Lawton, J. H., Beddington, J. R., and Bonser, R. 1974. "Switching in invertebrate predators," in M. B. Usher and M. H. Williamson, eds., *Ecological Stability.* London, Chapman and Hall.

Lechleitner, R. R. 1969. *Wild Mammals of Colorado.* Boulder, Colo., Pruett.

Leak, W. B. 1970. "Successional change in northern hardwoods predicted by birth and death simulation." *Ecology,* 51: 794–801.

Leck, C. F. 1970. "The seasonal ecology of fruit and nectar eating birds in lower middle America." Unpublished Ph.D. thesis, Cornell University.

LeCren, E. D., and Holdgate, M. W., eds. 1962. *The Exploitation of Natural Animal Populations.* New York, Wiley.

Lederberg, J., and Lederberg, E. M. 1952. "Replica plating and indirect selection of bacterial mutants." *J. Bact.,* 63: 399–406.

Lees, A. D. 1966. "The control of polymorphism in aphids." *Adv. Insect. Physiol.,* 3: 207–277.

Lehrman, D. S. 1961. "Hormonal regulation of parental behavior in birds and infra-human mammals," in W. C. Young, ed., *Sex and Internal Secretion,* 3d ed. Baltimore, Williams and Wilkins.

—— 1964a. "Control of behavioral cycles in reproduction," in W. Etkin, ed., *Social Behavior and Organization Among Vertebrates.* Chicago, Univ. of Chicago Press.

—— 1964b. "The reproductive behavior of ring doves." *Sci. Amer.,* 211: 48–54.

Leigh, E. G. 1968. "The ecological role of Volterra's equations," in M. Gerstenhaber, ed., *Some Mathematical Problems in Biology.* Providence, American Mathematical Society.

—— 1970. "Sex ratio and differential mortality between the sexes." *Amer.Nat.,* 104: 205–210.

Leopold, L. B., Wolman, M. G., and Miller, J. P. 1964. *Fluvial Processes in Geomorphology.* San Francisco, W. H. Freeman.

Lerner, I. M. 1968. *Heredity, Evolution, and Society.* San Francisco, W. H. Freeman.

—— and Dempster, E. M. 1962. "Indeterminism in interspecific competition." *Proc. Natl. Acad. Sci.,* 48: 821–826.

Lerner, I. M., and Ho, F. K. 1961. "Genotype and competitive ability of *Tribolium* species." *Amer. Nat.,* 95: 329–343.

Leslie, P. H., and Park, T. 1949. "The intrinsic rate of natural increase of *Tribolium castaneum* Herbst." *Ecology,* 30: 469–477.

Leslie, P. H., and Ranson, R. M. 1940. "Mortality, fertility and rate of natural increase in the vole (*Microtus agrestis*) in the laboratory." *J. Anim. Ecol.,* 9: 27–52.

Leuthold, W. 1966. "Variations in territorial behavior of Uganda kob *Adenota kob thomasi* (Neumann 1896)." *Behaviour,* 27: 214–257.

Levene, H. 1953. "Genetic equilibrium when more than one ecological niche is available." *Amer. Nat.,* 87: 331–333.

Levin, D. A. 1975. "Pest pressure and recombination systems in plants." *Amer. Nat.,* 109: 437–451.

Levin, S. A., and Paine, R. T. 1974. "Disturbance, patch formation, and community structure," *Proc. Nat. Acad. Sci.,* 71: 2744–2747.

Levins, R. 1968. *Evolution in Changing Environments. Some Theoretical Explorations.* Princeton, Princeton Univ. Press.

—— 1970. "Extinction," in *Some Mathematical Questions in Biology,* Vol. 2. Providence, *Amer. Math. Soc.*

—— and Culver, D. 1971. "Regional coexistence of species and competition between rare species." *Proc. Nat. Acad. Sci.,* 68: 1246–1248.

Levins, R., and MacArthur, R. H. 1966. "The maintenance of genetic polymorphism in a spatially heterogeneous environment: Variations on a theme by Howard Levene." *Amer. Nat.,* 100: 585–590.

Levins, R., Pressick, M. L., and Heatwole, H. 1973. "Coexistence patterns in insular ants." *Amer. Sci.,* 61: 463–472.

Lewert, R. M. 1970. "Schistosomes," in G. J. Jackson, R. Herman, and I. Singer, eds., *Immunity to Parasitic Animals,* Vol. 2. New York, Appleton-Century-Crofts.

Lewis, D. 1942. "The evolution of sex in flowering plants." *Biol. Rev.,* 17: 46–67.

—— 1954. "Comparative incompatibility in angiosperms and fungi." *Adv. Genet.,* 6: 235–287.

Lewis, H. 1953. "The mechanism of evolution in the genus *Clarkia*." *Evolution,* 7: 1–20.

Lewontin, R. C. 1967. "An estimate of average heterozygosity in man." *Amer. J. Human Genet.,* 19: 681–685.

—— 1974. *The Genetic Basis of Evolutionary Change.* New York, Columbia Univ. Press.

—— and Birch, L. C. 1966. "Hybridization as a source of variation for adaptation to new environments." *Evolution,* 20: 315–336.

Lewontin, R. C., and Hubby, J. L. 1966. "A molecular approach to the study of genic heterozygosity in natural populations, II: Amount of variation and degree of heterozygosity in natural populations of *Drosophila pseudoobscura*." *Genetics,* 54: 595–609.

Leyton, L., and Rousseau, L. Z. 1958. "Root growth of tree seedlings in relation to aeration," in K. V. Thimann, ed., *The Physiology of Forest Trees.* New York, Ronald.

Li, C. C. 1955. *Population Genetics.* Chicago, Univ. of Chicago Press.

—— 1967. "Genetic equilibrium under selection." *Biometrics,* 23: 397–484.

Li, H.-L. 1952. "Floristic relationships between eastern Asia and eastern North America." *Trans. Amer. Phil. Soc., N. S.,* 42, part 2.

Lidicker, W. Z. 1975. "The role of dispersal in the demography of small mammals," in F. B. Golley, K. Petrusewicz, and L. Ryszkowski, eds., *Small Mammals: Their Productivity and Population Dynamics.* Cambridge, Cambridge Univ. Press.

Lieth, H. 1973. "Primary production: Terrestrial ecosystems." *Human Ecol.*, 1: 303–332.

Light, S. F. 1942–43. "The determination of castes in social insects." *Quart. Rev. Biol.*, 17: 312–326; 18: 46–63.

Likens, G. E. 1972. "Eutrophication and aquatic ecosystems." *Amer. Soc. Limnol. Oceanogr., Spec. Symp.*, 1: 3–13.

—— Bormann, F. H., Johnson, N. M., Fisher, D. W., and Pierce, R. S. 1970. "Effects of forest cutting and herbicide treatment on nutrient budgets in the Hubbard Brook watershed-ecosystem." *Ecol. Monogr.*, 40: 23–47.

Likens, G. E., Bormann, F. H., Johnson, N. M., and Pierce, R. S. 1967. "The calcium magnesium, potassium, and sodium budgets for a small forested ecosystem." *Ecology*, 48: 772–785.

Likens, G. E., Bormann, F. H., Pierce, R. S., Eaton, J. S., and Johnson, N. M. 1977. *Biogeochemistry of a Forested Ecosystem*. New York, Springer-Verlag.

Likens, G. E., Bormann, F. H., Pierce, R. S., and Fisher, D. W. 1971. "Nutrient-hydrologic cycle interaction in small forested watershed-ecosystems." *Proc. Brussels Symp. Prod. For. Ecosyst., UNESCO 1969*: 553–563.

Lill, A. 1974. "Social organization and space utilization in the lek-forming white-bearded manakin, *M. manacus trinitatis* Hartert." *Z. Tierpsych.*, 36: 513–530.

Limbaugh, C. 1961. "Cleaning symbiosis." *Sci. Amer.*, 205: 42–49.

Lind, H. 1961. *Studies on the Behaviour of the Black-tailed Godwit* (Limosa limosa [L.]). Copenhagen, Munksgaard.

Lindeman, R. L. 1941. "Seasonal food cycle dynamics in a senescent lake." *Amer. Midl. Nat.*, 26: 636–673.

—— 1942. "The trophic-dynamic aspect of ecology." *Ecology*, 23: 399–418.

Lindgren, D. 1975. "Sensitivity of premeiotic and meiotic stages to spontaneous and induced mutations in barley and maize." *Hereditas*, 79: 227–238.

Linhart, Y. P. 1974. "Intra-population differentiation in annual plants, I: *Veronica peregrina* L. raised under non-competitive conditions." *Evolution*, 28: 232–243.

Lipps, J. H., and Mitchell, E. 1976. "Trophic model for the adaptive radiations and extinctions of pelagic marine mammals." *Paleobiol.*, 2: 147–155.

List, R. J. 1966. *Smithsonian Meterological Tables*, 6th rev. ed. *Smithsonian Misc. Coll.*, 114: 1–527.

Littlejohn, M. J. 1965. "Premating isolation in the *Hyla ewingi* complex (Anura: Hylidae)." *Evolution*, 19: 234–243.

—— and Loftus-Hills, J. J. 1968. "An experimental evaluation of premating isolation in the *Hyla ewingi* complex (Anura: Hylidae)." *Evolution*, 22: 659–663.

Lloyd, D. 1965. "Evolution of self-compatibility and racial differentiation in *Leavenworthia* (Cruciferae)." *Contr. Gray Herb.*, 195: 1–134.

Lloyd, J. A. 1975. "Social structure and reproduction in two freely-growing populations of house mice (*Mus musculus* L.)." *Anim. Behav.*, 23: 413–424.

—— and Christian, J. J. 1969. "Reproductive activity of individual females in three experimental freely growing populations of house mice (*Mus musculus*)." *J. Mammal.*, 50: 49–59.

Lloyd, J. E. 1975. "Aggressive mimicry in *Photuris* fireflies: Signal repertoires by *femmes fatales*." *Science*, 187: 452–453.

Lloyd, M., and Ghelardi, R. J. 1964. "A table for calculating the 'equitability' components of species diversity." *J. Anim. Ecol.*, 33: 217–225.

Lofts, B. 1970. *Animal Photoperiodism*. London, Arnold.

—— and Murton, R. K. 1968. "Photoperiodic and physiological adaptations regulating avian breeding cycles and their ecological significance." *J. Zool.*, 155: 327–394.

Longhurst, W. M., Leopold, A. S., and Dasmann, R. F. 1952. "A survey of California deer herds, their ranges and management problems." *Calif. Dept. Fish and Game, Bull. No. 6.*

Lorenz, K. Z. 1935. "Der Kumpan in der Umwelt des Vogels." *J. Ornith.*, 83: 137–213, 289–413.

—— 1966. *Evolution and Modification of Behaviour.* London, Methuen.

Lotka, A. J. 1922. "The stability of the normal age distribution." *Proc. Nat. Acad. Sci.*, 8: 339–345.

—— 1925. *Elements of Physical Biology.* Baltimore, Williams and Wilkins.

Loucks, O. L. 1962. "Ordinating forest communities by means of environmental scalars and phytosociological indices." *Ecol. Monogr.*, 32: 137–166.

Low, B. S. 1976. "The evolution of amphibian life-histories in the desert," in D. W. Goodall, ed., *Evolution of Desert Biota.* Austin, Univ. of Texas Press.

Lowrie, D. C., and Miller, R. S. 1973. "The myth of the Kaibab deer population." *Bio. Science*, 23: 458.

Lowry, W. P. 1969. *Weather and Life.* New York, Academic Press.

Luck, R. F. 1971. "An appraisal of two methods of analyzing insect life tables." *Can. Entomol.*, 103: 1261–1271.

Lucotte, G. 1975. "Le fardeau génétique de la caille Japonaise." *Ann. Biol.*, 14: 3–4.

Luttich, S., Rusch, D. H., Meslow, E. C., and Keith, L. B. 1970. "Ecology of red-tailed hawk predation in Alberta." *Ecology*, 51: 190–203.

Lynch, J. F., and Johnson, N. K. 1974. "Turnover and equilibria in insular avifaunas, with special reference to the California Channel Islands." *Condor*, 76: 370–384.

MacArthur, R. H. 1955. "Fluctuations of animal populations and a measure of community stability." *Ecology*, 36: 533–536.

—— 1957. "On the relative abundance of bird species." *Proc. Nat. Acad. Sci.*, 43: 293–295.

—— 1958. "Population ecology of some warblers of northeastern coniferous forests." *Ecology*, 39: 599–619.

—— 1960. "On the relative abundance of species." *Amer. Nat.*, 94: 25–36.

—— 1964. "Environmental factors affecting bird species diversity." *Amer. Nat.*, 98: 387–397.

—— 1965. "Patterns of species diversity." *Biol. Rev.*, 40: 510–533.

—— 1968. "The theory of the niche," in R. C. Lewontin, ed., *Population Biology and Evolution.* Syracuse, N.Y., Syracuse Univ. Press.

—— 1969. "Patterns of communities in the tropics." *Biol. J. Linnean Soc.*, 1: 19–30.

—— Diamond, J. M., and Karr, J. R. 1972. "Density compensation in island faunas." *Ecology*, 53: 330–342.

MacArthur, R. H., and Levins, R. 1967. "The limiting similarity, convergence, and divergence of coexisting species." *Amer. Nat.*, 101: 377–385.

MacArthur, R. H., and MacArthur, J. 1961. "On bird species diversity." *Ecology*, 42: 594–598.

MacArthur, R. H., and Pianka, E. R. 1966. "On optimal use of a patchy environment." *Amer. Nat.*, 100: 603–609.

MacArthur, R. H., Recher, H., and Cody, M. 1966. "On the relation between habitat selection and species diversity." *Amer. Nat.*, 100: 319–332.

MacArthur, R. H., and Wilson, E. O. 1963. "An equilibrium theory of insular zoogeography." *Evolution*, 17: 373–387.

—— 1967. *The Theory of Island Biogeography.* Princeton, Princeton Univ. Press.

MacLulich, D. A. 1937. "Fluctuations in the numbers of the varying hare (*Lepus americanus*)." *Univ. Toronto Studies, Biol. Ser. No. 43.*

Maelzer, D. A. 1970. "The regression of log N_{n+1} on log N_n as a test of density dependence: An exercise with computer constructed density-independent populations." *Ecology*, 51: 810–822.

Magistad, O. C. 1945. "Plant growth on saline and alkali soils." *Bot. Rev.,* 11: 181–230.

Maguire, B. 1963. "The passive dispersal of small aquatic organisms and their colonization of isolated bodies of water." *Ecol. Monogr.,* 33: 161–185.

Maher, W. J. 1970. "The pomarine jaeger as a brown lemming predator in northern Alaska." *Wilson Bull.,* 82: 130–157.

Maldonado, J. F. 1967. *Schistosomiasis in America.* Barcelona, Editorial Cientifico-Medica.

Malthus, R. T. 1798. *An Essay on the Principle of Population as it Affects the Future Improvement of Society.* London, Johnson.

Maly, E. J. 1969. "A laboratory study of the interaction between the predatory rotifer *Asplanchna* and *Paramecium.*" *Ecology,* 50: 59–73.

Mann, K. H. 1964. "The pattern of energy flow in the fish and invertebrate fauna of the River Thames." *Verh. Inter. Verein. Limnol.,* 15: 485–495.

―――― 1973. "Seaweeds: Their productivity and strategy for growth." *Science,* 182: 975–981.

Manning, A. 1961. "The effects of artificial selection for mating speed in *Drosophila melanogaster.*" *Anim. Behav.,* 9: 82–92.

―――― 1967. "Pre-imaginal conditioning in *Drosophila.*" *Nature,* 216: 338–340.

Manuwal, D. A. 1974. "Effects of territoriality on breeding in a population of Cassin's auklet." *Ecology,* 55: 1399–1406.

Manwell, C., and Baker, C. M. A. 1970. *Molecular Biology and the Origin of Species.* Seattle, Univ. of Washington Press.

Marchand, D. E. 1973. "Edaphic control of plant distribution in the White Mountains, eastern California." *Ecology,* 54: 233–250.

Mares, M. A. 1976. "Convergent evolution of desert rodents: Multivariate analysis and zoogeographic implications." *Paleobiol.,* 2: 39–63.

Margalef, R. 1958. "Information theory in ecology." *Gen. Syst.,* 3: 36–71.

―――― 1963. "On certain unifying principles in ecology." *Amer. Nat.,* 97: 357–374.

―――― 1968. *Perspectives in Ecological Theory.* Chicago, Univ. of Chicago Press.

Marks, G. F., and Kozlowski, T. T., eds. 1973. *Physiology and Ecology of Ectomycorrhizae.* New York, Academic Press.

Marshall, A. J., and Disney, H. S. de S. 1957. "Experimental induction of the breeding seasons in a xerophilous bird." *Nature,* 180: 647.

Marshall, D. R., and Jain, S. K. 1969. "Interference in pure and mixed populations of *Avena fatua* and *A. barbata.*" *J. Ecol.,* 57: 251–270.

Marshall, J. S. 1962. "The effects of continuous gamma radiation on the intrinsic rate of national increase of *Daphnia pulex.*" *Ecology,* 43: 598–607.

Marshall, J. T., Jr. 1948. "Ecologic races of song sparrows in the San Francisco Bay region, Part II: Geographic variation." *Condor,* 50: 233–256.

Masaki, S. 1967. "Geographic variation and climatic adaptation in a field cricket (Orthoptera: Gryllidae)." *Evolution,* 21: 725–741.

Massey, A. B. 1925. "Antagonism of the walnuts (*Juglans nigra* L. and *J. cinerea* L.) in certain plant associations." *Phytopathology,* 15: 773–784.

Mather, K. 1953. "The genetical structure of populations." *Symp. Soc. Exptl. Biol.,* 7: 66–95.

―――― 1955. "Polymorphism as an outcome of disruptive selection." *Evolution,* 9: 52–61.

―――― and Harrison, B. J. 1949. "The manifold effect of selection." *Heredity,* 3: 1–52, 129–162.

Mattingly, P. F. 1969. *The Biology of Mosquito-Borne Disease.* New York, American Elsevier.

May, J. M. 1961. *Studies in Disease Ecology.* New York, Hafner.

May, R. M. 1973. *Stability and Complexity in Model Ecosystems.* Princeton, Princeton Univ. Press.

—— 1975. "Patterns of species abundance and diversity," in M. L. Cody and J. M. Diamond, eds., *Ecology and Evolution of Communities*. Cambridge, Mass., Harvard Univ. Press.

—— and MacArthur, R. H. 1972. "Niche overlap as a function of environmental variability." *Proc. Nat. Acad. Sci.,* 69: 1109–1113.

Maynard-Smith, J. 1964. "Group selection and kin selection." *Nature,* 201: 1145–1147.

—— 1966. "Sympatric speciation." *Amer. Nat.,* 100: 637–650.

—— 1968a. "Evolution in sexual and asexual populations." *Amer. Nat.,* 102: 469–473.

—— 1968b. "'Haldane's dilemma' and the rate of evolution." *Nature,* 219: 1114–1116.

—— 1970. "The origin and maintenance of sex," in G. C. Williams, ed., *Group Selection.* Chicago, Aldine-Atherton.

—— 1971. "What use is sex?" *J. Theoret. Biol.,* 30: 319–335.

—— 1974. *Models in Ecology.* Cambridge, Cambridge Univ. Press.

—— 1976. "What determines the rate of evolution?" *Amer. Nat.,* 110: 331–338.

Mayr, E. 1942. *Systematics and the Origin of Species.* New York, Columbia Univ. Press.

—— 1956. "Geographical character gradients and climatic adaptation." *Evolution,* 10: 105–108.

—— 1963. *Animal Species and Evolution.* Cambridge, Mass., Harvard Univ. Press.

—— 1969. "Bird speciation in the tropics." *Biol. J. Linnean Soc.,* 1: 1–17.

—— and Phelps, W. H. 1967. "The origin of the bird fauna of the south Venezuelan highlands." *Bull. Amer. Mus. Nat. Hist.,* 136: 269–328.

McClelland, W. J. 1965. "The production of cercariae by *S. mansoni* and *S. haematobium* and methods for estimating the numbers of cercariae in suspension." *Bull. World Health Org.,* 33: 270–275.

McCormick, J. 1970. *The Pine Barrens: A Preliminary Ecological Inventory.* Trenton, New Jersey State Museum.

McGinnis, S. M., and Dickson, L. L. 1967. "Thermoregulation in the desert iguana *Dipsosaurus dorsalis.*" *Science,* 156: 1757–1759.

McIntosh, R. P. 1967. "The continuum concept of vegetation." *Bot. Rev.,* 33: 130–187.

—— 1974. "Plant ecology 1947–1972." *Ann. Missouri Bot. Gard.,* 61: 132–165.

McKay, F. E. 1971. "Behavioral aspects of population dynamics in unisexual-bisexual *Poeciliopsis* (Pisces: Poeciliidae)." *Ecology,* 52: 778–790.

McKenzie, J. A. 1975. "Gene flow and selection in a natural population of *Drosophila melanogaster.*" *Genetics,* 80: 349–361.

McLaren, I. A. 1962. "Population dynamics and exploitation of seals in the eastern Canadian arctic," in E. D. LeCren and M. W. Holdgate, eds., *Exploitation of Natural Animal Populations.* New York, Wiley.

—— 1969. "Primary production of nutrients in Ogac Lake, a landlocked fiord on Baffin Island." *J. Fish. Res. Bd. Can.,* 26: 1562–1576.

McLeese, D. W. 1956. "Effects of temperature, salinity and oxygen on the survival of the American lobster." *J. Fish. Res. Bd. Can.,* 13: 247–272.

McMillan, C. 1956. "The edaphic restriction of *Cupressus* and *Pinus* in the coast ranges of central California." *Ecol. Monogr.,* 26: 177–212.

—— 1959. "The role of ecotypic variation in the distribution of the central grassland of North America." *Ecol. Monogr.,* 29: 285–308.

McNab, B. K. 1963. "Bioenergetics and the determination of home range size." *Amer. Nat.,* 97: 133–140.

—— 1966. "An analysis of the body temperatures of birds." *Condor,* 68: 47–55.

McNaughton, S. J. 1976. "Serengeti migratory wildebeest: Facilitation of energy flow by grazing." *Science,* 191: 92–94.

—— and Wolf, L. L. 1970. "Dominance and the niche in ecological systems." *Science,* 167: 131–139.

McPhail, J. D. 1969. "Predation and the evolution of a stickleback (*Gasterosteus*)." *J. Fish. Res. Bd. Can.,* 26: 3183–3208.

McPhee, J. 1968. *The Pine Barrens.* New York, Farrar, Straus & Giroux.

McVay, S. 1966. "The last of the great whales." *Sci. Amer.,* 215: 13–21.

Mech, L. D. 1966. "The wolves of the Isle Royale." *U.S. Natl. Park Serv., Fauna Ser.* No. 7.

——— 1972. *The Wolf: The Ecology and Behavior of an Endangered Species.* New York, Natural Hist. Press.

Medawar, P. B. 1957. *The Uniqueness of the Individual.* London, Methuen.

Medina, E. 1974. "Dark CO_2 fixation, habitat preference and evolution within the Bromeliaceae." *Evolution,* 28: 677–686.

Menhinick, E. F. 1963. "Insect species in the herb stratum of a sericea lespedeza stand, AEC Savannah River Project, Aiken, South Carolina." *U.S. Atomic Energy Comm., Div. of Tech. Info.* (TID-19136).

——— 1967. "Structure, stability, and energy flow in plants and arthropods in sericea lespedeza stand." *Ecol. Monogr.,* 37: 255–272.

Merrell, D. J. 1951. "Interspecific competition between *Drosophila funebris* and *Drosophila melanogaster.*" *Amer. Nat.,* 85: 159–169.

Merriam, C. H. 1894. "Laws of temperature control of the geographic distribution of terrestrial animals and plants." *Nat. Geogr. Mag.,* 6: 229–238.

Merritt, R. B. 1972. "Geographic distribution and enzymatic properties of lactate dehydrogenase in the fathead minnow, *Pimephales promelas.*" *Amer. Nat.,* 106: 173–184.

Mettler, L. E., and Gregg, T. G. 1969. *Population Genetics and Evolution.* Englewood Cliffs, N.J. Prentice-Hall.

Michener, C. D. 1958. "The evolution of social behavior in bees." *Proc. 10th Intern. Congr. Entomol.,* 2: 441–447.

——— 1969. "Comparative social behavior of bees." *Ann. Rev. Entomol.,* 14: 299–342.

Mikola, P. 1960. "Comparative experiment on decomposition rates of forest litter in southern and northern Finland." *Oikos,* 11: 161–166.

Mildvan, A. S., and Strehler, B. L. 1960. "A critique of theories of mortality," in B. L. Strehler, ed., *The Biology of Aging.* Washington, D.C., Amer. Inst. Biol. Sci.

Milham, S., Jr., and Gittelsohn, A. M. 1965. "Parental age and malformations." *Human Biol.,* 3: 13–22.

Millbank, J. W., and Kershaw, K. A. 1969. "Nitrogen metabolism in lichens, I: Nitrogen fixation in cephalodia of *Peltigera aphthosa.*" *New Phytol.,* 68: 721–729.

Miller, A. H. 1956. "Ecologic factors that accelerate formation of races and species of terrestrial vertebrates." *Evolution,* 10: 262–277.

Miller, L. K. 1969. "Freezing tolerance in an adult insect." *Science,* 166: 105–106.

Miller, R. S. 1967. "Pattern and process in competition." *Adv. Ecol. Res.,* 4: 1–74.

——— 1969. "Competition and species diversity." *Brookhaven Symp. Biol.,* 22: 63–70.

Miller, S. L. 1953. "Production of amino acids under possible primitive earth conditions." *Science,* 117: 528–529.

——— and Orgel, L. E. 1974. *The Origins of Life on the Earth.* Englewood Cliffs, N.J. Prentice-Hall.

Milner, C., and Hughes, R. E. 1968. *Methods for the Measurement of the Primary Production of Grassland.* Oxford, Blackwell.

Minderman, G. 1968. "Addition, decomposition and accumulation of organic matter in forests." *J. Ecol.,* 56: 355–362.

Minshall, G. W. 1967. "Role of allochthonous detritus in the trophic structure of a woodland stream." *Ecology,* 48: 139–149.

Mirsky, A. E., and Ris, H. 1950. "The DNA content of animal cells and its evolutionary significance." *J. Gen. Physiol.,* 34: 451–462.

Misra, R., and Gopal, B. eds. 1968. *Proc. Symp. Recent Adv. Trop. Ecol.* Part I. Varansi, India, International Society for Tropical Ecology.

Mitchell, B. L., Shenton, J. B., and Uys, J. C. M. 1965. "Predation on large mammals in the Kafue National Park, Zambia." *Zool. Africana,* 1: 297–318.

Mohr, E. C. J., and Van Baren, F. A. 1954. *Tropical Soils.* New York, Interscience.

Möller, C. M., Müller, D., and Nielson, J. 1954. "Graphic presentation of dry matter production of European beech." *Forst. Fors Vaes. Danm.,* 21: 327–335.

Monk, C. 1966. "Ecological importance of root/shoot ratios." *Bull. Torrey Bot. Club,* 93: 402–406.

Mook, J. H., Mook, L. J., and Heikens, H. S. 1960. "Further evidence for the role of 'searching images' in the hunting behavior of titmice." *Arch. Néerl. Zool.,* 13: 448–465.

Mooney, H. A., and Dunn, E. L. 1970. "Photosynthetic systems of Mediterranean-climate shrubs and trees of California and Chile." *Amer. Nat.,* 104: 447–453.

—— 1970a. "Convergent evolution of Mediterranean-climate evergreen schlerophyll shrubs." *Evolution,* 24: 292–303.

—— Wright, R. D., and Strain, B. R. 1964. "The gas exchange capacity of plants in relation to vegetation zonation in the White Mountains of California." *Amer. Midl. Nat.,* 72: 281–297.

Moore, J. A. 1952. "Competition between *Drosophila melanogaster* and *Drosophila simulans,* II: The improvement of competitive ability through selection." *Proc. Natl. Acad. Sci.,* 38: 813–817.

Moore, N. W. 1964. "Intra- and inter-specific competition among dragonflies." *J. Anim. Ecol.,* 33: 49–71.

Moore, W. S., and McKay, F. E. 1971. "Coexistence in unisexual-bisexual species complexes of *Poeciliopsis* (Pisces: Poeciliidae)." *Ecology,* 52: 791–799.

Moran, P. A. P. 1952. "The statistical analysis of game-bird records." *J. Anim. Ecol.,* 21: 154–158.

—— 1953. "The statistical analysis of the Canadian lynx cycle, I: Structure and prediction." *Australian J. Zool.,* 1: 163–173.

—— 1970. "'Haldane's dilemma' and the rate of evolution." *Ann. Human Genet.,* 33: 245–249.

Morris, R. F. 1959. "Single factor analysis in population dynamics." *Ecology,* 40: 580–588.

—— 1963a. "The development of predictive equations for the spruce budworm based on key factor analysis," in R. F. Morris, ed., *The Dynamics of Epidemic Spruce Budworm Populations. Mem. Entomol. Soc. Can.* 32: 1–332.

—— 1963b. "The dynamics of epidemic spruce budworm populations." *Mem. Entomol. Soc. Can.,* 31: 1–332.

—— Chesire, W. F., Miller, C. A., and Mott, D. G. 1958. "Numerical responses of avian and mammalian predators during a gradation of the spruce budworm." *Ecology,* 39: 487–494.

Morris, R. F., Miller, C. A., Greenbank, D. O., and Mott, D. G. 1958. "The population dynamics of the spruce budworm in eastern Canada." *Proc. Tenth Inter. Congr. Entomol.,* 4: 137–149.

Morse, D. H. 1970. "Ecological aspects of some mixed-species foraging flocks of birds." *Ecol. Monogr.,* 40: 119–168.

Morton, E. S. 1973. "On the evolutionary advantages of fruit eating in tropical birds." *Amer. Nat.,* 107: 8–22.

Morton, J. K. 1966. "The role of polyploidy in the evolution of a tropical flora," in C. D. Darlington and K. R. Lewis, eds., *Chromosomes Today,* Vol. I. Edinburgh, Oliver and Boyd, and New York, Plenum Press.

Mosquin, T. 1964. "Chromosomal repatterning in *Clarkia rhomboidea* as evidence for post-Pleistocene changes in distribution." *Evolution,* 18: 12–25.

Moynihan, M. H. 1960. "Some adaptations which help to promote gregariousness." *Proc. XIIth Inter. Ornith. Congr.,* 523–541.

—— 1963a. "Two papers on the evolution of communal displays." *Auk,* 80: 381–394.

—— 1963b. "Interspecific relations between some Andean birds." *Ibis,* 105: 327–339.

—— 1968. "Social mimicry: Character convergence versus character displacement." *Evolution,* 22: 315–331.

Mueller-Dombois, D., and Ellenberg, H. 1974. *Aims and Methods of Vegetation Ecology.* New York, Wiley.

Mukai, T. 1969. "The genetic structure of natural populations of *Drosophila melanogaster,* VII: Synergistic interaction of spontaneous mutant polygenes controlling viability." *Genetics,* 61: 149–161.

Mullen, D. A. 1968. "Reproduction in brown lemmings (*Lemmus trimucronatus*) and its relevance to their cycle of abundance." *Univ. Calif. Publ. Zool.,* 85: 1–24.

Muller, C. H. 1966. "The role of chemical inhibition (allelopathy) in vegetational composition." *Bull. Torrey Bot. Club,* 93: 332–351.

—— Hanawalt, R. B., and McPherson, J. K. 1968. "Allelopathic control of herb growth in the fire cycle of California chaparral." *Bull. Torrey Bot. Club,* 95: 225–231.

Muller, H. J. 1932. "Some genetic aspects of sex." *Amer. Nat.,* 8: 118–138.

Murdoch, W. W. 1966. "Population stability and life history phenomena." *Amer. Nat.,* 100: 5–11.

—— 1966. "'Community structure, population control, and competition'—A critique." *Amer. Nat.,* 100: 219–226.

—— 1969. "Switching in general predators: Experiments on predator specificity and stability of prey populations." *Ecol. Monogr.,* 39: 335–354.

—— and Marks, J. R. 1973. "Predation by coccinellid beetles: Experiments on switching." *Ecology,* 54: 160–167.

Murie, A. 1944. *The Wolves of Mt. McKinley.* Fauna of the National Parks of the U.S., Fauna Series No. 5. Washington, D.C., U.S. Dept. Int., Nat. Park Ser.

Murphy, G. I. 1968. "Pattern in life history and the environment." *Amer. Nat.,* 102: 391–403.

Murphy, R. C. 1936. *Oceanic Birds of South America.* New York, Amer. Mus. Nat. Hist.

Murton, R. K. 1967. "The significance of endocrine stress in population control." *Ibis,* 109: 622–623.

—— 1970. "Why do some bird species feed in flocks?" *Ibis,* 113: 534–535.

—— 1971. "The significance of a specific search image in the feeding behaviour of the wood-pigeon." *Behaviour,* 40: 10–42.

—— Isaacson, A. J., and Westwood, N. J. 1971. "The significance of gregarious feeding behaviour and adrenal stress in a population of wood-pigeons *Columba palumbus.*" *J. Zool.,* 165: 53–84.

Mutch, W. R. 1970. "Wildland fires and ecosystems—a hypothesis." *Ecology,* 51: 1046–1051.

Myers, K. 1970. "The rabbit in Australia," in P. J. den Boer and G. R. Gradwell, eds., *Dynamics of Populations.* Wageningen, The Netherlands, Centre Agric. Publ. Documentation.

Mykytowycz, R. 1967. "Communication by smell in the wild rabbit." *Proc. Ecol. Soc. Australia,* 2: 125–131.

Nagy, K. A., Odell, D. K., and Seymour, R. S. 1972. "Temperature regulation by the inflorescence of *Philodendron.*" *Science,* 178: 1196–1197.

National Academy of Sciences. 1969. *Eutrophication: Causes, Consequences, Correctives.* Washington, D.C., National Academy of Sciences.

Neales, T. F., Patterson, A. A., and Hartney, W. J. 1968. "Physiological adaptation to drought in the carbon assimilation and water loss of xerophytes." *Nature,* 219: 469–472.

Neilson, M. M., and Morris, R. F. 1964. "The regulation of European spruce sawfly numbers in the Maritime Provinces of Canada from 1937 to 1963." *Canad. Entomol.,* 96: 773–784.

Nelson, E. W. 1918. *Wild Animals of North America.* Washington, D.C., National Geographic Society.

Nelson, J. B. 1966. "The behavior of the young gannet." *Brit. Birds,* 59: 393–419.

Newbould, P. J. 1967. *Methods for Estimating the Primary Production of Forests.* Oxford, Blackwell.

Newell, N. D. 1949. "Phyletic size increase, an important trend illustrated by fossil invertebrates." *Evolution,* 3: 103–124.

——— 1963. "Crises in the history of life." *Sci. Amer.,* 208: 77–92.

——— 1967. "Revolutions in the history of life." *Geol. Soc. Amer. Spec. Pap.,* 89: 63–91.

Newsome, A. E. 1969. "A population study of house-mice permanently inhabiting a reed-bed in South Australia." *J. Anim. Ecol.,* 38: 361–377.

Nice, M. M. 1937. "Studies in the life history of the song sparrow, I: A population study of the song sparrow." *Trans. Linnean Soc. N.Y.,* 4: 1–247.

——— 1941. "The role of territory in bird life." *Amer. Midl. Nat.,* 26: 441–487.

Nicholson, A. J. 1933. "The balance of animal populations." *J. Anim. Ecol.,* 2: 132–178.

——— 1955. "Compensatory reactions of populations to stresses, and their evolutionary significance." *Australian J. Zool.,* 2: 1–8.

——— 1958. "The self-adjustment of populations to change." *Cold Spring Harbor Symp. Quant. Biol.,* 22: 153–173.

——— 1958. "Dynamics of insect populations." *Ann. Rev. Entomol.,* 3: 107–136.

Nicol, J. A. C. 1967. *The Biology of Marine Animals,* 2d ed. New York, Wiley.

Nikolsky, G. V. 1963. *The Ecology of Fishes,* trans. from Russian by L. Birkett. New York and London, Academic Press.

Nixon, C. M. 1965. "White-tailed deer growth and productivity in eastern Ohio." *Game Res. Ohio,* 3: 123–136.

Norman, A. G. 1957. "Soil-plant relationships and plant nutrition." *Amer. J. Bot.,* 44: 67–73.

Norris, K. S. 1953. "The ecology of the desert iguana, *Dipsosaurus dorsalis.*" *Ecology,* 34: 265–287.

Norris, R. T. 1947. "The cowbirds of Preston Frith." *Wilson Bull.,* 59: 83–103.

North, W. J. 1970. "Kelp habitat improvement project." *W. M. Keck Lab. Environ. Health Eng., Calif. Inst. Tech., Ann. Rept.*

——— 1972. "Kelp habitat improvement project." *W. M. Keck Lab. Environ. Health Eng., Calif. Inst. Tech., Ann. Rept.*

Norton-Griffiths, M. 1967. "Some ecological aspects of the feeding behaviour of the oystercatcher *Haematopus ostralegus* on the edible mussel *Mytilus edulis.*" *Ibis,* 109: 412–424.

Novitski, E. 1947. "Genetic analysis of an anomalous sex ratio condition in *Drosophila affinis.*" *Genetics,* 32: 526–534.

O'Connor, F. B. 1963. "Oxygen consumption and population metabolism of Enchytraeidae," in J. Doeksen and J. Van Der Drift, eds., *Soil Organisms.* Amsterdam, North-Holland Publ. Co.

O'Donald, P. 1969. "'Haldane's dilemma' and the rate of natural selection." *Nature,* 221: 815–816.

—— 1973. "A further analysis of Bumpus's data: The intensity of natural selection." *Evolution,* 27: 398–404.

—— Wedd, N. S., and Davis, J. W. F. 1974. "Mating preferences and sexual selection in the arctic skua." *Heredity,* 33: 1–16.

Odum, E. P. 1958. "The fat deposition picture in the white-throated sparrow in comparison with that in long-range migrants." *Bird-Banding,* 29: 105–108.

—— 1959. *Fundamentals of Ecology,* 2d ed. Philadelphia, Saunders.

—— 1960. "Lipid deposition in nocturnal migrant birds." *Proc. XIIth Inter. Ornith. Congr.,* 563–576.

—— 1962. "Relationships between structure and function in the ecosystem." *Jap. J. Ecol.,* 12: 108–118.

—— 1969. "The strategy of ecosystem development." *Science,* 164: 262–270.

—— 1971. *Fundamentals of Ecology,* 3d ed. Philadelphia, Saunders.

—— Connell, C. E., and Davenport, L. B. 1962. "Population energy flow of three primary consumer components of old-field ecosystems." *Ecology,* 43: 88–96.

Odum, E. P., Connell, C. E., and Stoddard, H. L. 1961. "Flight energy and estimated flight ranges of some migratory birds." *Auk,* 78: 515–527.

Odum, E. P., and Kuenzler, E. J. 1963. "Experimental isolation of food chains in an old-field ecosystem with use of phosphorus-32," in V. Schultz and A. W. Klement, eds., *Radioecology.* New York, Reinhold.

Odum, E. P., and Smalley, E. A. 1959. "Comparison of population energy flow of a herbivorous and a deposit-feeding invertebrate in a salt marsh ecosystem." *Proc. Natl. Acad. Sci.,* 45: 617–622.

Odum, H. T. 1956. "Primary production in flowing waters." *Limnol. Oceanogr.,* 1: 102–117.

—— 1957a. "Trophic structure and productivity of Silver Springs, Florida." *Ecol. Monogr.,* 27: 55–112.

—— 1957b. "Primary production measurements in eleven Florida springs and a marine turtle-grass community." *Limnol. Oceanogr.,* 2: 85–97.

—— 1971. *Environment, Power, and Society.* New York, Wiley.

—— and Jordan, C. F. 1970. "Metabolism and evapotranspiration of the lower forest in a giant plastic cylinder," in H. T. Odum and R. F. Pigeon, eds., *A Tropical Rainforest. A Study of Irradiation and Ecology at El Verde, Puerto Rico.* Washington, D.C., U.S. Atomic Energy Comm.

Ogawa, H., Yoda, K., and Kira, T. 1961. "A preliminary survey on the vegetation of Thailand." *Nature Life S.E. Asia,* 1: 1–157.

Ohno, S. 1970. *Evolution by Gene Duplication.* New York, Springer-Verlag.

Olsen, C. 1948. "The mineral, nitrogen and sugar content of beech leaves and beech sap at various times." *Compt. Rend. Trav. Lab. Carlsberg, Ser. Chim.,* 26: 197–230.

Olson, J. S. 1958. "Rates of succession and soil changes on southern Lake Michigan sand dunes." *Bot. Gaz.,* 119: 125–170.

—— 1963. "Energy storage and the balance of producers and decomposers in ecological systems." *Ecology,* 44: 322–331.

O'Neill, R. V. 1968. "Population energetics of the millipede, *Narceus americanus* (Beauvois)." *Ecology,* 49: 803–809.

Oosting, H. J. 1942. "An ecological analysis of the plant communities of Piedmont, North Carolina." *Amer. Midl. Nat.,* 28: 1–126.

—— 1954. "Ecological processes and vegetation of the maritime strand in the southeastern United States." *Bot. Rev.,* 20: 226–262.

—— 1956. *The Study of Plant Communities,* 2d ed. San Francisco, W. H. Freeman.

Oparin, A. I. 1953. *The Origin of Life,* 2d ed. New York, Dover.

Opler, P. A. 1974. "Oaks as evolutionary islands for leaf-mining insects." *Amer. Sci.,* 62: 67–73.

Organ, J. A. 1961. "Studies of the population dynamics of the salamander genus *Desmognathus* in Virginia." *Ecol. Monogr.*, 31: 189–220.

Orians, G. H. 1961. "The ecology of blackbird (*Agelaius*) social systems." *Ecol. Monogr.*, 31: 285–312.

——— 1969a. "Age and hunting success in the brown pelican (*Pelecanus occidentalis*)." *Anim. Behav.*, 17: 316–319.

——— 1969b. "The number of bird species in some tropical forests." *Ecology*, 50: 783–801.

——— 1969c. "On the evolution of mating systems in birds and mammals." *Amer. Nat.*, 103: 589–603.

——— and Willson, M. F. 1964. "Interspecific territories of birds." *Ecology*, 45: 735–745.

Ovington, J. D. 1957. "Dry-matter production by *Pinus sylvestris* L." *Ann. Bot. N. S.*, 21: 287–314.

——— 1962. "Quantitative ecology and the woodland ecosystem concept," in J. C. Cragg, ed., *Advances in Ecological Research I.* New York and London, Academic Press.

——— 1965. "Organic production, turnover, and mineral cycling in woodlands." *Biol. Rev.*, 40: 295–336.

——— Heitkamp, D., and Lawrence, D. B. 1963. "Plant biomass and productivity of prairie, savanna, oakwood, and maize field ecosystems in central Minnesota." *Ecology*, 44: 52–63.

Ovington, J. D., and Madgwick, H. A. I. 1959. "The growth and composition of natural stands of birch, I: Dry-matter production." *Pl. Soil*, 10: 271–283.

Owen, D. F. 1963. "Polymorphism and population density in the African snail *Limicolaria martensiana*." *Science*, 140: 616–617.

——— 1966. *Animal Ecology in Tropical Africa.* San Francisco, W. H. Freeman.

Paine, R. T. 1963. "Trophic relationships of eight sympatric predatory gastropods." *Ecology*, 44: 63–73.

——— 1965. "Natural history, limiting factors and energetics of the opisthobranch *Navanax inermis*." *Ecology*, 46: 603–619.

——— 1966. "Food web complexity and species diversity." *Amer. Nat.*, 100: 65–75.

——— 1971. "Energy flow in a natural population of the herbivorous gastropod *Tegula funebralis*." *Limnol. Oceanogr.*, 16: 86–98.

——— 1971a. "The measurement and application of the calorie to ecological problems." *Ann. Rev. Ecol. Syst.*, 2: 145–164.

——— and Vadas, R. 1969. "The effects of grazing by sea urchins, *Strongylocentrotus* spp., on benthic algal populations." *Limnol. Oceanogr.*, 14: 710–719.

Pajunen, V. I. 1966. "The influence of population density on the territorial behavior of *Leucorrhinia rubicunda* L. (Odon., Libellulidae)." *Ann. Zool. Fenn.*, 3: 40–52.

Palmblad, I. G. 1968. "Competition in experimental populations of weeds with emphasis on the regulation of population size." *Ecology*, 49: 26–34.

Pandian, T. J. 1967a. "Intake, digestion, absorption and conversion of food in the fishes *Megalops cyprinoides* and *Ophiocephalus striatus*." *Marine Biol.*, 1: 16–32.

——— 1967b. "Transformation of food in the fish *Megalops cyprinoides*." *Marine Biol.*, 1: 60–64.

Paranjape, M. A. 1967. "Molting and respiration of euphausiids." *J. Fish. Res. Bd. Can.*, 24: 1229–1240.

Park, T. 1954. "Experimental studies of interspecific competition, II: Temperature, humidity and competition in two species of *Tribolium*." *Physiol. Zool.*, 27: 177–238.

—— 1962. "Beetles, competition, and populations." *Science,* 138: 1369–1375.

—— Leslie, P. H., and Mertz, D. B. 1964. "Genetic strains and competition in populations of *Tribolium.*" *Physiol. Zool.,* 38: 289–321.

Parr, J. C. 1968. "Toxicity of *Nicotiana* and *Petunia* species to larvae of the tobacco hornworm." *J. Econ. Entomol.,* 61: 1525–1531.

Parrish, J. D., and Saila, S. B. 1970. "Interspecific competition, predation, and species diversity." *J. Theoret. Biol.,* 27: 207–220.

Patrick, R. 1963. "The structures of diatom communities under varying ecological conditions." *Ann. N. Y. Acad. Sci.,* 108: 353–358.

—— 1966. "The Catherwood Foundation Peruvian-Amazon Expedition. Introduction, I: Limnological observations and discussion of results." *Monogr. Acad. Nat. Sci. Phila.,* No. 14: 1–4, 5–40.

—— 1967. "The effect of invasion rate, species pool, and size of area on the structure of the diatom community." *Proc. Natl. Acad. Sci.,* 58: 1335–1342.

—— Hohn, M. H., and Wallace, J. H. 1954. "A new method for determining the pattern of the diatom flora." *Notulae Naturae,* No. 259.

Patton, J. L. 1969. "Chromosome evolution in the pocket mouse, *Perognathus goldmani* Osgood." *Evolution,* 23: 645–662.

Payne, R. B. 1967. "Interspecific communication signals in parasitic birds." *Amer. Nat.,* 101: 363–375.

Peaker, M., and Linzell, J. L. 1975. *Salt Glands in Birds and Reptiles.* Cambridge, Cambridge Univ. Press.

Pearl, R. 1925. *The Biology of Population Growth.* New York, Knopf.

—— and Reed, L. J. 1920. "On the rate of growth of the population of the United States since 1790 and its mathematical representation." *Proc. Natl. Acad. Sci.,* 6: 275–288.

Pearson, O. P. 1954. "Oxygen consumption and bioenergetics of harvest mice." *Physiol. Zool.,* 33: 152–160.

—— 1966. "The prey of carnivores during one cycle of mouse abundance." *J. Anim. Ecol.,* 35: 217–233.

Peet, R. K. 1975. "Relative diversity indices." *Ecology,* 56: 496–498.

—— and Loucks, O. L. 1977. "A gradient analysis of southern Wisconsin forests." *Ecology,* 58: 485–499.

Perttunen, V. 1972. "Humidity and light reactions of *Sitophilus granarius* L., *S. oryzae* L. (Col., Curculionidae), *Rhizopertha dominica* F. (Bostry-chidae), and *Acanthoscelides obtectus* Say (Bruchidae)." *Ann. Ent. Fenn.,* 38: 161–176.

Petrides, G. A., and Swank, W. G. 1966. "Estimating the productivity and energy relations of an African elephant population." *Proc. Ninth Int. Grasslands Congr., São Paulo,* 831–842.

Petrusewicz, K., and MacFadyen, A. 1970. *Productivity of Terrestrial Animals. Principles and Methods. IBP Handbook* No. 13. Oxford, Blackwell.

Pfeiffer, W. 1963a. "Alarm substances." *Experientia,* 19: 113–168.

—— 1963b. "The fright reaction in North American fish." *Canad. J. Zool.,* 41: 69–77.

Phillipson, J. 1966. *Ecological Energetics.* London, Edward Arnold.

Pianka, E. R. 1966. "Latitude gradients in species diversity: A review of concepts." *Amer. Nat.,* 100: 33–46.

—— 1970a. "Comparative autecology of the lizard *Cnemidophorus tigris* in different parts of its geographic range." *Ecology,* 51: 703–720.

—— 1970b. "On *r*- and *K*-selection." *Amer. Nat.,* 104: 592–597.

—— 1971. "Lizard species density in the Kalahari Desert." *Ecology,* 52: 1024–1029.

—— 1971a. "Comparative ecology of two lizards." *Copeia,* 1971: 129–138.

—— 1971b. "Species diversity," in A. Kramer, ed., *Topics in the Study of Life: The Bio Source Book.* New York, Harper and Row.

——— 1972. "*r* and *K* selection or *b* and *d* selection?" *Amer. Nat.*, 106: 581–588.

——— 1973. "The structure of lizard communities." *Ann. Rev. Ecol. Syst.*, 4: 53–74.

——— 1974. "Niche overlap and diffuse competition." *Proc. Nat. Acad. Sci.*, 71: 2141–2145.

——— and Parker, W. S. 1975. "Age-specific reproductive tactics." *Amer. Nat.*, 109: 453–464.

Pielou, E. C. 1966. "Comment on a report by J. H. Vandermeer and R. H. MacArthur concerning the broken stick model of species abundance." *Ecology*, 47: 1073–1074.

——— 1966a. "The measurement of diversity in different types of biological collections." *J. Theoret. Biol.*, 13: 131–144.

——— 1977. *Mathematical Ecology.* New York, Wiley.

Pielou, E. C., and Arnason, A. N. 1966. "Correction to one of MacArthur's species-abundance formulas." *Science*, 151: 592.

Pierce, R. S. 1969. "Forest transpiration reduction by clear cutting and chemical treatment." *Proc. N. E. Weed Contr. Conf.*, 23: 344–349.

Pietsch, T. W. 1975. "Precocious sexual parasitism in the deep sea ceratioid anglerfish, *Cryptosaras couesi* Gill." *Nature*, 256: 38–40.

Pigott, C. D., and Taylor, K. 1964. "The distribution of some woodland herbs in relation to the supply of nitrogen and phosphorus in the soil." *J. Ecol.*, 52 (Suppl.): 175–185.

Pimentel, D. 1961. "Species diversity and insect population outbreaks." *Ann. Ent. Soc. Amer.*, 54: 76–86.

——— 1968. "Population regulation and genetic feedback." *Science*, 159: 1432–1437.

——— and Al-Hafidh, R. 1963. "The coexistence of insect parasites and hosts in laboratory populations." *Ann. Entomol. Soc. Amer.*, 56: 676–678.

Pimentel, D., Feinberg, E. H., Wood, P. W., and Hayes, J. T. 1965. "Selection, special distribution, and the coexistence of competing fly species." *Amer. Nat.*, 99: 97–109.

Pimentel, D., Nagel, W. P., and Madden, J. L. 1963. "Space-time structure of the environment and the survival of parasite-host systems." *Amer. Nat.*, 97: 141–167.

Pimlott, D. H. 1967. "Wolf predation and ungulate populations." *Amer. Zool.*, 7: 267–278.

Pitelka, F. A. 1957a. "Some characteristics of microtine cycles in the Arctic," in H. P. Hansen, ed., *Arctic Biology.* Corvallis, Oregon St. Univ. Press.

——— 1957b. "Some aspects of population structure in the short-term cycle of the brown lemming in northern Alaska." *Cold Spring Harbor Symp. Quant. Biol.*, 22: 237–251.

——— 1973. "Cyclic pattern in lemming populations near Barrow, Alaska," in M. E. Britton, ed., *Alaskan Arctic Tundra.* Arctic Inst. North Amer., Tech. Pap. No. 25.

——— Holmes, R. T., and MacLean, S. F. 1974. "Ecology and evolution of social organization in arctic sandpipers." *Amer. Zool.*, 14: 185–204.

Pitelka, F. A., Tomich, P. O., and Treichel, G. W. 1955. "Ecological relations of jaegers and owls as lemming predators near Barrow, Alaska." *Ecol. Monogr.*, 25: 85–117.

Plank, J. E. van der. 1968. *Disease Resistance in Plants.* New York, Academic Press.

Poe, S. L., and Enns, W. R. 1970. "Effects of inbreeding on closed populations of predaceous mites (Acarina: Phytoseiidae)." *Canad. Entomol.*, 102: 1222–1229.

Pond, C. M. 1977. "The significance of lactation in the evolution of mammals." *Evolution*, 31: 177–199.

Poole, R. W. 1974. *An Introduction to Quantitative Ecology.* New York, McGraw-Hill.

Popham, E. J. 1941. "The variation in the colour of certain species *Arctocorisa* (Hemiptera, Corixidae) and its significance." *Proc. Zool. Soc. London, A*, 111: 135–172.

———— 1942. "Further experimental studies on the selective action of predators." *Proc. Zool. Soc. London, A,* 112: 105–117.

Porter, J. W. 1972a. "Predation by *Acanthaster* and its effect on coral species diversity." *Amer. Nat.,* 106: 487–492.

———— 1972b. "Ecology and species diversity of coral reefs on opposite sides of the isthmus of Panama." *Bull. Biol. Soc. Wash.,* 2: 89–116.

———— 1974. "Community structure of coral reefs on opposite sides of the Isthmus of Panama." *Science,* 186: 543–545.

Porter, W. P., and Gates, D. M. 1969. "Thermodynamic equilibria of animals with the environment." *Ecol. Monogr.,* 39: 245–270.

Porter, W. P., Mitchell, J. W., Beckman, W. A., and deWitt, C. B. 1973. "Behavioral implications of mechanistic ecology. Thermal and behavioral modeling of desert ectotherms and their microenvironment." *Oecologia,* 13: 1–54.

Portmann, A. 1959. *Animal Camouflage.* Ann Arbor, Univ. of Michigan Press.

———— 1961. *Animals as Social Beings,* trans. by O. Coburn. New York, Viking.

———— 1967. *Animal Forms and Patterns.* New York, Schocken Books.

Potts, W. T. W., and Parry, G. 1964. *Osmotic and Ionic Regulation in Animals.* Oxford, Pergamon.

Powell, J. A., and Mackie, R. A. 1966. "Biological interrelationships of moths and *Yucca whipplei* (Lepidoptera: Gelechiidae, Blastobasidae, Prodoxidae)." *Univ. of Calif. Publ. Entomol.,* 42: 1–46.

Prakash, S. 1969. "Genetic variation in a natural population of *Drosophila persimilis.*" *Proc. Natl. Acad. Sci.,* 62: 778–784.

———— 1973. "Patterns of gene variation in central and marginal populations of *Drosophila robusta.*" *Genetics,* 75: 347–369.

———— Lewontin, R. C., and Hubby, J. L. 1969. "A molecular approach to the study of genic heterozygosity in natural populations, IV: Patterns of genic variation in central, marginal, and isolated populations of *Drosophila pseudoobscura.*" *Genetics,* 61: 841–858.

Prance, G. T. 1973. "Phytogeographic support for the theory of Pleistocene forest refuges in the Amazon Basin, based on evidence from distribution patterns in Caryocaraceae, Chrysobalanaceae, Dichapetalaceae and Lecythidaceae." *Acta Amazonica,* 3: 5–28.

Preston, F. W. 1948. "The commonness, and rarity, of species." *Ecology,* 29: 254–283.

———— 1960. "Time and space and the variation of species." *Ecology,* 41: 611–627.

———— 1962. "The canonical distribution of commonness and rarity." *Ecology,* 43: 185–215, 410–432.

———— 1969. "Diversity and stability in the biological world." *Brookhaven Symp. Biol.,* No. 22: 1–12.

Price, P. W. 1975. *Insect Ecology.* New York, Wiley.

Prosser, C. L. 1973. *Comparative Animal Physiology,* 3d ed. Philadelphia, Saunders.

———— and Brown, F. A. 1961. *Comparative Animal Physiology,* 2d ed. Philadelphia, Saunders.

Pulliam, H. R. 1973. "On the advantages of flocking." *J. Theoret. Biol.,* 38: 419–422.

———— 1975. "Coexistence of sparrows: A test of community theory." *Science,* 189: 474–476.

———— and Enders, F. 1971. "The feeding ecology of five sympatric finch species." *Ecology,* 52: 557–566.

Putwain, P. D., Machin, D., and Harper, J. L. 1968. "Studies in the dynamics of plant populations, II: Components and regulation of a natural population of *Rumex acetosella* L." *J. Ecol.,* 56: 421–431.

Pyke, G. H., Pulliam, H. R., and Charnov, E. L. 1977. "Optimal foraging: A selective review of theory and tests." *Quart. Rev. Biol.,* 52: 137–154.

Quayle, H. J. 1938. "The development of resistance in certain insects to hydrocyanic gas." *Hilgardia,* 11: 183–225.

Quispel, A., ed. 1974. *The Biology of Nitrogen Fixation.* Amsterdam, North-Holland Publ. Co.

Rabb, R. L., and Bradley, J. R. 1968. "The influence of host plants on parasitism of eggs of the tobacco hornworm." *J. Econ. Entomol.,* 61: 1249–1252.

Rabinowitch, E., and Govindjee. 1969. *Photosynthesis.* New York, Wiley.

Race, R. R., and Sanger, R. 1962. *Blood Groups in Man.* Philadelphia, F. A. Davis.

Rand, A. S. 1967. "Predator-prey interactions and the evolution of aspect diversity." *Atas do Simpósio sôbre a Biota Amazonica,* 5: 73–83.

Randall, D. J. 1968. "Fish physiology." *Amer. Zool.,* 8: 179–189.

Rapport, D. J. 1971. "An optimization model of food selection." *Amer. Nat.,* 105: 575–587.

Rasmussen, D. I. 1941. "Biotic communities of Kaibab Plateau, Arizona." *Ecol. Monogr.,* 11: 230–275.

Rathcke, B. J. 1976. "Competition and coexistence within a guild of herbivorous insects." *Ecology,* 57: 76–87.

Raunkiaer, C. 1934. *The Life Forms of Plants and Statistical Plant Geography.* Clarendon Press. Oxford, Oxford Univ. Press.

———— 1937. *Plant Life Forms,* trans. by H. Gilbert-Carter. Clarendon Press. Oxford, Oxford Univ. Press.

Raup, D. M. 1975. "Taxonomic survivorship curves and Van Valen's law." *Paleobiol.,* 1: 82–96.

Recher, H. 1969. "Bird species diversity and habitat diversity in Australia and North America." *Amer. Nat.,* 103: 75–80.

Recher, H. F., and Recher, J. A. 1969. "Some aspects of the ecology of migrant shorebirds, II: Aggression." *Wilson Bull.,* 81: 141–154.

Redfield, A. C. 1972. "Development of a New England salt marsh." *Ecol. Monogr.,* 42: 201–237.

Reeve, M. R. 1963. "The filter feeding of *Artemia,* I: In pure cultures of plant cells." *J. Exptl. Biol.,* 40: 195–205.

Rehr, S. S., Feeny, P. P., and Janzen, D. H. 1973. "Chemical defense in Central American non-ant-acacias." *J. Anim. Ecol.,* 42: 405–416.

Reichle, D. E., ed. 1970. *Analysis of Temperate Forest Ecosystems.* New York, Springer-Verlag.

Reid, G. K. 1961. *Ecology of Inland Waters and Estuaries.* New York, Reinhold.

Rensch, B. 1948. "Histological changes correlated with evolutionary changes of body size." *Evolution,* 2: 218–230.

Rettenmeyer, C. W. 1963. "Behavioral studies of army ants." *Univ. of Kans. Sci. Bull.,* 44: 281–465.

———— 1970. "Insect mimicry." *Ann. Rev. Entomol.,* 15: 43–74.

Rhijn, J. G. Van. 1969. "Behavioural dimorphism in male ruffs, *Philomachus pugnax* (L.)." *Behaviour,* 47: 153–229.

Rice, E. L. 1975. *Allelopathy.* London, Academic Press.

Richards, P. W. 1952. *The Tropical Rainforest.* London and New York, Cambridge Univ. Press.

Richman, S. 1958. "The transformation of energy by *Daphnia pulex.*" *Ecol. Monogr.,* 28: 273–291.

Ricker, W. E. 1954. "Stock and recruitment." *J. Fish. Res. Bd. Can.,* 11: 559–623.

Ricketts, E. F., and Calvin, J. 1952. *Between Pacific Tides,* 3d ed., rev. by J. Hedgpeth. Stanford, Calif., Stanford Univ. Press.

Ricklefs, R. E. 1965. "Brood reduction in the curve-billed thrasher." *Condor,* 67: 505–510.

———— 1966. "The temporal component of diversity among species of birds." *Evolution,* 20: 235–242.

———— 1968. "Patterns of growth in birds." *Ibis,* 110: 419–451.

———— 1969a. "Preliminary models for growth rates of altricial birds." *Ecology,* 50: 1031–1039.

———— 1969b. "An analysis of nesting mortality in birds." *Smithson. Contrib. Zool.,* 9: 1–48.

———— 1969c. "The nesting cycle of songbirds in tropical and temperate regions." *Living Bird,* 8: 165–175.

———— 1969d. "Natural selection and the development of mortality rates in young birds." *Nature,* 233: 922–925.

———— 1970a. "Clutch size in birds: Outcome of opposing predator and prey adaptations." *Science,* 168: 599–600.

———— 1970b. "Stage of taxon cycle and distribution of birds on Jamaica, Greater Antilles." *Evolution,* 24: 475–477.

———— 1972a. "A study in population ecology. Review of David Lack, *Ecological Isolation in Birds,* Harvard Univ. Press (1971)." *Science,* 175: 288–289.

———— 1972b. "Dominance and the niche in bird communities." *Amer. Nat.,* 106: 538–545.

———— 1973. "Fecundity, mortality, and avian demography," in D. S. Farner, ed., *Breeding Biology of Birds.* Washington, D.C., Nat. Res. Council.

———— 1973a. *Ecology* (1st ed.). Portland, Ore., Chiron Press.

———— 1973b. "Patterns of growth in birds, II: Growth rate and mode of development." *Ibis,* 115: 177–201.

———— 1974. "Energetics of reproduction in birds," in R. A. Paynter, ed., *Avian Energetics.* Cambridge, Mass., Nuttall Ornithol. Club.

———— 1975. "The evolution of co-operative breeding in birds." *Ibis,* 117: 531–534.

———— 1977. "On the evolution of reproductive strategies in birds: Reproductive effort." *Amer. Nat.,* 111: 453–478.

———— 1977a. "Environmental heterogeneity and plant species diversity: An hypothesis." *Amer. Nat.,* 111: 376–381.

———— Adams, R. M., and Dressler, R. L. 1969. "Species diversity of *Euglossa* in Panama." *Ecology,* 50: 713–716.

Ricklefs, R. E., and Cox, G. W. 1972. "Taxon cycles in the West Indian avifauna." *Amer. Nat.,* 106: 195–219.

———— 1978. "Stage of taxon cycle, habitat distribution, and population density in the avifauna of the West Indies." *Amer. Nat.,* in press.

Ricklefs, R. E., and Cullen, J. 1973. "Embryonic growth of the Green Iguana, *Iguana iguana.*" *Copeia,* 1973: 296–305.

Ricklefs, R. E., and Hainsworth, F. R. 1968. "Temperature dependent behavior of the Cactus Wren." *Ecology,* 49: 227–233.

———— 1969. "Temperature regulation in nestling Cactus Wrens. The nest environment." *Condor,* 71: 32–37.

Ricklefs, R. E., and O'Rourke, K. 1975. "Aspect diversity in moths: A temperate-tropical comparison." *Evolution,* 29: 313–324.

Riggs, A. 1960. "The nature and significance of the Bohr effect in mammalian hemoglobins." *J. Gen. Physiol.,* 43: 737–752.

Riley, C. V. 1892. "The yucca moth and yucca pollination." *Third Ann. Rept. Mo. Bot. Garden,* 99–159.

Riley, G. A. 1938–39. "Plankton studies, I: A preliminary investigation of the plankton of the Tortugas Region, II: The western North Atlantic, May–June, 1939." *J. Marine Res.,* I: 335–352, II: 145–162.

Ritter, W. E. 1938. *The California Woodpecker and I. A Study in Comparative Zoology.* Berkeley, Univ. of California Press.

Robbins, C. T., and Moen, A. N. 1975. "Composition and digestibility of several deciduous browses in the Northeast." *J. Wildl. Mgmt.,* 39: 337–341.

Robbins, J. D. 1971. "Differential niche utilization in a grassland sparrow." *Ecology,* 52: 1065–1070.

Roberts, M. R., and Lewis, H. 1955. "Subspeciation in *Clarkia biloba.*" *Evolution,* 9: 445–454.

Robertson, F. W. 1957. "Studies in quantitative inheritance, XI: Genetic and environmental correlation between body size and egg production in *Drosophila melanogaster.*" *J. Genet.,* 55: 428–443.

Robinson, M. H. 1969a. "Predatory behavior of *Argiope argentata* (Fabricius)." *Amer. Zool.,* 9: 161–174.

—— 1969b. "Defenses against visually hunting predators." *Evol. Biol.,* 3: 225–259.

Rockstein, M., ed. 1974. *Theoretical Aspects of Aging.* New York, Academic Press.

Rodin, L. E., and Bazilevich, N. I. 1964. "Biological productivity of the fundamental types of vegetation of the Northern Hemisphere of the Old World." *Doklady Akademii Nauk SSSR,* 157: 215–218.

Rohlf, F. J., and Schnell, G. D. 1971. "An investigation of the isolation-by-distance model." *Amer. Nat.,* 105: 295–324.

Rokop, F. J. 1974. "Reproductive patterns in the deep-sea benthos." *Science,* 186: 743–745.

Romanoff, A. L. 1960. *The Avian Embryo: Structural and Functional Development.* New York, Macmillan.

Romer, A. S. 1966. *Vertebrate Paleontology,* 3d ed. Chicago, Univ. of Chicago Press.

Rommell, L. G. 1930. "Comments on Raunkiaer's law and similar methods of vegetation analysis and the law of frequency." *Evolution,* 11: 589–596.

Rosenzweig, M. L. 1968. "The strategy of body size in mammalian carnivores." *Amer. Midl. Nat.,* 80: 299–315.

—— 1969. "Why the prey curve has a hump." *Amer. Nat.,* 103: 81–87.

—— 1973. "Evolution of the predator isocline." *Evolution,* 27: 84–94.

—— 1975. "On continental steady states of species diversity," in M. L. Cody and J. M. Diamond, eds., *Ecology and Evolution of Communities.* Cambridge, Mass., Harvard Univ. Press.

—— and MacArthur, R. H. 1963. "Graphical representation and stability conditions of predator-prey interactions." *Amer. Nat.,* 97: 209–223.

Rothenbuhler, W. C. 1964a. "Behaviour genetics of nest cleaning in honey-bees, I: Responses of four inbred lines to disease-killed brood." *Anim. Behav.,* 12: 578–583.

—— 1964b. "Behavior genetics of nest cleaning in honeybees, IV: Responses of F_1 and backcross generations to disease-killed brood." *Amer. Zool.,* 4: 111–123.

Rothschild, M. 1965a. "The rabbit flea and hormones." *Endeavor,* 24: 162–168.

—— 1965b. "Fleas." *Sci. Amer.,* 213: 44–53.

Rothstein, S. I. 1971. "Observation and experiment in the analysis of interactions between brood parasites and their hosts." *Amer. Nat.,* 105: 71–74.

—— 1975. "Evolutionary rates and host defenses against avian brood parasitism." *Amer. Nat.,* 109: 161–176.

Roughgarden, J. 1974. "Niche width: Biogeographic patterns among *Anolis* lizard populations." *Amer. Nat.,* 108: 429–442.

Rovira, A. D. 1965. "Interaction between plant roots and soil microorganisms." *Ann. Rev. Microbiol.,* 19: 241–266.

Rowell, C. H. F. 1970. "Environmental control of coloration in an acridid, *Gastrimargus africanus* (Saussure)." *Anti-Locust Bull.,* 47: 1–48.

Rowell, T. E. 1967. "Variability in the social organization of primates," in D. Morris, ed., *Primate Ethology.* Chicago, Aldine.

Royama, T. 1970. "Factors governing the hunting behaviour and selection of food by the great tit *(Parus major* L.)." *J. Anim. Ecol.,* 39: 619–659.

Ruano, R. G., Orozco, F., and López-Fanjul, C. 1975. "The effect of different selection intensities on selection response in egg-laying of *Tribolium castaneum.*" *Genet. Res.,* 25: 17–27.

Rubinoff, I. 1968. "Central America sea-level canal: Possible biological effects." *Science,* 161: 857–861.

Russell, E. W. 1961. *Soil Conditions and Plant Growth,* 9th ed. New York, Wiley.

Ruttner, F. 1963. *Fundamentals of Limnology,* 3d ed. Toronto, Univ. of Toronto Press.

Ryther, J. H. 1956. "Photosynthesis in the ocean as a function of light intensity." *Limnol. Oceanogr.,* 1: 61–70.

———— 1969. "Photosynthesis and fish production in the sea." *Science,* 166: 72–76.

———— and Yentsch, C. S. 1957. "The estimation of phytoplankton production in the ocean from chlorophyll and light data." *Limnol. Oceanogr.,* 2: 281–286.

Sabine, W. F. 1959. "The winter society of the Oregon junco: Intolerance, dominance and the pecking order." *Condor,* 61: 110–135.

Sadleir, R. M. F. S. 1969. *The Ecology of Reproduction in Wild and Domestic Mammals.* London, Methuen.

Sakai, K. 1961. "Competitive ability in plants: Its inheritance and some related problems." *Symp. Soc. Exptl. Biol.,* 15: 245–263.

Salisbury, E. J. 1942. *The Reproductive Capacity of Plants, Studies in Quantitative Biology.* London, G. Bell and Sons.

Salthe, S. N. 1969. "Reproductive modes and the number and sizes of ova in the urodeles." *Amer. Midl. Nat.,* 81: 467–490.

———— 1975. "Some comments on Van Valen's law of extinction." *Paleobiol.,* 1: 356–358.

———— and Duellman, W. E. 1973. "Quantitative constraints associated with reproductive mode in anurans," in J. L. Vial, ed., *Evolutionary Biology of the Anurans.* Columbia, Univ. of Missouri Press.

Sanchez, P. A., ed. 1973. *A Review of Soils Research in Tropical Latin America.* North Carolina Agric. Expt. Sta., Tech. Bull. No. 219.

Sanders, H. L. 1968. "Marine benthic diversity: A comparative study." *Amer. Nat.,* 102: 243–282.

———— 1969. "Benthic marine diversity and the stability-time hypothesis." *Brookhaven Symp. Biol.,* 22: 71–81.

Sandler, L., and Novitski, E. 1957. "Meiotic drive as an evolutionary force." *Amer. Nat.,* 91: 105–110.

Sarvella, P. 1970. "Sporadic occurrence of parthenogenesis in poultry." *I. Heredity,* 61: 215–219.

Saunders, P. T., and Bazin, M. J. 1975a. "On the stability of food chains." *J. Theoret. Biol.,* 52: 121–142.

———— 1975b. "Stability of complex ecosystems." *Nature,* 256: 120.

Saura, A., Halkka, O., and Lokki, J. 1973. "Enzyme gene heterozygosity in small island populations of *Philaenus spumarius* L. (Homoptera)." *Genetica,* 44: 459–473.

Sawyer, C. N. 1946. "Fertilization of lakes by agricultural and urban drainage." *J. New England Water Works Assoc.,* 61: 109–127.

Schad, G. A. 1966. "Immunity, competition, and natural regulation of helminth populations." *Amer. Nat.,* 100: 359–364.

Schaeffer, B. 1952. "Rates of evolution in the coelacanth and dipnoan fishes." *Evolution,* 6: 101–111.

Schaffer, W. M. 1968. "Intraspecific combat and the evolution of the Caprini." *Evolution,* 22: 817–825.

——— 1974. "Optimal reproductive effort in fluctuating environments." *Amer. Nat.,* 108: 783–790.

——— and Reed, C. A. 1972. "The co-evolution of social behavior and cranial morphology in sheep and goats (Bovidae, Caprini)." *Fieldiana Zool.,* 61: 1–62.

Scharloo, W. 1971. "Reproductive isolation by disruptive selection: Did it occur?" *Amer. Nat.,* 105: 83–86.

Scheffer, V. B. 1951. "The rise and fall of a reindeer herd." *Science Monthly,* 75: 356–362.

Schiess, L. R. 1963. "Die postembryonale Ausbildung der Körperproportionen bei Vögeln." *Rev. Suisse Zool.,* 70: 689–740.

Schindler, D. W. 1973. "Experimental approaches to limnology—an overview." *J. Fish. Res. Bd. Can.,* 30: 1409–1552.

——— 1974. "Eutrophication and recovery in experimental lakes: Implications for lake management." *Science,* 184: 897–899.

——— and Fee, E. J. 1974. "Experimental lakes area: Whole-lake experiments in eutrophication." *J. Fish. Res. Bd. Can.,* 31: 937–953.

Schmidt-Nielsen, B. 1972. "Mechanisms of urea excretion by the vertebrate kidney," in J. W. Campbell and L. Goldstein, eds., *Nitrogen Metabolism and the Environment.* London, Academic Press.

Schmidt-Nielsen, K. 1959. "Physiology of the camel." *Sci. Amer.,* 201: 140–151.

——— 1964. *Desert Animals: Physiological Problems of Heat and Water.* Oxford, Clarendon Press.

——— 1970. "Energy metabolism, body size, and problems of scaling." *Federation Proc.,* 29: 1524–1532.

——— 1972. *How Animals Work.* Cambridge, Cambridge Univ. Press.

——— 1975. *Animal Physiology. Adaptation and Environment.* London and New York, Cambridge Univ. Press.

——— 1975a. "Scaling in biology: The consequences of size." *J. Exptl. Zool.,* 194: 287–307.

——— Hainsworth, F. R., and Murrish, D. E. 1970. "Counter-current heat exchange in the respiratory passages: Effect on water and heat balance." *Resp. Physiol.,* 9: 263–276.

Schmidt-Nielsen, K., and Schmidt-Nielsen, B. 1952. "Water metabolism of desert mammals." *Physiol. Rev.,* 32: 135–166.

——— 1953. "The desert rat." *Sci. Amer.,* 189: 73–78.

Schmidt-Nielsen, K., and Taylor, C. R. 1968. "Red blood cells: Why or why not?" *Science,* 162: 274–275.

Schneirla, T. C. 1956a. "A preliminary survey of colony division and related processes in two species of terrestrial army ants." *Insectes Sociaux,* 3: 49–69.

——— 1956b. "The army ants." *Smithson. Inst. Publs., Rept. No. 4203:* 379–406.

Schoener, T. W. 1965. "The evolution of bill size differences among sympatric congeneric species of birds." *Evolution,* 19: 189–213.

——— 1967. "The ecological significance of sexual dimorphism in size in the lizard *Anolis conspersus.*" *Science,* 155: 474–477.

——— 1968. "The *Anolis* lizards of Bimini: Resource partitioning in a complex fauna." *Ecology,* 49: 704–726.

——— 1969a. "Models of optimal size for solitary predators." *Amer. Nat.,* 103: 277–313.

——— 1969b. "Optimal size and specialization in constant and fluctuating environments: An energy time approach." *Brookhaven Symp. Biol.,* 22: 103–114.

—— 1971. "Theory of feeding strategies." *Ann. Rev. Ecol. Syst.,* 2: 369–404.

—— 1974. "Resource partitioning in ecological communities." *Science,* 185: 27–39.

Scholander, P. F. 1955. "Evolution of climatic adaptation in homeotherms." *Evolution,* 9: 15–26.

—— 1956. "Climatic rules." *Evolution,* 10: 339–340.

—— van Dam, L., and Scholander, S. I. 1955. "Gas exchange in the roots of mangroves." *Amer. J. Bot.,* 42: 92–98.

Schopf, J. W. 1970. "Pre-Cambrian micro-fossils and evolutionary events prior to the origin of vascular plants." *Biol. Rev.,* 45: 319–352.

Schopf, T. J., and Gooch, J. L. 1971. "Gene frequencies in a marine ectoproct. A cline in natural populations related to sea temperature." *Evolution,* 25: 286–289.

—— 1972. "A natural experiment to test the hypothesis that loss of genetic variability was responsible for mass extinctions of the fossil record." *J. Geol.,* 80: 481–483.

Schreider, E. 1964. "Ecological rules, body-heat regulation, and human evolution." *Evolution,* 18: 1–9.

Schroder, G. D., and Rosensweig, M. L. 1975. "Perturbation analysis of competition and overlap in habitat utilization between *Dipodomys ordii* and *Dipodomys merriami.*" *Oecologia,* 19: 9–28.

Schroeder, H. 1919. "Quantitatives über die Verwendung der solaren Energie auf Erden." *Naturwiss.,* 7: 976–981.

Schultz, A. M. 1964. "The nutrient-recovery hypothesis for arctic microtine cycles, II: Ecosystem variables in relation to arctic microtine cycles," in D. J. Crisp, ed., *Grazing in Terrestrial and Marine Environments.* Oxford, Blackwell.

—— Launchbaugh, J. L., and Biswell, H. H. 1955. "Relationship between grass diversity and brush seedling survival." *Ecology,* 36: 226–238.

Schultz, R. J. 1961. "Reproductive mechanisms of unisexual and bisexual strains of the viviparous fish *Poeciliopsis.*" *Evolution,* 15: 302–325.

—— 1967. "Gynogenesis and triploidy in the viviparous fish *Poeciliopsis.*" *Science,* 157: 1564–1567.

—— 1969. "Hybridization, unisexuality, and polyploidy in the teleost *Poeciliopsis* (Poeciliidae) and other vertebrates." *Amer. Nat.,* 103: 605–619.

—— 1973. "Unisexual fish: Laboratory synthesis of a 'species.'" *Science,* 179: 180–181.

Schultz, V., and Klement, A. W., Jr., eds. 1963. *Radioecology.* New York, Reinhold.

Schulz, J. P. 1960. "Ecological studies on rain forest in northern Suriname." Amsterdam, North-Holland Publ. Co.

Schwab, J. H. 1975. "Suppression of the immune response by microorganisms." *Bact. Rev.,* 39: 121–143.

Scriber, J. M. 1973. "Latitudinal gradients in larval feeding specialization of the world Papilionidae (Lepidoptera)." *Psyche,* 80: 355–373.

Selander, R. K. 1965. "On mating systems and sexual selection." *Amer. Nat.,* 99: 129–144.

—— 1966. "Sexual dimorphism and differential niche utilization in birds." *Condor,* 68: 113–151.

—— 1970. "Behavior and genetic variation in natural populations." *Amer. Zool.,* 10: 53–66.

—— Hunt, W. G., and Yang, S. Y. 1969. "Protein polymorphism and genic heterozygosity in two European subspecies of the house mouse." *Evolution,* 23: 379–390.

Selander, R. K., and Johnson, W. E. 1973. "Genetic variation among vertebrate species." *Ann. Rev. Ecol. Syst.,* 4: 75–91.

Selander, R. K., Yang, S. Y., and Hunt, W. G. 1969. "Studies in Genetics." *Univ. of Texas Publ. No. 6918:* 271–338.

Selander, R. K., Yang, S. Y., Lewontin, R. C., and Johnson, W. E. 1970. "Genetic variation in the horseshoe crab (*Limulus polyphemus*), a phylogenetic 'relic.'" *Evolution*, 24: 402–414.

Serruya, C., and Berman, T. 1975. "Phosphorus, nitrogen and the growth of algae in Lake Kinneret." *J. Phycol.*, 11: 155–162.

Sette, O. E. 1943. "Biology of the Atlantic mackerel (*Scomber scombrus*) of North America, Part I: Early life history, including the growth, drift and mortality of egg and larval populations." *U.S. Fish Wildl. Ser., Fish. Bull. 50*, No. 38: 129–234.

Shan, R. K. 1969. "Life cycle of a chydorid cladoceran, *Pleuroxus denticulatus* Birge." *Hydrobiologia*, 34: 513–523.

Shanks, R. E. 1954. "Climates of the Great Smoky Mountains." *Ecology*, 35: 354–361.
——— and Olson, J. S. 1961. "First-year breakdown of leaf litter in southern Appalachian forests." *Science*, 134: 194–195.

Shantz, H. L. 1917. "Plant succession on abandoned roads in eastern Colorado." *J. Ecol.*, 5: 19–42.

Shaw, E. 1970. "Schooling in fishes: Critique and review," in L. R. Aronson, E. Tobach, D. S. Lehrman, and J. S. Rosenblatt, eds., *Development and Evolution of Behavior. Essays in Memory of T. C. Schneirla*. San Francisco, W. H. Freeman.

Sheldon, A. L. 1968. "Species diversity and longitudinal succession in stream fishes." *Ecology*, 49: 193–198.
——— 1969. "Equitability indices: Dependence on the species count." *Ecology*, 50: 466–467.

Shelford, V. E. 1911a. "Ecological succession: Stream fishes and the method of physiographic analysis." *Biol. Bull.*, 21: 9–34.
——— 1911b. "Ecological succession: Pond fishes." *Biol. Bull.*, 21: 127–151.

Shepard, M. P. 1955. "Resistance and tolerance of young speckled trout (*Salvelinus fontinalis*) to oxygen lack, with special reference to low oxygen acclimation." *J. Fish. Res. Bd. Canad.*, 12: 387–446.

Sheppard, P. M. 1951. "Fluctuation in the selective value of certain phenotypes in the polymorphic land snail *Cepaea nemoralis* (L.)." *Heredity*, 5: 125–134.
——— 1958. *Natural Selection and Heredity*. New York, Harper and Row.

Sheppe, W. A., and Adams, J. R. 1957. "The pathogenic effect of *Trypanosome duttoni* in hosts under stress conditions." *J. Parasitology*, 43: 55–59.

Shimwell, D. W. 1971. *Description and Classification of Vegetation*. Seattle, Univ. of Washington Press.

Shorrocks, B. 1970. "The distribution of gene frequencies in polymorphic populations of *Drosophila melanogaster*." *Evolution*, 24: 660–669.

Short, H. L. 1971. "Forage digestibility and diet of deer on southern upland range." *J. Wildl. Mgmt.*, 35: 698–706.

Shreve, F. 1915. "The vegetation of a desert mountain range as conditioned by climatic factors." *Publs. Carnegie Inst.*, 217: 1–112.
——— 1936. "The transition from desert to chaparral in Baja California." *Madroño*, 3: 257–264.

Sibley, C. G. 1954. "Hybridization in the red-eyed towhees of Mexico." *Evolution*, 8: 252–290.
——— and Short, L. L. Jr. 1959. "Hybridization in the buntings (*Passerina*) of the Great Plains." *Auk*, 76: 443–463.

Silberglied, R. E., and Eisner, T. 1969. "Mimicry of hymenoptera by beetles with unconventional flight." *Science*, 163: 486–488.

Silliman, R. P., and Gutsell, J. S. 1957. "Response of laboratory fish populations to fishing rates." *Trans. 22d North Amer. Wildl. Conf.*, 464–471.
——— 1958. "Experimental exploitation of fish populations." *U.S. Fish Wildl. Ser. Fish. Bull.* 58: 215–252.

Simberloff, D. S. 1969. "Experimental zoogeography of islands: A model for insular colonization." *Ecology,* 50: 296–314.

—— 1976. "Species turnover and equilibrium island biogeography." *Science,* 194: 572–578.

—— and Wilson, E. O. 1969. "Experimental zoogeography of islands: The colonization of empty islands." *Ecology,* 50: 278–296.

Simpson, E. H. 1949. "Measurement of diversity." *Nature,* 163: 688.

Simpson, G. G. 1949. *The Meaning of Evolution.* New Haven, Yale Univ. Press.

—— 1950. "History of the fauna of Latin America." *Amer. Sci.,* 38: 361–389.

—— 1953. *The Major Features of Evolution.* New York, Columbia Univ. Press.

—— 1964. "Species density of North American recent mammals." *Syst. Zool.,* 13: 57–73.

—— 1969. "The first three billion years of community evolution." *Brookhaven Symp. Biol.,* 22: 162–177.

Singh, J. S. 1968. "Net aboveground community productivity in the grasslands at Varanasi," in R. Misra and B. Gopal, eds., *Proc. Symp. Recent Adv. Trop. Ecol.* Part I. Varanasi, India, International Society for Tropical Ecology.

Sinnott, E. W. 1960. *Plant Morphogenesis.* New York, McGraw-Hill.

Skutch, A. F. 1949. "Do tropical birds rear as many young as they can nourish?" *Ibis,* 91: 430–455.

—— 1961. "Helpers among birds." *Condor,* 63: 198–226.

—— 1967. "Adaptive limitation of the reproductive rate of birds." *Ibis,* 109: 579–599.

Slatkin, M. 1973. "Gene flow and selection in a cline." *Genetics,* 75: 733–756.

Slobodkin, L. B. 1954. "Population dynamics in *Daphnia obtusa* Kurz." *Ecol. Monogr.,* 24: 69–88.

—— 1960. "Ecological energy relationships at the population level." *Amer. Nat.,* 95: 213–236.

—— and Richman, S. 1956. "The effect of removal of fixed percentages of the newborn on size and variability in populations of *Daphnia pulicaria.*" *Limnol. Oceanogr.,* 1: 209–237.

Slobodkin, L. B., Smith, F. E., and Hairston, N. G. 1967. "Regulation in terrestrial ecosystems, and the implied balance of nature." *Amer. Nat.,* 101: 109–124.

Smalley, A. E. 1959. "The growth cycle of *Spartina* and its relation to the insect population in the marsh." *Proc. Salt Marsh Conf.,* Mar. Inst., Univ. of Georgia, Athens: 96–100.

—— 1960. "Energy flow of a salt marsh grasshopper population." *Ecology,* 41: 672–677.

Smith, F. E. 1954. "Quantitative aspects of population growth," in E. J. Boel, ed., *Dynamics of Growth Processes, Growth Symposium II.* Princeton, Princeton Univ. Press.

—— 1961. "Density dependence in the Australian thrips." *Ecology,* 42: 403–407.

—— 1963. "Density-dependence." *Ecology,* 44: 220.

Smith, H. S. 1935. "The rôle of biotic factors in the determination of population densities." *J. Econ. Entomol.,* 28: 873–898.

—— 1941. "Racial segregation in insect populations and its significance in applied entomology." *J. Econ. Entomol.,* 34: 1–12.

Smith, N. G. 1967. "Visual isolation in gulls." *Sci. Amer.,* 217: 94–102.

—— 1968. "The advantage of being parasitized." *Nature,* 219: 690–694.

—— 1972. "Migrations of the day-flying moth *Urania* in Central and South America." *Carib. J. Sci.,* 12: 45–58.

Smith, W. G. 1975. "Dynamics of pure and mixed populations of *Desmodium glutinosum* and *D. nudiflorum* in natural oak-forest communities." *Amer. Midl. Nat.,* 94: 99–107.

Smith, W. J. 1969a. "Displays of *Sayornis phoebe* (Aves, Tyrannidae)." *Behaviour,* 33: 283–322.

——— 1969b. "Messages of vertebrate communication." *Science,* 165: 145–150.

Smithers, S. R., Terry, R. J., and Hockley, D. J. 1969. "Host antigens in schistosomiasis." *Proc. Roy. Soc. B.,* 171: 483–494.

Smyth, M. 1968. "The effects of removal of individuals from a population of bank voles (*Clethrionomys glareolus*)." *J. Anim. Ecol.,* 37: 167–183.

Snaydon, R. W., and Davies, M. S. 1972. "Rapid population differentiation in a mosaic environment, II: Morphological variation in *Anthoxanthum odoratum.*" *Evolution,* 26: 390–405.

Snow, B. K., and Snow, D. W. 1972. "Feeding niches of hummingbirds in a Trinidad valley." *J. Anim. Ecol.,* 41: 471–485.

Snow, D. W. 1962. "A field study of the black-and-white manakin, *Manacus manacus,* in Trinidad." *Zoologica,* 47: 65–104.

——— 1966. "A possible selective factor in the evolution of fruiting seasons in tropical forests." *Oikos,* 15: 274–281.

——— 1971. "Evolutionary aspects of fruit-eating by birds." *Ibis,* 113: 194–202.

Snyder, L. L. 1947. "The snowy owl migration of 1945–46: Second report of the Snowy Owl Committee." *Wilson Bull.,* 59: 74–78.

Sokal, R. R., and Rohlf, F. J. 1969. *Biometry.* San Francisco, Freeman.

Solbrig, O. T. 1972. "Breeding system and genetic variation in *Leavenworthia.*" *Evolution,* 26: 155–160.

Soulsby, E. J. L. 1968. *Helminths, Arthropods, and Protozoa of Domestic Animals.* Baltimore, Williams and Wilkins.

Southwick, C. H. 1955a. "The population dynamics of confined house mice supplied with unlimited food." *Ecology,* 36: 212–225.

——— 1955b. "Regulatory mechanisms of house mouse populations: Social behavior affecting litter survival." *Ecology,* 36: 627–634.

Southwood, T. R. E. 1960. "The abundance of the Hawaiian trees and the number of their associated insect species." *Proc. Hawaiian Entomol. Soc.,* 17: 299–303.

——— 1961. "The number of species of insects associated with various trees." *J. Anim. Ecol.,* 30: 1–8.

Spiess, E. B. 1968. "Low frequency advantage in mating of *Drosophila pseudoobscura* karyotypes." *Amer. Nat.,* 102: 363–379.

Spinage, C. A. 1972. "African ungulate life tables." *Ecology,* 53: 645–652.

Sprugel, D. G. 1976. "Dynamic structure of wave-regenerated *Abies balsamea* forests in the north-eastern United States." *J. Ecol.,* 64: 889–911.

Stanhill, G. 1970. "The water flux in temperate forests: Precipitation and evapotranspiration," in D. E. Reichle, ed., *Analysis of Temperate Forest Ecosystems.* New York, Springer-Verlag.

Stanley, S. M. 1975. "Clades versus clones in evolution: Why we have sex." *Science,* 190: 382–383.

Stearns, S. C. 1976. "Life-history tactics: A review of the ideas." *Quart. Rev. Biol.,* 51: 3–47.

Stebbins, G. L., Jr. 1950. *Variation and Evolution in Plants.* New York, Columbia Univ. Press.

——— 1957. "Self-fertilization and population variability in higher plants." *Amer. Nat.,* 91: 337–354.

——— 1971. *Chromosomal Evolution in Higher Plants.* Reading, Mass., Addison-Wesley.

Stebbins, R. C. 1949. "Speciation in salamanders of the Plethodontid genus *Ensatina.*" *Univ. of Calif. Publ. Zool.,* 48: 377–526.

Steel, R. G. D., and Torrie, J. H. 1960. *Principles and Procedures in Statistics.* New York, McGraw-Hill.

Steeman-Nielsen, A. 1960. "Productivity of the oceans." *Ann. Rev. Plant Physiol.*, 11: 341–362.

Steen, J. B. 1970. "The swimbladder as a hydrostatic organ," in W. S. Hoar and D. J. Randall, eds., *Fish Physiology*, Vol. 4. New York, Academic Press.

Steffan, W. A. 1973. "Polymorphism in *Plastosciara perniciosa.*" *Science*, 182: 1265–1266.

Stehli, F. G. 1968. "Taxonomic gradients in pole location: The recent model," in E. T. Drake, ed., *Evolution and Environment.* New Haven, Yale Univ. Press.

———— Douglas, R. G., and Newell, N. D. 1969. "Generation and maintenance of gradients in taxonomic diversity." *Science,* 164: 947–949.

Stenger, J. 1958. "Food habits and available food of ovenbirds in relation to territory." *Auk,* 75: 335–346.

———— and Falls, J. B. 1959. "The utilized territory of the ovenbird." *Wilson Bull.,* 71: 125–140.

Stephenson, T. A., and Stephenson, A. 1972. *Life Between Tidemarks on Rocky Shores.* San Francisco, W. H. Freeman.

Stewart, R. E., and Aldrich, J. W. 1951. "Removal and repopulation of breeding birds in a spruce-fir community." *Auk,* 75: 471–482.

Stickle, W. B. 1975. "The reproductive physiology of the intertidal prosobranch *Thais lamellosa* (Gmelin), II: Seasonal changes in biochemical composition." *Biol. Bull.,* 148: 448–460.

Stimson, J. 1970. "Territorial behavior of the owl limpet, *Lottia gigantea.*" *Ecology,* 51: 113–118.

Stoecker, R. E. 1972. "Competitive relations between sympatric populations of voles (*Microtus montanus* and *M. pennsylvanicus*)." *J. Anim. Ecol.,* 41: 311–329.

Storer, R. W. 1966. "Sexual dimorphism and food habits in three North American accipiters." *Auk,* 83: 423–436.

Strecker, R. L. 1954. "Regulatory mechanisms in house mouse populations: The effect of limited food supply on an unconfined population." *Ecology,* 35: 249–253.

———— and Emlen, J. T., Jr. 1953. "Regulatory mechanisms in house mouse populations: The effect of limited food supply on a confined population." *Ecology,* 34: 375–385.

Strehler, B. L., ed. 1960. *The Biology of Aging.* Washington, D.C., Amer. Inst. Biol. Sci.

Strickland, J. D. H. 1960. "Measuring the production of marine phytoplankton." *Bull. Fish. Res. Bd. Can.,* 12: 1–172.

Strong, D. R. 1974. "Rapid asymptotic species accumulation in phytophagous insect communities: The pests of cacao." *Science,* 185: 1064–1066.

———— and Levin, D. A. 1975. "Species richness of the parasitic fungi of British trees." *Proc. Natl. Acad. Sci.,* 72: 2116–2119.

Stross, R. G. 1969. "Photoperiod control of diapause in *Daphnia,* II: Induction of winter diapause in the arctic." *Biol. Bull.,* 136: 264–273.

———— and Hill, V. C. 1965. "Diapause induction in *Daphnia* requires two stimuli." *Science,* 150: 1462–1464.

Strum, S. C. 1975. "Primate predation and bioenergetics." *Science,* 191: 314–317.

Suthers, R. A. 1960. "Measurement of some lake-shore territories of the song sparrow." *Wilson Bull.,* 72: 232–237.

Svärdson, G. 1957. "The 'invasion' type of bird migration." *Brit. Birds,* 50: 314–343.

Sved, J. A. 1968. "Possible rates of gene substitution in evolution." *Amer. Nat.,* 102: 283–293.

———— Reed, T. E., and Bodmer, W. F. 1967. "The number of balanced polymorphisms that can be maintained in natural population." *Genetics,* 55: 469–481.

Sverdrup, H. V. 1945. *Oceanography for Meteorologists.* London, Allen & Unwin.

———— Johnson, M. W., and Fleming, R. H. 1942. *The Oceans; Their Physics, Chemistry and General Biology.* New York, Prentice-Hall.

Svihla, A., Bowman, H. R., and Ritenour, R. 1951. "Prolongation of clotting time in dormant estivating mammals." *Science,* 114: 298–299.

Swanberg, O. 1951. "Food storage, territory and song in the thick-billed nutcracker." *Proc. Tenth Inter. Congr. Ornithol.,* 545–554.

Swank, W. T., and Douglas, J. E. 1974. "Streamflow greatly reduced by converting deciduous hardwood stands to pine." *Science,* 185: 857–859.

Swanson, C. R. 1940. "The distribution of inversions in *Tradescantia.*" *Genetics,* 25: 348–365.

Taber, R. D. 1956. "Deer nutrition and population dynamics in the North Coast Range of California." *Trans. North Amer. Wildl. Conf.,* 21: 159–172.

—— and Dasmann, R. F. 1957. "The dynamics of three natural populations of the deer *Odocoileus hemionus columbianus.*" *Ecology,* 38: 233–246.

Tamarin, R. H., and Krebs, C. J. 1969. "*Microtus* population biology, II: Genetic changes at the transferring locus in fluctuating populations of two vole species." *Evolution,* 23: 183–211.

Tamm, C. O. 1951. "Seasonal variation in composition of birch leaves." *Physiol. Plant.,* 4: 461–469.

Tanner, J. T. 1966. "Effects of population density on growth rates of animal populations." *Ecology,* 47: 733–745.

Tansley, A. G. 1917. "On competition between *Galium saxatile* L. (*G. hercynicum* Weig.) and *Galium sylvestre* Poll. (*G. asperum* Schreb.) on different types of soil." *J. Ecol.,* 5: 173–179.

—— 1935. "The use and abuse of vegetational concepts and terms." *Ecology,* 16: 284–307.

Taylor, H. M., Gourley, R. S., Lawrence, C. E., and Kaplan, R. S. 1974. "Natural selection of life history attributes: An analytical approach." *Theoret. Pop. Biol.,* 5: 104–122.

Teal, J. M. 1957. "Community metabolism in a temperate cold spring." *Ecol. Monogr.,* 27: 283–302.

—— 1962. "Energy flow in the salt marsh ecosystem of Georgia." *Ecology,* 43: 614–624.

Teeri, J., and Stowe, L. 1976. "Climatic patterns and the distribution of C_4 grasses in North America." *Oecologia,* 23: 1–12.

Terborgh, J. 1971. "Distribution on environmental gradients: Theory and a preliminary interpretation of distributional patterns in the avifauna of the Cordillera Vilcabamba, Peru." *Ecology,* 52: 23–40.

—— and Faaborg, J. 1973. "Turnover and ecological release in the avifauna of Mona Island, Puerto Rico." *Auk,* 90: 759–779.

Tevis, L., Jr., and Newell, I. M. 1962. "Studies on the biology and seasonal cycle of the giant red velvet mite, *Dinothrombium pandorae* (Acari, Thrombidiidae)." *Ecology,* 43: 497–505.

Thayer, C. W. 1974. "Environmental and evolutionary stability in bivalve mollusks." *Science,* 186: 828–830.

Thoday, J. M. 1958. "Effects of disruptive selection: The experimental production of a polymorphic population." *Nature,* 181: 1124–1125.

—— 1959. "Effects of disruptive selection, I: Genetic flexibility." *Heredity,* 13: 187–203.

—— 1963. "Correlation between gene frequency and population size." *Amer. Nat.,* 97: 409.

—— and Boam, T. B. 1959. "Effects of disruptive selection, II: Polymorphism and divergence without isolation." *Heredity,* 13: 205–218.

Thoday, J. M., and Gibson, J. B. 1962. "Isolation by disruptive selection." *Nature,* 193: 1164–1166.

Thomas, M. D. 1955. "Effect of ecological factors on photosynthesis." *Ann. Rev. Plant Physiol.,* 6: 135–156.

Thompson, D. A. 1961. *On Growth and Form,* abridged edition edited by J. T. Bonner. London and New York, Cambridge Univ. Press.

Thompson, L. M. 1975. "Weather variability, climatic change, and grain production." *Science,* 188: 535–543.

Thorpe, W. H. 1963. *Learning and Instinct in Animals,* 2d ed. London, Methuen.

Thorson, G. 1950. "Reproduction and larval ecology of marine bottom invertebrates." *Biol. Rev.,* 25: 1–45.

Tinbergen, L. 1960. "The natural control of insects in pinewoods, I: Factors influencing the intensity of predation by songbirds." *Arch. Néerl. Zool.,* 13: 266–336.

Tinbergen, N. 1951. *The Study of Instinct.* London, Oxford Univ. Press.

—— 1952. "The curious behavior of the stickleback." *Sci. Amer.,* 187: 22–26.

—— 1953. *Social Life in Animals.* London, Methuen and New York, Wiley.

—— 1959. "Comparative studies of the behaviour of gulls (*Laridae*): A progress report." *Behaviour,* 15: 1–70.

—— Broekhysen, G. J., Feekes, F., Houghton, J. C. W., Kruuk, H., and Szuk, E. 1962. "Egg shell removal by the black-headed gull, *Larus ribundus* L.: A behaviour component of camouflage." *Behaviour,* 19: 74–117.

Tinkle, D. W. 1969. "The concept of reproductive effort and its relation to the evolution of life histories of lizards." *Amer. Nat.,* 103: 501–516.

—— and Ballinger, R. E. 1972. "*Sceloporus undulatus:* A study of the intraspecific comparative demography of a lizard." *Ecology,* 53: 570–584.

Tinkle, D. W., and Hadley, N. F. 1975. "Lizard reproductive effort: Caloric estimates and comments on its evolution." *Ecology,* 56: 427–434.

Tinkle, D. W., Wilbur, H. M., and Tilley, S. G. 1970. "Evolutionary strategies in lizard reproduction." *Evolution,* 24: 55–74.

Tompa, F. S. 1964. "Factors determining the numbers of song sparrows, *Melospiza melodia* (Wilson), on Mandarte Island, B.C., Canada." *Acta. Zool. Fenn.,* 109: 1–73.

Townsend, J. I., Jr., 1952. "Genetics of marginal populations of *Drosophila willistoni.*" *Evolution,* 6: 428–442.

Tracey, M. L., and Ayala, F. J. 1974. "Genetic load in natural populations: Is it compatible with the hypothesis that many polymorphisms are maintained by natural selection?" *Genetics,* 77: 569–589.

Tramer, E. J. 1969. "Bird species diversity: Components of Shannon's formula." *Ecology,* 50: 927–929.

Transeau, E. N. 1926. "The accumulation of energy by plants." *Ohio J. Sci.,* 26: 1–10.

Treshow, M. 1970. *Environment and Plant Response.* New York, McGraw-Hill.

Trewartha, G. T. 1954. *An Introduction to Climate.* New York, McGraw-Hill.

Trivers, R. L. 1972. "Parental investment and sexual selection," in B. Campbell, ed., *Sexual Selection and the Descent of Man 1871–1971.* Chicago, Aldine.

—— 1974. "Parent-offspring conflict." *Amer. Zool.,* 14: 249–264.

—— and Willard, D. E. 1973. "Natural selection of parental ability to vary the sex ratio of offspring." *Science,* 179: 90–92.

Turesson, G. 1922. "The genotypic response of the plant species to the habitat." *Hereditas,* 3: 211–350.

Turnage, W. V., and Hinckley, A. L. 1938. "Freezing weather in relation to plant distribution in the Sonoran Desert." *Ecol. Monogr.,* 8: 529–550.

Turner, B. J. 1974. "Genetic divergence of Death Valley pupfish species: Biochemical versus morphological evidence." *Evolution,* 28: 281–294.

Utida, S. 1957. "Population fluctuation, an experimental and theoretical approach." *Cold Spring Harbor Symp. Quant. Biol.*, 22: 139–151.

Uvarov, B. P. 1961. "Quantity and quality in insect populations." *Proc. Roy. Entomol. Soc.*, London, C 25: 52–59.

Uzzell, T. 1970. "Meiotic mechanisms of naturally occurring unisexual vertebrates." *Amer. Nat.*, 104: 433–445.

—— and Corbin, K. W. 1972. "Evolutionary rates in cistrons specifying mammalian hemoglobin α- and β-chains: Phenetic versus patristic measurements." *Amer. Nat.*, 106: 555–573.

Valentin, J. 1973. "Selection for altered recombination frequency in *Drosophila melanogaster*." *Hereditas*, 74: 295–297.

Valentine, J. W., and Ayala, F. J. 1974. "On scientific hypotheses, killer clams and extinctions." *Geology*, 2: 69–70.

Van Cleve, K. 1974. "Organic matter quality in relation to decomposition," in A. J. Holding *et al.*, eds., *Soil Organisms and Decomposition in Tundra.* Stockholm, Tundra Biome Steering Committee.

Vandenbergh, J. G., and Vessey, S. 1968. "Seasonal breeding of free-ranging rhesus monkeys and related ecological factors." *J. Reprod. Fert.*, 15: 71–79.

Vandermeer, J. H. 1969. "The competitive structure of communities: An experimental approach with protozoa." *Ecology*, 50: 362–371.

—— 1972. "Niche theory." *Ann. Rev. Ecol. Syst.*, 3: 107–132.

Van Emden, H. F., and Williams, G. F. 1974. "Insect stability and diversity in agroecosystems." *Ann. Rev. Entomol.*, 19: 455–475.

Van Valen, L. 1960. "Nonadaptive aspects of evolution." *Amer. Nat.*, 94: 305–306.

—— 1971. "The history and stability of atmospheric oxygen." *Science*, 171: 439–443.

—— 1973. "A new evolutionary law." *Evol. Theory*, 1: 1–30.

—— and Levins, R. 1968. "The origins of inversion polymorphisms." *Amer. Nat.*, 102: 5–24.

Van Valen, L., and Sloan, R. E. 1966. "The extinction of the multituberculates." *Syst. Zool.*, 15: 261–278.

Varley, G. C. 1947. "The natural control of population balance in the knapweed gallfly (*Urophora jaceana*)." *J. Anim. Ecol.*, 16: 139–187.

—— 1949. "Population changes in German forest pests." *J. Anim. Ecol.*, 18: 117–122.

—— 1975. "Should we control the use of the word control?" *Bull. Brit. Ecol. Soc.*, 6: 7.

—— and Gradwell, G. R. 1960. "Key factors in population studies." *J. Anim. Ecol.*, 29: 399–401.

Vaurie, C. 1950. "Adaptive differences between two sympatric species of nuthatches (Sitta)." *Proc. Tenth Inter. Ornithol. Congr.*, 163–166.

Vegis, A. 1964. "Dormancy in higher plants." *Ann. Rev. Plant. Physiol.*, 15: 185–215.

Vepsalainen, K. 1971. "The role of gradually changing daylength in determination of winglength, alary polymorphism and diapause in a *Gerris odontogaster* (Zett.) population (Gerridae, Heteroptera) in South Finland." *Ann. Acad. Sci. Fenn. Ser. A. IV*, 183: 1–25.

—— 1973. "The distribution and habitats of *Gerris* Fabr. species (Heteroptera, Gerridae) in Finland." *Ann. Zool. Fenn.*, 10: 419–444.

—— 1974a. "The life cycles and wing lengths of Finnish *Gerris* Fabr. species (Heteroptera, Gerridae)." *Acta Zool. Fenn.*, 141: 1–73.

—— 1974b. "Determination of wing length and diapause in water-striders (*Gerris* Fabr., Heteroptera)." *Hereditas*, 77: 163–176.

———— 1974c. "The winglengths, reproductive stages and habitats of Hungarian *Gerris* Fabr. species (Heteroptera, Gerridae)." *Ann. Acad. Sci. Fenn.*, Ser. A. IV, 202: 1–18.

Verghesi, M. W., and Nordskog, A. W. 1968. "Correlated responses in reproductive fitness to selection in chickens." *Genet. Res.*, 11: 221–238.

Verhulst, P. F. 1838. "Notice sur la loi que la population pursuit dans son accroissement." *Corresp. Math. Phys.*, 10: 113–121.

Verner, J. 1964. "Evolution of polygamy in the long-billed marsh wren." *Evolution*, 18: 252–261.

———— and Willson, M. F. 1966. "The influence of habitats on mating systems of North American passerine birds." *Ecology*, 47: 143–147.

———— 1969. "Mating systems, sexual dimorphism, and the role of male North American passerine birds in the nesting cycle." *Ornithol. Monogr.*, No. 9.

Vesey-Fitzgerald, D. F. 1960. "Grazing succession among East African game animals." *J. Mammol.*, 41: 161–172.

Vinegar, A., Hutchinson, V. H., and Dowling, H. G. 1970. "Metabolism, energetics, and thermoregulation during brooding of snakes of the genus *Python* (Reptilia, Boidae)." *Zoologica*, 55: 19–50.

Vogl, R. J. 1973. "Ecology of knobcone pine in the Santa Ana Mountains of California." *Ecol. Monogr.*, 43: 125–143.

———— and Schorr, P. K. 1972. "Fire and manzanita chaparral in the San Jacinto Mountains, California." *Ecology*, 53: 1179–1188.

Volterra, V. 1926. "Variations and fluctuations of the number of individuals of animal species living together," in R. N. Chapman, ed., *Animal Ecology*. New York, McGraw-Hill.

von Frisch, K. 1950. *Bees. Their Vision, Chemical Senses, and Language*. Ithaca, Cornell Univ. Press.

———— 1967. *The Dance Language and Orientation of Bees*. Cambridge, Mass., Harvard Univ. Press.

Vuilleumier, F. 1970. "Insular biogeography in continental regions, I: The northern Andes of South America." *Amer. Nat.*, 104: 373–388.

Waggoner, P. E., and Stephens, G. R. 1970. "Transition probabilities for a forest." *Nature*, 255: 1160–1161.

Wagner, F. H., and Stoddart, L. C. 1972. "Influence of coyote predation on black-tailed jackrabbit populations in Utah." *J. Wildl. Mgmt.*, 36: 329–343.

Waldron, I., and Ricklefs, R. 1973. *Environment and Population. Problems and Solutions*. New York: Holt.

Wali, M. K., and Krajina, V. J. 1973. "Vegetation-environment relationships of some sub-boreal spruce zone ecosystems in British Columbia." *Vegetatio*, 26: 237–381.

Walker, R. B. 1954. "The ecology of serpentine soils, II: Factors affecting plant growth on serpentine soils." *Ecology*, 35: 259–266.

Wallace, A. R. 1878. *Tropical Nature and Other Essays*. London and New York, Macmillan.

———— 1889. *Darwinism: An Exposition of the Theory of Natural Selection with Some of its Applications*. London, Macmillan.

Wallace, B. 1968. *Topics in Population Genetics*. New York, Norton.

———— and Srb, A. M. 1964. *Adaptation*, 2d ed. Englewood Cliffs, N.J., Prentice-Hall.

Waloff, N. 1941. "The mechanism of humidity reactions of terrestrial isopods." *J. Expt. Biol.*, 18: 115–135.

Waloff, Z. 1966. "The upsurges and recessions of the desert locust plague: An historical survey." *Anti-Locust Mem.*, 8: 1–111.

Walters, J. L. 1942. "Distribution of structural hybrids in *Paeonia californica*." *Amer. J. Bot.*, 29: 270–275.

Ward, P. 1963. *Contributions to the Ecology of the Weaver Bird,* Quelea Quelea, *Linnaeus, in Nigeria.* Unpublished Ph.D. thesis, Univ. College, Ibadan.

—— 1969a. "Seasonal and diurnal changes in the fat content of an equatorial bird." *Physiol. Zool.,* 42: 85–95.

—— 1969b. "The annual cycle of the yellow-vented bulbul *Pycnonotus goiavier* in a humid equatorial environment." *J. Zool.,* 157: 25–45.

—— and Zahavi, A. 1973. "The importance of certain assemblages of birds as 'information-centres' for food-finding." *Ibis,* 115: 517–534.

Ward, R. C. 1967. *Principles of Hydrology.* London and New York, McGraw-Hill.

Wareing, P. F. 1966. "Ecological aspects of seed dormancy and germination," in J. G. Hawkes, ed., *Reproductive Biology and Taxonomy of Vascular Plants.* Oxford, Pergamon.

Waring, R. H., and Major, J. 1964. "Some vegetation of the California coastal region in relation to gradients of moisture, nutrients, light, and temperature." *Ecol. Monogr.,* 34: 167–215.

Warner, R. E. 1968. "The role of introduced diseases in the extinction of the endemic Hawaiian avifauna." *Condor,* 70: 101–120.

Warner, R. R. 1975. "The adaptive significance of sequential hermaphroditism in animals." *Amer. Nat.,* 109: 61–82.

—— Robertson, D. R., and Leigh, E. G., Jr. 1975. "Sex change and sexual selection." *Science,* 190: 633–638.

Warren, C. E. 1971. *Biology and Water Pollution Control.* Philadelphia, Saunders.

Washburn, S. L., and DeVore, I. 1961. "The social life of baboons." *Sci. Amer.,* 204: 62–71.

Watson, A. 1965. "A population study of ptarmigan (*Lagopus mutus*) in Scotland." *J. Anim. Ecol.,* 34: 135–172.

—— 1967. "Population control by territorial behavior in red grouse." *Nature,* 215: 1274–1275.

—— and Jenkins, D. 1968. "Experiments on population control by territorial behaviour in red grouse." *J. Anim. Ecol.,* 37: 595–614.

Watson, A., and Moss, R. 1970. "Dominance, spacing behaviour and aggression in relation to population limitation in vertebrates," in A. Watson, ed., *Animal Populations in Relation to Their Food Resources.* Oxford, Blackwell.

Watson, G. F., and Martin, A. A. 1968. "Postmating isolation in the *Hyla ewingi* complex (Anura: Hylidae)." *Evolution,* 22: 664–666.

Watson, I. A. 1970. "Changes in virulence and population shifts in plant pathogens." *Ann. Rev. Phytopathol.,* 8: 209–230.

Watt, A. S. 1947. "Pattern and process in the plant community." *J. Ecol.,* 35: 1–22.

Watt, K. E. F. 1955. "Studies on population productivity, I: Three approaches to the optimum yield problem in populations of *Tribolium confusum.*" *Ecol. Monogr.,* 25: 269–290.

—— 1962. "The conceptual formulation and mathematical solution of practical problems in population input-output dynamics." in E. D. LeCren and M. W. Holdgate, eds., *The Exploitation of Natural Animal Populations.* New York, Wiley.

—— 1964. "Density dependence in population fluctuations." *Canad. Entomol.,* 96: 1147–1148.

—— 1964a. "Comments on fluctuations of animal populations and measures of community stability." *Canad. Entomol.,* 96: 1434–1442.

—— 1965. "Community stability and the strategy of biological control." *Canad. Entomol.,* 97: 887–895.

—— 1968. *Ecology and Resource Management.* New York, McGraw-Hill.

Weast, R. C., ed. 1969. *Handbook of Chemistry and Physics,* 50th ed. Cleveland, Chemical Rubber Co.

Webb, S. D. 1969. "Extinction—origination equilibria in late Cenozoic land mammals of North America." *Evolution,* 23: 688–702.

——— 1976. "Mammalian faunal dynamics of the great American interchange." *Paleobiology,* 2: 220–234.

Wecker, S. C. 1963. "The role of early experience in habitat selection by the prairie deer mouse, *Peromyscus maniculatus bairidi.*" *Ecol. Monogr.,* 33: 307–325.

——— 1964. "Habitat selection." *Sci. Amer.,* 211: 109–116.

Weeden, J. S. 1965. "Territorial behavior of the tree sparrow." *Condor,* 67: 193–209.

Weir, J. S. 1958. "Polyethism in workers of the ant *Myrmica.*" *Insectes Sociaux,* 5: 97–128, 315–339.

Weisbrot, D. R. 1966. "Genotypic interactions among competing strains and species of *Drosophila.*" *Genetics,* 53: 427–435.

Weiss, H. B. 1924. "Ratios between the food habits of insects." *Entomol. News,* 35: 362–364.

Weisskopf, V. F. 1968. "How light interacts with matter." *Sci. Amer.,* 219: 60–71.

Welch, H. 1968. "Relationships between assimilation efficiencies and growth efficiencies for aquatic consumers." *Ecology,* 49: 755–759.

Wellington, W. G. 1948. "The light reactions of the spruce budworm, *Choristoneura fumiferana* Clemans (Lepidoptera: Tortricidae)." *Canad. Entomol.,* 80: 56–82.

——— 1960. "Qualitative changes in natural populations during changes in abundance." *Canad. J. Zool.,* 38: 289–314.

——— 1965. "Some maternal influences on progeny quality in the western tent caterpillar, *Malacosoma pluviale* (Dyar)." *Canad. Entomol.,* 97: 1–14.

Welty, J. C. 1963. *The Life of Birds.* New York, Knopf.

Werner, E. E. 1977. "Species packing and niche complementarity in three sunfishes." *Amer. Nat.,* 111: 553–578.

West, G. C. 1972. "Seasonal differences in resting metabolic rate of Alaskan ptarmigan." *Comp. Biochem. Physiol.,* 42A: 867–876.

Westlake, D. F. 1963. "Comparisons of plant productivity." *Biol. Rev.,* 38: 385–429.

Westoll, T. S. 1949. "On the evolution of the Dipnoi," in G. L. Jepsen, E. Mayr, and G. G. Simpson, eds., *Genetics, Paleontology, and Evolution.* Princeton, N.J., Princeton Univ. Press.

Wheeler, W. M. 1923. *Social Life Among the Insects.* New York, Harcourt, Brace.

——— 1928. *The Social Insects, Their Origin and Evolution.* New York, Harcourt, Brace.

——— 1933. *Colony Founding Among Ants.* Cambridge, Mass., Harvard Univ. Press.

White, C. D. 1971. *Vegetation—Soil Chemistry Correlations in Serpentine Ecosystems.* Unpublished Ph.D. diss., Univ. of Oregon.

White, J. J. 1968. "Bioenergetics of the woodlouse *Tracheoniscus rathkei* Brandt in relation to litter decomposition in a deciduous forest." *Ecology,* 49: 694–704.

White, M. J. D. 1968. "Models of speciation." *Science,* 159: 1065–1070.

——— 1973a. *The Chromosomes.* London, Chapman and Hall.

——— 1973b. *Animal Cytology and Evolution,* 3d ed. Cambridge, Cambridge Univ. Press.

Whittaker, R. H. 1952. "A study of summer foliage insect communities in the Great Smoky Mountains." *Ecol. Mongr.,* 22: 1–44.

——— 1954. "The ecology of serpentine soils, I: Introduction." *Ecology,* 35: 258–259.

——— 1956. "Vegetation of the Great Smoky Mountains." *Ecol. Monogr.,* 26: 1–80.

——— 1960. "Vegetation of the Siskiyou Mountains, Oregon and California." *Ecol. Monogr.,* 30: 279–338.

——— 1961. "Experiments with radiophosphorus tracer in aquarium microcosms." *Ecol. Monogr.,* 31: 157–188.

——— 1965. "Dominance and diversity in land plant communities." *Science,* 147: 250–260.

——— 1966. "Forest dimensions and production in the Great Smoky Mountains." *Ecology*, 47: 103–121.

——— 1967. "Gradient analysis of vegetation." *Biol. Rev.*, 42: 207–264.

——— 1970. *Communities and Ecosystems.* New York, Macmillan.

——— 1972. "Evolution and measurement of species diversity." *Taxon*, 21: 213–251.

——— 1975. *Communities and Ecosystems,* 2d ed. New York, Macmillan.

——— and Feeny, P. P. 1971. "Allelochemics: Chemical interactions between species." *Science*, 171: 757–770.

Whittaker, R. H., and Likens, G. E. 1973. "Primary production: The biosphere and man." *Human Ecol.*, 1: 357–369.

Whittaker, R. H., and Niering, W. A. 1965. "Vegetation of the Santa Catalina Mountains, Arizona: A gradient analysis of the south slope." *Ecology*, 46: 429–452.

Whittaker, R. H., and Woodwell, G. M. 1968. "Dimension and production relations of trees and shrubs in the Brookhaven Forest, New York." *J. Ecol.*, 56: 1–25.

——— 1969. "Structure, production and diversity of the oak-pine forest at Brookhaven, New York." *J. Ecol.*, 57: 155–174.

Wickler, W. 1968. *Mimicry in Plants and Animals.* London, World Univ. Library.

Wiebe, W. J. 1975. "Nitrogen fixation in a coral reef community." *Science*, 188: 257–259.

Wiegert, R. G. 1964. "Population energetics of meadow spittlebugs (*Philaenus spumarius* L.) as affected by migration and habitat." *Ecol. Monogr.*, 34: 217–241.

——— 1965. "Energy dynamics of the grasshopper populations in old-field and alfalfa field ecosystems." *Oikos*, 16: 161–176.

——— and Owen, D. F. 1971. "Trophic structure, available resources and population density in terrestrial vs. aquatic ecosystems." *J. Theoret. Biol.*, 30: 69–81.

Wiens, J. A. 1966. "On group selection and Wynne-Edwards' hypothesis." *Amer. Sci.*, 54: 273–287.

Wilbur, H. M. 1972. "Competition, predation, and the structure of the *Ambystoma-Rana sylvatica* community." *Ecology*, 53: 3–21.

——— and Collins, J. P. 1973. "Ecological aspects of amphibian metamorphosis." *Science*, 182: 1305–1314.

Wilbur, H. M., Tinkle, D. W., and Collins, J. P. 1974. "Environmental certainty, trophic level, and resource availability in life history evolution." *Amer. Nat.*, 108: 805–817.

Wilde, S. A. 1968. "Mycorrhizae and tree nutrition." *Bio. Science*, 18: 482–484.

Wiley, R. H. 1974. "Evolution of social organization and life-history patterns among grouse." *Quart. Rev. Biol.*, 49: 201–227.

Williams, C. B. 1964. *Patterns in the Balance of Nature and Related Problems in Quantitative Ecology.* New York, Academic Press.

Williams, G. C. 1957. "Pleiotropy, natural selection, and the evolution of senescence." *Evolution*, 11: 398–411.

——— 1966a. "Natural selection, the costs of reproduction, and a refinement of Lack's principle." *Amer. Nat.*, 100: 687–690.

——— 1966b. *Adaptation and Natural Selection.* Princeton, N.J., Princeton Univ. Press.

——— 1975. *Sex and Evolution.* Princeton, N.J., Princeton Univ. Press.

——— and Mitton, J. B. 1973. "Why produce sexually?" *J. Theoret. Biol.*, 39: 545–554.

Williams, G. D. V., Joynt, M. I., and McCormick, P. A. 1975. "Regression analyses of Canadian prairie crop-district cereal yields, 1961–1972, in relation to weather, soil and trend." *Can. J. Soil Sci.*, 55: 43–53.

Williams, M. B. 1975. "Darwinian selection for self-limiting populations." *J. Theoret. Biol.*, 55: 415–430.

Williams, R. J. 1970. "Freezing tolerance in *Mytilus edulis.*" *Comp. Biochem. Physiol.*, 35: 145–161.

Williamson, G. B. 1975. "Pattern and seral composition in an old-growth beech-maple forest." *Ecology,* 56: 727–731.

Willis, E. O. 1963. "Is the zone-tailed hawk a mimic of the turkey vulture?" *Condor,* 65: 313–317.

——— 1967. "The behavior of bicolored antbirds." *Univ. of Calif. Publ. Zool.,* 79: 1–127.

Wilson, E. O. 1959. "Source and possible nature of the odor trail of the fire ant *Solenopsis saevissima* (Fr. Smith)." *Science,* 129: 643–644.

——— 1961. "The nature of the taxon cycle in the Melanesian ant fauna." *Amer. Nat.,* 95: 169–193.

——— 1962. "Chemical communication among workers of the fire ant *Solenopsis saevissima* (Fr. Smith): 1. The organization of mass-foraging; 2. An information analysis of the odour trail; 3. The experimental induction of social responses." *Anim. Behav.,* 10: 134–164.

——— 1963a. "Pheromones." *Sci. Amer.,* 208: 100–114.

——— 1963b. "The social biology of ants." *Ann. Rev. Entomol.,* 8: 345–368.

——— 1969. "The species equilibrium." *Brookhaven Symp. Biol.,* 22: 38–47.

——— 1975. *Sociobiology.* Cambridge, Mass., Harvard Univ. Press.

——— and Taylor, R. W. 1964. "A fossil ant colony: New evidence of social antiquity." *Psyche,* 71: 93–103.

Wilson, J. W. 1974. "Analytical zoogeography of North American mammals." *Evolution,* 28: 124–140.

Wimpenny, R. S. 1966. *The Plankton of the Sea.* London, Faber and Faber.

Witkamp, M. 1966. "Decomposition of leaf litter in relation to environment, microflora, and microbial respiration." *Ecology,* 47: 194–201.

——— 1969. "Cycles of temperature and carbon dioxide evolution from litter and soil." *Ecology,* 50: 922–924.

——— and van der Drift, J. 1961. "Breakdown of forest litter in relation to environmental factors." *Plant and Soil,* 15: 295–311.

Witter, J. A., Kulman, H. M., and Hodson, A. C. 1975. "Life tables for the forest tent caterpillar." *Ann. Entomol. Soc. Amer.,* 65: 25–31.

Wittwer, S. H. 1974. "Maximum production capacity of food crops." *Bio. Science,* 24: 216–224.

——— 1975. "Food production: Technology and the resource base." *Science,* 188: 579–584.

Wolf, L. L. 1969. "Female territoriality in a tropical hummingbird." *Auk,* 86: 490–504.

Wolfenbarger, D. O. 1946. "Dispersion of small organisms." *Amer. Midl. Nat.,* 35: 1–152.

Woodwell, G. M. 1970. "The energy cycle of the biosphere." *Sci. Amer.,* 223: 64–74.

Wright, H. E., Jr., 1964. "Aspects of the early postglacial forest succession in the Great Lakes Region." *Ecology,* 45: 439–448.

Wright, S. 1934. "Physiological and evolutionary theories of dominance." *Amer. Nat.,* 68: 24–53.

——— 1939. "The distribution of self-sterility alleles in populations." *Genetics,* 24: 538–552.

——— 1943a. "An analysis of local variability of flower color in *Linanthus parryae.*" *Genetics,* 28: 139–156.

——— 1943b. "Isolation by distance." *Genetics,* 28: 114–138.

——— 1951. "Fisher and Ford on 'the Sewall Wright effect.'" *Amer. Sci.,* 39: 452–479.

Wynne Edwards, V. C. 1962. *Animal Dispersion in Relation to Social Behavior.* Edinburgh, Oliver and Boyd.

——— 1963. "Intergroup selection in the evolution of social systems." *Nature,* 200: 623–628.

—— 1964. "Population control in animals." *Sci. Amer.*, 211: 68–74.

—— 1965. "Self-regulating systems in populations of animals." *Science,* 147: 1543–1548.

—— 1970. "Feedback from food resources to population regulation," in A. Watson, ed., *Animal Populations in Relation to Their Food Resources.* Oxford, Blackwell.

Yamazaki, T. 1971. "Measurement of fitness at the esterase-5 locus in *Drosophila pseudoobscura.*" *Genetics,* 67: 579–603.

Yang, R. S. H., and Guthrie, F. E. 1969. "Physiological responses of insects to nicotine." *Ann. Entomol. Soc. Amer.*, 62: 141–146.

Young, E. C. 1965. "Flight muscle polymorphism in British Corixidae: Ecological observations." *J. Anim. Ecol.*, 34: 353–390.

Zaret, T. M., and Rand, A. S. 1971. "Competition in tropical stream fishes: Support for the competitive exclusion principle." *Ecology,* 52: 336–342.

Zwölfer, H. 1963. "Untersuchungen uber die Struktur von Parasitenkomplexen bei einigen Lepidopteren." *Z. Angew. Entomol.*, 51: 346–357.

Index